P9-EED-159

# *Encyclopedia of* ENVIRONMENTAL BIOLOGY

## VOLUME 2

## F – N

# *Encyclopedia of* ENVIRONMENTAL BIOLOGY

## VOLUME 2
## F – N

*Editor-in-Chief*
***William A. Nierenberg***
*Scripps Institution of Oceanography*
*University of California, San Diego*

**ACADEMIC PRESS**
San Diego New York Boston London Sydney Tokyo Toronto

This book is printed on acid-free paper. ♾



Academic Press, Inc.
A Division of Harcourt Brace & Company
525 B Street, Suite 1900, San Diego, California 92101-4495

*United Kingdom Edition published by*
Academic Press Limited
24-28 Oval Road, London NW1 7DX

Library of Congress Cataloging-in-Publication Data

Encyclopedia of environmental biology / edited by William A. Nierenberg.
p. cm.
Includes bibliographical references and index.
ISBN 0-12-226730-3 (set). -- ISBN 0-12-226731-1 (v. 1). -- ISBN 0-12-226732-X (v. 2) -- ISBN 0-12-226733-8 (v. 3)
1. Ecology--Encyclopedias. 2. Environmental sciences--Encyclopedias. I. Nierenberg, William Aaron, date.
QH540.4.552 1995
574.5'03--dc20 94-24917
CIP

PRINTED IN THE UNITED STATES OF AMERICA
95 96 97 98 99 00 EB 9 8 7 6 5 4 3 2 1

# Contents

## VOLUME 1

## C



## D



## E

## VOLUME 2



## F

## W



## Z

# How to Use the Encyclopedia

The *Encyclopedia of Environmental Biology* is intended for use by both students and research professionals. Articles have been chosen to reflect important areas in the study of the environment, topics of public and research interest, and coverage of environmental issues vital to lawyers. Each article provides a comprehensive overview of the selected topic to satisfy readers from students to professionals.

The *Encyclopedia* is designed with the following features to allow maximum accessibility. Articles are arranged in alphabetical order by subject. A complete table of contents appears in the front matter of each volume. This list of titles represents topics carefully selected by our esteemed editorial board. Here, such general topics as "Acid Rain" and "Speciation" are listed.

A 10,000 entry Subject Index is located at the end of Volume 3. The index is the most direct way to access the *Encyclopedia* to find specific information. Although the article "Evolution and Extinction" is thorough and complete, additional text on evolution can be found in articles such as "Galápagos Islands" and "Bird Communities." Subjects in the index are listed alphabetically and indicate the volume and page number where corresponding information can be found.

The Index of Related Titles appears in Volume 3 following the Subject Index. This index presents an alphabetical list of the articles as they appear in the *Encyclopedia*. Following each article title is a group of related articles that appear in the encyclopedia.

Articles contain an outline, a glossary, cross-references, and a bibliography. The outline allows a quick scan of the major areas discussed within the article. The glossary contains terms that may be unfamiliar to the reader, with each term defined in the *context of its use in that particular article*. Thus, a term may appear in the glossary for another article defined in a slightly different manner or with a subtle nuance specific to that article. For clarity, we have allowed these differences in definition to remain so that the terms are defined relative to the context of each article.

Each article contains cross-references to other *Encyclopedia* articles. Cross-references are found at the end of the paragraph containing the first mention of a subject covered more fully elsewhere in the *Encyclopedia*. By using the cross-references, the reader gains the opportunity to find additional information on a given topic.

The bibliography lists recent secondary sources to aid the reader in locating more detailed or technical information. Review articles and research articles that are considered of primary importance to the understanding of a given subject area are also listed. Bibliographies are not intended to provide a full reference listing of all material covered in the context of a given article, but are provided as guides to further reading.

# Farming, Dryland

**V. R. Squires**
*University of Adelaide, Australia*



Dryland farming is rainfed agriculture principally concerned with raising crops and livestock in semiarid regions of the world. Recurrent drought and the potential for wind and water erosion are major constraints. Dryland farming therefore emphasizes water conservation and sustainable crop yields despite limited chemical inputs for soil fertility maintenance.

## I. WHAT IS DRYLAND FARMING?

The distribution of drylands is governed by the global circulation of the atmosphere. A high-pressure zone of sinking dry air around 30° latitude in each hemisphere is warmed by compression. The warm, dry air provides cloudless skies and exposes the land to the full effect of the sun. Brief seasonal invasions of moist air produce little rain, which is erratic in both area and time. The result is two series of deserts surrounded by drylands: in the northern hemisphere lie the Saharan, North American (Colorado, Gila, and Mexican), Arabian, Thar, and Gobi deserts; in the Southern Hemisphere are the Patagonian, Kalahari, and Australian deserts. All merge gradually into the habitable drylands that border them. [*See* DESERTS.]

The majority of the world's dryland areas lie within the developing, and mostly tropical and Mediterranean, regions (Fig. 1). Drylands embrace both arid and semiarid lands, as well as the more desertic (hyperarid) areas, but dryland farming is restricted to the wetter end of the spectrum. Several definitions of dryland have been proposed. The validity of the definition depends on its purpose. A multitude of factors, such as moisture, radiation, temperature, and nutrient cycling, may be included. Generally, drylands are characterized by low, erratic, and highly inconsistent rainfall levels. These are reflected in fluctuating and unpredictable levels of crop and livestock production.

A succinct definition of dryland farming is "husbandry under conditions of moderate to severe water stress during a substantial part of the year, which require special cultural techniques and adapted crops and systems for successful and stable agricultural production." Two factors—a limited supply of water and a serious erosion potential—are common to all dryland areas.

The main feature for "dryness"—extremely arid and desert areas excluded—is the negative balance between the annual rainfall and evapotranspiration. Where dryland farming is the objective, drylands have thus been defined as areas where mean annual precipitation is less than half the potential evapo-

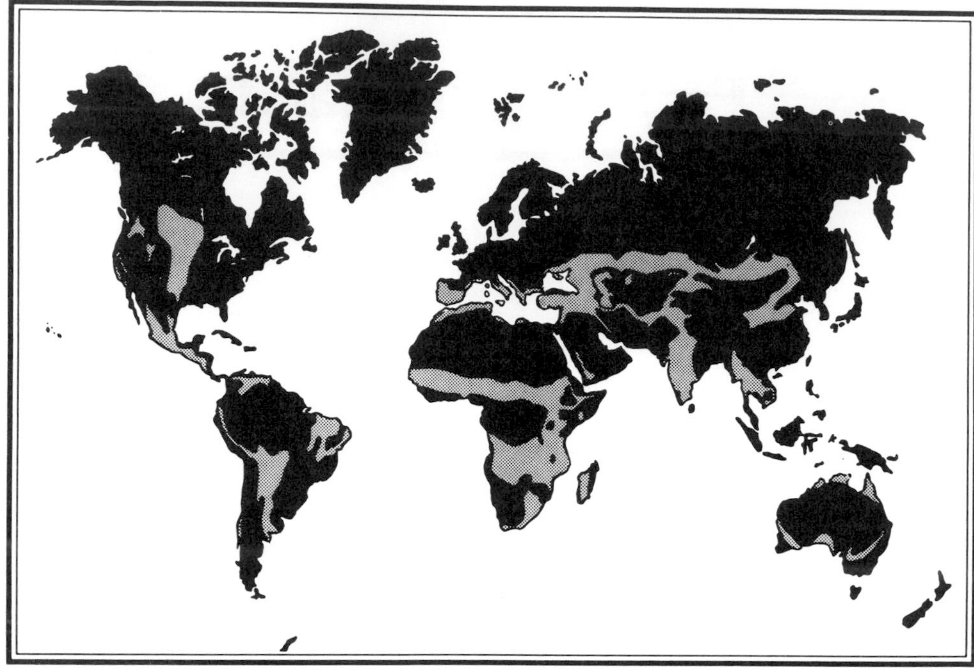

**FIGURE 1** Dryland farming is practiced in many parts of the world, principally in the midlatitude regions in each hemisphere.

transpiration. This in turn is reflected in the number of growing days that constitute the length of the growing period of the crop. Drylands are those areas with a growing period of less than 120 days. Within this range, aridlands have less than 75 growing days and semiarid lands have 75 days or more. This reflects a climatic situation where, in the absence of irrigation, rainfall is suboptimal to a greater or lesser extent, that is, it is inadequate to achieve the potential set by solar radiation.

Dryland farming and dryland agriculture are terms that can be used interchangeably. Strictly speaking, "agriculture" is a more comprehensive expression recognizing that there is more to it than just farming. Because dryland farming is practiced where the moisture supply is a limiting factor and irrigation is not used, it depends on the current season's precipitation and perhaps also on stored soil moisture trapped during a previous fallow period. It is estimated that the world's arable cropland is about 1470 million hectares (ha). Of this area, about 86% is dependent for its moisture supply on rain and/or snow (the remaining 14% is totally dependent on irrigation). This large area is suitable for "rainfed agriculture." Not all of this could be used for dryland farming as the two terms are not synonymous. Some rainfed areas (about 900 million ha) have no serious moisture limitation at any time of year. Dryland farming is rainfed agriculture that emphasizes water conservation, sustainable crop yields, limited inputs for soil fertility maintenance, and wind and water erosion constraints.

At best, dryland farming is a risky business with drought, hail, high winds, and destructive storms being ever-present threats. The vagaries of the weather still count heavily among the factors that control crop yields. Dryland farming can be thought of as a dynamic and highly complex production system in which the major limitation on food and fiber production is a deficiency of water for at least part of the year. But what exactly is this highly complex system? Where is it practiced and what are its attributes?

## II. WHERE IS DRYLAND FARMING PRACTICED?

Worldwide there is a lot of dryland farming, because 43% of the earth's surface receives insufficient rain to grow crops year-round. Some of it is too dry to grow crops at all. These are the rangelands. And of course it is not just how much rain falls but when it falls. The amount of moisture stored in the soil at the time of sowing is the most important measurable factor affecting potential production of the crop. It is axiomatic to say that distribution of rainfall is more important than the total recorded. However, good distribution does not mean, as is usually assumed, relatively light but frequent rains distributed regularly throughout the growing season. In such a case, much of the rainfall is dissipated by evaporation, moisture does not penetrate deeply, the root system remains shallow, and a relative or absolute crop failure is inevitable. A few heavy rains are more efficient, notwithstanding their poor distribution, for then the soil is moistened to a fair depth and losses due to evaporation are relatively slight. Spring wheat, for example, can be grown commercially in southern Australia on as little as 250 mm of rain. [*See* RANGE ECOLOGY AND GRAZING.]

Crops are grown in many parts of the world where precipitation is below so-called "safe" limits. Therefore, the first distinction to make is between subsistence and commercial agriculture. If both categories are combined, the proportion of the earth's surface that supports dryland farming is 460 million ha (Table I). Many areas of Africa, South America, the Middle East, southwestern Asia, the Indian subcontinent, China, and Central Asia support subsistence farmers on drylands.

**TABLE I**
Area of Arable Dryland by Continent[a]

| Continent | Area (thousands of ha) | Percent of world's dry farming land |
|---|---|---|
| Africa | 79,822 | 17.4 |
| Asia | 218,174 | 47.7 |
| Australia | 42,120 | 9.2 |
| Europe | 22,106 | 4.8 |
| North America | 74,169 | 16.2 |
| South America | 21,346 | 4.7 |
| Total | 457,737 | 100.0 |

[a] From H. Dregne, M. Kassas, and B. Rosanov, (1991). "Desertification Control Bulletin No. 20." United Nations Environment Program, Nairobi.

Much of this cropping is opportunistic and land is often in long fallow or used only occasionally. Crops planted include maize, sorghum, millet, upland rice, chickpeas (and other grain legumes), barley, and wheat. Collectively, the production from subsistence farmers represents a huge aggregate, but production statistics are not too reliable.

Crop production in semiarid and subhumid drylands of the world is dominated by three crops: wheat, grain sorghum, and millet. Maize, cotton, barley, pea, and bean are important crops in some regions. If we add the crops grown by subsistence farmers, the list grows longer. Climate and soil differences account for some of the regional variation in choice of crops and methods of crop management. There are also variations in the degree to which livestock are integrated with cropping.

The major producers of grain for international trade are the United States, Canada, France, Australia, and Argentina (Table II). Wheat dominates world grain trade, but grain sorghum is quite important too. Cotton is the other major dryland crop in international trade. It is significant that the New World lands of the United States, Canada, Australia, and Argentina should be the major exporters of commodities derived from dryland farming. Apart from the obvious benefits of soil and climate, which

**TABLE II**
Major Net Cereal[a] Importers and Exporters[b]

| Importers | | Exporters | |
|---|---|---|---|
| Country | Million metric tons | Country | Million metric tons |
| Soviet Union | 29 | United States | 83 |
| Japan | 27 | Canada | 28 |
| China | 16 | France | 26 |
| Egypt | 9 | Australia | 18 |
| Korea | 9 | Argentina | 9 |
| Saudi Arabia | 8 | Thailand | 6 |
| Iran | 6 | United Kingdom | 4 |
| Italy | 5 | South Africa | 2 |
| Mexico | 5 | Denmark | 1 |
| Iraq | 4 | New Zealand | 0.2 |

[a] Includes rice.
[b] From FAO (1987). "Trade Yearbook 1987," Vol. 41, pp. 114–116. FAO, Rome.

conferred considerable advantage, the remoteness and high transport costs were a major barrier to overcome. Efficiency in production and large-scale mechanization (Fig. 2) led to the lowering of production costs and made for price competitiveness. Significant too is the fact that the crops now grown so successfully in those New World regions are nonnative. No native crop plant (with the exception of maize) has been of significance in drylands of the New World. The success of agronomists in devising crop management strategies and of plant breeders in developing new and better varieties are testimony to the ingenuity of agricultural research and technology. Diseases, pests, and soil problems (including nutrient deficiencies) had to be overcome.

## III. DRYLAND FARMING SYSTEMS

Dryland farming involves both livestock (principally sheep, cattle, and goats) and crops. Farming systems evolve under a complex set of physical and cultural controls that ensure that successful systems survive while less suitable solutions become extinct. Climate and soil resources determine the range of possible crops, their productivity, and thus the possible land uses. But the particular systems that develop; the relative emphasis given to food, animal, or industrial products; and the degree of environmental control achieved are strongly determined by social factors. The reverse, that society is shaped by its agriculture, is also true. The course of urbanization and industrialization has been strongly controlled by the agricultural base that supported it. The development of a particular farming system is a response to a very wide range of factors—environmental (including climate), technological, economic, political and sociological opportunities and constraints, as well as personal and family goals.

Available soil moisture is normally an important limiting factor in choice of crop in any dryland

**FIGURE 2** Large-scale farming in dry regions depends on a high degree of mechanization.

farming system and the degree of moisture deficiency is a means of categorizing regions and the farming systems they support.

The principal groups of crops grown under dryland farming are grain crops, industrial crops (for fiber, oil, and sugar), and forages. The grain cereals belong to two main groups: the temperate cereals (called "small grains" in the United States), namely, wheat, barley, rye, and oats, and the predominantly tropical cereals, such as maize, sorghum, millets, and rice. The grain group that predominates in a particular region helps to define the farming system used.

On a world scale, classification of farming systems has been based not only on the climate but also on:

- purpose (types of outputs);
- structure (types of enterprises, rotations, equipment, farm size, land tenure, etc.);
- types and intensity of inputs (labor, machinery, chemicals, management, etc.); and
- arrangement of the components in space and time.

There are three major dryland farming systems:

- Mediterranean agriculture;
- mixed farming of western Europe and North America; and
- large-scale grain production.

In addition, there are specialized systems such as ley farming systems and crop/livestock systems.

### A. Mediterranean Agriculture

The Mediterranean-type climate is primarily characterized by a concentration of precipitation in the cooler half of the year (November to April in the Northern Hemisphere and May to October in the Southern Hemisphere) and drought in the warmer half. It is characteristic of the Mediterranean Basin, southern Australia, the Southwest Coast of the United States, the Cape region of South Africa, and central Chile. (Fig. 3).

A figure of 65% of the annual precipitation in winter has been used to define a boundary of this climatic type. Regions like California and central Chile can receive up to 80% of the annual precipitation in winter. Monthly rainfall for a range of stations with a Mediterranean-type climate is shown in Table III. Mediterranean-type climates are flanked by both wetter and drier regions in all areas where they occur.

The primary feature of agricultural land use in the Mediterranean Basin is the major importance of cereal production. Overall, about half of the arable land is under cereals, about 40% is under fallow, and the remainder is devoted to fruits, vegetables, and other crops, particularly the pulses (grain legumes). Wheat and barley are the predominant winter cereals in dryland farming areas in the Mediterranean Basin, especially in the WANA region (West Asia and North Africa). The main summer cereals are sorghum and maize. Other crops include lentils, chickpeas, and faba beans.

Agriculturally, wheat and barley, the predominant winter cereals, are quite different crops, in their growth requirements, in how they are usually managed, and in their utilization. Wheat is preferred and is grown whenever conditions permit, solely as a food grain. Barley has possibly the widest range of climatic adaptation worldwide of any cereal, but is valued in the WANA region mainly for its versatility as a feed source. It tends to be relegated to situations where wheat would often fail, for example, to dry areas and shallow or salty soils.

A crop's relationship with rainfall is particularly important. Wheat is more productive than barley in the wetter areas. The crossover point lies between 300 and 400 mm per annum depending on rainfall distribution, temperature regime, and cultivar planted. A full understanding of barley's superiority to wheat in harsh environments remains elusive, but important features of barley varieties adapted to dry areas include: rapid early growth during winter, which maximizes water-use efficiency; earliness, which promotes drought avoidance; and a modified carbon economy by which substantial amounts of soluble carbohydrate laid down before flowering may be translocated to the grain.

***FIGURE 3*** The world distribution of regions with Mediterranean-type climates.

These differences between crops and crop preferences are reflected in a predominance of wheat in the farming systems in the wetter areas and barley in the drier areas. Barley is of course grown in wetter areas too—on shallow soils, on weedier fields, or perhaps because the market price is attractive—and some wheat is grown in drier areas for home consumption. Nonetheless, a broad distinction between "wheat-based" systems in wetter areas and "barley-based" systems in drier areas remains valid.

Where mean annual rainfall is less than 250 mm, there are few suitable crop species and not even barley gives reliable yields. This is why small ruminant livestock become so important. Grazing animals are found in all dryland farming systems in the WANA region and generally their contribution to farm income increases as rainfall decreases. Small ruminant production is especially suited to the Mediterranean-type climate because livestock show patterns of reproduction periodicity that confine lambing/kidding activity to the spring months when availability of forages and survival of newborns are likely to be high.

In the traditional mode, Mediterranean livestock is complementary to arable agriculture. Sheep and goat production systems in the cereal zones are based on extensive grazing of the available resources: stubble in summer, re-shooting cereals in autumn (fall), clipping of cereals in winter, and rangeland and/or sown pastures in spring. In short, the livestock consume the by-products of the arable crops and graze land that is unsuitable for cultivation. Supplementation is supplied if pasture is not adequate to meet livestock needs, nevertheless it is more common to let the animals run down their body reserves.

Throughout the Mediterranean Basin, however, modes of livestock production are evolving away from what were predominantly nomadic, transhuman, and subsistence systems toward more commercial enterprises.

Farming systems are not static but are evolving in different ways according to a variety of pressures, particularly related to human population increase, land shortage, increased mechanization, changing market forces, and a variety of social and political forces. In the higher rainfall areas, wheat-based agricultural systems will predominate in the future. However, there will be increasing diversification of cropping systems and rotations in response to adoption of technical innovations and to changing

**TABLE III**

Examples of Monthly Rainfall Data in Countries with Mediterranean-Type Climates

| | | Monthly Rainfall (mm) | | | | | | | | | | | | |
|---|---|---|---|---|---|---|---|---|---|---|---|---|---|---|
| | | Northern Hemisphere | | | | | | | | | | | | |
| Place | Country | Jan. | Feb. | Mar. | Apr. | May | June | July | Aug. | Sep. | Oct. | Nov. | Dec. | Annual total |
| Marrakesh | Morocco | 28 | 29 | 32 | 31 | 17 | 7 | 2 | 3 | 10 | 21 | 28 | 33 | 241 |
| Oran | Algeria | 70 | 54 | 35 | 33 | 19 | 7 | 1 | 3 | 16 | 43 | 46 | 67 | 394 |
| Sfax | Tunisia | 18 | 18 | 25 | 21 | 12 | 5 | 1 | 5 | 26 | 38 | 26 | 15 | 210 |
| Tripoli | Libya | 62 | 38 | 19 | 14 | 3 | 1 | 1 | 1 | 10 | 32 | 41 | 65 | 287 |
| Aleppo | Syria | 68 | 54 | 42 | 36 | 19 | 3 | 1 | 1 | 1 | 19 | 25 | 72 | 341 |
| Amman (Marqa) | Jordan | 64 | 67 | 37 | 15 | 4 | tr | tr | tr | tr | 5 | 30 | 48 | 272 |
| Mosul | Iraq | 67 | 63 | 69 | 51 | 25 | 1 | tr | tr | tr | 10 | 36 | 65 | 388 |
| Shiraz | Iran | 144 | 50 | 4 | 26 | 1 | 0 | 0 | 0 | 0 | 1 | 46 | 24 | 296 |
| San Diego | California | 51 | 55 | 40 | 20 | 4 | 1 | tr | 2 | 4 | 12 | 23 | 52 | 264 |
| | | Southern Hemisphere | | | | | | | | | | | | |
| | | July | Aug. | Sep. | Oct. | Nov. | Dec. | Jan. | Feb. | Mar. | Apr. | May | June | Annual total |
| Merredin | Western Australia | 55 | 43 | 23 | 19 | 15 | 12 | 8 | 13 | 17 | 22 | 43 | 50 | 320 |
| Kadina | South Australia | 49 | 46 | 37 | 33 | 22 | 18 | 15 | 20 | 20 | 36 | 49 | 51 | 396 |
| Valparaiso | Chile | 72 | 50 | 19 | 4 | 0 | 4 | 2 | 0 | 3 | 18 | 82 | 96 | 349 |

economic conditions. In the lower rainfall wheat-based systems and the barley-based systems, animal production is assuming greater importance. Barley is replacing both wheat and fallow to satisfy the increased need for animal feeds. Similarly, sown fodder and forage legumes are now being used more to fill the feed gap.

An important feature of the farms in a region, whether wheat or barley based, is that cropping and livestock are not generally integrated. Cropping land is predominantly owned and/or worked by farmers who own very few livestock and livestock owners do not usually crop the land.

The degree of industrialization accounts for differences between the farming systems of the Mediterranean-climate countries within the Mediterranean Basin. It also accounts for differences between countries in other parts of the world, although influences of climate, geographical location, and history of settlement are also evident. Broadly speaking, the higher the national income per capita, the more intensive and specialized the farming patterns; thus the regions with Mediterranean-type climates may be ranked in an ascending order from North Africa to western Asia, Chile, the northwestern Mediterranean Basin, southern Australia, the southwestern Cape region, and California.

In general, cereal production is important in all four regions outside the Mediterranean Basin, (Fig. 3) where the same range of crops are grown. However, the mix of crops and animals is different. In Chile, cereals occupy three-fourths of the cropping area with wheat being four-fifths of the cereal area. Some fallowing is practiced and yields are generally low and variable as fertilizer input is low and little machinery is used even on larger holdings. In southern Australia and in the Cape region, wheat and barley together occupy about one-third of the cropped area. Although yields are low and variable, primarily due to rainfall deficiency, farms are large, highly mechanized, and use little farm labor. Cereal production in California, predominantly barley-grown dryland, occupies about one-quarter of the arable land. Farm operations are highly mechanized with minimal labor input.

Livestock production is generally as important in the New World Mediterranean-climate countries as in the Mediterranean Basin. However, the role of domestic livestock in the development of agricultural systems in the New World and their status are different. Mostly, this is due to the different pattern of settlement, the geographic location, and the level of economic development. There is no general integration of cropping and livestock in any region, other than southern Australia, where the formerly two separate enterprises became integrated through the use of pasture legumes in a ley farming system (see Section III, D).

### B. Mixed Farming of Western Europe and North America

Mixed farming in western Europe and North America occurs where there is a temperate climate and a predominance of summer rain. It is characterized by:

- a high degree of commercialization (little subsistence farming);
- dependence on urbanized markets and industrial input;
- farms are mainly owned and run by families that are faced with a declining labor force;
- farms produce both crops and livestock and the two enterprises are integrated; sown grassland occupies a substantial proportion of the arable land and is carefully managed; and
- a range of crops is grown in rotation.

The higher rainfall leads to smaller farm size and a greater use of crops for livestock feeding. There is commonly a high level of government subsidy, especially in western Europe, and this is related, in part, to small farm size. In North America, the farm size tends to be bigger and there is a high level of mechanization, because of the initial abundant supply of new (fertile) land and the relative scarcity of labor. At present, North America generally operates a land-extensive dryland agriculture with inputs and yields generally less than optimal relative to climate and soil potentials. In part this reflects the degree to which moisture deficits commonly or occasionally limit yields. The principal reason is economics: the low returns on agricultural produce fail to justify the costs of higher levels of input.

Although animals (e.g., beef cattle and pigs) are important in this system, they are fed on maize and other cereal grains rather than on pasture. Nitrogen is supplied by fertilizers rather than by legumes. A healthy national economy has been able to support this degree of intensification.

The principal crops are wheat and other small grains, maize, sorghum, and oilseeds (safflower, sunflower, oilseed rape, and mustard). Flax and cotton are also grown over large areas. Alfalfa and sweet clover are important forage crops. Strip cropping (Fig. 4) of compatible crops is widely practiced to reduce soil erosion and spread risks.

The production of cereals, especially wheat and grain sorghum, has always been the main occupation of farmers in the semiarid and subhumid regions. Under the erratic climatic conditions of the dryland farming regions, the choice of crops that can be grown is limited and none has yet challenged the supremacy of wheat as the most remunerative crop of these regions. Wheat, because of its wide adaptation and high favor with consumers, is an important grain in all regions. Rye is important only in Europe and the region covered by the former Soviet Union, where the combination of infertile soils and cool climate restricts the choice of crops. Oat production is confined to Europe, North America, and Oceania; the remaining regions are too hot and dry for it.

Summer fallow is widely practiced where warm-season crops are grown; this term is used to describe a system in which a crop is grown once in two seasons. Fallowing is a key feature in the development of flexible cropping systems, the so-called "opportunity rotations." The farmer assesses the available soil water and expected growing season precipitation at seeding time before deciding to either crop or summer fallow. If the farmer decides to crop, a crop is selected that should yield well on the available water supply. There is a considerable body of knowledge on the optimum crop sequences (expressed in terms of nutrient depletion,

***FIGURE 4*** Strip cropping can reduce erosion and break the cycle of pests and diseases.

disease, and pest control options) and other considerations such as crop quality and nutritive value.

Specialist grain crop/livestock combinations such as the corn/hogs (maize and pigs) system (see Section III, E) are a variant on this farming system, although the latter show more evidence of vertical integration with cropping, feed processing, and feedlotting as part of one enterprise.

### C. Large-Scale Cereal Grain Production

The large-scale cereal grain farming system has some features in common with the mixed farming systems, especially regarding choice of crop species. Large-scale cereal production has been conducted in parts of the United States, Canada, Argentina, the former Soviet Union, and Australia (Fig. 5). It is commonly a feature of the exploitation of large areas of land newly brought under cultivation. Because of initial high fertility, inputs of fertilizer are low for many years. The main aim is to maximize output per person, as labor is expensive and land is relatively cheap. The farms are highly mechanized.

Rainfall in such areas is generally variable, making average yields moderate to low (Table IV). Most of the cereal grains in the world are grown commercially in semiarid and subhumid regions. Here the length of the growing season varies from 90 to 179 days (where the growing season is defined as the number of days when both temperature and moisture permit crop growth). Days with mean temperatures of at least 5°C and with soil moisture resulting from rainfall at least equivalent to half the potential evapotranspiration are considered favorable for growth. Large areas are also sown to cereals in very low rainfall areas by accumulating 2 years of rainfall for a single crop (e.g., alternate wheat and fallow systems). Table V shows the effects of seasonal variation in precipitation on grain yields of three important dryland crops.

In North America (principally the United States and Canada) the dryland farming areas are clustered within six geographical regions lying in the 17 western states of the United States and the three prairie provinces of Canada (Fig. 6). These two countries share the Great Plains, which extend for over 2 million km$^2$ from Texas to Alberta. They were known a century ago as the "Great American Desert." Rainfall varies from 300 mm in the rainshadow of the Rocky Mountains to 600 mm, with 100–120 frost-free days. The U.S. part of the Great Plains has three subdivisions (southern, central, and

***FIGURE 5*** Many farms are highly mechanized to maximize output per person.

**TABLE IV**
Average Production Level and Yield of Cereals in the Major Dryland Cropping Regions of the World[a]

| Region | Ave. production of cereals (thousands of metric tons) | Ave. yield of cereals (kg/ha) |
|---|---|---|
| Africa | 84,225 | 1148 |
| North and Central America | 343,453 | 3600 |
| South America | 80,232 | 2034 |
| Asia | 782,291 | 2572 |
| Europe | 293,126 | 4280 |
| Soviet Union | 196,906 | 1806 |
| Oceania | 23,095 | 1557 |
| World | 1,803,327 | 2561 |

[a] From FAO of the United Nations (1988 data).

northern) and the Canadian part is called the Canadian Prairies.

Dryland farming in Mexico is practiced on about 11.3 million ha but Mexico is still an importer of cereal grains. The semiarid highlands (>1800 m elevation) are the principal dryland farming regions, encompassing three extensive plateaus characterized by sediment-filled basins surrounded by high mountains or lower hills. Sorghum and maize are the principal crops. Maize is the most important crop, occupying 2.9 million ha and giving an average yield of 1.05 to 2.2 tons/ha. But frost risk and dry seasons lead many farmers to switch to sorghum. In the north, the climate is too dry for commercial cropping without irrigation.

Although wheat is grown in many parts of North America, the biggest proportion is grown in regions where rainfall is insufficient for heavy maize or row-crop yields. The southern wheat area of the United States (the hard wheat area) of central and western Kansas, southwestern Nebraska, eastern Colorado, and northwestern Texas and Oklahoma surpasses the others in terms of area sown and total production.

**TABLE V**
Average Percentage Change in Yield for Various Dryland Crops for Normal, Dry, and Wet Years with a 10% Increase in Precipitation[a]

| Crop | Dry years[b] | Normal years[b] | Wet years[b] |
|---|---|---|---|
| Maize[c] | 2.8 | 2.2 | 1.1 |
| Wheat[c] | 2.3 | 2.3 | 2.5 |
| Soybeans[c] | 3.3 | 3.0 | 3.6 |

[a] The National Research Council (1976). "Climate and Food." Washington, D.C.: NRC.
[b] Each category consists of one-third of total sample.
[c] Main production areas.

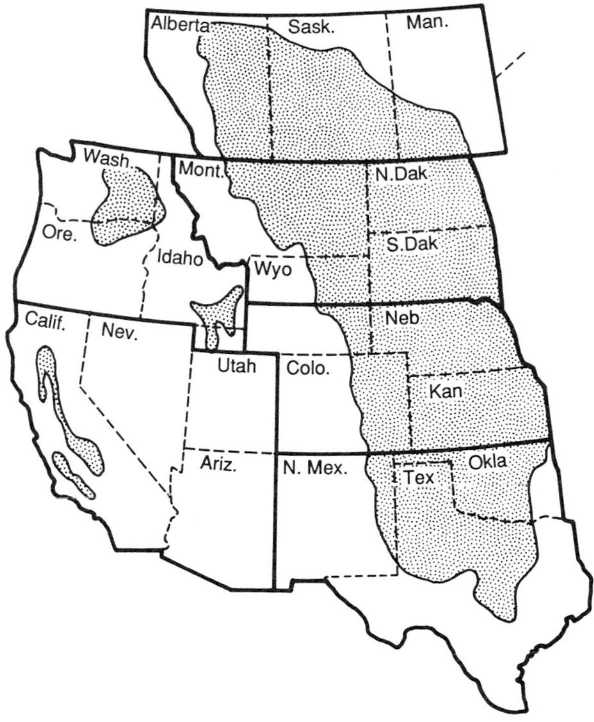


**FIGURE 6** Major dryland farming areas of the United States and Canada.

The amount of precipitation required to successfully produce a crop will depend on whether there is winter or summer dominance regarding rainfall. In the winter-rainfall areas of southern Australia, California, and Argentina, nearly all the small grains are grown in regions that have an average rainfall of between 250 and 375 mm during the active growth of the crop. Although successful crops can be grown in areas where rainfall is as low as 200 mm provided that it is received between late autumn and spring. In the Mediterranean Basin, 250 mm is considered as the minimum for producing a crop. (Fig. 7).

In the dryland farming areas with summer rainfall, the amount and distribution of rainfall that occurs during the growing season are usually en-

***FIGURE 7*** Dryland farming near the equator in Kenya's Rift Valley, where a tropical tree (*Acacia tortillis*) is the overstory for a temperate crop (wheat).

tirely inadequate; hence reserve supplies of stored soil moisture prior to sowing are essential for satisfactory grain production. Generally, if the average annual precipitation is 375–500 mm, there is likely to be only 150–200 mm of it falling during the average 85- to 90-day growing season. A minimum of 125 mm of available water is regarded as essential before any grain can be produced. For each additional 25 mm, the yields of winter wheat will increase by about 230–260 kg/ha. There seems to be no real difference between equal amounts of stored moisture or precipitation during the growth period of the crop.

In regions with nonseasonal rainfall and an average annual precipitation of 317 mm, with extremes of 150 mm and 475 mm, it was found that high yields of dryland wheat were directly related to the crop's emergence and early establishment, which in turn depended on favorable soil conditions at the time of sowing.

Winter temperatures determine to a large extent the time of sowing and choice of both crop type and variety. In regions with mild, moist winters and dry summers, wheat is grown at the beginning of the rainy season, but only spring types can be sown, because the temperatures are not sufficiently low for floral induction of winter types. In regions with moderately cold, dry winters, winter types of wheat are sown in autumn. They germinate and tiller before the onset of the winter's cold, and the plants remain prostrate during the winter months, sometimes under a snow cover. With the advent of spring and long days, growth resumes, the plant produces ears, and the grain mature in summer. In regions where the winter is so severe that few plants survive, spring types are sown in spring and mature in late summer. Similar constraints influence choice of type and variety of barley, oats, and rye. The number of frost-free days is the greatest influence, after available soil moisture, on the choice of type and variety of maize, sorghum, and millet.

Climate has largely determined the relatively stable pattern of specialized crop regions that, despite rapid technological evolution, has emerged in North America in this century. This pattern, characterized by the corn and wheat belts of the midwestern regions, follows the lines of the original climate–vegetation–soils regions, with moisture and temperature regimes and length of the growing season as key factors. A well-developed transportation system has enhanced the regional emphasis on the most economically rewarding crops. The dominant crop systems, however, are not necessarily the best adapted, nor the ones that exhibit the least year-to-year variation due to weather.

The tendency toward monoculture over extensive areas, which goes with dominance, is moderated by other factors. When the economic advantages of a particular crop are not great, the advantages of diversification receive more emphasis. For example, the wheat production centers (Palouse area, Kansas winter wheat area, and the Red River durum–spring wheat area) were well developed by 1910. Originally, there was emphasis on dryland crop–fallow techniques, and it was not until after the disastrous droughts of the 1930s that the emphasis on wheat was reduced in marginal-rainfall areas. Diversification of some of these plowlands into livestock production brought a measure of stability to the whole region.

Diversification to include livestock production usually allows more on-farm management alternatives than are possible with cash grain systems. Weather-damaged crops are more easily used and a wider array of crops with less weather vulnerability can be employed. However, livestock are generally not integrated into the cropping system, partly because of lack of farmer interest in livestock and partly because of the lack of suitable pasture legumes.

The emphasis in dryland farming regions on environmental stress (chilling, frost, heat, drought, soil salinity, and poor soil drainage) points to the extent to which many crops are cultivated in marginal environments. In the margins of the major cropping areas where one region gives way to another, several kinds of major crops and cropping systems coexist.

Because of low and variable precipitation and high machinery costs, the farmer is particularly vulnerable to fluctuating grain prices and to high costs of transport. Choice of species and cultivars that are well suited to and relatively safe in the local climate is perhaps the best form of insurance. Other management options include the timing and type of operations (fall versus spring plowing; stubble mulching versus clean tillage) and even the amount of labor and equipment available for performing operations in a timely manner.

The size of the operation is also important. Family-sized farms can operate with individually tailored decisions, whereas large-scale units usually must depend more on blanket assessments. The manager with a relatively small area has less inertia built into the production system and can respond more rapidly to expected short-term weather changes or to market fluctuations than a manager of a large unit who is unlikely to be able to engage or disengage resources in the short run.

The large-scale grain production system is characterized by a lack of flexibility, risk of soil erosion, and buildup of disease organisms and is subject to periodic low commodity prices. The long-term sustainability of large-scale grain production systems has been questioned. The chief concerns regard the cost and continued availability of cheap inputs (fuel, fertilizer, and pesticides) and the adverse environmental consequences of large-scale monocultures.

### D. Ley Farming Systems

The ley farming system involves integrated livestock and cropping, which has been used in rainfed agriculture in a number of geographic regions, for example, in the United Kingdom and New Zealand. The Australian dryland farming system is a unique subset of the Mediterranean-climate dryland farming system.

Integration of crops and livestock has been conducted most effectively in the cereal–livestock zone of southern Australia, but some progress has been made in transferring the technology to the WANA region of the Mediterranean Basin, whence many of the annual legumes came. Yet this has not been without its problems. In most areas the pasture phase is an important component of the system and, if managed properly, promotes high levels of both crop and livestock production (Fig. 8).

The rotation of cereals and annual pasture legumes became known as the southern Australian ley farming system, and the complementarity of cereal and sheep is an important feature. The main integrating factor is the legume-based pasture,

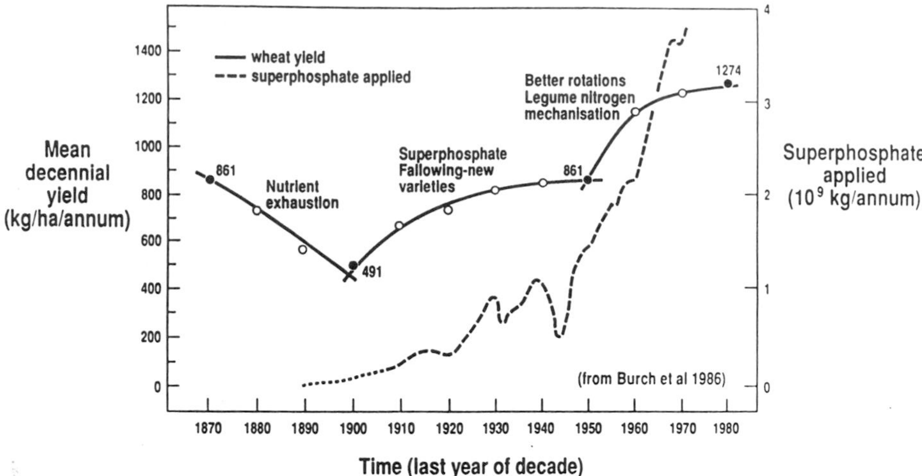


***FIGURE 8*** Ley farming, involving pasture legumes, crops in rotation, and grazing sheep, transformed Australia's dryland farming from about the 1950s onward.

which provides high-quality feed for sheep and extra soil nitrogen for the following crop. Another integrating feature is the grazing of the crop and pasture residues—the sheep obtain feed in the dry half of the year and reduce the volume of residues and also weed growth to assist in land preparation. The annual legumes (chiefly *Trifolium subterraneum* and *Medicago* species) are self-regenerating from seed set each year and deposited in the soil. A high proportion of the seeds (especially those of annual *Medicago* species) remain impermeable and viable for a varying number of years before germinating. This self-generation readily occurs after an intervening cereal crop.

A program of management procedures is followed to ensure the continued presence of the pasture species in the rotation, for example, shallow tillage to avoid deep burial of the legume seed, herbicides for weed control only in the cereal phase, phosphatic fertilizer application to both crop and pasture, and management to avoid both overgrazing at critical development stages (flowering and early seed-set) and consumption of the pods on the soil surface. Apart from forage supplied from pastures and crop residues, additional or supplementary feeding may be required to meet livestock requirements for survival, growth, and reproduction. The ley farming system is further characterized by relatively large farm size and the use of large machinery for timeliness of operations. Overall, the aim is to exploit the initial high fertility so characteristic of many of the newly cultivated soils.

Alternatives to cereal/pasture rotations have been sought and cereal/grain legume rotations have been used successfully. The choice between cereal/grain legume and cereal/pasture legume rotations depends on tradition and the relative prices for crops and livestock products, as well as rainfall and the availability of suitable cultivars that are adapted to the local area. Good results have been achieved on alkaline soils with several species of annual *Medicago*. On neutral or acid soils it is better to grow cultivars of *Trifolium subterraneum* or *T. balansa*. Hard seededness in these species generally breaks down quickly and consequently they do not regenerate as well after a period cropping.

The ley farming system is not without its problems. The incidence of serious plant diseases and insect pests of annual pasture legumes has increased. These and other problems of environmental adaptation have stimulated a pasture/legume breeding program as well as a stepped-up campaign of selection of suitable strains of other annual legumes such as *Trifolium balansa* and *Serradella*.

The integration of crops and livestock in a changing economic environment requires an understanding of the interrelationships in the system and how they can best be utilized. To maximize returns and to operate the system within sustainable limits, farmers must:

- devise livestock management strategies to achieve this objective, including matching of feed (forage and fodder) supply to livestock requirements, timing of operations, grazing management plans, and provision for seasonal differences and recurrent drought;
- efficiently utilize on-farm feed resources and supplements if required; and
- adopt appropriate cropping practices.

The main problem in the livestock/crop system arises when a cropping program is so intensive as to prevent improvement in soil structure and fertility via a pasture phase from taking place.

### E. Feed Grains and Livestock System

The feed grain/livestock system is best exemplified by the Corn Belt of the United States. It involves an extraordinarily close relationship between the growing of a crop and the utilization of the grain for livestock feeding—principally pigs, although poultry and beef cattle are also fed.

Grain/livestock farms in North America, Australia, and southern Africa are usually family-operated commercial enterprises, averaging less than 100 ha in area. Many farming enterprises display vertical integration, with cropping, feed processing, and feedlotting as part of one operation. Maize is the dominant crop of the so-called U.S. Corn Belt, where level, warm, and fertile land is typical. Much of it is farmed without livestock as the preoccupation is with grain crops. On the more rolling areas, where there is considerable permanent pasture, livestock production is combined with grain farming. Pigs, which efficiently convert concentrated feeds into meat, get most of the maize fed to livestock.

Although maize had its origin in a semiarid area, it is not a reliable crop for growing under dryland conditions with limited or erratic rainfall; under these conditions it cannot compete with sorghum or millets. Increasingly, the Corn Belt farmers are using supplementary irrigation to ensure a good crop. From this point of view the Corn Belt probably falls outside the scope of dryland farming, but similar grain/livestock systems would qualify in other geographic regions, such as southern Africa and Australia.

## IV. INPUTS, OUTPUTS, AND SUSTAINABILITY

Modern dryland farming has become more mechanized and has greater regional and on-farm specialization (often characterized by monocultures). A conflict has arisen between maintaining productivity and achieving sustainability.

Agricultural production depends on a variety of factors and begins with a natural resource base: the water, soil, and solar energy necessary for production (Fig. 9). Primitive agriculture used only these natural resources to produce food, but modern dryland agriculture increasingly employs management strategies and tactics developed through agricultural research to increase output and to ameliorate threats of food production. Some of these options involve an unlimited supply of cheap energy.

By increasing its reliance on mechanization and technology, developing a heavy dependence on fossil fuels, exploiting the productivity of the soil, and enhancing crop yields with fertilizers and pesticides, dryland agriculture has evolved to a state where short-term benefits have been gained at the expense of long-term costs. Such actions ignore the interdependence between agriculture and the environment. Most people accept that inevitably there is some trade-off between environmental quality and otherwise essential and beneficial agricultural activity. However, questions are being raised about the logic of putting sustained pressure on the environment to produce food and fiber from the marginal lands—the dryland farming regions of the world.

An agricultural system that relies heavily on energy-intensive, purchased inputs may be a success story in terms of traditional measurements of output and productivity. But times and economics are changing. The present mix of basic resources of labor and energy in North American agriculture, in particular, has developed because of the high cost and low availability of labor relative to energy.

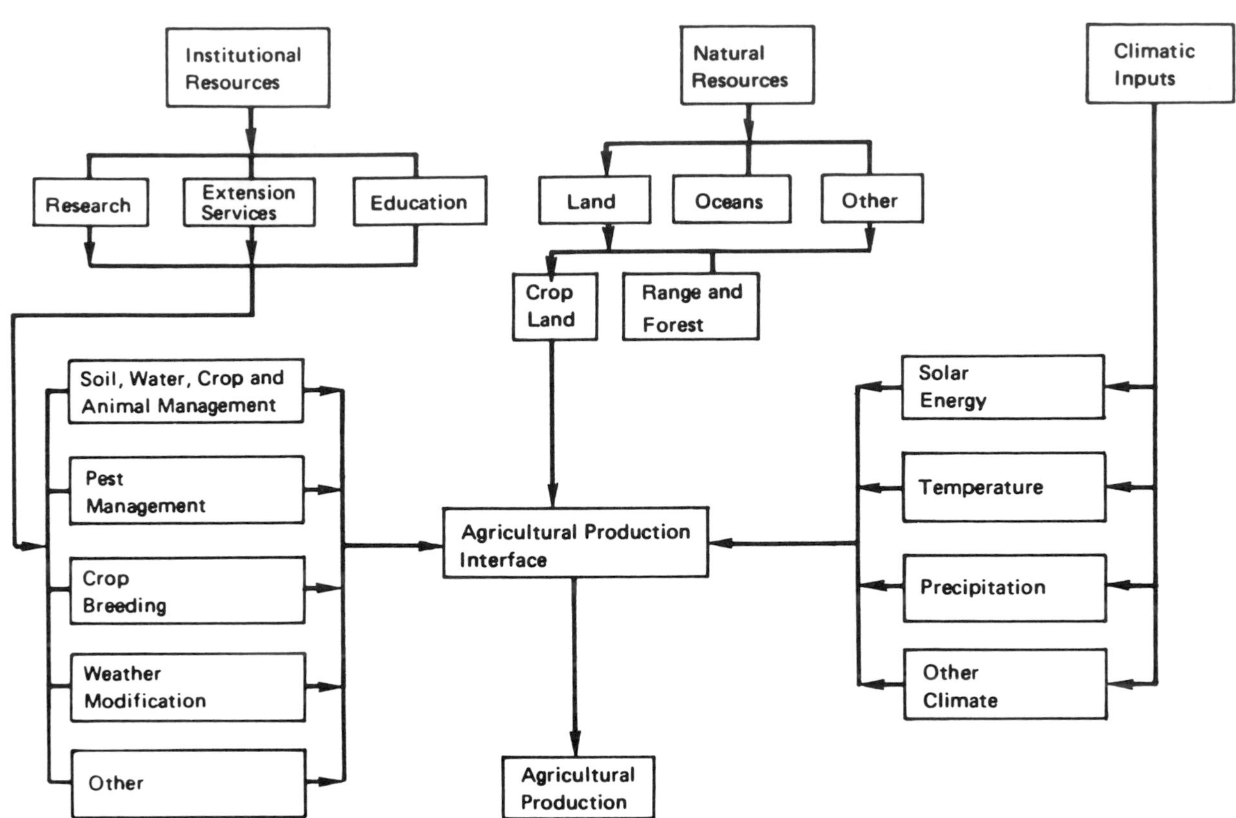


***FIGURE 9*** Flow of resources and information to support the food production system under the constraints of climatic fluctuation that are typical of dryland farming systems.

The impact on agriculture of declining availability of and increasing cost of energy is thus great.

Government programs that provide incentives for high-input farming were devised in a cheap-energy era and remain largely intact. Most require serious rethinking. Changes in dryland agriculture seem inevitable in the long run, particularly because petroleum (a depletable resource) is the major fuel used in dryland agriculture and because the pattern of world trade in food grains and other commodities is being drastically altered.

In industrialized nations, energy inputs do not end at the farm sector. Far more energy is now used in postfarm aspects such as transporting, processing, packaging, and cooking basic foods. About eight times more energy, in the form of fossil fuels, is required to process, distribute, and cook the food consumed than to produce it. A mere 29% of total farm inputs is involved in getting produce to the farm gate. In the course of getting this material to the dining table almost 50% of its total energy value is lost in various ways. Thus for each joule (J) of prepared food that makes it to the dining table, over 5 J of primary fuel is used to put it there, 11% of it at the farm gate, 38% takes it from the gate to the retail store, and 51% takes it from the store to the table. At least 40% of the total energy consumed in getting the food to the dining table is used in machines powered by internal combustion engines and a large part is consumed in cooking it. Clearly, improvements are needed at both of these stages. Any increase in the efficiency of such technologies could have a large effect on the overall energy costs. But there is scope for on-farm efficiency to be improved as well.

Despite sizable expenditures of fossil fuel, dryland agriculture is overall a mechanism for obtaining a net harvest of solar energy for human purposes. The fuel value of the primary plant material harvested by people and their livestock is about 1.5 times the energy value of primary fuel burned for all purposes (tillage, sowing, harvesting, etc.).

The major requirements for crop production on dryland farms are energy, labor, and land. Within limits, the three inputs are interchangeable. Industrialized agriculture relies heavily on energy inputs to substitute for labor and land. The question is, how long can this continue?

## V. FUTURE TRENDS

In a situation of generally suboptimal rainfall, future changes in dryland farming systems may lean less toward intensification of material inputs and more toward greater biological efficiency, scientific innovation, and technical managerial skill, together with conservation of soil and other resources.

Many of the new techniques for dryland farming call for bigger farms and less labor. However, the demand for food is price inelastic and income inelastic, that is, it does not respond very much to changes in price and income. As growth occurs and people's incomes rise, they spend a smaller proportion of their incomes on food and a larger proportion on secondary and tertiary industry. Thus the rapidly increasing supplies of agricultural commodities encounter a demand that is increasing only slowly. This will inevitably lead to the demise of many farm enterprises as farmers move out of dryland agriculture.

Dryland agriculture has been notoriously unstable with regard to prices received, output levels, and incomes. This instability has worsened over the past several decades because of the volatility of world trade and farmers' inability to influence prices. Fluctuations are cyclic, but the long-term trend is downward. This downward trend in rural commodity prices is a worldwide trend and is called by some commentators the "farm problem" as it reflects a fall in farm incomes relative to those enjoyed by other sectors of the economy. Productivity, growth, and viability of the farming sector have been achieved through considerable structural adjustment (Fig. 10).

There are now many fewer farmers and those that remain have:

- increased their productive base;
- substituted capital for labor;
- increased the scale or intensity of their operation;
- shown marked increases in output, especially crops; and
- developed new land areas, for example, by clearing native vegetation and by draining wetlands.

In some cases, farmers have increased their long-term viability by pressuring the authorities to develop complex infrastructures for support, for example, marketing boards, government-funded research, and extension services, and have led the way in seeking microeconomic reforms that affect transport, marketing, and trade agreements. There is a clear recognition that agricultural industries will need to adopt a marketing approach to business and shift away from the traditional production orientation. This change of emphasis will require producers to meet specific demands from consumers. In other words, if customers want assured quality (e.g., high-protein wheat), and continuity of supply, the farmer must deliver to retain market share. Achievement of these new objectives will require the development of new farming systems. Pressure for new systems will come from two main directions, namely, market pressures to produce what the consumer wants and pressure from society at large for agriculture to be environmentally responsible and sustainable.

These forces, as well as social, economic, and political pressures, may increase the need for dryland farmers to design better farming systems sooner rather than later. New systems, along with appropriate technology, would allow dryland farmers to have more price competitiveness and sustainability on their own farms.

Other pressures also mount. There is a worldwide shift toward alternative farming methods—organic, biodynamic, and variants of both. So far, they have made little impact on the broadscale dryland farm, but long-term field experiments are under way at the University of Adelaide and at the University of California to compare conventional farming with some alternative farming systems in integrated "farms" where crop, livestock,

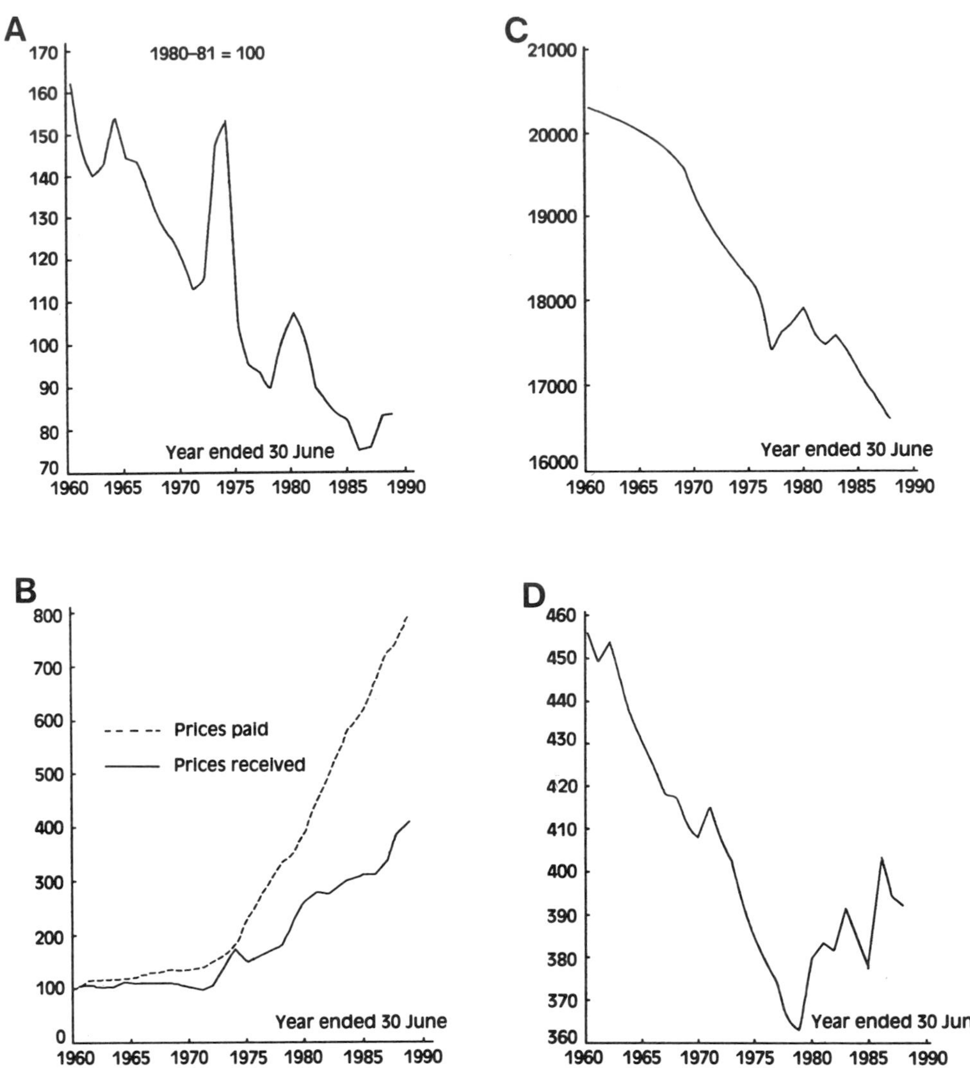


***FIGURE 10*** Structural adjustments in Australian dryland agriculture from 1960 to 1990 in response to (A) changing terms of trade and (B) prices received and paid. Principal changes are in (C) a reduction in the number of farms and (D) a decline in the rural workforce.

and pastures are the three principal components. Financial incentives (price premiums for organically grown grains and livestock) may hasten the acceptance of the alternative farming methods by the few, but for the many the imperative will be to produce more at a lower price to feed a hungry world.

## Glossary

**Agronomist** Specialist in crop and pasture production systems to meet human needs for food, fiber, and other materials.

**Biodynamic** Approach to agriculture characterized by the use of organic growth promoters in association with timing of planting and other operations.

**Crop rotation** Series of different crops planted in an annual sequence or alternated with a perennial crop or pasture phase to combat pests and improve soil fertility.

**Cultivar** Named variety selected within a plant species; derived from a cultivated variety.

**Evapotranspiration** Water loss from the soil by the combined effects of evaporation and transpiration through the plant.

**Fallowing** Practice of maintaining land free from plant growth to conserve moisture. This can be achieved by cultivation or by herbicides.

**Feed gap** Shortfall in the annual cycle of production of forage or fodder such that demand exceeds supply.

**Joule** Unit of energy in the metric system.
**Ley farming** Cropping systems that integrate livestock and include several years of pasture in the rotation.
**Monoculture** Cultivation of a single crop such as wheat or cotton to the exclusion of other uses of land.
**Organic agriculture** System of growing crops without the addition of artificial fertilizers or pesticides.
**Self-regenerating** Annual pastures that reappear after rain from seed reserves in the soil.
**Stubble mulching** Working of crop residues into the soil surface as a protection against the erosive forces of wind and water.
**Subsistence agriculture** Systems where crops and livestock are raised to provide for the immediate needs of the family.
**Sustainable agriculture** Set of goals or objectives for agricultural systems that integrates managing land with a healthy ecological balance, a sensitivity to the land's capabilities, and using technologies and practices that have minimal long-term impact while maintaining production.
**Technology transfer** Attempt to pass on to another group the technologies and practices that have proven successful elsewhere; it may involve both hardware (equipment) and software (know-how).
**Tillage** Cultivation of the soil, associated with the sowing of crops, and other treatments, including weed control and the handling of crop residues.

## Bibliography

Arnon, I. (1992). "Crop Production in Dry Regions. Vol. II. Systematic Treatment of the Principal Crops." New York: Barnes & Noble.

Dregne, H. E., and Willis, W. O., eds. (1983). "Dryland Agriculture," Agronomy No. 23. Madison, Wisc.: American Society of Agronomy, Crop Science Society of America and Soil Science Society of America.

Guessous, F., Kabbali, A., and Narjisse, H., eds. (1992). "Livestock in the Mediterranean Cereal Production Systems." Wageningen, The Netherlands: Pudoc Scientific.

Plucknett, D. L., and Smith, N. J. H. (1986). Sustaining agricultural yields. *BioScience* **36**(1), 40–45.

Reganold, J. P., Papendick, R. I., and Parr, J. F. (1990). Sustainable agriculture. *Sci. Amer.* **262**(6), 112–120.

Salman, A., ed. (1990). "Agriculture in the Middle East." New York: Paragon House.

Squires, V., and Tow, P., eds. (1991). "Dryland Farming: A Systems Approach." Melbourne: Sydney University Press.

Squires, V. (1991). "A Glossary of Terms in Dryland Agriculture." Roseworthy: University of Adelaide.

Unger, P. W., Sneed, T. V., Jordan, W. R., and Jensen, R., eds. (1988). "Challenges in Dryland Agriculture: A Global Perspective," Proceedings, International Conference on Dryland Farming, Amarillo/Bushland, Texas, August 15–19, 1988. Texas Agric. Exp. Station.

# Fire Ecology

**Norman L. Christensen**
*Duke University*



Naturally occurring fires affect virtually all terrestrial ecosystems. Climate and patterns of fuel accumulation make fire a regularly recurring phenomenon in many ecosystems. Fire regimes vary among landscapes and ecosystem types with regard to: frequency or return time, the average time between successive fires; intensity, the energy released during a fire; severity, the impact of the fire on the ecosystem; and predictability, the range of variation in the preceding features. These variations in fire regime are determined by climate and patterns of fuel accumulation. Millions of years of natural selection have produced a wide array of plant adaptations that include fire-resistant vegetative structures, fire-stimulated reproduction, seed release and germination, and perhaps even flammability. Patterns of postfire response vary among ecosystems as a consequence of species' life-history characteristics, previous site history, fire severity, landscape mosaic, and seasonality. In recognition of its importance to the preservation of biodiversity and health of many ecosystems, fire has become an integral tool in natural resource and ecosystem management.

## I. INTRODUCTION

Early in the twentieth century, it was argued that the vast majority of the world's ecosystems existed prior to human intervention as stable climax communities, perpetuated by small-scale disturbances such as individual tree deaths and replacements. For the most part, wildfire was thought to be a consequence of human activities and, therefore, not a "natural" ecosystem process. This worldview, along with perceived threats to life and property, led land managers to implement the strict fire suppression policies associated with the Smokey Bear campaign. [*See* BIOMASS BURNING.]

By midcentury, it was clear that fire suppression was resulting in unexpected and sometimes undesirable changes in many ecosystems. In the absence of fire, flammable woody debris was accumulating in some forests and shrublands, increasing both the likelihood and severity of fires when they occurred. For example, in forests of the mountains of western North America, fire suppression and exclusion by land managers resulted in diminished reproduction of many of the dominant tree species and invasion of shade-tolerant tree species. Such suppression had the effect of diminishing fire frequency, but increasing the intensity of fires when they did occur. Ecologists discovered that many species in fire-prone ecosystems were actually dependent on fire for their long-term survival.

The abundance of charcoal in sediments ranging in age from thousands to millions of years and the variety of organismal adaptations to fire clearly indicate that fire has played a significant role in ecosystems over evolutionary time scales. Furthermore, the patterning of ecosystems across landscapes as different as tropical savannas, the prairie steppe, and coastal wetlands has been shown to be a consequence of the long-term spatial distribution of fires.

Relatively few terrestrial ecosystems operate totally free of fire, however, the patterns of fire behavior, fire effects on the environment, and ecosystem response to burning vary enormously among ecosystems. We are just beginning to understand the basis for and consequences of these differences.

## II. FIRE REGIMES

The range of variation in fire behavior within a particular ecosystem constitutes its fire regime. The components of fire regime include intensity, severity, frequency, spatial extent, seasonality, and predictability. Fire intensity refers to energy released per unit area or per unit length along the fire's burning front. Severity, though often correlated with intensity, refers specifically to the ecological impact of the fire. Because of differences in ecosystem structure and organismal adaptations, fires of equal intensity may vary considerably in severity among ecosystems. Fire frequency for a particular ecosystem is typically expressed in terms of the average return time between fires. However, the number of fires occurring within a given area during a specified time, or regional frequency, is also an important fire regime feature. In general, fire return time is determined by the regional frequency of ignitions and the average spatial extent of fires.

In some circumstances, fire regimes are quite regular and predictable, but they are more often highly variable. The variance in or predictability of these characteristics may be especially important with respect to the character of fire effects on the environment and the nature of ecosystem recovery.

### A. Climatic Effects on Fire Regimes

In the absence of human intervention, lightning is by far the most common source of ignition for wildfires. Thus, one might expect that fire frequency would be highly correlated with the distribution of thunderstorms. In some landscapes, there is a rough correlation between the frequency of lightning strikes and regional fire frequency, however, fire return intervals for most ecosystems tend to be determined less by the frequency of ignition events and more by the nature of moisture content of fuels and by the landscape conditions that influence fire spread.

Figure 1 displays typical fire return intervals for different ecosystem types in western North America in relation to moisture conditions. At the very dry end of the moisture gradient, represented by deserts and desert grasslands, very low rates of production and fuel accumulation (perhaps coupled with limited ignition frequency) result in extended fire-free periods. Fires are most frequent under dry-mesic conditions that favor fuel accumulation. These conditions often favor comparatively high rates of biomass production and/or slow rates of

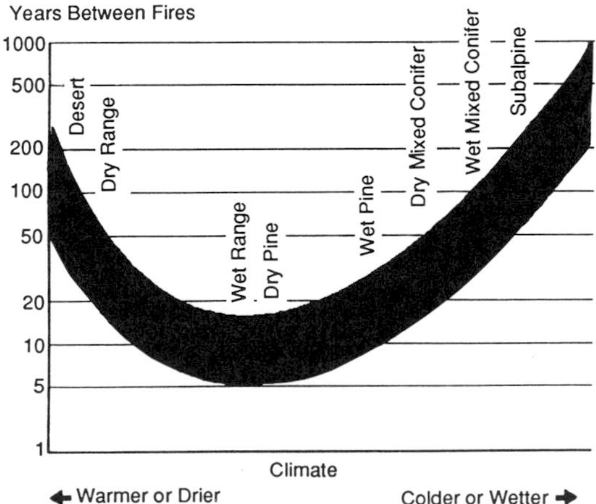


***FIGURE 1*** Fire return intervals in ecosystems of the western United States in relation to site moisture conditions. [After R. E. Martin (1982). *In* "Forest Succession and Stand Development in the Northwest" (J. E. Means, ed.), pp. 92–99. Forest Research Laboratory, Oregon State University, Corvallis.]

decomposition. The production of fuels tends to increase with increasing moisture, however, high fuel moisture content limits fire occurrence.

Climatic patterns have a major influence on the season of fire occurrence with a given region. For example, fires are most frequent and intense in Mediterranean-climate ecosystems late in the summer, when fuels are driest. In other regions, seasonal variation in fire behavior results from changes in both fuels and climate. In the savanna ecosystems of the United States, fires are most frequent in the fall and spring owing to a combination of common 3- to 4-week droughts and an accumulation of dry fuels during this period. Fires may also occur during the summer, however, the fuels tend to be much greener and less conducive to spread to high-intensity burning.

Long-term variation in climate has influenced historic patterns of fire behavior. Dendrochronological (tree ring) studies in the sequoia forests of California provide a measure of both climate change and fire frequency over the past two millennia. Fire return times were short (8–10 years) during the warm and comparatively dry conditions that prevailed during the period AD 800–1300. Fires were less frequent (every 16–24 years) during the so-called Little Ice Age, AD 1400–1800, when the climate in this region was cooler and moister. This same pattern of change in fire behavior has also been documented for the forests of the Great Lake States and southern Canada.

Shorter-term climatic variation accounts for much of the year-to-year variation in fire regimes. Extensive fires, such as the 1988 fire in Yellowstone National Park that burned 1.2 million acres of forest, are most common during periods of extreme drought. Such drought periods appear to occur in cycles. For example, fire frequency in the conifer forests of the southwestern United States is correlated with periods of drought associated with the so-called southern oscillation or "El-Niño."

### B. Fuels

Where net primary production is limited or decomposition of litter and woody debris is especially rapid, fire frequency may be limited by the rate of fuel accumulation. However, properties of the ecosystem biomass other than mere amount typically have a greater influence on fire regimes.

Plant materials vary considerably in their inorganic and organic chemical characteristics, and these variations influence flammability. In general, tissue minerals such as nitrogen, phosphorus, and calcium diminish flammability. Microbial decomposition is also slower in plant tissues with a low nutrient content, thus resulting in fuel accumulation. These relationships may contribute to the fact that fires are often more frequent in ecosystems with low soil fertility.

Organic compounds also influence fuel flammability. For example, flammable oils and lipids may account for over 30% of the dry weight of leaves in some fire-prone ecosystems. Although these chemicals are thought to be important in warding off herbivores, some ecologists have suggested that natural selection has actually favored such traits to increase flammability and guarantee conditions necessary for survival of some species (see the following).

As succession following fire proceeds, the ratio of dead to live biomass generally increases, making fuels more flammable. Variations in the size of such woody debris in forests and shrublands influence the rate at which it comes to equilibrium with atmospheric moisture conditions and thereby influence its ignitability. Larger-diameter fuels generally retain moisture longer and are extremely flammable for shorter periods than fine fuels. These fuels are classified by the fuel moisture lag times, or the time it takes for a branch or log to come into equilibrium with the moisture in the atmosphere. Small twigs and branches (<1 cm diameter) are classified as 10-hr fuels, meaning that they come into equilibrium in less than 10 hr. Large logs may require thousands of hours to attain equilibrium.

Where 10- to 100-hr fuels dominate, fuels quickly dry during drought periods and fires are more likely. Where larger fuel materials constitute most of the woody debris, long drought periods may be required before fuels will easily ignite.

The distribution of fuels in space and time influences fire regime characteristics. For example, early-successional even-aged forests are often vertically stratified, with the highly flammable fine fuels (e.g., leaves and small branches) of the canopy separated from the fuels on the forest floor. Such conditions favor light surface fires and provide little opportunity for fire to move into the forest canopy. Later in succession, mortality opens the canopy and dead snags and invading transgressive trees provide "ladder fuels" that favor high-intensity crown fires. Thus, such forests become more flammable as succession proceeds. Local variations in the horizontal distribution and accumulation of fuels are responsible for much of the spatial variation in fire behavior and postfire ecosystem response.

### C. Landscape Structure

The chance of a particular location being struck by lightning is infinitesimally small, thus the fire return interval for such a location is greatly influenced by the character of the vegetation that surrounds it. For example, fire return times are shortest in Arctic heathlands near their transition to the more easily ignited boreal forests. Similarly, southern California chaparral stands adjacent to more flammable coastal sage scrub have shorter fire-return intervals than more isolated stands. The scale of this spatial dependency varies widely in relation to landscape structure, terrain, frequency of ignition events, and climatic conditions.

Where fuels were once extensive and continuous, such as chaparral shrublands or some grasslands, fires of large extent were quite typical. In such situations, relatively infrequent ignition events still resulted in relatively short fire return intervals. Human activities have generally increased the number of ignition events on landscapes, however, fragmentation by roads, urban development, and agricultural activity have altered patterns of fire spread in many areas. As a result, the regional fire frequency (number of fires per area) has increased in many areas, but fire size is greatly diminished and return intervals have actually lengthened.

## III. HEAT RELEASE AND POSTFIRE MICROCLIMATE

Heat release during fire varies enormously. In grasslands and light surface fires in forests, fires consume approximately 8000–12,000 cal $\cdot$ $m^{-2}$, whereas crown fires in mature western conifer forests may consume 100 times that amount. Although temperatures exceeding 1000°C occur commonly in the burning fuels and flames, temperatures at the soil surface rarely exceed 300°C. Heat transfer down the soil profile is quickly attenuated so that temperatures rarely rise above 100°C below 3 cm.

The duration of elevated temperatures, which depends on the rate of fire spread and the burning of residual debris, may greatly influence survival of belowground plant parts. Thus, while unburned litter and humus may insulate soils and decrease soil heating, the ignition of litter and humus layers can result in elevated temperatures over a longer period of time.

Fire generally results in profound changes in microclimate at the soil surface. The removal of litter and understory plants increases light availability. However, this increased insolation coupled with reduced soil albedo (reflectance) may also cause considerable soil heating. Soil temperatures in excess of 70°C have been measured on recently burned exposed slopes in some shrublands and forests. In recently burned prairies, accelerated soil heating early in the growing season results in faster plant growth and longer growing seasons compared to unburned areas.

Increased soil exposure following fire often results in increased evaporation from the soil surface. Thus, surface soil horizons in recent burns may be quite dry, influencing patterns of postfire seedling establishment.

## IV. BIOGEOCHEMICAL EFFECTS OF FIRE

Fire often has profound effects on the storage, distribution, and processing of nutrients in ecosystems. Through several mechanisms, fire inevitably results in a loss of nutrient capital, however, it also often increases nutrient availability to plants. In many ecosystems, long-term nutrient budgets and patterns of productivity depend on repeated burning. [*See* Soil Ecosystems.]

### A. Nutrient Losses

Fire-caused losses of nutrients result from several processes, including vaporization, gasification (oxidation), ash convection, soil leaching, and accelerated erosion. Vaporization refers to losses of chemicals that are not chemically altered during a fire. For example, nitrate ($NO_3$) in organic matter volatilizes at temperatures as low as 80°C and some organic compounds vaporize at temperatures well below temperatures at which they might oxidize. Gasification, in contrast, refers to fire-caused chemical changes in nutrients that convert them to a gaseous phase. The oxidation of organic (reduced) carbon during fire, releasing gaseous $CO_2$, is the most obvious example of gasification. The importance of fire-caused carbon oxidation to the total carbon budget varies considerably among ecosystems (Fig.2). In deserts, where fires are infrequent and fuel consumption is modest, long-term annualized losses are generally less than 0.01 Mg · $ha^{-1}$ · $yr^{-1}$, or considerably less than 1% of annual carbon fixation. However, where fire return times are comparatively short and fires are intense, long-term losses may average over 1 Mg · $ha^{-1}$ · $yr^{-1}$, or more than 20% of annual carbon fixation. In such ecosystems, fire is clearly an important agent of organic decomposition.

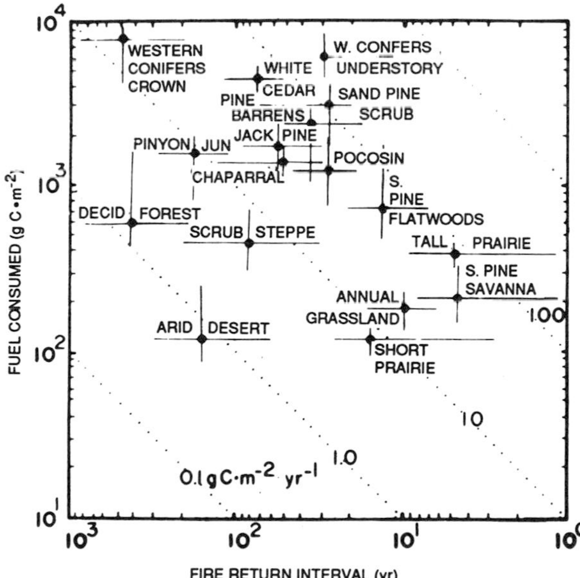


***FIGURE 2*** A comparison of averages and ranges of fire intensity (as fuel consumed) and fire return intervals in several North American ecosystems. Dotted diagonal lines indicate grams of carbon lost per square meter each year averaged over the fire cycle. [After N. L. Christensen (1993).]

Gasification is a major mechanism of nutrient loss for other nutrients as well. For example, oxidation of organic nitrogen or sulfur compounds during fire releases $N_2$ and $SO_2$, respectively. Fire may gasify 30–70% of the nitrogen contained in fuels, resulting in losses of 0.12–12 kg · $ha^{-1}$ · $yr^{-1}$. Such losses are roughly on the same order as nitrogen inputs in rainfall, which are 1–20 kg · $ha^{-1}$ · $yr^{-1}$.

The extent of volatilization and gasification losses depends in part on the nutrient content of fuels. For example, where nitrogen-fixing shrubs with high tissue nitrogen concentrations represent a major vegetation component, fire-caused losses of nitrogen capital are greater than where non-nitrogen-fixing shrubs predominate.

Relative to other pathways of loss and to total nutrient capital, nutrient loss from leaching is generally small. The leaching of nutrients from soil is influenced by the quantity of mineral made available, patterns of plant nutrient uptake and retention, adsorptive properties of the forest floor and soil, and patterns of rainfall and evapotranspiration. Because soil ion-exchange sites are predominantly positively charged, negatively charged ions are considerably more susceptible to such leaching. For example, where fire results in accelerated rates of nitrification and higher availability of nitrate ($NO_3^-$), nitrate may increase in stream waters draining burned watersheds. However, even in these situations leaching losses rarely exceed 0.5% of total soil nitrogen capital.

Fire often results in increased soil erosion owing to changes in vegetation, forest floor and soil properties, hydrology, and geomorphic processes. Increased overland flow and loss of soil binding by root systems result in increased rill and sheet erosion. The extent of such increases depends on fire intensity, soil drainage, topography, climate, and patterns of vegetation recovery. Removal of plant cover and litter exposes soil to increased kinetic energy of raindrops, causing soil movement. Where soils are porous, infiltration of water diminishes surface flow and erosion. However, ash may plug surface pores and fire may render some soils water repellent, causing increased overland flow and erosion.

### B. Nutrient Availability and Nutrient Cycles

Despite the temporary loss of total nutrient capital, fire often results in increased soil fertility. The increase in both plant diversity and production observed in many ecosystems following fire is often attributed to increased nutrient availability. This assertion is supported by the fact that concentrations of many essential nutrients in plant tissues are increased following fire.

The increase in nutrients associated with fire is often a "pulsed" event. Plant production may be enhanced in the first or second growing season following fire, but soil fertility quickly returns to prefire levels. Indeed, where overall losses of nutrient capital are high, nutrient availability to plants may decrease below prefire concentrations after a couple of growing seasons.

The direct addition of ash and its nutrients to soil is often cited as the major source of postfire nutrient pulses, and in shrublands and forests where ashfall is considerable, this is certainly the case. However, in prairies and savannas or light surface fires in forests, where total fuel consumption is relatively low, ashfall may not be as important as changes in soil microclimate in effecting nutrient pulses.

Fire may significantly alter organic matter decomposition and microbial activity. Increased soil temperatures and surface soil moisture accelerate microbial decomposition. Furthermore, fire may render some organic materials more decomposable and increased soil pH may also favor bacterial and fungal growth.

Fire effects on nitrogen cycling have been especially thoroughly studied. Immediately following fire, concentrations of ammonium often increase. Increased nitrification often results in a subsequent flush of nitrate production. Because nitrogen losses during burning may be substantial, considerable attention has been given to the impact of fire on nitrogen fixation. There is no evidence that fire accelerates rates of nonsymbiotic nitrogen fixation in soils, however, burned areas are often colonized by plants capable of symbiotic nitrogen fixation. Annual rates of fixation for such plants typically range from 10–30 kg $N \cdot ha^{-1}$ for herbaceous legumes and shrubs in shrublands to over 100 kg $N \cdot ha^{-1}$ for such pioneer trees and shrubs as *Alnus* and *Ceanothus*. Increased availability of phosphorus and other minerals immediately following fire may be important to the success of such nitrogen-fixing plants. Where nitrogen losses are significant, such increased nitrogen fixation may be important to the long-term balance of nitrogen budgets.

## V. PLANT ADAPTATIONS TO FIRE

Plant adaptations to fire range from structural or morphological characteristics that impart resistance to heat and thereby increase survivorship to life-history characteristics that take advantage of environmental changes associated with fire to improve reproductive success. In some cases, such adaptations merely increase plant tolerance of fire, but for other species these adaptations have resulted in fire dependence. Indeed, it has been proposed that where reproductive success of certain species is enhanced by fire, natural selection may have even favored plant characteristics that influence flammability.

### A. Adaptations That Improve Survivorship

Where fires are a recurrent feature, plants often possess the ability to sprout from belowground buds. Because the heat is attenuated rapidly in upper soil layers, such buds are killed only during the most intense fires. In prairies or shrublands where fires occur at regular intervals, the vast majority of postfire plant production is from resprouting plants.

In the case of many grasses or grasslike plants, buds or basal meristems are protected by leaf bases that are often relatively nonflammable. This is especially true of so-called bunch or tussock-forming species. Photosynthate and nutrients stored in extensive belowground root systems provide the resources for rapid resprouting following fire in such species. In many shrub species, the buds at the base of stems may multiply to form a swollen woody burl or lignotuber. The buds on such burls are often protected by fire-resistant coverings. Such burls store considerable quantities of nutrients and carbohydrates. [*See* GRASSLAND ECOLOGY.]

Many trees possess thick, fire-resistant bark that insulates the meristem from the heat of light- to moderate-intensity fires. In some cases, trees may possess epicormic or stem buds along the main bole, from which new branches may sprout following fire. Thus, even though many of the branches may be killed, the tree is able to retain its place in the forest canopy if the fire is not severe.

Ecologists have pointed out that the ability to sprout following damage to aboveground parts is a characteristic shared among many plants, regardless of the importance of fire. Sprouting may have evolved originally as a mechanism to survive herbivory. In some shrublands where the fire return intervals are long and fires quite intense, natural selection may actually have favored the loss of the ability to sprout. These species depend entirely on reproduction from seed following fires. Here, intense fires may make survival of belowground buds unlikely and favor investment of plant resources into seeds rather than burls.

Where frequent, light surface fires are the norm, natural selection has sometimes favored the development of unique patterns of plant growth. This is certainly the case of longleaf pine *(Pinus palustris)*, which often grows in savanna habitats of the southeastern United States where light surface fires occur with a return time of 4–6 years. Fire creates a mineral seed bed that is ideal for germination of its seeds. Following germination, seedlings allocate most of their resources to the development of an extensive tap root system and aboveground growth is limited to a tuft of needles surrounding a fire-resistant apical bud. This so-called "grass stage" is quite resistant to fire. After approximately 6 years, presumably after the next fire, the seedling begins rapid apical growth. During this stage, the apical bud is vulnerable to heat damage, but fire is relatively unlikely. After a couple of years, the apical bud has grown well above the zone of potential heat damage. Where this fire regime has been altered by human activities, longleaf pine populations have declined (Fig. 3).

### B. Reproductive Adaptations

The full array of environmental factors that influence plant reproductive success is affected by fire, and patterns of flowering, seed production and release, and seed germination may be dependent on fire regime patterns. In many ecosystems, seed dormancy results in the development of a soil seed bank. Fire causes extensive germination of these seeds. In such ecosystems, the probability of successful germination and growth in interfire years is very low owing to low soil fertility, competition for light or water from established plants, the accumulation of toxic chemicals in the soil, and/or herbivory. Burning ameliorates these negative effects.

Several cues have been identified that "inform" dormant seeds that a fire has occurred. In many shrubland species, germination is stimulated by a short burst of heat that may rupture impermeable seed coats or melt seed coat waxes. In other cases, chemicals leached from charred plant material appear to stimulate germination. In some shrub species, such as the California chamise *(Adenostoma fasciculatum)*, a portion of any year's seed crop germinates after a brief ripening period, and the remainder requires heat treatment to germinate. This "bet hedging" strategy allows the plant to take advantage of reproductive opportunities that may occur with or without fire.

Many species of trees, especially in the Pinaceae and Myrtaceae, retain their seeds in fire-resistant cones or fruits that are opened by the heat of a fire. Such reproductive structures are said to be "closed" or "serotinous" and are especially common among coniferous trees and members of the Myrtaceae in Australia (Fig. 4). In these trees, a seed pool is maintained in the forest canopy and seeds are dispersed following fire. Successful germination of these species often depends on the exposed mineral soil typical of postfire environments.

For some plant species, flower production is stimulated by fire. This pattern of increased reproduction in immediate postfire years is, in some

**FIGURE 3** (a) The "grass stage" of longleaf pine, *Pinus palustris*. During this stage, seedlings devote the majority of their growth to the development of an extensive root system. Aboveground plant parts are quite resistant to fire. (b) The sprouting stage of longleaf pine. Five to six years following germination, usually following a fire, seedlings sprout and put on rapid apical growth. The apical bud grows quickly above fire flame height. (c) A "pole stage" longleaf pine. With buds well above the flames of light surface fires, this tree is now quite resistant to fire.

cases, simply a result of increased resource availability. However, some plants will only flower when stimulated by burning or removal of aboveground parts.

As indicated earlier, germination, growth, or reproduction of some plant species is not strictly fire dependent. However, in other plants fire provides specific cues for these processes. The physiological mechanisms for such processes as heat-stimulated germination or flowering are obviously complex and highly evolved. In such cases, natural selection for mechanisms to concentrate reproduction in immediate postfire years has obviously been strong. It is no surprise that strict fire-dependent reproduc-

tive mechanisms are most common in situations where competition for limited resources or herbivory severely limits the probability of successful reproduction in interfire years.

### C. Flammability

In has been proposed that where plant fitness, for whatever reason, is positively correlated with fire, natural selection might favor plant traits that increase flammability, or influence fire behavior. Plants in fire-prone ecosystems often do possess traits such as high concentrations of oils in leaves or low tissue mineral content that do indeed increase flammability. Such circumstantial evidence, plus the appealing notion that disturbance cycles in some ecosystems might be under genetic control, makes this hypothesis tempting.

However, many of the traits influencing flammability have other functions such as repelling herbivores or increasing drought tolerance. Their impact on flammability may thus be incidental to the forces that selected for them. Natural selection operates at the level of individuals or groups of genetically related individuals. In most ecosystems, the behavior of fire is influenced relatively little by the

***FIGURE 4*** Serotinous or closed cones of pond pine, *Pinus serotina,* found in shrub bogs of the southeastern United States. The heat of a fire melts waxes that seal cone scales, causing the cone to open and release its seeds.

characteristics of individual plants and more by the collective properties of the ecosystem and climate. During drought periods, chemical differences among individual plants have comparatively little effect on fire behavior.

## VI. FIRE AND PATTERNS OF ECOSYSTEM SUCCESSION

Patterns of ecosystem response to fire vary widely, not only among fuel types and landscapes, but also within an ecosystem from location to location and fire to fire. Among the determinants of such patterns are life-history characteristics of constituent species, previous site history, fire severity, landscape mosaic, and seasonality.

Even where species must reproduce from seeds following fire, immigration from nearby unburned areas may be relatively unimportant to successional patterns. Widespread weedy plants are typically found in naturally burned ecosystems only adjacent to roads or in association with human-caused disturbances such as plow lines.

Where fire intensity is low, ecosystem recovery is quite rapid. Savannas and grasslands return to their prefire status within a year or two. However, severe, crown-killing fires in forests and shrublands often result in a succession of predictable stages. The open postfire environment initiates the *establishment stage,* during which conditions are favorable for vegetative and sexual reproduction. With the growth of early pioneers, competition increases and the *thinning phase* is initiated. During this stage, competition is intense, establishment from seed is unlikely, and overall plant density may decrease while total biomass and fuel increase. In many shrublands and forests, fires are likely during this stage and the process is reset. Where fires do not occur, plant density decreases to the point where new individuals may become established in gaps left by the death of pioneer plants, resulting in a sometimes prolonged *transition stage.* Invaders during this time are often more shade tolerant than pioneer species. In the continued absence of fire, a *steady-state stage* may be initiated in which the processes of establishment, thinning, and gap formation produce a mosaic of relatively small patches.

The actual time frame over which these various changes occur varies considerably among ecosystems. Furthermore, changes in fuels during this process result in changes in fire probability and behavior. In shrublands and certain forest types where the distribution of fine fuels such as branches and needle leaves facilitates movement of fire into the canopy, fire is likely to interrupt this process of change during thinning. However, in many pine forests, these fine fuels are concentrated in a compact canopy that is separated from the forest floor during the thinning stage, and crown-killing fires are less likely. In such forests, flammability increases sharply with fuel changes associated with the invasion of shade-tolerant trees in the transition stage.

Different stages of this process may result in different fire regimes. For example, highly stratified thinning forests may experience repeated light surface fires, which in turn may intensify the vertical stratification and prevent invasion of shade-tolerant species even as gaps open in the canopy. The complex vertical structure of transition and steady-state forests often produces high-intensity but heterogeneous fires. All of these factors influence patterns of postfire ecosystem development.

It is far too simple to view the dynamics of fire occurrence and ecosystem recovery as an endless set of relatively deterministic cycles. Ecologists prefer to interpret landscapes as being a mosaic of patches undergoing constant successional change. The character of this "patch-mosaic" is determined by the frequency, intensity, severity, and spatial extent of disturbances such as fire that create patches as well as the rate and nature of the process of patch succession.

The periodicity of fires is only partially determined by successional changes in fuels in a particular patch. Variations in climate, the vagaries of ignition events, and the character of the surrounding landscape patches influence the likelihood of fire in a particular area. Climatic conditions that influence fire behavior vary on scales of decades to millennia and have undoubtedly determined the relative importance of different ecosystem types on landscapes through time.

## VII. FIRE AND ECOSYSTEM MANAGEMENT

Conventional ecological wisdom during the first half of the twentieth century held that, prior to human settlement, most landscapes were dominated by climax ecosystems whose composition and structure were determined by regional climate patterns. The important biodiversity of any region was viewed to be contained in such climax ecosystems, and major disturbances such as fire were thought to be largely the result of human activities. From the standpoint of natural resource management, fire was viewed as an undesirable process, and fire prevention and suppression were the cornerstones of land management.

By midcentury, ecologists and managers began to appreciate the importance and inevitability of fire in many ecosystems. Fire suppression in many forests and shrublands resulted in successional changes that actually increased the likelihood of severe fires and diminished growth and survival of keystone species. Studies of lake sediments and tree ring data demonstrated that wildfires occurred often in many ecosystems. Studies of species life histories revealed that many species not only were tolerant of fire, but depended on it for their survival. In many cases, early stages of the fire cycle described here contain unique species and provide an important component of landscape biodiversity in many regions.

The prevention of fire in the giant sequoia *(Sequoiadendron giganteum)* forests of the Sierra Nevada provides an excellent case study. Tree ring studies demonstrate conclusively that, prior to settlement by Europeans, light surface fires occurred in these ecosystems every 8–16 years. The absence of fire allowed invasion of shade-tolerant conifers, which increased the chance of severe crown fires. Furthermore, because the giant sequoia has serotinous cones and its seeds require bare mineral soil to germinate and grow, exclusion of such light surface fires resulted in diminished reproduction of this important species. Clearly, management strategies aimed at preserving this majestic ecosystem were having the opposite effect.

Beginning in the 1960s, the National Park Service and other land management agencies initiated programs to reintroduce fire into such ecosystems. Prescribed fire has since become an important management tool. In highly managed forests, such as pine plantations, fires are set in the understory to control growth of competing trees and to prepare the forest for harvest. In wilderness preserves, fire management has focused on restoring the natural effects of fire to landscapes.

Restoration of natural fire regimes has proven to be a challenge. Where human land use has been limited and there has been little landscape fragmentation, lightning-set fires can be allowed to burn so long as the fire behaves within preset prescriptions of intensity and size. However, on many landscapes, human activities have created a variety of obstacles (e.g., roads, urban development, and agricultural activities) to fire spread and wildfires pose a serious threat to human life and property. In such cases, human-set prescribed fires may be used to simulate natural fire regimes. Nevertheless, public misunderstanding of the importance of fire, issues of public liability, and impacts of prescribed burning on air quality near some urban areas have limited the application of such management strategies.

The appreciation of the importance of fire in many ecosystems has significantly changed resource management policies. Nevertheless, we have much to learn. Although we understand in broad terms that fire regimes are determined by interactions of climate and fuels, our ability to predict with high precision the behavior of specific fire events is quite limited, especially on complex landscapes. The general range of potential fire effects and responses is understood, however, we are just beginning to recognize the tremendous variability in such effects and responses. Much of the diversity generated by fire is a consequence of such variability. Finally, we live in a world where human impacts on ecosystems are ubiquitous. We have modified the structure of landscapes, altered climates, and changed the chemistry of our atmosphere. Our understanding of the impacts of these changes on fire regimes is primitive indeed. These uncertainties will occupy fire ecologists for decades to come.

## Glossary

**Epicormic bud** Meristems or growth points along the main bole or branches of trees and shrubs that may sprout following a disturbance such as fire.

**Gasification** Chemical oxidation of minerals often associated with heating that results in their conversion from a solid to gaseous phase, for example, the conversion of organic carbon to carbon dioxide during burning.

**Fire intensity** Energy released by a fire as a function of its area or length of burning front.

**Fire return interval** Average time between successive fires at a particular location.

**Fuel moisture lag time** Time in hours required for the moisture content of fuel material such as litter or woody debris to equilibrate with the atmosphere.

**Lignotuber** Swollen root or stem base in some shrubs and trees from which new sprouts emerge following disturbance such as fire.

**Nutrient leaching** Removal of nutrients from soil solution in water.

**Regional fire frequency** Number of fires occurring within a specified area and time.

**Serotinous cones or fruits** Cones and fruits that remain closed after seed maturation and are opened by the heat of a fire, releasing their seeds.

**Vaporization** Conversion of chemicals from a solid to gaseous phase without chemical alteration or oxidation.

## Bibliography

Agee, J. K., (1993). "Fire Ecology of Pacific Northwest Forests." Washington, DC: Island Press.

Crutzen, P. J., and Goldammer, J. G. (1993). "Fire and the Environment: The Ecological, Atmospheric, and Climatic Importance of Vegetation Fires." New York: John Wiley & Sons.

Fuller, M., (1991). "Forest Fires: An Introduction to Wildland Fire Behavior, Management, Firefighting, and Prevention." New York: John Wiley & Sons.

Goldammer, J. G., and Jenkins, M. J., eds. (1990). "Fire in Ecosystem Dynamics." The Hague: SBP Academic Publishers.

Johnson, E. A., ed., (1992). "Fire and Vegetation Dynamics: Studies from the North American Boreal Forest." New York: Cambridge University Press.

Martin, R. E. (1982). Fire history and its role in succession. *In* "Forest Succession and Stand Development in the Northwest." (J. E. Means, ed.), pp. 92–99. Corvallis: Forest Research Laboratory, Oregon State University.

Trabaud, L. V., Christensen, N. L., and Gill, A. M. (1993). "Historical biogeography of fire in temperate and Mediterranean ecosystems. *In* "Fire and the Environment: The Ecological, Atmospheric, and Climatic Importance of Vegetation Fires." (P. J. Crutzen and J. G. Goldammer, eds.), pp. 277–295. New York: John Wiley & Sons.

Wallace, L. L. (1990). "Fire in North American Tallgrass Prairies." Norman, Oklahoma: University of Oklahoma Press.

# Fish Conservation

***Jack E. Williams***
*Bureau of Land Management*



Fish diversity, from individual populations to whole species and even complex communities, has been declining in response to the reduced health of aquatic ecosystems and invasions by nonnative species. During the past decade, the number of North American freshwater fishes requiring special protection has increased by 45%. Marine fishes are also threatened, primarily by overfishing and pollution, as well as habitat loss. Overall, the status of aquatic species appears to be deteriorating faster than that of mammals, birds, reptiles, and other terrestrial animal groups. Recent improvements in scientific knowledge and public education provide some reason for optimism. This article outlines fundamental improvements that are needed in fish conservation, which include greater attention to ecosystem health and community conservation.

## I. FISH DIVERSITY

Worldwide there are approximately 25,000 species of fishes. Nearly half of these occur in fresh waters, which is surprisingly large considering the small amount of fresh water compared to salt water, and this indicates the amount of isolation and speciation that has occurred in our rivers, lakes, and springs. Within fresh waters, riverine habitats typically harbor a greater fish diversity than do reservoirs, lakes, and springs. [*See* LIMNOLOGY, INLAND AQUATIC ECOSYSTEMS.]

The fresh waters of North America provide habitat for approximately 1033 species of native fishes in 51 families. The most diverse families are the minnow nows (Cyprinidae: 272 species), perches (Percidae: 146),suckers (Catostomidae: 68), killifishes (Cyprinodontidae: 65), and livebearers (Poeciliidae: 63). About 790 of these species occur in the United States.

Total biodiversity, however, includes more than species. Genes, populations, subspecies, and communities as well as species are primary components of diversity. Maintenance of diversity below the species level often is overlooked for those fishes organized more or less into isolated populations, such as anadromous fishes or desert fishes. Conservation strategies for salmon, for instance, should maintain individual stocks because they are the evolutionary building blocks of anadromous fishes. Stocks are adapted to local conditions, possess specializations for timing of spawning, and in other ways possess unique genetic adaptations to certain streams. [See BIODIVERSITY.]

Conspecific populations may evolve unique genetic characteristics when isolated in small habitats. Classic examples of such diversity occur in desert fishes. The Amargosa pupfish (*Cyprinodon nevadensis*) of the Death Valley region contains three recognized and described subspecies. Other species may not contain recognized subspecies, but still contain substantial genetic variation among populations. Researchers from Oklahoma State University, for example, found that 52% of the total genetic variability of the Pecos gambusia (*Gambusia nobilis*) occurs separately among the numerous isolated populations along the Pecos River, as compared to 48% of variation that is found within single populations. Although no subspecies of Pecos gambusia are recognized, two populations are at the extreme of genetic differentiation in the species and are critical to maintaining genetic diversity. Populations of another desert fish, the Gila topminnow (*Poeciliopsis o. occidentalis*), exhibit substantial genetic variation within a single subspecies. In general, populations at the periphery of the subspecies' range are lower in genetic diversity. Also, populations from thermally constant spring systems contain little genetic variability compared to those from variable, desert stream habitats. Clearly, for fishes with multiple populations, resource managers should seek to maintain the viability of existing populations rather than assume that one population contains all or most of the genetic diversity of recognized taxa.

## II. FISHES AS INDICATORS OF ENVIRONMENTAL QUALITY

Flowing waters integrate the landscape. As water flows from mountaintops to valleys and ultimately to lakes and oceans, it collects runoff, sediments, nutrients, and pollutants. Point and nonpoint sources of pollution are synergized and often form complex and lethal compounds as they accumulate downstream. Fishes, as the best-known species of the aquatic world, reflect the health or dysfunction of aquatic habitats. As a result, biological monitoring of fishes can be utilized to assess overall environmental degradation of the landscape. [See FISH ECOLOGY.]

Certain fish or fish guilds may be utilized as "indicators" of environmental degradation. The presence of certain introduced fishes, such as carp (*Cyprinus carpio*) and mosquitofish (*Gambusia affinis*), may indicate poor-quality habitats because of their broad tolerance to degraded environmental conditions. Green sunfish (*Lepomis cyanellus*) and black bullhead (*Ictalurus melas*) also are broadly-tolerant species that may become abundant in degraded habitats. Other species, such as bull trout (*Salvelinus confluentus*), are highly sensitive to siltation or warm temperature, and are present only in high-quality habitats. Increasing incidence of hybridization, diseases, and parasites also often indicate degraded habitat conditions.

The use of single species as indicators can be misleading. For that reason, broader measures of fish community diversity have been developed to more clearly qualify and quantify habitat degradation. For example, the Index of Biotic Integrity (IBI) combines attributes of communities, populations, and species to assess biological integrity by making comparisons between disturbed and relatively undisturbed habitats of the same region. Factors considered in determining the IBI in a particular stream might include fish species richness and community composition, ratios of native versus nonnative species, trophic composition, and overall fish abundance and condition. In regions of the country where natural fish diversity is relatively low, aquatic macroinvertebrates can be used as partial or total substitutes for fish community data.

## III. FACTORS IN THE DECLINE OF FISHES

Causes of declines in fish diversity are numerous and complex. In a survey on the extinctions of North American fishes during the past century, more than one factor contributed to 82% of the extinctions. Physical habitat alteration was cited as a causal factor in 73% of the extinctions, and detrimental effects associated with nonnative introductions were cited in 68%. Pollution and hybridization each were cited in 38% of the extinctions, and overharvesting was a factor in 15%. The complexity in the decline of a wide-ranging fish is typified by that of the Grande Ronde River coho salmon (*Oncorhynchus kisutch*) in Oregon. The Grande Ronde historically supported a coho stock of 2000 to 4000 spawning adults. Initial declines were caused by the cumulative effects of livestock grazing, timber harvest and associated road construction, and agriculture. In the 1960s and 1970s, dams constructed on the migratory path of the coho in the Snake River further reduced the size of spawning runs to the point that fishery managers mistakenly considered the stock to be too weak to be worth protecting from overfishing in mixed-stock fisheries. By 1980, only about 50 spawners returned to the Grande Ronde.

Adverse anthropogenic impacts on fish diversity can be described in the following four categories: (1) physical habitat alteration, (2) pollution, (3) fishery mismanagement (including overfishing and poor hatchery practices), and (4) nonnative species.

### A. Physical Habitat Alteration

Common activities that physically alter the habitat of fish include dam construction, water diversion, stream channelization, bank riprap, dredging, and other projects that change the natural course or flow of aquatic habitats. Dams alter stream habitats by inundating riverine areas and change downstream habitats by altering natural flow, temperature, and turbidity regimes. Negative impacts also occur to fish communities in remnant riverine habitats because dams fragment the remaining high-quality areas and inhibit migrations. Even when dams are designed with fish ladders and other structures to facilitate fish movement, mortality and delays in migrations may cause severe problems. Each dam on the Northwest's Columbia River, for example, causes up to 20% mortality of juvenile salmon as they move to the Pacific Ocean. Other habitat changes such as dredging and channelization degrade habitats and result in reduced diversity of macroinvertebrates and fishes by reducing habitat complexity and eliminating backwaters that are critical for many species.

According to a Nationwide Rivers Inventory (NRI) completed in 1982, only 1.9% of the 5.2 million km of streams in the contiguous United States remain in a high-quality condition, free of dams or other major structural changes. Only 42 rivers longer than 200 km were found by the NRI to be free-flowing (Fig. 1).

### B. Pollution

Most fish communities, perhaps as many as 81% according to one survey, are adversely affected by environmental degradation. Pollution, in the form of point sources from industrial outfalls and sewage treatment plants, as well as nonpoint sources such as runoff from agriculture, rangelands, and urban areas, is a serious problem. The U.S. Environmental Protection Agency reported that 602 stream segments in the Pacific Northwest have impaired water quality because of chemical contamination. In the Midwest, widespread agricultural development during the 1800s and resulting pollution and increased stream turbidity resulted in the extinction of the harelip sucker (*Lagochila lacera*), once regarded as "abundant" or "common" in Alabama, Indiana, Ohio, Kentucky, and Tennessee. Many Northeast streams and lakes have been adversely affected by acid precipitation. Depending on the level of increased acidity, some fish communities have lost diversity, and individual populations have decreased in size or been eliminated altogether. As many as 100,000 lakes in Ontario and Quebec have undergone chemical change as a result of acid precipitation. Within the contiguous 48 states, approximately 30% of rivers, lakes, and estuaries are impacted by pollution to the point that beneficial uses to human society are lost or significantly impaired (Table I). Many point pollution sources have been eliminated or improved through enforcement of the Clean Water Act, but the subtle, chronic impacts of nonpoint source pollution have proved more difficult to overcome. [*See* ACID RAIN.]

### C. Fishery Mismanagement

Certain fishery management practices are now known to have adverse side effects on fish diversity.

***FIGURE 1*** Free-flowing rivers of the contiguous 48 states that are longer than 200 km in length and do not have major physical habitat alterations according to the Nationwide Rivers Inventory.

Some practices that have been questioned in recent years include stream and lake poisoning projects designed to eliminate "rough" or "trash" fish in favor of certain "more desirable" fishes. In addition, some hatchery practices, and poorly managed commercial and sport fishery harvests, have had detrimental effects on fish diversity.

Rotenone, Antimycin, and other chemicals are sometimes used to rid waters of unwanted species in an effort to improve growth and survival of sport fishes. These chemicals prevent the uptake of oxygen by gill-breathing species and kill a wide range of fish and invertebrates. The extinction of the Miller Lake lamprey (*Lampetra minima*), endemic to a single

***TABLE I***

**Water Quality in the Contiguous 48 States as Compiled by the Environmental Protection Agency in 1990**

| Habitat type | Beneficial uses (%) | Leading causes of degradation (%) | Leading sources of degradation (%) |
|---|---|---|---|
| Rivers | Full (70) | Siltation (42) | Agriculture (55) |
| | Partial (20) | Nutrients (27) | Municipal discharge (16) |
| | None (10) | Fecal coliforms (19) | Habitat modification (13) |
| Lakes | Full (74) | Nutrients (49) | Agriculture (58) |
| | Partial (17) | Siltation (25) | Habitat modification (32) |
| | None (9) | Organics/low DO (25) | Storm sewers/runoff (28) |
| Estuaries | Full (72) | Nutrients (50) | Municipal discharge (953) |
| | Partial (22) | Pathogens (48) | Resource extraction (34) |
| | None (6) | Organics/low DO (29) | Storm sewers/runoff (28) |

lake in Oregon, is attributed to a purposeful fish poisoning project during the 1950s that was undertaken in an effort to improve sport fishing. The realization of the negative impacts of such practices on native species has reduced the use of such chemicals.

The potential for hatchery practices to negatively impact native fishes also has become clear in recent years. Hatchery fish may act as vectors for diseases, which are common in densely populated hatchery environments, into wild populations. Artificial propagation may alter the genetic composition of species by unwittingly or purposefully selecting for certain life-history traits in brood stock, or through the inevitable production of generations of fish selected for the unnatural conditions common in hatchery ponds or raceways. One common problem is the selection of brood stock from the beginning or end of spawning runs, which eventually changes the timing of the spawning run as only those fish that return early or late are selected to produce next year's progeny. Other problems include the release of large numbers of artificially propagated fish that compete or hybridize with remaining wild fish. The large numbers of fish typically produced in hatcheries can swamp the natural genetic structure of remaining wild fish by overwhelming spawning grounds. Some of the negative consequences of artificial hatchery selection may go unnoticed until it is too late for the remaining wild fish. As an example,. many of the native coastal coho salmon stocks in the Pacific Northwest have been replaced by hatchery-produced fish. Fishery managers were surprised by how poorly the hatchery fish survived when ocean currents changed during El Niño events. The native stocks, which were well adapted to dealing with periodic changes in ocean upwelling cycles, have largely been replaced by stocks of hatchery coho that nearly collapse during such events.

Mixed-stock commercial and sport fisheries often lead to overfishing that has been detrimental to fish populations and communities, especially anadromous and marine species. Many fisheries catch mixes of stocks and species that spawn in different times and places but combine into larger intermingled groups in the ocean. "Weaker" stocks often are caught at unsustainable levels that were established for larger stocks and species that are the targets upon which fishing quotas are established. Some coastal fisheries deplete one stock and then move on to another. Such "serial depletions" have been common in certain ocean areas, where fish population data often are lacking and enforcing quotas among several nations is difficult. Still another problem is large ocean by-catch, or the capture of nontarget fish, that are collected in trawl fisheries and discarded. For example, approximately 12.4 million juvenile red snapper (*Lutjanus campechanus*) are lost annually when they are caught and discarded in the Gulf of Mexico shrimp fishery. For every 1 lb. of shrimp caught, 9 lbs. of fish are thrown overboard.

### D. Nonnative Species

The introduction and establishment of nonnative species is a pervasive problem for freshwater fish conservation. Introductions occur through both authorized and unauthorized means and can include intentional stockings of forage and predatory species to provide angling opportunities and stocking of species for biocontrol (e.g., mosquitofish to control mosquito larvae), as well as unauthorized transfer of baitfish by anglers, release of aquarium fishes, and escapes from aquaculture facilities. Release of ship ballast water also has been documented as the source of many introduced fish larvae and invertebrates in places like San Francisco Bay and the Great Lakes. Many species react in unexpected ways when introduced into established communities and novel environments. Often, the positive results predicted prior to intentional introductions are replaced by unexpected problems, resulting in a "Frankenstein effect" as the complete impact of the introductions becomes clear. Problems resulting from introductions include competition with, predation upon, or hybridization with native fishes and a disruption of food webs that can affect even terrestrial species. Introduced species also may carry nonnative parasites and diseases and spread these vectors to native species.

Introduced species now dominate many aquatic habitats of western North America, where native fish communities are less diverse than in the East. Clear Lake, California, once harbored 11 native fishes of which 3 were endemic to the lake. Sixteen nonnative fishes, primarily sunfishes (family Centrarchidae), have been introduced into Clear Lake. As a result, 6 of the native fishes, including the endemic Clear Lake splittail (*Pogonichthys ciscoides*), have been eliminated.

Once nonnative species are introduced into aquatic habitats, floods or the numerous interconnections provided by rivers and canals make their control or containment very unlikely. Most introductions should be viewed as irreversible and any such proposals should be subjected to a rigorous environmental analysis.

## IV. STATUS OF FISHERY RESOURCES

Within the United States, 260 (33%) of the 790 fish species are considered to be endangered, threatened, of special concern, or extinct. The number of North American freshwater fish species and subspecies considered by the American Fisheries Society to be endangered, threatened, or of special concern increased 45% between 1979 and 1989. The most recent compilation, published in 1989, found 364 taxa at risk. Only 7 of the 251 fish taxa included in the previous 1979 list had improved enough in status by 1989 to warrant reclassification to an improved listing (e.g., "endangered" to "threatened," or to "of special concern"), and none of the taxa had recovered enough to warrant removal from the list. In general, states in the Southwest and Southeast contain the highest numbers of rare fish taxa, although as a percentage of the total fauna, Southwest states clearly have a larger proportion of the fish fauna that is endangered, threatened, or of special concern (Fig. 2).

The status of aquatic species in the United States appears to be deteriorating much faster than that of terrestrial species. Thirty-three percent of fish species in the United States qualify for special management classification because of rarity compared to 13% of mammal species and 11% of birds (Table II). Comparisons of the recovery status of federally listed endangered or threatened species also reveal similar disparity between the status of aquatic versus terrestrial species. According to the U.S. Fish and Wildlife Service, only 4% of the federally listed endangered and threatened aquatic species exhibit an improving trend compared to 20% of the terrestrial species (Fig. 3).

The status of diversity, of course, includes more than just measures of endangered species. Surveys conducted by the Environmental Protection Agency and the U.S. Fish and Wildlife Service found that 81% of fish communities in streams of the continental United States are affected by environmental degradation. Although no endangered species occur in many of these streams, declines of fish populations and local extirpations are widespread, indicating a loss of biological integrity of aquatic systems.

The following sections contain detailed information on the conservation status of selected fish groups in North America and the world.

### A. Western Trouts

Native trouts of western North America include cutthroat, rainbow, redband, golden, Gila, and Apache trouts (genus *Oncorhynchus*), and Dolly Varden and bull trouts (genus *Salvelinus*). Many of the species are polytypic and contain numerous subspecies. Depending on the taxonomic authority cited, there may be as many as 14 subspecies of cutthroat trout (*O. clarki*), each native to a particular drainage basin. Still others, not formally recognized by taxonomic authorities, may contain unique adaptations to specialized stream or lake environments.

Many trouts native to interior basins of the West are recently evolved (i.e., within the last 40,000 years) and lack reproductive isolating mechanisms that would allow them to coexist with introduced trouts. As a result, the indiscriminate and widespread introductions of hatchery rainbow trout have resulted in hybridization with and the eventual loss of many native forms. In Montana, for instance, genetically pure Westslope cutthroat trout (*O. clarki lewisi*) have been eliminated from 97.5% of their historic range because of stockings of nonnative trouts. The Alvord cutthroat trout, an undescribed subspecies from the high desert of southeastern Oregon and northwestern Nevada, is now extinct largely because of the introductions of nonnative trout.

***FIGURE 2*** Number of native freshwater fish species in each of the contiguous 48 states (upper number) and comparison to the number of fish species and subspecies that are recognized by the American Fisheries Society as endangered, threatened, or of special concern (lower number). Note that because the lower number includes both subspecies and species, it may exceed the upper number in rare cases.

In some cases, hybridization with introduced trouts has combined with habitat degradation to jeopardize even the most widespread native trouts. Cutthroat trout of the Humboldt River drainage of northeastern Nevada now occupy only about 12% of their historic habitat, which may have included up to 300 small streams. Surveys of the Nevada Department of Wildlife in the 1980s documented that 72% of streams were in fair or poor condition as a result of water diversions, mining, and overgrazing by livestock. In at least one subbasin of the Humboldt, the Marys River in northeastern Nevada, the native trout is being brought back to its native range through acquisition and restoration of riparian areas and improved livestock management.

**TABLE II**
Relative Rarity among Selected Animal Groups in the United States[a]

| Animal group | Total species | Percentage rare or extinct |
|---|---|---|
| Mammals | 442 | 13 |
| Birds | 762 | 11 |
| Reptiles | 301 | 14 |
| Amphibians | 226 | 28 |
| Fishes | 790 | 33 |
| Freshwater mussels | 297 | 72 |

[a] Data for fishes and freshwater mussels from the American Fisheries Society; all others from The Nature Conservancy Natural Heritage Database.

Even national parks have not remained free of nonnatives. Unauthorized introduction of lake trout (*Salvelinus namaycush*) into Yellowstone Lake threatens the largest wild populations of Yellowstone cutthroat trout (*O. c. bouvieri*) and indirectly, many other species that feed on the cutthroat during their spring spawning runs up tributary streams. Unlike the native cutthroat, lake trout spawn in lakes, often at 40 or more feet in depth, during the autumn.

Because of their aesthetic and recreational values, active recovery programs exist for the greenback cutthroat (*O. clarki stomias*), Apache (*O. apache*),

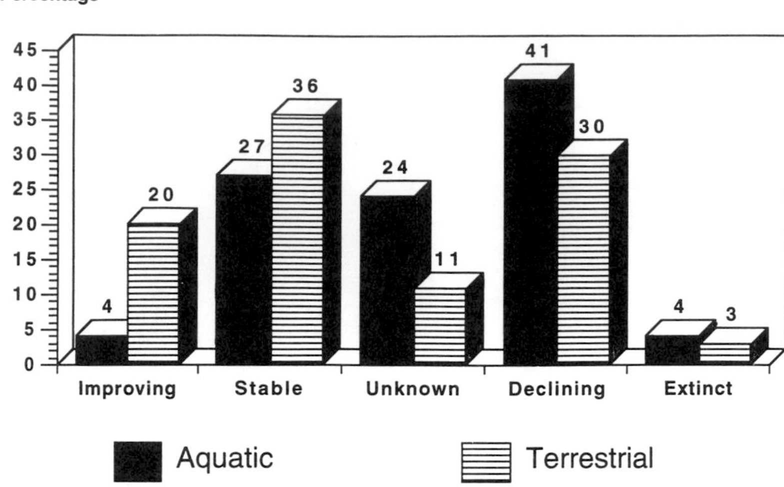


***FIGURE 3*** Recovery status of terrestrial ($n$ = 117) versus aquatic ($n$ = 104) species listed as threatened or endangered pursuant to the Endangered Species Act. Data from the U.S. Fish and Wildlife Service 1990 Report to Congress.

and other trouts. Many recovery programs attempt to reintroduce the native trout into their historic range through removal of nonnative trouts by poisoning or repeated stream electroshocking, construction of instream barriers to help prevent reinvasion, conducting habitat restoration, stocking of the native trout from a secure site, and repeated monitoring to ensure that genetic integrity is maintained.

### B. Appalachian Stream Fishes

Streams draining the Appalachian Mountains of the eastern United States contain the richest fish fauna of any temperate river systems. The Tennessee River contains 224 fish taxa, including 25 endemic species, in 25 families. Even smaller river systems, such as the Holston, Clinch, and French Broad in Virginia and Tennessee, contain 96, 118, and 80 native fish species, respectively. The most diverse families are the riffle-dwelling darters (Percidae: *Etheostoma, Percina*) and minnows (Cyprinidae: *Hybopsis, Nocomis, Notropis*).

Dam construction and associated reservoirs have inundated large portions of riverine habitats. The Tennessee Valley Authority has constructed 36 flood control and hydroelectric dams in the Tennessee River system. The dams and impounded waters cause the loss of many species that require shoal and riffle habitats and prevent movement of migratory species. The Norris Reservoir site on the Clinch River, for example, supported 65 fish species from 17 families prior to dam construction, but only 30 species immediately after. Habitat for native fishes in tailwaters below the dams also was decreased because of changes in flow rates, water temperature, and turbidity. The combined effects of dams, impounded waters, and tailwaters have eliminated natural habitats in more than 1100 km of rivers out of 2800 km in the Clinch, Holston, and French Broad systems. General pollution and industrial chemical spills periodically result in large fish kills in the remaining riverine areas.

Historically, riverine habitats in the Appalachian river systems also supported diverse but equally endangered assemblages of freshwater mussels. Fishes and mussels have coevolved and are dependent on each other. Mussels filter impurities from the water and the fish serve as hosts for the larval mussels, which act as benign parasites as they are dispersed to new riffles by the fish. Of 297 native freshwater mussel taxa (Margaritiferidae and Unionidae) in the United States and Canada, many of which are restricted to Appalachian stream systems, the American Fisheries Society estimated that 72% are endangered, threatened, or of special concern. A new and potentially devastating threat to the biological integrity of aquatic stream communities is the spread of the nonnative zebra mussel (*Dreissena polymorpha*), which encrust on and smother the native mussels.

### C. Anadromous Fishes

Many anadromous fishes, including salmon, steelhead, striped bass, and sturgeons, are of particular interest because of their great economic and cultural importance. Because of their complex life-history migrations and organization into many discrete stocks, however, they pose significant problems for fishery managers and conservationists. In their long migrations, anadromous fish are confronted with a variety of threats, such as overfishing in oceans, pollution of coastal areas, hydroelectric dams and altered riverine flows during freshwater migration, and degraded spawning and rearing habitat in fresh waters. In addition, hatchery operations designed to supplement dwindling populations or to mitigate for habitat losses actually have been detrimental to survival of wild stocks because of competition between hatchery and wild fish and because of the loss of genetic integrity resulting from hybridization between hatchery and wild fish.

Despite major efforts to improve the management of ocean fisheries, freshwater habitats, and artificial propagation, many stocks of salmon, steelhead, and sea-run cutthroat trout continue to decline and many have become extinct. The Columbia River Basin supported salmon runs estimated at 10 to 16 million adults prior to 1850. In the early 1990s, run sizes had declined to 2.5 million adults, of which about 75% are artificially spawned and reared in about 100 hatcheries. Of 192 identifiable stocks of anadromous salmonids in the Columbia River Basin, 35% are extinct, 19% are at high risk of extinction, 7% are at moderate risk of extinction, 13% are of special concern, and only 26% are considered to be secure. The American Fisheries Society has documented 214 stocks of anadromous salmonids from California, Idaho, Oregon, and Washington that are at risk of extinction or of special concern, plus another 106 stocks that already have become extinct. Several stocks, such as the Sacramento River winter chinook salmon (*Oncorhynchus tshawytscha*) and the Snake River sockeye salmon (*O. nerka*), are now listed as endangered or threatened species pursuant to the Endangered Species Act.

The increasing number of declining anadromous fish stocks has brought into question which ones qualify as "species" pursuant to the Endangered Species Act and, therefore, which qualilfy for special legal protection. The law clearly defines "species" as "species, subspecies, or any distinct population segment of any species of vertebrate fish or wildlife which interbreeds when mature." The concept of evolutionarily significant units (ESUs) has been coined to determine which stocks or combination of stocks meet the criteria of "distinct population segment" pursuant to the Endangered Species Act. The National Marine Fisheries Service has determined that a stock must satisfy two criteria to be considered an ESU: (1) it must be substantially reproductively isolated from other conspecific population units and (2) it must represent an important component in the evolutionary legacy of the species. The key is that isolation by itself is not adequate to qualify a stock as an ESU, but the stock also should be ecologically or genetically important to the species as demonstrated, for instance, by occupying unusual or distinctive habitat or otherwise displaying a distinctive adaptation to its environment.

Worldwide there are about 25 species of sturgeons. All are "big water" fishes, occurring in oceans and/or large lakes and rivers, but all species spawn in fresh waters. They are long-lived, with members of some species not attaining sexual maturity until 15 or 20 years of age. This life-history pattern provides substantial prereproduction vulnerability to fishing or other unnatural mortality factors. Many species range across numerous international jurisdictions, further complicating conservation efforts.

The fate of many sturgeons can be illustrated by the endangered European Atlantic sturgeon (*Acipenser sturio*). Historically, this long-lived, large species (length of 3 m, weight of 200 kg) was abundant throughout much of the northeastern Atlantic Ocean and supported commercial fisheries in many parts of western Europe. By the 1970s, overfishing, dams, and pollution had reduced populations until only occasional individuals could be collected from the Rhine, Po, Gironde, Danube, and Douro rivers. By the early 1990s, the only viable population of European Atlantic sturgeon remaining consists of 300–1000 individuals in the Black Sea. Five of eight North American species of sturgeons, including the Pallid sturgeon (*Scaphirhynchus albus*) of the Missouri and Mississippi rivers and the shortnose sturgeon (*Acipenser brevirostrum*) of the Atlantic Coast, are classified as threatened or endangered by the American Fisheries Society. In central Asia, three shovelnose sturgeons (*Pseudoscaphirhynchus fedtschenkoi, P. hermanni,* and *P. kaufmanni*) are critically endangered or extinct.

### D. Colorado River Fishes

The Colorado River and most of its larger tributaries are characterized by high turbidity and great fluctuations in daily and seasonal flows. The native fishes are well adapted to the floods, droughts, and waterborne sand and silt. Seventy-four percent of the native fishes are endemic to the basin. Species such as the woundfin minnow (*Plagopterus argentissimus*) possess a scaleless, torpedo-shaped body with large, falcate fins supported by spinelike rays as a brace against the fast flows and shifting sands of tributary streams. In the mainstream Colorado and Green rivers, the razorback sucker (*Xyrauchen texanus*), bonytail chub (*Gila elegans*), and humpback chub (*G. cypha*) are known by their specialized body shapes that work with the current to keep the fish oriented toward the river bottom and away from higher flows (Fig. 4).

**FIGURE 4** Endangered fishes of the Colorado and Green rivers. These species are adapted to the fast-flowing, turbid habitats typical of these rivers prior to construction of mainstem dams. From top, bonytail chub, humpback chub, the Colorado squawfish, and razorback sucker. (Courtesy of the Colorado Division of Wildlife.)

Habitat modification and fragmentation caused by dams, coupled with a myriad of nonnative fish introductions, have brought nearly the entire native fish fauna of the Colorado River to the edge of extinction. One of the most costly and devastating efforts to harness the waters of the Colorado was the Flaming Gorge Dam, constructed on the Green River in Utah, just downstream of the Utah-Wyoming border. Associated with dam construction was the poisoning of 715 km of the Green River and its tributaries in Wyoming and Utah during 1962 in an effort to destroy the native fish fauna and assist the soon to be introduced rainbow trout in establishing a sport fishery. The dam was designed to impound 145 km of the Green River, but the dam's cold, clear tailwaters also degrade natural habitat for more than 100 km downstream of the dam. The river reaches that are inundated or otherwise modified by dams typically are stocked with large numbers of sport fishes such as black basses and crappies (Centrarchidae), and channel catfish and bullheads (Ictaluridae). The native species, if not poisoned, often slowly become extirpated in the now foriegn habitats. Unfortunately, the problems caused by loss of habitat are often exacerbated by nonnative species that invade riverine habitats between the reservoirs. Surveys of the Colorado River in Cataract Canyon (Canyonlands National Park) found 28 fish species, of which only 8 were native. Many of the invading species originated by stockings into Lake Powell, located downstream of the national park. As a result of such changes in natural habitats and faunal composition, nearly all the "big river" fishes of the Colorado and Green rivers, including the Colorado squawfish, razorback sucker, bonytail chub, and humpback chub, are listed as endangered species pursuant to the Endangered Species Act.

### E. Fishes of the Great Lakes of Africa

The African Great Lakes of Victoria, Tanganyika, and Malawi contain the richest diversity of fishes of any of the world's lakes. Of particular interest in these lakes is the rich and highly endemic cichlid fish fauna. The fish fauna of Lake Tanganyika consists of 287 species in 18 families, of which about 220 (76%) are endemic to the lake. Of the endemic forms, 176 are members of the Cichlidae. Lake Malawi contains more than 500 cichlid species, of which all but 4 are endemic to the lake. Similarly, approximately 300 cichlid species, many of which are endemic, occur in Lake Victoria. Smaller, but still substantial numbers of locally endemic species occur in nearby lakes Kioga, Albert, Kivu, Kariba, and others. Each species occupies a specialized niche and many possess highly evolved characteristics, such as mouthparts specialized to remove scales from other fishes.

The lakes and their watersheds have suffered from a burgeoning human population and associated lake sedimentation resulting from clearing of nearby forests and increased pollution. More devastating than even these problems has been the introduction of nonnative fishes, particularly the Nile perch (*Lates*) and tilapia (*Oreochromis* and *Tilapia*). During the early 1980s, the combination of widespread nonnative fish introductions, an abundance of foods, and modifications to the natural habitats resulted in an explosion of Nile perch and other invading species. Introductions of Nile perch, a large predator, are responsible for the extinction of an estimated 200 of Lake Victoria's endemic species. Losses of endemic species are occurring throughout the African Great Lakes. Large components of the faunas, estimated to have taken 750,000 years to evolve, have been decimated during the past two decades.

### F. Marine Fishes

Many countries depend on viable marine fisheries as a primary food source. Yet, because of the large expanses of oceans and multinational aspects of fisheries, marine fish resources are among the most difficult to properly manage. Incomplete life-history and population data for many marine fishes complicate management problems.

Because of unregulated fishing pressure and improved technology, many marine fishes are caught beyond sustainable levels. Off the coasts of the United States, including Alaska, approximately 28% of fish stocks are overfished, 26% are being fished at maximum sustainable levels, 12% are fished below sustainable levels, and 34% are of unknown status. Atlantic Ocean fisheries experience the highest fishing pressure, with 45% of stocks of the northeast coast and 33% off the Southeast being caught at levels that are not sustainable. Worldwide, although fishing pressure has increased, fish catches peaked at 100 million tons in 1989, and have declined since then. [*See* MARINE PRODUCTIVITY.]

By-catch, or nontarget fishes collected in commercial fisheries, is another significant source of marine fish mortality. Between 300 and 400 million pounds of fish were discarded each year between 1985 and 1989 by commercial shrimp fisheries in the Gulf of Mexico. In New England, the discarded by-catch of fish exceeds the annual catch of shrimp landed in the commercial shrimp fishery. Improvements in regulations, fishing gear, and methods of use could substantially decrease losses.

Overfishing and mortality from by-catch are not the only problems facing marine fishes. About 90% of marine species depend on coastal wetlands, bays, and estuaries for spawning and rearing sites. Many wetlands have been diked and filled, and remaining areas often are highly polluted by coastal industries and urban areas.

## V. FISH CONSERVATION STRATEGIES

The development of sound and measurable goals is critical to conservation work. But there is debate regarding what level management should focus on: the gene, population, species, or community? The appropriate answer to this question may depend on the issue, geography, or other factor. For many species, it is important to maintain the diversity both within populations and among populations. As shown with desert fishes, the genetic diversity of one population may differ greatly from another population of the same species. Similarly, Pacific and Atlantic salmon, striped bass, and many other anadromous fishes are organized into discrete stocks as a result of homing instincts that lead adults to return to the streams from which they were spawned. Each stock may have distinct genetic qualities worthy of preservation.

Many recovery efforts for fish focus on the species level and include plans for habitat protection and reintroduction. Recovery Plans are mandated for each species listed as threatened or endangered pursuant to the Endangered Species Act. Typically, the plan details specific needs for habitat protection, research, reintroductions, monitoring, and public education. The U.S. Fish and Wildlife Service maintains the Dexter National Fish Hatchery in New Mexico solely for the culture of threatened and endangered fishes. Dozens of rare fishes are reared at Dexter and used for reintroduction efforts or to maintain genetic insurance against loss of remaining wild stock. Success of such single-species programs has been slow, but such efforts have been invaluable for a wide variety of fishes.

### A. Ecosystem-Based Conservation Efforts

The increasing rate of endangerment among species has recently led conservationists and the public to question the efficacy of traditional species-based approaches to management. As a result, ecosystem-based approaches to management are being developed that focus on critical ecosystem processes and functions that maintain communities. For example, floods often serve to control riparian plant succession, maintain aquatic habitat complexity, and remove nonnative fishes from stream systems. In this way, floods are becoming viewed as beneficial, much like the role of fires in maintaining the vigor of terrestrial ecosystems. Further, with the emphasis on broad ecosystem processes, protection or restoration of entire communities can be achieved.

For fish conservation, fundamental areas of focus for ecosystem-based management are riparian areas and watersheds. Riparian areas are the interface between terrestrial and aquatic habitats, and as such, they are one of the most dynamic and critical components of the landscape. Healthy riparian areas dissipate flood flows, moderate drought, store surface waters, and reduce erosion. Riparian areas also directly influence the quality of aquatic habitats by reducing sediments, modifying water temperature, and contributing woody materials that are critical for maintaining habitat complexity and for pool development. Watersheds are a logical focus for ecosystem-based management of both terrestrial and aquatic species because (1) their boundaries can be easily determined on topological maps and in the field, (2) they possess a hierarchical organization by aggregating into larger basins or subdividing into smaller watersheds, and (3) their rivers provide focal points for cumulative effects analyses. For these reasons and others, riparian areas and watersheds have been used as the foundation for management strategies developed for late-successional old-growth forests in the Pacific Northwest.

An example of ecosystem-based management for aquatic and riparian habitats is the strategy of the USDA Forest Service and USDI Bureau of Land Management for habitat needs of hundreds of declining stocks of anadromous salmonids in the West. The strategy consists of the following elements: (1) broad riparian goals; (2) measurable Riparian Management Objectives that define landscape-scale desired future conditions based on thorough studies of historical stream conditions;

(3) Riparian Habitat Conservation Areas (called Riparian Reserves in some strategies) that greatly expand the concept of riparian areas to include intermittent streams and headwall areas that often are critical in sediment delivery and geomorphic processes that influence stream condition; (4) standards and guidelines for all kinds of management activities within riparian zones; (5) Key Watersheds to provide extra protection to remaining areas of high quality and identify high-priority areas for restoration efforts; (6) Watershed Analysis to better understand cumulative effects and to tailor the strategy to local watershed conditions; (7) watershed restoration; and (8) monitoring. The overall goal is to restore the biological integrity and health of watersheds and thereby provide the habitat necessary for the natural production and restoration of salmon, steelhead, sea-run cutthroat, and other aquatic species.

## VI. PROSPECTS FOR THE FUTURE

The decline of fish diversity in recent decades is severe and provides a call to action for fishery biologists, land managers, and the general public. Fortunately, there are reasons for some optimism. First, the decline of fishes and their habitats is becoming well known and documented. In the past, conservationists and the public have been slow to recognize problems in aquatic systems. People are more familiar with terrestrial systems and have been less likely to recognize degradation of aquatic habitats and aquatic communities. Recent studies by the American Fisheries Society and others have increased our understanding of the scope and causes of the decline. This insight provides the first step toward a solution. Second, in response to the decline of fish diversity, public aquaria, scientific societies, and conservation groups are focusing on the decline of our aquatic habitats. Third, our scientific understanding of fishes and ecosystem processes has substantially improved during the past decade and now provides a firmer foundation for proper management of fish, aquatic communities, watersheds, and oceans. As noted by J. E. Deacon and W. L. Minckley in "Battle against Extinction: Native Fish Management in the West:" "During the past two decades it has become evident that knowledge no longer limits our ability to protect native fishes. Most endangered species can be recovered, if we choose."

### A. Improving Aquatic Resource Management

Implementation of the following management principles would substantially improve the status of fish conservation and would help ensure a sustainable supply of fish resources for future generations.

1. Fishery management practices such as harvest quotas and artificial manipulation should be cautious and conservative in the face of uncertainty.
2. Remaining aquatic habitats of high environmental quality should be preserved to conserve fish diversity and demonstrate the components of healthy ecosystems.
3. Management efforts should focus on maintaining biodiversity, which provides stability and resilience to communities.
4. Genetic studies need to be integrated into management programs to ensure that the full array of fish diversity is identified and maintained.
5. Although efforts should continue for species-based conservation, new efforts should focus on community conservation, including the restoration and maintenance of ecosystem processes and functions, such as natural flow regimes, that maintain healthy communities.
6. Fishery conservation should include a focus on entire watersheds, particularly riparian areas, which provide the interface between actions on land and the quality of aquatic habitats.
7. Artificial management tools, such as hatcheries, barging fish past dams, lake fertilization, and so on, should be viewed as methods of last resort and should not substitute for habitat restoration or improved regulation.

8. Management efforts should emphasize monitoring and control of nonnative species.

Advances in our scientific understanding of aquatic and riparian systems generally have outpaced our willingness to put the needed principles into practice. Improved public understanding and political leadership are needed to affect desired changes in aquatic resource management.

## B. Public Aquaria

Most of the major aquaria provide a stimulating source for more information about fishes, their habitats, and their conservation. In particular, the following public aquaria feature fish and habitat conservation as central themes: Dallas Aquarium in Dallas, Texas; Monterey Bay Aquarium in Monterey, California; National Aquarium in Baltimore, Maryland; New England Aquarium in Boston, Massachusetts; New York Aquarium in Brooklyn, New York; Newport Bay Aquarium in Newport, Oregon; Seattle Aquarium in Seattle, Washington; John G. Shedd Aquarium in Chicago, Illinois; Steinhart Aquarium in San Francisco, California; Tennessee Aquarium in Chattanooga, Tennessee; and Vancouver Public Aquarium in Vancouver, British Columbia.

## C. Fish and Aquatic Conservation Organizations

Many conservation groups work to preserve and protect habitats that fishes depend on. The following conservation groups and professional societies specialize in aquatic systems and can provide more information.

**American Fisheries Society**
5410 Grosvenor Lane, Suite 110
Bethesda, MD 20814
Oldest organization for professional fisheries biologists. Provides numerous publications on fisheries topics.

**American Rivers**
801 Pennsylvania Avenue, S.E., Suite 303
Washington, D.C. 20003
General membership organization seeking to preserve and restore America's river systems.

**Aquatic Conservation Network**
540 Roosevelt Avenue
Ottawa, Ontaria K2A 1Z8, Canada
Aquarium hobbyists, professional biologists, and conservationists working to preserve aquatic life.

**Atlantic Salmon Federation**
P.O. Box 807
Calais, ME 04619 *or*
P.O. Box 429
St. Andrews, New Brunswick, EOG 2X0, Canada
General membership organization promoting the conservation and wise management of Atlantic salmon and its habitat.

**Desert Fishes Council**
P.O. Box 337
Bishop, CA 93515
Professional and general membership organization dedicated to the preservation of native fishes and their ecosystems in arid and semiarid lands.

**Izaak Walton League of America**
1401 Bishop Boulevard, Level B
Arlington, VA 22209
General membership organization promoting resource education. Operates Save Our Streams programs.

**Living Oceans Program/National Audubon Society**
550 South Bay Avenue
Islip, NY 11751
General membership organization promoting the sustainable management of marine fisheries, the protection and restoration of ocean and coastal habitats, and science-based education for marine conservation issues.

**Ocean Voice International**
2883 Otterson Drive, Head Office
Ottawa, Ontario K1V 7B2, Canada
General membership organization promoting harmony among people, marine life, and the environment.

**Pacific Rivers Council**
P.O. Box 10798
Eugene, OR 97440
General membership organization promoting the improved management of riverine ecosystems in the West.

**Southeastern Fishes Council**
Tulane Museum of Natural History
Tulane University
Belle Chasse, LA 70037
Professional membership organization dedicated to the preservation of fishes of the southeastern United States.

**Trout Unlimited**
1500 Wilson Blvd., Suite 310
Arlington, VA 22209
General membership organization dedicated to conserving, protecting, and restoring wild trout and salmon.

## Glossary

**Anadromous** Species that migrate from marine to freshwater environments to spawn.

**Biological diversity** Variety and variability among living organisms and the ecological processes in which they occur; often shortened to "biodiversity."

**Biological integrity** Capacity of habitat to support and maintain a balanced, integrated, adaptive community of organisms having a composition, diversity, and functional organization comparable to that of natural habitat of the region.

**Ecosystem-based management** Approach to management that emphasizes ecological processes and functions in such a way as to maintain the integrity of natural communities; as opposed to species-based management.

**Endemic** Naturally-occurring in a restricted, geographical area.

**Environmental Stochasticity** Random or chance changes in the environment, such as caused by floods, drought, or fire.

**Evolutionary significant unit (ESU)** Reproductively isolated population that represents an important component in the evolutionary legacy of the species.

**Extinction** Termination of a species or subspecies by death of all remaining members of that taxon.

**Extirpation** Local extinction; a species or subspecies disappearing from a locality or region without becoming extinct throughout its entire range.

**Frankenstein effect** Intentional introduction of nonnative species to "improve" fisheries, but that often has the opposite effect, causing declines in fisheries and threats to native species (term coined by Peter Moyle of the University of California, Davis).

**Habitat fragmentation** Reduction in total habitat area with remaining habitat being redistributed into isolated segments.

**Indicator species** Particularly sensitive species to pollutants, human disturbance, or ecological instability.

**Resilience** Speed at which a population recovers when disturbed.

**Riparian** Strip of moist soils bordering watercourses, seeps, springs, and lakes that typically supports an abundance of plant and animal life.

**Species-based management** Approach to management that emphasizes individual species rather than groups of species or communities; contrasts with ecosystem-based management.

**Stock** Group of fish that spawn in a particular river system (or portion of it) during a particular season, and do not interbreed to any substantial degree with any other group of the same species. Term commonly used in reference to anadromous fishes.

**Taxon (plural, taxa)** Term for any category used in classification, such as species or subspecies.

**Viable population** Population with sufficient numbers and distribution to provide a high likelihood that the population will continue to grow, develop, and be well distributed throughout its range.

**Watershed** Drainage basin contributing water, organic matter, dissolved nutrients, and sediments to a stream or lake.

## Bibliography


Allan, J. D., and Flecker, A. S. (1993). Biodiversity conservation in running waters. *BioScience* **43**(1), 32–43.

Allendorf, F. W. (1988). Conservation biology of fishes. *Conservation Biol.* **2**(2), 145–148.

Benke A. C. (1990). A perspective on America's vanishing streams. *J. No. Amer. Benth. Soc.* **9**(1), 77–88.

Fausch, K. D., Lyons, J., Karr, J. R., and Angermeier, P. L. (1990). Fish communities as indicators of environmental change. *Amer. Fisheries Soc. Sympos.* **8,** 123–144.

Forest Ecosystem Management Assessment Team (1993). "Forest Ecosystem Management: An Ecological, Economic, and Social Assessment." July 1993 Report to President Clinton.

Gilbert, C. R., ed. (1992). "Rare and Endangered Biota of Florida." Vol. II, "Fishes." Gainesville: University Press of Florida.

Hilborn, R. (1992). Hatcheries and the future of salmon in the Northwest. *Fisheries* **17**(1), 5–8.

Lowe-McConnell, R. H. (1993). Fish faunas of the African Great Lakes: Origins, diversity, and vulnerability. *Conservation Biol.* **7**(3), 634–643.

McAllister, D. E., Parker, B. J., and McKee, P. M. (1985). "Rare, Endangered and Extinct Fishes in Canada," Syllogeus No. 54. National Museums of Canada, Ottawa, Ontario.

Miller, R. R., Williams, J. D., and Williams, J. E. (1989). Extinctions of North American fishes during the past century. *Fisheries* **14**(6), 22–38.

Minckley, W. L., and Deacon, J. E., eds. (1991). "Battle against Extinction: Native Fish Management in the American West." Tucson: University of Arizona Press.

Naiman, R. J., ed. (1992). "Watershed management: Balancing Sustainability and Environmental Change." New York: Springer-Verlag.

Nehlsen, W., Williams, J. E., and Lichatowich, J. A. (1991). Pacific salmon at the crossroads: Stocks at risk from California, Oregon, Idaho, and Washington. *Fisheries* **16**(2), 4–21.

Palmer, T. (1994). Lifelines: the Case for River Conservation. Washington, D.C.: Island Press.

Reisner, M. (1987). "Cadillac Desert: The American West and Its Disappearing Water." New York: Viking Penguin.

Wheeler, A., and Sutcliffe, D., eds. (1990). The biology and conservation of rare fish. *J. Fish Biol.* **37**(Suppl. A), 1–269.

Williams, J. E., Johnson, J. E., Hendrickson, D. A., Contreras-Balderas, S., Williams, J. D., Mendoza-Navarro-M., McAllister, D. E., and Deacon, J. E. (1989). Fishes of North America endangered, threatened, or of special concern: 1989. *Fisheries* **4**(6), 2–20.

# Fish Ecology

**K. O. Winemiller**
*Texas A&M University*



Fish ecology is the study of fishes and their interactions with the physical and living components of their environments. Because fishes are the oldest and most species-rich lineage of living vertebrates, a particularly diverse variety of adaptations and ecological relationships are observed among different species and even among different populations within some species. Fishes are found in a great diversity of habitats, including temporary ponds in arid environments, subterranean waters, mountain streams, lakes, rivers, estuaries, coral reefs, pelagic ocean waters, and oceanic abysses. On a global basis, the greatest number of living fishes occupy marine environments, but the highest species densities (number of species per unit area) tend to occur in fresh waters. These differences in species densities may be related to greater habitat complexity and opportunities for geographical isolation of populations in freshwater ecosystems. Aquatic and estuarine habitats are very sensitive to pollution and human-induced landscape alterations, and as a result the abundance and diversity of fishes have declined in many regions of the world. Despite increased fishing effort and advances in the technology of fishing, the collective annual catch of the world's major marine fisheries has leveled off over the past two decades.

## I. INTRODUCTION

Water covers over 70% of the earth's surface, so it is perhaps not surprising that fishes are the most abundant and speciose group of living vertebrates (estimates of the number of living species range between 20,000 and 30,000). Fishes occupy aquatic habitats ranging in elevation from mountain lakes and streams, approximately 5 km above sea level, to deep ocean trenches 11 km below sea level. Fishes are also the oldest vertebrate lineage; the first agnathan fishes originated some 500 million years ago. The long evolutionary history of fishes, coupled with the large quantity and diversity of aquatic and marine habitats, has resulted in tremendous variation in the ecological characteristics of modern species of fishes. This article reviews the relationship between fishes and the living and nonliving components of their environments. Because it is impossible to survey the full range of ecological characteristics of such a large and diverse group of organisms, select species and habitats will be used to illustrate major similarities and differences among taxa and ecological settings.

Fish ecology is greatly influenced by the physical properties of water: density, heat capacity, viscosity, and miscibility as a solvent. Because water is more than 750 times denser than air, fishes need not invest great amounts of matter and energy into the development and maintenance of a massive skeleton to support the body. Most fishes possess a gas-filled swim bladder that allows them to maintain neutral buoyancy over a range of water depths with low energetic cost. A number of benthic fishes that lack functional swim bladders [e.g., sculpins (Cottidae), many darters (Percidae), and gobies (Gobiidae)] use fin movements and body undulations to rise above the substrate. These forays into the water column require large energy expenditures, and as a result they tend to be of short duration, as when feeding or changing locations on the substrate. The high density of water also causes sound to be transmitted much more efficiently than through air. Sound attenuates slowly in water, so that sonic signals can be transmitted over very long distances. Fishes perceive sound with the inner ear and the acousto-lateralis system. The acousto-lateralis system consists of pressure receptors housed in tubes embedded in scales of the head and body flanks. In minnows and suckers (Cypriniformes), tetras (Characiformes), and catfishes (Siluriformes), a chain of small bones, called the Weberian ossicles, transmits sound pressure from the swim bladder to the inner ear. Sound pressure first received by the body wall is transmitted to the swim bladder, which functions as an ear drum. Sounds are produced for intraspecific communication by many minnows and catfishes, plus a variety of fishes from families lacking Weberian ossicles.

Water achieves its maximum density at 4°C and can stratify over relatively shallow depth gradients. Nearly all fishes are ectothermic and can exploit temperature gradients to locate their thermal optima or to increase their efficiency of energy utilization. By reducing vertical or horizontal mixing of water, thermal and density gradients also affect nutrient dynamics and ecosystem productivity. In cold climates, lakes may experience vertical mixing (turnover) when they become isothermal during the spring and fall. This vertical mixing can release nutrients locked in deeper waters into the light penetration layer (epilimnion), where photosynthesis takes place. Blooms of phytoplankton production often follow these seasonal lake turnovers.

Light attenuation occurs much more rapidly in water than in air. Long wavelengths are the first to be absorbed (within the first 25 m), so that objects viewed at great depths appear blue due to reflectance of shorter light wavelengths. Fishes are assumed to have color vision, especially those diurnal species that have flashy species-specific or sex-specific color patterns. The retinal cone receptors have been investigated in several fish species, such as the walleye (*Stizostedion vitreum*) and goldfish (*Carassius auratus*). Many fishes have countershading, darker pigmentation in the dorsal region fading to a light-colored ventral surface. With illumination from above, countershading results in uniform reflectance that masks the normal shading cues that reveal object depth. Countershading is a particularly important adaptation for concealment in pelagic fishes.

Water has a high specific heat, meaning that it must receive or lose more energy than air or rock to change the same number of degrees. Therefore, aquatic and marine habitats are more thermally buffered against sudden climatic changes than adjacent terrestrial environments. Because of its higher density, water also conducts heat more rapidly than air. Fishes are ectotherms, and the time lag between a change in ambient temperature and subsequent change in body temperature is very short.

Because water is more viscous than air, fishes can propel themselves by swimming. Drag influences the energetics of alternative methods of locomotion, with a fusiform body producing less drag than stout or compressed body shapes. Water's high viscosity enhances the efficiency of suction feeding and, to a lesser extent, hinders grasping. Water's high density and viscosity facilitate passive dispersal of buoyant gametes, larvae, and even some adult fishes. The high viscosity of water also affects the architecture of aquatic and marine landscapes via sediment erosion, transport, and substrate scouring.

Water has been called the universal solvent, because it dissolves a great variety of chemical substances, including gases, salts, acids, bases, nutrients (nitrates, phosphates), and biological waste compounds (carbon dioxide, ammonia). Because of the solvency of nutrients in water, the dynamics of nutrient cycling and primary production tend to be faster in aquatic ecosystems compared with terrestrial systems. Partial pressures and diffusion rates of oxygen and carbon dioxide are lower for water than for air. Even so, high $CO_2$ concentrations can cause severe physiological stress or death in many fishes. Acute or chronic hypoxia (low concentrations of dissolved oxygen) can pose a serious problem for fishes in swamp, lake, and estuarine habitats. Fishes from these habitats have evolved a wide variety of respiratory adaptations, including aerial respiration with lungs (lungfishes), swim bladder (loricariid and clariid catfishes), gut (callichthyid catfishes), skin (some blennies and anguillid eels), gill chamber (synbranchid eels), and suprabranchial chambers (snakeheads and anabantids). During periods of acute hypoxia, many freshwater fishes rise to the surface and skim the oxygen-rich surface film (aquatic surface respiration). The electrical conductivity of water permits several groups of fishes (African mormyriforms, neotropical gymnotiforms) to utilize electrogeneration and reception for navigation, prey detection, and communication. At least three groups of fishes (torpedinid rays, electrophorid eels, and malapterurid catfish) have evolved the ability to produce electrical shocks as mechanisms for subduing prey and for defense.

## II. EVOLUTION OF FISHES

### A. Phylogenetic Diversity

Fishes belong to the phylum Chordata and are derived from a common ancestor with the protochordates (acorn worms, sea squirts). The three groups of living fishes are the Agnatha (jawless cartilaginous fishes), the Chondrichthyes (jawed cartilaginous fishes), and Osteichthyes (bony fishes). Agnatha has by far the fewest living taxa: about 50 species of hagfishes and lampreys of the classes Pteraspidomorphi and Cephalaspidomorphi. The class Chondrichthyes contains about 800 living species of sharks, skates, rays, and chimaeras. The vast majority of Chondrichthyes are restricted to marine habitats, although some species, such as bull sharks (*Carcharinus leucus*) and sawfishes (*Pristis*), regularly enter freshwater habitats where they may reside for extended periods. All the tetrapod vertebrates (amphibians, reptiles, birds, mammals) are derived from a primitive lineage within the Osteichthyes. If all the known undescribed species and estimated undiscovered species were tallied, the total number of bony fishes might be well over 25,000. Of the many dozens of new fish species described each year, the vast majority come from tropical latitudes, especially from freshwater habitats.

Bony fishes inhabit an incredibly wide range of aquatic habitats, including marginal aquatic habitats like thermal springs (North American killifishes of the genus *Cyprinodon,* African cichlids of the genus *Sarotherodon*), underground waters [cavefishes (Amblyopsidae), North American catfish genera *Satan* and *Trogloglanis* (Ictaluridae)], torrential mountain streams [Asian loaches (Cobitidae), South American hillstream catfishes (Astroblepidae)], ephemeral savanna pools [African and South American killifishes (Rivuliidae), lungfishes (Ceratodontidae, Lepidosirenidae)], and tidal mudflats [mudskippers (Periophthalmidae)]. Except for a few special cases, the generation of new fish species (speciation) is believed to have occurred by way of geographical isolation and genetic divergence of populations (allopatric speciation). Freshwater habitats provide numerous opportunities for geographical isolation of populations, and on a global basis the density of fish species in fresh water is much greater than in marine habitats. About 40% of all fishes live in fresh water, which comprises less than 0.01% of the earth's total volume of surface water. Populations can be split into isolated subunits when river drainages are divided by geological and climatic changes, river captures (anastomoses), confinement in separate lake basins, or formation of habitats hostile to dispersal between separate river drainage basins. Species flocks occur in a number of lake basins (cyprinodontids in Lake Titicaca, Peru; atherinids in central Mexico; salmonids in the North American Great Lakes; cichlids in the African Rift Lakes). Evolutionary biologists debate the potential mechanisms of sympatric speciation for generating these species flocks, especially the African cichlids. Most evidence seems to support a hypothesis of allopatric speciation within isolated or partially isolated lake subbasins, or within preferred habitats separated by shoreline regions containing hostile habitat. Geographical isolation of marine populations occurs by a variety of geographical and climatic changes, including formation of isthmuses and islands, tectonic movements of landmasses, zones of freshwater intrusion, and a host of oceanic currents that influence patterns of dispersal.

## B. Ecological Diversity

Regions at higher latitudes generally contain fewer fish species than tropical areas, and this pattern holds for both marine and freshwater ecosystems. Likewise, fish diversity tends to be lower at higher altitudes compared with lowland areas in the same biogeographic region. A variety of historical, climatic, and ecological factors interact to produce these gradients of fish species diversity. The basic mechanisms influencing fish biodiversity probably do not differ significantly from models proposed for terrestrial vertebrates. Compared with the tropics, regions at high latitudes and altitudes suffered more species extinctions during glacial epochs, experience more frequent and more severe climatic changes over evolutionary time scales, and generally experience greater habitat changes with season. Consequently, coevolution, ecological specialization, and adaptive radiations are much more apparent in tropical freshwater fish communities than in arctic or temperate communities. Research has documented more morphological variation and ecological strategies in tropical fish assemblages relative to fish assemblages in comparable physical habitats at higher latitudes. For example, arctic and temperate freshwater fish assemblages are dominated by invertebrate-eating and fish-eating fishes, whereas tropical fishes in the same kinds of habitats contain a number of specialized invertebrate- and fish-eating fishes, plus fishes that specialize on algae, detritus, wood, macrophytes, seeds, fruit, fish fins, fish scales, and the external mucus slime layer of other fishes.

Fishes have proven to be a particularly good group for the study of the relationship between functional morphology and ecology (ecomorphology). Body form and the sizes, shapes, and positions of fins determine swimming performance in fishes (Fig. 1). The bodies of benthic fishes tend to be cylindrical or stocky with a flattened belly region, or their bodies are entirely depressed in the dorsoventral plane, as in skates, rays, and many catfishes. Soles, flounders, and other flatfishes are compressed in the lateral

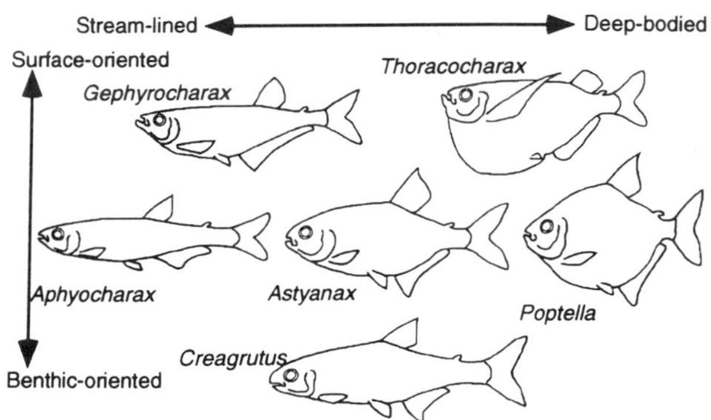


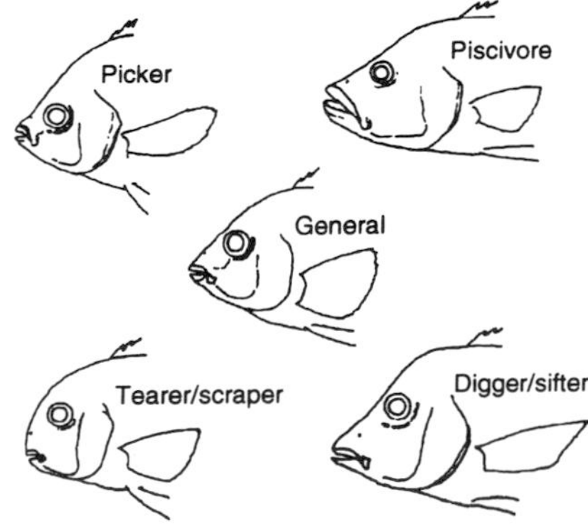


**FIGURE 1** Ecological and morphological diversification of body form and swimming performance in South American characiform fishes (above) and head form and feeding in Central American cichlid fishes (below). (With permission from the National Geographic Society and Kluwer Academic Publishers.)

plane and rest on their sides on the substrate. Midwater fishes that are compressed in the lateral plane can maneuver well in three dimensions. A laterally compressed body and broad fins permit greater stability in water, which is important for fishes that suction feed on small food items in the water column or glean prey from surfaces. North American sunfishes (*Lepomis*), many tropical cichlids, and coral reef damselfishes and surgeonfishes illustrate this ecomorphological strategy. These deep-bodied fishes achieve slow but stable locomotion by sculling their broad pectoral fins. They can also perform rapid swimming by folding their broad fins close to the body (reducing drag) and using undulatory propulsion by laterally flexing the body and caudal (tail) fin.

Tunas, mackerels, salmon, and other fishes that rely on rapid and sustained swimming generally possess fusiform bodies with relatively narrow dorsal and ventral fins. The more elongate, fusiform body reduces drag in these strong swimmers. Drag is further reduced by a narrow caudal peduncle (region between the body and the caudal fin) and a caudal fin that is tall and forked to some degree (high aspect ratio). Pike, barracuda, and other fishes that rely on stealth and a rapid swimming burst to capture prey often have very elongate bodies and medial fins positioned posteriorly. These fishes can accelerate rapidly by bending the body in an s-shape, then shooting off like an uncoiling spring. These burst predators perform sustained swimming by the normal undulating mechanism of lateral flexion of the body and tail. Eels have extremely elongate bodies and reduced fins that permit swimming or burrowing by undulatory locomotion (lateral body flexion). The elongate eel-like morphology has evolved independently in a number of taxa, including anadromous eels, moray eels, and snipe eels (Anguilliformes), neotropical knife fishes (Gymnotiformes), paleotropical spiny eels (Mastacembelidae), and swamp eels (Synbranchidae).

Much can be determined about a fish's feeding behavior by examining its jaw morphology (Fig. 1). Predatory fishes generally have large mouths armed with jaw teeth for grasping, piercing, or cutting the flesh of prey. Some piscivores (fish eaters), like sculpins (Cottidae) and scorpion fishes (Scorpaenidae), use suction feeding rather than grasping. Others, like the tarpon (*Megalops*), use an engulfing mode of prey capture in which the predator passes over its prey with jaws and gill covers (opercula) flared. Some large-mouthed plankton feeders, such as paddlefish (*Polyodon spathula*) and whale sharks (*Rhincodon typus*), use this engulfing feeding mode, but in this case small food items are strained from the water passing through the gill openings using comblike gill rakers. In modern bony fishes like the North American crappie (*Pomoxis*), the Central American cichlid (*Petenia*), the tarpon snook (*Centropomus pectinatus*), and the African thin-face cichlid (*Serranochromis angusticeps*), suction feeding is enhanced by highly protrusible jaws. During jaw protrusion, the premaxillary and maxillary bones of the upper jaw swing forward as the mouth opens and, with the mandible, form a tubular gape. Fishes that primarily use biting or grasping often have less protrusible jaws, and some, like piranhas (*Pygocentrus*) and barracudas (*Sphyraena*), lack significant protrusibility altogether.

Depending on the ecological setting, piscivores that feed on a wide variety of prey often can adopt either grasping, sucking, or biting prey capture modes. North American bass (*Micropterus*) and marine groupers (*Epinephelus*) feed on a wide variety of invertebrate and fish prey using all three feeding modes. Many small-mouthed invertebrate-feeding fishes use a picking (biting) mode to glean insects and small crustaceans from the substrate or vegetation. The same species may also use suction feeding to capture small food items from the water column or to glean surfaces. The jaw elements of sea horses, pipefishes, and trumpet fishes are fused to form a long tubular snout used for suction feeding. In habitats with soft bottom sediments, some fish forage by grabbing a mouthful of sediment and sifting immature aquatic insects from the sediments with their gill rakers or pharyngeal teeth (the latter located on the throat region). These digger/sifters, like "earth-eating" cichlids (*Geophagus, Satanoperca*) and marine mojarras (*Diapterus, Eucinostomus*), usually have long snouts or highly protrusible jaws to reach deep into soft sediments. Cypriniform fishes lack jaw teeth and manipulate prey primarily with their pharyngeal teeth prior to ingestion. Fishes that feed on hard-bodied prey, like corals and molluscs, usually have massive flattened pharyngeal teeth used for crushing.

Herbivorous and detritivorous fishes generally have relatively compact jaws for grasping and tearing plant material. Seed- and fruit-eating fishes usually have multicuspid teeth for tearing and crushing. The relationship between powerful compact jaws, multicuspid teeth, and a frugivorous diet is well illustrated by a large number of neotropical characids, including species of the genera *Astyanax, Brycon,* and *Colossoma*. Grazers have flat unicuspid or bicuspid teeth used for scraping algae from the surfaces of rocks, wood, or vegetation. Great diversity in tooth morphology is often observed among sympatric grazers, and different tooth morphologies seem to select for different algae taxa in their diets. Herbivorous and detritivorous fishes usually have long, coiled alimentary canals, and detritivores sometimes have crop and gizzard stomach chambers and pyloric ceca. The muscular gizzard contains sand particles ingested along with detritus that aid the mechanical breakdown of plant cell walls. A long gut and pyloric ceca provide more surface area for digestion of tough plant tissues and absorption of nutrients. Some herbivorous and detritivorous species, like the wood-eating South American catfishes (*Panaque*), have coevolved gut faunas (microbial organisms) that aid the digestion of plant tissues.

## III. PHYSIOLOGICAL ECOLOGY OF FISHES

Growth in fishes has been called indeterminate because they exhibit some growth even after attainment of sexual maturation, even though the rate of growth usually declines with size and age. As in any organism, a fish's growth potential is determined genetically, and its realized growth is derived from the interaction between genes and environment (temperature, salinity, food quality and quantity). Fisheries ecologists frequently employ the von Bertalanffy curve to describe growth rates in the following manner:

$$L_t = L_\infty(1-e^{-K(t-t_0)})$$

where $L_t$ is length at age $t$, $L_\infty$ is the asymptotic length, $t_0$ is the hypothetical time when length is zero, and $K$ is the rate of approach to $L_\infty$. Some piscivorous fishes show an accelerated growth rate during the juvenile stage that coincides with an ontogenetic shift from invertebrates to larger, more energy-rich fish prey. Because fishes are ectotherms, each individual has an optimal temperature for food assimilation and growth. The biotic environment can also affect growth, with crowding reducing growth via exploitation competition for limited food or via interference competition for limited space. The relationship between fish weight ($W$) and length ($L$) can be described by the equation $W = aL^b$, where $a$ and $b$ are constants. If growth is isometric, with no change in shape with change in size, then $b$ is approximately 3.0. The relationship between fish length and fish size has been used as an index of condition of plumpness in fisheries management. Species-specific standards that describe average or good condition have been derived empirically by regression methods. Fish condition can be influenced by foraging rate, environmental stress, reproductive state, and habitat characteristics.

Fishes can be aged by a variety of methods, all of which involve counting concentric bands that are deposited on hard structures during their growth process. Spines, scales, and otoliths (mineral grains in the inner ear) grow radially at different rates depending on seasonal ecological conditions, maturation, or reproduction. Minerals are deposited as daily layers (circuli) that appear as rings in cross section. During periods of rapid growth, the layers are thicker and the more opaque boundaries between layers are spaced farther apart. During periods of slow growth, as in winter or spawning seasons, the daily rings are tightly compacted and appear as bands. The bands on scales, spines, and otoliths correspond to annual age increments (annuli) in most temperate-zone fishes and many tropical fishes from seasonal habitats. In young fishes, otolith circuli often can be counted under a microscope to estimate their age in days.

Freshwater fishes are hypertonic to their aquatic environment, and marine fishes are hypotonic to marine waters that typically range between 32 and 35 ppt (parts per thousand) solute concentration. Because the epithelia of the skin, mouth, and gills of fishes are permeable to ions and water molecules, freshwater fishes' ions tend to diffuse out and water tends to diffuse into their bodies. Likewise, ions passively diffuse into the body and water diffuses out of the bodies of marine fishes. Freshwater fishes use their kidneys to eliminate excess water in dilute urine and gain lost salts in their diets and by active ion transport in branchial cells of the gill membranes. Marine fishes produce concentrated urine and eliminate as little as 3 ml of urine per kilogram of body weight per day. Up to 90% of nitrogenous wastes may be eliminated through the gills of marine fishes. Cartilaginous fishes in marine habitats have elevated levels of urea in their body fluids, which reduces the diffusion gradient of water. Most fish species live out their entire lives in habitats that experience little variation in salinity. These species are generally stressed by salinity changes of only a few ppt and are intolerant of changes greater than 15–20 ppt. Salmon (*Oncorhynchus, Salmo*), American shad (*Alosa*), striped bass (*Morone*), American eels (*Anguilla*), and other diadromous fishes have the ability to make rapid osmotic adjustments by changing urine volumes and the uptake or secretion of ions against diffusion gradients by the gills.

Sex in most fishes is genetically or chromosomally determined, but sexual development in some species is influenced by environmental conditions. Sex ratio in the Siamese fighting fish (*Betta splendens*) is strongly influenced by temperature. A number of marine fishes and a few freshwater species exhibit sex switching with age and size or in response to their social environment. Some groupers (Serranidae) and wrasses (Labridae) mature as males and become females when they are older and larger. This protandrous strategy maximizes lifetime fitness because mating is not highly selective, and a small male can produce large quantities of sperm, but female fecundity is strongly size-dependent. Some reef-dwelling wrasses, such as the cleaner fish *Labroides dimidiatus,* mature as females and later change into males (protogyny), depending on their status in the dominance hierarchy of a social group inhabiting a territory. When the dominant male is removed, the most dominant female rapidly begins to show hormonal and behavioral changes followed by physiological and morphological changes to male characteristics.

Several all-female species (gynogenetic species) have been identified within the live-bearing poeciliid genera *Poeciliopsis* and *Poecilia,* and these appear to have orginated sympatrically via reproductive isolation from genetic mechanisms following hybridization. Some of these all-female species must mate with males of heterospecific species in order for the egg to develop into a genetic clone of its mother. In other species, male genetic material is incorporated in offspring and then subsequently lost during alternate generations, and no significant amount of the male's genetic material enters into the gene pool of the all-female species (hybridogenesis).

## IV. LIFE-HISTORY VARIATION

Fishes demonstrate more variation in life-history traits than any other comparable taxonomic group (Table I). For example, clutch sizes range from 1–2 in the longfin mako (*Isurus paucus*), thresher (*Alopias*), and sandtiger sharks (*Eugomphodus taurus*) to over $6 \times 10^8$ in the ocean sunfish (*Mola mola*). Even within species, individuals may vary considerably in their sizes and ages at maturity, growth rates, fecundities, and longevities. Fishes have been grouped based on their spawning habitat and method of egg deposition, parental care, or brooding. At one end of the spectrum are nonguarding egg scatterers that usually have external fertilization and group spawning. Pelagic spawners scatter their eggs into the water column. Eggs and early larvae of pelagic egg scatters usually contain oil droplets in the yolk that enhance buoyancy. In freshwater habitats, many nonguarding species scatter their eggs over vegetation, gravel, or stones. Some nonguarding fishes hide their broods. Trout and

**TABLE I**
Interspecific Variation in North American Fish Life-History Traits

| | Age of maturity[a] (years) | Length of maturity[a] (mm) | Longevity (years) | Clutch size[b] ($N_{eggs}$) | Egg diameter (mm) |
|---|---|---|---|---|---|
| Freshwater Species | | | | | |
| *Polyodon spathula* (Polyodontidae) | 9–10 | 2235 | 30 | 141,531 | 3.35 |
| *Lepisosteus osseus* (Lepisosteidae) | 6 | 1370 | 30 | 59,422 | 2.65 |
| *Dorosoma cepedianum* (Clupeidae) | 2 | 486 | 14 | 543,912 | 0.75 |
| *Salmo clarki* (Salmonidae) | 4 | 663 | 7 | 4,420 | 4.70 |
| *Salvelinus namaycush* (Salmonidae) | 8 | 1240 | 41 | 18,051 | 5.50 |
| *Umbra limi* (Umbridae) | 1–2 | 85 | 4 | 1,489 | 1.60 |
| *Esox lucius* (Esocidae) | 3 | 1296 | 24 | 226,000 | 2.80 |
| *Notemigonus crysoleucas* (Cyprinidae) | 1–2 | 259 | 9 | 200,000 | 1.25 |
| *Pimephales promelas* (Cyprinidae) | 1 | 74 | 2 | 1,136 | 1.30 |
| *Ictiobus bubalus* (Catostomidae) | 3–4 | 909 | 19 | 427,880 | 5.00 |
| *Ictalurus punctatus* (Ictaluridae) | 4 | 976 | 14 | 70,000 | 3.75 |
| *Amblyopsis spelaea* (Amblyopsidae) | 3–4 | 113 | 7 | 70 | 2.15 |
| *Lepomis macrochirus* (Centrarchidae) | 2 | 384 | 9 | 81,104 | 1.20 |
| *Pomoxis annularis* (Centrarachidae) | 2 | 505 | 8 | 213,000 | 0.89 |
| *Etheostoma spectabile* (Percidae) | 1 | 80 | 3 | 320 | 1.42 |
| *Stizostedion vitreum* (Percidae) | 5 | 790 | 14 | 400,000 | 1.75 |
| Marine Species | | | | | |
| *Acipenser oxyrhynchus* (Acipenseridae) | 20 | 2743 | 60 | 3,755,745 | 2.55 |
| *Anguilla rostrata* (Anguillidae) | 8 | 740 | 40 | 2,561,000 | 1.00 |
| *Alosa pseudoharengus* (Clupeidae) | 3–4 | 352 | 9 | 466,701 | 0.90 |
| *Anchoa mitchilli* (Engraulidae) | <1 | 100 | 2 | 2,100 | 0.80 |
| *Oncorhynchus tshawytscha* (Salmonidae) | 4–5 | 1490 | 5 | 13,619 | 6.50 |
| *Osmerus mordax* (Osmeridae) | 2 | 310 | 5 | 69,600 | 0.95 |
| *Arius felis* (Ariidae) | ? | 355 | 8 | 68 | 16.0 |
| *Merluccius productus* (Gadidae) | 4 | 910 | 11 | 496,000 | 1.12 |
| *Lycodopsis pacifica* (Zoarcidae) | ? | 460 | 5 | 52 | 2.00 |
| *Gasterosteus aculeatus* (Gasterosteidae) | 1 | 102 | 3 | 150 | 1.70 |
| *Katsuwonus pelamis* (Scombridae) | 2 | 870 | 7 | 1,900,000 | 1.05 |
| *Ammodytes americanus* (Ammodytidae) | 2 | 210 | 9 | 1,313 | 0.83 |
| *Pomatomus saltatrix* (Pomatomidae) | 2 | 710 | 14 | 195,000 | 1.05 |
| *Haemulon aurolineatum* (Pomadasyidae) | 3 | 289 | 9 | 83,000 | 0.93 |
| *Scianops ocellatus* (Scianidae) | 4–5 | 1986 | 33 | 3,500,000 | 0.93 |
| *Lagodon rhomboides* (Sparidae) | 1 | 437 | 7 | 39,200 | 1.02 |
| *Centropristis striata* (Serranidae) | 1–2 | 550 | 11 | 1,050,000 | 0.95 |
| *Lutjanus campechanus* (Lutjanidae) | 2 | 906 | 13 | 9,300,000 | 0.82 |
| *Morone saxatilis* (Percichthyidae) | 4–5 | 1245 | 17 | 4,010,325 | 1.78 |
| *Microgobius gulosus* (Gobiidae) | 1 | 71 | 3 | 567 | 1.14 |
| *Sebastes caurinus* (Scorpaenidae) | 4 | 570 | 20 | 640,000 | 0.95 |
| *Ophiodon elongatus* (Hexagrammidae) | 5 | 1520 | 16 | 500,000 | 3.50 |
| *Paralichthys dentatus* (Bothidae) | 2 | 445 | 10 | 4,190,000 | 0.98 |
| *Pseudopleuronectes americanus* (Pleuronectidae) | 4–5 | 570 | 12 | 3,329,000 | 0.81 |

[a] Average for a population near the center of a species range.
[b] Maximum clutch size reported for a population near the center of a species range.

salmon use their caudal fins to bury their eggs in gravel (create spawning redds), and European bitterling (*Rhodeus sericeus*) deposits their eggs in the gills of unionid clams. Brood-guarding occurs in a variety of forms. Ovoviviparity (live-bearing without maternal nutritional contribution during gestation) and viviparity (live-bearing with maternal nutritional contribution during gestation) are common in a variety of fish taxa, including many sharks, skates, rays, guppies and other poeciliids

(Poeciliidae), surfperches (Embiotocidae), and the coelacanth *Latimeria chalumnae*. The coelacanth gives birth to 1–5 very large, advanced offspring that gain nutrition during gestation within the oviduct by consuming unfertilized trophic eggs and perhaps even smaller less-advanced siblings.

Male guarding of the nest and brood is perhaps the most common form of parental care in fishes. Most cichlids have biparental care, but a few species, like the dwarf cichlids (*Apistogramma*), have maternal care. External-bearing and mouth and gill chamber brooding are other forms of parental care in fishes. Males of several South American aspredinid and loricariid catfishes brood their embryos on the surface of their bellies, and male pipefishes and sea horses (Syngnathidae) carry their eggs and larvae in a belly brood pouch. Males of the Australian nursery fishes (*Kurtus*) carry their developing embryos on a hook that protrudes from their foreheads. The developing embryos hang on either side of the head in small clusters suspended by a twisted cord made from egg membranes looped through the eye of the hook. Oral brooding occurs in a number of freshwater and marine families, including the ariid catfishes (Ariidae), bony tongues (Osteoglossidae), cichlids (Cichlidae), and cardinal fishes (Apogonidae) [*See* LIFE HISTORIES.]

*Life-history strategies* result from trade-offs among attributes, such as clutch size and egg size, that have either direct or indirect effects on reproduction and fitness. For example, one way to achieve a larger clutch is to partition the biomass available for reproduction into smaller eggs. This functional constraint contributes to the negative correlation between clutch size and egg size frequently observed among fish species. Larger clutches can also be attained by delaying reproduction and growing to a larger body size. Comparing both within and across species, larger fishes tend to produce more eggs per batch than do smaller fishes. Yet, exceptions to this rule are common. Many fishes with parental care, like the mouth-brooding gaftopsail catfish (*Bagrus marinus*), have much smaller clutches than those of smaller species that scatter their eggs, like anchovies (*Anchoa mitchilli*) and dace (*Rhinichthys atratulus*).

Comparative life-history studies of fishes have repeatedly identified associations of large adult body size with delayed maturation, long life span, large clutches, small eggs, and few spawning bouts during a short reproductive season. Fishes with well-developed parental care tend to have larger eggs and often have longer reproductive seasons and serial spawning. Three primary life-history strategies define the end points of a continuum derived from comparisons of diverse ecological and taxonomic groupings of fishes worldwide (Fig. 2). At one end point, *periodic-strategists* have delayed maturation, mature at intermediate or large body sizes, produce large clutches of small eggs, tend to spawn in annual episodes, and tend to exhibit rapid growth during the first year of life. *Opportunistic-strategists* mature early at small sizes, produce small- to medium-sized clutches of small eggs, have multiple spawning bouts each year, and grow rapidly as larvae. *Equilibrium-strategists* may be any size, but tend to produce small- or medium-sized clutches of relatively large eggs, and often have brood guarding or maternal provisioning of nutrients to developing embryos.

Fishes near the periodic region of the life-history continuum probably reap two major benefits from delayed maturation and large adult body size: capacity to produce large clutches and enhanced adult survival during periods of suboptimal environmental conditions, like winter and periods of reduced food availability. Fishes with large clutches frequently spawn in synchronous bursts that coincide

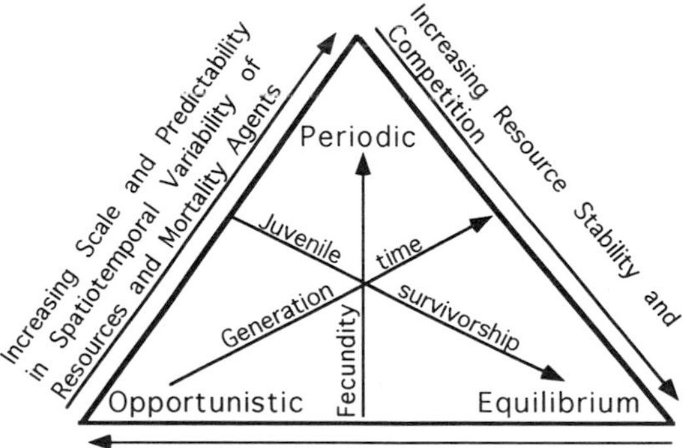


**FIGURE 2** Schematic representation of the triangular continuum of primary life-history strategies in fishes and environmental factors associated with selection gradients. [Based on K. O. Winemiller and K. A. Rose (1992). *Canad. J. Fish. Aquat. Sci.* **49**(10), 2196–2218.]

either with migration into favorable habitats or with favorable periods within the temporal cycle of the environment, like the spring or rainy season. Cod, cobia, mackerels, tunas, and other marine fishes with tiny pelagic eggs and larvae also have among the highest fecundities. These species cope with large-scale spatial variation in the marine pelagic environment by producing huge numbers of tiny offspring, at least some of which are bound to thrive once they encounter favorable strata or patches. On average, larval survivorship is extremely low among highly fecund fishes in the marine environment, and the average larval fish probably dies during the first week of life. Fast larval growth rates reflect successful exogenous feeding by the lucky survivors that encounter areas of relatively high prey density.

At higher latitudes, large-scale and cyclic temporal variation in environmental conditions is a major factor selecting for the timing of reproduction. Highly fecund fishes can exploit predictable patterns in time or space by releasing massive numbers of progeny in phase with periods in which environmental conditions are most favorable for larval growth and survival. Selection favors physiological mechanisms that enhance a fish's ability to detect cues that predict the periodic cycle (photoperiod, ambient temperature, solute concentrations). In tropical marine pelagic environments, large-scale variation in space may represent a periodic signal as strong as the seasonal variation at temperate latitudes. Research in physical oceanography has shown patchy distributions for a variety of physical parameters (salinity, temperature), primary production, and zooplankton due to upwellings, gyres, convergence zones, and other currents.

Many periodic-type species are migratory. Anadromous American shad (*Alosa sapidissima*) show more repeat spawning and devote a greater portion of energy to migration at higher latitudes where environments are more variable and less predictable. Anadromous sticklebacks (*Gasterosteus aculeatus*) have more periodic-type life-history traits (larger clutches, larger size at maturity) compared with conspecific freshwater populations. By adopting anadromy, adult fishes can find favorable environments for the development and survival of their larvae. In contrast, the reproductive success of marine broadcast spawners depends on rates of larval encounters with suitable zones or patches. Massive clutches of small pelagic eggs undoubtedly enhance dispersal capabilities of wide-ranging marine fishes during the early life stages. In a stable population, losses due to settlement in hostile habitats (called *advection*) ultimately are balanced by the survival benefits derived from the passage of some fraction of larval cohorts into suitable regions or habitats.

The opportunistic life-history strategy is associated with rapid population turnover rates and a high intrinsic rate of population increase (*r*, an index of the potential for exponential growth). By having among the smallest rather than largest clutches, opportunistic-type fishes differ markedly from the traditional model of *r*-strategists. Yet because of their small size, the *relative* reproductive effort of opportunistic strategists is actually high, despite the fact that *absolute* clutch size and egg size are small. In these small species, serial spawning sometimes results in an annual reproductive biomass (i.e., annual fecundity) that exceeds the female's body mass. Small fishes with early maturation and frequent spawning are well equipped to repopulate habitats following disturbances and to sustain their numbers when faced with continuously high mortality during the adult stage. Given their high intrinsic rates of increase, opportunistic-strategists are relatively efficient colonizers of disturbed habitats. A relative opportunistic-strategy is observed in the bay anchovy (Engraulidae), silversides (Atherinidae), annual killifishes (Rivuliidae), marsh killifishes (Cyprinodontidae), and mosquito fishes and other freshwater live-bearers (Poeciliidae, Goodiidae). These small fishes often maintain dense populations in marginal or constantly changing habitats and persist in the face of high predation mortality during the adult stage. Extreme examples of the opportunistic-strategy appear to be more common in tropical fresh waters than in the temperate zone, and in shallow marginal habitats than in deeper freshwater and marine habitats.

An equilibrium life-history strategy in fishes corresponds largely with the suite of traits associated with the traditional *K*-strategy of adaptation to life in resource-limited or density-dependent environments (delayed maturation, large body size, clutches containing few large offspring). Large eggs and parental care yield larger or more developmentally advanced juveniles at the onset of independent life. Live-bearing sharks and rays that bear relatively large, advanced offspring would lie near the equilibrium end point of the life-history spectrum. Among bony fishes, marine ariid catfishes (egg diameters 16–20 mm, oral brooding of eggs and larvae) and amblyopsid cave fishes (branchial brooding of small clutches of relatively large eggs) illustrate relatively extreme forms of this equilibrium-strategy. Cave fishes inhabit among the most stable and resource-limited of aquatic habitats. Parental care tactics (including long gestation in live-bearers) tend to be more highly developed and widespread in tropical fresh waters and among certain reef-dwelling marine fishes, such as sea horses, surfperches, and sharks.

Of course most fishes are associated not with a particular end point strategy, but rather with intermediate strategies within the triangular gradient of primary life histories. Some of the largest periodic-type fishes, the sturgeons (Acipenseridae) and paddlefish (Polyodontidae), have relatively large eggs, which reduces their theoretical maximum clutch size. Salmon and trout possess even larger eggs (4–6.5 mm diameter) and smaller clutches than fishes exhibiting the extreme periodic-strategy. Yet among populations of coho salmon (*Oncorhynchus kisutch*), egg size declines and clutch size increases with increasing latitude. Selection seems to favor local optima in egg size with clutch size adjustments resulting from physiological constraints and ecological performance. Relative to periodic-strategists with larger clutches and smaller eggs, salmon and trout apparently have evolved a more equilibrium-strategy of fewer but larger offspring at the onset of independent life. In eutrophic ecosystems, a pulse of primary and secondary production during short summers probably favors a periodic-type strategy at high latitudes. Studies of several fish species with large ranges along the eastern North American coast show that juvenile growth rates are actually faster for fishes at higher latitudes where the growing season is shorter. Yet, the growing season at high latitudes may be so short that brood guarding is not a viable tactic for large fishes in oligotrophic systems. Data from arctic char (*Salvelinus alpinus*) indicated that the growing season at high latitudes probably constrains age at maturity and the frequency of spawning. Migration to special spawning habitats and burial of fertilized eggs (brood hiding) by salmon, char, and trout are forms of parental investment that carry large energetic and survival costs in relation to future reproductive effort. Senescence associated with semelparity in Pacific salmon might have evolved as a consequence of the survival cost of returning to the sea after energetically costly upstream runs to fluvial habitats that enhance larval survivorship. By comparison, freshwater whitefishes (*Coregonus, Prosopium*) exhibit a perennial periodic-strategy that involves large clutches, small eggs, and annual spawning bouts.

A number of intermediate-sized fishes have seasonal spawning, moderate clutch sizes, and nest guarding (North American ictalurid catfishes and sunfishes, *Lepomis*). Rockfishes of the eastern Pacific (Scorpaenidae) have large clutches and small eggs and bear living young. All of these fishes lie between periodic and equilibrium end points of a triangular gradient. Small fishes with rapid maturation, small clutches, large eggs relative to body size, and a degree of parental care (minnows of the genus *Pimephales,* madtoms, darters, sticklebacks, pipefishes, and sculpins) lie between opportunistic- and equilibrium-strategists. Similarly, small fishes with seasonal spawning, moderately large clutches, small eggs, and only one or a few bouts of reproduction per season lie between opportunistic and periodic extremes of the gradient. It is important to note that fishes with divergent life-history strategies frequently coexist in the same habitats. Each species' morphology and feeding niche determine the nature of the resource variation and predation that it experiences; and morphological constraints, including features involved in feeding within particular microhabitats, restrict the evolution of life-history features. In addition, phylogenetic constraints will result in varying degrees of adaptive divergence or evolutionary convergence toward a given adaptive suite of life-history traits.

## V. POPULATION REGULATION

The study of population regulation in fishes has been dominated by the search for density-dependent recruitment in commercial stocks. Several simple models describe density-dependent relationships between stock abundance (expressed as either spawning adults or egg cohorts) and the abundance of recruits (expressed as age 1 fishes, or cohorts when they enter the fishery or become vulnerable to sampling gear). These include the well-known Ricker and Beverton–Holt stock/recruitment models. By and large, data from large marine and lake fisheries conform very poorly to density-dependent, stock-recruit models. This has spurred greater examination of potential density-independent factors that influence recruitment, such as climatic fluctuations and seasonal changes in marine currents. Evidence for density-dependent population dynamics has been obtained from a variety of sources, including studies of local population dynamics, niche relationships, and predator- or resource-regulated community dynamics. [*See* Population Regulation.]

In terms of life-history strategies, fitness can be estimated by either $V_x$, the reproductive value of an individual or age-class, or by $r$, the intrinsic rate of natural increase of a population or genotype. Each of these fitness measures can be expressed as a function of three essential components: survivorship, fecundity, and the onset and duration of reproductive life. In the case of reproductive value,

$$V_x = m_x + \sum_{t=x+1}^{\omega} (l_t m_t)/l_x,$$

where for a stable population $m_x$ is age-specific fecundity, $l_x$ is age-specific survivorship, and $\omega$ is the last age-class of active reproduction. When $x$ is equal to $\alpha$, the age of first reproduction, reproductive value is equivalent to the lifetime expectation of offspring, and contains survivorship, fecundity, and timing components. The intrinsic rate of population increase can be approximated as $r \simeq \ln R_0/T$, where $R_0$ is the net replacement rate, $T$ is the mean generation time, and $R_0 = \Sigma\, l_x\, m_x$, resulting in $r \simeq \ln (\Sigma\, l_x\, m_x)/T$. The relative rate of population increase is directly dependent on fecundity, timing of reproduction, and survivorship during both immature and adult stages. Averaged over many generations, the three parameters ($l_x$, $m_x$, $T$) must balance, or populations would decline to extinction or would grow to precariously high densities and eventually crash.

The three end point life-history strategies of fishes referred to previously are associated with trade-offs among age at maturation ($\alpha$ positively correlated with $T$), fecundity, and survivorship. The periodic-strategy corresponds to high values on fecundity and age at maturity axes (the latter is a correlate of population turnover rate) and a low value on the juvenile survivorship axis. The opportunistic-strategy of high population turnover rate via rapid maturation corresponds to low values on all three axes. The equilibrium-strategy corresponds to low values on the fecundity axis and high values on the age of maturity and juvenile survivorship axes.

Clearly, the periodic-strategy maximizes age-specific fecundity (clutch size) at the expense of optimizing turnover time (turnover times are lengthened by delayed maturation) and juvenile survivorship (maximum fecundities are attained by producing smaller eggs and larvae). Large body size enhances adult survivorship during suboptimal conditions and permits storage of energy and biomass for future reproduction. Iteroparity permits a fish to sample its environment several times until, sooner or later, reproduction coincides with favorable conditions and strong recruitment occurs. Virtually all ecosystems exhibit either spatial or temporal variation that is to some degree predictable. This may be especially true in freshwater and marine environments, because batch spawning of large clutches is predominant among bony fishes worldwide. Spawning by these periodic-type fishes is usually annual and synchronous, so that generations are often recognized as discrete annual cohorts. Yet, correlations between parental stock densities and densities of young-of-the-year recruits have been shown to be negligible in these fishes. Recruitment often depends on climatic conditions that influence water currents, larval retention zones, productivity of prey patches, and a host of other environmental factors that determine early growth and survival. Accurate predictions of recruitment by periodic-type species in large marine ecosystems require understanding of physical oceanography plus the ability to forecast weather conditions. Because weather cannot be predicted over long time intervals, fisheries projections often rely on short-term estimates of juvenile cohort strength several weeks or months following spawning, rather than long-term estimates based on parental stocks.

Many periodic-type fishes spread their reproductive effort over many years (or over large areas), so that high larval/juvenile survivorship during one year (or in one area) compensates for the many bad years (or areas). For example, anadromous female striped bass (*Morone saxatilis*) live up to 17 years on average and produce an average clutch of $4 \times 10^6$ eggs every year or two. This requires survivorship of roughly $3 \times 10^{-8}$ during the egg to maturation interval to maintain a stable population. Most years probably result in a larval survivorship approaching zero for most females. For an individual female, the fitness payoff comes only during one or two spawning acts over the course of a normal life span. In species like striped bass, the variance in larval survivorship that serves as input for population projections lies well beyond our ability to measure differences in the field. Management of exploited populations of long-lived, highly fecund fishes requires the maintenance of critical densities of adult stocks and the protection of spawning habitats during the short spawning season. Because recruitment is largely determined by unpredictable interannual environmental variation, this critical density is impossible to determine with any degree of precision. And because most larvae never recruit into the adult population even under pristine conditions, it follows that some spawning must proceed unimpeded each year if strong recruitment is to occur during the exceptional year.

The opportunistic life history maximizes the intrinsic rate of population growth ($r$) through a reduction in the mean generation time ($T$). Fishes exhibiting an opportunistic-type strategy are often associated with shallow marginal habitats. These edge habitats are the kinds of environments that experience the largest and most unpredictable changes on small temporal and spatial scales. Changes in precipitation and temperature induce major alterations in water depth, substrate characteristics, and productivity in shallow aquatic habitats. Population density estimates for small fishes in shallow marginal habitats, like headwater streams and salt marshes, has shown large monthly variation. In the absence of chronic intense predation and resource limitation, opportunistic-type populations can quickly rebound from localized disturbances.

Because they tend to be small and occur in shallow marginal habitats, opportunistic-type fishes are not usually exploited commercially. Some important commercial species, like gulf menhaden (*Brevoortia patronus*), are intermediate between opportunistic- and periodic-strategies. Yet, small fishes are often the most important food resources for larger piscivorous species. Bay anchovies (*Anchoa mitchilli*) and silversides (*Menidia menidia*) inhabit relatively stable habitats, yet they suffer high adult mortality from predation. Given the capacity for opportunistic-strategists to sustain losses during all stages of life spans that are typically rather short, one of the keys to their management is protection from large-scale or chronic perturbations that eliminate important refugia.

The equilibrium-strategy in fishes is roughly equivalent to the traditional $K$-selection model of evolution in density-dependent and resource-limited environments. For example, some stream-dwelling darters (*Etheostoma*) and madtoms (*Noturus*) probably have fewer refuges in shallow riffles during periods of reduced stream flow. If refuges are limited, individuals may be forced to compete for depleted food supplies in areas immediately surrounding refuges. In the East African rift lakes, many brood-guarding and mouth-brooding cichlids have home ranges that cover only a few square meters. In the marine environment, parental care is most frequently seen in small fishes associated with the benthos or structure (damselfishes, pipefishes, sea horses, eelpouts, some gobies). Compared with opportunistic- and periodic-strategists, equilibrium-strategists ought to experience lower temporal variation in population density and conform better to stock-recruit models.

Because equilibrium-strategists produce small numbers of offspring, early survivorship must be relatively high for these populations to avoid local extinction. When parental care is involved, survivorship during early life stages depends on both the condition of adults and their nesting habitat. Relatively few equilibrium-type fishes are commercially exploited on a massive scale. Sev-eral species exploited by sport fisheries exhibit brood guarding and are intermediate between equilibrium- and periodic-strategies, including lingcod (*Ophiodon elongatus*) and the North American sunfishes. Management of exploited stocks of equilibrium-type fishes requires the maintenance of undegraded habitats and adult stock densities that promote surplus yields that can be fished and replaced via natural compensatory mechanisms.

## VI. ECOLOGICAL INTERACTIONS

### A. Predation

Predation has a major influence on fish evolution, population dynamics, and community structure. The diversity of feeding mechanisms of fishes is derived from the mechanical trade-offs involved in harvesting different kinds of food resources. In the absence of stable food resources and interspecific competition, generalized feeding is often favored. Many fishes show seasonal diet shifts, in which a greater variety of prey is consumed during periods of prey abundance. In diverse fish assemblages, predatory fishes usually exhibit divergence in feeding tactics and diet. Piscivores may feed by a sit-and-wait tactic (flounders, scorpion fishes), stealth and rapid pursuit (barracudas, pikes), solitary active searching (snappers, black bass), or actively searching in schools (tunas, piranhas). Divergence in piscivore foraging tactics has coevolved with an equivalent amount of evolutionary and ecological divergence in prey escape tactics. Prey species avoid predation by crypsis, mimicry, hiding, schooling, spines, bony plates, and aggression. Small fishes, like anchovies and killifishes, often have few specialized defenses against predators (other than hiding and schooling), yet their populations thrive owing to their rapid population turnover times and high growth potentials.

Piscivores can have major effects on the density, population structure, and behavior of their prey. The addition of piscivores (largemouth bass, *Micropterus*) to North American lakes has been shown to decrease the abundance of small planktivorous fishes (minnows), which in turn leads to wholesale changes in zooplankton, phytoplankton, and nutrient dynamics. Because mouth gape and the diameter of the throat limit the size of prey ingested, piscivores that swallow their prey whole are often very size selective. Size selectivity can alter the size and age structure of prey populations. For example, small gizzard shad are more vulnerable than larger conspecifics to predation by largemouth bass and other piscivorous fishes. Shad populations in predator-dense habitats tend to be dominated by older age-classes comprising larger individuals. In Europe, the size/age structure of roach (*Rutilus*) populations is often influenced by predatory perch (*Perca*). Piscivores can also influence the competitive interactions between prey species. Walleye (*Stizostedion*) are efficient predators of both young-of-the-year perch (*Perca*) and gizzard shad (*Dorosoma*). In temperate lake ecosystems, young perch compete with shad for zooplankton. If shad densities are high, competition can cause a competitive bottleneck for perch recruitment into older age-classes. At high densities of gizzard shad, young perch grow more slowly and remain vulnerable to walleye for a longer period of time.

Effects of piscivores on aquatic ecosystems are clearly demonstrated by introductions of exotic predators. The predatory sea lamprey has contributed to the decline of a number of fishes of the Laurentian Great Lakes, including lake trout and endemic whitefishes (*Coregonus*). The introduction of peacock bass (*Cichla*) into Lake Gatún, Panama, resulted in major changes in the aquatic food web and local extirpation of several native fish species. In Africa, the introduction of predatory Nile perch (*Lates*) into Lake Victoria ranks among the most devastating exotic species introductions in history. Prior to the introduction of Nile perch, the lake's fish community was dominated by more than 400 species of endemic haplochromine cichlids. Lake Victoria is now dominated by Nile perch, a native minnow, and an introduced species of tilapia (a detritivorous/herbivorous cichlid), and a great number of haplochromine cichlids are now extinct. The complex food web supported by the ecologically diverse haplochromines has been replaced by a simple food web linking detritus–shrimp, tilapia, minnows–Nile perch. Recent evidence indicates that the entire lake ecosystem's nutrient dynamics may have been destablized by the replacement of haplochromine biomass and diversity with shrimp, minnow, and tilapia biomass.

By influencing the behavior of their prey, piscivorous fishes can have subtle effects on ecosystem structure and function. Recent studies have demonstrated that fishes are able to perceive predation risks in different habitats, and that they alter their use of habitats in accordance with the relative costs and benefits of foraging gains versus predation risks. Experiments have shown that algae-feeding stone rollers (*Campostoma*) will avoid stream pools containing largemouth bass. As a consequence, pools that contain bass develop and maintain more benthic algae than pools without bass. In North American streams, dace (*Rhinichthys*) will avoid pools that containing large piscivorous creek chubs (*Semotilus*). Stream areas containing higher concentrations of food induce dace to take greater predation risks. Similarly, in the presence of largemouth bass (*Micropterus*), bluegill sunfish (*Lepomis*) spend more time foraging in vegetation than in open water areas of ponds where foraging is more profitable.

### B. Competition

Competition occurs when the supply of resources is insufficient to meet the demand of two or more conspecific consumers (intraspecific competition) or consumer populations (interspecific competition). Food and habitat have been identified as the resources that most often limit individual fitness and population densities. Historical competition is frequently inferred from comparisons of species characteristics within natural assemblages. Diverse fish assemblages in the tropics show more interspecific variation in ecomorphology than assemblages from comparable habitats at higher latitudes. Resource segregation in response to interspecific competition is a likely causal factor for this pattern. In a study of four tropical fish assemblages, fishes were clustered into feeding guilds and species within guilds were segregated in their use of food resources more than would be expected by chance. Many studies have shown high levels of resource segregation, particularly in fishes' use of habitat. Considerable diet and habitat overlap have been observed during early stages of ontogeny and during seasonal increases in resources in many different aquatic and marine ecosystems. Declines in fish diet or habitat overlap during periods of natural resource depression provide comparative evidence for interspecific competition.

Interspecific competition has been inferred from observed niche shifts in response to species introductions in natural and experimental systems. The introduction of planktivorous alewife (*Alosa pseudoharengus*) into Lake Michigan resulted in native bloaters (*Coregonus hoyi*) shifting to forage in deeper waters on larger and more benthic invertebrate prey. The bloater's shift in ecology was accompanied by a reduction in the length and number of gill rakers formerly used for straining plankton from the water column. In experimental ponds, bluegill sunfish (*Lepomis macrochirus*), green sunfish (*L. cyanellus*), and pumpkinseed sunfish (*L. gibbosus*) use a similar range of habitats and food resources when stocked as single species. When all three are stocked together, bluegills feed more in open water on zooplankton, green sunfish feed more in the vegetation on aquatic insects, and pumpkinseeds feed more on benthic invertebrates. The growth rates of each species are slower in multispecies ponds compared with similar single-species ponds, offering further evidence of interspecific competition for resources.

## VII. COMMUNITY PATTERNS AND BIODIVERSITY

In general, regional fish diversity is positively correlated with drainage basin area and negatively correlated with latitude. South American fresh waters contain the highest regional freshwater fish diversity, and the southwestern Pacific Ocean contains the greatest diversity of marine fishes. Within freshwater families having little or no salinity tolerance, species and populations show conservative biogeographic patterns in which recently divergent taxa are restricted to disjunct drainage basins. On a longer time scale, distributions of fish families reflect dispersal across continental landmasses via stream captures and freshwater intrusions into coastal marine habitats. For example, catastomid suckers range from eastern Asia across the Bering Strait into North America and south to central Mexico; and centrarchid sunfishes are restricted to eastern, central, and southern portions of North America. Minnows, carp, and other cyprinid fishes are found on all continents except South America, Australia, and Antarctica. Marine taxa have distributional patterns that reflect biogeographical processes on different scales. The Central American isthmus separates recently divergent sister species of numerous coastal marine taxa. Several marine families have circumglobal distributions that are restricted by latitude, such as the anadromous salmonids in the Northern Hemisphere and their sister clade, the anadromous argentiniids in the Southern Hemisphere [*See* BIODIVERSITY.]

At a local level, fish species composition results from both historical regional processes and contemporary ecological processes. Analyses of lake assemblages in Canada showed that geographical distance explained the level of fish assemblage similarity more than local habitat conditions. In contrast, streams located only a few kilometers apart, but with different habitats, may share only a very small fraction of their species. For example, some tropical freshwater fishes are restricted to "blackwater" streams that flow through regions containing nutrient-poor soils. Other species inhabit only nutrient-rich "whitewater" rivers that support high aquatic primary production and biomass of invertebrates and fishes. Species composition of these assemblages correlates more with habitat conditions than geographical distance. Several studies have used multivariate statistical methods to examine fish assemblage composition in relation to habitat gradients. Stream systems in many different regions of the world show clear patterns of faunal change and species turnover in relation to longitudinal fluvial gradients. Many small species are restricted to headwater tributary streams, where stream flow is highly variable and habitats are very dynamic. As one moves downstream into larger, more stable, and more productive habitats, a few of the headwater species are retained and new species are added.

Fish biodiversity is threatened by a number of factors. The introduction of exotic species can eliminate native species directly via competition and predation or indirectly via alteration of the food web. On a global basis, the tiny North American mosquito fish (*Gambusia affinis*) and herbivorous African tilapias (*Tilapia, Sarotherodon, Oreochromis*) may be the most widespread and damaging exotic species. Carp (*Cyprinus carpio*), largemouth bass (*Micropterus salmoides*), rainbow trout (*Oncorhynchus mykiss*), and brown trout (*Salmo salar*) have been introduced worldwide for sportfishing. The effects of these introductions on native species have varied greatly, but in most areas the full impact on the native fauna is unknown. Exotic introductions can result in genetic contamination of locally adapted fish populations, as when hatchery salmon are introduced to rivers containing native stocks. Overfishing and environmental degradation from a variety of agents have caused the decline of fish populations and species diversity in virtually all major aquatic and marine ecosystems. Fluvial and estuarine ecosystems are among these most affected by pollution and other anthropogenic perturbations in regional landscapes. Deforestation and plowing cause increased soil erosion, runoff, and sedimentation, all of which degrade fish habitat. Trace contaminants like lead and PCBs can attain high concentrations in fish tissues through the process of bioaccumulation. Elevated nutrient levels from pollution and agricultural runoff increase primary productivity of aquatic ecosystems, which frequently leads to higher biomass of less desirable species like carp and gar and declines in game fishes like largemouth bass. In arid regions, fragile aquatic habitats have been completely destroyed by removal of subsurface water for irrigation, industry, and municipalities. When they block movement to critical spawning habitats, dams can eliminate local populations of migratory fishes in little more than a single generation. [*See* MARINE BIOLOGY, HUMAN IMPACTS.]

Fishes can serve as sensitive indicators of water quality and ecosystem health. Small species that adjust well to laboratory conditions, such as the fathead minnow (*Pimephales promelas*) and the medaka (*Orizias latipes*), are used extensively in toxicological research. Species that have broad geographical ranges and are highly sensitive to habitat degradation are useful as bioindicators. For example, many North American stream-dwelling darters (Percidae) require oxygen-rich water flowing over hard substrates. These species are among the first eliminated from habitats degraded by pollution, siltation, and other landscape perturbations. In addition, the structure of local fish communities has been used as an indicator of ecosystem stress. Under normal conditions, diverse fish faunas will contain a mixture of detritivorous, insectivorous, omnivorous, and piscivorous fish species. Oftentimes, piscivores will be supported by greater numbers and a greater variety of fishes at lower trophic levels. An index of biotic integrity was developed to evaluate ecosystem change as a result of anthropogenic perturbations. This index compares the structure of fish communities in degraded habitats with the structure of communities in similar but unperturbed habitats from the same biogeographic region.

## Glossary

**Anadromous** Of or pertaining to species that spend most of their lives in the sea and migrate to fresh water to breed.

**Assemblage** Group of species populations coexisting in a local area, usually pertaining to species belonging to the same higher taxonomic grouping (e.g., fish assemblage, minnow assemblage, trout assemblage).

**Benthic** Of or pertaining to the bottom or substrate region of freshwater and marine environments. Category of fishes that inhabit the bottom region (e.g., many catfishes, sculpins, darters).

**Catadromous** Of or pertaining to species that spend most of their lives in fresh water and migrate to the sea to breed.

**Demersal** Of or pertaining to the deep-water regions near the bottom or substrate of aquatic and marine environments.

**Diadromous** Of or pertaining to species that migrate between fresh water and the sea.

**Fluvial** Of or pertaining to flowing freshwater ecosystems such as streams and rivers.

**Iteroparity** Condition of performing repeated episodes of reproduction over the lifetime.

**Marine** Of or pertaining to oceanic or coastal environments, typically containing solute concentrations greater than 30 parts per thousand.

**Pelagic** Of or pertaining to the open-water region of freshwater and marine environments.

**Pharyngeal teeth** Throat teeth that occur on pads on various gill arch elements in many fishes.

**Semelparity** Condition of performing a single bout of reproduction during the lifetime.

**Stock** Local population of fishes, usually referring to one that is exploited as a commercial or recreational resource.

## Bibliography

Breder, C. M., Jr., and Rosen, D. E. (1966). "Modes of Reproduction in Fishes." Garden City, N.Y.: Natural History Press.

Hilborn, R., and Walters, C. J. (1992). "Quantitative Fisheries Stock Assessment." New York: Chapman & Hall.

Lowe-McConnell, R. H. (1987). "Ecological Studies in Tropical Fish Communities." Cambridge, England: Cambridge University Press.

Marshall, N. B. (1971). "Explorations in the Life of Fishes," Harvard Books in Biology, No. 7. Cambridge, Mass.: Harvard University Press.

Matthews, W. J., and Heins, D. C., eds. (1987). "Community and Evolutionary Ecology of North American Stream Fishes." Norman: University of Oklahoma Press.

Moyle, P. B., and Cech, J. J., Jr. (1988). "Fishes: An Introduction to Ichthyology." Englewood Cliffs, N.J.: Prentice-Hall.

Nelson, J. S. (1984). "Fishes of the World," 2nd ed. New York: John Wiley & Sons.

Sale, P. F., ed. (1991). "The Ecology of Fishes on Coral Reefs." New York: Academic Press.

Winemiller, K. O., and Rose, K. A. (1992). Patterns of life-history diversification in North American fishes: Implications for population regulation. *Canad. J. Fish. Aquat. Sci.* **49**(10), 2196–2218.

Wooton, R. J. (1990). "Ecology of Teleost Fishes." New York: Chapman & Hall.

# Foraging by Ungulate Herbivores

**John F. Vallentine**
*Brigham Young University*



Foraging is a complex process by which ungulate herbivores search for and consume forage from a standing forage crop. In foraging, a hierarchy of instinctive responses and behavioral actions are included within the following steps: (1) site selection for grazing/browsing; (2) selection of plant species, individual plants, and plant parts on that site; and (3) prehending and ingesting the selected forage. In all aspects of foraging, there are substantial differences between animal species but also between individuals within species caused in part by different past experience and training.

## I. SITE SELECTION IN FORAGING

Free-ranging wild or domestic animals often exhibit extreme nonrandomness in the use of grazing lands. When environmental resources are heterogeneous and patchy, both spatially and temporally, the herbivores are likely to strongly select against some sites and congregate on others. Site preference for foraging (and also for resting and ruminating, drinking, bedding, and other activities) results from complex interactions of both abiotic and biotic factors as follows:

*Abiotic*—weather (temperature, precipitation, wind, storminess, etc.), soil characteristics, topography and landform features, elevation, aspect, water availability, and fencing and other barriers.

*Biotic*—plant communities, botanical composition, quantity of forage, quality/palatability of plants, shade and shelter, escape cover, inter- and intraspecific animal behavior, insect pests, and human activity.

When animals forage, they narrow down the alternatives to a specific location and then search for desirable forage within that location. The site selection decisions required of large generalist herbivores can be generalized into the following:

1. geographical region (determined by environment, kind of animal, evolution, or management restrictions);
2. landscape (bounded by fence, other management control, or home range determinants);
3. plant community or terrain type (affected by many factors contributing to the favored habitat);

4. feeding area (influenced by terrain, proximity of cover, physical boundaries, animal sociality, established trails, etc.);
5. feeding station (the area immediately available for grazing when the forefeet are stationary).

The daily search area may comprise most or all of the landscape area or may be restricted to only a portion thereof by size of area, shape of area, topography, or distance from water. After full utilization has been made of forage on selected sites by the foraging animal, the search for new sites may be accelerated. The seasonal migration of mule deer is, in part, a response to seasonal change in forage quality as well as quantity in and between different vegetation types. [*See* FORAGING STRATEGIES.]

### A. Forage Factors

Forage factors play a prominent role in plant community selection by foraging animals. An abundance of palatable plants attracts grazing animals into the communities in which the plants are found. Nevertheless, animal preferences for plant species and plant communities are interrelated: animal preferences for plant species greatly influence selection of the grazing site, but the site being grazed then influences the plant composition of the diets. [*See* PLANT–ANIMAL INTERACTIONS.]

Riparian sites are favored areas for many species of foraging animals, including cattle, sheep, elk, moose, and often deer. A major attractant is the relatively high palatability, quality, and variety of forage on riparian sites; the availability of water, shade, thermal cover, and ease of accessibility are other attractants. In addition, drainage ways passing through grasslands or other seasonally dry sites often receive additional run-in water from surrounding slopes, permitting the forage to remain green longer than on adjoining upland sites. However, upland sites periodically offer attractive forage conditions, and management tools are frequently used to induce fuller use of less attractive areas while reducing stress on more attractive areas. Even the minimally attractive forage areas can play an important role in providing a reserve forage supply.

### B. Nonforage Vegetation Factors

Nonforage vegetation factors also influence the sites that animals select to graze, rest, and bed. These nonforage vegetation factors include timber harvesting, edge effects, shade, and shelter. Extensive clearings in wooded or brush areas (up to several hundred acres) may greatly increase forage production and be highly attractive to livestock and bison, but small patch cuttings (45 acres or less) are often more readily used by deer and elk. The willingness of big game animals to enter openings is influenced by a requirement for security during the feeding period but is locally modified by the past experience of the animals in the available environment.

An *edge* is the place where plant communities meet and often overlap; such transition areas are often richer in forage plant species and provide forage and cover in close proximity. Such areas are particularly attractive to big game animals, and special consideration is often given to create or maintain a mosaic of vegetation types thereby maximizing edge effects. The natural cover of trees and large shrubs provide shade in very hot weather, thermal protection in very cold weather, and possibly security from real or imagined enemies and can increase foraging in their vicinity.

Natural shade can be either a help or a hindrance to optimal livestock foraging depending on where it is located—good if located in otherwise less preferred areas but bad if located in overused areas. Anything that induces foraging animals to forage radially from some more or less fixed attraction point (water, salt, shade, bedding area), particularly on otherwise homogeneous areas, results in a heavily exploited zone nearest that point and a gradient of decreasing resource exploitation that diminishes with distance from that point. During cold, stormy weather, free-ranging livestock and big game seek shelter under natural tree cover, in tall grass or brush, behind ridges or planted tree windbreaks, in gullies and other depressions, or in selected man-made structures; this, in turn, tends to modify where herbivores select to forage. During the winter season, snow depth may forestall foraging on some areas by either covering the forage supply or restricting access, or both, while leaving adjoining, wind-swept areas open to grazing. Deep snow impairs search capability, animal mobility, and site selectivity and can greatly modify plant species selection.

### C. Slope and Other Physical Factors

Slope is an important factor affecting site selection and thus foraging distribution in hilly or mountainous areas, but its effect varies greatly among kinds of grazing animals. Cattle, horses, bison, and antelope generally prefer areas such as flatlands and rolling lands, valley bottoms, low saddles, level benches, and mesas but are probably more unwilling than incapable of grazing steeper areas. At the other extreme in animal habits, deer, domestic goats, domestic sheep, mountain goats, and bighorn sheep are more willing to utilize steeper terrain.

The availability of drinking water is an important limiting factor with all ungulate herbivores, particularly under conditions of dry or salty forage, high temperature, and low humidity. Open water for drinking should normally be provided yearround for both livestock and big game animals. When other factors do not limit foraging distribution, distance from drinking water ultimately controls the limit of vegetation utilization. Although forage factors play a major role in determining where herbivores forage, that is, set the inner boundary, distance from water is apt to set the outer boundary within which animals will forage.

Management tools are commonly employed to reduce site selectivity in order to disperse foraging animals and associated forage utilization over a greater part of the management unit. When foraging animals are not kept well distributed, the areas foraged too heavily as well as those foraged too lightly expand in size while those areas receiving optimal use become smaller. Primary tools for encouraging greater foraging in underused areas include providing additional watering places, utilizing fences, utilizing herding, placing salt and supplement attractants, and improving access. [*See* RANGE ECOLOGY AND GRAZING.]

## II. INGESTIVE BEHAVIOR

After the desired forage is selected by the foraging herbivore, it must be prehended (grasped) and taken into the mouth. The forage is then chewed, mixed with saliva, and prepared for swallowing. Jaw activity during foraging is complex, as it involves initial movements to arrange the herbage in the mouth, gripping the herbage with mouthparts, severing it from the plant by biting or jerking the head, and masticating and arranging the herbage for swallowing.

The anatomy of the jaw, teeth, and other mouthparts results in differences between animal species in how the herbage is grasped and severed from the plant. Cattle depend on their mobile tongue to encircle a mouthful of forage and draw it into the mouth, unless the vegetation is very short. The forage is then gripped between the upper and lower molars or between the incisor teeth in the lower jaw and the muscular pad in the upper jaw and severed from the plant by a backwards jerk of the head.

Bison ingest forage in a manner similar to cattle, but seldom prehend forage with their tongues in a horizontal plane as do cattle. The large, flat muzzles of cattle and bison allow relatively large clumps of vegetation to be drawn into the mouth at one time. Sheep either bite the foliage off the plant or break it off as they grip and jerk their heads backward or less commonly forward. Sheep are similar to cattle in having only molar teeth in the upper jaw, the incisors being replaced by a muscular pad. On the other hand, horses are able to bite close to the ground, having the advantage over sheep and cattle of both upper and lower sets of incisors. Because sheep have smaller mouths, they can take smaller bites and so are able to be more selective of plant species and plant parts. Goats have mouthparts and ingestive technique similar to those of sheep but are noted for mobile upper lips and prehensile tongues that permit them to eat tiny leaves of browse even from thorny species.

Camels are like sheep and goats in having mouthparts adapted for browsing on woody plants. The muzzle of the pronghorn is long and narrow, its mouth is small, and its upper lip is cleft like that of sheep, giving it a great deal of manipulative ability. The anatomy of the deer mouth is similar, and forage is either gripped between the molars and severed by biting action or seized between the incisors and the upper dental pad and sheared off with an upward or downward jerking action. Ungulate herbivores exhibit considerable plasticity in feeding behavior; this is necessary for animals that feed on plants that vary greatly in structure. The biting mechanics are varied somewhat—this within the limitations of mouth structural limitations—as the animal attempts to achieve its intake potential.

The typical activity of a grazing animal has been described as interrupted forward movements in which moving intervals and feeding intervals are interspersed. The pause for feeding has been referred to as the *feeding station interval* and the location as the *feeding station,* the latter more fully defined as the area available in a half-circle shape in front of and to each side of the foraging animal while its front feet are stationary. Feeding station intervals are normally short, seldom more than a few seconds unless animals are feeding selectively on a large plant such as a shrub, and are often not fully differentiated from the moving intervals. Besides interrupting its biting activity to move to a new feeding station, a foraging animal may be forced to pause to move out of the way of another animal or in response to any one of a number of disturbance factors.

Walking locomotion is an inherent part of foraging, and both free-ranging livestock and big game animals only occasionally travel without foraging. When high-quality forage is limited or the forage stand is heterogeneous and foraging animals are being highly selective, additional travel time and distance often result. Daily travel distances may be greatly increased when grazing animals must travel longer distances for adequate food and water. The energy cost of increased travel—this can result from animals being driven or chased by predators or insects, when responding to outside disruption, or when responding to unusual weather conditions—may be substantial, particularly on upslopes and difficult terrain. Herbivores often are more restless, graze less intensely, and cover more ground when foraging in stormy and unsettled weather.

The mechanical task that is presented to large herbivores in biting off their daily requirements of green forage (154 to 209 pounds for cattle) appears almost formidable. Even under optimum forage conditions, it has been calculated that the cow must take about 80 bites per minute through an 8-hour grazing day to harvest 198 pounds of green material. Biting rates of 30 to 50 bites per minute (50 to 60 bites under good forage conditions) are common in both cattle and sheep. Bites per day for adult foraging cattle range from 12,000 to 36,000, and bite size varies greatly, ranging from 0.05 to 8 g of organic matter. Although bite rate, bite size, and time spent grazing each day theoretically determine daily forage intake, this formula is useful only in showing relationships and not in determining actual forage intake.

Foraging animals vary bite size, biting rate, and grazing time to deal with a variable and changing environment, but their ability to effectively adjust is limited. As available forage increases, bite size usually increases also. Biting rate usually declines as sward height or herbage mass increase and as intake per bite increases, principally because the ratio of manipulative to biting jaw movements increases as intake per bite and the size of individual plant parts prehended increase. Reciprocal changes in intake per bite and bite rate may balance to maintain a roughly constant rate of intake on relatively tall (or abundant) forage stands; but on shorter (i.e., limiting) stands any increase in biting rate is inadequate to balance the decline in intake per bite, and rate of intake declines. It is generally concluded that bite size has the greatest influence on forage intake, with rate of biting and grazing/foraging time being compensatory variables.

Ease of prehension affects forage intake rates, and either exceedingly long or very short leaves delay grasping and ingesting by the animal and generally reduces intake rate. The presence of thorns and spines restricts bite size, and foraging animals are often unable to increase their biting rates to compensate for the smaller bite size. In has been widely observed that ingestion rate decreases with decreasing forage availability while both biting rate and grazing time increase. It is apparent that the necessity or urge to sort green material from standing dead material reduces biting rate. A high degree of plant selectivity compared to essentially nonselective foraging can also be expected to cause declines in the rate of biting and the amount of herbage intake per bite. Foraging tends to become more selective and more casual as hunger is alleviated. Reduced forage intake rate after consuming substantial amounts of forage may result from rumen fill and satiety but could also result from fatigue.

Daily forage dry matter intake by foraging animals is determined by a large number of animal factors (physical, physiological, and psychogenic), dietary/forage factors, weather factors, and management factors. The control of forage intake is apparently multifactorial, because for any single treatment to suppress intake, it has to be administered at an artificially high level. Although diet quality is obviously important, variation in voluntary forage intake has been deemed the most urgent factor determing level and efficiency of production in foraging animals. (For detailed discussions of forage intake factors and opportunities for favorably manipulating these factors, refer to appropriate sections in "Grazing Management" and "Forage in Ruminant Nutrition" as listed in the bibliography.)

## III. FORAGING TIME AND DIURNAL PATTERNS

Domestic livestock commonly spend 7 to 12 hours per day foraging, including time spent searching for as well as consuming forage. As grazing time increases, more energy is used for activity and less for production. Thus, the minimum grazing time that results in adequate dry matter intake is considered optimal. Foraging time depends on ease of ingestion, which varies with accessibility of preferred plant parts and availability of total forage.

Foraging time is generally lowest when forage is abundant, of good quality, and readily accessible and highest when forage is of low quality or availability is limited. Intake can be maintained for a time when forage and thus bite size are limited by compensating with increased grazing time and number of bites per minute. However, this compensation is seldom adequate to prevent a fall in daily intake once the short-term rate of intake starts to decline; grazing time itself may fall under severely limited forage availability, thus contributing further to decline in forage intake. Low forage digestibility in ruminants is directly related to slow rates of ingesta passage rate, which in turn is apt to limit both intake and foraging time.

High selectivity for limited green material may result in the herbivore having difficulty in harvesting enough forage for a complete fill, even when foraging time increases. Foraging animals selectively searching for palatable new green growth in short supply or when palatable plant species make up a minor portion of the standing crop are apt to increase their grazing time. Ruminants respond to short-term thermal stress, either hot or cold, by reducing grazing activity. Adverse winter weather often reduces both grazing activity and forage intake in the short term, but animals subsequently readjust somewhat to running mean temperatures. Reduced foraging activity tends to have a sparing effect on energy expenditures but compensates only in part for reduced forage and energy consumption.

Foraging herbivores exhibit a daily grazing cycle that is remarkably consistent and reoccurs each day with minimal change. Most studies in temperate environments have shown that a major grazing period begins around sunrise and another in late afternoon, with shorter and less regular periods during midday and at night. Current weather conditions may modify the time when animals graze during the 24-hour day. During hot summer days, livestock reduce or eliminate midday grazing bouts, seek shade or water sites, and spend time remaining idle or ruminating. In the tropics and subtropics and during prolonged periods of hot weather in temperate zones, night grazing may account for substantial portions of total time spent foraging during each 24-hour day. During hot weather, herbivores tend to forage more at breezy points, but seek protection from the wind on very cold days.

## IV. SELECTIVE GRAZING: PALATABILITY AND PREFERENCE

Foraging animals are always selective in what they eat, that is, they choose or harvest plant species, individual plants, and plant parts differently from random removal or from the average of what is available. Herbivores range from generalists to specialists in their diet selection. However, there are no obligatory ungulate herbivores, that is, restricted to a single species or genera of plants as there are in insects. What grazing animals actually ingest is a complex phenomenon determined by the animal, by the plants being offered to the animal, and by the environment in which the selection occurs. Selectivity can be diminished by grazing techniques such as heavy stocking and high animal density to more fully utilize the standing crop, but eliminating all selectivity generally lowers nutritive quality of the ingesta and associated animal performance.

When given an opportunity to do so, foraging animals select the greener, finer, leafier, and thus more nutritious plants and plant parts. In the short run, foraging herbivores continuously "high-grade" the forage supply by eating the most preferred portions—and generally the most nutritious portions. As a result, relative to the forage rejected and even the average in the total standing crop, the ingesta typically has enhanced levels of nitrogen, phosphorus, carotene, cell contents, and generally useful energy and lower levels of fiber, lignin, cell-wall materials, and silica. Nevertheless, there seems to be no scientific basis to substantiate the hypothesis that foraging animals knowingly or instinctively select for protein, energy, or other nutrient requirements, and the existence of generalized *nutritional wisdom* in foraging animals about the forage they consume is now widely rejected.

Although both aspects are involved in selectivity, evaluating the extent and cause of selectivity is enhanced by restricting the term *palatability* to plant characteristics or conditions and the term *preference* to the reactions of the grazing animal to these differences. Selectivity by one foraging animal may be influenced by the presence, either concurrently or previously, of one or more other animal species in the area. This results from changing the short-term relative availability of the different plant species or differentially affecting the palatability of the remaining forage. As a result, under mixed grazing the timing of when an animal utilizes the forage in relation to access by another animal species may determine if the quality and probably the palatability of the remaining forage have been reduced or, in some cases, actually enhanced.

Palatability and preference are always relative to the variety in the vegetation and alternatives available for selection by the foraging herbivore. Palatability is subject to seasonal changes in forage plant species and in associated plant species; preference also varies between animal species and even individual animals. Forage species of mediocre palatability are often consumed well if they are available as the sole choice, but may be discriminated against or even rejected if more palatable plant species are offered simultaneously. The proportion of a given plant species in the diet will depend on both relative abundance and palatability of the plant. It has also been observed that less palatable species may be taken to a greater degree where they make up a smaller proportion of the total vegetation, possibly having a novelty attraction.

### A. Plant Physical Factors

Several plant physical or morphological factors are known to improve or lower palatability. Succulence appears to be an important if not the major plant characteristic sought by grazing animals. Animals prefer young plants to older plants or the young and actively growing leaves, twigs, and stems of older plants to their mature counterparts. A high preference for living over dead plant material is generally exhibited by foraging animals; an almost exclusive preference for green materials often persists down to very low proportions in the stand. A long growing season favors a prolonged period of green, succulent forage of higher palatability. In grasses and grasslike plants, a high leaf to stem ratio is also related to higher palatability.

Leaves that are fine, tender, and of low tensile strength are generally preferred. The ability of animals to select depends to some extent on the accessibility of the preferred forages. Selection is enhanced when the green material is physically more separated from the dead material and depressed when the green material is growing up through an overburden of standing dead materials. Cattle grazing crested wheatgrass have been shown to prefer medium-sized plants over either the smallest or the largest, most robust plants.

Any growth form or anatomical characteristic that enhances avoidance of grazing will effectively reduce relative use. The barrier effect of thorns and spines of some woody plants and cacti are readily apparent. Thick, stubby twigs, particularly when compacted and sharp pointed at the ends, may also prove to be effective barriers against utilization of leaves within the protected area. Stiff, spiky leaves, feathery inflorescences, or sticky foliage may have similar herbivory-inhibiting effects. Utilization of plants may be sharply reduced when shielded by overstory plants, when occurring in the presence of other highly unacceptable species, when avoiding grazing by prostrate or elevated growth, or when growing principally on steep slopes, rough terrain, or unstable soils of reduced accessibility to herbivores.

### B. Environmental Factors

Excessive growth rate early in the season may promote coarse, stemmy growth of both the primary forage crops and weeds. Slow growth, as long as growth remains active, is often associated with increased nutritive levels and palatability. Although forage quantity may be higher in high-rainfall years, forage quality in terms of nutritive value and leafiness may suffer. Moisture in the form of rain, dew, or light melting snow on semiarid rangelands commonly increases palatability by "softening up" the matured forage. Weather conditions that force forage plants into rapid dormancy may preserve higher palatability, whereas those conditions that expedite rapid weathering following maturity will reduce palatability. Also, the forage of herbaceous plant species produced on localized unfavorable sites (shallow, gravelly, low fertility, low moisture) is commonly of greater palatability than when produced on adjoining but favorable sites because of lower stature and finer plant parts.

### C. Plant Chemical Factors

The chemical composition of forage is a major palatability factor, although the effective constituents are not well known and their associations with palatability have sometimes only been situation specific rather than universal. High nitrogen levels are often correlated with high palatability but without necessarily demonstrating a cause and effect relationship. High levels of sugars, fat, and cellular contents improve palatability, whereas high levels of fiber, lignin, and silica and low levels of magnesium and phosphorus typically lower palatability.

Antipalatability factors existing in plants as secondary plant metabolites may provide defense against insect and large animal herbivory. Such compounds include phenols, tannins, monoterpenes, and alkaloids. In addition to (1) causing animal avoidance because of low palatability, they may serve as defensive mechanisms by (2) causing aversive conditioning (animal learns that the plant makes it ill and so avoids it) or (3) having extreme toxicity (kills outright and thus prevents further attack).

### D. Animal Preference Factors

Animal factors that influence food preference can be partitioned into three major categories: (1) use of the senses, (2) previous experience or adaptation of the animals, and (3) variations between species, breeds, and individuals. All five senses—taste, smell, touch, sight, and even hearing—appear to be involved in forage preference behavior by foraging animals, but their interactions are complex, and no one sense seems to predominate in every situation.

It is commonly accepted that taste is the most important sense in forage selection and that the other senses serve primarily to supplement or modify its expression. Smell is commonly intricately involved with taste in determining whether the herbivore will select against or for certain plants. Smell seems to be a mechanism by which sheep discriminate against high-alkaloid varieties of forage plants and against high selenium levels in plants and by which cattle avoid forage contaminated with or even growing near their own dung pats. Even after being positively sensed by receptors in the nose and throat, reappraisal by taste receptors of old or new flavors arising from the masticated and broken cells may be responsible for secondary rejection of some forages.

The sense of touch is important in that grazing animals generally select against rough, harsh, or spiny plant material. The sense of hearing is least used in forage selection, but it is used by mast feeders in targeting the sound of falling acorns striking the ground. Sight is probably the next least important sense in determining forage preference as animals can be effective foragers even in total darkness. It is apparent that sight mostly aids grazing animals in spatial orientation in relation to vegetation, that is, in the discovery of forage areas, whereas other senses primarily determine actual plant selection. Sight does appear useful in distinguishing between widely differing plant forms.

Previous experience of the foraging animal, particularly that obtained in early life, affects its behavior in forage selection. It can render a particular forage species more or less acceptable; it can also result in preferences for or aversions to available plants and can aid in acquiring the motor skills necessary to harvest and ingest the preferred forages. Failure to utilize a newly found or offered plant species may result from lack of previous experience with the species, lack of prehension skills, or both. However, individual animal variation in forage selection may be inherited as well as acquired. And lastly, the physiological condition of the foraging animal can alter forage preferences. Animals with voracious appetites may discriminate less than finicky eaters with lower demands. Hunger lowers the threshold of forage acceptance, a fact of particular significance when grazing animals are stressed.

## V. SELECTIVE GRAZING: KIND OF HERBIVORE

### A. Body Size and Rumen Capacity

Except for horses and burros (and also tapirs, elephants, and rhinoceroses outside North America), the most common ungulate herbivores are ruminants. Ruminants are unique in that they rechew most of the forage consumed (i.e., chew the cud). Also, the ruminant stomach is unique in that the large anterior part, referred to collectively as the reticulo-rumen, functions as a holding tank where fermentation can occur and from which the ingesta is regurgitated for rumination.

Forages ingested by ruminants are fragmented into various sizes as the result of ingestive mastication (initial chewing) but subsequently by ruminative mastication (rechewing the cud after regurgitation). Ingestive mastication reduces the ingesta to sizes that can be incorporated into a bolus and swallowed, but rumination results in further particle size reduction and exposure to microbial attack. Ruminative mastication appears to be the major factor in decreasing the size of forage particles and increasing the probability of passing out of the reticulo-rumen into the small intestines and being digested.

Greater efficiency in energy digestion of fibrous material and lower basal metabolism requirement per unit of body size favor the larger herbivore species. In the ruminant, high reticulo-rumen volume to body size in cattle and bison is considered an adaptation to exploiting thick-cell-walled, high-cellulose diets, that is, graminoids. High reticulo-rumen volume is associated with higher consumption rate, more time spent in rumination, and longer ingesta retention time.

By contrast, low reticulo-rumen volume to body size in deer and pronghorn antelope is considered an adaptation for exploiting high-cell-soluble and thin but lignified cell-walled diets, that is, browse; and this is associated with lower consumption rates, more selective grazing, less time spent in rumination, and faster ingesta passage. Animal species with very small rumen capacity relative to weight have been referred to as "concentrate selectors" based on their reliance on fruits, forb leaves, and tree and shrub foliage (but not excluding immature grass forage) to provide higher levels of crude protein, phosphorus, and energy.

It is generally concluded that small ruminants with high metabolic requirement relative to digestive tract capacity must eat food composed largely of rapidly digestible fractions, because rumen function does not permit more rapid ingesta passage rate and thus higher intake to counter low digestibility. In ruminant species of large body size, the relatively large reticulo-rumen capacity can take advantage of the longer retention time to more fully digest the slowly digesting fraction of the forages. Within ruminants, feeding time decreases with large size while rumination time increases. If rumination time is considered feeding activity, then small and large ruminant feeding time may be similar.

## B. Herbivore Diets by Forage Class

Prominent differences in forage preference are exhibited between animal species groups and even between related species. In addition to body size and digestive tract capacity, inherited differences in anatomy (teeth, lips, and mouth structure), foraging and prehending ability, agility, and kind of digestive systems may account for some of the differences. Ruminants with small mouthparts can more effectively utilize many shrubs while selecting against woody material. Narrow mouthparts and thus the ability to be highly selective may permit such animals to be more efficient browsers. Animal species with broad mouthparts may be inhibited from feeding on spinescent species where small, individual leaves must be selected but are at an advantage in rapidly consuming large quantities of accessible forage where high levels of mechanical discrimination are not required.

On the basis of this discussion, ungulate herbivores have been divided into three feeding types: (1) the bulk and roughage eaters, (2) the concentrate selectors, and (3) the intermediate feeders. An equivalent classification system, based more directly on forage class consumed (i.e., grasses, forbs, and shrubs), is (1) grazers, (2) browsers, and (3) intermediate feeders. Both classifications have obvious limitations, such as failure to adequately treat forbs and implying invariable rigidity in dietary selection. Also, classification of an ungulate as a grazer or a browser rather than an intermediate feeder suggests forage preference, but it tends to ignore seasonal differences and implies a rigidity in dietary selection that is seldom experienced. It should also be noted that a breakdown of diets by forage classes is probably never totally dependent on animal preferences because availability by forage class will also affect foraging animal diets.

*Grazers* include bison, musk-oxen, horses, cattle, elk, mountain bighorns, and Dalles sheep. They (1) mostly consume diets dominated by grasses and grasslike plants and (2) may locally consume substantial amounts of forbs and shrubs (except possibly bison and musk-oxen), particularly when graminoids are not available. They also (3) show a strong avoidance of browse high in volatile oils and appear to lack mechanisms to effectively reduce the toxic efforts of these substances.

*Browsers* are best represented by moose, pronghorn antelope, and deer; the domestic goat is also placed here but is quite versatile in its dietary selection. Browsers (1) are noted primarily for consuming large amounts of forbs and shrubs, (2) commonly consume substantial amounts of green grass during rapid growth stages but avoid dry, mature grass; and (3) often experience digestive upsets if forced to consume diets dominated by mature grass. The smaller browsers are also (4) better able to use plants high in volatile oils such as the monoterpenes.

*Intermediate feeders* include domestic sheep, burros, caribou, desert bighorn sheep, and mountain goats. They (1) use large amounts of grasses, forbs, and shrubs and (2) have substantial capability to adjust their feeding habits to whatever forage is available. Because of the versatility in their diets, the domestic goat (forb and browse preferring) could also be included here.

## C. Mixed Species Grazing

Combining grazing by two or more animal species under mixed grazing can be advantageous if they have different diet and/or site and terrain preferences and habitat requirements. The advantage of mixing kinds of grazing animals in terms of total foraging capacity and stability of vegetation composition increases as the diversity of vegetation and site and terrain within a foraging unit increases. Although the principle of mixed grazing conceivably applies to every situation where grazeable vegetation grows, even homogeneous improved cropland pasture, limitations including that of management will continue to prevent its universal application. Although grazing more kinds of animals in common will increase the likelihood that more plant species will be utilized and a greater portion of the grazing unit will be covered, the greater number of animal species will increase management requirements and make even more urgent the setting of proper stocking rates.

Under mixed grazing there are social interactions between the different species of foraging animals. These can be negative when two or more species come together, particularly under conditions of crowding such as at watering places, on common favored sites, or where supplemental feed is available. In considering negative social interactions, distinction has been made between *interference competition* (when one animal species takes aggressive defense of territory) and *disturbance competition* (when one animal species seemingly voluntarily leaves the vicinity where other animal species are found). It is often impossible to determine whether the exit of one species is resulting from interference or the mere presence of the prevailing species and/or from the latter's effects on the forage supply. However, negative social interactions are seldom a major problem, and management techniques are available to minimize such instances.

## Glossary

**Browse** (n) Leaf and twig growth of shrubs, woody vines, and trees available and acceptable for animal consumption; (v) to consume browse; synonymous with *browsing*.

**Browser** Herbivore that concentrates its plant consumption from woody plants rather than herbaceous plants.

**Forage** That part of vegetation that is available and acceptable for animal consumption, whether considered for grazing or mechanical harvesting; includes herbaceous plants in mostly whole-plant form; generally extended to include browse.

**Foraging** Complex process by which ungulate herbivores search for and consume forage from a standing forage crop; includes *browsing* and *grazing*.

**Grazer** Grazing animal; (specific) any herbivore that concentrates its plant consumption from standing herbaceous plants.

**Grazing** Act of consuming a standing forage crop by ungulate herbivores; often expanded to include browsing.

**Mixed grazing** Grazing the current year's forage production by two or more kinds of foraging animals, either at the same time or at different seasons of the year; synonymous with *common use* or *multispecies grazing*.

**Palatability** Summation of plant characteristics that determine the relish with which a particular plant species or plant part is consumed by an animal.

**Preference** (in foraging) Selective response made by the herbivore between alternative plant species or plant parts.

**Ruminant** Even-toed, hoofed mammal that chews the cud and has a four-chambered stomach, i.e., *Ruminantia*.

**Ungulate herbivore** Hoofed animal, including ruminants, that subsists principally or entirely on plants or plant materials.

## Bibliography


Arnold, G. W., and Dudzinski, M. L. (1978). "Ethology of Free-ranging Domestic Animals." Amsterdam: Elsevier.

Frame, J., ed. (1986). "Grazing," British Grassland Society Occasional Symposium 19. Berkshire, UK: Herley, Maidenhead.

Heitschmidt, K., and Stuth, J. W., eds. (1991). "Grazing Management: An Ecological Perspective." Portland, Ore.: Timber Press.

Horn, F. P., Hodgson, J., Mott, J. J., and Brougham, R. W., eds. (1987). "Grazing-Lands Research at the Plant–Animal Interfaces." Morrilton, Ark.: Winrock International.

Minson, D. J. (1990). "Forage in Ruminant Nutrition." San Diego: Academic Press.

Morley, F. H. W., ed. (1980). "Grazing Animals." Amsterdam: Elsevier.

Provenza, F. D., Flinders, J. T., and McArthur, E. D. (1987). "Proceedings, Symposium on Plant–Herbivore Interactions," Forest Service General Technical Report INT-222. Washington, D.C.: USDA.

Senft, R. L., Coughenour, M. B., Bailey, D. W., Rittenouse, L. R., Sala, O. E., and Swift, D. M. (1987). Large Herbivore Foraging and Ecological Hierarchies. *BioScience* **37,**(11), 789–799.

Vallentine, J. F. (1990). "Grazing Management." San Diego: Academic Press.

# Foraging Strategies

**D. W. Stephens**
*University of Nebraska, Lincoln*



The phrases "foraging strategies," "foraging theory," and "optimal foraging theory" describe a body of ideas and observations that provide an economic interpretation of animal feeding behavior. Students of "foraging" make no distinction between "foraging behavior" and "feeding behavior." Foraging theorists use economic ideas, in the sense that they try to understand why it is better to feed in a particular way in a particular set of environmental conditions, and they invoke the honing action of natural selection to justify this point of view.

## I. THE APPROACH

Most foraging models are optimization models, like the calculus student's problem of choosing the diameter of a tin can to maximize the can's volume given a specified surface area. Foraging models imagine that some attribute of feeding behavior can be adjusted, like the can's diameter (except that the adjuster is not a human designer, but the process of natural selection). They also suppose that these adjustments will be made so as to maximize some measurable quantity, like the can's volume. This maximized quantity should correlate with Darwinian fitness. Finally, optimization problems have many constraints, like the can's fixed surface area, that limit and define the relationship between the quantity that is maximized and the quantity that is adjusted.

## II. THE PREY AND PATCH MODELS

### A. The Patch Model

Consider an animal whose foraging consists of traveling between patches and feeding within patches. We assume (1) it takes $\tau$ seconds to travel from one patch to next, and (2) that a function $g(t)$ describes the relationship between food acquired in a patch and the time $t$ spent searching for food within the patch. The mathematical problem is to "adjust" the *patch residence time* $t$ to maximize the long-term rate of food acquisition. We express this long-term rate as

$$\frac{g(t)}{\tau + t} \qquad (1)$$

We suppose that the *gain function* $g(t)$ increases at a decreasing rate with the *patch residence time* $t$. Figure 1 shows a "typical" $g(t)$ function on the right, with patch residence time $t$ increasing as we move

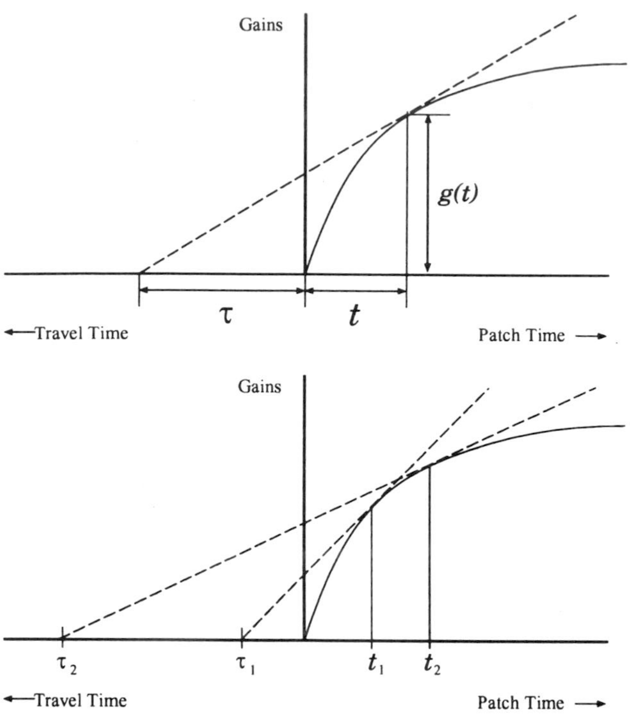


**FIGURE 1** Graphical solution of the patch model, or marginal-value theorem. Top: the slope of the dashed line represents the rate of energy intake, since the slope is simply the "rise" $g(t)$ over the "run" $\tau + t$. The model chooses a rate-maximizing $t$ given a fixed $\tau$. Evidently, the rate-maximizing $t$ occurs where the dashed line is just tangent to the curve. Bottom: the main qualitative prediction of the model is that a longer travel time $\tau$ leads to a longer patch residence time $t$. Using the tangent construct method, this panel shows how the longer travel time $\tau_2$ leads to a longer patch residence time $t_2$.

along the $x$ axis to the right of the origin. To the left of the origin, we plot the *travel time* $\tau$ increasing as we move to the left away from the origin. On this graph, the rate of intake as given by Equation (1) is the slope of a line with its left end point at $(\tau, 0)$ and its right end point *anywhere* on the $g(t)$ curve. If you imagine this right end point sliding up and down the $g(t)$ curve, then you can see that the slope is maximized when the dashed line is *exactly* tangent to the $g(t)$ curve. This "tangent" result leads to the most significant prediction of the model: we expect foragers that must spend more time traveling between patches to spend more time exploiting each patch (see the lower panel of Fig. 1). This qualitative prediction is, as far as I am aware, universally supported in both field and laboratory situations. The situation is muddier when one examines tests of quantitative predictions. Behavioral ecologists usually call this model the *marginal-value theorem*. [*See* FORAGING BY UNGULATE HERBIVORES.]

### B. The Prey Model

Consider a different foraging situation in which an animal periodically encounters prey items that differ in quality. Which prey types should it attack, and which should it ignore? We can associate three quantities with the $i$th potential prey type: an *encounter rate*, $\lambda_i$, an *energy value* $e_i$, and a *handling time*, $h_i$. Let $p_i$ be the probability that our forager will attack an item of type $i$ upon encounter. For a case with two potential prey types, we write the long-term rate to energy intake as

$$\frac{p_1\lambda_1 e_1 + p_1\lambda_2 e_2}{1 + p_1\lambda_1 h_1 + p_2\lambda_2 h_2} \quad (2)$$

The mathematical problem is to choose values of $p_1$ and $p_2$ to maximize expression (2). One can use calculus to show that the best $p$ is either zero or one, and never in between. Moreover, if $e_1/h_1 > e_2/h_2$, then a diet of type 2 alone can never be best, so we only need to consider diets of 1 alone and of 1 and 2. We call the $e$ over $h$ quotients the *profitabilities*. So, our question becomes "when should the type of lower profitability be attacked?" A little algebra leads to the condition

$$\frac{\lambda_1 e_1}{1 + \lambda_1 h_1} < \frac{e_2}{h_2} \quad (3)$$

In words, the predator should attack type 2 if its profitability exceeds the rate of gain achieved by a diet of type 1 alone. The most surprising result here is that no matter how abundant type 2 is, it should not be attacked if the better prey type is abundant enough [notice that $\lambda_2$ is not part of the inclusion condition, Equation (3)]. This model is not as well supported as the patch model, neither qualitatively nor quantitatively. Despite its shortcomings, most "foraging" practitioners agree that the model plays an important role as a starting point for asking questions about animal prey choice.

## III. MODIFICATIONS OF THE BASIC MODELS

The two models described here are the basic models of foraging theory, because they represent the core from which new models and ideas have radiated. Both models have, for example, been extended to consider the specific problems of returning food to a central place, such as a nest or cache. Many authors have been concerned with the abilities of animals to adjust their behavior to natural changes in the encounter rate or profitability of a prey item, and they have tried to incorporate information acquisition and cognitive processes (such as learning and memory) into the basic models. Two observations suggest difficulties with the premise of rate maximization. First, the variability of food reward affects animal feeding preferences. Animals sometimes prefer and sometimes avoid variability, and these preferences are said to be *risk-sensitive*. Second, animals have strong and consistent preferences for immediate reward. Foraging theorists sometimes label this preference for immediacy as *time-discounting*.

## IV. LIMITATIONS OF THE APPROACH

The assumption that natural selection optimizes phenotypes implies several tenuous assumptions. The difficulty is that when a single prediction fails, we do not know whether something about the model is incorrect or whether the premise of optimization by natural selection has failed. Advocates of foraging models have tended to patch up the model and soldier on, and critics have accused them of post hoc prediction fudging. Foraging theory is idiosyncratic, for example, it offers almost nothing about how animals search for food, or about sit-and-wait foragers. It has been applied primarily to insectivorous, carnivorous, and granivorous animals, but is consistently dismissed by students of folivorous animals.

## V. IMPLICATIONS FOR ENVIRONMENTAL BIOLOGY

Community ecologists believe that when a predator preferentially consumes the "competitive dominants" in a community, it can increase the community's species diversity, but other preferences can lead to very different outcomes. Similarly, population biologists argue that the availability of refugia can stabilize predator–prey population dynamics. The patch model presented earlier gives an economic explanation of refugia, because it predicts that predators should leave some prey behind: more in some cases and less in others. An explanatory and predictive theory of animal feeding behavior would provide an invaluable tool to environmental biologists, because many of the key interspecific interactions in the natural environment are feeding interactions.

## Glossary

**Dynamic optimization** Method of solving for an optimal sequence of behaviors, usually dependent on some internal property of the animal (e.g., hunger). In contrast, conventional or static optimization solves for the best value of a small number of values (usually one) with no temporal order.

**Gain function** Relationship between the time spent hunting within a patch and amount of food extracted from the patch. Gain functions usually deal with the net amount of energy extracted, but in special circumstances a gross energy or another measure, such as protein, may be used.

**Marginal-value theorem** Model of patch exploitation originally attributed to E. L. Charnov. The model predicts that feeding animals will spend more time exploiting each patch when they must travel farther between patches. The term "marginal" derives from economics, where "marginal rate" loosely translates to the derivative or tangent line of a function.

**Profitability** Extractable energy (*e*) divided by the time required to attack and consume (i.e., to "handle," *h*) a given prey type.

**Rate maximization** The basic premise of earlier foraging models was that natural selection had "designed" animal feeding behavior so that they could obtain the most food (usually measured as energy) in the least time. This, in turn, suggests a rate-maximizing approach.

**Risk sensitivity** A feeding animal is said to have a risk-sensitive preference if it has a preference when offered a choice between (1) a probability distribution and (2) the mean of the probability distribution offered with certainty.

**Time discounting** Phenomenon in which animals prefer small immediate food rewards over larger delayed rewards. The delayed reward is said to be discounted by the intervening delay.

## Bibliography

Kamil, A. C., Krebs, J. R., and Pulliam, H. R., eds. (1987). "Foraging Behavior." New York: Plenum Press.

Kamil, A. C., and Sargent, T. D., eds. (1981). "Foraging Behavior: Ecological, Ethological and Psychological Approaches." New York: Garland STPM Press.

Mangel, M., and Clark, C. W. (1988). "Dynamic Modeling in Behavioral Ecology." Princeton, N.J.: Princeton University Press.

Stephens, D. W. (1990). Foraging theory: Up, down, and sideways. *Studies in Avian Biol.* **13,** 444–454.

Stephens, D. W., and Krebs, J. R. (1986). "Foraging Theory. Monographs in Behavior and Ecology." Princeton, N.J.: Princeton University Press.

# Forest Canopies

**J. J. Landsberg**
*CSIRO, Australia*



The cover of leaves and branches that forms the upper layer of a forest is called the canopy. Canopy architecture, described by the vertical and horizontal distributions of foliage, determines the amount of light absorbed by a stand, which drives the process of photosynthesis. The transfer processes that determine the fluxes of carbon dioxide and water vapor to and from the foliage surfaces are strongly influenced by canopy architecture and the structure of the stand as a whole. These fluxes, integrated over time, reflect the carbon and water balances of forest stands. Stand structure and canopy architecture are strongly influenced by the availability of water and nutrients, which affect the distribution of carbohydrates between the component parts of trees as well as growth patterns of the foliage, and litterfall. Canopy architecture varies spatially as a result of variation in soil properties and differences in development patterns among trees. Spatial variation increases with stand age. The influence of disturbance on forest ecosystems can be assessed by evaluating the way the disturbance affects the physiological processes of the canopy, its physical structure, and its interactions with the atmospheric environment.

## I. INTRODUCTION

At one time forests covered much of the surface of the earth in areas with reasonably abundant rainfall. Even now, after centuries of clearing, forests and woodlands cover about one-quarter to one-third of the earth's surface. Natural forests may consist of a wide range of species or they may be virtually monospecific. They range from the often stunted conifers of the northern boreal zone, which merges into the tundra, to temperate coniferous forests that include some of the largest trees in the world—for example, in the North American coastal forests of spruce, Douglas-fir, and cedar—to highly diverse tropical rain forests. Temperate deciduous forests include species such as oak and ash and beech, associated with conifers in some areas. There are mixed evergreen forests in the temperate parts of the Southern Hemisphere, dominated by beech and kauri pine in New Zealand and eucalyptus species in Australia. In South America, Africa, and Southeast Asia, tropical rain forests contain a bewildering array of species. We must also recognize the plantations and managed forests that, although relatively small in total area, are important in terms of commercial forest products. These are invariably single-species stands, the species used, in some cases, being exotic to the area in which they grow.

Natural forest types tend to be categorized according to climatic regions: the commonly used descriptions such as boreal, cool temperate, warm temperate, and tropical imply relationships between climate and forest type, but even though such relationships exist, in very broad terms there is enormous variation caused by local factors such as slope, aspect, soil type, and disturbance regime.

Forests may be regarded as wood-producing resources, as interesting and important plant communities of great ecological significance, as significant factors ameliorating local or global climate change and buffering the effects of weather on the surface of the earth, and as areas of recreational and often spiritual value. They are also home for myriads of fauna, from insects to mammals. Forests affect the weather in their vicinity directly by modifying air flow and temperature, intercepting precipitation, and altering the amount and pattern of water flow across a region. They thus play a major role in governing stream flow and floods and the development of wetlands. Forests are communities dominated by large plants in which considerable quantities of carbon dioxide ($CO_2$) are converted into carbohydrates by the process of photosynthesis, so they are important factors in the global carbon balance, although their relative importance in this respect (in relation to other vegetation types and the oceans) is not unequivocally established. [*See* Global Carbon Cycle.]

Whatever view is taken of the significance of forests, it is clear that the extent to which they interact with their environment, their rate of growth, and their visual and aesthetic impacts depend on the size, architecture, and health of their canopies. The canopy of a forest may be described as the continuous, or almost continuous, cover of leaves and branches forming the upper layers of the forest. Canopies are formed by the crowns of trees: stand structure—the number, spacing, height, and size distribution of the trees—is obviously a major determinant of canopy architecture. As a general rule we can expect to find trees with narrow columnar crowns associated with high latitudes and more xeric sites and broad or spherical crowns associated with humid sites and moist environments. This is because narrow, columnar crowns (large height to crown radius ratio) result in greater energy interception in areas where the average sun angle above the horizon is small. Such crown shapes also reduce damage from snow and ice loading. Where sun angles are larger, low densities (small number of trees per unit area) of narrow-crowned trees with relatively dense foliage will use less water than trees with the same leaf area exposed in broad, spherical crowns (smaller height to radius ratio). The trees in tropical forests may cover the range from tall, large-leaved species with broad crowns to short, small-leaved forms in drier areas. Many forests are multilayered, with understories of shrub and herb species, whereas others are essentially single layered.

Forest canopies vary in their depth, density, uniformity, and surface (aerodynamic) roughness: they are, to some extent, characteristic of particular forest types but there will generally be considerable variation in canopy architecture within a forest type, induced by soil fertility, precipitation and the water balance, the stand structure, and the understory species present. Forests may be destroyed by fire, wind, human activities, and insects or disease: replanted, or left to regenerate naturally, they move through phases of development during which canopies vary in their architecture.

The nature of their canopies provides a further means of classifying forests, which may be deciduous or evergreen, broad-leaved or needle-leaved conifers. Most, but not all, deciduous trees are broad-leaved, and most, but not all, conifers are evergreen. There are also evergreen broad-leaved species, including eucalypts and many species in tropical forests. The implications of being deciduous or evergreen are enormous. Deciduous canopies lose all their leaves each year, so the canopy has to be replaced for the growing season. Foliage production occurs largely in spring, when there is massive production of leaves, which fall in autumn. The pattern of leaf production varies with species, some producing virtually all the leaves for a particular year simultaneously, others producing an initial "crop" of leaves on the existing branch and shoot structure, then producing new leaves on summer-growing shoots. Evergreen canopies retain their foliage, enabling them to take advantage of favorable growing conditions whenever they occur. There are metabolic energy costs and savings associated with both strategies. For example, foliage longevity is positively correlated to initial construction costs, specific leaf area ($m^2/kg$) and photosynthesis rates are negatively correlated to leaf longevity, and species with greater leaf longevity tend to support greater foliage mass. [*See* DECIDUOUS FORESTS.]

The interactions between forests and their aerial environments take place in and through the canopies. These interactions include the interception of shortwave, visible radiant energy (light) and the exchange of gases (primarily $CO_2$ and water vapor) by diffusion and turbulent transfer processes between leaf surfaces and the air. The canopy is the site of three fundamental physiological processes:

- photosynthesis—driven by light and mediated by the exchange of carbon dioxide through the stomata;
- transpiration—water vapor loss through the stomata in response to the energy balance of the foliage and the transfer processes within the canopy;
- respiration associated with photosynthesis (photorespiration) and the metabolic processes that release the energy needed for the construction of new tissue and for tissue maintenance.

In addition, nutrients, originally derived from the soil, are translocated between the foliage, branches, and stems. In this article forest canopies are considered in terms of these physiological processes and physical interactions: in view of the enormous variation in canopies, across the range of forest types and the environments in which forests grow, we must understand them in terms of these processes and interactions if we are to understand the way canopies determine both the growth rates and growth patterns of the forests and the effects of the forests on their local environments. Because function depends on canopy architecture, the next section outlines the parameters of canopy architecture, followed by consideration of the factors that influence canopy development and hence final form. Later sections discuss canopy–environment interactions, light interception and the carbon balance, evaporation and the water balance, and the effects of disturbance on forest canopies.

In all considerations of forest canopies and their functions, we have to recognize that canopies are not discrete, independently functioning entities. They are integral parts of trees, affecting and affected by the roots and conditions in the soil as well as by conditions in the atmosphere. This principle of integration will be observed throughout this article.

## II. CANOPY ARCHITECTURE

The distribution and arrangement of foliage through the canopy space is called canopy architecture, which may be formally described in terms of the vertical distribution of foliage area density [$D_f(z)$—$m^2$ foliage/$m^3$ canopy volume, where $z$ denotes height]. There are a number of useful mathematical descriptions of $D_f(z)$, although in most cases there is probably not much practical difference between the results obtained with the various formulae used. These have included the Weibull distribution, the normal curve, and triangular approximations. These descriptions are only likely to be useful for stands with relatively simple structure, such as even-aged, single-species stands. Integration in terms of height gives the leaf area index ($L^\star$)—the surface area of foliage per unit land area. Figure 1 provides some examples of profiles of $D_f$, including the type of pattern that would be observed in a multilayered canopy.

From the point of view of physical and physiological processes, the leaf area index is the most important parameter describing the canopy of any plant community. Information about $L^\star$ is needed for the study of the interception and exchange of radiation, sensible heat, mass ($CO_2$ and water vapor), and momentum in canopies. But information about $L^\star$ alone is inadequate; for a complete description of radiation absorption by canopies, and energy transfers within them, information about

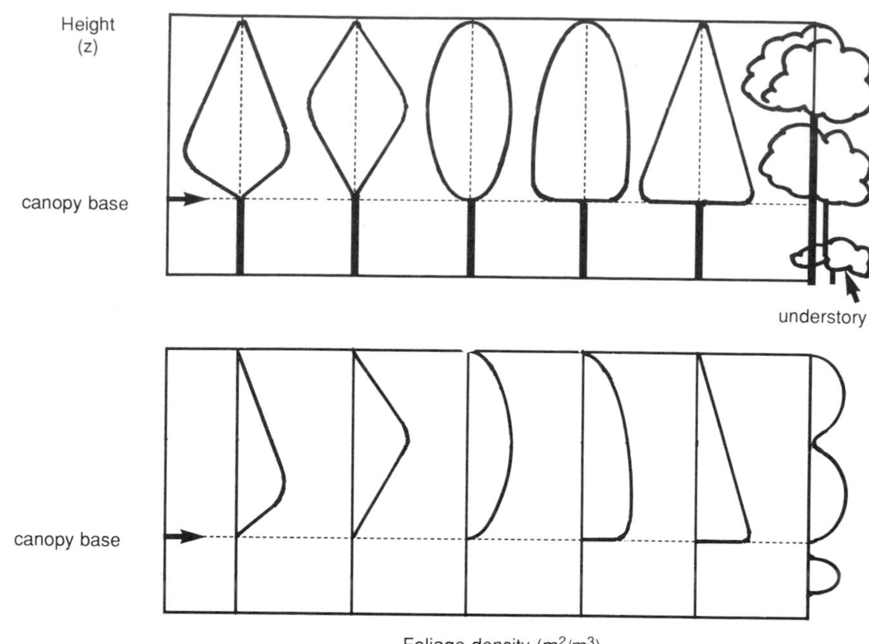


***FIGURE 1*** Geometrical approximation of tree shapes (top) and profiles of the vertical distribution of foliage density, $D(z)$ (bottom). The shapes are (from left to right): trapezoid, the profile of which can be approximated by a triangle or Weibull distribution; normal curve rotated through 360°; ellipsoid; truncated ellipsoid; linear decrease in $D(z)$; layered canopy, not conveniently described mathematically. Maximum values of $D(z)$ may be 1.5–2 $m^2/m^3$.

the distribution of leaf area with height is also essential, and information about leaf size, shape, and angle distribution is very useful.

Maximum $L^\star$ values attained by forest canopies appear to be about 10–12. The optimum $L^\star$ value from the point of view of light interception is about 5–6. This is demonstrated by using the well-established equation for the interception of solar radiation by plant canopies:

$$I(z)/I_0 = \exp(-kL^\star), \tag{1}$$

where $I(z)$ is the average, radiation flux density at level $z$, below $L^\star$, and $I_0$ is radiation incident on the top of the canopy. A value of 0.5 for the extinction coefficient ($k$) is representative for forest canopies; exp(-3) = 0.05, indicating that about 95% of energy is intercepted by a canopy with $L^\star = 6$. More detail is provided in Section IV,A.

Canopy depth is the distance between tree height and the average height of the lowest green foliage on the trees. In mixed-species forests with significant understory, the canopy includes the understory, so the lowest green foliage may be almost at ground level. Foliage may be distributed randomly within the canopy or it may be clumped or layered. The assumptions underlying Equation (1)—that foliage is randomly distributed through the canopy space, and that the leaf angle distribution is spherical—are seldom fulfilled, but for the purposes of analyzing and modeling radiation interception and absorption by forest stands the relationship generally yields acceptable results for relatively uniform canopies. In situations where the tree crowns are widely separated it may be necessary to consider the forest as an array of individual objects. For modeling purposes, tree crowns may be assumed identical in shape and represented by truncated ellipsoids or cones, with the upper and lower limits set by tree height and canopy depth (see Fig. 1). Dimensions can be varied between specified size limits.

In trees with conical or ellipsoid crowns the foliage is often arranged in a shell around the outside of the canopy volume; the inner, nonfoliated zone is called the "bare core." As crown size increases, the bare core tends to occupy an increasing proportion of total canopy volume, although this will of course depend on the overall structure of the stand. Average crown size will decrease as stand density increases, so that a larger proportion of the canopy will be foliated at high stand population densities. Trees with flat or umbrella-shaped crowns may change their shape during their development cycle, with crown diameter increasing proportional to stem diameter.

The concept of canopy depth is particularly useful in describing relatively simple canopies; in mixed-species forests with layered canopies and significant understory it is more useful to consider the depth of each layer. Empirical relationships have been established between canopy depth, tree density, and mean tree height for some species; in other cases where estimates of canopy depth have been required for the calculation of radiation interception, canopy depth has been estimated simply by assuming that crown height was, for example, half the tree height. Such assumptions are probably adequate as a basis for relatively simple calculations such as estimates of stand productivity in relation to annual radiant energy income or regional comparisons between forest types. However, where variations in canopy depth are expected to have significant consequences for some process of interest, such as tree growth patterns or responses to thinning, detailed empirical studies may be necessary.

For a long time, foresters have used stem diameter as a measure of tree size. In general, the foliage mass carried by individual trees is proportional to stem diameter. It follows that it should be possible to estimate foliage mass from stem diameter and relationships between these parameters have been established for many species. However, the relationship is not linear as tree foliage amounts increase at a greater rate than diameter: sapwood cross-sectional area is a better predictor of foliage mass than stem diameter, as it does produce a linear relationship. Information about sapwood cross-sectional area can be obtained from stem cores. Foliage surface area can be estimated from foliage mass through the parameter specific leaf area ($F_A$, $m^2/kg$), although this varies between species and indeed within a single species, depending on leaf age and growing conditions. Typical values of $F_A$ (total leaf surface areas) are 8–10 $m^2/kg$ for temperate deciduous trees, 12–16 $m^2/kg$ for a pine species (*Pinus radiata*) widely used in plantation forestry in the southern hemisphere, and 6–9 $m^2/kg$ for eucalypt species. Hence, given the appropriate relationships between foliage mass and stem diameter or sapwood cross-sectional area for the species of a particular forest type, and given an appropriate value for $F_A$ and information about the stand population and stem sizes, then $L^\star$ can be calculated.

Observed ratios of projected leaf area to sapwood cross-sectional area ($m^2/m^2$) range from about 1000 to 7500. These relationships have generally been established for trees from mature stands, but the coefficients of the equations can be expected to vary from site to site, and there have been few rigorous studies of how they vary with stand density. Clearly the trees of a very dense stand ("overstocked" in forestry terms) will not carry as much foliage as trees from open stands, although stem diameters in dense stands, and hence foliage mass per stem, would be smaller than in low-density stands. Despite the establishment of such relationships, it must be recognized that the foliage mass of trees is not constant over time, even in mature evergreen forests (see Section III). The relationships between foliage surface area and stem diameters, or sapwood cross-sectional area, are average relationships and may be significantly in error for any particular tree at any particular time. Foliage on deciduous trees, by definition, has a life of one growing season, and relationships between foliage mass and some measure of tree size are usually derived when the maximum amount of foliage is on the trees, in midsummer. In evergreen forests foliage may be held on the trees for a number of seasons—as long as 20 years in cold climates (up to 40 years on bristlecone pine)—but even within seasons the foliage mass of a canopy is dynamic, the amount present at any particular time being dependent on the balance between the (integrals of) foliage production and loss rates (litterfall).

Forest canopies are also not spatially homogeneous, and statements about canopy architecture in relation to particular forest types must be accepted as approximations. Even in a relatively young, even-aged, single-species forest on a site with uniform soils there will be differences from point to point. This variability increases as the forest ages, the difference in size between dominant trees and their smaller neighbors increases, and gaps are formed as trees fall. The canopies of mixed-species forests, particularly those on sites where soils are highly variable, may vary considerably in height and in their vertical structure across relatively small distances. A study of the aboveground dry weight of trees in a tropical rain forest gave values ranging from 2.5 to more than 480 tons/ha within a few hundred meters. This may be an extreme example but it reflects the influence of gaps, and the development of new growth within them, on forest structure.

Gaps caused by tree-fall, fire, wind, or human activities are an intrinsic part of the structure of major forest stands. They are obvious and major causes of horizontal nonuniformity as well as being important ecologically: species regenerating in them may not be the same as the dominant tree species of a stand, so gaps can exert significant influence on the species composition of stands as well as on the population dynamics of the dominant species. The influence of gaps is exerted through their effect on canopy microclimate, and hence on the conditions for growth for the seedlings that develop in them and the growth of the trees surrounding them. Gap effects therefore depend on their size, age, and position in terms of slope, aspect, and frequency of occurrence.

## III. FACTORS DETERMINING CANOPY ARCHITECTURE

### A. Tree Density, Age, and Site Properties

As noted earlier, stand structure is a major determinant of canopy architecture, and stem density is a major factor in stand structure. If the stand density is very low, so that the trees do not affect one another except in relation to air flow and light at very low sun angles, the structure of the tree crowns will be determined largely by genetic differences and growing conditions. This is particularly the case with young stands. With increasing stand density the influence of the trees on one another increases; in a very dense stand of young trees, mutual interference in terms of light interception and water use will begin at an early stage and growth patterns will be affected. In general, height growth rates of closely spaced trees tend to be higher at an early stage than those of trees at wider spacings, which will produce more branches and tend to more rounded crown forms. $L^*$ for a close-spaced stand will increase more rapidly than that of a widely spaced stand, although the resultant leaf area per tree will be lower.

The time required to form a closed canopy, or to reach the point where $L^*$ does not continue to increase, will vary with species, location, and spacing. With some species—for example, Ponderosa pine, which is a dry area tree—canopy closure will seldom, if ever, occur. Some species, such as eucalypts in temperate areas of Australia, regenerating after fire or clearance in areas with prolific seed sources, will close canopy within three to five years, but this will be followed by rapid and massive mortality (self-thinning) and a long period of slower mortality as the stand matures. During this period, leaf area per tree increases as the number of trees decreases; height growth is relatively rapid and $L^*$ may decline slightly or remain more or less unchanged, stabilizing (within the limits of annual variation) as the stand reaches maturity. Canopy development in plantation eucalypts planted at spacings of 2–3 m in favorable locations in places such as Australia, South Africa, South America, or Portugal, will follow the same developmental path. The canopies formed by such single-species communities would be even and relatively homogeneous in appearance and in terms of aerodynamic roughness.

There is little information available about the development patterns of regenerating stands that are not dominated by one species. Anecdotal observation indicates that such stands will regenerate much less evenly, because different species do not grow at the same rate or develop their canopies in the same way. The spatial distribution of individuals within mixed-species stands is also unlikely to be even, so the rate and pattern of canopy development will vary from place to place; some patches will close more quickly than others, so there may be a phase during which parts of the forest will have closed canopy while other parts remain relatively open. In terms of biomass and the vertical distribution of leaf area density, the structure and canopy architecture of such a stand may appear similar to that of an old stand with large gaps, but the species composition and morphology of the trees would be quite different. As mixed-species canopies mature, they will develop layers with dominant and subdominant species.

Mature canopies are highly variable in their architecture, depending on the ratios of the different species and crown types, their relative heights and spatial distributions, and site properties. Clearly the canopy architecture of a mixed deciduous forest in Europe or North America, with about 10 species per hectare, will be different from that of a single-species coniferous forest, which in turn will be vastly different from the structure of a tropical rain forest with up to 400 tree species per hectare.

The tree species that will grow in an area are determined largely by "macroclimate" factors, such as the temperature patterns of the area, annual rainfall, and solar energy income. In many cases the presence or absence of species is determined by the extremes of temperature or drought. Given a range of tree species, the size and density of the forest canopy at a site depend on the growing conditions there. On a fertile soil, well supplied with water, where there are no extremes of temperature or persistent strong wind, canopies will tend to be luxuriant, with high $L^\star$, supporting rapidly growing stands of trees that will tend to become large. Adverse conditions of any sort are reflected in reduced $L^\star$ and reduced growth.

Our capacity to make quantitative estimates of the influence of site factors on canopy architecture derives from knowledge about the influence of those factors on, and their interaction with, the processes that govern canopy development and the growth of trees. These influences are determined by the interactions between site characteristics and macroclimate. The relevant site characteristics include soil depth, fertility, and the water-holding capacity of the soil—determined by its depth and hydraulic properties—and the slope and aspect of the site. Temperature patterns are also influenced by local topography, which can cause cold hollows or, as a result of the influence of aspect on energy interception, warmer or cooler slopes. The development and maintenance of forest canopies are strongly influenced by the amount of water held in the soil and available to plants. This is dependent on the amount and temporal distribution of rainfall, the water storage capacity of the soil, and the rate at which water is transpired by the vegetation. The influence of water and nutrition on canopy development and architecture is considered in the following sections.

### B. Water

Water constitutes a large proportion of living tissue and if the water status of tissue is unfavorable, growth will be checked or halted entirely. However, the amount of water held in plant tissues is small relative to the amount transpired. Most of the water that passes through plants plays no part in physiological processes, but is lost by transpiration through the leaves, where the wet cell surfaces behind the stomatal pores that provide the avenue of entry for the $CO_2$ essential for photosynthesis make water loss inevitable.

Favorable tissue water status requires a favorable soil water balance. The water balance of any land surface can be described by the hydrologic equation, which is essentially an expression of the conservation of water:

$$P - E_t - q_D - q_I = \Delta S, \qquad (2)$$

in which $P$ is the rainfall amount in the time interval of concern, $E_t$ denotes the amount of water lost by evapotranspiration, $q_R$ is the water lost by surface runoff, $q_D$ in the water that drains out of the root zone, $q_I$ is the water intercepted by the canopy and lost by evaporation from there, and $\Delta S$ is the change in the water content of the soil. The depth of soil that can be exploited by roots is largely a function of the soil type, but the extent to which the soil is actually exploited depends on the species present on a site and the carbon allocation relationships between the component parts of the trees: there are strong interactions between canopy size, stand structure, and root mass. The amount of water ($E_t$) extracted from the soil in any interval depends on the (average) rate of transpiration, which is directly dependent on $L^\star$ and atmospheric conditions (see Section IV,B).

If precipitation exceeds the sum of the loss terms, $\Delta S$ will be positive. When the soil is saturated, additional precipitation is ineffective and is all lost through drainage or runoff. During periods when $P = 0$, $q_R$ and $q_t = 0$ and $q_D$ tends to zero as the larger soil pores empty, but plants continue to extract the water stored in the root zone. As soil water content falls, the resistances in the flow pathways from the soil to the roots through the stems and leaves increase, leaf water potentials fall, and water stress develops.

Water balance can be considered over any time scale, from minutes to long-term averages applying over years; their effect on forest canopies are generally important over seasonal time scales. The seasonal water balance of forests varies widely. Lowland wet tropical forests, almost universally, occur in areas where rainfall exceeds evaporation throughout the year, so the water balance is usually favorable, although periods of water stress do occur. However, most tropical forests experience dry seasons and in many regions where plantation forestry is important, such as parts of South America, South Africa, and Australia, the southeast United States, and some Mediterranean countries, evaporation during summer normally exceeds rainfall by significant amounts and endemic summer water stress is the norm. Shorter and less predictable periods of stress occur in many areas and are caused, not by a shortage of water in the root zone, but by high rates of evaporative demand, when rates of water movement through the soil–root–plant system cannot meet the potential loss rates (see Section IV,B). The effects will be exacerbated by high $L^\star$, although such periods of atmospherically induced stress will seldom lead to foliage shedding (see also a later comment). [*See* FOREST HYDROLOGY.]

In considering the effects of seasonal water deficits we have to recognize that precipitation is a highly variable factor; the amount of rain (or snow) received in any particular month may vary considerably from the long-term monthly mean, particularly in drier areas, whereas a given amount of rain falling in a short period has a very different effect from the same amount received as small falls over an extended period (see remarks in Section IV,B about intercepted water). Nevertheless, the long-term average water balance of an area provides a firm basis for estimating $L^\star$. This has been demonstrated quantitatively for forests in Oregon, where there is a strong climatic gradient from the coast to the inland ranges, characterized by coniferous species that range from spruce to juniper. A linear relationship was established between $L^\star$, calculated from regressions of foliage biomass on stem diameter and foliage dry weight/area relationships, and average annual site water balance.

Water stress generally affects foliage through its effects on leaf expansion; these can be described quantitatively by a parameter called the stress integral—a measure of water stress obtained by summing the predawn values of leaf water potential over any chosen period. Reductions in leaf expansion have been shown to be linearly related to this parameter.

Prolonged severe water stress may also cause reduction in $L^\star$ by foliage shedding, which results in improved water relations for the trees as the leaf surface area, and hence the amount of water lost by transpiration, is reduced. In a study on coastal eucalypts in Australia, the $L^\star$ fell from 3 to less than 1 in a severe drought, and in an experiment with plantation-grown radiata (Monterey) pine, foliage senescence was observed to increase rapidly as a result of periods of severe water stress. This experiment provided an interesting insight into one of the possible consequences of fertilization in some areas: foliage mass increased rapidly in trees that were well fertilized before a wet growing season, but in a following drought period there was massive foliage loss caused by the high rates of soil water depletion by these trees and the consequent development of serious water stress. Foliage loss in response to water stress has also been observed in some tropical forests.

### C. Nutrition

Plant nutrients are taken up from the soil by the roots and translocated mainly to the foliage. The amount and availability of nutrients impose a major control on tree growth: basically, there must be a certain minimum amount of nutrient available to achieve some minimum growth increment. But that should not be taken to imply a simple, proportional relationship between nutrient uptake and growth. This may be achieved for young seedlings under carefully controlled experimental conditions, but for forests as a whole the amount of nutrient taken up is likely to be proportional to biomass production: the dynamics of the process are complex. [*See* Nutrient Cycling in Forests; Nutrient Cycling in Tropical Forests.]

In general, canopy mass will increase with increasing nutrient availability, assuming soil water is adequate to support the additional $L^*$. Furthermore, nutrient uptake is dependent on favorable soil water conditions. As an indication of the amounts of nutrients incorporated in biomass of forests, we may note that in a pine stand with about 110 tons per hectare (tons/ha) aboveground biomass, of which about 10 tons would be foliage, there would be about 200 kg/ha N, 25 kg/ha P, 200 kg/ha K, 140 kg/ha CA, and 50 kg/ha Mg in the wood and bark, with about 120 kg N, 10 kg P, 70 kg K, 40 kg Ca, and 20 kg Mg in the foliage. The equivalent figures for a temperate hardwood stand, which would have only 3–4 tons/ha foliage, would be 270 kg N, 40 kg P, 350 kg K, 78 kg Ca, and 60 kg Mg in the wood and bark, with 50 kg N, 5–7 kg P, 25 kg K, 35 kg Ca, and 5 kg Mg in the leaves.

These figures are clearly only indicative. The amounts of nutrients in standing biomass will vary with the age, stand density, time of year, soil fertility, and water status. In mature stands nutrient uptake is likely to be more or less balanced by losses in litterfall (foliage and small branches), tree-fall, and soil losses. The annual nutrient requirements of forests can be estimated by multiplying annual foliage production by the average nutrient concentrations in foliage. Indicative figures for these are (% dry weight) 1–2% N, 0.05–0.1% P, 0.5% K, and 0.1% Mg, and indicative figures for total nutrient uptake by forests are 50–100 kg N, 3–8 kg P, 20–40 kg K, and about 10 kg Mg per hectare per year. Foliar efficiency, in terms of the amount of dry matter produced by unit mass of foliage over any given period, is related to foliage nutrient concentrations. However, there are upper and lower limits to the nutrient concentrations of foliage, so that the use of those concentrations at any particular time is a poor guide to the nutrient status of the trees, as they vary seasonally in response to the requirements of adjacent tissue for nutrients and in relation to their age and position in the canopy.

During the period of canopy development, in a regenerating stand or young plantation, nutrient uptake is rapid, because enough nutrient must be taken up to supply the requirements of all the new foliage and the living cells of the wood. During this period the amount of nutrient taken up will be proportional to the dry weight increment or, conversely, the amount of dry matter that trees can produce in a given period will be limited by the amount of nutrient they can obtain. [*See* Forest Stand Regeneration, Natural and Artificial.]

As trees grow the amount of nutrients retranslocated internally increases. Studies on hardwoods have indicated that nutrients may be withdrawn from heartwood, which is not living tissue, as it is formed, becoming available for growth by retranslocation to actively growing tissues. Although retranslocation generally refers to the movement of nutrients from old to new tissue, it has been shown that the nutrient concentrations of the foliage of coniferous trees are subject to continual change, with considerable evidence that retranslocation may take place from young foliage to active growing areas, such as new shoots. For example, nitrogen in the needles of radiata pine fluctuated from 130 to 380 $\mu$g/needle, and back, over several seasons, whereas P varied from 15 to 35 $\mu$g/needle. Some of the variation occurred in response to rain, which stimulated a flush of growth and hence nutrient retranslocation to meet the nutrient requirements. Typical concentrations (% dry weight) of these elements in the foliage of these species are 0.15% N and 0.02% P.

Retranslocation of nutrients during growth influences the efficiency with which forest canopies use nutrients to produce dry matter. When foliage senesces, prior to its abscission and fall (litterfall), whether at the end of the season on deciduous trees or at the end of the life of foliage on evergreens, there is also substantial withdrawal of mobile nutrient elements, especially N, P, and S, which can account for about 50% of the nutrient requirements for new growth. Nitrogen use efficiency by evergreen trees is about 20% higher than that by deciduous trees. The remaining nutrients in litterfall are recycled through decomposition processes, becoming available again through the soil. [*See* SOIL ECOSYSTEMS.]

As a result of the combination of retranslocation and recycling through litter and the soil, net nutrient uptake from the soil by a mature forest where the foliage mass is effectively constant may be very low, so that nutrients are used very effectively and efficiently. This highlights an important difference between water and nutrients in their effects on forest canopies and their functions. Shortage of water constitutes a fundamental limitation: forests do not grow in dry areas. But they may grow, and produce dense canopies, in nutrient-poor, wet areas, because of retranslocation and recycling.

## IV. INTERACTIONS BETWEEN FOREST CANOPIES AND WEATHER

Normal weather measurements, made in open spaces, do not reflect the conditions in forest canopies, which create their own microclimates where conditions reflect the interactions between the physical structure of the forest and the physical and physiological processes taking place within it. The most fundamental of those processes is energy interception, which determines the amount of energy available for photosynthesis and transpiration. Both of these processes also depend on the aerodynamic exchange processes that transport $CO_2$ to the canopy and to the leaf surfaces within it, and transport water vapor away.

### A. Energy Interception and the Carbon Balance

The growth patterns and productivity of forests depend on the interception of shortwave visible radiant energy by the foliage, which drives the photosynthetic process. Photosynthesis involves the absorption of $CO_2$ by the wet mesophyll cells subtending the stomatal pores on the foliage, and the conversion of that $CO_2$, by complex photochemical and biochemical processes, to carbohydrates. A proportion of the carbohydrates produced by photosynthesis is lost by respiration: the difference (photosynthesis minus respiration) is net primary production (NPP). Of the carbon fixed by the trees, some is lost in fine-root turnover (see later), and over a growing season there are losses such as litterfall and branch abscission. The end result is net biomass accumulation.

The interception of radiant energy by canopies is a purely physical process depending on the canopy architecture, the spectral properties of the foliage, and the intensity and distribution of the incoming energy. The most important foliage spectral properties are transmissivity and reflectance. Radiation may be diffuse—coming from all directions—or strongly directional, which is the case when the sky is clear. The angle of incidence of direct beam radiation depends on the solar elevation, which is a function of latitude, season, and time of day.

There has been an enormous amount of research into energy interception by canopies, much of it concerned with detailed mathematical descriptions of the path of beams and scattered radiation through canopies of specified structure. The objective is to be able to describe the light regime of the foliage and to use that information, in association with models describing photosynthesis in terms of light climate, to calculate the amount of $CO_2$ that will be fixed by a canopy in a specified interval. The simplest formulation, the extinction coefficient model [Equation (1)] mentioned earlier in relation to optimum $L^\star$, only provides accurate results for short time intervals (minutes to hours) if the assumption that foliage within canopies is randomly distributed in space, with spherical leaf angle distribution, is not seriously violated. If this assumption is invalid because foliage is clumped, more complex models may have to be used.

However, to obtain useful average values of the energy intercepted by a forest canopy over periods longer than a few days, simple formulations such as Equation (1) will provide good results. Strong linear relationships have been established between aboveground net biomass accumulation and estimates of the energy absorbed by stands over seasons: the relationship generally has a slope of about 1–1.5 g dry matter/MJ for total solar radiation, or about double that for energy in the visible range—usually called photosynthetically active radiation (PAR).

As noted earlier, few forest stands are uniform in the horizontal plane, and spatial variations in canopy structure will be reflected in variations in photosynthesis and dry matter production. Estimates of these processes based on detailed information about canopy structure collected at only a few locations will not necessarily provide an accurate mean value for the whole stand. The spatial variability of the radiation environment within a forest canopy is matched by temporal variability. On days with broken clouds, the flux density of radiant energy incident on canopies may vary by more than an order of magnitude over intervals of minutes, and even on days when incident energy varies smoothly the movement of foliage caused by wind, over short intervals, causes continual variation at leaf level.

Gaps are an important factor in spatial variability and the most important aspect of the climate in gaps is the light climate. This is determined by the size of the gap relative to the height of the surrounding trees and the density of the surrounding foliage. As gap size increases the average energy flux density in the gap increases, and the probability of short-period, high-intensity sunflecks on the forest floor increases. These are important for the growth of understory vegetation and canopy species recolonizing gaps. Large gaps also affect the light regime of the forest floor in the immediate vicinity of the gap.

It has been established, for temperate forests, that productivity is generally proportional to $L^\star$, although forest canopies where $L^\star$ is greater than 5–6 will absorb virtually all the incident radiant energy and can be expected to be producing carbohydrate at the maximum possible rate, commensurate with foliage condition and nutrient status (see comments in Section III,B). Maximum foliage efficiency, in terms of wood productivity per unit area of foliage, is reached at $L^\star$ values of 2–4, after which efficiency declines, but the decline in efficiency is not sufficient to counteract the increase in total carbon fixed at larger values of $L^\star$. Seasonal NPP in deciduous forests depends on the length of time the foliage is functional, as well as the values of $L^\star$ reached in the forests. In evergreen forests the length of the growing season, determined by temperature, tends to be a determining factor.

The allocation of the carbohydrates produced in the canopy to the various parts of the trees is an important determinant of canopy structure. It is generally possible to describe the relationships between the component parts of trees by allometric equations, which provide a useful basis for estimating the size and structure of a stand. The relationships between leaf mass and stem diameter are of this type. All allometric equations suffer from the fact that they are static, empirical descriptions of dynamic systems. The proportions of carbohydrates allocated to the component parts of the trees vary, depending on nutrition and the water balance. Experimental data for actively growing radiata pine indicate that about 0.1–0.2 NPP goes to foliage growth, depending on fertility (higher fertility yields greater foliage mass); stem growth takes about 0.4 NPP and root growth from 0.2 to 0.4 NPP. There are indications from some experimental work that the proportion of carbon allocated by trees to belowground growth is higher under poor growing conditions than under favorable conditions. This is not universally accepted. If it is the case it implies that, for every unit of carbohydrate produced on a poor site, more will go to support fine root production than foliage, generating a feedback effect—the canopy will have less foliage, intercept less energy, and produce less carbohydrate. The converse holds.

### B. Evaporation and Water Balance

The rate of water loss by transpiration from forest canopies depends on the foliage surface area and how it interacts with the solar radiation incident upon it and with the air flowing over and through it.

The energy balance of a canopy depends on the relative rates of input and loss of radiant energy, and the rate of dissipation of absorbed energy. The net radiation retained below any surface is

$$Q_a = (a - \alpha)Q_S + Q_L, \qquad (3)$$

where $\alpha$, the albedo or reflectivity of the surface, is the fraction of incident shortwave energy ($Q_S$) reflected from the surface and $Q_L$ is the longwave balance.

Equation (3) shows that the canopy has to dissipate an amount of energy $Q_n$, which may be as much as 15–20 MJ/day in some areas and seasons. The main mechanism for dissipation is conversion to latent heat through the process of transpiration. This process has been intensively studied and is well understood. The rate of transpiration depends not only on the physical interactions between the canopy and its environment, but also on the physiological controls exerted by stomata. The energy balance of individual leaves can be written in terms of the fluxes of sensible and latent heat; if stomata are wide open and the leaf to air transport processes are efficient, the process will be dominated by the latent heat term.

Stomata are the key control points in the gas exchange pathways between the atmosphere and the internal cell of leaves. We know a great deal about how they function and the factors that affect them, but it is sufficient, in the present context, to note that stomata are generally sensitive to light, to the water status of leaves, and to atmospheric humidity. Stomata usually open during the day and close at night. They tend to close in response to relatively severe water stress, which reduces water loss, although it also reduces the rate of photosynthesis. The tendency of stomata to close at low atmospheric humidity reduces water loss under conditions of high evaporative demand. As in almost all such matters, there is considerable variation between species in the size, number, and distribution of stomata and in their responses to environmental changes.

Values of $\alpha$ for forests range from about 0.1 to 0.2. Estimates of the proportion of net radiation converted into sensible and latent heat may be made through the so-called Bowen ratio or, in experimental situations, by using aerodynamic transfer theory. Forests are large, aerodynamically rough communities, generating a high degree of turbulence above their canopies, so that transfer of heat and water vapor from the upper layers of canopies to the atmosphere tends to be highly effective unless wind speeds are very low. The elements of forest canopies are also highly effective in the absorption of the momentum transferred downward by wind, with the consequence that, in closed canopies, wind speeds drop quickly through the canopy and tend to be very low at most levels. Consequently leaf–air and bulk aerodynamic transfer processes in the canopy may be relatively ineffective, water vapor and $CO_2$ are not transported to the free air above the canopy and the concentrations of both rise. Because wind speeds in tropical regions are usually low, air humidities within the canopies and trunk space of such forests tend to be very high.

For general purposes of estimating transpiration from forests, the most common procedure is to use estimates of $Q_n$ in an equation that combines energy balance and mass transfer considerations and includes a term accounting for the resistance to vapor flow from leaf to air imposed by stomata—the stomatal resistance term. The aerodynamic term in the commonly used combination equation is derived from theory that is strictly only applicable to flat, uniform forests, but the equation suffices for most purposes; appropriate values of the resistance term are available. Maximum rates of transpiration from forests (measured from Australian plantations during hot periods in summer) are about 5–8 mm (depth equivalent) per day, which may include a period when the loss rate is up to 1 mm/hr. In most temperate areas summer transpiration rates would be less than half these values.

The water balance equation [Equation (2)] presented earlier contains a term for interception of water by canopies, from where it is lost by evaporation without contributing to the soil water balance.

Interception may be a significant factor in the overall water balance of forests, which have a canopy storage capacity usually equivalent to about 1–2 mm rainfall. In areas where rainfall is evenly distributed, and much of it comes in small showers, up to 50% may be intercepted and lost by evaporation. Where rainfall is heavy, storage capacity is rapidly saturated and most of the rain reaches the soil as direct throughfall, stemflow, or drip from the foliage. Up to a limit value of $L^\star$ about 5 interception losses will tend to increase with increasing foliage mass.

## V. EFFECTS OF DISTURBANCE ON FOREST CANOPIES

Canopy disturbance may be caused by wind, insects, disease, fire, pollution, or the direct activities of humans. Space limitations preclude a thorough treatment, but the following comments outline some of the effects of disturbance.

Most insect damage is physical: many species of insects eat leaves, causing partial, and in some cases complete, defoliation. In most years, where there has been no pest outbreak, insects consume less than 10% of forest foliage. Others feed by leaf sucking, such as the scale insects that attach themselves to leaves and extract the contents of the cells. Yet others may be borers, damaging the conducting tissue of branches or stems. In each case the consequences of the damage can be assessed—at least in principle—on the basis of the type of injury inflicted on the trees. Defoliation reduces foliage area and hence the capacity of a tree to photosynthesize and produce carbohydrates. This may not be serious at low damage levels: all healthy trees have reserve carbohydrates, in the form of stored starches, and even a single complete defoliation will seldom cause lasting damage, although it would certainly reduce growth. Partial defoliation will also alter the water balance of the tree, which may or may not be important depending on its environment and the season. Repeated defoliation may, in the end, cause death as a consequence of the exhaustion of the reserves, with resultant deterioration of root systems and inability to produce new leaves. The massive nutrient losses associated with repeated defoliation may also have serious effects.

Disease may cause defoliation by damaging leaves and leading to their abscission. The effects would be the same as those produced by insects, or they may lead to foliage malfunction and loss of photosynthetic capacity. Diseases may disrupt transport processes in the trees, physically preventing the movement of sap, or they may cause the transport of toxins that lead to the death of large parts of the foliage. [*See* FOREST PATHOLOGY.]

Air pollution may exert its effects through the soil—the acid rain problem affects the acidity of soil and hence the availability of nutrients—or directly on leaves by disrupting metabolic processes. Photosynthesis appears to be particularly sensitive to air pollutants, with the degree of susceptibility proportional to the average stomatal conductance of the species. [*See* AIR POLLUTION AND FORESTS.]

The effects of fire may range from slight scorching, in the case of low-intensity fire, to complete destruction of a stand. Similarly, clearing by humans may range from selective logging, or light thinning, to clear felling a stand. The effects will vary depending on subsequent treatment. However, the germination and early growth of some species are heavily dependent on fire, and fire suppression in forests over recent decades has resulted in the decline of these species.

In all cases, the effects of damage can be assessed in terms of the processes disrupted and their effects, not only on the physiological processes but also on the physical environment that influences the functioning of those processes. For example, any reduction in $L^\star$ will alter the energy interception characteristics of the canopy; reduced $L^\star$ will result in reduced momentum absorption and higher wind speeds at lower levels in the canopy. The magnitude of such effects will depend on the degree of damage to the canopy and their importance will vary from place to place, depending on the environmental characteristics of the area—whether it is dry or wet, whether changes in energy interception will affect the dominant trees of the stand, whether more air movement in the lower levels of the stand will make a significant difference to disease organisms. Whatever the results, the forest canopy is inextricably coupled to its environment and disturbance may trigger a chain of events with complex consequences.

## Glossary

**Canopy** Cover of leaves and branches forming the upper layer of forests.

**Canopy architecture** Distribution and arrangement of foliage through the canopy space.

**Canopy depth** Distance from the top of the canopy to the average level of the lowest foliage.

**Foliage area density** Surface area of foliage ($m^2$) per unit volume of canopy ($m^3$).

**Foliar efficiency** Dry matter produced per unit mass of foliage per unit time.

**Leaf area index** Total surface area of foliage per unit land area.

**Net primary production** Difference between the total amount of $CO_2$ fixed by the canopy and that lost by respiration. Clear definition of the term is essential to clarify whether litterfall is included. Commonly called aboveground net primary production.

**Specific foliage area** Area of foliage per unit mass ($m^2/kg$).

**Spherical leaf angle distribution** Leaf angle distribution such that, whatever the angle of a beam of radiation, there is always an equal area of leaf normal to the beam.

**Stand structure** Stem population and size distribution, average height, tree crown shape(s), and the presence or absence of understory species. Species composition and stand age exert strong influence on stand structure.

## Bibliography

Fujimori, T., and Whitehead, D., eds. (1986). "Crown and Canopy Structure in Relation to Productivity." Japan: International Union of Forest Research Organisations.

Gholz, H. L., Ewel, K. D., and Teskey, R. O. (1990). Water and forest productivity. *Forest Ecol. Management* **30,** 1–18.

Jarvis, P. G., and Leverenz, J. W. (1983). Productivity of temperate, deciduous and evergreen forests. *In* "Encyclopedia of Plant Physiology. Vol. 12D. Physiological Plant Ecology IV. Ecosystem Processes: Mineral Cycling, Productivity and Man's Influence" (O. L. Lange, P. S. Nobel, C. B. Osmond, and H. Ziegler, eds.). Berlin: Springer-Verlag.

Landsberg, J. J. (1986). "Physiological Ecology of Forest Production." London/Orlando, Fla.: Academic Press.

Myers, B. J. (1988). Water stress integrals. *Tree Physiol.* **4,** 315–323.

Nambiar, E. K. S. (1991). Nutrient translocation in temperate forests. *Tree Physiol.* **9,** 185–207.

Vitousek, P. M. (1984). Litterfall, nutrient cycling and nutrient limitation in tropical forest. *Ecology* **65**(1), 285–298.

Waring, R. H., and Schlesinger, W. H. (1985). "Forest Ecosystems: Concepts and Management." London/Orlando, Fla.: Academic Press.

# Forest Clear-Cutting, Soil Response

J. G. McColl
*University of California*



Forest clear-cutting refers to the cutting of all trees over a relatively large area and the removal of merchantable portions. The effects of clear-cutting on soil differ between forests, their age and degree of development, species diversity, the time and frequency of clear-cutting, the soil conditions, and the climate. Soil responses are also determined by the method of clear-cutting, the degree of utilization of merchantable biomass, and what is left and what is done at the site following harvesting. Postharvest treatments interact with the direct effects of clear-cutting; such treatments include slash removal or burning, ploughing, chipping and spreading of slash, and herbicide and fertilizer applications. The crop that follows the clear-cutting also interacts with the effects of clear-cutting on the soil.

Conventional clear-cutting, when applied correctly, does not generally have lasting effects on soil. However, there are many actions that could lead to soil deterioration. If roads cause erosion, if the litter layer is removed, if postharvest practices are too severe, or if conventional harvesting is replaced by whole-tree harvesting, there could be deleterious effects on physical, nutritional, and biological properties of soil.

## I. INTRODUCTION

The soil is a complex, dynamic medium, involving physical, chemical, and biological processes that are interlinked. It is difficult to generalize about the effects of forest clear-cutting on soil responses, as effects differ owing to the interactions of the controlling soil factors and the clear-cutting and postharvest practices. Long-term effects are more difficult to predict than short-term effects, because there are only a few, well-documented studies over years or decades, and because the time required for measurable changes in mineral weathering and soil development is usually very long. There is concern about effects of clear-cutting on soil properties and the potential decrease that may subsequently occur in forest productivity, particularly if whole-t ree harvesting widely replaces conventional harvesting.

## II. SOIL PHYSICAL STATUS

Clear-cutting removes the forest canopy, and the mechanical operation itself often disturbs the forest floor and upper mineral soil horizons and changes the soil bulk density. Thus, the soil microclimate of a clear-cut is different from that within a forest. The main physical effects of harvest practices on the soil are changes in temperature regime, compaction and reduced aeration, altered soil moisture regimes and changes in water yield. [*See* SOIL ECOSYSTEMS.]

### A. Soil Temperature

In clear-cuts, variations in daily temperatures are wider and snow accumulation is usually greater, causing lower soil temperatures. In very cold climates this may lead to changes in depth of permafrost. The albedo (reflectivity of solar radiation) of a forest canopy is usually less than that of many other types of vegetation, and much less than that of snow in a clear-cut. The local climate is thus altered, particularly if such factors lead to a change in the time of spring snow-melt. In hot climates, soil surface temperatures in clear-cuts can soar in summer and be lethal to regenerating vegetation.

### B. Soil Compaction and Aeration

Soil compaction by heavy machinery and dragging logs can cause soil porosity to decrease and soil bulk density to increase, and can result in decreased aeration and decreased hydraulic conductivity. In fine- and medium-textured soils, increased surface runoff may result. Together with decrease soil moisture utilization following clear-cutting, decreased porosity may result in anaerobic soil conditions, thus retarding soil microbial decomposition of organic matter and resulting in increases in acidity and facilitating chemical valence reductions in iron and manganese oxides. Waterlogged conditions and increased soil strength due to compaction may inhibit root penetration and growth of the subsequent forest stand.

To minimize compaction problems in clear-cuts, there should be little disturbance of litter and upper mineral soil that contains organic matter. Carefully planned trafficking to avoid wet conditions should be exercised, and logs should be lifted off the ground rather than across the soil surface. A high level of soil organic matter should also be maintained by avoidance of severe slash burning to reduce the risk of deleterious soil compaction.

### C. Soil Erosion

In most undisturbed forests, surface erosion by wind or water is negligible, and even in clear-cuts such erosion is small if the litter layer is left intact and unburned, and if logging access roads are revegetated. However, erosion by mass movement, especially on steep slopes in wet climates, can be a serious problem. The cause of this kind of erosion is primarily related to the undercutting of slopes in road construction rather than the removal of the forest stand per se. A study in the United States Pacific Northwest showed that nearly all the sediment in a stream after clear-cutting was due to roading, even though roads covered only a very small area of the clear-cut watershed.

If clear-cuts on steep slopes are not quickly regenerated with trees, or if they are turned into pasture, there is a risk of slope failure and mass movement of soil. This is because deeply penetrating tree roots contribute to the shear strength of the soil, much like reinforcing roads do in concrete. In one experimental study, the shear stress of soil with actively growing pine roots was double that of the same soil without roots. When tree roots are removed or die and decompose, the soil shear strength decreases and may result in slope failure for several years after clear-cutting. [*See* RHIZOSPHERE ECOPHYSIOLOGY.]

Increased soil moisture due to less evapotranspiration also adds to the probability of mass movement following clear-cutting. Greater moisture increases the downslope force and reduces soil shear strength because of reduced internal friction and cohesion of soil particles.

### D. Soil Moisture Regime

Clear-cutting generally reduces transpirational use of soil water by the forest stand, and if the soil is compacted, deep drainage may also be impeded. This usually results in more moisture being retained in the soil and in more water yield to streams as runoff, at least until the subsequent stand regenerates and water use by the vegetation reaches the level prior to clear-cutting.

In low-lying areas where the water table is usually high, clear-cutting may result in saturation of the soil for long periods, making it almost impossible to regenerate the site. In upland sites, soil moisture levels are also increased, and moisture retention in the soil can be doubled even during summers in Mediterranean-type climates where there is no summer rainfall.

Direct effects of the different soil moisture regime on soil properties are difficult to determine, but at least temporary changes in soil leaching, decomposition of organic matter, and other processes, should be expected.

### E. Water Yield

There have been many studies of water yield from the soil to streams following clear-cutting, conducted in both small plots and whole watersheds. Generally in the temperate climatic zone, a 10% reduction in vegetation cover yields 40 mm more annual water yield in coniferous forest, 25 mm more in hardwood forest, and 10 mm more in scrubland. Clear-cutting and conversion of one type of forest to another, or to pasture, has been done to increase water yields in areas where water catchment is the primary objective. The main reason for the increase in water yield is that a new forest (or pasture) usually transpires less water than the previous forest of larger trees. In other words, younger, smaller trees in the *aggradation phase* of growth (which have not yet achieved a *steady state* as in most mature forests) usually require less water.

This principle usually applies but there are notable exceptions, for example, in *Eucalyptus* forest in a water-catchment area of southeastern Australia. Here, logging or burning old-growth forest followed by extremely rapid regrowth and transpiration of tree seedlings results in *reduction* of water yields measured as decreased stream flow. This revelation has influenced local land-management strategies that aim to maximize water yield.

Clear-cutting may also reduce the net precipitation reaching the soil by the removal of tree crowns that intercept fog and cause it to precipitate. Such fog-drip can be a significant contribution to total precipitation in foggy coastal areas of relatively low rainfall. [*See* FOREST CANOPIES.]

Deforestation on a large scale may even change patterns of precipitation and energy balance at local, continential, and perhaps global scales. In the humid tropical forests of equatorial Amazonia, which use about 50% of the total rainfall, large-scale deforestation may not only increase total runoff, but may lower the total rainfall and cause more pronounced seasonality. [*See* DEFORESTATION.]

## III. SOIL NUTRIENT STATUS

In undisturbed forests, chemical nutrients cycle between the soil and plants, and soil development proceeds over time as a function of the interactions of the soil-forming factors at the site, namely, the climate, topography, soil parent material, and organisms, including the forest trees themselves. The outputs of nutrients are generally balanced or slightly exceeded by the inputs over the life span of the forest (Table I).

Clear-cutting disrupts the natural cycling of nutrients in forest ecosystems by removing nutrient elements that the forest has taken up from the soil. Nutrients can also be removed by leaching, that is, by movement offsite in solution, and in particulates by erosion. Thus the nutrient capital of the soil can be reduced, unless there are equivalent inputs of nutrients from rock weathering, the atmosphere, nitrogen-fixing plants, or addition of fertilizer salts. [*See* NUTRIENT CYCLING IN FORESTS.]

Clear-cutting and post-clear-cutting treatments, such as the burning of slash, can also change the *availability* of soil nutrients, often manifested as a flush of soluble nutrients that plants can use or that is lost from the site in drainage waters. Important nutrient elements and soil processes that are most affected by clear-cutting are discussed below as nutrient removals and nutrient additions, followed by a discussion on the resultant effects on soil nutrient characteristics.

### A. Nutrient Removals

#### *1. In Biomass*

Harvesting the stem and bark in conventional harvesting removes the following approximate

***TABLE I***

**General Nutrient Budget for Mature, Temperate-Climate Forests in North America over 50- to 100-Year Rotations, Showing Ranges of Natural Inputs and Outputs and Additional Removals in Biomass by Conventional Clear-Cut Harvesting of Stemwood (in kg/hectare/year)**

| | Nutrient element | | | | |
|---|---|---|---|---|---|
| Inputs and outputs | K | Ca | Mg | P | N |
| Inputs | | | | | |
| Weathering | tr[a]–11 | tr–100 | tr[a]–52 | 0.2–3.0 | |
| Atmosphere | 0.5–12 | 1–21 | 0.7–3.9 | tr–1.3 | 2–22 |
| N-fixation | | | | | 0–7.5,300[b] |
| Outputs | | | | | |
| Leaching and erosion | 2–13 | 8–72 | 3–20 | 0.01–0.1 | tr–10,22[b] |
| Biomass | 2–10 | 2–20 | 0.5–10 | 0.2–0.3 | 1–6,202[b] |

[a] tr = trace amount.
[b] Nodulated actinorrhizal and leguminous plants are the larger amounts.

amounts of nitrogen (N), phosphorus (P), calcium (Ca), magnesium (Mg), and potassium (K) in kilograms per hectare: 50–300 of N, 10–150 of P, 100–1000 of Ca, 25–500 of Mg, and 100–500 of K. These are broad ranges for mature, temperate North America forests, over typical 50- to 100-year rotations. Amounts vary greatly with species, stand development, and the site quality, which is largely a function of the soil fertility and climate. For forests with N-fixing trees, N removals will be much higher than the range given. Approximate yearly removals are given in Table I.

If conventional harvesting is compared to whole-tree harvesting, in which even roots may be taken, these quantities of nutrients removed may be multiplied by two to four times. The effect of these nutrient removals on the soil will be determined by many factors, especially the total and available nutrient capital of the soil.

The amount of nutrient elements taken up by forests from the soil is largely a function of the net primary productivity of the forest. Thus the sum of nutrient elements in the biomass of humid tropical forests is about twice that of broad-leaved temperate forests and triple that of conifer forests. The relative proportion of nutrients in biomass above ground is also highest in the humid tropics, and the proportion of nutrients stored in the forest floor is very low. Thus, whole-tree harvesting in the humid tropics will remove more nutrients, and soil fertility reductions would be greater, than in the temperate zone. Similarly, harvesting of a deciduous forest would generally remove more nutrients than harvesting of Northern Hemisphere evergreen forests of similar biomass, unless the harvest occurs after leaf fall.

Rotation length, which is the time between successive forest harvests, is also a factor determining nutrient removal. With shorter rotations and greater harvest utilization, a site is in disturbed state for longer periods, and the potential for nutrient removals in erosion and through leaching becomes greater. Nutrient deficiencies may result on sites with low-nutrient soils, and addition of fertilizers then becomes a necessity to offset decreased soil fertility and possible decline in forest productivity.

Nutrient removals in biomass compared to total and available reserves of nutrients in the soil and forest floor will be discussed for the individual nutrients N, P, K, and Ca.

There is about 8 to 11 times as much N in the forest floor and mineral soil than there is in the forest biomass. The forest floor is a particularly important (but vulnerable) reserve for N, as it is easily subject to loss from mineralization and leaching and from fire. In Canadian spruce forests, for example, about three-quarters of the N reserve is in the forest floor. In such northern climate forests, relatively long rotation times may prevent excessive N losses. By contrast, in warmer temperature and tropical climates there is greater potential for soil N decline because there is less forest floor and relatively more N is contained in the biomass. This N removal may offset the potential for greater biomass production on shorter rotations that appears as an attractive option because of the more rapid nutrient cycling in the warmer climates.

The P removal in biomass as a proportion of the total P in forest soil is usually less than that for N, except for very poor soils low in P, such as those of the Southern Coastal Plain of the southeastern United States. Here, soils may become P-deficient before two 20-year rotations of pine. (Considerable analytical problems remain in defining what fraction of the total soil P is available, particularly because much of the P is organic, and meaningful fractionation of organic soil P is difficult.)

Potassium is not usually deficient in forests, although red pine growing on poor sandy soils in the northeastern United States showed deficiencies. Biomass removal of K is generally small compared to total soil K, but may be quite large compared to the exchangeable, and therefore available, soil K.

Calcium removal in biomass is high compared to other nutrients (Table I), and Ca removal with whole-tree harvesting is greater than the Ca amount in the forest floor in some species. In such cases, Ca replenishment following clear-cutting must depend on increased weathering. In low-Ca soils, the weathering rate may not maintain an adequate Ca supply for subsequent rotations. Similarly, biomass losses of Mg from the soil exchange complex must be replaced by weathering if the soil Mg availability is to be maintained.

### 2. By Leaching and Erosion

Losses from leaching and erosion following clear-cutting are usually small and comprise a negligible fraction of the total or available nutrients in the soil. However, leaching may cause undesirable changes in composition of water draining the clear-cut area. Such is the case in some northeastern United States hardwood forests, where excessive nitrate ($NO_3^-$) concentrations result from increased soil nitrification and reduced uptake following clear-cutting.

The production and leaching of $NO_3^-$ also affect the leaching of cations from the soil. Anions, such as $NO_3^-$ and bicarbonate ($HCO_3^-$), are produced from the decomposition of harvest residues and organic matter in the soil. They are very mobile in solution, not being readily adsorbed by mineral soil, and they induce an equivalent amount of cations to be leached with them, the main ones being $Ca^{2+}$, $Mg^{2+}$, $K^+$, and $Na^+$. Measurable flushes of soluble nutrients may last for years following clear-cutting, but the magnitude decreases rapidly after the first year or two owing to uptake by the regenerating forest and absorption lower in the soil profile. Loss of P from clear-cut areas is usually restricted to that from soil erosion, as most soil P is bound in organic forms, and inorganic forms of P are very insoluble in the acidity range of most forest soils.

Under certain conditions though, clear-cutting can actually result in a *decrease* in nutrients leached from the soil. Clear-cutting a *Eucalyptus* forest in California resulted in a decreases in soil solution concentrations of $K^+$, $Ca^{2+}$, $Mg^{2+}$, $NO_3^-$ and $HCO_3^-$, at least during the first year. This case is an example in which particular soil and climatic conditions interacted to create exceptional results. The clayey soil had a high cation-exchange capacity, rainfall was low and limited seasonally to December through March, and nearly all organic material was removed from the soil surface. Mineralization of organic matter was minimal and cations were retained on the soil exchange complex, $NO_3^-$ production from nitrification was reduced, and there was no longer any cycling of soluble nutrients between the forest canopy and the soil.

Accelerated leaching of organic as well as ionic substances can occur after clear-cutting, especially if slash is left on the ground, and more so if it is chipped and respread across the clear-cut. For example, where wood and bark residues were chipped and spread following clear-cutting of a pine forest in Wyoming, phenol concentrations greatly exceeded allowable aquatic standards for at least two years. Also of interest were the low $NO_3^-$ concentrations under the chipped slash, suggesting that harvest residues may be manipulated to control nitrification rates and thus ionic leaching losses. However, soluble organic substances may also enhance mobilization of soil metals such as aluminum, which may be deleterious to plant roots and aquatic organisms.

### B. Nutrient Additions

#### 1. By Weathering

Nutrient elements, other than N, C, and O are primarily supplied to the soil by the weathering of rock minerals. Because there is no reliable way to measure weathering rates directly, mass-balance studies provide the best way to calculate weathering rates. The mass-balance approach involves detailed information about the complex geochemistry of weathering reactions plus data on the intricate cycling of nutrients between plants, soil, microbes, and the atmosphere. Because of this complexity, little reliable information about weathering rates has been collected over long time periods under real field conditions. From the handful of comprehensive studies available, ranges of elemental nutrient release from rock weathering in forested ecosystems are given in Table I. Clear-cutting effects on weathering rates have not been measured, but there may be short-term, subtle increases in weathering rates due to increased moisture and/or temperature of the soil profile and underlying rock, and due to changes in soil microbes and fauna.

#### 2. From the Atmosphere

Nutrients added to forest ecosystems from the atmosphere (Table I) as wet precipitation in rain, snow, hail, and fog, and as dry precipitation in small particulate matter of salts, windblown soil, and dust, include both organic and inorganic components. Near the ocean, sea salt additions predominate, with large quantities of sulfur (S), Na, and chloride (Cl). Terrestrial dust primarily contributes Ca and K. Air pollution and "acid rain" from automobiles and industry also provide inputs, especially of N, S, and heavy metals.

Tree crowns intercept both wet and dry atmospheric depositions, resulting in generally greater inputs in forests than would occur in an open field. Atmospheric precipitation also leaches soluble nutrients from tree crowns as it falls through the forest to the ground, and throughfall (precipitation beneath the canopy) typically contains more N, K, and other relatively soluble nutrients. Thus removal of the forest canopy by clear-cutting generally reduces these throughfall inputs to the soil, often by about one-half.

#### 3. By Nitrogen Fixation

Nitrogen, one of the most important plant nutrients, is often growth-limiting in forests, and maybe leached from soil particularly after clear-cutting. Nitrogen fixation is the biological conversion of elemental N in the atmosphere to organic forms utilized in biological processes, and it is the primary source of N to forest ecosystems. In forests, N-fixation occurs primarily in nodules of leguminous trees mediated by bacteria (e.g., black locust with *Rhizobium*), and in other root-nodulated trees mediated by actinomycetes (e.g., alder with *Frankia*), but also occurs in free-living soil micoorganisms (including bacteria and cyanobacteria) associated with roots of many other forest trees without nodules (e.g., pine and fir). Fixation of N can also occur in epiphytes in forest canopies. By far the largest inputs of N to forest soils are from symbiotic N-fixing organisms in nodulated roots (Table I).

Therefore, the magnitude of clear-cutting effects on N-fixation in soil largely depends on the species being clear-cut and on the nature of the following regenerating vegetation, in terms of both the species of trees and the understory that invade or are planted in the clear-cut area.

### C. Resultant Effects on Soil Nutrient Status

The inputs and outputs of major nutrients to forest soil ecosystems and the effects of clear-cutting on these fluxes have been discussed, but what is the resultant effect on the soil chemical status, and is it significant to the long-term nutrient status of the soil and to the maintenance of forest productivity?

First, it is difficult to generalize about effects of clear-cutting on soil nutrients because of the complexity of factors that differ between sites. However, the general conclusion is that removal of nutrients in a conventional harvest of stemwood does not deplete soil nutrients that are usually replaced by inputs from weathering and the atmosphere, at least on the basis of amounts expressed on a yearly basis over typical rotations (Table I). Where there is serious nutrient depletion by harvesting, it is often shown first by Ca.

But if the rotation time is shortened and/or if whole-tree harvesting is implemented, nutrient removals in biomass may be two to four times greater than the ranges in Table I, and the ability of the soil to supply N, P, K, Ca, and other elements for succeeding rotations may be seriously decreased. Forest productivity may decline after one or more rotations of such intensive harvesting unless fertilizers are added and/or unless N-fixing plants are included in the silvicultural practices employed in regenerating the site after clear-cutting. Potential declines in forest productivity due to whole-tree harvesting were generally thought to be greatest on the lower-productivity sites, but there is now evidence that declines may be proportionally greater on the better sites.

It may be misleading to express nutrient losses in biomass on a yearly basis as shown in Table I. Effects of clear-cutting are not equally distributed over a rotation, but peak immediately following clear-cutting and then rapidly decrease as the stand regenerates.

What is sometimes overlooked is that removal of what seems to be a very small proportion of the total soil nutrient supply may actually be a large proportion of the *available* nutrient supply and may reduce the available supply below the level required for adequate tree nutrition. The replacement of the available supply is probably the most important issue for sustained soil fertility and forest productivity.

The issue of availability versus total amount is also pertinent for nutrients that are primarily organically bound (rather than occurring as simple inorganic ions or compounds) and whose availability is primarily biologically controlled; N and P are such elements. This point highlights the importance of the forest floor and the dangers of disturbing or removing it during clear-cutting, because it is a major source of available nutrients, especially those organically bound.

Even if there is no net loss or gain of nutrients from the soil at the time of clear-cutting or thereafter, changes may be induced *within* the soil that affect important soil processes and alter the course of soil development, at least temporarily. Such was the case in some northeastern United States hardwood stands after whole-tree harvesting. Exchangeable cations were retained in the soil profile, but there was movement of cations from the upper to the lower horizons in the soil profile, which readsorbed them. Although there were accelerated leaching losses of nutrient cations following clear-cutting, they were not the result of depletion of exchangeable cation supplies. However, the acidification of the upper soil horizons that resulted from the downward movement of cations would affect microbial activity, organic matter decomposition, and mineral weathering rates. Growth and species composition of the following stand could also be affected; acidification of the surface soil may favor regeneration of conifers over hardwoods.

## IV. SOIL BIOLOGICAL STATUS

The disturbance of the forest floor (and often the upper mineral soil horizons) and the severing of tree roots by clear-cutting affect the diversity, numbers, and activities of microbes that are important to the maintenance of soil fertility, for example, the processes of nutrient mineralization, especially that of N, and decomposition of organic matter mediated by soil microorganisms. Decomposition changes induced by clear-cutting may be manifest as changes in the carbon dioxide ($CO_2$) evolved to the atmosphere and in the amount and distribution of soil C.

### A. Soil Microbe and Mesofauna Populations and Processes

Changes noted in soil microbial status following clear-cutting include increases in bacterial populations and numbers of nitrifying bacteria (i.e., those oxidizing ammonium to $NO_3^-$). Microbial uptake of N during decay of residual organic matter is a process that retains much of the soluble N that would otherwise be leached following clear-cutting. Reductions in fungal biomass and mycorrhizae (symbiotic fungal/plant root associations), and decreases in numbers of soil mesofauna (including arthropod populations of springtails and mites), have also been documented after clear-cutting.

Effects vary and largely depend on the degree of disturbance and/or removal of the litter layer. Whole-tree harvesting and some postharvest practices that lead to almost complete removal of organic debris (including severe burning) have more drastic effects on microbial populations and activities in the soil than does conventional clear-cutting alone. In some cases, reduced microbial decomposition of litter results from excessive drying and high temperatures of the forest floor following clear-cutting. In other cases, particularly for subsurface soil, clear-cutting may lead to greater soil moisture contents and enhanced microbial decomposition.

Reducing the size of clear-cut areas so that microbial and mesofaunal populations can reinvade from adjoining forest, leaving some mature trees, and avoiding use of machinery that causes soil compaction are ways to help maintain below-ground biological systems.

### B. Carbon Dioxide Evolution

Carbon dioxide is an end product of the decomposition of organic matter, and measurement of $CO_2$ evolution is used as an index of the rate of decomposition and the activity of decomposer organisms in the soil. However, $CO_2$ is also produced by the respiration of roots, and it is extremely difficult to separate these two main sources of $CO_2$ in soil. The concern over global warming that is possibly caused by increased $CO_2$ levels in the atmosphere has piqued interest in the possible effects of forest clear-cutting on evolution of $CO_2$ from soil. But recent studies have shown that there is no significant increase in atmospheric $CO_2$ from clear-cutting that has been followed by immediate regrowth of the forest. However, some *post*-harvesting practices, such as combustion of logging debris, would produce much more $CO_2$ and may result in elevated levels in the atmosphere. [*See* GREENHOUSE GASES IN THE EARTH'S ATMOSPHERE.]

If there are declines in productivity of next-rotation forest following clear-cutting, there would be less C fixed in biomass from the atmosphere. This may be one *indirect* way that clear-cutting may result in elevated $CO_2$ levels in the atmosphere.

### C. Soil Carbon

Forests accumulate tremendous quantities of C in organic matter compared to other plant ecosystems, and consequently the soils that support them also contain comparatively large quantities of C. Generally, the cold-climate forest ecosystems toward the poles have less C than do humid, tropical forest ecosystems, reflecting the major control by climate on primary productivity. However, the proportion of the C accumulated in the soil decreases from the poles to the tropics, as decomposition is greater with increasing temperature.

Nutrients that are available in the soil, and nutrients taken up in biomass, are largely proportional to the C stored in the soil and biomass, respectively. This generalization is particularly relevant for N, where soil N limits forest productivity, as nearly all N in forest ecosystems is bound with C in organic forms. Thus the decomposition of organic matter to yield available plant nutrients through mineralization is a focal point of nutrient cycling in forests, and changes in the C cycle provide an overall picture of changes in nutrient cycles.

It follows that clear-cutting effects on soil C should provide an integrated index of such effects on the status of most soil nutrients. This generalization is confirmed by a recent comprehensive review of research studies documenting C storage in forest soils, which revealed that losses of C following conventional harvesting were negligible in most cases. Although there may be no net differences in soil C before and after clear-cutting, the activity of logging equipment can result in *redistribution* of C within the soil profile.

Results markedly differ, however, if the clear-cut site is revegetated with different species, or if clear-cutting is followed by intense burning of slash. When forests are converted to pasture following clear-cutting, large losses of soil C occur, on the order of 30–50% over a period of several decades.

The review paper of C in forest soils, referred to previously, concluded that there are good opportunities to increase soil C content in forest ecosystems by management of the nutrient regime, as soil C generally increases when N-fixing plants are present and when nutrient fertilizers are applied.

## V. CONCLUSIONS

Conventional harvesting of tree stems generally has negligible effects on soil nutrients and does not cause major deleterious physical effects on soil, if the litter layer is not seriously disturbed and if logging roads are correctly built to minimize erosion.

Deleterious effects on the soil are sometimes associated with the postharvest treatments, including removal of the litter layer, severe burning of slash, conversion to pasture, or where whole-tree harvesting removes nutrients in excess of those available in the soil.

### Glossary

**Clear-cutting** Forest harvest practice of cutting and removing all trees over a large area.

**Conventional harvesting** Harvesting the merchantable stems of trees.

**Exchangeable cations** Positively charged ions (mainly those of calcium, magnesium, potassium, and sodium) adsorbed on surfaces of clay and organic matter in soil, which are generally available for plant uptake.

**Leaching** Removal of nutrients and other soluble materials from soil by percolating water.

**Litter layer** Relatively fresh plant debris on the soil surface.

**Mineral soil** Soil body that is primarily mineral in composition (rather than organic), reflecting many properties of the rock or mineral parent material below.

**Nutrients** Chemical elements in soil that are plant nutrients.

**Rotation** Forest harvest of a sequence; rotation length is the time between harvests.

**Soil** Thin surface layer of the earth's crust that is a mixture of mineral and organic matter and that supports vegetation.

**Whole-tree harvesting** Harvesting the stems and crowns of trees, and possibly the roots.

### Bibliography

Attiwill, P. M. (1991). The disturbance of forested watersheds. *In* "Ecosystem Experiments" (H. A. Mooney, E. Medina, D. W. Schindler, E. D. Schulze, and B. H. Walker, eds.), Scope 45, Chap. 11, pp. 193–213. New York: John Wiley & Sons.

Binkley, D. (1986). "Forest Nutrition Management." New York: Wiley–Interscience.

Fernandez, I. J., Son, Y., Kraske, C. R., Rustad, L. E., and David, M. B. (1993). Soil carbon dioxide characteristics under different forest types and after harvest. *Soil Sci. Soc. Am. J.* **57,** 111–1121.

Frazer, D. W., McColl, J. G., and Powers, R. F. (1990). Soil nitrogen mineralization in a clearcutting chronosequence in a northern California conifer forest. *Soil Sci. Soc. Am. J.* **54,** 1145–1152.

Gessel, S. P., Lacate, D. S., Weetman, G. F., and Powers, R. F., eds. (1990). "Sustained Productivity of Forest Soils." Vancouver: Faculty of Forestry, University of British Columbia.

Johnson, C. E., Johnson, A. H., and Siccama, T. G. (1991). Whole-tree clear-cutting effects on exchangeable cations and soil acidity. *Soil Sci. Soc. Am. J.* **55,** 502–508.

Johnson, D. W. (1992). Effects of forest management on soil carbon storage. *Water, Air, Soil Pollut.* **64,** 83–120.

Leaf, A. L., ed. (1979). "Impact of Intensive Harvesting on Forest Nutrient Cycling." Syracuse: College of Environmental Science and Forestry, State University of New York.

McColl, J. G., and Powers, R. F. (1984). Consequences of forest management on soil–tree relationships. *In* "Nutrition of Plantation Forests" (G. D. Bowen and E. K. S. Nambiar, eds.), Chap. 14, pp. 379–412. New York: Academic Press.

Mroz, G. D., Jurgensen, M. F., and Frederick, D. J. (1985). Soil nutrient changes following whole tree harvesting on three northern hardwood sites. *Soil Sci. Soc. Am. J.* **49,** 1552–1557.

# Forest Genetics

**Raymond P. Guries**
*University of Wisconsin—Madison*



Forest genetics is concerned with the conservation, study, and utilization of the genetic resources contained in forests. Forests vary in space and time, whether they are natural or man-made, and whether they are subject to forest management practices or left in a wild condition. Humans have exploited forests for millenia, harvesting a diverse array of products, including food, fuel, building materials, gums and resins, and medicines. Forests also provide other goods and services, such as wildlife habitat, watershed protection, and esthetic and spiritual values. In recent years, scientists have begun to explore the ecological and genetic factors responsible for creating and maintaining these vast resources. Forest trees appear to contain enormous stores of genetic variation, and many national, state, and industrial organizations have begun to exploit this natural variation in tree improvement programs. Our efforts are propelled by a sense of urgency as deforestation increases in many parts of the world.

## I. INTRODUCTION

Forest genetics is the study of heritable variation (variation due to genetic rather than environmental factors) in forest trees. Trees, like people, begin life as embryos, each with a different genetic endowment and therefore a different potential for development. Owing to the segregation and assortment of chromosomes (and genes) at the time pollen and eggs are formed, virtually every seed represents a unique combination of genes. And as with people, the growth and development of trees are controlled by the unique genetic constitution of each individual, subject to modification by the environment in which it develops. Though the exact details of inheritance may vary slightly from organism to organism, the underlying principles transcend species lines and permit us to study the genetics of trees as we would almost any other organism. In addition to individual differences, variation among populations of trees is also of interest, because different geographic regions may give rise to populations with unique characteristics affecting survival, growth, pest resistance, and other traits. Thus, forest genetics is concerned with both individuals and populations, the genetic characteristics of each, and their role in the conservation and utilization of genetic resources.

## II. MEASURING GENETIC VARIATION

Much of what we know about genetic variation in forest trees has been learned from provenance tests.

In such tests, seeds of a species are collected from many different geographic locations (or provenances) and then grown together at one or more locations. Because conditions for growth are relatively uniform at any one location, any observed differences can be attributed largely to genetic differences among the original populations. Provenance testing originated in the nineteenth century as a way to identify superior seed sources for reforestation purposes. Since the 1930s, it has become an important tool in forest genetics research to examine the patterning of genetic variation that occurs over large geographic areas, or across markedly different elevations or soil types. Climatic factors such as temperature, precipitation, or day length, which vary in a continuous fashion, can give rise to patterns of genetic variation that also vary continuously (clinal variation). Results for a number of forest trees indicate that clinal patterns of variation are common for metric traits (traits that vary more or less continuously and can be quantified statistically) related to survival or growth. These patterns are established and maintained in response to environmental factors, and reflect the evolution of adaptations governing survival, growth, reproduction, and other life-history variables. Patterns of variation formed in response to more abrupt environmental or physical features such as soil types or large bodies of water may be discontinuous (or ecotypic) and change sharply over relatively short distances. [*See* Species Diversity.]

The large stores of variation present in wild populations are attributed to the complex interaction of various evolutionary forces that create and modify genetic variation. No attempt will be made here to detail the main features of evolutionary biology. However, it is important to note that evolutionary mechanisms that give rise to genetic variation, such as mutation, recombination, and "gene flow," together with forces that modify this variation, especially natural selection and "random genetic drift," give rise to the rich and complex patterns of genetic variation observed in wild populations of trees. Opportunities to examine genetic variation using molecular approaches as opposed to field studies have confirmed that trees are among the most genetically diverse organisms on earth. In addition, genetic variation in forest trees has become better understood via the use of molecular characteristics. Our ability to explore genetic variation at the molecular level permits us to better assess, for example, taxonomic relationships among species, genera, and families using large numbers of traits, variation in the organization of local "neighborhoods" of trees (groups of related individuals), or the physiological, biochemical, and molecular bases for various processes during growth and development that are difficult or impossible to measure with intact organisms. Such techniques offer new ways to study, conserve, and exploit genetic variation in trees.

## III. DOMESTICATION PRACTICES

The enormous stores of genetic variation contained in most forest trees have served as a starting point for forest tree breeding programs around the world. Tree breeding is the application of genetic principles to the domestication of forest trees. It is a process that parallels most earlier plant and animal domestication programs, and for which the goals are to manipulate the production of various tree products to human use. Even without a knowledge of genetic principles, our ancestors recognized individual variation in their crops and livestock and favored the better individuals to generate breeds and races. Typically, the domestication process begins with the identification of a need for a product such as lumber or pulp and a species capable of yielding this product. The favored species may have a variety of desirable traits and also some undesirable features (e.g., susceptibility to a pest), but may possess sufficient variation among individuals or populations to permit "improvement," the breeding of better performing individuals or populations for some human use.

Genetic improvement programs have two components: a short-term component in which rapid progress may be made by intensifying selection in relatively small populations to maximize the rate of improvement for one or a few characters, and a long-term component in which the evolutionary potential of a population or species is maintained by more conservative breeding practices. Selection of superior phenotypes (the morphological aspect of an individual) within different forests, followed by propagation by seed or rooting of cuttings, permits convenient development of seed orchards or similar collections for testing, breeding, and seed production. Testing is normally conducted to determine whether superior phenotypes selected in the field are actually superior genotypes (the genetic constitution of an individual). Such testing evaluates the genetic quality of a parent based on its ability to produce superior progeny, the "improved" trees actually used in reforestation programs. Normally, only those individuals whose progeny are deemed to be genetically superior are retained to produce seed. For short-term development of seed orchards, only a small number of elite parents are employed in order to maximize genetic gain (the difference in performance between progeny of the original population and progeny from the elite parents). For long-term conservation, larger numbers of selected parents are mated in different combinations to create populations for subsequent evaluation and to permit natural selection to operate in concert with domestication practices.

## IV. GENETIC RESOURCE CONSERVATION

Maintaining wild forests *in situ* is one obvious way to conserve genetic resources. Wild forests will continue to evolve in response to evolutionary forces with little or no human influence. Whether human-induced global change via pollutants or climate modification will strongly impact genetic resources is the subject of some debate in forestry circles, and various models of long-term genetic change in response to anthropogenic factors are being studied. Forest genetic resources can also be maintained *ex situ* in various breeding populations, seed or clone banks (repositories of genetic materials), or provenance tests. This practice may be necessary for species or populations whose existence is threatened because of deforestation or development, especially in the tropics. Such maintenance is not without cost, and future technologies such as cryopreservation (long-term subzero storage) may provide alternatives to maintaining field collections. A more formidable challenge concerns prioritizing genetic resources for *in situ* conservation using forests, protected reserves, or parks, and *ex situ* conservation using more expensive technologies. Striking a balance between conservation and utilization requires a better understanding of present and future values and needs, but limitations of time and other resources make it important that the choices made for current use of forest genetic resources do not foreclose other options for future generations. [*See* SEED BANKS.]

## Glossary

**Adaptation** Trait or combination of traits under genetic control that permits an organism to survive and reproduce under a particular set of environmental conditions.

**Clinal variation** Pattern of genetic variation in which individuals and populations vary in continuous fashion in parallel with a geographic, climatic, or edaphic features.

**Ecotypic variation** Pattern of genetic variation in which individuals and populations vary in discontinuous fashion, typically resulting from fragmented or abrupt discontinuities in geographic distribution.

**Ex situ** Out of or away from a region of natural or historic occurrence, such as plants grown in a region to which they are not native.

**Gene flow** Exchange of genes between neighboring populations, usually at low levels, and typically involving migration of individuals or pollen and seed.

**Genetic gain** Difference in performance between the average of an original population and that of a population derived from mating among elite individuals following selection.

**Heritable variation** Proportion of variation among a group of individuals that can be explained by genetic effects as opposed to environmental modifications.

**In situ** Growing in a native range of occurrence or habitat.

**Natural selection** Any natural process that leads to a differential contribution of genes from some individuals in succeeding generations, normally through differential reproductive success.

**Provenance test** Procedure to evaluate the genetic differences among a group of seeds or progenies collected in different geographic regions (provenances) and raised in a common environment, usually a single locale.

**Random genetic drift** Random fluctuations in gene frequencies in effectively small populations frequently attributed to selectively neutral mutations.

## Bibliography

Burley, J., and Wood, P. J. (1976). "A Manual on Species and Provenance Research with Special Reference to the Tropics." Oxford, England: Commonwealth Forestry Institute.

Guries, R. P. (1990). Forest genetics and forest tree breeding. *In* "Introduction to Forest Science" (R. Young and R. Giese, eds.), 2nd ed. New York: John Wiley & Sons.

Namkoong, G., Kang, H. C., and Brouard, J. S. (1988). "Tree Breeding: Principles and Strategies." New York: Springer-Verlag.

National Research Council (1991). "Managing Global Genetic Resources. Forest Trees." Washington, D.C.: National Academy Press.

Zobel, B., and Talbert, J. B. (1986). "Applied Forest Tree Improvement." New York: John Wiley & Sons.

# Forest Hydrology

**David R. DeWalle**
*The Pennsylvania State University*



Forest hydrology is a branch of the general field of hydrologic science that deals with forest effects on the water and energy cycles, especially as it relates to the quantity, timing, and quality of water yielded as streamflow and groundwater.

## I. WATER AND ENERGY CYCLES AND FORESTS

The water cycle or budget for a forested land area over a specific period of time is generally represented in equation form as

$$P + E_t + Q + dS = 0, \quad (1)$$

in which $P$ is precipitation, $Q$ is water yield generally represented by streamflow, $E_t$ is evapotranspiration or total evaporation, and $dS$ is the change in water storage in soil, rock layers, snowpacks, lakes, and so on. The October 1 to September 30 budget period is the standard "water year" used to minimize storage changes in Equation (1). All quantities are generally expressed as volumes of water per unit land area or depth of water over the land surface. Land areas typically represent watersheds, defined by surface topography, which contribute water to streamflow. The water cycle is diagrammed in Fig. 1, showing the various flows and storages of water within a watershed.

### A. Precipitation

Forest canopies intercept water from rain and snow and affect the pathways and amounts of precipitation reaching the ground. Once intercepted water satisfies an initial canopy storage capacity, water from precipitation reaches the ground in forests as throughfall and stemflow. Throughfall ($T$) is water that drips from the forest canopy or falls directly through gaps within the forest canopy, and most water reaches the ground via this pathway. Small amounts of precipitation water may also reach the ground in forests as stemflow ($S$) or water flowing downward along branches and tree trunks. Even while throughfall and stemflow are occurring, evaporation losses are taking place. Canopy interception losses ($I_c$) represent intercepted water that is ultimately evaporated from the canopy back to the atmosphere. $I_c$ can be represented as the difference between total precipitation above the forest ($P$) and the sum of water reaching the ground as throughfall and stemflow:

$$I_c = P - (T + S). \quad (2)$$

Canopy interception losses can be appreciable, ranging from about 10% of precipitation in decidu-

Encyclopedia of Environmental Biology
Volume 2

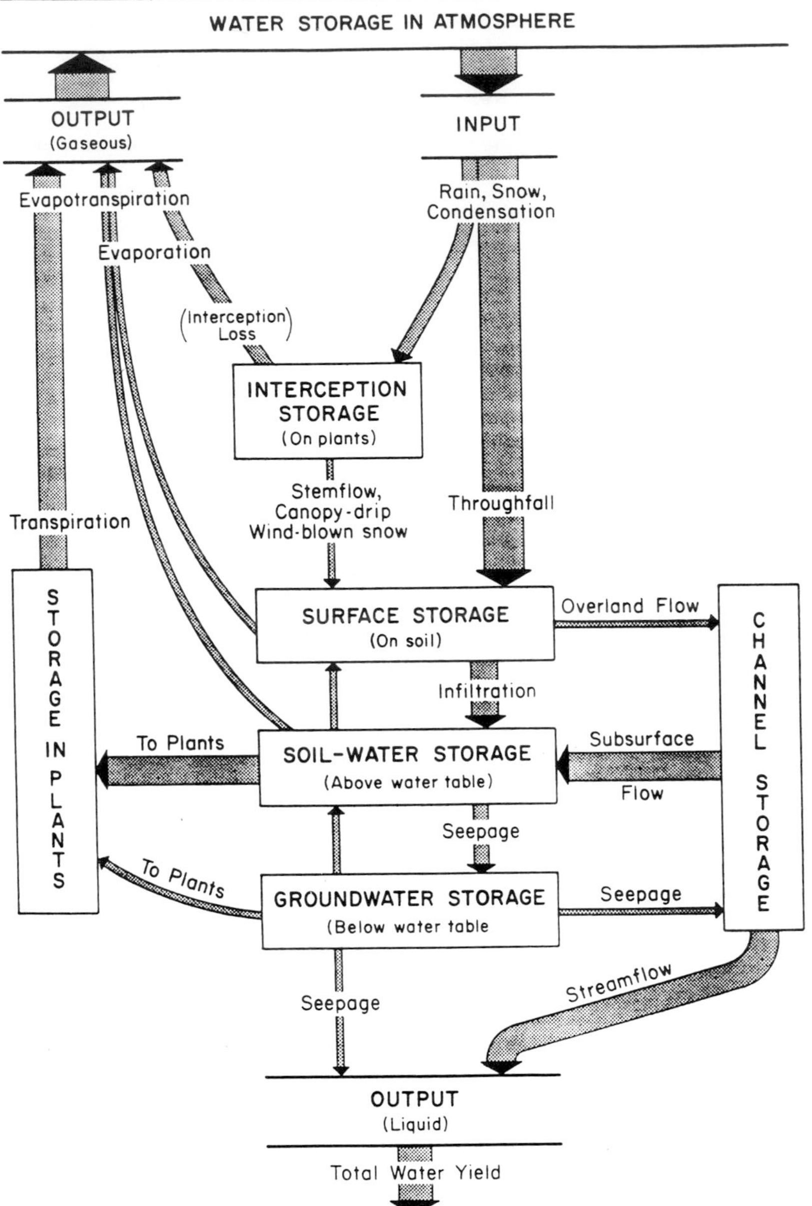


***FIGURE 1*** The hydrologic cycle showing the input, internal storage and flows, and the output of water on a watershed. [Courtesy of H. W. Anderson, M. D. Hoover, and K. G. Reinhart (1976). "Forests and Water: Effects of Forest Management on Floods, Sedimentation, and Water Supply," Gen. Tech. Rpt. PSW-18/1976. USDA Forest Service, Washington, D.C.]

ous hardwood forests to about 30% of precipitation in dense spruce–fir–hemlock forests. Interception losses for snowfall events are usually even greater than that for rainfall events, because snow near the freezing point is cohesive and tends to be retained within the forest canopy, where evaporation/sublimation can occur for longer periods of time. [*See* Forest Canopies].

In coastal or high-elevation regions, where fog or low-elevation clouds occur, forests may augment the water reaching the ground via fog drip. Fog drip, or occult precipitation as it is sometimes called, comes from fog or cloud water droplets that are intercepted onto the forest canopy and eventually grow large enough to drip to the ground. Fog drip represents an input to the hydrologic cycle that would not exist without vegetation.

The question of whether the presence of forests increases regional precipitation remains controversial in forest hydrology. Although forests do evaporate more water than many types of land cover types and also cause minor uplifting of air masses owing to the height of the trees, both of which could lead to some augmentation of precipitation, the effect (perhaps only a 1–2% increase) appears to be within the measurement error for precipitation amounts. Loss of rain forest (Amazon) and desert shrubs (Sahel Africa) have been linked to reductions in precipitation in regions where much of the precipitation is derived from local evaporation, but other factors are also at work. Definitive studies on this issue are needed to determine if precipitation reductions could be linked to vegetation loss.

### B. Snowpacks

Snowfall accumulates as a snowpack on the ground surface within the forest until liquid water is released by melt or water is evaporated or sublimated back into the atmosphere. The water equivalent of the snowpack is of primary interest to hydrologists as it represents the total amount of water stored in the snowpack as ice and liquid water in transit. Forests influence both the accumulation patterns and melt rates of snow. The importance of snow in the hydrologic cycle obviously varies regionally, but in some locations where precipitation is dominated by snowfall the influence of forests on snow is paramount.

Snowflakes are easily transported by the wind owing to their high surface area to mass ratio and effects of trees on wind circulation patterns can affect snowpack distribution. Snow is relatively uniformly distributed in homogeneous forest stands; however, forest openings alter wind circulation patterns such that greater amounts of snow water (often 30–40% more) can accumulate in small openings (Fig. 2). Part of the increase in snowpack water equivalent in forest openings is

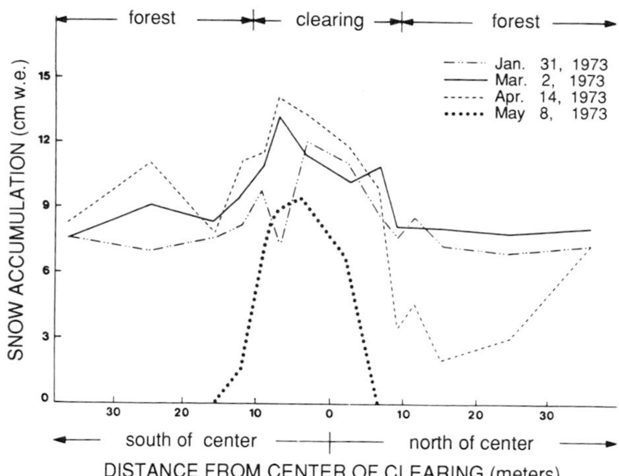


***FIGURE 2*** The distribution of snowpack water equivalent on a north–south transect across a small forest opening, showing increased snowpack accumulations in the opening in winter (Jan. 31) and the effects of faster melt rates along the warmer forest edge facing to the south later in spring (May 8). [From D. L. Golding and R. H. Swanson (1986). *Water Resources Res.* **22**(13), 1931–1940. Used with permission.]

due to lack of canopy interception losses. The relative roles of reduced canopy interception losses and wind redistribution of snow in producing increased snowpacks in small openings have not been totally resolved.

Once snow has accumulated to form a snowpack, the rates of melt and evaporation/sublimation are also influenced by the presence of the forest. The energy supply for snowmelt ($Q_m$) can be expressed as

$$Q_m = R_n + H + L_vE + G + P, \qquad (3)$$

where $R_n$ is net allwave radiation exchange, $H$ is sensible heat exchange between the snowpack and air, $L_vE$ is the energy equivalent of snowpack water vapor exchange with the atmosphere, $G$ is soil heat conduction, and $P$ is energy inputs due to rainfall. Energy gains to the snowpack on the right side of Equation (3) are considered positive and energy losses are negative. Generally, energy for snowmelt is primarily derived from net allwave radiation and sensible heat convection, both of which are affected by the forest.

Net allwave radiation ($R_n$) is reduced in the forest owing to the reduction of incoming solar radiation by the forest canopy. Solar radiation reaching the snowpack can be reduced by 50% in leafless deciduous forests and by 80–90% in dense conifer forests. Increased longwave radiation emitted by the forest canopy reaches the snowpack surface and partially compensates for the reduction in solar radiation in the forest, but the net radiant energy supply for melt is still reduced overall in the forest compared to open areas.

Sensible heat exchange between the snowpack and atmosphere ($H$) is also reduced in the forest compared to open areas due primarily to the reduction in wind speeds near ground level. Wind speeds above the snowpack in the forest can be reduced by as much as 70–80% in conifer forests and roughly 50% in leafless deciduous forest, although density and structure of the forest and distance from the upwind edge of the forest are important variables. Regardless, wind speed reductions in forests usually give rise to reduced transport of sensible heat from the air to the snowpack. Wind speed reductions in the forest can also reduce the evaporation/sublimation losses from snowpacks, but losses of water from snowpacks by evaporation/sublimation are generally small in humid zones.

Reductions in net allwave radiation and sensible heat exchange in the forest generally result in lower snowmelt rates in the forest compared to in the open. Snowmelt rates for open, deciduous forest and conifer forest sites generally are in the ratio of 3:2:1, respectively. Melt rates in the open are also influenced by the surrounding forest. Snow accumulating in openings along forest edges that face the sun and warm during midday usually melts faster than snow that accumulates along edges facing away from the sun (Fig. 2).

In mountainous terrain, where deep snowpacks can lead to snow avalanching, timberline forests may help retard the initiation of snow avalanches. Forests in high-elevation basins, where extremely large amounts of snow tend to accumulate from drifting, can reduce the initiation of avalanches by anchoring the snowpack in place. Forests in high-elevation avalanche start-up zones deserve special protection, because these forests exist in a very harsh environment and are nearly impossible to regenerate. Once avalanches are initiated, forests at lower elevation in avalanche tracks provide little or no protection and can be severely damaged or destroyed by snow avalanches.

### C. Evapotranspiration

Evapotranspiration ($E_t$) represents the total evaporation from a forest and can be thought of as the sum of canopy interception losses ($I_c$), transpiration ($T$), and evaporation from the ground surface ($E_s$):

$$E_t = T + I_c + E_s. \tag{4}$$

Transpiration represents the water evaporated from leaves through stomatal openings and to a lesser extent water evaporated from plant stems through lenticels. Transpiration losses from the forest canopy generally represent the majority of evapotranspiration (roughly 60%), whereas canopy interception losses may represent about 30% and ground evaporation about 10% of total evapotranspiration losses. Ground evaporation in forests is generally due to evaporation of throughfall and stemflow intercepted on the organic matter accumulated on the surface of the forest soil or forest floor. Forest floor evaporation losses are generally low in forests, less than 5% of precipitation in the summer, owing to the shading of the ground surface by the forest canopy.

Evapotranspiration rates from forests are relatively high compared to those of other vegetation types because of the unique features of forest vegetation. Deep root systems, the aerodynamically rough forest canopy, the relatively low solar reflectivity coefficient or albedo of forests (especially in conifer forests), the relatively large canopy interception losses, and the high water-holding capacities of forest soil all contribute to high evapotranspiration rates in forests. Unfortunately, measurements of evapotranspiration rates from forests for comparison with other vegetation types are difficult to obtain and indirect methods of comparison must be used.

The best long-term average estimates of forest evapotranspiration losses probably come from the water budget [Equation (1)]. The average difference between precipitation and streamflow for forested watersheds from each water year can be used to estimate evapotranspiration. In the eastern United States, the long-term difference between precipitation and streamflow for deciduous forest land equals about 40–60% of annual precipitation. Thus, for a typical annual precipitation of 120 cm in this region, 48–72 cm of evapotranspiration loss is expected with the remainder of the water yielded as streamflow. Comparable percentages for other cover types or surfaces are shown in the Table I.

In an energy budget context, the energy equivalent of evapotranspiration can be expressed as:

$$L_v E_t = -(R_n + G)/(1 + B), \qquad (5)$$

where $R_n$ and $G$ have been previously identified and $B$ represents the Bowen ratio, or ratio of energy exchange by sensible heat convection to energy exchange by vapor exchange processes. Just as evapotranspiration losses can be related to precipitation using the water balance [Equation (1)], Bowen ratios allow evaporation to be related to the energy supply using Equation (5). Bowen ratios are typically low (e.g., $B = 0.1–0.2$) for open-water surfaces, which dissipate net radiation primarily by evaporation, and high (e.g., $B > 2$) for dry semiarid lands, where most radiation is dissipated by convection of sensible heat into the air. Long-term average forest Bowen ratios range from about 0.6 to 0.8. Thus, as evaporators, forests are intermediate between dry and wet surfaces, but much more similar to wet surfaces.

**TABLE I**
Annual Average Evapotranspiration Relative to Precipitation in Midlatitudes, Northern Hemisphere

| Cover type | $E_t/P$ (%) |
|---|---|
| Bare soil | 30 |
| Cultivated land | 40 |
| Meadow | 65 |
| Deciduous forest | 50 |
| Conifer forest | 70 |
| Open water | 75 |
| Wet pasture | 100 |

Estimates of annual average forest evapotranspiration using the water balance equation (1) can be used to illustrate the typical forest Bowen ratio for the eastern United States. An annual average $E_t$ loss of 60 cm per year would be equivalent to an average energy loss of about $-4.07$ MJ m$^{-2}$ day$^{-1}$. If annual average $R_n$ is about $+6.91$ MJ m$^{-2}$ day$^{-1}$ (typical of the eastern United States), then the long-term annual average Bowen ratio in Equation (4) would be 0.7. Note that soil heat conduction in Equation (4) can be neglected for annual average computations. An average Bowen ratio of 0.7 implies that forests would dissipate about $1/(1 + B)$ or 59% of net radiant energy supply by evapotranspiration and the remainder of 41% by sensible heat convection.

Evapotranspiration rates from forests for short time periods can vary widely from these annual average estimates. During and immediately after rainstorms, evaporation of intercepted water in the forest canopy can proceed at exceptionally high rates of more than 0.5 mm per hour, with energy derived from net allwave radiation and sensible heat extracted from the air. Transpiration rates when the canopy is not wetted by rain typically range from about 2 to 4 mm per day in the summer, and in winter they may range between 0 and 1 mm per day. Species differences in transpiration rates exist because of differences in stomate characteristics. They are extremely difficult to generalize to forest stands owing to the complexities of stand age and size structure, rooting depth, varying site and climatic characteristics, forest species composition, and so on.

### D. Streamflow

Total amounts of streamflow and the timing of flows from forested watersheds generally differ from that on agricultural and urban/suburban watersheds, due largely to the effects of the forests on the soil. Comparisons of watersheds with differing land use are complicated because forests are often found on steep, upland watersheds, whereas agricultural and urban/suburban lands are more common in the gently sloping lowlands.

Forests promote high soil water infiltration rates and high soil water-holding capacity, which in turn lead to larger evapotranspiration losses from stored soil water and higher streamflow rates during summer low-flow periods. High infiltration rates on forest land are attributable to:

1. organic debris from the forest canopy accumulating on the soil surface, which absorbs the kinetic energy of raindrops and retards compaction and soil freezing;
2. soil organisms and tree roots, which promote incorporation of organic matter and increase soil porosity; and
3. high forest evapotranspiration rates, which increase the amount of water storage capacity in the soil.

When forest soils are dry, infiltration capacities of over 20 cm per hour are common. If land with compacted soil and little vegetation cover is replanted to forest, it is common for infiltration capacities of the soil to increase several orders of magnitude, but the changes are gradual and may take several decades to reach conditions equal to that in natural forest. High soil infiltration rates in forests increase the amount of water that can move through subsurface soil and rock materials and provide more substantial groundwater and streamflow water during dry periods.

Although flows from forested basins during summer low-flows generally are higher owing to higher infiltration, total annual volumes of water yielded as streamflow often are less owing to high forest evapotranspiration loss. If forests evaporate 50–70% of precipitation annually, then only the remaining 30–50% of precipitation is available to support streamflow. During storms, high infiltration rates on forestland also lead to less overland flow, less surface soil erosion, and greater subsurface stormflow. Subsurface stormflow produces a large fraction of stormflows on forested basins. Peak flows in streams are generally reduced on forestland during summer owing to the drier soils, but even the effects of forests cannot counteract the effects of uncommonly large storms that completely saturate the soil and lead to floods. Thus, maximum peak flows from forestland are probably as great as those from other lands, but peak flows from forestland are not accompanied by as much eroded soil material.

## II. WATER QUALITY AND FORESTS

Forests have long been associated with excellent quality water. Water quality relates to the physical, chemical, microbiological, and radiological properties of water. Forests influence the physical properties of water in streams and lakes, especially the water temperature, dissolved oxygen concentration, and sediment concentration. The movement, uptake, and storage of nutrients and pollutants dissolved in water are also influenced by the forest.

### A. Water Temperature and Dissolved Oxygen

Shade from the forest canopy maintains a moderate temperature environment in small, shallow streams on forested uplands. Maximum daily water temperatures in headwater streams are commonly reduced by 5–10°C by forest shade compared to streams exposed to full summer sunlight, and reductions up to 15°C have been found (Fig. 3). Minimum daily water temperatures may also be slightly increased by the protective influence of the canopy against longwave radiation loss at night. Aquatic organisms dependent on low water temperature are often stressed and do not survive when the protective influence of the forest canopy is removed by forest clearing. Salmon and trout experience stress at temperatures above 20°C. For this reason, foresters commonly leave buffer zones of vegetation to protect the stream against temperature increases during and after timber harvesting operations. Shading by vegetation will have a lesser effect on the temperature regime of deep and wide streams and lakes that are only partially or minimally shaded.

Because the solubility of dissolved oxygen increases as water temperature decreases, cool forest streams generally also exhibit high dissolved oxygen levels that are favorable to many aquatic organ-

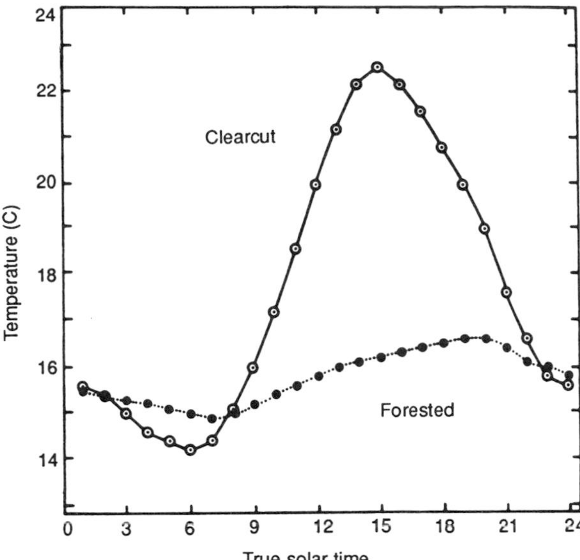


***FIGURE 3*** Daily variations in stream temperature on a forested watershed and a cleared basin, showing the effects of shading by the forest canopy in maintaining low maximum daily water temperatures. [From R. Lee and D. E. Samuel (1976). *J. Environ. Qual.* **5**(4), 362–366. Used with permission.]

isms (Table II). Organic debris from the forest that falls into water bodies can cause reductions in dissolved oxygen levels due to the chemical oxidation and decomposition of this debris by microorganisms. Buffer zones of vegetation along shorelines are also used for reducing the amount of this organic debris that reaches water bodies following forest clearing.

## B. Sedimentation

The major impact of forests on water quality is to limit the sediment load in streams and rivers. As

***TABLE II***
Water Temperature–Oxygen Solubility Relationship

| Water temperature (°C) | Oxygen solubility[a] (mg liter$^{-1}$) |
|---|---|
| 0 | 14.6 |
| 5 | 12.8 |
| 10 | 11.3 |
| 15 | 10.0 |
| 20 | 9.0 |
| 25 | 8.2 |

[a] Atmospheric pressure of 1000 mbar.

previously mentioned, forests promote infiltration of water into the soil and thereby reduce overland flow and soil surface erosion. Organic debris accumulated on the soil surface in forests absorbs the kinetic energy of raindrops and protects against surface erosion. Root systems of trees along shorelines help stabilize banks and reduce channel erosion. In regions with very steep slopes, tree roots even lessen the occurrence of soil slumping and avalanching by increasing soil cohesion.

The total load of sediment transported in a stream channel is thus reduced on forestland. Forestland typically yields less than 20 metric tons of sediment per km$^2$ per year, whereas acceptable erosion losses from agricultural land typically range from 200 to 1000 metric tons per km$^2$ per year. Because of low sediment levels, forest-land often produces water that can be used as a domestic water supply without filtration. Low sediment levels from forestland also promote high quality aquatic habitat for fish and reduce sediment accumulations in reservoirs. In fact, maintaining forestland with low erosion rates in the headwaters of large watersheds can significantly prolong the useful life of large downstream reservoirs by keeping reservoir sedimentation rates low.

## C. Nutrient and Pollutant Movement

Pathways for water movement and hence movement of substances dissolved in the water are also influenced by the forest. Precipitation reaching the ground as throughfall and stemflow is greatly enriched in dissolved substances due to contact with the forest canopy. High soil infiltration capacities bring more water into contact with the subsurface mineral soil and tree root systems in forests than in many other land-use types. Forest uptake, accumulation, and recycling of chemical elements in organic matter can create a significant dynamic reservoir of stored chemical elements (Fig. 4). Forests depend on the cycling of nutrients in water for sustained growth, but pollutants introduced into the forest can move with nutrients in a dissolved state through the forest hydrologic system. [*See* NUTRIENT CYCLING IN FORESTS.]

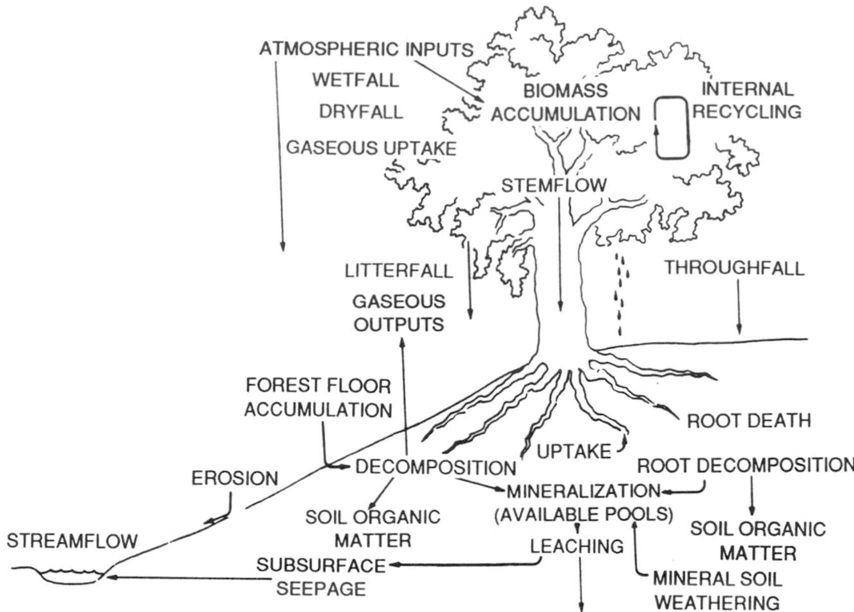


***FIGURE 4*** Nutrient cycle in a forest showing fluxes, storages, and processes affecting natural water quality. [From B. Binkley (1986). "Forest Nutrition Management." John Wiley & Sons, New York. Used with permission.]

Throughfall and stemflow contain elevated concentrations of dissolved substances compared to precipitation. Washoff of substances deposited on the forest canopy from the atmosphere during dry periods, leaching of plant tissues, and microorganisms in the canopy all contribute to the enrichment of throughfall and stemflow. Concentrations of nearly all dissolved substances increase in throughfall compared to open precipitation and stemflow is even more enriched than throughfall. Concentrations of biologically available elements, such as potassium, are especially increased as they are easily leached from the canopy. This enriched throughfall and stemflow generally is available to infiltrate the soil.

The concentration of elements in water moving downward and laterally in the forest soil is the net result of many important processes, including:

1. the concentrating effects of transpiration and direct soil evaporation;
2. release of dissolved substances due to decomposition of organic matter, including soil organisms;
3. Uptake and release of dissolved substances by roots and soil organisms;
4. mineral weathering and formation in the soil; and
5. soil anion adsorption and cation exchange.

Because many of these processes are related to biological activity, the age and condition of the forest play roles in controlling soil water, groundwater, and ultimately streamflow concentrations of dissolved substances. Young, healthy, actively growing forests will generally be net accumulators of nutrients and pollutants, whereas ageing forests or forests in poor health may exhibit net release of dissolved substances.

Pollutants often follow the same pathways as do nutrients through the forest; however, the specific chemical nature of the pollutant will determine its ultimate fate. Chemical substances may volatilize, biologically or chemically degrade, adsorb onto organic or inorganic soil material, or remain in a dissolved mobile state. Forests may be able to accumulate pollutants in organic matter without harm to the plants initially, but long-term pollutant retention or health of the forest cannot be assured because of the natural recycling of organic material in the forest. A complete understanding of the properties of each pollutant and its interaction with the natural processes in the forest is needed to predict long-term behavior of pollutants in forests.

## Glossary

**Bowen ratio** Ratio of convective sensible heat exchange between a surface and the atmosphere to the latent heat exchange by evaporation and sublimation of water.

**Canopy interception loss** Precipitation water intercepted on the forest canopy that evaporates back into the atmosphere during and after a storm.

**Energy budget** Account of the various sources, sinks, and storages of energy within a system, which considers radiant, sensible, latent, and chemical energy exchanges and transformations.

**Evapotranspiration** Total evaporation from the land surface, including transpiration by plants, canopy interception loss, and ground evaporation.

**Infiltration capacity** Maximum rate at which water can enter the soil.

**Stemflow** Precipitation water that reaches the ground by flowing down the branches and trunks of trees.

**Subsurface stormflow** Streamflow derived from subsurface sources that arrives in the stream channel quickly enough to be part of a particular rain or snowmelt runoff event.

**Throughfall** Precipitation water that falls through or drips from the forest canopy.

**Water budget** Account of the various sources, sinks, and storages for the mass of water on a watershed over a specified period of time.

**Water year** The 12-month period used in water budgeting to give the smallest storage changes from year to year; typically October 1 to September 30 of the following year.

## Bibliography

Black, P. E. (1991). "Watershed Hydrology." Englewood Cliffs, N.J.: Prentice–Hall.

Brooks, K. N., Ffolliott, P. F., Gregersen, H. M., and Thames, J. L. (1991). "Hydrology and the Management of Watersheds." Ames: Iowa State University Press.

Hewlett, J. D. (1982). "Principles of Forest Hydrology." Athens: University of Georgia Press.

Lee, R. (1980). "Forest Hydrology." New York: Columbia University Press.

Tomlinson, G. H., and Tomlinson, F. L. (1990). "Effects of Acid Deposition on the Forests of Europe and North America." Boca Raton, Fla.: CRC Press.

# Forest Insect Control

**S. C. Pathak**
*Rani Durgavati University, India*

**N. Kulkarni and P. B. Meshram**
*Tropical Forest Research Institute, India*



Human existence is inconceivable on a planet that has no forests. Enormous animal populations (including the ever-increasing human population) exhale huge quantities of $CO_2$ (carbon dioxide), which is balanced by green plants that can absorb the $CO_2$ and provide $O_2$ (oxygen) for the respiratory needs of animals. Forests are nature's answer to this problem. In addition to providing $O_2$ for billions of animals, forests also help maintain the delicate equilibrium of atmospheric gases. Forests have rightly been called the lungs of the world.

Insects as a group affect the forests, though to a lesser extent than they do agroecosystems. In fact, insects and forest trees may actually have a mutual relationship in which exposure to insect pests increases host resistance and decreases susceptibility to some extent and may even contribute to slightly increased host vigor. But large-scale damage to forest ecosystems occurs every now and then, due to abnormally high populations of specific insect species. Whenever the damage from forest insect pests is high, forest trees start dying. This may lead to higher $CO_2$ content in the atmosphere, leading to what is referred to as the global warming or greenhouse effect. Thus, insect pest populations on forest trees need to be managed so that their numbers are never so high that they kill the hosts or cause irreparable damage to the forest ecosystem. At the same time, no insect species need be eliminated from an ecosystem, as insect/tree interrelationships are complex and integral parts of a healthy ecosystem.

## I. INTRODUCTION

Insects constitute the largest single group of animals. The number of insect species is believed to be nine times that of all the other species put together. With the enormous reproductive potential of most of these species, actual population figures would defy imagination. Fortunately, various natural factors, including unfavorable conditions, disease, and inter- and intraspecific competition for food and shelter, keep the populations within moderate limits. Although insects may be generally considered as "pests" or irritants to many humans, there are many economically useful species, for example, lac insects, honeybees, and silkworms. But it is generally forgotten that insects are one of the major agents of pollination in plants. Biodiversity in plants would be much poorer without the active help of insects, even those that may be irritants to humans. Because of the numerous and essential roles played by insects in most ecosystems, we need to preserve the biodiversity of insect species. [*See*

INSECT DEMOGRAPHY; INSECT INTERACTIONS WITH TREES.]

## II. TYPES OF FORESTS

Forests are categorized on the basis of geographical distribution (1) with respect to latitude, such as alpine, temperate, subtemperate, and tropical, and (2) with respect to altitudes. Generally alpine forests are found in mountains, but alpine forest patches do occur on hilly tracts at latitudes where tropical forests are generally found.

Another way to categorize forests is by the type of vegetation, for example, evergreen, wet deciduous, or dry deciduous (Fig. 1). Forests located along a coastline are exposed to swampy conditions due to the tides. These forests are called littoral or tidal swampy forests, also known as mangroves. Other types are freshwater swamps and seasonal swampy forests on wetlands.

Natural mixed forests are composed of a variety of trees and countless smaller plants and animal life of a very diverse nature. With the precipitous decline in natural forests, foresters have been trying to grow forests that may have only one predominant tree species, known as "monoculture" forests. Man-made forests can, however, be of polyculture or mixed type as well, though these are less common. Monoculture forests meet a variety of demands, for example, for timber, industrial, and furniture wood and for pulps (industrial forestry) and fuelwood (social forestry). Other benefits of forests include habitat for a variety of wildlife, soil management, water recycling, and recreation.

## III. ECONOMIC DAMAGE CAUSED BY INSECTS

Insects have a close relationship with the plant kingdom. They act as pollinating agents for forest trees, yet are also pests. This article concentrates on the negative relationship—as pests of forest trees. Table I lists important forest trees, their pests, and the type of damage caused by each. [*See* FOREST PATHOLOGY.]

Insects cause damage to trees in various ways, including defoliation, skeletonization, phloem and wood boring, damage to seeds, cones, buds, shoots, and roots, gall formation (Fig. 2 to 10),

***FIGURE 1*** Dry deciduous forest in Madhya Pradesh, India. The predominant tree species is *Shorea robusta* (The Hindi name is *sal*).

**TABLE I**
Some Major Insect Pests of Economically Important Tree Species and the Nature of Damage

| Host | Insect pests | Nature of damage |
|---|---|---|
| 1. *Acacia catechu* (L.F.) Willd. | *Celosterna scabrator* Fabr. (Coleoptera: Cerambycidae) | Stem borer |
| | *Sternocera diardi* Gray (Coleoptera: Buprestidae) | Defoliator |
| | *Hyposidra talaca* Walk. (Lepidoptera: Geometridae) | Inflorescence feeder |
| | *Indarbela quadrinotata* Walk. (Lepidoptera: Indarbelidae) | Bark borer |
| 2. *Acacia nilotica* (L.) Del. | *Bruchus chinensis* (Coleoptera: Bruchidae) | Seed borer |
| | *Celosterna scabrator* Fabr. (Coleoptera: Cerambycidae) | Stem and root borer |
| | *Indarbela quadrinotata* Walk. (Lepidoptera: Indarbelidae) | Bark-eating caterpillars |
| | *Selepa celtis* Moore (Lepidoptera: Noctuidae) | Defoliator |
| | *Sternocera diardi* Gray (Coleoptera: Buprestidae) | Defoliator |
| 3. *Ailanthus excelsa* Roxb. | *Atteva fabriciella* Swed. (Lepidoptera: Yponomeutidae) | Webworm (defoliator) |
| 4. *Albizia* spp. | *Eurema* spp. (Lepidoptera: Pieridae) | Defoliator |
| | *Hyposidra successaria* Walk. (Lepidoptera: Geometridae) | Defoliator |
| 5. *Anogeissus latifolia* Wall. | *Henicolabus discolor* Fahraeus (Coleoptera: Curculionidae) | Defoliator |
| | *Platypeplus aprobola* Meyr. (Lepidoptera: Tortricidae) | Defoliator |
| 6. *Anthocephalus cadamba* Roxby. | *Arthroschista hilaralis* Walk. (Lepidoptera: Lymantriidae) | Defoliator |
| 7. Ash | *Lepicerinus aculeatus* (Coleoptera: Scolytidae) | Bark borer |
| | Fall webworm, *Hyphantria cunea* Drury (Lepidoptera: Arctiidae) | Defoliator |
| 8. *Azadirachta indica* A. Juss. (*Neem* in Hindi) | *Anoplophora chinensis* Forst. (Coleoptera: Cerambycidae) | Wood borer |
| | *Apate monachus* Fabr. (Coleoptera: Bostrychidae) | Powder post beetle |
| | *Boarmia variegata* Moore (Lepidoptera: Geometridae) | Defoliator |
| | *Bostrychopsis jesuita* Fabr. (Coleoptera: Bostrychidae) | Sapwood borer |
| | *Aspidiotus hederae* Vallot (Homoptera: Coccidae) | Sap sucker |
| | *Pseudococcus marieimus* Ehrhoen (Homoptera: Pseudococcidae) | Shoot/fruit sap sucker |
| | *Pulvinaria maxima* Green (Homoptera: Coccidae) | Sap sucker |
| 9. *Bambusa* spp. and *Dendrocalamus strictus* Nees. | *Chlorophorus annularis* Fabr. (Coleoptera: Cerambycidae) | Dry bamboo borer |
| | *Heterobostrychus aequalis* Waterh. (Coleoptera: Bostrychidae) | Shoot hole borer |
| | *Holotrichia consanguinea* Blanch. (Coleoptera: Scarabaeidae) | Root damage in nursery |
| | White grub, *Holotrichia* spp. (Coleoptera: Scarabaeidae) | Damage rhizomes and roots in plantations |
| | *Estigmena chinensis* Hope (Coleoptera: Chrysomelidae) | Culm and shoot borer |
| | Bamboo weevils, *Cyrtotrachelus dux* Boh. (Coleoptera: Curculionidae) | Young culm and shoot borer |
| | Bamboo leaf roller, *Pyrausta bambucivora* Moore (Lepidoptera: Pyralidae) | Defoliator |
| | *Sitotroga cereallela* Olive (Lepidoptera: Gelechiidae) | Stored seed borer |
| | Bamboo aphids, *Oregma bambusae* Buckton (Homoptera: Aphididae) | Sap sucker |
| | Termites (Isoptera: Rhinotermitidae) | Damage dead or green culms |
| 10. Beech spp. | *Cryptococcus fagi* Baer. (Homoptera: Coccidae) | Sap sucker |
| | *Phyllaphis fagi* Linn. (Homoptera: Aphididae) | Sap sucker |
| 11. Birch spp. | *Agrilus anxius* Gory (Coleoptera: Buprestidae) | Bark borer |
| 12. *Bombax cieba* | *Batocera rufomaculata* DeGeer (Coleoptera: Cerambycidae) | Sapwood borer |
| | *Tonica niviferana* Walk. (Lepidoptera: Oecophoridae) | Shoot borer |
| | *Dysdercus cingulatus* Fabr. (Hemiptera: Pyrrhocoridae) | Sap sucker |
| 13. *Butea monosperma* (Lam.) Taub. | *Anomala dimidiata* Hope (Coleoptera: Scarabaeidae) | Adult is a defoliator and grub is a root feeder |
| | *Holotrichia serrata* Fabr. (Coleoptera: Scarabaeidae) | Adult is a defoliator and grub is a root feeder |
| 14. *Cassia* spp. | Amaltas defoliator, *Catopsilia* spp. (Lepidoptera: Pieridae) | Defoliator |
| 15. *Casuarina* spp. | *Celosterna scabrator* Fabr. (Coleoptera: Cerambycidae) | Root and shoot borer |
| | *Myllocerus undecimpustulatus marmoratus* Desbro. (Coleoptera: Circulionidae) | Young shoot borer |

*continues*

*Continued*

| | | |
|---|---|---|
| | *Indarbela tetraonis* Moore (Lepidoptera: Indarbelidae) | Bark and stem borer |
| 16. *Cedrus deodara* (Roxb.) Loud. (*Deodar* in Hindi) | *Lophosternes hugelii* Redt. (Coleoptera: Cerambycidae) | Stem and root borer |
| | *Ectropis deodarae* Prout. (Lepidoptera: Geometridae) | Defoliator |
| | *Euzophera cedrella* Hamp. (Lepidoptera: Pyralidae) | Cone borer |
| | *Dioryctria abietella* De & Schf. (Lepidoptera: Pyralidae) | Cone and shoot borer |
| 17. *Dalbergia sissoo* Roxb. | *Batocera rufomaculata* DeGeer (Coleoptera: Cerambycidae) | Meristem and sapwood borer |
| | *Leucoptera sphenograpta* Meyrick (Lepidoptera: Lyonetiidae) | Leaf miner |
| 18. *Diospyros melanoxylon* Roxb. | *Trioza obsoleta* Buckton (Homoptera: Psyllidae) | Leaf gall maker |
| | *Hypocala rostrata* Fab. (Lepidoptera: Noctuidae) | Defoliator |
| 19. Elm spp. | *Sapedra tridentata* Oliv. (Coleoptera: Scolytidae) | Bark and sap wood borer |
| | The gypsy moth *Porthetria dispar* Linn. (Lepidoptera: Lymantriidae) | Defoliator |
| 20. *Eucalyptus* spp. | *Celosterna scabrator* Fabr. (Coleoptera: Cerambycidae) | Root and stem borer |
| | *Agrotis ipsilon* Huf. (Lepidoptera: Noctuidae) | Cutworm |
| | *Odontotermes* spp. (Isoptera: Termitidae) | Root/bark feeding |
| 21. *Gmelina arborea* Roxb. | *Diahammus cervinus* Hope (Coleoptera: Cerambycidae) | Canker grub |
| | *Indarbela quadrinotata* Walk. (Lepidoptera: Indarbelidae) | Bark-eating caterpillar |
| 22. *Leucaena leucocephala* (Lam.) | *Heteropsylla cubana* Crawf. (Homoptera: Psyllidae) | Sap sucker |
| 23. *Madhuca indica* Roxb. | *Achaea janata* Linn. (Lepidoptera: Noctuidae) | Defoliator |
| | *Acrocercops* spp. (Lepidoptera: Lithocolletidae) | Leaf miner |
| | *Unaspis acuminata* Green (Hemiptera: Diaspididae) | Sap sucker |
| 24. *Michelia champaca* L. | *Urostylis punctigera* Westwood (Hemiptera: Urostylidae) | Sap sucker |
| 25. *Morus* spp. | *Apriona germari* Hope (Coleoptera: Cerambycidae) | Shoot girdler |
| | *Glyphodes pyloalis* Walk. (Lepidoptera: Pyralidae) | Defoliator |
| 26. *Pongamia pinnata* (L.) Merr. (Indian Beech) | *Acrocercops anthracuria* Meyrick (Lepidoptera: Lithocolletidae) | Leaf miner |
| 27. *Populus* spp. (Poplar) | *Phalantha phalantha* Drury (Lepidoptera: Nymphalidae) | Defoliator |
| | *Apriona cinerea* Chevr. (Coleoptera: Cerambycidae) | Stem and root borer |
| | *Amsacta moorei* Butl. (Lepidoptera: Arctiidae) | Defoliator |
| | *Clostera* (*Pygaera*) *fulgurita* Walk. (Lepidoptera: Notodontidae) | Defoliator |
| | *Prodenia litura* Fabr. (Lepidoptera: Noctuidae) | Defoliator |
| | *Agonoscelis nubila* Fabr. (Hemiptera: Pentatomidae) | Sap sucker |
| | *Odontotermes* spp. and *Microtermes obesi* Holm. (Isoptera: Termitidae) | Root-feeding termites |
| 28. *Prosopis spicigera* L. Jhand. | *Caryedon gonagra* Fabr. (Coleoptera: Bruchidae) | Seed weevil |
| 29. *Salix* spp. | *Aeolesthes sarta* Solsky (Coleoptera: Cerambycidae) | Sapwood and heartwood borer |
| | *Batocera horsfieldi* Hope (Coleoptera: Cerambycidae) | Sapwood and heartwood borer |
| | *Clostera* (*Pygaera*) *cupreata* Butler (Lepidoptera: Notodontidae) | Defoliator |
| | *Clostera* (*Pygaera*) *fulgurita* Walk. (Lepidoptera: Notodontidae) | Defoliator |
| | *Lymantria obfuscata* Walk. (Lepidoptera: Lymantriidae) | Defoliator |
| | *Malacosoma indica* Walk. (Lepidoptera: Lasiocampidae) | Defoliator |
| | *Tuberolachnus salignus* Gmelin (Hemiptera: Aphididae) | Sap sucker |
| 30. *Santalum album* L. | *Ascotis selenaria imparata* Walk. (Lepidoptera: Geometridae) | Defoliator |
| | *Sahyadrassus malabaricus* Moore (Lepidoptera: Hepialidae) | Stem and root borer |
| | *Zeuzera coffeae* Niet. and *Z. multistrigata* Moore (Lepidoptera: Cossidae) | Shoot borer |
| 31. *Sesbania* spp. | *Alcides bubo* Fabr. (Coleoptera: Curculionidae) | Shoot borer |
| | *Amsacta moorei* Butl. (Lepidoptera: Arctiidae) | Shoot borer |
| | *Eurema hecabe* Linn (Lepidoptera: Pieridae) | Defoliator |
| | *Prodenia litura* Fabr. (Lepidoptera: Noctuidae) | Defoliator |

*continues*

*Continued*

| | | |
|---|---|---|
| 32. *Shorea robusta* Gaertn. f. | *Batocera rufomaculata* DeGeer (Coleoptera: Cerambycidae) | Meristem and sapwood borer |
| | *Hoplocerambyx spinicornis* Newn. (Coleoptera: Cerambycidae) | Bark, sapwood, and heartwood borer |
| | *Sitophilus rugicollis* Casey (Coleoptera: Curculionidae) | Seed borer |
| | *Antheraea paphia* Linn. (Lepidoptera: Saturniidae) | Defoliator |
| | *Indarbela quadrinotata* Walk. (Lepidoptera: Indarbelidae) | Bark-eating caterpillars |
| | *Pammene therisis* Meyr. (Lepidoptera: Tortricidae) | Seed borer |
| | *Trabala uishnou* Lef. (Lepidoptera: Lasiocampidae) | Seed borer |
| | *Kerria lacca* Kerr. (Homoptera: Coccidae) | Sap sucker |
| 33. *Sterculia urens* | *Batocera rufomaculata* DeGeer (Coleoptera: Cerambycidae) | Wood borer |
| | *Indarbela quadrinotata* Walk. (Lepidoptera: Indarbelidae) | Bark-eating caterpillars |
| | *Oglassa separata* Walk. (Lepidoptera: Noctuidae) | Defoliator |
| | *Pingasa tephrosia* Guen. (Lepidoptera: Geometridae) | Defoliator, inflorescence feeder, and fruit borer |
| 34. *Swietenia* spp. and *Toona ciliata* M.J. Roem. | *Pagiophloeus longiclavis* Marsh. (Coleoptera: Curculionidae) | Collar borer |
| | *Diacrisia obliqua* Walk. (Lepidoptera: Arctiidae) | Defoliator |
| | *Hypsipyla robusta* Moore (Lepidoptera: Pyralidae) | Shoot inflorescence and fruit borer |
| 35. *Syzygium cumini* Skiels. (Jambolan) | *Batocera rufomaculata* DeGeer (Coleoptera: Cerambycidae) | Bark borer |
| | *Sitophilus rugicollis* Casey (Coleoptera: Curculionidae) | Bark borer |
| | *Trioza jambolanae* Craw. (Hemiptera: Psyllidae) | Sap sucker |
| 36. *Tectona grandis* L.F. (Teak) | *Dihammus cervinus* Hope (Coleoptera: Cerambycidae) | Canker grub |
| | *Holotrichia* spp. (Coleoptera: Scarabaeidae) | Root grub |
| | *Dichocrocis punctiferalis* Guenee (Lepidoptera: Pyralidae) | Seed borer |
| | *Hyblaea puera* Cram. (Lepidoptera: Hyblaeidae) | Defoliator |
| | *Indarbela quadrinotata* Walk. (Lepidoptera: Indarbelidae) | Bark-eating caterpillars |
| | *Pagyda salvalis* Walk. (Lepidoptera: Pyralidae) | Seed borer |
| | *Sahyadrassus malabaricus* Moore (Lepidoptera: Hepialidae) | Phassus borer |
| | *Xyleutes ceramicas* Walk. (Lepidoptera: Cossidae) | Beehole borer |
| | *Zeuzera coffeae* Nietner (Lepidoptera: Cossidae) | Shoot borer |
| 37. *Terminalia* spp. | *Amblyrhinus poricollis* Shoen. (Coleoptera: Curculionidae) | Defoliator |
| | *Apoderus tranquebaricus* Fabr. (Coleoptera: Curculionidae) | Leaf roller |
| | *Araecerus fasciculatus* DeGeer (Coleoptera: Anthribidae) | Stored seed borer |
| | *Celosterna scabrator* Fabr. (Coleoptera: Cerambycidae) | Bark borer |
| | *Dasychira mendosa* Hueb. (Lepidoptera: Lymantriidae) | Defoliator |
| | *Trioza fletcheri* Crwf. (Homoptera: Psyllidae) | Leaf gall makers |

phytotoxemia, and gout disease. All of these separately and cumulatively affect forest productivity and cause different degrees of economic losses. One estimate of annual damage to timber in the United States from the pest complex (which includes insects, diseases, and weeds) is 15 billion board feet. This represents one-third of the U.S. total, so one can easily visualize how great the total global damage must be considering the vast forest tracts of South America, Australia, Asia, and Africa. In India, the damage to stored wood from insects and diseases is estimated to be in the range of 13 to 40% per year, and timber is only one benefit derived from forests. Forests are equally important for soil management, water recycling in the environment, a diverse forest flora and fauna, and of course recreation.

In India, it has been estimated that during a short period of heavy and complete defoliation of teak, caused by *Hyblaea puera,* the loss was

***FIGURE 2*** A mahaneem (*Ailanthus excelsa*) tree completely defoliated by *Atteva fabriciella*. (Photograph courtesy of S. S. Sandhu.)

13% of the normal annual yield. In Myanmar (formerly Burma), the loss in a teak forest due to the same pest was estimated to be 25% of the annual yield.

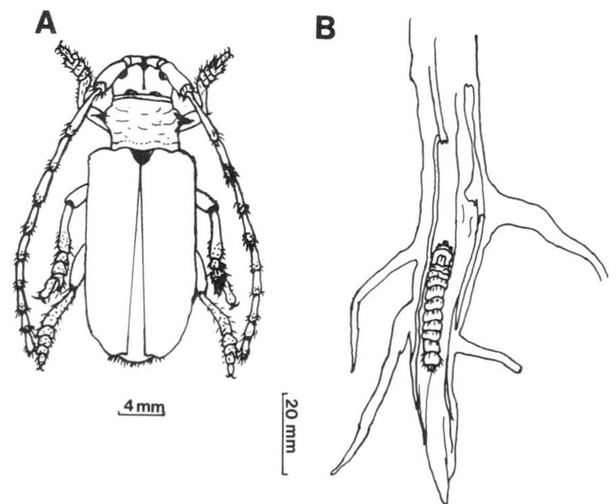


***FIGURE 3*** The wood borer coleopteran *Celosterna scabrator* (A) and its larva inside the root of a *Eucalyptus* sapling (B).

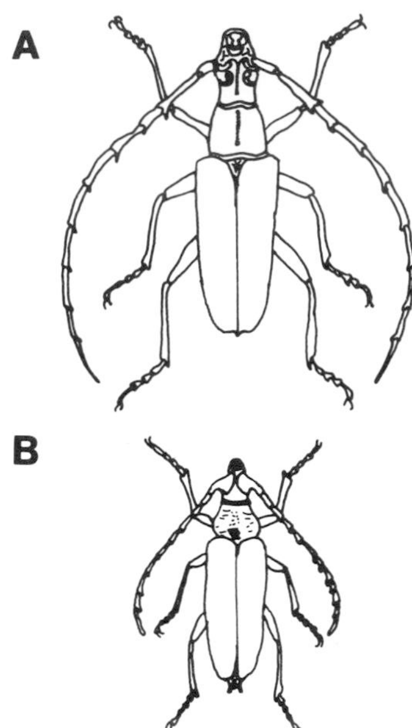


***FIGURE 4*** The wood borer *Haplocerambyx spinicornis* male (A) and female (B).

## IV. MANAGEMENT PRACTICES

Factors governing the population of any organism include availability of food, suitability of climatic conditions, intraspecific and interspecific competition, and the presence of predators, parasites, and diseases. In forests, both natural and man-made, conditions may become conducive for a sharp rise in populations of some pests, resulting in large-

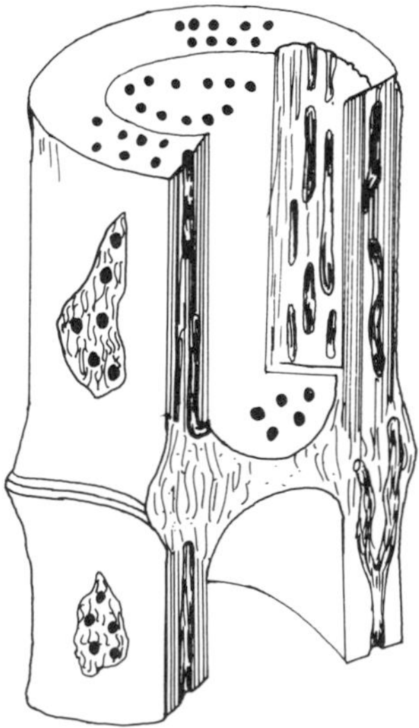

***FIGURE 5*** Bamboo shoot damaged by *Dinoderus* sp.

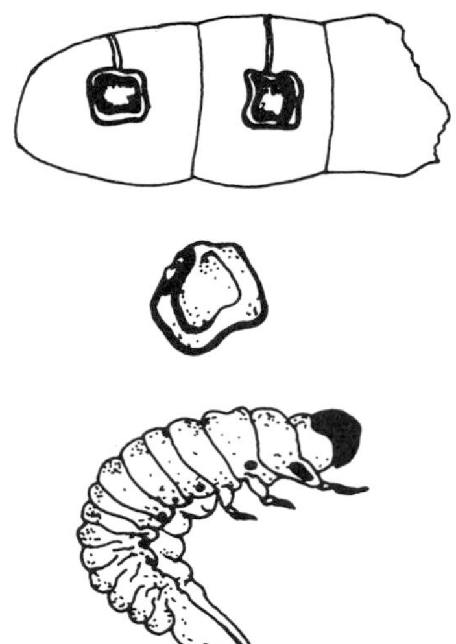

**FIGURE 6** Seed borer grub causing damage to the seeds of *Albizzia lebbek*.

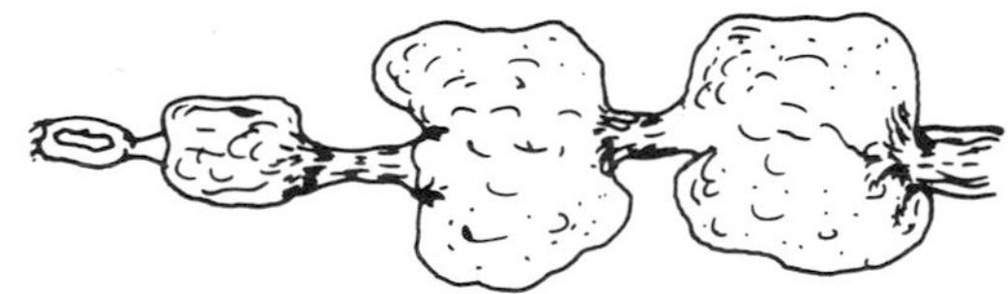

**FIGURE 7** Multilocular galls of the gall midge *Asphondylia tectonae* on teak.

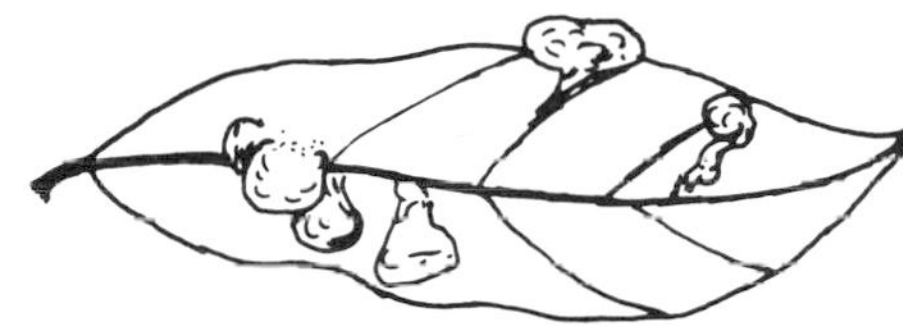

**FIGURE 8** Leaf galls caused by *Trioza fletcheri* on leaves of *Terminalia arjuna*.

**FIGURE 9** A leaf damaged by the action of a skeletonizer. (Photograph courtesy of S. S. Sandhu.)

**FIGURE 10** A leaf damaged by the action of a defoliator. (Photograph courtesy of S. S. Sandhu.)

scale damage. When this occurs, it is imperative that the situation be immediately assessed and prompt action taken.

A few decades ago, pest management placed no emphasis on preventive measures and steps were taken to control pest populations only when an epidemic was noted. However, there is now a growing acceptance to also employ preventive practices so that chances of a pest outbreak are minimized. But such advance planning is not always noticed in many areas of the world, partly because of lack of funds and partly because of the slow percolation of modern practices of pest control to the worker in the field. As a result, pest outbreaks are still quite common, particularly in monocultures with a large number of even-aged trees. Although there is a growing realization of the hazards of producing and using chemical insecticides, this method continues to be the most commonly used one. Several other methods are gaining effective employment, but no single method assures a foolproof success against the variety of insect pests. It is ultimately necessary to devise a strategy of using many methods in a definite sequence to gain effective control over the pests. Such a strategy is often referred as integrated pest management. Let us examine the impact of various methods of pest control before discussing integrated pest management.

## A. Chemical Control

An insecticide to be used in a forest should fulfill the following conditions:

(1) be easy to formulate and apply;
(2) be effective in small amounts;
(3) be long-lasting;
(4) be relatively safe in the forest environment; and
(5) be relatively inexpensive.

All insecticides have to be tested for their toxicity against other animal groups, particularly mammals, birds, fishes, and other wildlife. The rate of breakdown of these chemicals and the toxicity of various metabolites or intermediate products also influence the decision of foresters to use a particular formulation. Insecticides act on insects in a variety of manners, such as stomach poisons, contact poisons, systemic insecticides, fumigants, repellents/antifeedants, deterrents, or attractants. Based on their origin, insecticides may be inorganic or organic. Inorganic compounds in extensive use are based on arsenic, fluorine, sulfur, lime sulfur, phosphorous, aluminum, mercury, borax or boric acid, and zinc. Organic insecticides may be of either plant or animal origin. Insecticides of plant origin may act as repellents and/or be capable of disrupting the normal life cycle of the insect pests. Nicotine sulfate has been used as a contact insecticide. Pyrethrum and pyrethrin have been effective repellents. Other plants that provide effective insecticides include *Acorus calamus, Labelia excelsa, Croton oblongifolium, Randia dumetorum,* and of course *Azadirachta indica* or *neem* (in Hindi), which is now being considered as a primary agent for the future management of insect pests.

Among the organic insecticides of animal origin that have a high possibility of success against pest populations are juvenile hormone/JH analogues, which influence the growth, metamorphosis, and reproductive cycle of adults. Another effective method of killing a pest population is to first bring it together at one place by using pheromones. Cantharidine, extracted from meloid beetles, has traditionally been used against Orthoptera and Lepidoptera. Oils (vegetable, animal, petroleum) and soaps have also been used in a variety of ways, particularly in protecting stored grain from insect pests.

Besides the foregoing insecticides, a number of synthetic organic compounds effective against a variety of insect pests have been developed. These include thiocyanates, organochlorides, dichlorodiphenyl dichloroethane (DDD), dichlorodiphenyl trichloroethane (DDT), methoxychlor, and kelthane.

BHC (benzene hexachloride) is an example of a synthesized monocyclic organic compound used against a large number of pests. Cyclodiene or polycyclic insecticides represent a large number of chemicals presently in use. These include aldrin, chlordane, and heptachlor. Dieldrin and endosulfan (Thiodan) are heterocyclic synthetic compounds. Organophosphate compounds are either contact type or systemic poisons. The former include dichlorvos (Nuvan), diazinon (Basudin), trichlorphos, fenthion (Bay Fex), malathion, parathion (Folidol), and so on. The latter include demeton (Systox), dimethoate (Roger), formothione (Anthio), methyl demeton, phosphamidon (Dimecron), monocrotophos (Nuvacron), phorate (Thimet), Phosdrin, and so on. Carbamates used include carbonyl compounds, carbofuron (Furadon), and aldicarb (Temik). Synthetic pyrethroids include cypermethrin (Cyperguard, Cymbush), deltamethrin (Decis), and fenvalerate (Sumicidin).

### *1. Formulation*

Insecticides are applied either as liquids or as dust. Liquid formulations may have water or oil as the base, and some are in the form of emulsions. Liquid insecticides are applied from the ground or air. Dust formulations are bulky and subject to drift from winds and so are less effective. Additives are often added to liquid insecticides for specific purposes, for example, controlling viscosity, antievaporants, wetting, spreading, antifoaming, ultraviolet protecting, and reducing photodecomposition. Assessment of the success of application of specific insecticides is done by adding tracers or dyes to a spray formulation for qualitative and quantitative spray deposit assessment.

### 2. Manner of Application

In nurseries, smaller plantations, and small woodlands, ground spraying equipment is exclusively used. Also, when combating bark beetles, this type of equipment is of considerable value because aerial spray is not effective against insects hiding under the bark of trees. The equipment consists of a tank, pump, regulator for spray pressure, strainers, hose, nozzles, and hand gun. In specific cases, ultra low volume (ULV) applications have to be made, for which ground equipment is used. This equipment may be hand operated (e.g., backpack or knapsack sprayers, mist blowers, wheelbarrow models, etc.), may have a hydraulic pressure system, or may be power operated. The latter include mist-blowing machines mounted on vehicles that target large trees.

Aerial spraying of insecticides is a far more successful method where large forest areas are involved. Aircraft are selected on the basis of (1) size and nature of treatment area, (2) proximity of landing, refueling, and loading facilities, (3) altitude, topography, and meteorological data of the treatment area, (4) carrying capacity, maneuverability, speed, and aerodynamics of the aircraft, and (5) proximity of sensitive nontarget areas. Spray equipment is fixed on the aircraft and operated manually. Although aerial spraying of insecticides is more successful, there are problems regarding the winds and their direction, as drifting of the spray to nontarget areas is possible, not only resulting in a total loss of the insecticide sprayed and the cost of operating an aircraft at high costs, but also unnecessarily introducing hazardous chemicals into a nontarget area.

Different claims have been made about the success of various chemical insecticides. However, the success rate can never be equal everywhere and always depends on many factors besides the training of the personnel involved in a particular method of application. It is well known that target insects develop resistance to insecticides over a period of time. Furthermore, one has to choose a different insecticide for every pest. Effective concentration and the form of application have to be determined in the field, depending on the extent of infestation. The timing of application to coincide with a vulnerable stage of the pest is also crucial. But despite claims of preventing annual losses running to billions of dollars, long-term changes in the soil and environment from the continuous addition of chemicals have been documented. That most of the changes will be detrimental is also clear, because every application alters the soil environment.

## B. Biological Control

The use of one organism against another, which may be a pest from the human point of view, is called biological control. This phenomenon was not invented by humans as it already existed in nature. Biological control has been used extensively in agriculture and horticulture, as well as in forestry, where a number of applications have yielded excellent results. [*See* BIOLOGICAL CONTROL.]

### 1. Biological Control by Predators

Biological control operates by either predation or parasitization of the pests. Both phenomena maintain a state of equilibrium between predator/prey and parasite/host populations. However, sometimes and in specific regions, some pest populations appear to increase unchecked due to either a lack of natural predators or the existence of other conditions that prevent effective predation by locally available predator species. In these cases, humans have sometimes introduced exotic species that prey on the pest or altered the conditions that prevent the effectiveness of these local predators.

However, the processes underlying this strategy are not simple. Predator–prey relationships and interactions depend on a number of factors, including (1) density of prey and predator populations, (2) behavioral and physiological characteristics of the prey and predators, and (3) density and quality of alternate food available in the vicinity. The success of a species in a particular region also depends on a number of other physicochemical and biotic factors. An in-depth study of these is necessary before the final choice of a local or exotic predator can be made.

Among the successful predators of forest insect pests are birds and small animals, including reptiles and mammals. The former operate at middle and upper levels of the tree and the latter on the ground-inhabiting stages of the insects. A common example of a bird predator is that of a woodpecker that feeds on cerambycid beetles, which damage timber. Bird predation on the jack pine budworm accounts for 40–65% of the mortality of late-instar larvae and pupae of this species. Insectivorous mammals are known to play a key role in checking the population of the European spruce sawfly by consuming the cocoons of this pest, which are laid in the soil. However, no pest can be completely controlled by the activity of a single predator, although the contribution of this predator may remain central to the effort. It is the cumulative effect of not only the predators themselves but also the sequence of predators that determines the success in checking a pest population. Local predators are preferred in habitats that can be manipulated (e.g., plantations and nurseries), but exotic predator species ensure a greater success rate in large habitats in which manipulation is less feasible. Native species are sometimes encouraged by offering special inducements like providing nest boxes.

Predator control of insect pests has been most successful when insect predators are used (Fig. 11).

**FIGURE 11** *Hyblaea puera*, a pest of teak, being preyed on by a coleopteran.

Coccinellid beetles, syrphid flies, and mites are well known as predators of a number of insect pests with piercing and sucking mouthparts (e.g., aphids). One of the more recent success studies is provided by ants that have successfully controlled a variety of forest pests (at least 17 well-known pest species populations are known to be controlled by ants). Only ants remain active from the ground level up to the tips of the uppermost branches. Furthermore, this predator is able to invade and exploit almost all habitats. The density of the red ant *Formica rufa* reaches approximately 4237 ants per square meter or 180 ant nests per hectare. It is estimated that ants from a million nests in the Italian Alps consume as much as 14,000 tons of live insects in a 200-day growing season, and one ant nest effectively protects an area up to a radius of 35 m. The activity and consumption of such predator ants are expected to be many times higher in the warmer climates of Asia and Africa. Another advantage of using ants is that they attack the eggs, larvae, pupae, and adults with equal fervor. The introduction of *Formica lugubris* in Canada has been very successful against insect pests of mixed conifer plantations.

Apart from ants, clerid beetles are of special interest to humans, as they prey on various bark beetles that cause much damage to the timber. Carabid beetles rid the forest floor of various soil insects, caterpillars, pupae, and other phytophagous beetles. A number of other insects from the orders Diptera, Hemiptera, and Neuroptera also prey on specific insects; spiders and mites are effective predators as well. Much work is still needed to understand the ecological, physiological, and behavioral aspects of these species before they can be safely introduced in new habitats to control specific pests.

### 2. Biological Control by Parasites and Parasitoids

Insect parasitoids (which are parasites as larvae but free-living as adults) also help to keep some insect pest populations in check. Insect parasitoids that prey on insect pests of forest trees belong to the families Tachinidae, Chalcidae (Hymenoptera), and Ichneumonidae (Hymenoptera). Specific variants include hyperparasitoids (i.e., parasitoids of parasitoids), multiple parasitism (in which two species lay their eggs on or in the same host individual, with only one being generally successful), and superparasitism (in which many eggs or larvae are placed on or in the same species, of which one species survives in solitary hosts but many may survive in gregarious species). [*See* PARASITISM, ECOLOGY.]

### *3. Biological Control by Microbes*

By far the most promising method of insect control is microbial control. A number of microbes belonging to diverse groups have been and are being tested, but basic biological and ecological studies are still required before field tests are possible. The potential of microbial control is high, as the pathogens are relatively host specific and this mode is compatible with other control procedures. Among the pathogens currently being tested against forest insect pests are nuclear polyhedral virus (NPV) and the bacteria *Bacillus thuringiensis* (against the basin tent caterpillars *Malacosoma, Thaumetopoea, Neodiprion,* gypsy moth, and tussock moth), which provide foliage protection up to 25% and larval mortality of 100%.

In Canada, the European spruce sawfly was effectively controlled by a polyhedrosis virus that was accidently introduced. This virus has also controlled sawfly of jack pine and redheaded pine sawfly in Canada and the United States. Application of the pathogen is comparatively simple. Once applied, the pathogen spreads by (1) the movements of the infected hosts, (2) contact with the fecal matter of the insects or (3) animals that feed on diseased hosts, (4) physical factors like wind and rain, (5) contaminated eggs, and (6) dead hosts. Other pathogens include cytoplasmic viruses (CPV), of which NPV is the most effective one. In addition, there are more than a hundred species of fungi, the more effective ones being *Beauveria bassiana, Entomophthora, Empusa, Metarrhizium,* and *Isana*. Of these, much work has been done on *Beauveria bassiana* and *Metarrhizium anisopliae*.

## C. Other Methods of Insect Control

Besides the chemical and biological methods, a number of other control measures have been effectively used in many cases. These include screening the germplasm of forest trees for susceptible and resistant varieties. Augmentation of the saplings to be planted is done by tissue culture/cloning methods. This practice minimizes the need for insect pest control at a later stage.

Silvicultural practices include crop rotation, lower planting density, and planting of mixed forests in place of pure stands. Use of fire prior to planting eliminates many insects, and felled trees must be removed promptly. Adult insects can be removed by employing light traps (Fig. 12). The reproduction of pests can be disrupted by using pheromone traps to catch adults of one sex.

Integrated pest management (IPM) entails a judicious combination of all or some of the control measures to keep the population density of the pest at a level where it cannot cause much damage.

IPM is defined as "the maintenance of destructive agents (including insects) at tolerable levels by using a variety of preventive, suppressive or regulatory tactics in a planned manner. The tactics used must be ecologically and economically efficient and socially acceptable."

Availability of ecological and biological data of the interacting insects and trees is a prerequisite for

***FIGURE 12*** A light trap extensively used for trapping adults, particularly lepidopterans, in both agroecosystems and forests. (Photograph courtesy of S. S. Sandhu.)

effective pest management. The data provide an insight into the population dynamics of a given species, the key triggers or suppressors of population fluctuations, relative consistency of population trends, impact of insect populations on the hosts, and interdependence of various dendrophagous species on a given host. Based on this information of the forest insect community and its various components (the system), suitable models are developed to reflect the real ecosystem. These models are then used for predicting the effectiveness of specific preventive/suppressive/regulatory tactics at precise junctures. Predictive and information-handling capacities of the computer are thus put to a very effective use.

When these measures are applied in the field conditions, the models get tested and shortcomings, if any, are revealed. These are then attended to with the help of further research. A number of IPM models have been successfully employed as in the case of spruce budworm population model in forest insect management.

## V. CONCLUSION

The selection of pest-resistant varieties of forest trees is perhaps the most important and essential prerequisite for a vigorous and pest-free stand. It has been found that mixed forests are less prone to insect pest epidemics as compared to pure or monoculture plantations. In case of pest outbreak, it is often necessary to employ chemical control as a short-term measure, but because the addition of chemicals to the environment has a negative impact in the long and short run, their use should be kept to a minimum. Biological control is a viable alternative, but a thorough study of the ecology of predator–prey or parasite–host interactions has to be done before introducing an exotic species to a new environment. Cultural or silvicultural practices help in achieving the ideal equilibrium. Integrated pest management requires flexible and sophisticated strategies that are appropriate to a particular situation.

## Glossary

**Antifeedant** Chemical that causes cessation of feeding activity in insects.

**Defoliation** Loss of all leaves of a tree.

**Fumigant** Insecticide applied by fumigation.

**Phytotoxemia** Poisoning of a tree.

**Repellent** Chemical used to repel insect pests.

**Skeletonization** Process in which leaf veins are left intact while the soft tissue is consumed.

## Bibliography

Barbosa, P., and Wagner, M. R. (1989). "Introduction to Forest and Shade Tree Insects." New York: Academic Press.

Browne, F. G. (1968). Pests and diseases of forest plantation trees. Oxford: Clarendon Press.

Coulson, R. N., and Witter, J. A. (1984). "Forest Entomology, Ecology and Management." New York: John Wiley & Sons.

Dent, D. (1991). "Insect Pest Management." Wallingford, U.K.: C.A.B. International.

Garner, W. Y., and Harvey, J., Jr. (1984). "Chemical and Biological Controls in Forestry," American Chemical Society Sympos. Ser. 238. Washington, D.C.: ACS.

Joshi, K. C. (1992). "Handbook of Forest Zoology and Entomology." Dehradun: Oriental Enterprise.

# Forest Pathology

**Robert A. Schmidt**
*University of Florida*



Forest pathology is the science that studies diseases (maladies) of trees in natural and planted forests. In its broadest sense, forest pathology includes diseases of trees wherever they grow, including urban settings, and extends to the deterioration of wood products. The practical goal of forest pathology is the proper management of the forest and trees in other landscapes to prevent or mitigate the undesired impacts of the agents that cause disease.

## I. SCOPE OF FOREST PATHOLOGY

Forest pathology is a young discipline, substantially begun in Germany in the mid-nineteenth century and initiated in the United States in the beginning of the twentieth century. Early studies were primarily of decay in living mature trees in natural forests. Contemporary studies have expanded to include plantations, nurseries, seed production orchards, and shade and ornamental trees, the latter becoming more important with increased urbanization. Also included in forest pathology is wood products pathology, the study of the cause and prevention of deterioration of wood during storage and use. Forest pathologists also study microorganisms that are beneficial to trees, for example, mycorrhizal fungi, which are important root symbionts.

Tree diseases are caused by climatic, edaphic and biotic factors in the environment, the latter, i.e. organisms, being the most numerous causes of tree diseases. Forest pathology is multi- and interdisciplinary, encompassing (1) many botanical sciences, for example, physiology, ecology, genetics, mycology, microbiology, and silviculture, (2) physical sciences, for example, chemistry and climatology, and (3) hybrid sciences, for example, biochemistry and aerobiology. Forest pathology spans all levels of the biological sciences from the molecular to the ecosystem or landscape level. [*See* FOREST INSECT CONTROL.]

Historically, research and extension (public service) in forest pathology were crisis-driven by catastrophic epidemics, often caused by the introduction of exotic plant pathogens into indigenous forests. Recent trends are toward preventing diseases in individual trees and managing forests to minimize disease impacts. As such, forest pathology has evolved along with forest entomology into forest protection, integrated pest management, and most recently forest health. Contemporary concerns in forest pathology have addressed biological control, molecular biology, genetic diversity, and global warming.

## II. CAUSES OF TREE DISEASES

Plant disease can be conceptualized as a triad interaction among a plant (host), a causal agent (pathogen), and contributing environment factors, including time as an important dimension. The causal agents of forest tree disease are biotic (fungi, bacteria and phytoplasma, viruses, nematodes, and parasitic higher plants) and abiotic (climate and edaphic factors, e.g., drought, nutrition, and air pollutants). The causes of forest tree diseases are often complex, involving one or more predisposing factors and secondary factors acting in sequence, or concert. Predisposing and secondary causal agents can be biotic or abiotic. Biotic pathogens can be parasites or saprophytes. Saprophytes play an important role in forests because they decay wood and decompose dead organic matter, returning nutrients to the soil. Some pathogens are highly specialized, occurring only on a few tree species or genotypes and/or only on or in certain host tissues, for example, poplar leaf diseases, pine stem rusts, and root rots of conifers. Other pathogens attack a broad spectrum of hosts in different regions and environments. There are hundreds of forest tree pathogens and resulting diseases, each with its unique life cycle (disease cycle). Fortunately, only a small portion of these diseases result in significant impacts that are important for forest management. Traditionally, diagnosis of tree diseases is accomplished from morphological evidence of the pathogen (signs) and/or host abnormalities (symptoms). Often the patterns of disease incidence and spread are clues for diagnosis of cause. Recent advances in biotechnology are aiding the diagnosis of pathogens and plant diseases.

## III. EPIDEMIOLOGY OF TREE DISEASES

Environmental factors (climatic, biotic, edaphic, and cultural) that are favorable or unfavorable for disease development can profoundly affect disease occurrence and severity, temporally and spatially, both in individual trees and, most importantly, in populations of trees. These interactions determine the dynamics of disease development, that is, the rate and magnitude of increase and the rate and extent of spread of the epidemic. Some forest tree pathogens produce prodigious amounts of inoculum (e.g., fungal spores) and with favorable conditions for spread and infection result in explosive epidemics. Other forest tree pathogens reproduce and spread slowly or may be slowed by environmental conditions, low pathogen aggressiveness, or host resistance, resulting in endemic disease conditions. However, even endemic disease can result in a significant impact in a long-lived, perennial tree crop.

In natural forests, pathogens and hosts have coevolved into a dynamic balance. Local epidemics may occur in response to environmental disturbances such as wildfires or weather events, but normally diverse homeostatic mechanisms (functional diversity) serve to limit disease incidence, both temporally and spatially. One exception is the deterioration of our mature forests caused by root and stem decay, which, although of natural occurrence, can result in significant impacts (degrade and mortality) in overmature or ancient forests.

Human activities, including silvicultural practices, can profoundly affect the natural balance of host and pathogen and thereby the incidence of disease in forests. The introductions of exotic tree pathogens into indigenous forests have resulted in disastrous epidemics: Chestnut blight, white pine blister rust, and Dutch elm disease are classic examples. The chestnut blight epidemic virtually eliminated the valuable American chestnut (*Castanea dentata*) from the forests of the eastern United States. The causal fungus (*Cryphonectria parasitica*) spread rapidly and extensively in the absence of limiting factors—especially host resistance—in an otherwise diverse ecosystem. Wounding caused by silvicultural practices, such as harvesting or thinning, also can lead to severe losses from pathogens that normally do not infect unwounded host tissues. A more recent anthropogenic cause of disease increase is intensive plantation culture, when rapidly growing monocultures, devoid of functional diversity, are intensively managed. One such example is the current epidemic of fusiform rust on indigenous slash (*Pinus elliottii* var. *elliottii*) and loblolly (*Pinus taeda*) pines intensively managed for fiber production in the southern United States. This disease, rare in natural forests at the turn of the twentieth century, has reached epidemic proportions in intensively managed plantations. Losses to this disease are being mitigated through the selection, breeding, and planting of genetically improved disease-resistant pine genotypes.

## IV. TREE DISEASE IMPACTS AND MANAGEMENT

The financial losses caused by the direct impacts of forest tree diseases (tree mortality, growth loss, decay, and defects) are substantial. It is estimated that one in three trees succumbs to forest health problems. In addition, there are indirect impacts including delayed regeneration, reduced stocking, changes in species composition, deterioration of site quality, reduced management opportunities, and pest management costs. Economic losses caused by pests are greatly magnified in shade and ornamental trees and on some forest trees (e.g., walnut and veneer logs) owing to the high value of individual trees. Structural damage resulting from decay of wood in use is also very costly.

The ultimate applied goal of forest pathology is the biologically sound, environmentally safe, and economically feasible control or management of tree diseases. In attempting to reach this goal, forest pathology has evolved from reactive crisis management of epidemics (often caused by introduced pathogens) to proactive and integrated pest management strategies. A recent strategy emphasizing the crop, in contrast to the pest, and integrated management has led to the concept of forest health and resource management. Forest health is a proactive management strategy in that it embodies periodic monitoring to detect real and potential pest problems. Survey and detection of forest tree diseases will be greatly aided by GIS (geographic information systems) monitoring, and also perhaps satellite imagery, although at present the latter lacks the resolution for many forest disease studies.

In concept, plant disease control principles (strategies) include exclusion, eradication, protection, resistance, avoidance, and therapy, and the control methods (tactics) include regulation, cultural, biological, physical, and chemical. In practice, exclusion (prevention) and avoidance are cornerstones of forest disease management. It is clearly the case that in long-lived perennial forest crops, "an ounce of prevention is worth a pound of cure." Sanitation of diseased trees (reduction of inoculum) is sometimes useful, especially during forest thinning operations. Also, genetically controlled disease resistance is gaining in importance, especially in intensively managed forest plantations. Chemical control is most often not economically feasible in forests and has associated environmental risks; however, pesticides are useful in nurseries and seed production orchards. Protection of wood-in-use with chemical preservatives is effective and may increase the life of wood products from 5 to 30 years. Biological control of forest tree diseases has great intrinsic appeal, but has proven effective in only a few instances. There is a hope that new biotechnologies will lead to better control of forest tree diseases. For example, through genetic engineering of biocontrol organisms to enhance their ability as competitors or antagonists.

## Glossary

**Anthropogenic** Pertaining to the impact of humans.

**Disease cycle** Chain of events in the interaction of a pathogen and a host resulting in disease, including infection, colonization, sporulatron, secondary cycles and spread of the pathogen.

**Edaphic** Pertaining to the soil.

**Endemic** Native or indigenous to a certain region; also used to imply a balanced low amount of disease.

**Epidemiology** Ecology of disease; the study of epidemics.

**Homeostatic** Stable state between interdependent organisms.

**Inoculum** Propagules of pathogens that initiate infection of the host.

**Mycorrhizae** Symbiotic association between a fungus and a plant root.

**Parasite** Organism living upon another living organism and deriving food from it.

**Pathogen** Causal agent of disease.

**Saprophyte** Organism that lives upon dead organic matter.

**Symbiont** Organism that lives in a state of symbiosis, that is, dissimilar organisms living together to their mutual benefit.

## Bibliography

Blanchard, R. O., and Tattar, T. A. (1981). "Field and Laboratory Guide to Tree Pathology." New York: Academic Press.

Hepting, G. H. (1971). "Diseases of Forest and Shade Trees of the United States." U.S. Dept. Agric. For. Serv., Handb. No. 386. Washington, D.C.: USDA Forest Service.

Manion, P. D. (1991). "Tree Disease Concepts." Englewood Cliffs, N.J.: Prentice–Hall.

Phillips, D. H., and Burdekin, D. A. (1982). Diseases of Forest and Ornamental trees. MacMillian, London.

Schmidt, R. A. (1978). Disease in forest ecosystems: The importance of functional diversity. *In* "Plant Disease: An Advanced Treatise. Vol. II. How Disease Develops in Populations of Plants." (J. E. Horsefall and E. B. Cowling, eds.), Chap. 14, pp. 287–315. New York: Academic Press.

Sinclair, W. A., Lyon, H. H., and Johnson, W. T. (1987). "Diseases of Trees and Shrubs." Ithaca, N.Y.: Cornell University Press.

Smith, W. H. (1970). "Tree Pathology: A Short Introduction." New York: Academic Press.

# Forests, Competition and Succession[1]

**David A. Perry**
*Oregon State University*



Competition, the struggle for limited resources, and succession, the sequence of change in dominant organisms following colonization, have long been key concepts employed by ecologists to understand and organize the patterns of nature. Although competition and succession are distinct processes, they are closely related for at least two reasons. First, successional trajectories are largely driven by interactions among organisms, including (but not restricted to) competition. Second, both are intimately related to the degree of equilibrium or disequilibrium in ecosystems and landscapes. Ecologists once believed that succession led inexorably to a stable equilibrium within a given community of organisms, the composition of which was determined in large part by who won the struggle for limited resources. Although that view has not been totally discarded, most ecologists now recognize that change is the rule rather than the exception in nature, with few if any ecological communities achieving a long-lasting equilibrium in species composition. Disturbances at many spatial and temporal scales create "shifting mosaics" of communities in different stages of succession, resulting in diverse niches that allow more species to coexist than would be possible if all were competing for the same set of resources.

This article first discusses competition: why it occurs, why it does not occur, and how it shapes the structure of communities. It then turns to the patterns and mechanisms of succession, many of which turn on the nature of both competitive and cooperative interactions among species.

## I. COMPETITION

> "The inhabitants of the world at each successive period in its history have beaten their predecessors in the race for life." (Charles Darwin, "The Origin of Species," 1859)

One of the oldest ideas in ecology is that individuals utilizing the same resource will compete if that resource is in short supply. For many years ecologists assumed that the sizes of all populations within a given community were ultimately limited by resources, hence competition was believed to be an inevitable consequence of making a living and the major determinant of community structure (i.e., the number of species and size of each population). Constant struggle is not necessarily implied; over many generations species may evolve ways to avoid competition through allocating resources. Nevertheless, communities are ultimately struc-

[1] Portions excerpted from D. A. Perry (1995). "Forest Ecosystems." Copyright Johns Hopkins University Press, with permission.

tured by competition, be it ongoing or, as a popular phrase puts it, the "ghost of competition past."

Although competition for resources undoubtedly occurs (at least in some trophic levels), the notion that it is the major organizing force in nature is overly simplistic. Most ecologists now recognize that ecological communities are complicated and variable, with their structure shaped by many interacting environmental and biotic factors in addition to resource competition. Species may not compete for resources at all or they may simultaneously compete and cooperate. For example, multispecies flocks of insectivorous birds commonly follow ant swarms in the tropics, feeding on insects flushed by the ants. In 1983, F. Bourliere listed two possible advantages to such mixed flocking: more efficient hunting in a patchy environment and more eyes and ears for protection against predators. Different species of monkeys also frequently forage together, even when they are quite similar in body size and diet. Once again, Bourliere (1983, p. 87): ". . . the main advantage for mixed groups (of monkeys) appears to be to make easier the location of patchily distributed food sources and the detection of predators. A monospecific group of similar size might well offer the same benefits to its members, but at the cost of a stringent social hierarchy which would imply a greater energy expenditure for its enforcement." A. F. Hunter and L. W. Aarssen list the following ways in which coexisting plant species have been demonstrated to help one another: ". . . improving the soil or microclimate, providing physical support, transferring nutrients, distracting or deterring predators or parasites, reducing the impact of other competitors, encouraging beneficial rhizosphere components or discouraging detrimental ones, and attracting pollinators or dispersal agents." [*See* Foraging Strategies.]

At least three things come into play to modify the importance of competition within ecological communities.

1. Species within a given trophic level may be limited by factors other than resources, hence seldom or never have to compete. It has been suggested that the importance of competition alternates up the trophic ladder: plants compete; herbivores are held down by predation, hence do not compete; and carnivores compete because they are at the top of the food chain and therefore have no predators. Evidence in support of this idea is equivocal. However, there is little doubt that predation, disturbance, or climatic fluctuations can and frequently do act to maintain populations below what their food supply would permit. This is often the case with herbivorous insects and its can also occur in plant communities when herbivores and pathogens increase the species richness of plant communities by reducing the ability of any one species to competitively dominate others. Higher order interactions other than predation may come into play to reduce competition. Such is the case with at least some types of mycorrhizal fungi, which by mediating a more equitable distribution of resources among individual plants reduce the ability of one or more species to dominate a community.

2. Species may compete for the same resources but also benefit one another in some way that tends to dilute or negate their negative effects. This undoubtedly occurs in ecosystems (as exemplified by the mixed foraging groups discussed earlier), but may not be readily apparent from casual observation or even from experiments unless conducted over many years (which seldom happens). It has been hypothesized that plant species participate in defense guilds, i.e., either directly or indirectly reduce herbivory and/or pathogenesis within the community. For example, flowering plants are common in young conifer forests, where they probably compete with the conifers for various resources. However, nectar produced by flowers of these plants is important in the diet of at least some insects that prey on defoliating insects. It has been documented that 148 species of parasitoids (important predators of defoliating insects) are associated with flowering plants in forests of northern Germany. What is the net effect of these plants on conifers? If they were not in the ecosystem would faster conifer growth eventually be negated by larger populations of defoliating insects made possible by lower populations of their predators? Questions such as these are seldom entertained in

competition studies. Potentially competing species have also been hypothesized to form cooperative guilds based on protecting and stabilizing ecological commons such as shared mutualists (e.g., pollinators, mycorrhizal fungi) and soils. "Facilitation," in which one plant species has some effect that benefits others, is commonly observed during succession. It does not follow that a "facilitator" does not also compete for resources; however, the net effect is positive rather than negative. On the other hand, there are clear instances in which one plant species has a net negative effect on others, and other cases in which the net effect of one species on another varies with time and environment. These points are followed in the section on succession.

3. Complexity at scales ranging from landscapes through individual trees to soil aggregates creates diverse niches that allow species to avoid competition through specializing.

One of the primary criticisms leveled against the idea that communities are structured primarily by competition is that few experiments have really tested this. If one wants to know the effect of species A on species B, then species A must be removed in a properly designed study, and the response of species B must be measured over a sufficiently long time to assess the fitness of B. Experiments in which one or more plant species are removed are common in forestry and frequently show that removing "competitors" improves the growth of those that remain. However, these are seldom followed long enough to tell whether long-term effects tell a different story. While competition is generally an ongoing process that is readily measured in short-term experiments, the beneficial effects of one species on another may be subtle, long term, or manifest only during certain critical periods. For example, nodulated nitrogen-fixing plants are common pioneers on disturbed sites, where they play an important ecological role by replenishing soil nitrogen and carbon. In the short term these plants compete with others for resources whereas in the longer term they benefit others by increasing soil fertility. To give another example, some hardwood species that grow intermixed with conifers are relatively resistant to fire. During much of the lifetime of a stand these may compete with conifers, but during wildfire they protect the conifers. Species-diverse grasslands are more stable during a severe drought than species-simple grasslands. [*See* FIRE ECOLOGY.]

Interactions among species requiring the same resources are complex, varying over time and with environmental conditions. In contrast, much of the traditional ecological thinking about the role of competition in structuring communities has been shaped by mathematical models that treat two interacting species as if they were in a constant environment and isolated from other species in the system. L. Stone and A. Roberts developed a more realistic model that evaluates the interactions between any two species ". . . within the framework of the community to which they belong." Their approach, which they refer to as the "inverse method," deals strictly with interactions *within* a given trophic level. In other words, these are interactions in which the participants potentially compete for resources. The criterion they use to determine whether a species benefits or suffers in interaction with another is population growth: in essence, they ask "what happens to numbers of species A if numbers of species B increase?" If A increases, it benefits from B (at least within the range of increase of B that is modeled), if A decreases, it suffers from B. Stone and Roberts conclude:

> "Remarkably, the 'inverse' method finds that generally a high proportion (20–40%) of interactions must be beneficial, or 'advantageous,' when not lifted out of the community context in which they actually occur. The contrary case, called here 'hypercompetitive,' in which each species suffers from every other species, can occur only if the environment is nearly constant, and the species closely akin to each other, with both of these conditions holding and persisting to a degree that must be considered implausible."

Given the current evidence for major extinctions caused by meteor impacts, Darwin's "survival of the fittest" may be appropriately modified to "survival of the luckiest." Nevertheless, while various factors reduce competition for limited resources, there seems little doubt that important aspects of community structure reflect the "ghost of competition past." Evidence suggests that through evolutionary time conflicts over food have frequently been resolved through specialization; this is thought by some to have been a significant diversifying force in nature. Examples of food-related niche diversification within animals are widespread in nature. In the neotropics, fruits compose the primary diet of 405 species of birds, 33 species of primates, and 96 species of bats. But the diverse array of fruits produced by the numerous plant species of these areas permits frugivores to specialize to some degree and thus reduce competition or avoid it altogether. In the dry tropical forests around Monte Verde, Costa Rica, it has been estimated that only 13 of the 169 species of fruit eaten by birds are also eaten by bats. Similar studies throughout the tropics have found little overlap in the diets of fruit-eating bats, birds, and primates. [*See* EVOLUTION AND EXTINCTION.]

Niche diversification may take numerous forms, including type of food, timing of feeding, place of feeding, and, in some instances, the ability of a species to capitalize on food provided by periodic extraordinary events. One of the better known examples is Robert MacArthur's study of five warbler species that coexist in conifer forests of New England. All are insect eaters and are about the same size. These species feed in different positions in the canopy, move in different directions through the trees, feed in different manners, and have slightly different nesting dates.

Numerous studies of niche diversification among animals have focused on the so-called *metric traits,* readily measurable attributes such as body size or, in birds, shape and size of the bill, that are assumed to reflect differences in diet. MacArthur discusses bird species diversity on the island of Puercos off the coast of Panama.

> "(T)here are . . . four species of interior forest flycatchers on Puercos. The smallest, the beardless tyrannulet, . . . has an average weight of 8 gm; the next smallest is the scrub flycatcher . . . with an average weight of 14.6 gm; then follows the short-crested flycatcher . . . with a mean weight of 33.3 gm. Each of these is about double the weight of the previous one. Finally, the largest is the streaked flycatcher . . . that weighs an average of 44.5 gm. . . . Other families do not seem to sort by size. For instance, there are two flower-feeding 'honeycreepers,' the bananaquit . . . and the red-legged honeycreeper. They feed together among the flowers in the canopy and their mean weights are 10.7 and 12.8 gm, respectively. There is a plausible explanation for the (fact that flycatchers sort by size while honey-creepers do not). Large flycatchers do eat larger foods than small ones . . . , whereas there is no simple way that a large honeycreeper could eat . . . different food than a small one. . . . Rather the bills are of different shape, and it is very likely that these species eat nectar from different flowers or eat different insects while feeding on nectar."

In summary, long-term, diffuse interactions are the rule in ecological communities instead of the exception. Species certainly compete among themselves for resources, and it will never be known how many are now extinct because they lost a struggle to a superior competitor. However, species also depend on one another in numerous ways that are not readily apparent. Those that compete most of the time might benefit one another during certain critical periods. When one reflects on the multiplicity of indirect, diffuse, and subtle interactions that are possible in ecosystems, it is apparent that the experiments necessary to truly grasp the patterns of nature will be formidably difficult at best, and maybe even impossible (which is not to say that they should not be attempted). Like physicists (who deal with much simpler systems), ecologists may have to accept an "uncertainty principle," i.e., we may never completely capture the richness of nature within the framework of scientific hypotheses and models.

## II. SUCCESSION

> "No matter what forms we observe, but particularly in the organic, we shall find no-

where anything enduring, resting, completed, but rather that everything is in continuous motion." (Goethe, 1790)

Envision a landscape of bare rocks exposed by a retreating glacier. This is an inhospitable environment for life. There is no soil, hence no water and nutrient storage capacity to support plants. Animals find little shelter and no food. Some life is adapted to such conditions, however. Lichens colonize the rocks, obtaining nitrogen directly from the air and other nutrients by releasing acids that break down the rock. They scrounge water from cracks in the rock. Lichens provide a food base for animals and, mixing their own organic matter with the products of rock weathering, slowly build a soil, which allows higher plants to establish. These, along with the set of animals and microbes that accompany them, further modify the environment, casting shade and building litter layers, resulting in yet another set of plants and animals—adapted to the new conditions—coming to occupy the site.

This sequence represents what is termed *primary succession;* the term "primary" is applied because there was no preexisting community on the site (or if there was all traces were obliterated). It is an idealized picture—some higher plants, including trees, are quite capable of colonizing fields of bare rock and do not have to wait on lichens—but nevertheless illustrates a common pattern in primary succession:

- A disturbance (glaciation in the example just given, but it could be something else, e.g., volcanic eruption, lava flows) wipes out life and most or all traces of life on a site.
- A set of organisms adapted to survive and reproduce in these "primary" conditions becomes established; the colonizing plants are often characterized by an ability to extract nitrogen (N) directly from the atmosphere and other nutrient elements directly from rock.
- Colonizing organisms modify the site, accumulating nutrients and building soil, thus creating the conditions that permit a second wave of organisms to establish.
- Over many years the site becomes increasingly modified by the biotic communities that occupy it: that combination of minerals, dead organic matter, complex biochemical molecules, and living organisms that is called soil continues to be built, litter accumulates, and, particularly in climates capable of supporting trees, the accumulation of leaf area increasingly shades and buffers the interior of the community from environmental extremes.
- As one set of organisms modifies the site, it is replaced by another set better adapted to the new conditions. Barring another disturbance, a relatively persistent community eventually comes to occupy the site, in forests often (but not always) dominated by tree species that are able to reproduce in shade. The qualifying term *relatively* must be taken seriously when applied to the persistence of late successional stages. Trees may live from hundreds to thousands of years, but they are not immortal. If fire, wind, insects, pathogens, chainsaws, or something else does not kill them, old age eventually will. Each death creates space for new individuals of the same or different species to grow, hence forests are dynamic rather than static. A given set of species virtually never persists indefinitely on a given piece of ground, although constancy in species composition does occur at regional scales (except during major changes in climate).

There are many variations on this theme, but one feature is common to all primary successional sequences: primary succession involves a progressive "imprinting" of biological features onto a physical landscape.

Perhaps the most important biological imprint is soil. Joan Ehrenfeld discusses soil development during primary succession on sand dunes:

> "The soil microflora interacts with plants in promoting soil development in dune ecosystems. Hyphae of both saprophytic and endomycorrhizal fungi help bind sand grains into aggregates through the excretion of amorphous polysaccharides which in turn serve as substrate for colonization by bacteria,

actinomycetes, and algae. The presence of a diverse microflora enhances the process of aggregation. The degree of soil aggregation increases (as succession proceeds) . . . soil aggregates (mg/kg soil) increase from 5 in the foredunes to 40 on the mobile dune slope, 300 on the dune crest, and 1260 on young fixed dunes. The aggregates contain a variety of fungal species, including mycorrhizal species, and various bacteria . . . thought to be nitrogen fixing. There is an interactive effect between plant root growth and aggregate formation . . . (the) total amount of aggregation, and concomitantly the abundance of all microfloral species . . . increases dramatically in the presence of roots."

Now envision the mature forest that is the "end point" of primary succession on that bare rock. In fact, it is not an end point at all, but one stage in a (more or less) cyclical alternation of communities that will dominate that site. At some future date the glaciers will probably return, but in the intervening period there will be many more disturbances such as fire or severe windstorms that will kill the trees and initiate the process called secondary succession, which occurs where disturbance has left biological imprints (or legacies) such as soil, surviving individuals, and dead wood. Virtually all ecosystems exist within a matrix of fluctuating environments punctuated by periodic disturbances ranging from mild to severe. The twin processes of disturbance and succession form the core of natural dynamics and create much of the variety that is seen in the natural world. On the other hand, disturbances that are too frequent, severe, or "foreign" (i.e., have characteristics to which the species composing the system are not adapted) can throw succession off track and lead to persistent changes that frequently include loss of diversity and productivity, a widespread phenomenon in today's world. Hence understanding the mechanisms of community response to and recovery from disturbance is more than an academic exercise, it yields insights into how humans can protect and sustainably utilize natural systems.

### A. Brief Historical Notes

Among American ecologists, two names stand out in the development of successional ideas during the early years of the 20th century: Frederick Clements and H. A. Gleason.[2] Clements believed that communities were superorganisms and that succession was a maturation of the community toward its most mature state, which he called the "climax:"

> "Succession must then be regarded as the development or life history of the climax formation. It is the basic organic process of vegetation, which results in the adult or final form of this complex organism. All the stages that precede the climax are stages of growth. They have the same essential relation to the final stable structure of the organisms that seedling and growing plant have to the adult individual." (Clements, 1916)

For Clements, the composition of the climax vegetation was uniquely determined by climate:

> "Such a climax is permanent because of its entire harmony with a stable habitat. It will persist just as long as the climate remains unchanged, always providing that migration does not bring a new dominant from another region."

Like Clements, Gleason recognized the importance of environment in determining the composition of plant communities; however, he rejected Clements' idea that a given community was a repeatable entity that occurred whenever a given set of environmental conditions occurred. In his 1926 paper, Gleason argued that two factors came into play to make each community distinct from every other. The first was the independent nature of plant species:

[2] The same contrasting viewpoints were developed in early 20th century Europe. There, the Russian Sukatchew and the Frenchman Braun-Blanquet argued that plant communities were repeatable entities, whereas the individualistic view was developed by the Russian, Ramensky, and the Frenchman, Lenoble.

> ". . . every species of plant is a law unto itself, the distribution of which in space depends upon its individual peculiarities of migration and environmental requirements. . . . The behavior of the plant offers in itself no reason at all for the segregation of definite communities." (Gleason, 1926)

The second factor was randomness, by which Gleason meant that the composition of a given community was not completely predetermined by environmental factors. Instead, any number of plant species might be able to occupy a given site, but may or may not depending on whether their seeds were dispersed into disturbed areas. Gleason argued that community composition might be repeatable in regions with few species, simply because there were few alternative communities that might develop. But, as species diversity increased within a region, so did the variety of community types that might develop on a given site during the course of succession.

As with most polar issues, the "truth," at least how it is perceived today, contains some elements of both Clements and Gleason, but is adequately captured by neither. Most modern ecologists reject Clements' idea of communities as superorganisms. But, if a community is not an organism in the same sense as an oak tree or a swallowtail butterfly, neither is an individual plant, as Gleason suggests, a law unto itself. Every organism is part of and in interaction with a larger community: feeding, being fed upon, competing, cooperating, and coexisting. Moreover, although randomness is clearly a powerful force in nature, species evolve strategies to reduce the uncertainty associated with randomness and to retain a presence, or a potential presence, on a site. Biological legacies, including (but not restricted to) buried seeds, live roots from which new tops sprout, and mycorrhizal fungi, are passed from the old community to the new, shaping the new in the image of the old, as Clements suggested. Finally, despite the undeniable (and not surprising) fact that different species differ in their environmental requirements (as Gleason argued), the Clementsian view of mutual dependence among members of a community also has validity.

Through the years there has been a shift in emphasis among ecologists who study succession. Clements, Gleason, and others who followed them focused on end points—of which Clements' climax was the archetype. Today, ecologists are more concerned with the mechanisms and processes that shape community dynamics. The following two sections explore successional patterns, then consider mechanisms behind those patterns. It must be kept in mind that what follows are generalizations that may or may not hold in a given situation. As Blaise Pascal observed, "Imagination tires before nature."

### B. Stages of Succession

Particularly in severe disturbances, rapidly growing, often short-lived species with widely dispersed seeds are often the most abundant early pioneers. These nomads, as Gomez-Pompa and Vasquez-Yanes call them, are frequently herbs but may also include some tree species. Nomads seldom dominate a site for long periods, generally being quickly replaced by shrubs and trees that were present in one form or another in the predisturbance community. These include plants that grow from buried seeds, sprout from roots, or that survive the disturbance unharmed. In many cases, early successional plants are intolerant of shade, hence their seedlings do not survive and grow beneath an established canopy; in the absence of disturbance the early successional community generally does not perpetuate itself.

Early successional stages are relatively short in both time and stature of the dominant vegetation, whereas the intermediate stages are increasingly lengthy and taller, culminating in Clements' climax community, one that, in theory, persists indefinitely, but in fact rarely does. (In fact, coniferous forests become highly susceptible to crown fires when the mid-successional trees are senescing and late successional conifers begin to grow taller.) Note that a successional sequence refers to changes in *dominance,* or the degree to which the site is occupied by canopies and roots, not to the presence or absence of a given life form. Shrubs and trees that sprout from roots or grow from buried seed, legacies of the old forest, are likely to be present during the earliest stages of secondary succession (unless the disturbance is so catastrophic that surface soils are lost). Many species that come to dominate forests in late successional stages are also able to pioneer newly disturbed sites. Western hemlock, a shade-tolerant, late seral tree of the Pacific Northwest, has light, readily dispersed seed and frequently pioneers clear-cuts in relatively moist habitats. In New England forests, both early and late-seral tree species establish soon after disturbance, with fast-growing pin cherry dominating early in succession and slower growing species emerging to dominate later. In general, throughout any given successional sequence, the community at any one point in time is likely to contain not only the dominants, but seedlings of future dominants. The period prior to complete canopy closure is a time of great species richness, containing mixtures of herbs, shrubs, and tree seedlings, which in turn create diverse habitats for animals.

Most, if not all, forest communities include species that are adapted to recover quickly from disturbances. For example, in the black spruce forests common to the interior of Alaska, burned sites are quickly occupied by sprouting grasses, shrubs, and small trees (willows, birch) and by numerous black spruce seedlings originating from seed stored in semiserotinous cones. In the Pacific Northwest, wildfires create a mosaic of species that either survive the fire through heat-resistant bark or regenerate through sprouting or from buried seed. The initial colonizers in areas where the overstory is killed are generally nomads that persist for a few months to a few years before being succeeded by former residents growing from sprouts or buried seed. [*See* Forest Stand Regeneration, Natural and Artificial.]

Disturbance severity acts as a filter on the available species pool, modifying the composition of the early successional community. Disturbances that preserve soil but destroy aboveground parts favor sprouters or species with seed stored in the soil. Species with serotinous cones are generally an exception, but not always. In the Rocky Mountains, very intense burns may consume the serotinous cones of lodgepole pine, favoring sprouting aspen. This may also occur in black spruce/aspen stands that occupy certain habitats in interior Alaska. On the other hand, fires that generate excessive heat in the soil either delay recovery by sprouting plants or kill the roots so no sprouting is possible. Tropical trees, many of which sprout prolifically following windthrow, are particularly vulnerable to roots being killed in fire. Without the sprouters, the composition of the early successional community depends on seeds stored in the soil or input to the site following disturbance. Disturbances severe enough to destroy soil (e.g., landslides) generally initiate primary succession, and sites must be colonized by seeds from elsewhere. However, biological legacies have a surprising ability to persist and shape early successional communities. Foreign disturbances—those for which species that comprise the system have no adaptations (e.g., herbicides, fire in some forests)—may eliminate biological legacies and open the site to colonization by nomads.

Along with disturbance severity, timing of a disturbance also filters the available species pool. Composition of the early successional community often depends on coincidences between the time at which a disturbance occurs and the natural rhythms of species within the colonizing pool. Three different time scales are important: time of year, the year itself, and the interval between disturbances. The first two time scales relate to coincidence between disturbance and the availability of propagules (seeds or sprouts) whereas the third relates to life span.

Plant species vary in their seasonal rhythms, hence a disturbance occurring at one time of the year may select for quite a different set of early successional plants than one occurring at another time of year. The ability of some species to sprout following the destruction of aboveground parts varies seasonally: destroyed at one time of year these recover vigorously, at another time of year not at all. Seeds of different species mature, hence are available for colonization, at different times of the year. For example, in interior Alaska most wildfires occur during June and July, coincident with the ripening and dispersal of aspen and balsam poplar seed, but before seeds of white spruce and paper birch ripen. In the tropics, where yearly rhythms of the biota are not constrained by low temperature for part of the year, species vary widely in phenology. Trees that produce animal-dispersed seeds tend to fruit year-round, while those producing wind-dispersed seeds fruit only during the dry season and, moreover, seeds are dispersed only on days with relatively low humidity. Other factors being equal, a disturbance coinciding with this dispersal would probably result in a relatively high proportion of wind-dispersed species in the pioneering community, whereas one that did not coincide would have more animal-dispersed species.

Many trees produce seeds at intervals of several years. In Alaska, for example, birch produces heavy seed crops at least once every 4 years, but white spruce only once every 10 to 12 years. Hence the capacity of a given species to deliver seed to a newly disturbed site depends, among other things, on whether the disturbance coincides with a good seed year. Such a coincidence is not totally random, however, because weather conditions that increase the probability of wildfire, such as hot, dry springs, also trigger seed production by white spruce.

Finally, the interval between disturbances can also influence the composition of the pioneering community. For example, an 80-year-old western hemlock forest in northern Montana was established following a fire without being preceded by lodgepole pine, which is the usual early successional tree of that area. In that case the interval between fires had been sufficiently long that shade-intolerant lodgepole pine had dropped out of the forest, leaving only hemlock to colonize. (Unlike many late-successional trees, hemlock produces light, widely dispersed seed, enabling them to play the role of the pioneer.)

### *The Steady-State Forest*

Once the forest becomes dominated by species that reproduce successfully under their own canopy or in gaps created by the death of old trees, community composition may become relatively stable. This is Clements' climax, and is also called the *steady-state* or *equilibrium* community. The dominant trees are called *climax species*. The steady-state forest is not static, but is composed of a dynamic mosaic of patches created by old trees dying and young trees filling the gaps that are left—a condition termed "shifting-mosaic steady state." Species composition may or may not change over time at any one point on the ground, but on the larger scale of the landscape it will remain constant. The age structure of the forest changes from the relatively even-aged condition of earlier successional stages to the many-aged condition. Biomass accumulation levels out to zero and total biomass remains relatively constant. When discussing steady states, it is important to distinguish between forests and forested landscapes. In theory, all forests attain a steady state as the end point of succession. In fact, however, many do not, or if they do, they do not stay there long because disturbance is always part of the scene. On the other hand, forested landscapes may maintain a relatively stable distribution of stands in different successional stages (the shifting mosaic), even when disturbances are frequent. The landscape area within such a relative steady state depends on the average scale of the disturbance: a regime dominated by small-scale disturbances, such as minor windthrow or low intensity fire, produces a steady state within relatively small areas. This seems to be the case in moist tropical forests that are not on hurricane tracks, mixed conifer hardwood forests of eastern and central North America, dry ponderosa pine forests of interior western North America, and dry miombo woodlands of southern Africa. In each of these, the steady state is characterized by frequent minor disturbances, such as the death of old trees, minor windthrow, or ground fires, that create space within which young trees can establish and grow. The steady state of both ponderosa pine forests and miombo woodland depends on frequent ground fires, in the absence of which new species invade and the character of the forest changes. When the disturbance regime is characterized by large events (e.g., high intensity crown fires), a steady-state may be found only within very large landscapes. Because the disturbance regime of many forest types is characterized by relatively frequent small-scale events punctuated by infrequent large-scale events, the scale at which constancy is found on the landscape varies over time. F. G. Hall and colleagues used remote imagery (Landsat) to document the nature of the shifting mosaic in the 900-$km^2$ area of the Superior National Forest in northeastern Minnesota, a little less than one-half of which was in wilderness (no logging allowed). Using spectral characteristics (light reflected from the surface) of both visible light and near-infrared, different tree species were distinguished along with the degree of crown closure. This information was used to identify five successional stages (from early to late) from Landsat photos: clearings, areas of regeneration (cleared areas covered by low shrubs and young trees), mature stands of deciduous trees, mixed deciduous–conifer stands, and closed canopy pure conifer stands. Photographs from 1983 were then compared to 1973 photos to determine the rate of change from one type to another. The landscapes of both the wilderness and nonwilderness were very dynamic. Over the 10-year period that was studied, about one-half of the stands changed from one successional stage to another. Despite the dynamism at the stand level, however, the proportion of different successional stages across the landscape remained relatively constant.

### C. Mechanisms of Succession

In brief, successional trajectories are influenced by two primary factors: which species colonize first (the so-called priority effects) and how the initial dominants (and each succeeding wave of dominants) influence what follows. Numerous factors combine to determine which species (or set of species) initially establish, although biological legacies provide threads of continuity that facilitate the recovery of species with prior history on a site. According to J. H. Lawton and V. K. Brown, "There is growing evidence for priority effects in community assembly (either species A or species B can establish in the habitat; which one actually does depends upon which arrives first). Priority effects may then lock community development into alternative pathways, generating different end points or alternative states." Through what mechanisms do the earliest arrivals shape subsequent patterns of community development?

In an influential paper in 1977, J. H. Connell and R. O. Slatyer proposed three ways in which a plant might influence a potential successor: *facilitation, tolerance,* and *inhibition*. These terms refer to the effect of environmental modification by early colonizers on the subsequent establishment of late successional species. In the facilitation model, only early successional species are able to colonize disturbed sites, and these modify the environment in such a way that it becomes *less* suitable for their own species and *more* suitable for others. In the tolerance and inhibition models, disturbed sites are potentially colonized by both early and late successional species (i.e., there is nothing inherent in the newly disturbed environment to prevent colonization by late successional species), which then modify the environment in such a way that new individuals of early successional species are unable to become established, with late successional species either unaffected by these modifications (tolerance model) or also inhibited (inhibition model).

J. H. Connell and R. O. Slatyer provided a valuable framework for thinking about species interactions during succession. However, except in a few cases, successional dynamics rarely fit neatly into one or another of the categories they proposed. The interaction between individuals of two different plant species during succession often contains elements of both inhibition (e.g., competition for resources) and facilitation, with the net effect varying depending on factors such as soil fertility, climate, and the relative stocking density of each species. The net effect may also vary over time, a relationship dominated by competition at one stage of stand development becoming predominantly facilitative later on or vice versa. Moreover, successional dynamics can rarely be reduced to interactions between two species: the nature of the relationship between any two individuals is conditioned by numerous other plants, animals, and microbes.

With this background in mind, the mechanisms of interaction will be explored in more detail, beginning with facilitation, then moving to inhibition, and closing with higher-order interactions.

### D. Facilitation

Primary successions probably always involve facilitation of one kind or another, most often related to soil building and nutrient accumulation. Soils do not develop without plants, and pioneers facilitate establishment of their successors by weathering rocks, accumulating nutrients and carbon, and providing the energy base that allows populations of soil microbes and animals to establish and grow. Facilitation commonly occurs during secondary as well as primary succession. For example, large amounts of nitrogen can be lost from ecosystems during fire, and nodulated plants are often among the earliest pioneers on burned sites. These plants are often rapid growers that initially compete with other trees and shrubs for water and nutrients, and foresters have often viewed these as undesirable competitors with crop trees. However, in the long run they facilitate the growth of other plants in the ecosystem by restoring soil fertility.

Early successional plants may create certain structures or habitats that facilitate the establishment of later successional species. Providing cover or perches for animals that disperse seeds is one way in which this happens: fruits are eaten and the seed is defecated and nuts are dispersed through the caching behavior of animals. Seeds buried by birds (particularly nutcrackers and jays) and mammals (e.g., squirrels, bears) are a primary avenue of establishment for heavy-seeded species such as oaks, beech, hickories, and some pines (whitebark, limber). It has been estimated that, in a good seed year, a single Clark's nutcracker may cache 100,000 whitebark and limber pine seeds and that jays are able to disperse 150,000 nuts from a beech woodlot. Since animals are exposed to predators when in the open, they frequently constrain their movements, including seed caching, to areas with cover. Jays, for example, avoid open fields when burying acorns. Hence the cover provided by early successional trees and shrubs facilitates the seed dispersal of late successional trees.

One of the more common examples of facilitation during early succession, at least in some environments, is the so-called "island effect," in which tree or shrub seedlings establish most readily in the vicinity of an already established tree or shrub (the *nurse plant*). (This should not be confused with facilitation by nodulated plants discussed earlier; nurse plants may or may not be nodulated.) Tree seedlings invading savannas in Belize, for instance, establish preferentially near other trees, and the same is true for tree seedlings establishing in savannas in the Philippines and in abandoned pastures in Amazonia. The island effect has been noted often in both forests and deserts of western North America. Both ponderosa and pinyon pines require nurse plants to establish on certain droughty and/or frosty sites. Live oak seedlings are strongly associated with some species of woody shrubs in central California, and Douglas fir seedlings establish preferentially beneath some species of oaks in northern California. One study of natural regeneration in Oregon found nearly five times more Douglas fir seedlings beneath Pacific madrone trees than in the open. Not all trees and shrubs necessarily act as nurse plants on a given site. For example, while abundant Douglas fir seedlings establish beneath canopies of Pacific madrone and some species of oaks, none establish beneath nearby Oregon white oak stands.

Reasons for the island effect are not always clear, but there are at least three plausible mechanisms, any or all of which could be operating in a given situation. Nurse plants might (a) shelter seedlings from environmental extremes, (b) act as foci for seed inputs, and (c) provide enriched soil microsites. Shelter can significantly improve survival in droughty sites as well as in cold environments. For example, in the droughty forests of southern Oregon and northern California, shade cast by early successional hardwood trees and shrubs may reduce the water use by conifer seedlings growing beneath them (less transpiration is needed to cool leaves). On high elevation or other frosty sites, nurse plants provide a relatively warm nighttime environment by preventing excessive loss of radiant heat. As discussed earlier, established trees and shrubs act as foci for seed inputs because they attract birds, and birds often leave behind seeds. Over one 6-month period in an abandoned pasture in Amazonia, nearly 400 times more tree seeds were dispersed beneath *Solanum crinitum* trees colonizing the pasture than fell in the open. Eighteen different tree species were represented in the seed rain beneath *Solanum*.

Plant islands may also facilitate the establishment of later-arriving species during secondary succession through soil chemistry, biology, or structure.

This is in some ways similar to, but in other ways quite different than, the facilitation that occurs through soil building during primary succession. Pioneers during secondary successions may restore soil carbon, nutrients, and organisms lost during disturbance; however, what is probably more common following natural disturbances is that the most resiliant members of the former community, species that are able to sprout from roots or grow from buried seeds, prevent soil degradation in the first place by preventing excessive nutrient loss and by maintaining critical elements of soil biology and structure.

Ecologists have known for some time that early successional plants prevent excessive nutrient loss after disturbance. A growing body of evidence suggests that islands of pioneering shrubs and trees, especially those that are legacies of the previous forest, also stabilize soil microbes that facilitate the reestablishment of later-arriving plants. The survival and growth of tree seedlings establishing in disturbed areas may depend on their ability to quickly reestablish links with their belowground microbial partners, especially on infertile soils or in climatically stressful environments. That would seem not to be a problem for plants that can sprout from roots because they presumably never lose contact with belowground partners. However, it could be a problem for trees that reestablish slowly because their seeds must be dispersed to disturbed sites from elsewhere. What happens to their microbial partners during the period the host plant is absent? One possibility is that the microbe simply goes dormant until its host plant reestablishes. Another possibility is that the microbe is flexible enough to utilize other food sources, perhaps by soil organic matter or a pioneering plant. The latter seems to be the case in tropical and at least some temperate forests. The most common mycorrhiza-forming fungal species in tropical forests are widely shared among different tree species, as are some of the fungi that form mycorrhizas with temperate trees and shrubs. An early successional plant that supports microbes needed by later-arriving plants effectively facilitates the reestablishment of the latter, although it may also compete with the late arrival for light, water, and nutrients. In southwest Oregon and northwest California, Douglas-fir and various hardwood trees and shrubs share some of the same mycorrhizal fungi. The hardwoods sprout from roots following disturbances whereas Douglas-fir must reestablish from seeds. Douglas-fir seedlings tend to survive and grow better in the proximity of at least some hardwood species than in the open; controlled studies indicate that the phenomenon is related to soil biology. The Douglas-fir are believed to "plug into" the network of hyphae extending from the hardwood mycorrhizae, which allows seedlings to rapidly develop their own water- and nutrient-gathering capacity. But the phenomenon is complex and appears to involve other factors as well, including nitrogen-fixing bacteria and perhaps bacteria that stimulate root tip production by seedlings. Nutrients also cycle faster in soils near hardwoods than in the open, a reflection of greater biological activity. The evidence amassed so far suggests that hardwoods of this area act as selective filters of soil biology, retaining beneficial soil microbes and inhibiting detrimental ones. Studies of unreforested clear-cuts have found that the inability of seedlings to establish may be related to the buildup of certain types of microbes that inhibit seedlings and their mycorrhizal fungi. In at least one instance, soils near sprouting hardwood islands within a clear-cut were relatively free of deleterious microbes. At present, ecologists have only a rudimentary understanding of the complex relationships among plants and soil organisms, and how these influence successional dynamics.

### I. Inhibition

Succession always implies a change in the availability of resources. The deepening canopy shades and alters the microclimate within a stand, favoring shade-tolerant species over those that need high light levels to establish. Nutrients become increasingly tied up in biomass. In what has been called the resource ratio hypothesis, David Tilman argues that relative change in the availability of different resources is generally an important mechanism for species change during both primary and secondary successions. In many cases early successional species create the conditions that inhibit their own progeny from succeeding them. For instance, many sites that had been clear-cut in the Oregon Cascades during the 1960s and early 1970s became dominated by nitrogen-fixing shrubs in the genus *Ceanothus*. Foresters and some scientists were concerned that ceanothus might exclude trees for many decades. Eventually, however, intermixed conifers began to grow above the shrub canopy, and most of those sites are now dominated by Douglas-fir. Inhibition was not permanent in these cases because of the inherently different growth rhythms of the species: ceanothus and alder grow fast and reach maximum heights at a relatively young age, whereas Douglas-fir grows more slowly but maintains growth, eventually becoming taller than the others.

On the other hand, inhibition of one species by another can be relatively long lasting when circumstances permit an aggressive early successional plant to form a dense, monospecific cover that effectively excludes other species. The pioneer could exclude other species by preempting site resources so fully and quickly that no other species can establish, or it might allelopathically inhibit other plants and/or their mycorrhizal fungi. In eastern North America, for example, the failure of trees to reestablish in old clear-cuts, abandoned agricultural fields, and areas burned by wildfire has been related to the allelopathic inhibition of tree seedlings by herbs, ferns, and grasses. In the Sierra Nevada of California, the herbaceous perennial *Wyethia mollis* has spread widely in old burns and allelopathically inhibits tree regeneration. Excluding wildfires from Swedish forests resulted in the spread of a dwarf shrub (*Empetrum hermaphroditum*) that allelopathically inhibits tree regeneration. The inhibitory plants in these examples are often natives that once had been present in relatively low numbers and that were apparently triggered into a more aggressive mode by some foreign disturbance; in other words, a balance was disrupted. In Pennsylvania, tree seedlings were originally eliminated from recovering clear-cuts by forest fires and exceptionally high populations of deer. In California, overgrazing allowed the unpalatable *Wyethia* to spread at the expense of more tasty plants. In Sweden and elsewhere, excluding wildfire has shifted a balance so as to favor the spread of plants previously limited by fire.

Woody plants can have particular difficulty getting a foothold within established grass communities. In western North America, annual grasses are often deliberately sown in recently burned forests to stabilize soils. However, the grasses can completely inhibit the recovery of native shrubs, at least for several years. In Central and South America, areas cleared of forest are frequently seeded to grasses to provide cattle pasture, then abandoned after a few years because they decline in productivity. Trees have great difficulty in reinvading abandoned pastures. According to D. Nepstad *et al.* (1990):

> "Directly or indirectly, grasses present barriers to tree seedlings at every step of establishment in abandoned pastures with histories of intensive use. Seed dispersal into grass-dominated vegetation is low because grasses do not attract birds and bats that eat fleshy fruits of forest trees. Grasses provide food and shelter for large populations of rodents that consume tree seeds and seedlings . . . . The dense root systems of grasses produce severe soil moisture deficits in the dry season and compete for available soil nutrients. Finally, grasses favor fire so that tree seedlings that do surmount the numerous obstacles to establishment are periodically burned."

If a pioneer successfully excludes other plants and is also capable of reproducing under its own canopy, it can, in theory at least, hold a site indefinitely. Such is the case with the Pacific coast shrub salmonberry *Rubus spectabilis,* which produces pure stands of 30,000 or more stems per ha following disturbance. By sprouting from basal buds and rhizomes, salmonberry quickly replaces old stems with new ones, thereby creating unevened periods. Once a pure stand (i.e., without intermixed tree seedlings) attains a sufficiently high density, plants such as salmonberry are likely to persist until weakened by pathogens or insects or until confronted with a disturbance to which they are not adapted. Because of their relative simplicity, species monocultures may be especially vulnerable to pests and pathogens; however, this remains to be seen.

### 2. *Higher-Order Interactions: Those Involving More Than Two Species*

In the past, ecologists thought of succession as a process driven primarily by plant–plant interactions. Studies have now shown that many other elements of the ecosystem can either directly or indirectly influence successional trajectories, including particular microbes, animals, and abiotic environmental factors. Community interactions invariably involve several species, not just two, hence complex relationships may develop during succession. In the North Carolina Piedmont, for example, early successional pines inhibit the establishment of fast-growing hardwoods such as Liquidambar, which, in the long term, facilitates the entry of slower-growing oaks and hickories. Animals often modulate plant–plant interactions during succession. In the Pacific Northwest, browsing elk and deer prefer hardwood shrubs over conifer seedlings, accelerating succession from shrubs to trees. This was demonstrated by a study that excluded elk and deer from a portion of a clear-cut in western Washington. In areas accessible to elk and deer there were 8.7 woody stems per $m^2$, one-half of which were Douglas-fir, whereas the area with no animals present had 19 woody stems per $m^2$, 11 of which were salmonberry, a particularly aggressive competitor with conifer seedlings. In eastern Oregon, early successional communities in which deer, elk, and cattle are excluded are dominated by *Ceanothus* sp. Where those animals are present, however, browsing limits the height growth of the shrub, and sites are dominated relatively quickly by conifers. On the other hand, where trees are favored food or, as is more often the case, when excessively high animal numbers result in a shortage of preferred food, animals will definitely retard succession and even jeopardize the existence of trees on a site.

Soil organisms play an important but poorly understood role in shaping the composition of successional plant communities. The belowground foodweb is critically important to the nutrient cycle, especially invertebrates and protists that graze microbes. Soils contain microbes such as mycorrhizal fungi and some types of bacteria that directly benefit plants, and microbes that are pathogenic or otherwise inhibitory toward plants. Particular microbial species within those broad groups seldom affect all plant species equally, e.g., a given species of mycorrhizal fungus may benefit some plants and not others; the same is true for the detrimental effect of pathogens. In some instances, a microbe that stimulates one plant species is pathogenic toward another. Because of the selectivity of their action, the composition of the microbial community on a site feeds back to affect the relative success of different plant species. The relationship is reciprocal because a microbe that depends on living plants for food—whether it is a mycorrhizal fungus or a pathogen—presumably cannot persist indefinitely in the absence of a host plant. As a result, feedback relationships develop between composition of the plant community and composition of the soil microbial community.

The availability of beneficial microbes in some instances determines whether their host plants can establish on a site or how well they grow once established. Mycorrhiza formation by plants may be reduced where host plants have been absent too long, on highly disturbed areas (e.g., where erosion is severe), and in some instances even with rather mild soil disturbance. Inoculating seedlings either with mycorrhizal fungi or with forest soils or litter has significantly improved the survival of trees planted in mine spoils, abandoned fields, old clear-cuts, and natural grasslands. Research in Canada indicates that inoculating with certain types of rhizosphere bacteria significantly improves the growth of outplanted tree seedlings. In one case, forest soil transfers enhanced tree seedling establishment in clear-cuts through reintroducing invertebrates and protists that are keystones in the nutrient cycle.

Inhibitory soil microbes include well-known pathogens, such as root rots and the so-called "damping-off" fungi, and less well-known groups, sometimes called "exopathogens," that can have sublethal inhibitory effects on plants and/or mycorrhizal fungi. Actinomycetes, a form of filamentous bacteria, have been implicated in reforestation failures in the Pacific Northwest. *Streptomyces,* from which the antibiotic streptomycin is derived, is a genetically diverse soil actinomycete that has complex effects on other organisms. Depending on the isolate, *Streptomyces* allelopathically inhibit plants, bacteria, and/or plant pathogens, and may either inhibit or stimulate mycorrhizal fungi. *Streptomyces* have been found to be higher in soils of unreforested clear-cuts than in forest soils and, within clear-cuts, higher in soils between islands of sprouting trees and shrubs than in soils beneath the islands.

Some of the more interesting research questions relating to successional dynamics relate to the role of the belowground community. What triggers the buildup of inhibitory microbes in some disturbed areas and how widespread is that phenomenon? How long can mycorrhizal fungi or beneficial rhizosphere bacteria persist in the absence of host plants? How does the composition of the early successional community influence the composition of the soil microbial community and how does that in turn influence the successional trajectory?

### E. Threads of Continuity: Legacies and Guilds

Scientists studying the recovery of plants and animals following the eruption of Mt. St. Helens in May 1980, found some surprises. Quoting from J. F. Franklin *et al:*

> "Successional theory traditionally emphasizes invading organisms or immigrants . . . but this script for ecosystem recovery could be played out at only a few sites (at Mt. St. Helens), as surviving organisms over most of the landscape provided a strong and widespread biological legacy from the preeruption ecosystem. In fact, essentially no posteruption environment outside the crater was completely free of preeruption biological influences, although there were substantial differences in the amounts of living and dead organic material that persisted."

Within 3 years after the eruption, 230 plant species—90% of those in preeruption communities—had been found within the area affected by the blast deposit and mudflows.

> "Plant and animal species that live belowground or that have reproductive structures belowground were the most likely survivors . . . Fossorial mammals, such as the pocket gopher (*Thomomys talpoides*), and subterranean and log-dwelling invertebrates, such as ants, survived in the areas of deepest deposits. The most obvious of surviving plants were 'weedy' species such as common firewood (*Epilobium angustifolium*), thistle (*Cirsium* sp.), pearly everlasting (*Anaphalis margaritacea*), various species of blackberry (*Rubus* sp.), and bracken fern (*Pteridium aquilinum*). These plants typically have perennating structures belowground and display vigorous shoot growth which can penetrate overlying deposits." (Franklin *et al.*, 1985)

Webster defines legacy as "anything handed down from . . . an ancestor." In an ecological context, legacies are anything handed down from a predisturbance ecosystem, including:

- surviving propagules and organisms, such as buried seeds, seeds stored in serotinous cones, surviving roots and basal buds, mycorrhizal fungi and other soil microbes, invertebrates, and mammals;
- dead wood; and
- certain aspects of soil chemistry and structure, such as soil organic matter, large soil aggregates, pH, and nutrient balances.

Most, if not all, legacies probably influence the successional trajectory of the recovering system to one degree or another. That is clearly the case with surviving plant propagules, which directly affect composition of the early successional community. Other legacies may shape successional patterns in more subtle ways or perhaps not at all—this is a relatively new area of ecology that needs much more research. Despite the uncertainties, however, a variety of plausible mechanisms exist through which legacies might influence succession.

### *1. Soil Biology*

As already discussed, the composition of the soil biological community following disturbance is a legacy that potentially influences the relative success of different plant species during succession.

### *2. Dead Wood*

Dead wood has the potential to influence system recovery in several ways. Standing dead snags mitigate environmental extremes within disturbed areas by shading and preventing excessive heat loss at night. Down logs within forests are centers of biological activity, including not only organisms of decay, but also roots, mycorrhizal hyphae, nitrogen-fixing bacteria, amphibians, and small mammals. As Franklin *et al.* noted at Mt. St. Helens, logs provide protective cover for their inhabitants during catastrophic disturbances. After disturbance, down logs reduce erosion by acting as physical barriers to soil movement and provide cover for small mammals that disseminate mycorrhizal spores from intact forest into the disturbed area. The sponge-like water-holding capacity of old decaying logs helps seedlings that are rooted in them survive drought.

### *3. Soil Aggregates and Soil Organic Matter*

Plants and associated microbes literally glue minerals together to form soil aggregates, which are intimate mixtures of minerals, organisms, and nutrients. These aggregates are essentially little packages of mycorrhizal propagules, other microbes, and nutrients that are passed from the old forest to the new. Soil organic matter in general, whether contained in aggregates or not (most is), provides a legacy of nutrients for the new stand. Depending on its origin and stage of decay, soil organic matter can either stimulate or inhibit plant pathogens.

### *4. Soil Chemistry*

Different plant species may affect soil chemistry quite differently: by the particular array of nutrients they accumulate, their effect on soil acidity, or allelochemicals they release. To the degree these chemical imprints persist after the plant is gone constitute legacies which, in theory at least, could influence the composition of the early successional community.

Legacies interact with one another, creating chains of direct and indirect influence. At Mt. St. Helens, for example, pocket gophers, which survived below ground, facilitated the establishment of some plant species by digging up soil buried beneath the ash. The exposed soil provided establishing plants with nutrients and mycorrhizal spores, and its organic matter retained water during drought. As discussed previously, sprouting plants, and those growing from buried seeds, often become foci for the recovery of other plants. Whatever the reason, whether providing shelter, perchs for birds that disseminate seeds, or food for mycorrhizal fungi, pioneers that sprout from roots or buried seed constitute legacies that influence the recovery of other species within the system. The legacies provided by pioneers do not necessarily affect other species uniformly, hence can shape successional trajectories by favoring the establishment of some species over others.

One hypothesis holds that species within a given community form into guilds based on common interests in mycorrhizal fungi and perhaps other beneficial soil organisms. According to this view, early colonizers during secondary succession facilitate subsequent colonization by members of the same guild by providing a legacy of mycorrhizal fungi (and perhaps other beneficial soil organisms) and inhibit colonization by members of other guilds because they provide no such legacy. The guild concept may be extended to include animals that are tied into a relational network with plants and their mycorrhizal fungi. For example, truffles, the belowground fruiting bodies of some species of mycorrhizal fungi, are the primary food source for some small mammals, and the small mammals spread spores of the fungi. In forests of the Pacific Northwest, the primary diet of the endangered northern spotted owl is the northern flying squirrel, whose primary diet is truffles. Hence the long-term welfare of both flying squirrels and spotted owls depends on successional trajectories that lead back to trees that support truffle-producing mycorrhizal fungi.

## G. Summary of Successional Mechanisms

To summarize this section, patterns of species establishment and changing dominance during succession are likely to result from a mixture of random factors and complex interactions among plants, animals, and microbes. To a certain degree, composition of a pioneer community is determined by which species arrive first, which is, in turn, a function of interactions between the nature and timing of disturbance on the one hand and the early successional environment, which filters colonizers according to their adaptations, on the other. Biological legacies facilitate recovery of the system and act to shape the new community in the image of the old. Dominance changes over time in part because developmental patterns differ among species—some are fast growing and short-lived, others are slow growing and long lived, and yet others are somewhere in between—and in part because, for various reasons, some species establish more successfully in a preexisting community than in a newly disturbed site. Interactions among plant species during succession often include elements of both facilitation and inhibition, and are influenced by complex interactions among climate, resource availability, and many nonplant species such as animals, pathogens, mycorrhizal fungi, and other soil microbes. Moreover, the nature of interaction may change with time: inhibition becoming facilitation or facilitation becoming inhibition. As a result, one must proceed cautiously when judging interactions among plants during succession.

## Glossary

**Competition** Interaction between individuals of the same species (intraspecific competition) or between different species (interspecific competition) at the same trophic level, in which the growth and survival of one or all species or individuals are affected adversely. The competitive mechanism may be direct (active), as in allelopathy and mutual inhibition, or indirect, as when a common resource is scarce. Competition leads either to the replacement of one species by another that has a competitive advantage or to the modification of the interacting species by selective adaptation (whereby competition is minimized by small behavioral differences, e.g., in feeding patterns). Competition thus favors the separation of closely related or otherwise similar species. Separation may be achieved spatially, temporally, or ecologically (i.e., by adaptations in behavior, morphology, etc.). The tendency of species to separate in this way is known as the competitive exclusion or Gause principle.

**Niche (ecological)** Functional position of an organism in its environment, comprising the habitat in which the organism lives, the periods of time during which it occurs and is active there, and the resources it obtains there.

**Seral stage** A phase in the sequential development of a climax community.

**Sere** Characteristic sequence of developmental stages occurring in plant succession.

**Succession** Sequential change in vegetation and the animals associated with it, either in response to an environmental change or induced by the intrinsic properties of the organisms themselves. Classically, the term refers to the colonization of a new physical environment by a series of vegetation communities until a final equilibrium state, the climax, is achieved. The presence of the colonizers, the pioneer plant species, modifies the environment so that new species can join or replace the initial colonizers. Changes are rapid at first but slow to a more or less imperceptible rate at the climax stage, composed of climax plant species. The term applies to animals (especially to sessile animals in aquatic ecosystems) as well as to plants. The characteristic sequence of developmental stages (i.e., nudation, migration, ecesis, competition, reaction, and stabilization) is termed a sere.

**Trophic level** A step in the transfer of food or energy within a chain. There may be several trophic levels within a system, for example, producers (autotrophs), primary consumers (herbivores), and secondary consumers (carnivores); further carnivores may form fourth and fifth levels. There are rarely more than five levels since usually by this stage the amount of food or energy is greatly reduced.

## Bibliography

Allaby, M., ed. (1994). "The Concise Oxford Dictionary of Ecology." Oxford: Oxford Univ. Press.

Amaranthus, M. P., and Perry, D. A. (1989). Interaction effects of vegetation type and Pacific madrone soil inocula on survival, growth, and mycorrhiza formation of Douglas-fir. *Can. J. For. Res.* **19,** 550–556.

Amaranthus, M. P., Li, C.-Y., and Perry, D. A. (1990). Influence of vegetation type and madrone soil inoculum on associative nitrogen fixation in Douglas-fir rhizospheres. *Can. J. For. Res.* **20,** 368–371.

Amaranthus, M. P., Trappe, J. M., and Perry, D. A. (1993). Soil moisture, native revegetation, and *Pinus lambertiana* seedling growth, and mycorrhiza formation following wildfire and grass seeding. *Restor. Ecol.* Sept. 188–195.

Atsatt, P. R., and O'Dowd, D. J. (1976). Plant defense guilds. *Science* **193,** 24–29.

Borchers, J. G., and Perry, D. A. (1992). The influence of soil texture and aggregation on carbon and nitrogen dynamics in southwest Oregon forests and clearcuts. *Can. J. For. Res.* **22,** 298–305.

Borchers, S. L., and Perry, D. A. (1990). Growth and ectomycorrhiza formation of Douglas-fir seedlings grown in soils collected at different distances from pioneering hardwoods in southwest Oregon. *Can. J. For. Res.* **20,** 712–721.

Bossema, I. (1979). Jays and oaks: An eco-ethological study of a symbiosis. *Behaviour* **70,** 1–117.

Bourliere, F. (1983). Animal specie diveristy in tropical forests. *In* "Tropical Rain Forest Ecosystems" (F. B. Golley, ed.), pp. 77–91. New York: Elsevier.

Chanway, C. P., Turkington, R., and Holl, F. B. (1991). Ecological implications of specificity between plants and rhizosphere micro-organisms. *Adv. Ecol. Res.* **21,** 121–169.

Clements, F. E. (1916). "Plant Succession, an Analysis of the Development of Vegetation," Pub. 242, pp. 1–512. Washington: Carnegie Institute.

Connell, J. H., and Slatyer, R. O. (1977). Mechanisms of succession in natural communities and their role in community stability and organization. *Am. Nat.* **111,** 1119–1144.

Denslow, J. S., and Gomez Diaz, A. E. (1990). Seed rain to tree-fall gaps in a Neotropical rain forest. *Can. J. For. Res.* **20,** 642–648.

Ehrenfeld, J. G. (1990). Dynamics and processes of barrier island vegetation. *Rev. Aquat. Sci.* **2**(3,4), 437–480.

Finegan, B. (1984). Forest succession. *Nature* **312,** 109–114.

Fleming, T. H., Breitwisch, R., and Whitesides, G. H. (1987). Patterns of tropical vertebrate frugivore diversity. *Annu. Rev. Ecol. Syst.* **18,** 111–136.

Franklin, J. F., MacMahon, J. A., Swanson, F. J., and Sedell, J. R. (1985). Ecosystem responses to the eruption of Mount St. Helens. *Nat. Geogr. Res.* Spring, 198–216.

Gleason, H. A. (1926). The individualistic concept of the plant association. *Bull. Torrey Bot. Club* **53,** 7–26.

Gomez-Pompa, A., and Vazquez-Yanes, C. (1981). Successional studies of a rain forest in Mexico. *In* "Forest Succession, Concepts and Applications." (D. C. West, H. H. Shugart, and D. B. Botkin, eds.), pp. 246–266. New York: Springer-Verlag.

Hairston, N. G., Smith, F. E., and Slobodkin, L. B. (1960). Community structure, population control, and competition. *Am. Nat.* **94,** 421–425.

Hall, F. G., Botkin, D. B., and Strebel, D. E. (1991). Large-scale patterns of forest succession as determined by remote sensing. *Ecology* **72,** 628–640.

Harvey, A. E., Jurgensen, M. F., Larsen, M. J., and Graham, R. T. (1987). Relationships among soil microsite, ectomycorrhizae, and natural conifer regeneration of old growth forests of western Montana. *Can. J. For. Res.* **17,** 58–62.

Hibbs, D. E. (1983). Forty years of forest succession in central New England. *Ecology* **64,** 1394–1401.

Howe, H. F., and Smallwood, J. (1982). Ecology of seed dispersal. *Annu. Rev. Ecol. Syst.* **13,** 201–228.

Hunter, A. F., and Aarssen, L. W. (1988). Plants helping plants. *BioScience* **38**(1), 34–40.

Janos, D. P. (1980). Vesicular-arbuscular mycorrhizae affect lowland tropical rain forest plant growth. *Ecology* **61,** 151–162.

Kauffman, J. B. (1991). Survival by sprouting following fire in tropical forests of the eastern Amazon. *Biotropica* **22,** 219–224.

Lanner, R. M. (1985). Some attributes of nut-bearing trees of temperate forest origin. *In* "Attributes of Trees as Crop Plants" (M. G. R. Cannell and J. E. Jackson, eds.), pp. 426–437. England: Institute of Terrestrial Ecology.

Lawton, J. H., and Strong, D. R., Jr. (1981). Community patterns and competition in folivorous insects. *Am. Nat.* **118,** 317–338.

Lawton, J. H., and Brown, V. K. (1992). Redundancy in ecosystems. *In* "Biodiversity and Ecosystem Function" (E.-D. Shulze and H. A. Mooney, eds.), Springer-Verlag.

MacArthur, R. H. (1964). Environmental Factors Affecting Bird Species Diversity. *American Naturalist* **98,** 387–397. 1972. Geographical Ecology. Harper & Rowe, New York. 269 pp.

Marx, D. H. (1975). Mycorrhiza and establishment of trees on strip-mined land. *Ohio J. Sci.* **75,** 288–297.

Maser, C., Trappe, J. M., and Nussbaum, R. A. (1978). Fungal-small mammal interrelationships with emphasis on Oregon coniferous forests. *Ecology* **59,** 799–809.

Nepstad, D., Uhl, C., and Serrao, E. A. (1990). Surmounting barriers to forest regeneration in abandoned, highly degraded pastures: A case study from Paragorninas, Para, Brazil. *In* "Alternatives to Deforestation" (A. B. Anderson, ed.), pp. 215–229. New York: Columbia Univ. Press.

Peet, R. K., and Christensen, N. L. (1980). A population process. *Vegetatio* **43,** 131–140.

Perry, D. A., Margolis, H., Choquette, C., Molina, R., and Trappe, J. M. (1989b). Ectomycorrhizal mediation of competition between coniferous tree species. *New Phytol.* **112,** 501–511.

Pickett, S. T. A., Collins, S. L., and Armesto, J. J. A. (1987a). A hierarchical consideration of causes and mechanisms of succession. *Vegetatio* **69,** 109–114.

Pickett, S. T. A., Collins, S. L., and Armesto, J. J. (1987b). Models, mechanisms and pathways of succession. *Bot. Rev.* **53**(3), 335–371.

Pickett, S. T. A., and Mconnell, M. J. (1989). Changing perspectives in community dynamics: A theory of successional forces. *TREE* **4**(8), 241–245.

Schisler, D. A., and Linderman, R. G. (1989). Influence of humic-rich organic amendments to coniferous nursery soils on Douglas-fir growth, damping-off and associated soil microorganisms. *Soil Biol. Biochem.* **21**(3), 403–408.

Schoener, T. W. (1989). Food webs from the small to the large. *Ecology* **70**(6), 1559–1589.

Stenstrom, E., Ek, M., and Unestam, T. (1990). Variation in field response of *Pinus sylvestris* to nursery inoculation with four different ectomycorrhizal fungi. *Can. J. For. Res.* **20,** 1796–1803.

Stone, L., and Roberts, A. (1991). Conditions for a species to gain advantage from the presence of competitors. *Ecology* **72**(6), 1964–1972.

Strong, D. R., Simberloff, D., Abele, L. G., and Thistle, A. B., eds. (1984). "Ecological Communities: Conceptual Issues and the Evidence." Princeton: Princeton Univ. Press.

Tilman, D. (1985). The resource-ratio hypothesis of plant succession. *Am. Nat.* **125**(6), 827–852.

Tilman, D., and Downing, J. A. (1994). Biodiversity and stability in grasslands. *Nature* **367,** 363–365.

Valdes, M. (1986). Survival and growth of pines with specific ectomycorrhizae after 3 years on a highly eroded site. *Can. J. Bot.* **64,** 885–888.

# Forest Stand Regeneration, Natural and Artificial

**David M. Smith**
*Yale University*



Trees can live longer than other organisms, but not forever. Sooner or later death creates vacant growing space for the regeneration of new life. The natural vegetation of each locality includes species adapted to colonize any kind of vacancy that may come into being as a result of any kind of lethal natural disturbance that typically occurs in nature. In the silvicultural practices of forestry, these natural disturbances are simulated in varying degrees to renew forest stands. When new plants or seeds are brought to the site and established there by humans the practice is called "artificial regeneration." When natural forces put the plants there the process is called "natural regeneration," even if it is deliberately induced by human treatments. Artificial regeneration can evade some of the limitations that restrict natural processes.

## I. ROLE OF LETHAL DISTURBANCE

Vegetation abhors any vacuum in the *growing space,* the space above and below the surface that can be successfully occupied by the roots and tops of plants. The roots quickly expand to occupy all of the soil space up to the limit of its capacity to supply water and chemical nutrients. The canopy of leaves fills the aboveground space to the extent allowed by the supply of light, water, carbon dioxide, and soil nutrients, but only as fast as the supportive woody stems can grow to elevate the leaves.

New plants can start only if older plants or parts thereof are killed and leave vacant growing space above and below ground. Woody plants usually die only when their roots are killed either directly or by being starved of sugar from the leaves. Such starvation occurs when sugars produced in the leaves cease to reach the roots. This may happen because of defoliation (as by insects or browsing animals) or when the living inner bark through which sugars move is interrupted by the heat of fire or physical severance of the bark.

Regenerative disturbances vary very widely in severity and magnitude. The most severe are geologic upheavals such as landslides, volcanic eruptions, and severe erosion. Such events expose bare rock or unconsolidated parent materials of soil that are devoid of the incorporated plant remains that are called *organic matter*. Woody plants seldom become established on these until algae, mosses, lichens, grasses, or herbaceous annuals have started to build true soil. Even then the first woody plants that are established are likely to be hardy, drought-resistant, shade-intolerant shrubs; the trees that follow are likely to have similar characteristics. [*See* SOIL ECOSYSTEMS.]

The second most severe kind of lethal disturbance is fire hot enough to kill the roots of preexisting vegetation. Such fires can occur only when some previous lethal event such as insect defoliation, windthrow, or an earlier fire has left large volumes of dead wood on the ground. The heat generated by burning such debris can be hot enough to kill the roots but usually does not destroy any organic matter incorporated in the soil except in spots where large fallen logs burn for hours. Such vacancies are devoid of plants, and the new tree seedlings and herbaceous plants that quickly recolonize them are of specialized species aptly called *pioneers*. These pioneers are capable of capturing light by building foliar canopies quickly and taking advantage of the large amount of nutrients suddenly made available by the burning of organic matter. The amount of available soil water is also suddenly increased by fire because the plants that once used large amounts in *transpiration* (evaporation from plant surfaces) are dead. [*See* FIRE ECOLOGY.]

Fires of lesser severity, which burn only the leaf litter of the forest floor, can also kill trees by overheating the living inner bark and cambium completely around the bases of the trees. This kind of killing or "girdling" results from cutting off the movement of sugar from leaves to roots so that the roots die. Oftentimes, however, such fires kill only the tops of plants but not their roots and, if the species are capable of resprouting from buds, the forest is renewed without any or many truly new woody plants becoming established. [*See* RHIZOSPHERE ECOPHYSIOLOGY.]

There are many kinds of quite different lesser disturbances that do not kill so much woody vegetation. The most important are those caused by insect attack, diseases, animal damage, wind, ice, and snow. Most of these are enemies of the larger trees so they tend to kill the old forest from the top down. Fires and landslides kill from the bottom up. [*See* FOREST PATHOLOGY.]

Lethal disturbances that kill from the top down tend to favor shade-tolerant species capable of starting as *advanced regeneration* in very small vacancies that are fully or partially shaded by remaining vegetation. Advanced regeneration can persist for many years until it receives additional light and freedom from root competition in some subsequent larger disturbance that also kills from the top down.

Natural disturbances are not uniform in severity or extent. Even a single hectare seemingly laid bare by a hot fire is actually a myriad of tiny spots, each a regeneration microenvironment different from adjacent ones. The chief differences are in degree of exposure of mineral soil, in extent of freedom from root competition, and in solar radiation. These factors are the main determinants of the kind of vegetation that appears on a given vacancy a few square centimeters in area. Therefore, on the one hand, the kind of vegetation that appears after disturbances that kill only some of the trees can be quite heterogeneous. On the other hand, the more severe the disturbance the more uniform and simpler will be the species composition of the resulting vegetation.

Disturbances severe enough to cause the establishment of entirely new vegetation begin a process called *natural succession* in which one category of species replaces another. As each stage develops, more shade-tolerant plants appear beneath it and take over when the taller plants succumb. This process, in which one stage replaces another, is sometimes called *relay floristics*. However, if the reinitiating disturbance kills only the taller plants and development proceeds from the established survivors, the resulting process is called *initial floristics*. Regardless of the starting situation, either process can theoretically lead to a *climax* vegetational stage in which species composition remains the same so long as lethal disturbances are either small or not severe.

## II. SIMULATION OF NATURAL DISTURBANCE IN SILVICULTURE

The practices followed in growing forests generally imitate those of natural forest development. Attention is given not only to the initiation of new stands but also to their structure and to their development over the decades.

Natural disturbances similar to landslides, in which the parent materials of true soils are exposed, are seldom if ever deliberately simulated in silvicultural practice. However, it is often necessary to establish trees on such areas to correct the effects of severe erosion or earth moving. Such efforts constitute *afforestation* in which forests are created where they could not or did not exist before.

The simplest kind of silvicultural techniques involve simulation of the effects of fires hot enough to kill most preexisting vegetation. In such cases, fire usually does not heat the soil enough to destroy the incorporated organic matter. The new stand is then created from trees that arrive on the site afterward, either from planting nursery-grown trees or from the germination of seeds applied naturally or artificially.

The preliminary reduction of natural vegetation does not necessarily come from use of fire. If fire is used it may be necessary to create substantial amounts of fuel if fires must be hot enough to kill the roots of sprouting woody plants. Mechanical equipment can also be used to rip the roots out of the soil. Herbicides capable of killing the roots of potential competitors are often the most effective means and also the least damaging to the soil. However, they do not expose the mineral topsoil which may be necessary for the establishment of small-seeded species.

Many species, mostly angiosperms, can resprout from the roots after the tops of the trees have been killed by cutting, light fires, or animal browsing. Few conifers are capable of such sprouting. No form of forest regeneration is more dependable, but sometimes the new trees are crooked or are subject to early attack by rot fungi.

Several different cutting patterns are employed to simulate the effect of disturbances such as wind that are more likely to kill large trees than small ones. Some involve making openings of various sizes and orientations with relation to the sun. Others may be designed to increase the amount of light uniformly throughout a stand by removing scattered trees or whole strata of the crown canopy. These patterns of partial cutting are usually also designed to leave chosen trees for additional growth.

## III. MICROENVIRONMENTAL CONTROL OF FOREST REGENERATION

When vegetation is established from seeds or spores the particular species that appear depend heavily on the physical characteristics of the very small units of surface on which they land. The most important determinants are the amount and kind of solar radiation that reaches the spots and the ability of the surface medium to supply water.

The solar radiation phenomena depend chiefly on shading. If there is no shade, surface temperatures become hot by day and cold by night. This is especially true of the surfaces of dead leaf litter because of the insulating effect of the included air. This effect reduces the downward conduction of heat to the soil when the sun shines and up from the soil to offset nocturnal radiational cooling. The extremes of surface temperature are accentuated in dry climates or at high elevations where there is not much water vapor above the surface to block incoming and outgoing radiation. Ventilation by wind somewhat reduces the extremes because it causes convective movement of heat upward by day and downward by night. The ability to withstand such extremes is one of the attributes of sun-living pioneers that can cope with exposed and difficult microenvironments.

Different kinds of shading (Fig. 1) provide mitigating effects and may also determine which species are successful. It does not require much shading to blunt the extremes of surface temperature.

Many plants grow very well in *side shade* where the slanting direct rays of the sun are blocked by crowns of adjacent trees. This condition is sometimes called "blue shade" because the light that is received is rich in the blue wavelengths of scattered or diffuse light from the blue sky. "Green shade" exists where much of the light must come through several layers of leaves in full shade. Since green light is photosynthetically useless, light rich in green supports only the most shade tolerant of species.

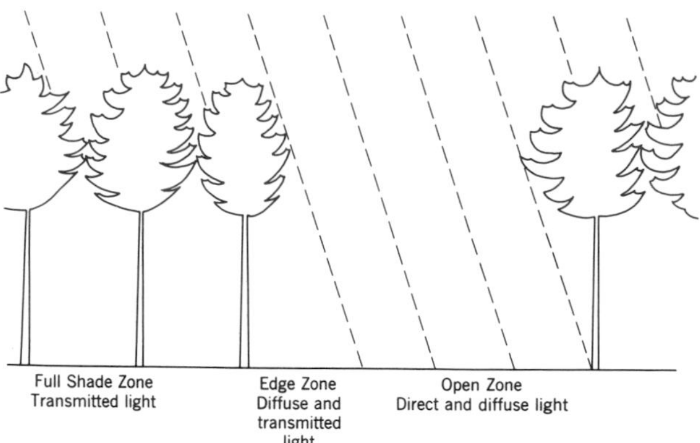


**FIGURE 1** The zonation of solar radiation across a forest opening showing the slanting rays of direct sunlight, the diffuse or scattered light that comes most abundantly from the sky straight above, and the transmitted light (including sun flecks) that comes through the leaves of the trees. [From Smith (1986). John Wiley & Sons, New York. Used with permission.]

Two other light conditions important for some species also exist. The *sun flecks* that come through small gaps between overhead leaves provide brief periods of intermittent direct sunlight and shade that are optimum for photosynthesis in some species. The other intermediate condition is that of the *side light* found just beyond the edges of forest openings where the light is a combination of sun flecks and diffuse radiation that is optimum for seedlings of some species.

Much also depends on the nature of the surface materials. All propagules must be moist to germinate. The denser the surface or soil medium the better is its ability to provide moisture to seeds or small plants. Dense media also conduct heat well both upward and downward.

Another factor that influences the regeneration microenvironment is the competition of preexisting roots for water and nutrients. This *root competition* is commonly exerted by the roots of older trees that extend into openings created for the establishment of regeneration. The roots of trees almost always extend farther than the crowns. Root competition from any preexisting vegetation, including herbaceous plants, sometimes hampers or kills new trees.

The pattern of arrangement of newly established species (Fig. 2) within forest openings often indicates the effect of controlling factors. The effects of direct solar radiation result in patterns shaped like crescents. Those of factors such as diffuse solar radiation, radiation frost, ventilation by wind, or the root competition of trees adjacent to the opening produce concentric patterns. Such patterns do not affect plants that were established before the openings were created, except to the extent that they control subsequent growth or health.

Natural disturbances or artificial treatments that simulate them can create many different combinations of light, shade, alteration of surfaces, and degree of root competition. The conditions that favor the establishment and height growth of one species may become fatal for another. Knowledge of these factors is used in silviculture to regulate the species composition of newly regenerated stands. However, with seedlings more is not always better. If all surfaces were rendered uniformly favorable to a given species, if the supply of seeds was sufficient, and if there was enough light, water, and nutrients to support height growth, a hopelessly overcrowded stand would result. The ideal condition is one in which uniformly distributed small spots aggregating somewhat less than 1% of the surface are favorable and the intervening surfaces are either unfavorable to other plants or favorable only to those that do not outgrow the desirable ones.

The most dangerous period during the life of the new seedling comes when it is very young. The successful seedling usually must quickly send its roots downward and its top upward through inhospitable strata that extend 2–3 cm above and below the soil surface. Just above the surface, air is so tightly held by friction that the turbulent transfer of heat (convection) is very sluggish. If the surface is exposed to sunlight, heat accumulates and it may reach temperatures in the lethal range above 50°C.

Even at lower temperatures the dangerous uppermost soil stratum may dry out from direct evaporation. Below that level, water is lost to the atmosphere only by transpiration through the roots and leaves of plants; surface tension effects are no longer capable of moving water to the surface by capillarity. Whether the water loss is by evaporation or by transpiration, it is imperative that the seedling

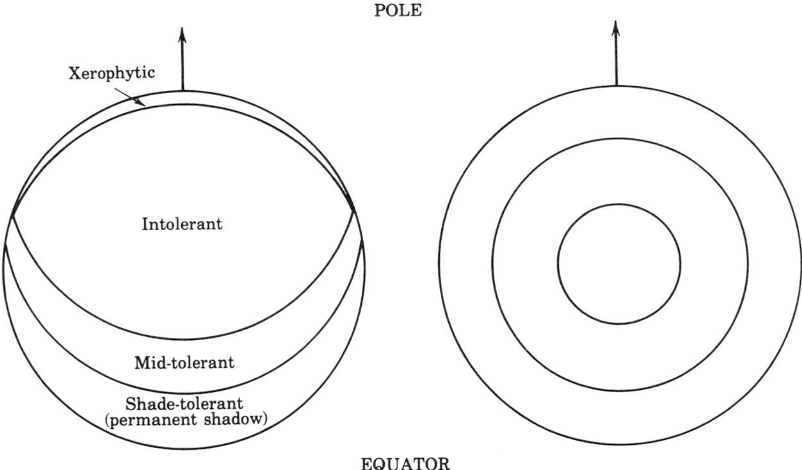


***FIGURE 2*** Crescentic (left) and concentric (right) patterns of variation in effects of microenvironmental factors in circular forest openings at middle latitudes. Shade tolerances of species favored by different kinds of shading and exposure are indicated in crescentic pattern. [From Smith (1986). John Wiley & Sons, New York. Used with permission.]

root penetrate the soil faster than it can dry out from top to bottom.

This condition is alleviated by partial shade, except that any living plants that provide the shade may also dry out the soil too much by transpiration. One of the fundamental advantages of planting nursery-grown seedlings is that both tops and roots extend beyond the dangerous zones above and below the air–soil interface right from the beginning.

## IV. STEPS IN THE REGENERATION PROCESS

The natural regeneration of forest trees from seed consists of a series of steps; failure at any one step defeats the whole process. Since trees live a long time the outcome of the process determines the composition and structure of at least part of the forest stand for decades or even centuries.

### A. Seed Production

The first step in the regeneration process is seed production. While trees that are almost moribund may produce many seeds, the most dependable producers are generally the most vigorous ones. Forests have so many rodents, birds, and other seed predators that only a very small fraction of seeds escape. Therefore, success is almost always after sporadic years of usually abundant seed production. During most years the seed-bearing fruits or cones start to form in abundance and then the activity of insects and other pests that feed on them halts the whole process. Cross-pollination is usually essential for production of seeds with high percentages of viability.

### B. Dissemination

Seeds are dispersed by many different agents such as wind, birds, bats, ants, rodents, flowing water, and gravity. When wind is the disseminating agent, sufficient amounts of seed are usually moved for several tree heights in the direction of movement of drying winds. Seeds with hard coats and edible flesh are designed to survive in the digestive tracts of the birds and other animals that move them. For almost every conceivable transportation mechanism there is a species to take advantage of it. For example, certain large-seeded species, such as oaks and walnuts, often depend on squirrels and birds for dispersal; not only do these animals transport the seeds but they also store them in the soil where they germinate if the animals do not return to eat them. Some species of mangroves have seeds that germinate on the trees; the seedlings fall into the sea, then float upright in search of some suitable strand on which to take root.

### C. Storage

Storage conditions, such as those of overwintering, can be limiting factors in seasonal climates. Seeds are usually produced during a favorable period but then must survive a dry or cold period and be ready to germinate during the next growing season. To do this they develop varying degrees of *dormancy,* a condition in which they do not grow and their physiological processes become very slow. This dormancy can become so pronounced that the seeds will not germinate until particular special conditions have been met. This phenomenon often prevents seeds from germinating prematurely as during abnormal warm periods during winter.

The required conditions are often those that prevail in the forest litter or soil during the periods unfavorable for germination. For example, many species will not germinate without a period of moist storage at low temperature. Some bird-disseminated species have seeds with hard coats that must be abraded by the sand in bird gizzards before water and oxygen can penetrate the seed and start germination. Where seeds are stored artificially it may be necessary to carry out treatments that imitate those that break dormancy in nature.

The seeds of many species of tropical rain forests do not develop dormancy because conditions are always favorable for growth and there is no benefit in delaying germination. The seeds must germinate in hours or days and usually cannot be stored artificially. At the other extreme, the seeds of some species can remain dormant yet survive for many years under natural or artificial conditions in which respiration is kept very slow (but not halted) by dryness, low temperature, or limited aeration. For example, the forms of jack and lodgepole pine that are adapted to regenerate after forest fires have seeds that are stored dry for decades in sealed cones on the trees. When hot fires melt the cone seals, the seeds are released onto the bare soil left by the fires.

### D. Germination

Germination, the start of development of the embryo, depends on having adequate amounts of liquid water, heat, and oxygen, although sometimes it also requires certain wavelengths of light. Any condition of dormancy must also have been broken. Water is the most common limiting factor; if it is too dry the seeds of some species may go back into dormancy for another year. The early germinators are the most likely to survive, except when there is a late frost.

### E. Succulent Stage

After germination the seedlings must survive a succulent stage in which their stems remain green, red, or white. During this period the seedlings are very subject to loss from the attacks of weakly saprophytic fungi that cause a disease misnamed "damping off." Insect larvae commonly feed on them too. Any seeds that remain attached to the seedling are still the prey of rodents and birds. This is also the period during which seedlings in the open are extremely subject to mortality from extreme surface heating or nocturnal frost as well as the time when the root must penetrate below the zone where water can be lost by direct evaporation. The succulent stage ends when a layer of insulating material, usually straw-colored, forms on the surface of the stem.

### F. Establishment Period

The new tree is regarded as established when its crown has reached a zone where there is enough light for it to survive indefinitely and the roots are in contact with a dependable supply of soil moisture. Ordinarily, mortality is much lower during the second year of life but it may take longer than that for the seedling to become established.

Since one large dead tree is replaced by dozens of plants, the next major source of mortality is the thinning that commences when the small trees start to compete with each other. This process continues throughout the life of the stand. In a managed stand the period between the regeneration of a stand and its replacement is called the *rotation*.

## V. SOURCES OF REGENERATION

Not all forest regeneration comes from new, currently produced seeds that fall on some vacant receptive surface after a disturbance. Many sources of regeneration are stored on the site in various latent forms or in arrested stages of development; they start or resume development when there is new space in which they can grow.

Seeds can be stored for long periods on the trees as in the closed cones of some pines and spruces or the fruits of some eucalypts. With some hard-coated seeds such storage is commonly in the soil and may continue for decades.

There are also many forms of small trees that represent quasi-dormant stages of slow or nearly arrested development. Some species do not start rapid height growth and the formation of large amounts of leaf surface until the root systems are large and deep enough to provide adequate water. Longleaf pine, for example, does not grow in height or form a stem until a large tap root develops. The tops of oak seedlings sometimes keep growing up and dying back until there is enough light for them to keep growing; it is possible to have a 1-year-old top and a 30-year-old root. Certain shade-tolerant species can persist for decades beneath taller trees and suddenly speed up in growth when something eliminates the taller ones. These opportunistic growth forms are called *advanced regeneration* in the sense of preceding whatever disturbance initiates rapid height growth.

There are several forms of asexual *vegetative regeneration* in which new trees develop from special parts of old ones instead of sexually from seeds. They represent adaptations to the killing of aboveground parts of trees by fire or other injury but are often deliberately secured by cutting.

The most common form of vegetative regeneration is sprouting from *dormant buds*. These buds start out as normal leaf buds, but instead of developing, they grow outward just under the bark but do not burst unless the crown of the parent tree suffers some sort of debilitation. These buds can form shoots at any level on the tree. New branches are called *epicormic shoots*; the basal shoots that may develop after a tree is cut or its top is killed by fire are called *stump sprouts*. Fires or cuttings that are repeated every few years may convert the stumps to bark-covered lumps of callus growth called *stools* and very dense, crooked, spindly shoots arise from them.

Buds that develop anew from cambial tissue (as distinct from dormant buds in the axils of leaves) are called *adventitious buds*. When they are damaged, some species develop such buds on root surfaces. The shoots that arise from such buds on the roots are called *root suckers*. For example, aspens are almost completely dependent on such regeneration; they commonly grow in clumps or natural *clones* in which all the trees are genetically identical.

A few species can regenerate vegetatively by the rooting of branch ends that droop down on the ground and are partially covered by litter or mosses. Such rooting is called *layering* and can lead to the development of independent trees.

## VI. ARTIFICIAL REGENERATION

Artificial regeneration is most commonly accomplished by planting nursery-grown seedlings. The sowing of seed out in the forest, i.e., *direct seeding*, is successful only when seed predation by rodents and birds is avoided and also when it rains often enough to keep the seeds wet. It is seldom feasible to apply the stupendous quantities of seeds that nature uses to overwhelm the appetites of the predators. Sometimes direct seeding works with species that have seeds so small that the predators ignore them. Otherwise direct seeding usually waits on the invention of environmentally acceptable repellents of predators.

Nursery-grown trees that are moved only once are called *seedlings*, but those which are replanted again one or more times in the nursery are called *transplants*. The purpose of such transplanting is to confine a large amount of root surface into a small volume and to allow the top to grow larger before the plant goes to the field. The chief advantage in planting large stock is that taller seedlings are more likely to overtop competing vegetation. There is, however, the risk that the roots, which are inevitably reduced during planting, may not provide enough water for the leaves.

Nursery stock is sometimes grown from rooted vegetative cuttings. Cuttings of some species, such as cottonwood poplars that normally grow along moist river banks, can be planted as unrooted cuttings directly in the field.

Nursery stock is planted either as *bare-rooted* or *containerized* plants. Bare-rooted plants are separated from the soil when lifted from the nursery beds while containerized plants are planted with their roots attached to the medium in which they grew.

### A. Bare-Root Planting

It is easy to transport large numbers of bare-rooted seedlings out into the field but success depends on meeting several restrictive factors. The most important factor is the need for the seedling to reestablish contact between the root and the soil. This can be accomplished initially by packing loose soil around the roots, but it is extremely important that the planting be done just before or during a period in which the roots grow rapidly. Such periods of rapid root growth usually occur at the beginning of the growing season and before the leaf buds burst. Therefore, it is best to lift and move the seedlings late in their dormant period; the roots will not grow unless the soil is moist. Sometimes there is a second period of root growth after the period during which leaves and stems grow, but success during that period depends on getting enough root growth to supply the water that is lost even in the dormant season.

Dormant bare-rooted seedlings can be kept in cold storage (above freezing) to postpone the time of maximum root growth and to extend the planting season. If there is no dormant season, as in the tropical rain forest, bare-rooted seedlings survive only if planted within hours of the lifting. The roots of any bare-rooted stock must be kept visibly moist at all times; freezing or waving them in the air before planting will desiccate them. The period during which this kind of planting can be done is often limited to a few frantic weeks.

Ordinarily, the plants should be reestablished with their root collars level with the soil surface as they were in the nursery, but it is better to be too deep than too shallow. The roots should extend downward vertically and not be bent upward into shapes like the letters J or L. Despite all these limitations, bare-root planting is the most common mode of artificial regeneration.

### B. Containerized Planting

This type of planting is expensive but more dependable. The main advantage is that contact between the roots and the medium in which they grow is not broken. This means that the planting can be done whenever the soil is moist and unfrozen. Containerized planting has always been the standard method of tree planting in the moist tropics where there is no dormant period and in arid regions where survival after planting is poor at best. The same is true for planting trees that are more than about half a meter tall.

Containerized planting is also adopted where labor costs are high and it is costly to dig holes. Tools are used which punch holes and "shoot" containers of uniform size with seedlings into the soil. One of the problems with the containers is that the roots are forced to spiral around the container walls. This "root-bound" condition may prevent normal radial growth. Therefore, plants should not be left too long in the containers and it is best that the containers be removed or ruptured severely before the plants are put into the soil. This problem can be avoided, however, if the container is a block of some organic material and has no walls.

## VII. SILVICULTURAL METHODS OF REGENERATION

In silvicultural practice, forest stands are usually regenerated by arranging cuttings and other operations in time and space according to several general patterns and programs that are called *methods of regeneration.* These patterns are part of the control of the microenvironmental factors and sources of seeds that determine the outcome of the regeneration process. It is crucial that these methods do not create conditions favorable to undesirable vegetation.

These patterns also govern the structure of the stand throughout its life. The various patterns that can be created not only govern regeneration but can be even more important for such objectives as securing a sustained yield of benefits, scheduling the tending of immature stands, facilitating timber harvesting, regulating stream runoff, manipulating wildlife habitat, and reducing losses to damaging agencies.

*Stand structure,* the physiognomy of the stand, is determined mainly by differences (or lack of difference) within the stand between species and classes of tree age (Fig. 3). The simplest kind of structure consists of aggregations of trees that are all of the same age and species. Each such aggregation has trees that are nearly the same height and form a single canopy stratum. If there are two or more tree species, there is usually a tendency for the species to become segregated into different strata even if all are of the same age. This is because no two species have precisely the same temporal pattern of height growth and because shade-tolerant species are adapted to grow beneath intolerant ones. If there are many species, some may occupy essentially the same stratum in these *stratified mixtures of species.*

The structures of stands can also become complicated if new regeneration develops in patch openings made at different times such that the stands ultimately consist of different intermingled age classes. Each new age unit can also be a stratified mixture. Another dimension of complexity is added if some trees from one rotation are carried into the next one. A stand is *even-aged* if all of its trees are nearly the same age and *uneven-aged* if it has at least three distinctly different age classes of trees. *Stands with two age classes* are recognized as such only if both age classes persist for a whole rotation; if the older age class is soon removed the stand is regarded as even-aged.

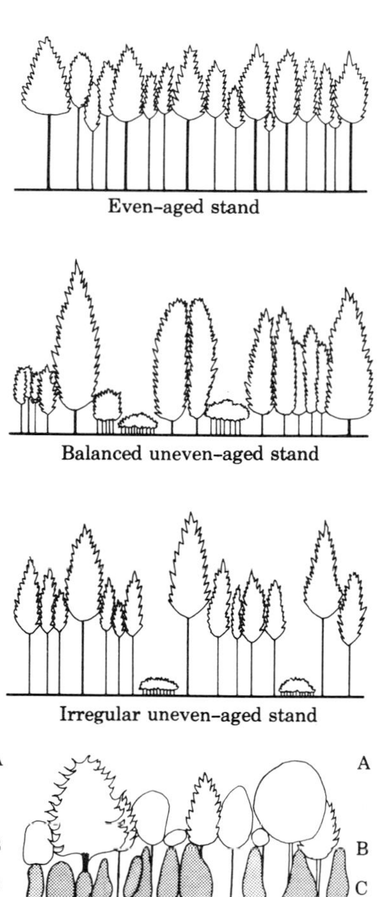


**FIGURE 3** Four kinds of forest stand structures. The trees of the first three stands are all of the same species but the fourth stand consists of several species all of the same age. [From Smith (1986). John Wiley & Sons, New York. Used with permission.]

It is not easy to alter the age–class structure of stands because it usually requires cutting some trees prematurely or carrying others beyond normal rotation age. Cuttings that are not severe enough to create vacant growing space for the regeneration or height growth of new age classes do not alter age–class structure. Figure 4 shows how some of the different methods (or patterns) of cutting affect the more important microenvironmental factors that regulate the species that regenerate after the cuttings.

The simplest pattern for creating a new stand is done by the *clear-cutting method* in which virtually all of the preexisting vegetation in the whole stand area is killed by cutting and site preparation. The

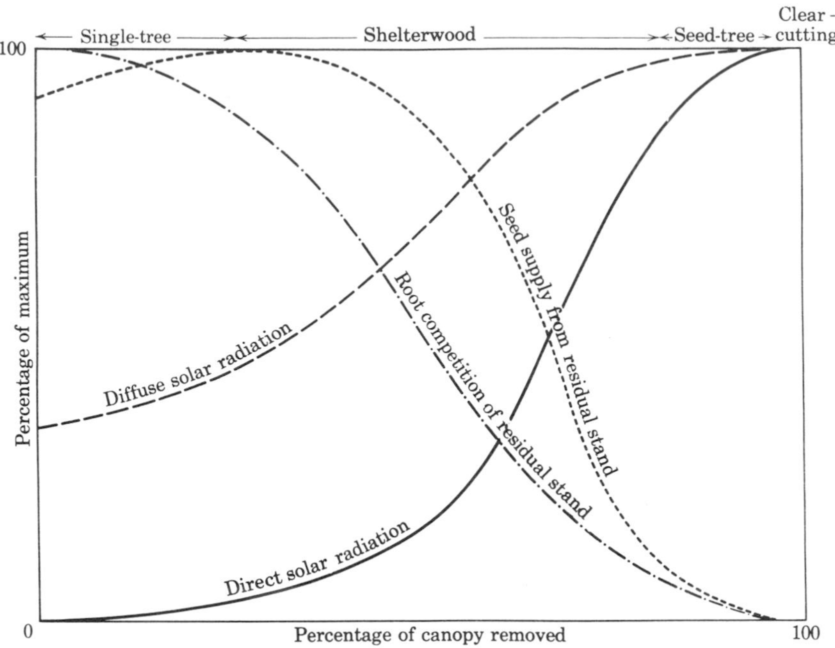


***FIGURE 4*** The effects of the initial cuttings of various regeneration methods applied uniformly over an area in a humid, temperate climate on several factors that affect the establishment of regeneration from seed. [From Smith (1986). John Wiley & Sons, New York. Used with permission.]

actual regeneration is usually done by planting or by direct seeding. Natural seeding on the exposed areas is usually successful only if the source of seed is nearby and the species is intolerant of shade. A similar method of creating even-aged stands is the *seed–tree method* which is generally the same as clear-cutting except that some scattered trees are left to provide seeds.

The *shelterwood method* also leads to even-aged stands and involves leaving more trees than seed–tree cutting. Often the purpose of these trees is to provide both seeds and shelter from excessive sun, frost, and other vicissitudes. Another purpose is to leave some chosen trees to grow larger. The old stand is removed in two or more separate stages with the final cutting not coming until the earlier cutting has induced adequate advanced regeneration. If such regeneration is present without any cutting, the old stand may be removed all at once in a so-called *one-cut shelterwood* cutting. The reliance on advanced regeneration makes the shelterwood method one of the safest ways to regenerate a stand because the old stand remains until there is a new one beneath it. The new stand also has a head start because it is taller at the time of the final cutting than it would be if it had been started after clear-cutting.

The *selection method* is used to create or maintain uneven-aged stands where there is some reason to have several different age classes in close proximity. New age classes are established by making small scattered openings. If the regeneration openings are large enough, even the most intolerant species can be regenerated in the sunny parts of the openings. However, this method is most often associated with seeking regeneration of the more tolerant species.

The *coppice method* involves regeneration from sprouts and is used mainly to grow fuelwood on short rotations. However, some of the regeneration secured by any of the other methods frequently comes from sprouts.

In any of these methods the openings or the arrangement of reserved trees can involve a uniform distribution or be in strips, circles, or groups. Sometimes the openings are gradually enlarged if it helps to keep moving a zone of partial shade conducive to regeneration.

During the rotation it may be desirable to conduct various *tending* or *intermediate* operations to regulate competition between the various plants. These include *release* operations to free desirable trees by killing undesirable overtopping vegetation; *thinning* may also be done to salvage trees destined to die from competition or to allocate more growing space to trees that are favored.

## Glossary

**Advanced regeneration** Trees that appear spontaneously or are induced to appear under existing stands.

**Forest stand** An aggregation of trees with sufficient uniformity of species composition, age(s), spatial arrangement, or condition as to be distinguishable from adjacent aggregations and large enough to be treated separately for purposes of forest management or study.

**Regeneration** or **reproduction** The renewal (v.) of aggregations of trees either naturally or artificially or the small trees (n.) resulting from the renewal.

**Regeneration** or **reproduction cuttings** Tree removal treatments made to create conditions favorable for establishment of regeneration.

**Microenvironment** A small space of a few cubic centimeters throughout which the physical, chemical, and biotic environmental factors are uniform in their ecological effect.

**Silviculture** The theory and practice of controlling forest establishment, species composition, structure, and growth.

**Site** or **habitat** An area considered in terms of its environment, particularly as this determines the type and quality of biota the area can carry.

**Tending** or **intermediate cutting** Treatments carried out during the life of a stand in order to improve it, regulate its growth, or obtain early financial returns *but not to regenerate it*.

## Bibliography

Barrett, J. W., ed. (1980). "Regional Silviculture of the United States," 2nd Ed. New York: John Wiley & Sons.

Kozlowski, T. T., Kramer, P. J., and Pallardy, S. G. (1991). "The Physiological Ecology of Woody Plants." San Diego: Academic Press.

Oliver, C. D., and Larson, B. C. (1990). "Forest Stand Dynamics." New York: McGraw-Hill.

Smith, D. M. (1986). "The Practice of Silviculture," 8th Ed. New York: John Wiley & Sons.

Spurr, S. H., and Barnes, B. V. (1980). "Forest Ecology," 3rd Ed. New York: John Wiley & Sons.

U.S. Forest Service. (1974). "Seeds of Woody Plants in the United States." U.S.D.A., Agriculture Handbook 450. Government Printing Office, Washington, D.C.

U.S. Forest Service. (1990). "Silvics of North America." U.S.D.A., Agriculture Handbook 654. Government Printing Office, Washington, D.C.

# Galápagos Islands

**George W. Cox**
*San Diego State University*



The Galápagos Islands, the largest oceanic archipelago uninhabited by humans prior to European discovery, have provided a rich location for studies of evolution, island biogeography, and marine ecology. The terrestrial flora and fauna of the islands, derived largely by rafting or aerial transport from Central and South America, provide outstanding examples of speciation and adaptive radiation, leading to an abundance of endemic forms adapted to insular conditions. The marine biota also shows a strong degree of endemism. Human impacts, beginning soon after discovery, have caused the extinction and endangerment of many species. The growing human population, intensified resource exploitation, and increased tourism now are creating heavy pressures on island ecosystems.

## I. GEOLOGICAL ORIGIN AND STRUCTURE

### A. Geography of the Archipelago

The Galápagos Islands, an oceanic archipelago of volcanic origin, lie on the equator 900–1000 km west of South America (Fig. 1). The archipelago comprises 13 islands larger than 10 km$^2$ in area, 6 islands between 1 and 5 km$^2$ in area, and many small islets and rocks. The total land area of the archipelago is 7882 km$^2$. Most of the large islands are formed of shield volcanos that rise to elevations of several hundred to more than a thousand meters, topped by craters or wide calderas. The largest island, Isabela, is 4670 km$^2$ in area and is formed of five shield volcanos, one of which reaches an elevation of 1707 m. [*See* ISLAND BIOGEOGRAPHY, THEORY AND APPLICATIONS.]

### B. Island Origins and Ages

The Galápagos archipelago is located near the junction of three crustal plates. The island platform lies about 30–300 km south of the northern edge of the Nazca Plate, which is separated from the Cocos Plate to the north by the Galápagos Rift. To the west, both the Nazca and Cocos plates meet the Pacific Plate along the East Pacific Rise. Seafloor spreading along the East Pacific Rise and the Galápagos Rift therefore tends to carry the islands eastward and southward at a rate of perhaps 7 cm per year.

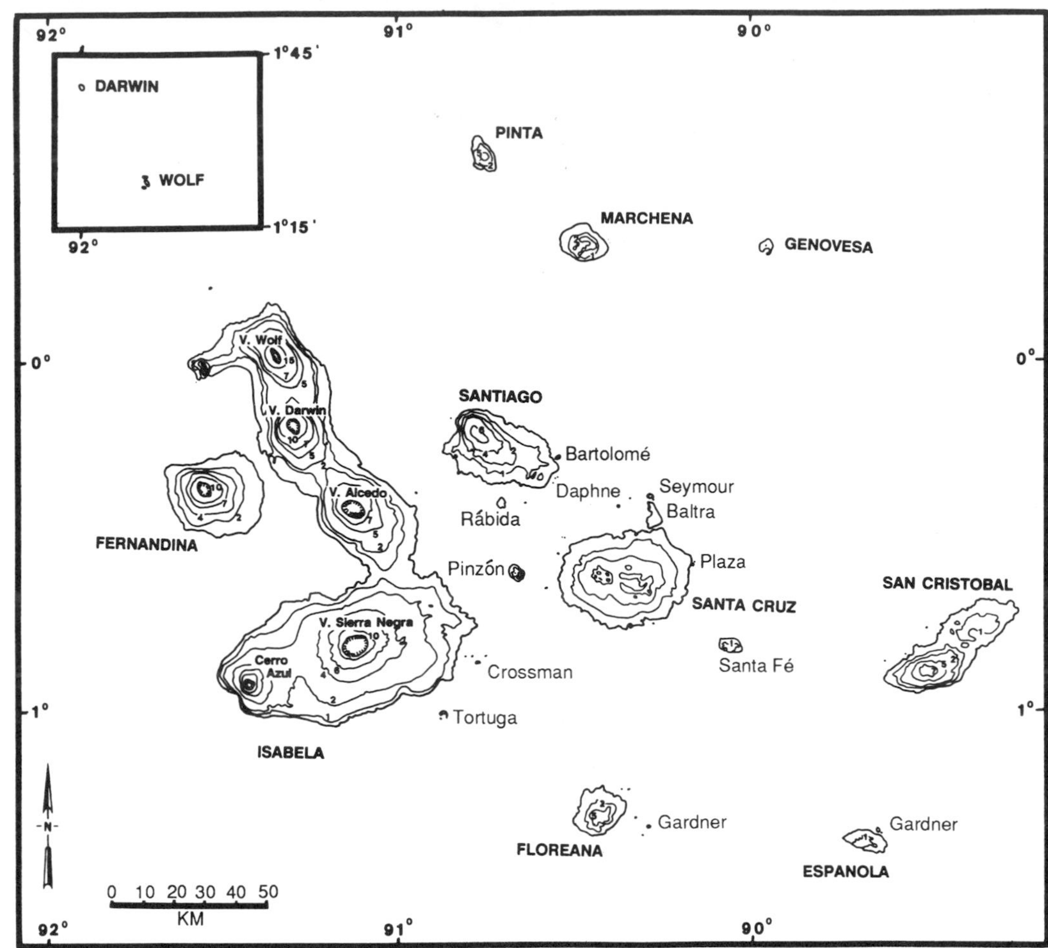


***FIGURE 1*** The Galápagos Archipelago. Numbers adjacent to contour lines indicate elevations in 100-m units. [From S. B. Peck (1991). The Galapagos Archipelago, Ecuador. *In* "The Unity of Evolutionary Biology" (E. C. Dudley, ed.), pp. 319–336. Dioscorides Press, Portland, Ore. Reprinted with permission.]

The islands range in age from about 1 to 3–5 million years. The oldest islands are those farthest east, San Cristobal and Española, and the youngest are those farthest west, Fernandina and Isabela. Island environments range from arid lowlands to humid highland forests. Recent volcanic activity has created landscapes of lava and ash in many places.

In all likelihood, the present archipelago is only the current generation of islands formed over a "hot spot" in the earth's mantle over which the Nazca Plate has been gradually moving. Eastward from the present islands, along the Carnegie Ridge, lie seamounts, their tops about 300–2200 m below the sea surface. Dredged material from these sites contains beach-worn cobbles, suggesting that these seamounts were once islands. This material has been dated at 5–9 million years B.P. Thus, an archipelago probably existed in the Galápagos region much earlier than suggested by the oldest existing islands. Based on the extent of the Cocos and Carnegie submarine ridges that diverge from the present Galápagos Islands, islands may well have been present in the general region for 15–20 million years, or even longer.

### C. Recent Volcanic Activity

Numerous eruptions have been recorded on Fernandina and Isabela. Fernandina is one of the most volcanically active islands on earth, and in 1825, a major eruption there generated enough heat to melt pitch on the decks of the sealing vessel *Tartar,* anchored 16 km to the north. Several expeditions in the early 1900s also noted volcanic activity in Fernandina and Isabela. In 1958, eruption of the Fernandina volcano evaporated the lake in its caldera, which re-formed only after two years. In 1968, another eruption led to subsidence of the caldera floor by about 300 m. Subsequent eruptions were noted on Fernandina in 1984 and 1990. A spectacular eruption of Cerro Azul Volcano on Isabela occurred in 1979, and in September 1991, an eruption occurred on Marchena Island, producing a lava flow that reached the ocean. Uplift and subsidence of areas of island coastlines have also been noted in several locations in the archipelago.

## II. TERRESTRIAL AND MARINE CLIMATE

### A. Winds, Ocean Currents, and Seasonal Weather Patterns

The climate of the islands is controlled by the seasonal patterns of trade winds and ocean currents, and by the El Niño–Southern Oscillation phenomenon. The weather pattern in most years is one of warm, clear weather from January through June, and cooler, cloudy weather from July through December. During the first half of the year, the influence of the southeast trade winds weakens, as this belt shifts southward during the Southern Hemisphere summer. As a result, the westward flow of the Humboldt Current from the South American coasts toward the islands weakens, and warmer water from the Gulf of Panama invades the region. Sea temperatures rise to 25–28°C, and sunny weather with occasional convectional rains prevails. As the southeast trade wind belt moves northward during the latter half of the year, the influence of the cool Humboldt Current becomes dominant, and sea temperatures fall to 16–22°C. The ocean cools the lower atmosphere, creating a temperature inversion and a dense cloud layer at the junction with warmer, overlying air. This cool, cloudy condition brings fog, mist, and drizzle, a weather condition known as *garúa,* to the higher elevations of the islands, while the low elevations remain cool and dry. During the warm season the mean daily range of air temperatures at sea level on Santa Cruz Island averages 24.1–27.9°C; during the cool season this range is 18.8–22.7°C.

### B. Altitudinal Climatic Belts

These weather patterns interact with the geography of the islands to create altitudinal climatic belts. The lowlands are arid, with annual rainfall of about 10–40 cm, mostly during the warm season from January through June. With increasing elevation on the volcanic mountains, rainfall increases, especially during the *garúa* season. At elevations of 300–600 m, annual rainfall reaches about 100–180 cm, enough to support moist forest vegetation.

### C. El Niño–Southern Oscillation Events

Every few years, typically at intervals of 3–7 years, the El Niño–Southern Oscillation (ENSO) phenomenon brings a pronounced change to weather conditions. In these years, warm ocean water that has accumulated in the western Pacific Basin floods eastward, sealing off the upwelling of deep, cold, nutrient-rich water along the western coast of South America and effectively eliminating the westward flow of the Humboldt Current. During ENSO years, warm, sunny weather with heavy convectional rainstorms dominates the islands. In 1982–1983, the longest and strongest ENSO event on record struck the archipelago. This ENSO began in April 1982 and lasted through September 1983, increasing rainfall 10- to 20-fold, cutting off the flow of cool, nutrient-rich water to the island region, and profoundly affecting vegetation and animal life throughout the archipelago.

## III. NATURE AND ORIGIN OF TERRESTRIAL BIOTA

The biota of the Galápagos has been derived largely from Central and South America and is strongly

**TABLE I**
Diversity and Endemicity of Major Biotic Groups in the Galápagos Archipelago

| Group | Total species | Native species | Percent endemic |
|---|---|---|---|
| Flowering plants | 736 | 540 | 42.4% |
| Marine algae | 333 | 333 | 34.8% |
| Corals | 44 | 44 | 22.7% |
| Marine annelids | 199 | 199 | 26.0% |
| Marine crustaceans | 285 | 285 | 26.0% |
| Marine mollusks | 652 | 652 | 22.7% |
| Echinoderms | 126 | 126 | 16.7% |
| Marine fish | 307 | 307 | 16.6% |
| Land snails | 83 | 80 | 96.4% |
| Land arthropods | | | |
| Noninsects | 296 | 279 | 25.8% |
| Insects | 1616 | ca. 1500 | ca. 50% |
| Reptiles | 22 | 21 | 95.2% |
| Seabirds | 17 | 17 | 29.4% |
| Land and freshwater birds | 41 | 38 | 57.9% |
| Mammals | 19 | 9 | 90.0% |

endemic (Table I). Over 42% of the native plants, 57% of the native land birds, and most of the native mammals and reptiles are endemic. Several genera of plants and animals have shown elaborate patterns of adaptive radiation in the archipelago, making it an important natural laboratory of evolution.

## A. Flora

### 1. Richness and Endemicity

The native flora of 540 species may have arisen from colonization of the islands by about 400 ancestral forms. Of these, 229 are endemic. About 196 other plant species have been introduced deliberately or inadvertantly to the islands. More than 92% of the native species, both endemic and nonendemic, are of South American or pantropical affinity. Most of the plant colonists have not shown extensive speciation in the islands. Some 18 genera, however, contain three or more endemic species, and these alone account for almost half of the plant endemism.

### 2. Evolution of Major Endemic Taxa

Most notable among the endemic plant genera are *Scalesia* (Asteraceae) and *Opuntia* (Cactaceae). The morphology and cytology of *Scalesia* species suggest that they were derived from an ancestor close to members of the genera *Helianthus* or *Viguiera,* which are widely distributed through western North, Central, and South America. All 20 species in the Galápagos are woody plants, ranging in size from low shrubs to trees about 15 m in height. Adaptive radiation in this genus rivals that of the more frequently discussed geospizine finches.

The genus *Opuntia,* which is widespread in the continental Americas, is represented by 14 endemic species that may have been derived from two colonization events. Nine of these species are treelike in growth form, reaching heights of up to 12 m. These species have well-developed trunks with red, flaky bark and well-developed spines. Browsing by tortoises may have been a major selective force in the evolution of these forms.

## B. Fauna

### 1. Land and Freshwater Birds

The Galápagos Islands are one of the few archipelagos of substantial size and diversity for which we have data on the full biota prior to disturbance by humans. Only 28 native land bird species breed on the Galápagos, the product of perhaps as few as 14 colonization events. Some of these species are undifferentiated from their relatives on the coast of Ecuador. Others, such as the finches of the subfamily Geospizinae, are forms that originated by speciation within the archipelago. An area of equivalent size and vegetational diversity on the South American continent, however, would probably have a bird fauna 10 or 20 times as rich.

Of the 28 native, breeding land bird species, all but 6 are endemic at the specific or generic level. The endemic forms include the Galápagos rail, Galápagos hawk, Galápagos dove, and 19 passerines. The passerines include the large-billed flycatcher, Galápagos martin, 4 endemic mockingbirds, and 13 endemic geospizine finches. The mockingbirds include the widespread Galápagos mockingbird, together with the Charles mockingbird, now extinct on Floreana but resident on small islets nearby; the Hood mockingbird, resident on Española and a nearby islet; and the Chatham mockingbird, resident on San Cristobal (the names of these birds are taken from former English names for the islands). The nonendemic land birds include the paint-billed crake, barn owl, short-eared owl, dark-billed cuckoo, vermilion flycatcher, and yellow warbler. The dark-billed cuckoo and the yellow warbler, which do not show endemicity even at the subspecific level, may have colonized the archipelago within the last few hundred years. Fossils of these species do not appear in lava-tube deposits of bone that contain many other endemic forms. One wild species, the groove-billed ani, was apparently introduced deliberately by cattle ranchers. Domestic fowl and rock doves are feral near some human settlements.

Ten species of waterbirds also breed in the archipelago. Only one, the lava heron, is an endemic species. The nonendemics include the great blue heron, cattle egret, great egret, yellow-crowned night heron, white-cheeked pintail, greater flamingo, common gallinule, oystercatcher, and common stilt. The cattle egret has become established naturally in recent times, the first breeding having been noted in 1986.

## 2. The Galápagos Finches and Their Evolution

The Galápagos finches, which usually are classified as the endemic subfamily Geospizinae, have diversified from an ancestral seed-eater to become fruit-eaters, wood-excavators, bark-scalers, and foliage-gleaners. This group consists of 13 species in the Galápagos and 1 species restricted to Cocos Island, 630 km to the north of the Galápagos archipelago. These species have probably been derived from one ancestral colonist species from Central or South America, which may have arrived in the Galápagos between 5 million and 0.5 million years ago. The blue-black grassquit, *Volatinia jacarina,* and the St. Lucia black finch, *Melanospiza richardsoni,* have been proposed as descendants of the same ancestor.

Speciation in the Galápagos has produced 13 species, grouped into 5 genera, with the Cocos Island finch being placed in the sixth genus. The species on the Galápagos show a striking pattern of adaptive radiation in feeding morphology and diet. Four species, the large, medium, small, and sharp-beaked ground finches, are adapted to seed-eating and show striking patterns of beak size differences related to seed-crushing ability. The sharp-beaked ground finch also pecks at the wings or tail of seabirds to drink the blood that appears and breaks seabird eggs to consume their contents. The cactus and large cactus finches feed extensively on both seeds and *Opuntia* fruits. Another, the vegetarian finch, feeds extensively on buds and soft plant tissues. Six species are insectivorous. The large, medium, and small tree finches, together with the woodpecker and mangrove finches, search for insects on the branches, trunks, and fallen parts of woody plants. The tree finches often excavate insects from bark and rotting wood. The woodpecker and mangrove finches are notable for their use of cactus spines or slender twigs as tools to tease insects out of crevices. One species, the warbler finch, feeds by foliage-gleaning for small insects. Coexistence of these species is extensive, and as many as 11 occurred together on individual islands before human disturbance led to local extinction of some island forms [*See* SPECIATION.]

Several zoologists have examined the evolution of this group. Analysis of evolutionary trends is complicated by high variability within and between populations, and by the tendency of some forms to hybridize. The current view is that speciation has occurred by divergence of allopatric populations on different islands, and that competition among species that have subsequently become sympatric has promoted adaptive radiation of the group in food resource use features. I have suggested that most of the speciation and divergence of species have occurred in the central, large-island region, and the colonization of the outer islands has been a one-way route. Adaptive radiation is exhibited most strikingly in body size and in the size and form of the beak.

## 3. Reptiles

Reptiles are the second most diverse group of terrestrial vertebrates, represented by 21 species. Three species of snakes of the genus *Dromicus* live on islands of the central and southern part of the archipelago. Each of these species is represented by two or three subspecies, and forms of two species occur on seven islands. Two families of lizards, Gekkonidae and Iguanidae, are native to the islands. Six species of geckos of the genus *Phyllodactylus* occur in the archipelago. Five of these are endemic, and *P. tuberculosis* is also found on the South American mainland. These species are thought to have been derived from three colonizations of the islands from South America, the earliest perhaps 8.9 million years before present. Three geckos, including *P. tuberculosis,* coexist on San Cristobal. The iguanid genus *Tropidurus* is represented by seven endemic species, known as lava lizards, with no more than one species per island. Lava lizards are abundant on most of the islands. This genus may have colonized the islands twice from South America, the earliest as early as 10.2 million years before present.

Of perhaps greater interest among the iguanid lizards are the land and marine iguanas, which are considered to belong to genera endemic to the archipelago. Two species of land iguanas occur, *Conolophus subcristatus* and *C. pallidus*. The former was originally found on four of the large central islands and is now extinct on one of these. The latter is restricted to Santa Fé. Both species have been greatly reduced in numbers.

The marine iguana, *Amblyrhynchus cristatus,* is one of the more remarkable evolutionary products of the islands. Classified into seven subspecies, it occurs on all major islands and most smaller ones. Marine iguanas are restricted to rocky shores, where they feed primarily on marine algae in intertidal and shallow subtidal habitats. They are capable swimmers and dive to depths of up to 15 m. They possess remarkable three-cuspate teeth that enable them to crop algae growing on rock surfaces.

The remaining group of reptiles is the giant tortoise, *Geochelone elephantopus,* represented by 11 subspecies, some now extinct. On Isabela, the largest island, five subspecies are recognized, one on each of the large volcanos. The populations on Floreana, Santa Fé, and Fernandina are now extinct, and that from Pinta is represented by a single captive individual, "Lonesome George." The remaining healthy populations of giant tortoises are in the highlands of northern Isabela and Santa Cruz.

The various subspecies differ in shape and other features of the carapace. On Española and Pinta, the populations show a strong "saddle-shaped" carapace, in which the front edge of the shell is vaulted to permit the head and neck to be raised higher than is possible for animals with symmetrically domed shells. The saddle-shaped carapace is interpreted as an adaptation to allow the tortoises to feed on pads of tree-form prickly pears in the xeric habitats of these islands. Animals with domed carapaces occur in more humid environments, where they graze on herbaceous plants and low shrubs.

### 4. Mammals

The native land mammal fauna is limited to two species of bats, one of which is endemic, and at least seven endemic species of rice rats. Both bats are members of the genus *Lasiurus:* the endemic Galápagos red bat, *L. brachyotis,* and the hoary bat, *L. cinereus*. All but two of the rice rats have become extinct within historic time. *Oryzomys bauri* is abundant on Santa Fé Island, and a closely related form occurred on San Cristobal. *Nesoryzomys narboroughii,* of Fernandina Island, is a generic endemic that represents a colonization event distinct from that of *Oryzomys.* Other forms of *Nesoryzomys,* now extinct, were collected in the 1800s and early 1900s on Santa Cruz, Baltra, and Santiago islands, and recent fossils suggest that a member of this genus occurred on Isabela. Remains of rice rats as large as a muskrat have been discovered in cave and lava-tube deposits on Santa Cruz and Isabela. These animals have been placed in the endemic genus *Megaoryzomys*.

### 5. Invertebrates

Terrestrial invertebrates are represented by land snails and arthropods, the vast majority of which are of Central or South American origin. About 80 species of terrestrial molluscs inhabit the archipelago, the product of about 16 colonization events. A single ancestral land snail of the genus *Bulimulus* has given rise to about 65 endemic species and subspecies.

Among arthropods, about 21 species of isopods and amphipods, 19 species of centipedes and millipedes 80 species of spiders, and 176 species of mites, scorpions, and their relatives have been recorded. More than 1600 insects are known, representing 24 orders. The orders Coleoptera, Lepidoptera, Diptera, Hemiptera, and Hymenoptera, and Homoptera are well represented. Most of the flightless and some weak-flying invertebrates are presumed to have reached the islands by rafting. Other weak fliers and all strong fliers such as dragonflies, butterflies, and moths presumably reached the islands by aerial means. The insect fauna is dominated by generalist species, some 40.1% of the beetles being scavengers, for example. Plant pollinators are poorly represented. Bees are represented by a single species of carpenter bee, which may be the most important insect pollinator species in the archipelago. The level of endemism varies greatly from group to group but is substantial in many cases. For example, about 52% of the 371 species of beetles are endemic.

#### *6. Plant and Animal Communities*

The vegetation ranges from arid littoral and lowland communities to moist highland forests, shrublands, and pampas. The vegetation is arrayed as a series of altitudinal zones, their altitudinal limits generally lower on southern slopes than on northern slopes. Along the coast, mangrove forests occur in sheltered coves. Sandy areas of coastline possess a flora of perennial grasses, succulent herbs, and vines. The arid zone, from sea level to an elevation of about 100 m, is characterized by a rich diversity of drought-tolerant shrubs, small trees, and cacti. Many of the woody plants of this zone are drought-deciduous. In the transition zone, between 100 and 180 m, different species of woody plants become dominant, and cacti decline in importance. From about 180 m to 500–600 m, *Scalesia* cloud forest, which is rich in epiphytes, was originally widespread. On Santa Cruz Island, a distinctive open forest of *Zanthoxylum* trees, thickly covered by epiphytic ferns, mosses, and liveworts, occurred on south-facing slopes above the *Scalesia* forest. Because of the color of the dry epiphytes during the dry season, this is sometimes called the "brown zone." Above the *Scalesia* and *Zanthoxylum* zones on Santa Cruz and San Cristobal, a zone dominated by shrubs of the genus *Miconia* also occurred at elevations of 400–700 m on south-facing slopes. Finally, at the highest elevations, meadows of sedges and ferns are widespread. Vegetation of the higher elevations has been greatly disturbed by human settlement, feral livestock, and introduced plants.

### C. Specialized Habitats

Three freshwater lakes occur in the archipelago. El Junco Lake lies at an elevation of 700 m in a volcanic crater on San Cristobal Island. This lake, about 6 m in depth, has a very low dissolved solute content. Frigate birds commonly use this lake as a source of drinking water, scooping water from the surface in flight. Two other lakes occur in the crater of Fernandina Island, at an elevation of 920 m. The larger of these lakes is 1.7 km in maximum diameter. This lake disappeared following the eruption of the volcano in 1958. The smaller lake, 200 m in maximum diameter, is over 15 m deep. None of these lakes contains fish.

Several saline lakes occur at low elevations. A crater in the center of Genovesa Island is occupied by a lake with twice the salinity of seawater and about 30 m deep. On Isabela Island, Beagle Crater contains a lake 2 km in diameter and 15 m deep. Although twice as saline as the ocean, a fish of the genus *Xystaema* lives in this lake. Tagus Crater, a short distance north of Beagle Crater, contains a smaller, more saline lake. Ephemeral saline lakes also occur at several locations on the western end of Santiago Island. Brine shrimp and notonectid hemipterans are abundant in most of the saline lakes. Coastal lagoons occur on several of the large islands, and these are home to the greater flamingo, white-cheeked pintail, and common gallinule.

At higher elevations on Santa Cruz, southern Isabela, and Floreana, *Sphagnum* bogs occur. The largest of these, at about 700 m elevation on Santa Cruz, is about 100 m in diameter, with a peat accumulation of about 5 m thickness at its center.

### D. Impacts of ENSO Events

The record rainfall during the 1982–1983 and 1992–1993 ENSO events created lush growth of vegetation in the arid lowlands and created shallow lakes in the craters of several volcanos. The rains in 1982–1983 may have triggered a severe dieback of *Scalesia* trees in the highlands of Santa Cruz Island, but the forest has reestablished itself from seedlings. The lush plant growth and the abundance of seeds stimulated breeding of the ground finches. Some pairs nested repeatedly over an 8-month period, fledging as many as 25 young. Some birds reared early in this period began nesting toward its end. After weather conditions returned to normal in late 1983, breeding largely ceased, and populations of these species declined until 1987.

The turbulent sea conditions during El Niño shortened the available feeding time for marine iguana populations, especially those that fed largely in intertidal areas. The red algal turf on which these animals preferentially graze was replaced in many areas by a brown alga never before recorded in the islands. This alga was poorly digested, and large numbers of iguanas starved. Very little reproduction occurred. Over a 6-month period, the marine iguana population declined by 60–70%. Recovery of the population after the end of El Niño was rapid, however.

## IV. NATURE AND ORIGIN OF MARINE BIOTA

### A. Marine Flora, Invertebrates, and Fish

The marine biota of the Galápagos archipelago is rich and varied. Most species are of Panamic or western Pacific affinity, but the level of endemism is relatively high in almost all groups.

Although lacking in macrophytes such as kelps, the marine flora is diverse (333 species) and strongly endemic (35%). Molluscs, crustaceans, and echinoderms are well represented and also strongly endemic. Corals, including reef-building species, are numerous, although reefs themselves are only moderately developed. Incipient reefs occur along most coastlines, except where upwelling is strongly developed. The best development of reefs is near Floreana in the southern part of the archipelago, and near Wolf and Darwin in the northernmost part of the island group.

The marine fish fauna is also diverse, compared to other islands of the tropical eastern Pacific, apparently due to the diversity of marine habitats and the biotic influence of several current systems that bring larvae to the archipelago. Some 307 species of 92 families have been identified. Most species are of Panamic affinity, with 60% being forms that also occur along the coasts of Central and South America. Nevertheless, about 17% are endemic to the archipelago.

### B. Seabirds

Some 17 species of seabirds (Galápagos penguin, waved albatross, Audubon's shearwater, three petrels, brown pelican, three boobies, flightless cormorant, two frigate birds, two gulls, and two terns) also breed in the archipelago. Of these, only five are endemic: the Galápagos penguin, waved albatross, flightless cormorant, lava gull, and swallow-tailed gull. Populations of most of the resident seabirds are in the thousands or tens of thousands. Among the endemic species, the flightless cormorant and lava gull are least abundant, with normal populations probably fewer than 1000 birds.

### C. Marine Mammals

Two marine mammals breed in the islands, the California sea lion, *Zalophus californicus,* and the Galápagos fur seal, *Arctocephalus galapagoensis*. The sea lion population is considered a distinct subspecies and the fur seal an endemic species. The sea lion population varies considerably but is believed to lie in the range of 20,000–50,000 animals. Sea lions feed diurnally and largely consume cephalapods and fish. The fur seal was nearly driven to extinction in the late 1800s and early 1900s. Fur seals now number 30,000–40,000 animals and are concentrated near upwelling areas in the western and northern portion of the archipelago, where they feed nocturnally on squid and small fish.

Sperm whales are relatively common in Galápagos waters, with the total population perhaps consisting of about 270 animals in 13 groups. These whales are most common in the upwelling zone west of Isabela Island.

### D. Sea Turtles

The green sea turtle, *Chelonia mydas,* is the only sea turtle to nest in the Galápagos, which is probably the major nesting area for this species in the central eastern Pacific. The most important nesting beach is Quinta Playa, on the southeast coast of Isabela, where 610 nesting females were recorded in 1978. Other important nesting beaches are on Santa Cruz and Santiago. Hawksbill and leatherback sea turtles are occasionally seen in Galápagos waters, but are not known to nest.

### E. Impacts of ENSO Events

ENSO events profoundly affect the marine ecosystems of the archipelago. During the 1982–1983 ENSO, for example, warm, nutrient-poor water low in salinity flooded into Galápagos waters from the Gulf of Panama, leading to a sudden, sharp decline in biological productivity. Almost all marine organisms were severely impacted. Corals with symbiotic zooxanthellae experienced severe bleaching and mortality. On many patch reefs, all corals died. High sea levels and altered swell direction led to extensive mechanical damage to reefs, especially during high spring tides. Deterioration of the physical framework of these reefs reduced the habitat available for reef fish and invertebrates.

The major fish-eating seabirds, including the Galápagos penguin, flightless cormorant, common frigate bird, all three booby species, and the waved albatross, experienced virtually complete reproductive failure and showed catastrophic declines in numbers. The Galápagos fur seal was severely affected, losing almost all individuals of the youngest four age classes and about 30% of older adults. The sea lion, which can forage at greater depths, experienced lesser effects.

Recovery of Galápagos marine ecosystems has been slow for some species and rapid for others. A fivefold increase in population densities of sea urchins caused severe erosion of the dead coral framework of reefs for years after the end of the ENSO, due to their rasping mode of feeding. Most seabirds have shown rapid population recovery, although the lush vegetation growth in some seabird colonies reduced suitable nesting habitat for several years. The Galápagos fur seals and Galápagos penguins have recovered more slowly than other species.

## V. HUMAN HISTORY AND IMPACT

### A. Presettlement Exploitation and Human Impact

The Galápagos archipelago is one of the few oceanic island groups that was not colonized by humans in prehistoric time. Thus, the biota present when the islands were first visited by European voyagers represented the result of evolution unaffected by human activity.

Human impacts date from the sixteenth century, when the islands were discovered in 1535 by Spanish voyagers from Panama. From the 1500s through the 1700s they were visited by whalers and privateers. During this period, populations of giant tortoises were hunted to extinction or severely reduced in numbers. These large tortoises were taken primarily by crews of ships, because the animals could be kept alive and in good condition (as sources of meat) for months. It was common for a single ship to take aboard 100–400 tortoises at one time. The total number of animals removed from the Galápagos probably numbers in the hundreds of thousands. Tortoises were also killed and rendered for oil. At least three races of the Galápagos tortoises were hunted to extinction, and a fourth survives as only a single individual.

### B. Early Agricultural Settlement Period

Settlement began in 1832, when an Ecuadorean colony was established in Floreana. Only four major islands—Isabela, Santa Cruz, San Cristobal, and Floreana—have permanent human settlements. This settlement, concentrated in the higher elevations where moisture permitted farming, resulted in the direct destruction of unique vegetation types. Domestic animals such as goats, pigs, cattle, donkeys, horses, dogs, and cats were introduced, leading in many instances to the establishment of feral populations. Black rats were also introduced accidentally to many islands. The *Scalesia* forests, in particular, were severely affected by land clearing and animal disturbance. Only fragments remain of these original highland forests.

### C. Darwin and Other Scientists in the Islands

The *H.M.S. Beagle,* carrying Charles Darwin as naturalist, arrived in the Galápagos on September 15, 1835, and remained until October 20 of the same year. Darwin spent 19 days on land in the islands. He visited only four islands, spending five days on San Cristobal, four days on Floreana, one day at Tagus Cove on Isabela, and nine days on Santiago, including three days in the highlands. He and several other shipmates collected specimens of a wide variety of plants and animals, and he recorded his impressions of the geology and biology of the islands. Unfortunately, the island sources of many of the specimens were not recorded and were labeled by guesswork after he returned to England. At the time, Darwin evidently still believed in creation and the fixity of species. Later, reflection on his observations and on the analyses of the collected materials by various specialists contributed to the development of his ideas on natural selection.

The first American scientist to visit the islands was Louis Agassiz, who led a 9-day expedition to the islands in 1872. The next major scientific expedition to the Galápagos was that of the California Academy of Sciences in 1905–1906. In 1923, William Beebe led an expedition of the New York Zoological Society to the islands. His popular account of this visit, "Galapagos: World's End," attracted considerable public attention to the biology of the islands. G. Allan Hancock led several expeditions to the islands in the late 1920s and 1930s. Studies of the Galápagos finches by David Lack, who visited the islands for four months in 1938–1939, stimulated much of the interest in competitive processes in evolution, as well as in island ecology in general, in the latter part of the twentieth century.

### D. Recent Economic Development and Tourism

The total Ecuadorean population on the Galápagos is now about 14,000. The major economic activities of these residents are tourism, farming, fishing, and employment in the Ecuadorean armed forces and government. Tourism, the largest component of the economy, is estimated to bring in about $180 million each year.

Since the establishment of the Galápagos National Park, tourism has boomed. In 1970, only about 4500 tourists visited the islands. By 1993, this had grown to about 40,000. Most visitors are housed on boats, which also provide transportation among the islands. About 80 boats, with a capacity of about 1070 persons, are licensed to transport visitors to park areas. Visitors are permitted access to designated areas and restricted to specific trails or observation areas in the most heavily utilized locations. Guides licensed by the park service must accompany visitors to park areas.

## VI. CONSERVATION ECOLOGY

### A. Extinctions and Their Causes

Hunting and killing by humans, together with the impacts of introduced exotic mammals, have reduced or eliminated populations of several Galápagos species. On Floreana, seven vertebrates have disappeared: the giant tortoise, the snake *Dromicus biserialis,* Galápagos hawk, barn owl, the mockingbird *Mimus trifasciatus,* and two species of ground finches. The giant tortoise population was probably eliminated by overcollecting by visiting ship crews by about 1840–1850. Killing of the Galápagos hawk and barn owl by settlers was likely the final cause of local extinction of these species, as both were very tame, and extinction of the remaining species was probably the result of predation or habitat disturbance by introduced mammals.

Overexploitation by humans has also led to extinction of giant tortoises on Santa Fé and the depletion of their numbers on most islands. On Fernandina, intense volcanic activity may have actually caused extinction. Today, predation and other impacts of introduced mammals are probably the greatest threat to their survival.

Introduced black rats probably led to the extinction of native rice rats on five islands. Rice rats survive only on Santa Fé and Fernandina, islands free of black rats. Extinction of the large *Megaoryzomys* rice rats also might have been caused by black rats and other human-induced changes, because these species may have survived into the very early human settlement period. [*See* INTRODUCED SPECIES.]

### B. Impacts of Introduced Exotics

The Galápagos Islands provide a good case study of the effects of both animal and plant introductions. Most of the larger islands, especially those with human settlements, have six to nine species of wild or feral introduced mammals. Goats were introduced by early buccaneers, and cattle, pigs, horses, and donkeys have become feral on some of the inhabited islands. On several of the smaller islands, goat introductions led to population explosions. On Pinta, for example, one male and two female goats were introduced in 1959. By 1970 the population on this island of only 60 km$^2$ was estimated to be between 5000 and 10,000. On Santiago Island, goats introduced in the early 1800s increased to 80,000–100,000 animals in the 1970s and 1980s. Goats consume a wide variety of plants, including ferns, herbaceous plants, shrubs, trees, and cacti. On several islands, goat browsing decimated the lowland vegetation and severely damaged the forest understory at higher elevations. On Floreana, more than 70% of the native plant species were reduced or eliminated by goat browsing. Cattle grazing in highland areas of Santa Cruz severely disturbed the native shrublands that occur above the *Scalesia* forests. In overgrazed areas of these shrublands, exotic grasses tend to invade.

Dogs and cats have also established feral populations on various islands. These animals proved themselves to be major predators on marine and land iguanas, as well as on seabirds and fur seals. Black rats, house mice, and most recently, the Norway rat are other exotic invaders. Rats are major predators on young iguanas on some islands. The fire ant *Wasmannia auropunctata* has become established on all four inhabited islands and Santiago, where it is causing major changes in native invertebrate communities. In 1989, the black fly *Simulium bipunctatum* became established in the highlands of San Cristobal, raising serious concern because these insects are potential vectors of several vertebrate diseases.

About 240 exotic plants have become naturalized, and about 15 of these are invasive species that are now problems. Guava, *Psidium guajava,* has invaded large areas of the highlands where the native vegetation has been overbrowsed by goats and other animals. Its seeds are dispersed by both introduced mammals and native birds that eat the fruits. Even where overbrowsing has been terminated, recovery of the native vegetation is now inhibited in many places by this weedy shrub. The quinine tree, *Cinchona succirubra,* a tree with wind-dispersed seeds, has invaded much of the high-elevation forest and shrubland of Santa Cruz. Lantana, *Lantana camara,* a weedy tropical shrub introduced for ornamental purposes, is another invasive exotic that has replaced *Scalesia* forest in many areas on Santa Cruz and San Cristobal.

Vegetation destruction has secondarily affected many animals. Damage and destruction of the *Scalesia* forests probably caused the extinction of four populations of the sharp-beaked ground finch, which has disappeared from all islands that are inhabited by humans. Browsing by goats, donkeys, and other large mammals has probably been responsible for several extinctions of birds. On Floreana Island, for example, destruction of the forests of tree prickly pears by these animals probably led to the extinction of the native mockingbird and the large-billed ground finch, both of which depended on this plant.

### C. The Charles Darwin Foundation and Research Station

Organized conservation efforts in the Galápagos date from the creation of the Charles Darwin Foundation in 1959, through the efforts of a group of scientists who had become concerned about the impacts of human activities on the biota of the archipelago. This foundation, supported by the Ecuadorean govenment and various international groups, sponsored the construction of the Charles Darwin Research Station at Academy Bay on Santa Cruz Island. This station, which opened in 1964, has become the center for research and conservation efforts in the islands.

### D. The Galápagos National Park

In 1959, the Ecuadorean government also formally established the Galápagos National Park. This action placed 97% of the island land into the park, protecting it against settlement, domestic animal grazing, and mining. Personnel of the Charles Darwin Research Station aided the creation of the Galápagos National Park Service, which was not formed until 1968. The park is now listed as a World Heritage Site by UNESCO. [*See* NATURE PRESERVES.]

### E. The Galápagos Marine Resources Reserve

In 1986 the Ecuadorean government established the Galápagos Marine Resources Reserve, which formalized Ecuadorean federal control over waters within the archipelago and to a distance of 24 km from the outermost islands, an area of 70,000 $km^2$. Initially, this decree protected the archipelago against commercial exploitation of biotic or mineral resources. Management of the reserve, however, is overseen by a commission made up of representatives of various national ministries, including those concerned with mining, fishing, commerce, and national development.

### F. Current Conservation Priorities and Programs

Major efforts are being made to control the spread of invasive plants and to eradicate introduced animals. Goats have been eliminated from South Plaza, Santa Fé, Española, Marchena, and Pinta. Substantial recovery of the vegetation has followed goat control. Feral cattle have been eliminated from San Cristobal, Santa Cruz, and Floreana and feral dogs from a number of areas.

The Charles Darwin Research Station is concentrating on protection and recovery programs for three species: the giant tortoise, the land iguana, and the dark-rumped petrel, which nests only in the Galápagos and Hawaiian islands. Captive rearing programs are maintained for tortoises and land iguanas, and large numbers of these species have been reintroduced to their native islands. Tortoises from Española are bred in captivity, whereas young of other subspecies are reared from eggs collected in the wild. Captive-reared tortoises have been returned to several islands. Successful breeding by repatriated tortoises was noted on Española in 1990. The total population of giant tortoises in the wild is now estimated to be about 13,000 animals. Captive-bred land iguanas have been reintroduced successfully to Baltra Island, where they had become extinct shortly after World War II.

The dark-rumped petrels nest in burrows in highland areas of four islands, and the adults leave their chick unattended for up to several days as they forage for food at sea. Thus protection of the dark-rumped petrel has concentrated on reducing predation in nesting areas, particularly by rats.

### G. Present and Future Threats to Biota

Under a policy of unrestricted immigration, the resident Ecuadorean population of the islands has grown from fewer than 3000 in the 1960s to 14,000 in 1993. This population is placing heavy demands on scarce resources, particularly fresh water, causing the Ecuadorean government to consider restricting immigration to the islands. The archipelago is now served by airports on three islands, however, and tourism is thought likely to increase to 100,000 visitors annually within a few years. Thus, it is likely that the resident human population will continue to grow rapidly.

The increased human influx has also increased the rate of invasion and spread of exotic species through the archipelago. Recently, for example, goats have expanded their range into the northern arm of Isabela, where some of the healthiest giant tortoise populations remain.

Exploitation of marine resources has also intensified. In spite of the protected status of Galápagos internal waters by the Galápagos Marine Reserve, illegal harvest of marine life is heavy. Black coral jewelry is sold openly in shops, and illegal fishing is extensive, particularly for sharks. In June of 1994, the Ecuadorean government announced plans to open the Marine Resources Reserve to commercial sea cucumber harvesting, a fishery notorious for its unsustainability, and to a large-scale shark fishery. The Darwin Foundation is presently contesting these actions.

## Glossary

**Caldera** Volcanic mountain in which the upper cone has collapsed into internal space created by withdrawal of lava, creating a wide, flat-bottomed crater.

**Endemic form** Species or other taxonomic group that has evolved its distinct characteristics in a particular region, to which it is restricted.

**Panamic** Having a biogeographic affinity to the biota of the tropical Pacific coastal waters of the Bay of Panama.

**World heritage site** Natural or cultural site judged to be of importance to all of humanity and listed by the World Heritage Program of the United Nations Educational, Scientific, and Cultural Organization (UNESCO).

## Bibliography


Cox, A. (1983). Ages of the Galapagos Island. "Patterns of Evolution in Galapagos Organisms" (R. I. Bowman, M. Berson, and A. E. Leviton, eds.), pp. 11–23. San Francisco: Pacific Division, American Association for the Advancement of Science.

Cox, G. W. (1990). Centers of speciation in the Galapagos land bird fauna. *Evol. Ecol.* **4,** 130–142.

Grant, P. R. (1986). "Ecology and Evolution of Darwin's Finches."Princeton, N.J.: Princeton University Press.

Hamman, O. (1981). Plant communities of the Galapagos Islands. *Dansk Botanisk Archiv* **34,** 1–163.

Peck, S. B. (1992). The Galapagos Archipelago, Ecuador. With a special emphasis on terrestrial invertebrates, especially insects; and an outline for research. *In* "The Unity of Evolutionary Biology" (E. C. Dudley, ed.), pp. 319–336. Portland, Ore.: Dioscorides Press.

Schofield, E. K. (1989). Effects of introduced plants and animals on island vegetation: Examples from the Galapagos Archipelago. *Cons. Biol.* **3,** 227–238.

Slevin, J. R. (1959). "The Galapagos Islands: A History of Their Exploration." San Francisco: California Academy of Science.

Smith, G. T. C. (1990). A brief history of the Charles Darwin Foundation for the Galapagos Islands 1959–1988. *Noticias de Galapagos* **49,** 1–36.

Steadman, D. W. (1986). Holocene vertebrate fossils from Isla Floreana, Galapagos. *Smithsonian Contrib. Zool.,* No. 413.

Wellington, G. M. (1984). Marine environment and protection. *In* "Galapagos" (R. Perry, ed.), pp. 247–263. Oxford, England: Pergamon Press.

# Gene Banks

**Garrison Wilkes**
*University of Massachusetts at Boston*



Gene banks are exactly what the name implies—where genes are stored for safekeeping. The first gene banks came into being early in this century from what were plant breeders' working collections, yet the majority have been created since the 1970s by national governments or international organizations. In theory, any live cell is a holder of genes, but in practical reality, gene banks usually store plant seeds. Purists use the phrase "seed gene banks" to distinguish them from germplasm repositories that preserve clonally reproduced material, sometimes referred to as cuttings, bud wood, or cell lines. The word seed also implies plants to distinguish them from animal genetic resources in a sperm gene bank (storing male gametes), egg gene bank (storing female gametes), or embryo gene bank. Seed bank, on the other hand, usually refers to the potential of a plant community to regenerate from seeds already present in the soil. The classic example would be the regeneration of a forest after a devastating fire that burned everything above ground level. In practice, the vast majority of the plant seed gene banks or just plain gene banks store seed of useful plants (food, fibers, drugs, and medicines) whose reproducible genetic potential is referred to as plant genetic resources.

## I. INTRODUCTION

A few botanical gardens have seed collections that represent a cross section of the world's flora biodiversity, but most gene banks hold only the seeds of plants useful to humans (referred to as economic plants), with about 500 species being of major importance. Most gene bank storage consists of the fifty or so widely planted food crops. Operationally, the term gene bank has come to mean live seed collections held under storage conditions, usually at refrigeration temperatures, for the major food crops such as wheat, corn, rice, potatoes, beans, and soybeans, although some timber trees, ornamentals, and medicinal plants can be found in gene banks. [*See* SEED BANKS.]

To be stored safely for a long time, living seeds and the genes they hold must be at a low moisture content, usually 4 to 9% depending on the chemical composition of the seed. A large number of seed crops may be stored for up to 25 years or more at refrigerated temperatures just above the freezing point at 25% relative humidity, or preferably in sealed containers. These are called "active storage conditions." Longer storage periods are maintained at subfreezing temperatures. At temperatures of

−18 to −20°C, safe storage of dry seed (4 to 9%) in sealed containers for 50 years or more is the norm. These are called "base storage conditions." At the extra cold temperatures of −150°C or lower of liquid nitrogen vapor, dry seed (4 to 9% moisture) is expected to remain viable for up to 500 years, and certainly safe storage for 100 years is a reasonable expectation. This cryopreservation is thought to bring cell activity to a complete stop and therefore is a standard, extra-secure preservation method for orthodox seed (seeds that can be frozen). Some (mostly tropical) seed do not survive low temperatures, and these recalcitrant seeds are a special problem for gene banks. Plant tissues such as meristems and excised embryo of recalcitrant seeds can be kept in cryopreservation and brought back to growth phase under appropriate, often taxa-specific, growth-stimulating conditions. The number of national and institutional gene banks with refrigerated active collection storage conditions exceeds 100. The number of gene banks with base collection storage facilities is close to 40, which is up from the 7 in operation in 1970. The number of gene banks using cryopreservation is between 7 and 10.

The International Board of Plant Genetic Resources (IBPGR), part of the International Agricultural Research Centers (IARC), was established in 1974 as an autonomous international scientific organization at the Food and Agriculture Organization (FAO) in Rome, with a mandate to promote a world wide network of genetic resource centers. By 1980, there were over 50 national, regional, and international institutions conserving and storing germplasm. In 1994, the IBPGR became the International Plant Genetic Resources Institute (IPGRI). [*See* Germplasm Conservation.]

There are no guidelines on how big, complete, or specialized a gene bank needs to be before it is considered a gene bank. Some national gene banks hold accessions numbering in a few hundred to several thousand, with almost all holdings coming from their own country or the geographic region. Other national collections, such as those of the former U.S.S.R. and the United States, number in the hundreds of thousands and have holdings from all parts of the world where crops are grown. The Russian geneticist Nikolai I. Vavilov established the first gene bank in the 1920s, the All Union Institute of Plant Industry, in Leningrad. This bank, now called the Russian Vavilov Institute (VIR), still exists in St. Petersburg and is one of the largest gene banks in the world.

Intermediate in size are regional gene banks such as the Nordic Gene Bank for the Scandinavian countries or the Tropical Agriculture Research and Training Center (CATIE) of the Central American countries. Contrasting with these broad-based banks are the crop-specific holdings of the International Agricultural Research Centers, such as the International Rice Research Institute (IRRI) world collection of rice and Centro Internacional de Mejorimento del Maíz y Trigo (CIMMYT) world collection for maize. There are about one hundred major and/or significant gene banks or gene bank systems (*See* Germplasm Conservation for a listing). Gene bank systems exist in some large nations, such as the United States, India, and China, that have central seed storage facilities and regional crop-specific and/or climate-specific gene bank repositories with active collections. The larger gene banks tend to be in the developed countries, and the holding of crops outside the basic cereals, legumes, and potatoes are not as well represented, especially if they are strictly tropical.

## II. CROP GENETIC RESOURCES OR GERMPLASM

Germplasm is the source of the genetic potential of living organisms. Among other things, diversified germplasm allows organisms to adapt to changing environmental conditions. No single individual of any species, however, contains all the genetic diversity of that species. This means that the total genetic potential is represented only in populations made up of many individuals. Such genetic potential is referred to as the gene pool. The potential represented in a gene pool is the foundation for crop plants in both agriculture and forestry.

Germplasm is maintained only in living tissue, most often the embryo of seeds. When the seed of an annual plant dies, the germplasm is lost. Extinction of a species, crop variety, or a genetic line represents the loss of a unique resource. This type of genetic and environmental impoverishment is irreversible and is called genetic erosion. When the seed of a unique variety is no longer planted, it is a silent loss. Like soil erosion, it is gone with no drama of disappearing; genetic erosion leaves a void and a diminished gene pool.

To the extent that we cannot be certain what needs may arise in the future (new plant diseases or pests, climatic change due to the greenhouse effect, and so forth), it makes sense to keep options open and maintain the most comprehensive gene pool possible for the future. This conservation rationale has led to the establishment of plant seed gene banks.

## III. USE OF CROP PLANT GERMPLASM

To better understand why crop plant genetic resources have been stored in gene banks we must consider how they are used. Certainly the most important use is for plant breeding. Plant breeding is a method of germplasm enhancement of already existing allelic variation, of creative recombinations through hybridization of differing genotypes and intense artificial selection of plant forms that probably would not survive in the wild. True domesticates have been so altered that they survive only under cultivation in agriculture or horticulture. One example of a characteristic that has been altered is seed germination. Farmers expect the seed of cultivated plants to germinate immediately after being placed in the soil. Such a characteristic usually would be lethal for a wild plant, which must have mechanisms to ensure germination only at the proper season, hence promoting survival. In cultivated plants, survival is keyed to the farmers' commitment for seedbed preparation, decreased competition with other plants, the sowing of seed, the protection of the plants during growth from pest and pathogens, and finally the collection of fruits, seeds, and other edible parts for human use, but always saving sufficient seed to sow and renew the crop the following year.

All major cultivated food plants have lost the ability to exist in the wild; in other words, they are fully domesticated. These plants have been selected to produce unusually large plant parts, low-fiber edible tissue, thick flesh with intense color, and fruits attached by tough stems. They are so altered genetically that they are dependent on humans to sow them in the proper season, to protect them from competition and predation, to supply them with water and nutrients, and to harvest their propagules in order to repeat the cycle. If we did not save the seed, we would lose the ability to repeat the cycle and feed ourselves. If we saved only one variety of every crop, we would lose the ability to promote the further evolution of that crop and thus run an increasing risk of crop failure due to changing pest pathogens and weather conditions. When a society can no longer change crops by plant breeding to meet changing conditions, it has lost the key to agricultural sustainability.

## IV. GERMPLASM AND THE HISTORY OF PLANT BREEDING

Plant breeding as a human activity can be viewed as developing historically through three phases, and we are currently on the threshold of a fourth. The process began about 8000 to 10,000 years ago when humans started to save seed. The earliest stages of domesticated crops were probably not much more productive than their wild progenitors, but the act of cultivation and saving seed to replant was a radical break with the past. We know that this environmental rearrangement can be traced back to at least five areas: China (rice, soybeans, Chinese cabbage, peach), South and Southeast Asia (banana, sugarcane, mango, rice), the Near East (wheat/barley, peas, apples/pears, grape), Mesoamerica of Mexico and Guatemala (maize, beans, squash, tomatoes, chilies), and the South American Andean highlands (potatoes, beans, maize). Probably agriculture also originated independently in different parts of the globe with native plants of the region. This was the first stage of plant breeding or human control over crop plant evolution. All the important world crops were developed in this first stage.

The second stage of plant breeding began with the European colonization of the Americas and the circumnavigation of the world, and the rapid diffusion of crops, livestock, and farming techniques that followed. The U.S. experience with agricultural productivity has been rich because of the worldwide genetic contributions to the nation's plant breeding and improvement process. The earliest European settlers brought with them from their homeland the native varieties or folk landraces. Thus, barley from England grew beside barley from Germany in the newly settled colonies. Ship captains often brought back cargoes of wheat grown in the Punjab of British India and purchased in the port of Calcutta, and rice from China, Madagascar, or Madras for a relative or friend to try on the farm. The Spanish missions of the U.S. West introduced arid land crops that were completely foreign to farmers of the East Coast. And later, with the large-scale immigration from central and southern Europe, new and distinct genetic diversity was added to the basic crops. These immigrants who brought seed with them helped to establish a broad genetic wealth for plant breeding and improvement. This migration and acclimatization of crops throughout the world, often in conjunction with hybridization between dissimilar varieties, was the basis of the tremendous variation generated in the second stage of plant breeding.

The third stage of plant breeding and improvement began with the rediscovery of Gregor Mendel's classic experiments on the heredity of garden peas at the beginning of this century. For the first time, the plant breeding community had a set of principles by which to proceed with the crop improvement process. Products of this era are hybrid corn, changes in the photoperiod response of soybeans, and the dwarf-stature wheat and rice from CIMMYT and IRRI, respectively. These late 1960s Green Revolution cereals (dwarf-stature wheat and rice) and the genes they hold now enter the food supply of 2 billion people and are directly responsible for feeding more than 800 million people by their increased yields alone.

The irony of using improved varieties is that they have a tendency to eliminate the resource upon which they are based and from which they have been derived. Current elite varieties yield better than their parents, and they displace them in farmers' fields. Once a displaced variety is no longer planted, its genes are lost to future generations unless it is conserved, usually in a gene bank collection.

The fourth era of plant breeding, genetic engineering, is dependent on the optional use of crop plant gene pools. Traditional plant breeding is, in fact, genetic engineering, but this term is now being limited to biotechnologies such as cell culture and recombinant DNA techniques. These biotechnologies can speed the breeding because genetic material is introduced directly into cell cultures (splicing), thus completely sidestepping the genetic recombination through the usual sexual processes or when cell cytoplasm is altered as in protoplast fusion. And, of course, all of these changes depend on cloning technologies in which hundreds of identical plants are grown from units as small as a single cell, which ultimately result in seed-producing plants.

The advent of these technologies has ramifications regarding legal and ethical issues of plant germplasm. Because these advances use manipulative and unnatural means, they can become intellectual property and thus are potentially protected by the laws of ownership (patents). The spector of patent protection has polarized the public view of germplasm as either public good (common heritage) or private good (rewards for value added). The developed world has the technology and financial resources to enter the global biotechnology market, but because most of the developing world lacks the technical resources, it is fearful of what these new technologies hold for it in the future. The dichotomy is further accentuated by the fact that the origins of most of the world's basic economic plants lie in the developing countries, yet they do not have clear title or recognized ownership claim as the original inventor or developer (farmer rights). Because most of the genetic variation of these crops is already present in the major gene banks of the developed nations, the developing nations feel fraudulently deprived. Much of this fear is reflected in an International Undertaking on Genetic Resources of the Food and Agriculture Organization of the United Nations, which aims for international sovereignty over gene banks. This has focused attention on plant genetic resources as a keystone in our food production system; but unfortunately, the discussions and solutions sought have been more political and have not led to accelerated gene bank activities to conserve plant genetic resources or to facilitate the availability of these resources to the plant breeding community.

## V. GENE BANKS AND CROP GENETIC RESOURCES

Gene banks are defined in part by the types of materials they hold. Each individual sample is referred to as an accession and came from a donor such as a plant collector, plant breeder, university researcher, or another gene bank through exchange. Each accession possesses or should possess basic passport information about its site of origin: country, exact location, elevation, soil type conditions where collected such as cultivated fields or native vegetation, date of collection, name of collector, and any other notes made by the collector. This passport information is extremely valuable to gene bank managers in the decision-making process of whether to enter the material into the bank (i.e., it may be a duplication of already held accessions) and where and under what conditions to grow out the collection for seed regeneration. The emphasis in most gene banks is to possess representational holding of the genetic diversity found in a crop or useful plant. Often this means that special emphasis is placed on a region of the world where a crop has originated and/or areas where farmers have developed a large number of native landraces or farmer-bred varieties. Ultimately, the decision of what is accessed into the bank is made by the curator. Once a collection is accessed, it becomes the obligation of the gene bank to store it safely, regenerate the accession to renew the viability of the seed, and make the seed available to the plant improvement community. Taking entries out of a gene bank by deaccessing a collection is not a casual activity, because once accessed they are usually held in trust for future human generations.

Once a sample is accessed, the curator assigns a permanent accession number unique to that bank and a storage address in the storage facility, much like the name of a book and its call letters for its position on a library's shelves. As a management tool, the curator will develop characterization data, such as seed characteristics of color, weight per 100 seeds, and type of storage tissue (i.e., large white seeded flour corn), which will help to confirm that the seed in the bottle, can, or foil/plastic laminate pouch does indeed match the accession information and confirm the labeling as correct. In addition, the curator needs to discover useful attributes about the accessions so it can be put in the hands of breeders and researchers. This information is called evaluation data and herein lies a major dilemma for gene bank curators. Their obvious obligation is safe storage and their job performance is measured by the number of accessions held at the least expense. Unfortunately, growing the accessions to develop evaluation data greatly increases the cost and liabilities for the gene bank and the curator. Therefore, few gene banks have evaluation/performance data and their holdings are largely unknown. Less than half of the world's gene banks have computer-accessible accession/passport records available for the public or user community. Only about one in ten gene banks has machine-readable evaluation data for the public on their accessions. Gene banks were initially designed to increase the availability of crop genetic diversity, but because of the lack of public interest and financial support, they focus on storing seed.

Clearly plant breeders need to establish the efficacy of gene banks. Knowledge of what is in the world collections has been the major stumbling block to their use. A great deal of new and novel genetic variation is locked up for major crops in the approximately 2 million gene bank accessions worldwide. If the function is solely storage, the germplasm will never be evaluated, and its potential will not be discovered. The evaluation of samples requires the growing and assessing of plants in the field to discover genes that control specific traits and to use that discovery to exploit that trait in genetic lines for tomorrow's high-yielding elite varieties. Breeders point out that without better passport and descriptor data, they have no rational entry into gene banks and must search more or less blindly for useful materials. These unknown materials are called exotic germplasm and are used by most breeders only as the last resort in choosing parental material. For the plant breeding community, the vast majority of the accessions in gene banks are considered to be "exotic" because so few performance data on productivity and adaptation are known.

Because most gene banks hold many more accessions than can be screened by a plant breeder in a single season, or even five seasons, for a needed trait there has developed the concept of a Core collection. The holdings of a gene bank are a sample of farmers' fields and the Core is a representative sample of the accessions of a gene bank (therefore, the Core is a representative sample of a sample with a suggested size of 10% of the accessions in the collection). For large gene banks, the Core should not exceed a maximum of 500. In theory, a Core should be constructed to form a representative collection of the different ecological habitats or farming regions with a minimum of repetitiveness of the genetic diversity of a crop species and its relatives. The Core is an outline of crop holdings for the gene bank and the accessions not in the Core are called the Reserve collection. If a Core accession proved of interest to a breeder, they would want to look at the Reserve collection from the same geographical area. The Core subset is an organized method to search the holdings of a gene bank.

Currently gene banks hold aging seed, yet many banks have no action plan to grow out their collection and rejuvenate the seed. A major concern is that some gene banks might be storing more dead than alive seed. This certainly is not true of well-managed gene banks, but not all banks worldwide are well managed.

Gene banks were relatively easy to establish in the 1970s, when the threat of genetic erosion compelled the creation of major collections. The real test of a gene bank comes when it is time to grow the seed. Most gene banks have neither adequate facilities nor funding to achieve this function. Some crops are relatively easy to regenerate: these are genetically homogeneous, inbreeding, short-cycle annual plants such as wheat, where an accession grow-out of about thirty plants would take the field space of a large dinner plate. At the other extreme is a genetically highly heterogeneous, outcrossing annual crop plant that exhibits considerable inbreeding depression, which might take the field space of approximately a basketball court (or one third the size of a soccer field). As an example of the latter, the recommendations to regenerate an accession of maize call for a regeneration packet of 512 seed planted two seeds per hill for 256 hills and thinned to one plant once they germinate. When the plants flower, each ear is pollinated by the male tassel of another plant, which is used only once. This plant by plant or chain crossing is more labor-intensive (because of hand pollenization of each plant individually) than bulking male pollen to pollenate all plants, but ensures greater probability that genes at low frequency will be captured and saved in the regeneration. Once harvested, a minimum of 100 ears are needed; if fewer than 100 good ears result, the grow-out is repeated. A balanced regeneration packet requires an equal number of seed from each ear. To be kept under base storage conditions, an adequate bank accession should have at least three balanced regeneration packets and from 1 to 5 kg of bulk seed for distribution from the active collection. Accessions that are in demand might require seed holdings of 25 kg to satisfy requests. Balanced regeneration seed packets should never be used for distribution—they are made specifically for the next bank grow-out to produce fresh seed. Many banks have a backup regeneration packet stored at another site or gene bank in case there is an unforeseen disaster that would cause their facility to fail. This is called secondary storage or an off-site backup collection.

If the crop is a long-lived perennial, it might be kept as a date or apple orchard, because individual varieties do not breed true and (1) are clonally propagated and/or (2) take a substantial length of time to grow from seed to flower. Plant germplasm that must be maintained vegetatively presents a special challenge for curators. Often these are called germplasm repositories, with the implication that a gene bank stores only seeds. By their very nature they hold fewer accessions than seed gene banks and are much more vulnerable to weather, diseases, and day-to-day problems of maintenance.

A sense of the breadth of activities involved in gene banking plant genetic resources becomes obvious from a review of the National Plant Germplasm System (NPGS) of the United States. The NPGS has evolved over a period considerably shorter than that of many of the plants held within it. The Department of Agriculture, begun in 1862, took over from the U.S. patent office the already existing and very popular free distribution of crop plant seed, most of them being garden vegetable seed. In 1898, the Section of Seed and Plant Introduction (PI) was founded to undertake plant exploration and introduction of new useful germplasm to the country. These collections were given PI numbers and distributed to the appropriate region of the country to test their agronomic potential. As late as the 1940s only a few institutions had the facilities to provide minimal refrigerated and

**TABLE I**

**Crops and Location for the U.S. National Plant Germplasm System's Regional Plant Introduction Stations**

| Ames, Iowa | Geneva, New York | Griffin, Georgia |
|---|---|---|
| Amaranth | Apples | Castor bean |
| Jerusalem artichokes | Crabapples | Bermuda grass |
| Asparagus | Broccoli | Blackeye peas |
| Beets | Brussels sprouts | Eggplant |
| Bent grass | Cabbage | Gourds (with Ames) |
| Buckwheat | Chinese cabbage | Guar |
| Cantaloupe | Cauliflower | Kenaf |
| Carrots | Celery | Lespedeza |
| Chicory | Grapes (cool season) | Luffa |
| Sweet clover | Onions (with Pullman) | Mungbean |
| Collards | Peas | Okra |
| Coriander | Pumpkins | Peanuts |
| Corn | Radish | Pearl millet |
| Crambe | Shallots | Pennisetum–sorghum |
| Cucumber | Squash (with Ames and Griffin) | Peppers |
| Dill | Tomatoes | Squash (with Ames and Geneva) |
| Endive | Cherry tomatoes | Pigeonpea |
| Gourds (with Griffin) | | Serradella |
| Honeydew melon–muskmelon | **Pullman, Washington** | Sesame |
| Horseradish | Alfalfa | Sweet potato (with Mayaguez) |
| Kale | Beans | Water chestnuts |
| Kohlrabi | Bluegrass-fescue | Watermelons |
| Mustard | Brome grass | Wingbeans |
| Parsley | Canary grass | Zoysia grass |
| Parsnips | Chickpeas | |
| Pawpaw | Chives | |
| Pumpkins (with Geneva and Griffin) | Garlic | |
| Rutabaga | Leeks | |
| Spinach | Lentils | |
| Squash (with Geneva and Griffin) | Lupine | |
| Sugar beets | Milkvetch | |
| Sunflowers | Orchard grass | |
| Turnips | Pak choi | |
| Zucchini | Ryegrass | |
| | Safflower | |
| | Sainfoin | |
| | Teff | |
| | Vetch (with Griffin) | |
| | Wheat grass | |
| | Wild rye | |

dehumidified conditions for preserving seed viability. The Research and Marketing Act of 1946 (Public Law 733) established the Regional Plant Introduction Stations, and the National Seed Storage Laboratory (NSSL) became the first U.S. facility specifically designed for long-term preservation of seed. It opened in 1958 in Fort Collins, Colorado, and had facilities for storing seed at 4°C and 32% humidity. Presently, seed is stored at −18°C with seed moisture around 5 to 10%, and some storage is cryogenic using the vapor of liquid nitrogen (−150 to −197°C). The NSSL stores more than 250,000 accessions, and in 1993 it opened expanded facilities with a storage capacity for more than a million accessions. The NSSL is only for long-term storage and the regional plant introduction stations, crop-specific seed collections, and national clonal germplasm repositories are responsible for active storage conditions, accession evaluation, and distribution of the crop under their mandate. A list of the facilities and their crop responsibilities is shown in Tables I and II and Fig. 1.

## VI. FUNCTIONS OF A GENE BANK

Gene banks have been the institutional solution to genetic erosion. Although a degree of conserva-

**TABLE II**
National Clonal Germplasm Repositories

| Corvallis, Oregon | Davis, California | Brownwood, Texas |
|---|---|---|
| Blackberries | Almonds | Chestnut |
| Blueberries | Apricots | Hickory |
| Boysenberries | Cherries | Pecan |
| Cranberries | Figs | |
| Currants | Warm-season grapes | |
| Filberts–hazelnuts | Cold-season grapes | |
| Gooseberries | Kiwifruit | |
| Hops | Nectarines | |
| Mint | Olives | |
| Pears | Peaches | |
| Raspberries | Persimmons | |
| Strawberries | Plumcots | |
| | Plums | |
| | Pomegranates | |
| | Tomato genetic stocks | |
| | Walnuts–pistachios | |
| Miami, Florida | Riverside/Brawley | Hilo, Hawaii |
| Avocados | Dates | Carambola rambutan |
| Cassava (with Mayaguez) | Grapefruit | Guava |
| Cocoa (with Mayaguez) | Lemons | Lychee |
| Coffee | Limes | Macadamia |
| Mangoes | Mandarin oranges | Papaya |
| Passionfruit (with Hilo) | Oranges | Passionfruit (with Miami) |
| Sugarcane | Tangerines | Pilis |
| | | Pineapple |
| Mayaguez, Puerto Rico | | Washington, D.C. |
| Bamboo | | Dogwoods (with Ames) |
| Bananas–plantains | | Holly |
| Brazilnuts–cashews | | Landscape ornamental plants |
| Cassava (with Miami) | | Magnolias |
| Cocoa (with Miami) | | Maple |
| Sweet potato (with Griffin) | | Oaks |
| Taniers | | Rhododendrons |
| Yams | | |

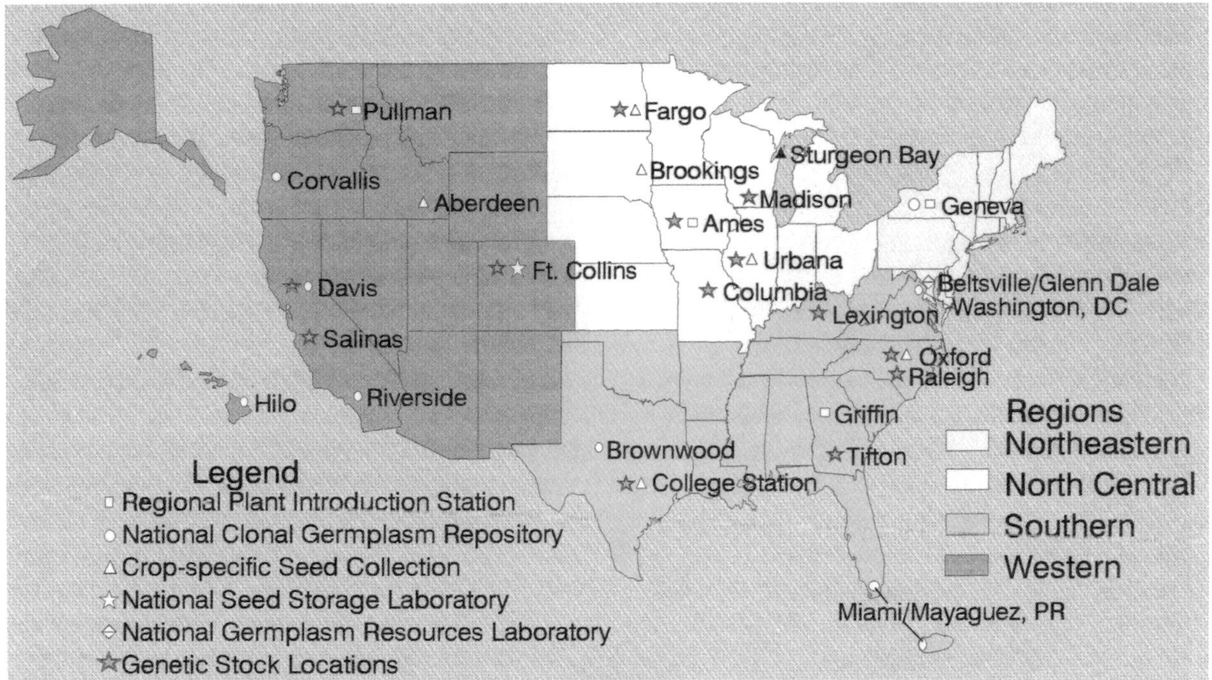


***FIGURE 1*** National Plant Germplasm System (U.S.) as of March 30, 1994.

tion can be achieved by farmers maintaining old varieties and home gardeners growing heirloom vegetable varieties, this cannot be as inclusive of the gene pool as gene banks. To be successful, gene banks must have four distinct functions in addition to preservation of seeds and sometimes clones to counter the loss of genetic erosion. They must:

1. Be proactive and linked to exploration of undercollected zones to increase and maintain representation of the gene pool as samples in the gene bank.
2. Maintain "in trust" the genetic integrity of samples in storage over time and regenerate stocks periodically.
3. Establish "utility" of the gene bank by actively undertaking to evaluate and document a useful data base to guide management and use of the holdings; also, effect the free and timely distribution of accessions to bona fide users so that "raw materials" will be in the hands of those that mold crop plant evolution.
4. Be linked to active evaluation prebreeding and early breeding for enhancement so that gene bank materials will be in a useable form for the plant community. Unless there are more plant breeders working with gene bank materials to evaluate and use the genes, agriculture with only current elite varieties will undermine itself and fail.

Because crop plant evolution is precluded for accessions in gene banks, both hybridization and segregating populations of a breeding process become necessary adjuncts to the well-functioning gene bank. A complete listing of the kinds of genetic materials that can be saved as seed is given in Tables I and II.

## Glossary

**Accession** Individual sample of seed or plants entered into a germplasm collection in a seed gene bank or clonal repository.

**Active collection** Collection of gene bank accessions maintained for medium-term viability (for most crops 20 to 40 years), usually stored at temperatures just above freezing (0°C) and at 4 to 9% moisture. Requests for seed are usually supplied from these storage conditions.

**Base collection** Collection of accessions kept for long-term secure storage (30 to 50 years) at below freezing temperatures (−18°C) and at 4 to 9% moisture. These are not routinely available for distribution.

**Cultivar** Variety (cultivated variety) of a plant produced by selective breeding, which has been specifically improved for agriculture or horticultural purposes and is grown in cultivated conditions.

**Germplasm** (1) Genetic material that forms the physical basis of heredity for a species and that is transmitted from one generation to the next by means of the germ cells. (2) Individual, clone, or cell line representing a type, species, or culture that may be held in a repository for agronomic, historic, or other reasons.

## Bibliography

CIMMYT (1986). "Seed Conservation and Distribution: The Dual Role of the CIMMYT." Mexico, D.F.: Maize Germplasm Bank. (Available from CIMMYT, A.P. 6-641 Mexico, D.F. 06600 Mexico.)

Holden, J., Peacock, J., and Williams, J. T. (1993). "Genes, Crops and the Environment." Cambridge, U.K.: Cambridge University Press.

Janick, J. ed. (1989). "The National Germplasm System of the United States," Plant Breeding Reviews, Vol. 7. Portland, Ore.: Timber Press.

Plucknett, D. L., Smith, N. J. H., Williams, J. T., and Anishetty, N. M. (1987). "Genebanks and the World's Food." Princeton, N.J.: Princeton University Press.

Skovmand, B., Varughese, G., and Hettel, G. (1992). "Wheat Genetic Resources at CIMMYT: Their Preservation, Enrichment and Distribution." Mexico, D.F.: (Available from CIMMYT.)

USDA (1990). "Seeds for Our Future: The U.S. National Plant Germplasm System." Washington, D.C.: USDA. (Available from National Plant Germplasm System, ARS/USDA, Room 140, Bldg. 005, BARC-West Beltsville, Md. 20705, U.S.A.)

White, G., Shands, H., and Lovell, G. (1989). History and operation of the National Plant Germplasm System. "Plant Breeding Reviews" (J. Janick, ed.), Vol. 7, pp. 5–56. Portland, Ore.: Timber Press.

Wilkes, G. (1988). Plant genetic resources over ten thousand years: From a handful of seed to the crop-specific mega-gene bank. *In* "Seeds and Sovereignty." (J. Koppenburg, ed.), pp. 68–89. Durham, N.C.: Duke University Press.

# Germplasm Conservation

**_Garrison Wilkes_**
_University of Massachusetts at Boston_



The imperative for germplasm conservation develops from three processes in our current world: (1) an increasing human population that leads to further alteration of natural ecosystems and the expansion of food-producing agriculture; (2) the widespread adoption of elite crop plant germplasm and agricultural technology that promotes genetic uniformity, sometimes worldwide; and (3) because of habitat displacement and genetic erosion, the centers of genetic variability are moving from natural systems and primitive agriculture to genebanks and breeders' working collections. Clearly, the future of plant breeding, and therefore agriculture and world food supply, rests on the conservation of plant genetic resources of our most productive crop plants held in international agricultural research centers and national program genebanks. The responsibility is double sided; both the holder of genetic diversity and the breeders that enhance the genetic architecture have an obligation to ensure that these resources are used to improve the human condition.

## I. INTRODUCTION

Germplasm is the hereditary message that dictates the development of plants and animals. This biological information found in the cytoplasm and more extensively on the chromosomes in the cell nucleus is passed from generation to generation in an unbroken chain of living cells. Unlike other resources, once cell life is lost, the germplasm is lost. Therefore, for all practical purposes, life's code is stored only in living cells. Biological and genetic diversity maintenance has been one of the unrecognized public service functions of natural ecosystems. Suddenly, with the rapid doubling of the human population and the disappearance of natural systems, germplasm conservation has become a concern of society.

Given the needs of the future, genetic resources can be reckoned among society's most valuable raw materials. Any reduction in the diversity of resources narrows society's capacity to respond to new problems and opportunities. To the extent that we cannot be certain what needs may arise in the future, it makes sense to keep our options open. This conservation rationale applies to the earth's endowment of useful plants more than to almost any other category of natural resource.

The world's human population could not survive at its present density were it not for cultivated crop plants and domesticated animals. Because approximately 80% of our calories are supplied by plants,

Encyclopedia of Environmental Biology
Volume 2

the primacy of plants in meeting the food needs is evident. The germplasm of crop plants is in a much more advanced state of conservation or gene banking than the other useful plants of forestry or microorganisms. For animals the situation is even less advanced, there being very limited knowledge of the genetic variation that exists worldwide in domesticated animals. In most cases, the genetic variation of plants can be conserved as living seed or cell cultures. The conservation problems with animals are much more extensive, difficult, and expensive because of the size and labor of maintaining herds and flocks, not to mention the problems of disease and quarantine. Because of the more advanced worldwide effort and planning for plant germplasm conservation, the rest of this article will focus on crop plants.

## II. THREATS TO GERMPLASM OF CROP PLANTS

### A. Genetic Erosion

A technological fact of improved varieties is that they have a tendency to eliminate the resource that they are based on and from which they have been derived by breeding. Current elite varieties yield better than the varieties they displace, and once a displaced variety is no longer planted, its genes are lost to future generations unless it is conserved. The gene-rich ancestral forms are also lost because of bad land use planning, environmental degradation, and urbanization. The key point is that when a unique variety is no longer planted, it is a silent loss. Like soil erosion, it is gone with no drama of disappearing; what remains is diminished genetic diversity. [*See* PLANT CONSERVATION.]

### B. Genetic Viewpoint

The wholesale loss of plant genetic resources is called genetic wipeout by some. Genetic erosion is a slow, gradual process based on the independent individual decisions of farmers, whereas genetic wipeout is the rapid and sometimes one-stroke destruction of genetic resources, usually by institutional failure. Social disruptions such as political instability or crop failure and famine owing to natural disasters can eliminate genetic resources rapidly and have done so repeatedly in the past. The process of erosion and genetic wipeout are not mutually exclusive but are, in fact, two ends of a spectrum for biodiversity impoverishment of plant genetic resources.

### C. Genetic Vulnerability

In the developing world, high-yielding varieties (HYV) for the major crops have come into dominance just within the last two decades. As a few varieties from the same breeding programs come to dominate world production, increased vulnerability is almost certain to emerge.

The southern corn blight of 1970 in the United States was a dramatic demonstration of the inherent danger of uniform genetic susceptibility to a new and unrecognized pathogen. Approximately 15% of the U.S. corn crop was destroyed and in some zones essentially the entire crop was lost. Similar disasters have occurred in the past. In the eighteenth century a new food plant from the Andes of South America, the potato, was introduced to Ireland. The genetic diversity within the potato was limited, but being geographically isolated from some of its New World diseases and pests, the potato yielded well and the Irish population increased. In the 1840s, after the population had increased threefold to eight million, a previously unknown disease, late blight, caused by the fungus *Phytophthora infestans*, appeared in Ireland.

Within ten years approximately two million Irish emigrated, two million died, and four million remained, many in abject poverty. The potential exists for a similar crisis with today's HYVs, yet this time the human population is much greater. This condition is called genetic vulnerability and it is promoted by widespread monocultures and genetic uniformity for susceptibility.

### D. Use of Germplasm

The twenty-first century is going to see many new medicinal and industrial uses for plant products and many high-technology substitutes for conventional plant products (lumber and fabrics, for instance), but the human dietary requirements are going to change little and be met by basically the same short list of crop plants that feed us now: wheat, corn, rice, potatoes, barley, millets, sorghum, sugar cane and sugar beets, soybeans, beans, peas, tomatoes, onions, carrots, cabbages, apples, grapes, oranges, peaches, mangoes, bananas, yams, sweet potatoes, coconuts, and peanuts. These are the crops that demand new germplasm for improvement and to which we commit our plant breeding energies. [*See* PLANT SOURCES OF NATURAL DRUGS AND COMPOUNDS.]

The process of plant breeding is a dynamic one of genetic selection in response to changing diseases, insects, parasites, agricultural techniques, and human use. Since the beginnings of agriculture about 8000 to 10,000 years ago, the growing of these crops has expanded into many different environments and an enormous wealth of genetic variation has been generated and preserved over the centuries in locally adapted races. Only a small sample of the variation is now employed in the elite breeding stocks of leading crop varieties and much of the rest is still located in native indigenous traditional agriculture, which is rapidly disappearing. Because of genetic erosion, extensive collecting begun in the 1950s and peaking in the 1970s and 1980s has moved the diversity from farmers' fields and seed stores to genebanks for genetic conservation to support the future plant breeding process.

## III. GENES AND GERMPLASM

Genebanks are an institutional solution to genetic erosion and genetic wipeout. Although a degree of *in situ* conservation can preserve parts of the gene pools of crop plants, to date this mode of conservation has not been implemented broadly enough to be as successful as *ex situ* genebanks. To be successful, genebanks must have four distinct functions in addition to preservation of seed and clones to counter loss due to genetic erosion. They must:

1. be linked to exploration of undercollected ecological zones to increase and maintain representation of the gene pool as samples in the genebank;
2. maintain the genetic integrity of samples, provide storage over time, and regenerate stocks periodically;
3. evaluate and document samples and maintain a useful data base to guide the management and use of the holdings;
4. be linked to active evaluation, prebreeding, and early breeding for enhancement so that genebank materials will be in a useful form for the plant community.

Seed is the most common form of genebank preservation. Seeds age in even the best genebank and ultimately die sooner or later depending on the species. Periodically the seed must be planted, populations maintained through controlled pollinations, and new, fresh seed stock returned to the genebank. A seed genebank is generally the easiest to establish and does not incur a significant cost or management crisis until years later when the need for scientifically based regeneration develops. [*See* SEED BANKS.]

A better understanding of genebanks requires an appreciation of the kinds of genetic materials that can be saved as seed. Germplasm can be organized into five distinct categories:

- *varieties or cultivars in current use,* often very elite varieties;
- *obsolete cultivars,* often the elite varieties of 20 to 50 years ago and usually found in the parentage of current cultivars;
- *primitive cultivars* and landraces of traditional agriculture;
- *wild and weedy taxa,* near relatives of crop plants;
- *special genetic stocks,* including induced mutants.

Varieties or cultivars in current use have generally undergone a vigorous selection process by plant breeders for plant type, response to input, and predictability of yield and are more or less genetically homogeneous. The released varieties possess a "highly tuned" set of elite genes but a considerably narrowed gene base against the foundation parents from which they have come. Ultimately these foundation stocks trace back to landraces in diverse parts of the world.

Obsolete cultivars are advanced cultivars from the most recent past that have been displaced by a newer release. Often specifically selected older materials appear in the pedigree of a wide variety of releases.

Primitive cultivars and landraces are the real treasure house because they are the largest repository of genes for a crop. They are also the largest unknown because they are usually genetically heterogeneous and few data exist on their morphological, biochemical, and genetic traits, or on their responses to pests or environmental stresses.

Landraces are a rich source for unidentified, uncharacterized traits but usually exhibit narrow local adaptation. Landraces seldom make ready-made, broadly adapted cultivars. Generally, landraces perform poorly under high inputs of fertilizer, water, and intensive cultivation and are replaced by new elite seed strains. On the other hand, there is a fairly wide variation in the ability of landraces to survive fluctuating environments and to withstand cold, drought, insect damage, and other such variables. After all, most landacres represent accumulated mutational events integrated and balanced over centuries in the real world. Using the genes that landraces possess, modern plant breeding has formed the current elite cultivars.

The wild relatives and weedy taxa are poorly represented in genebank collections and are the most difficult to regenerate. This is because these ancestral forms and wild relatives closely related to the no longer extant ancestral taxa do not come from cultivated fields but rather from unique and often highly specific wild habitats. These taxa self-sow their seeds and exhibit genetic adaptation in an often very narrowly defined environment for day length, soil, and water relationships. They possess many dominant wild-type alleles and carry rare deleterious recessive alleles at very low frequencies. It is almost impossible to regenerate the genetic integrity of the wild collected seed because the site of regeneration does not match the site of collection and there are site regeneration adaptation and selection forces. Also, it is difficult to regenerate a population of sufficient size to capture rare alleles with confidence. In most genebanks, the wild and weedy forms are small samples from wild populations and may not represent the genetic diversity of these unique genetic systems with genes compatible with, but not found in, the crop plant.

## IV. MODES OF CONSERVATION

The method employed to preserve crop plant genetic resources will influence both what is held in the conservation and how it is evaluated for the discovery of potentially useful traits or attributes. There are essentially three modes of increasing human management intensity to conserve in perpetuity these resources.

### A. Biome Conservation

The preservation of entire vast tracts represents *in situ* conservation of the animal and plant ecosystems. This level of preservation will be extremely important in slowing extinction rates, especially in the tropical forest, but will have little impact on genetic resources of useful plants with the exception of timber trees. Some wild relatives of food and fiber crops will be affected, but most of these can be accounted for in the next level of management.

### B. *In Situ* Conservation

Prime candidates for *in situ* conservation include preservation of the agroecosystems of village farming where both fields and forest are skillfully managed, of landraces and wild relatives of the crop in ecogeographical pockets of genetic diversity, and of natural vegetation where near relatives and weedy forms exist. Probably the best case for *in situ* agroecosystems rests in the ancient cradles of agriculture, the Vavilov centers, where tremendous potential exists to combine *in situ* preservation with tourism and demonstrating the best of local cultural heritage. A second type of *in situ* preservation is the evolutionary garden habitat, where crops and their wild relatives hybridize and the resulting variation is in dynamic evolution with evolving pests and pathogens. A third form is a land reserve set aside for the wild relatives and weedy primitive forms of crop plants that are losing habitat through more intensive land uses. The unique aspect of *in situ* preservation is that it does not freeze the seed in a genebank generation but permits grow-out as a living population. If plants are growing in their natural place there are no questions of sampling strategy of the population; if the population is big enough there is no question with holding allelic frequencies of 1 in a 1000, and no question about selection shift or genetic drift at the time of seed regeneration. All of these technical aspects pose problems for *ex situ* genebank collections. On the other hand, *in situ* genetic conservation has a lot of practical problems with management, economics, communications, and local public support, but it does conserve genes at the ideal population level.

### D. *Ex Situ* Conservation

Preservation of plants or seed in genebanks or *in vitro* cell lines or clones in tissue collections under appropriate conditions for long-term storage often seems to be the easiest and most preferred mode. The disadvantage with genebanks is that they are only as complete as the starting sample and seldom can alleles at a frequency of less than 1 in 100 be either originally collected or maintained in rejuvenated seed (a truer statement for outcrossers than inbreeders at regeneration). A more significant problem is that hidden variation in genebanks, sealed in airtight containers under low temperatures to promote longevity, is useless until discovered. Genebanks store dormant seed and thus draw genes out of circulation. Discovery requires growing the plant to evaluate and document its traits so that users will have enough information to know where to look among the vast array of stored seed. Of the estimated 2 to 2 1/2 million genebank accessions held in just over a hundred genebanks worldwide, about one-half lack basic passport information for site collected and ecogeographical conditions and only 100,000 have this passport on machine-driven management systems. Genebanks retard crop plant evolution so that the evaluation and plant breeding process become a necessary part in making *ex situ* conservation useful.

Genebank collections, and in fact many breeder's collections, are too large for scientists to intensively investigate and record all traits in the screening process. Recently the core collecting concept has become popular. In the core, a selected sample of approximately 10% of the holdings (not to exceed approximately 500 entries) is carefully chosen to represent the agroecogeographical variation or other aspects of the crop worldwide. When a specific accession in the core shows promise in evaluation for a desired attribute or trait, then the remainder of the geographical cluster in the bank from which that core sample came is focused on for further evaluation. Many single valuable genes are often present in regions at very low levels (below 0.05) and remain undetected except for a single accession or in a single population where the gene frequency may be very high (as high as 80 to 90%). Once discovered, this attribute can be sought for in other materials in trials designed to uncover very rare alleles. The core idea is as good as the completeness of the designated core in capturing genetic variation for a starting point in evaluation. The total collection of the genebank, not the core, remains the unit of conservation.

Not all genebank accessions are held as seed. Some collections are maintained in *ex situ* clonal depositories with the plants grown in the ground (*in vivo*) or in meristem tissue culture (*in vitro*). The large collections store seed at very low seed moisture. These collections (active genebank) can be maintained above freezing for short-term storage (approximately 20 years) or considerably below freezing (base genebank) with special drying for long-term storage (50 to 100 years and even longer periods in liquid nitrogen at −173°C). This method of conservation removes seed from use and evaluation and is fine for seed depositories, but for gene banking there must be an ability to regenerate fresh seed and supply living germplasm to the plant breeding process.

## V. STEPS IN GERMPLASM CONSERVATION

How do accessions get into a genebank and what are the steps from a farmer's field to the seed packet stored for up to 50 years? There is no single route for inclusion in a genebank. Some accessions have been specifically collected because a landrace or region was poorly represented in a genebank. Other accessions were once part of a breeder's collection that was used to create a genebank years ago and formed the nucleus of the collection. Some accessions are representatives of a landrace, others are parents that have been widely used in plant breeding. Some accessions are highly heterozygous both as a collection and as individual seed, whereas others, such as inbreds used in commercial hybrid production, are homozygous as individual seed and all the seed in the accession are genetically identical. The central criterion for inclusion in the genebank is the possibility that genes found in the seed might be useful in the future and that the same seed or its unique genes are not already included in the collection.

The following account is a construction of a "typical" accession based on synthesis of elements from hundreds of accessions with which I am familiar. The genebank is by its very nature a sample of genetic variation of a crop, and its quality parallels the quality of the accessions as they entered the collection. A good accession possesses unique genes for both a region and the farmer landrace. The collector's sample also must be large enough to capture alleles that occur at low frequency, hopefully those down to the 5% level.

To obtain a typical accession our plant collector has traveled to an isolated valley where a road entering the mountainous region has just been built. The crop has been grown in this valley for more than 500 years and the local landraces are recognized as unique and different from those found in the lowlands. The collector has gone during the harvest season when the crop can be best observed in the field. Both color variation in the seed and differences in the fruiting structure have been observed and included in the sample. The collection of 500 to 5000 seed from over a hundred plants is labeled for geographic location, altitude above mean sea level, collector, date, name of farmer, local name of landrace, growing season (planting to harvest), cultivation condition, rain pattern, and special qualities valued by the local people. This information is called the passport data and will follow the accession as long as it is in the genebank. Well-dried seeds are very stable and can be shipped internationally, but if the collection is of vegetative clonal material, such as potato tubers or apple budwood, then special packaging and handling will be necessary to ensure that it quickly arrives at the genebank alive and healthy. Recipient country rules for seed and plant health must be strictly followed. This often leads to postentry quarantine, a protocol that must be followed.

Universally recognized ethics of plant collecting ensures that a duplicate collection is left with the appropriate institution within the country where the collection was made. Of course, no collection trip is undertaken without permission of the host country and, hopefully, nationals will accompany collectors on the plant collecting trip. These guidelines have not always been followed in the past but they are now the recognized standard as elucidated in FAO's Code of Conduct on Plant Collecting and Transfer. A plant collector must have permission of both the farmer and the appropriate national institution to make any collection. In the past, many accessions have come from bulk samples taken from a market and these may represent a landrace but not the efforts of a single farmer. Obviously, farmer permission is not possible for these bulk samples purchased in the open market.

Interestingly, the commodity price paid in the market, which on average would be a fraction of a U.S. dollar per kilogram, is a fairly accurate value of the worth of a genebank accession. Another way to establish value is to sum the cost of collecting, international and local travel, and shipping costs of the collector, which could run to hundreds of dollars but more typically range between ten and a hundred dollars. For genebanks the real cost of an accession is not incurred until the seed needs to be regenerated and fresh seed put back into the bank to replace the aged seed. The cost of a regeneration sample is somewhere between ten and a hundred dollars (possibly more for some crops). A well-run genebank might require between one and five dollars (U.S.) per year per accession as maintenance cost (electricity, repairs, salary of management personnel), so the storage cost for 25 years approximates or equals the regeneration cost. This generalized statement says that effective conservation of seed in a genebank incurs a long-term cost that exceeds the collection cost of the original sample. For this reason it is absolutely necessary that the best representation of the genetic diversity be in the original collection, because it costs as much to store and maintain a poorly made collection as a well-made collection once the material is accessioned into the genebank.

Most genebanks do not automatically accept all samples sent to them. First, there are the problems of phytosanitary permits and the passing of quarantine for most seeds when they move internationally. Often this requires a grow-out of the plants to check for disease. This also becomes a regeneration cycle for the bank so absolutely fresh seed is put into storage. Assuming that the seed has passed quarantine, the genebank would then investigate the value of adding the sample to the collection. Generally accessions are not made of small samples of a few seed, say ten, seed with no or poor passport (origin of seed = Africa, which is too general to be useful) or materials that duplicate holdings already in the genebank.

After seed has been accessed by a genebank there is the commitment to maintain the accession and ensure that the passport data can be related back to the sample. Often the bank develops identifiers and characterization data during regeneration cycles that help bank personnel identify the seed as being the accession that the label indicates. For example, characterization traits for maize such as "white seed with soft floury endosperm and a weak dent" helps establish that the accession number in fact matches the seed in the envelope or container.

Both passport and characterization data are necessary for the sound management of the accession, but are of limited usefulness to the plant breeder. What interests plant breeders and helps them locate desired parents for their crosses are evaluation data: plant growth form, quality of root system, days to harvest, and response to fertilizer, diseases, pest, cold, and drought. Seldom are genebank accessions suitable for release directly as a food crop plant, although there are some forages and range forbs that have been direct introductions. The primary use and value of genebank accessions is really in expanding the gene base of breeders' plots. Generally it is a long process from plant collecting to breeders' plots to finished cultivars. Spans of 20 to 50 years are more the norm than the exception. Two examples of useful traits reaching cultivars via genebanks are high-lysine corn and high-protein barley.

Collections of Hiproly barley C.I. (cereal investigation) 3947 and CI 4362 were made in Ethiopia by H.V. Harlan and added to the United States Department of Agriculture (USDA) world collection of small grains in 1924. It was not until 46 years later (1970) that high lysine and total protein were discovered (evaluation date) in these accessions and they became valued as parents in commercial barley cultivar development. In maize, Opaque-2, a mutant found in an Enfield Connecticut, farmer's field (1922), was studied by D.F. Jones and R. Singleton (1938) as a new endosperm mutant and maintained in their collection of mutants. It was not valued until the 1960s, when it was discovered to possess a gene for high lysine, an essential amino acid for humans that is classically low in maize diets. Often the payoff from collecting and maintaining accessions in genebanks has come too late for personal satisfaction or public recognition for either the original collector or the genebank manager. This long time scale usually leads to a sense of impersonal achievement that makes it hard to recognize heroes. The exception was the Russian geneticist and plant breeder N.I. Vavilov. More often the conservation of plant genetic resources (the genes and allelic diversity of crop plants) is associated with national programs such as the USDA or specialized institutions such as the Centro Internacional de Mejormiento de Maiz y Trigo (CIMMYT). The major institutions holding plant genetic resources are listed in Table. I.

**TABLE I**

Partial List of Crops and Their World Network of Major Genebanks[a]

| Crop | Material | Genebank |
|---|---|---|
| **Cereals** | | |
| Rice | *Oryza sativa* | IRRI, Los Baños, Philippines |
| | *O. indica* | IRRI, Los Baños, Philippines |
| | *O. javanica* | NIAS, Tsukuba, Japan |
| | Mediterranean forms, temperate South American forms, and intermediate types from the United States (plus duplicates from other centers) | NSSL, Fort Collins, U.S.A. |
| | Wild species | IRRI, Los Baños, Philippines |
| | African forms | IITA, Ibadan, Nigeria |
| Wheat | Cultivated species | VIR, St. Petersburg, Russia CIS<br>CNR, Germplasm Institute, Bari, Italy<br>ICARDA, Syria<br>CIMMYT, El Batan, Mexico<br>NSSL, Fort Collins (each institute's collection duplicated at one of the others) |
| | Wild species of *Triticum* and *Aegilops* | Plant Germplasm Institute, University of Kyoto, Japan (duplicated in one of the above institutions and NIAS, Japan) |
| Maize | New World material | CIMMYT, El Batan, Mexico |
| | Asiatic material | NSSL, Fort Collins, U.S.A.<br>NIAS, Tsukuba, Japan<br>TISTR, Bangkok, Thailand |
| | European material | VIR, St. Petersburg, Russia CIA<br>Braga, Portugal (for Mediterranean material) |
| Sorghum | Cultivated and wild | NSSL, Fort Collins, U.S.A.<br>ICRISAT, Hyderabad, India |
| Millets | Cultivated and wild *Pennisetum* spp. | NSSL, Fort Collins, U.S.A.<br>PGR, Ottawa, Canada<br>ICRISAT, Hyderabad, India |
| | *Eleusine* spp. | ICRISAT, Hyderabad, India<br>PGRC, Addis Ababa, Ethiopia |
| | Minor Indian millets | ICAR, New Delhi, India |
| | *Eragrostis* spp. | PGRC, Addis Ababa, Ethiopia |
| | *Panicum miliaceum* | ICRISAT, Hyderabad, India |
| | *Setaria italica* | ICRISAT, Hyderabad, India |
| Barley | Cultivated and wild (global collection) | PGR, Ottawa, Canada |
| | European material | Nordic Genebank, Lund, Sweden |
| | African material | PGRC, Addis Ababa, Ethiopia |
| | Asian material | NIAS, Tsukuba, Japan<br>CIMMYT, Mexico<br>ICARDA, Syria |
| Oats | Cultivated and wild<br>Nordic Genebank, Lund, Sweden | PGR, Ottawa, Canada |
| **Sugar Crops** | | |
| Sugar beets and other beets | | German Genebank, Braunschweig, Germany |
| Sugarcane | | NIAR, Tsukuba, Japan<br>NPGS, U.S.A.<br>ICAR, New Delhi, India |
| **Legumes** | | |
| Phaseolus | New World material (all species but emphasis on *P. vulgaris, P. coccineus, P. Lunatus,* and *P. acutifolius*) | CIAT, Cali, Colombia (duplicated in NSSL, Fort Collins, U.S.A.) |
| | European material | German Genebank, Braunschweig-Germany |
| | Wild species | University of Gembloux, Belgium |

*continues*

*Continued*

| | | |
|---|---|---|
| Pigeon pea | | ICRISAT, Hyderabad, India |
| Groundnut | | INTA, Pergamino, Argentina<br>Cenargen, Brasilia, Brazil |
| Chickpea | | ICARDA, Syria<br>ICRISAT, Hyderabad, India |
| Cowpea | | ICARDA, Syria<br>IITA, Ibadan, Nigeria |
| Pea | | Nordic Genebank, Lund, Sweden |
| Soybean | | AVRDA, Taipei, Taiwan<br>NIAR, Tsukuba, Japan<br>NPGS, U.S.A. |
| Wild perennial forage crops | | CSIRO, Canberra, Australia |
| **Root Crops** | | |
| Potato | Wild and cultivated species | CIP, Lima, Peru |
| Sweet potato | | CIP, Lima, Peru<br>AVRDC, Taipei, Taiwan |
| **Vegetables** | | |
| Allium | Global collection | NVRS, Wellesbourne, U.K. |
| | Asian collection | NIAS, Tsukuba, Japan |
| Capsicum | Global collection | AVRDA, Taipei, Taiwan<br>CATIE, Turrialba, Costa Rica |
| | Global collection | IVT, Wageningen, Netherlands |
| Eggplant | Global collection | IVT, Wageningen, Netherlands |
| | New World collection | NSSL, Fort Collins, U.S.A. |
| | Southeast Asian collection | IPB, Los Baños, Philippines |
| Tomatoes | Global collection | CATIE, Turrialba, Costa Rica<br>ARVDA, Taipei, Taiwan<br>NSSL, Fort Collins, U.S.A. |
| | Asian collection | IPB, Los Baños, Philippines |
| Crucifers | *Brassica oleracea* | NVRS, Wellesbourne, U.K.<br>IVT, Wageningen, Netherlands |
| | Vegetable and fodder types: *B. campestris, B. juncea, B. napus* | NVRS, Wellesbourne, U.K. |
| | Vegetable and fodder types: *B. napus* | German Genebank, Braunschweig, Germany |
| | Oilseed and green manure crucifers: *B. campestris, B. juncea, B. napus, Sinapis alba, B. carinata* | PRG, Ottawa, Canada<br>German Genebank, Braunschweig, Germany<br>PGRC, Addis Ababa, Ethiopia<br>German Genebank, Braunschweig, Germany |
| | *Raphanus* species | NVRS, Wellesbourne, U.K. |
| | Wild relatives | Universidad Politecnica, Madrid, Spain<br>Tohoku University, Sendai, Japan |
| | East Asian collection | NIAS, Tsukuba, Japan |
| Other vegetables | Southeast Asian species | IPB, Los Baños, Philippines<br>ARVDA, Taipei, Taiwan |

[a] The following abbreviations are used: AVRDC, Asian Vegetable Research and Development Center; CATIE, Centro Agronomico Tropical de Investigación y Enseñanza; CENARGEN, Centro Nacional de Recursos Geneticos; CIAT, Centro Internacional de Agricultura Tropical; CIMMYT, Centro Internacional de Mejoramiento de Maiz y Trigo; CIP, Centro Internacional de la Papa; CNR, National Research Council; CSIRO, Commonwealth Science and Industry Research Organization; ICAR, Indian Council Agricultural Research; ICARDA, International Center for Agricultural Research in the Dry Areas; ICRISAT, International Crops Research Institute for the Semiarid Tropics; IITA, International Institute of Tropical Agriculture; INTA, Instituto National de Technología Agropecuaria; IPB, Institute of Plant Breeding; IRRI, International Rice Research Institute; IVT, Institute for Horticultural Plant Breeding; NIAS, National Institute of Agricultural Sciences, NPGS, National Plant Germplasm System; NSSL, National Seed Storage Laboratories, NVRS, National Vegetable Research Station; PGR, Plant Gene Resources of Canada; PGRC, Plant Genetic Resources Center; TISTR, Thailand Institute of Scientific and Technological Research; VIR, N.I. Vavilov Institute of Plant Industry. Table adapted from G. Wilkes (1983). "Current Status of Crop Plant Germplasm," CRC Critical Reviews in Plant Sciences, Vol. I, pp. 133–181. CRC Press, Boca Raton, Fla.

## VI. WORLD NETWORK OF CROP PLANT GERMPLASM

The development of a worldwide consciousness of the value of genetic resources formed between 1950 and 1975. Although the work of N.I. Vavilov and the All Union Institute of Plant Industry (St. Petersburg) in the 1920s anticipated the developments of the latter half of the century, most early work with assembling genetic diversity was either by individual breeders or through a national effort whose main function was not gene conservation but plant breeding self-sufficiency.

The 1960s saw a worldwide recognition of the threat of genetic erosion and the Food and Agriculture Organization became more active in exploration and conservation of crop plant germplasm. The 1960s was a decade for urgency—the human population was increasing and the world was running short of food. This condition was reversed in the last years of the decade by the successful spread from the International Agricultural Research Centers (IARCs) of the high-yielding, dwarf-stature wheat from CIMMYT and rice from the International Rice Research Institute (IRRI), subsequently called the Green Revolution. There was also an urgency to counteract the replacement of local varieties by new exotic varieties which often resulted in genetic erosion and widespread monocultures of elite cultivars, by the few workers that recognized the problem. At the time, there was little experience with gene banking and its methodology and no overall strategy for plant exploration and gene collecting.

The late 1960s and 1970s were a time of increasing environmental awareness and concern by the general public, especially in the developed nations. Considerable publicity focused on the United Nations Conference on the Human Environment held in Stockholm in 1972. The Stockholm meeting passed seven (Nos. 39–45) resolutions that helped focus world attention on the urgency of action to preserve genetic resources. The climate for further initiative was set when the Consultative Group on International Agricultural Research (CGIAR), the umbrella organization of the IARCs, created the International Board for Plant Genetic Resources (IBPGR) in 1974. IBPGR was to be an autonomous, international, scientific organization under the aegis of the CGIAR with its secretariat provided by the FAO in Rome. The basic function of IBPGR, as defined by the CGIAR, was to promote an international network of genetic resource centers to further the collection, conservation documentation, evaluation, and use of plant germplasm, and thereby contribute to raising the standard of living and welfare of people throughout the world. It proceeded to do this in the next 15 years with a success that exceeded expectation. By the 1980s there was a vague but established working network of genebanks conserving a vast storehouse of genetic variation of useful cultivated plants. The IBPGR has now become the International Plant Genetic Resources Institute (IPGRI) and is still a center of the CGIAR. [*See* CONSERVATION AGREEMENTS, INTERNATIONAL.]

The world was quick to rally around the concept of genebanks when the concern was genetic erosion, but it has been slow to take the next steps to sustain them so that they could effectively evaluate and regenerate conserved materials. In part this has been due to a decade (the 1980s) of political maneuvering over who owns germplasm: a struggle between farmers' rights (recognizing national origin of individual landraces and the farm families that still cultivate, or once cultivated, the landrace in the past) and the potential for breeders' rights for advanced breeding lines and other elite germplasm materials such as the inbred parents of hybrids. The controversy pitted activists urging establishment of the FAO Undertaking versus the IBPGR in the early part of the decade, and later it focused on the heated debate of who owned the intellectual property rights of germplasm (was it a private good, public good, or a mix in which developing countries could not lay claim to raw or primitive germplasm of landraces and developed countries would still be protected by the ownership of improved, elite cultivars). That decade ended with a mutually agreed upon consensus arrived at through a series of meetings sponsored by the Keystone Center. The Convention on Biological Diversity (Rio Convention) from the 1992 Earth Summit in Rio de Janeiro now appears to supersede and potentially resolve the issues of the decade-long dialogue.

## VII. FUTURE PROSPECTS

The evolution under domestication of our cultivated plants has taken place around the world, starting from a specific region where the plant was once part of the native vegetation and underwent its early evolution as a crop. Often major gene diversity and significant changes in productivity of a crop have taken place far from where the crop originated. The major food plants of the world are not owned by any one people and are quite literally a part of our human heritage from the past. Some cultures have influenced the development of specific crops more than others, but no one culture owns the invention of agriculture; this was a human discovery in several separate and distinct regions of the world, each contributing food plants of importance to the world collection that feeds humans today. Therefore, the conservation of genetic resources is a world problem because no one nation possesses all the genetic diversity and every nation is dependent on plant genetic resources from other nations. The consensus to arrive at this conclusion has been slow to develop, implementation has only recently taken place, and funding to sustain the collections is far from stable and assured.

## Glossary

**Accession** Individual sample of seed or plants entered into a germplasm collection in a seed bank or clonal repository.

**Active collection** Storage of germplasm (at nonfreezing temperature), as either seeds or clones, that breeders can regularly tap for use in the development of new crop varieties.

**Base collection** Long-term storage of crop plant germplasm usually as superdry seed held below the freezing point (sometimes in liquid nitrogen), not intended for everyday use but the ultimate backup for materials not found in active collections.

**Genetic resource** Germplasm, plants, animals, and microorganisms containing potentially useful characteristics under genetic control.

**Germplasm conservation** Planned management of a natural resource over time, allowing for the utilization of the resources while preventing exploitation, destruction, or neglect, which would limit the full spectra of genetic diversity under preservation.

**Landrace** Crop plant population developed *in situ* through continued farmer selection and not subject to formal breeding procedures.

**Vavilov centers of diversity** Recognized regions where the greatest diversity for plant genetic resources could be found, first identified and studied by the Russian plant breeder and geneticist N.I. Vavilov. These nine centers—Ethiopia, Mediterranean, Near East, Central Asia, South Asia, Malaysia—Indonesia, China, Mexico–Guatemala, and Andean South America—cover less than one-fortieth of the land area of the globe, but from them have come almost all of our food plants. These centers are located in mountainous regions, essentially immediate to the two topics (Cancer and Capricorn), and in regions long populated by agricultural people.

## Bibliography

Brown, A. H. D., Frankel, O. H., Marshall, D. R., and Williams, J. T., eds. (1991). "The Uses of Plant Genetic Resources." Cambridge, England: Cambridge University Press.

Hawkes, J. G. (1983). "The Diversity of Crop Plants." Cambridge, Mass.: Harvard University Press.

Juma, C. (1989). "The Gene Hunters: Biotechnology and the Scramble for Seeds." Princeton, NJ: Princeton University Press.

Keystone. (1991). "Oslo Report: Third Plenary Session Keystone International Dialogue Series on Plant Genetic Resources." Washington, D.C.: Genetic Resources Communication System.

Kloppenburg, J. R., ed. (1988). "Seeds and Sovereignty: The Use and Control of Plant Genetic Resources." Durham, N.C.: Duke University Press.

Plucknett, D. L., Smith, N. J. H., Williams, J. T., and Anishett, N. M. (1987). "Genebanks and the World's Food." Princeton, N.J.: Princeton University Press.

Wilkes, G. (1983). "Current Status of Crop Plant Germplasm," CRC Critical Reviews in Plant Sciences, Vol. I, pp. 133–181. Boca Raton, Fla.: CRC Press.

Witt, S. C. (1985). "Brief Book: Biotechnology and Genetic Diversity." San Francisco: Center for Science Information.

# Global Anthropogenic Influences

**W. S. Fyfe**
*University of Western Ontario*



Anthropogenic influences on the environment include all significant, observable changes on earth that have occurred because of the presence of our species, *Homo sapiens*. Many would agree that at this time humans are the dominant influence on change in the near-surface environments, including atmosphere, hydrosphere, biosphere, and lithosphere (the outer skin of the solid earth).

## I. INTRODUCTION

In the past two decades there has been a vast increase in awareness of anthropogenic forcing of change on our planet. In 1972, the first Earth Summit was held in Stockholm, Sweden, with its declaration "to bear a solemn responsibility to protect and improve the environment for present and future generations." There was concern that the support systems for the growing world population were a threat to our most basic life-support system. But the actions that followed Stockholm were minimal. Population grew from 3.8 billion in 1972 to 5.5 billion by 1992. The annual increase now approaches 100 million per year. There has been vast growth in the numbers living in urban centers: in 1972, there were 3 mega-cites (over 10 million inhabitants), there are now 14. In 1972 there were about 250 million motor vehicles producing their exhaust gases of $CO_2$, CO, and $NO_x$, and now there are over 600 million. Back then ozone depletion was not even recognized. In 1972, about 100,000 km$^2$ of forests were being destroyed annually, today the figure is 170,000 km$^2$. As Daniel Koshland, editor of *Science,* recently wrote (1993) in a discussion of clean air:

> First of all it is important to identify the main villain as overpopulation. In the good old days (viewed through the myopia of nostalgia), the water, air, flora and fauna existed in an idyllic utopia. But in truth, there were famine, starvation, horses and buggies that contributed to pollution, fireplaces that spewed forth soot from burning soft coal, and water contaminated with microorganisms. The humans

were so few, and the land so vast, that these insults to nature could be absorbed without serious consequence. That is no longer true.

Long before, Thomas Peacock of England wrote (1860): "They have poisoned the Thames and killed the fish in the river. A little further development of the same wisdom and science will complete the poisoning of the air, and kill the dwellers in the banks. I almost think it is the destiny of science to exterminate the human race."

There is no doubt that modern science and the associated technologies have allowed the spectacular expansion of our species. Without many such technologies our population would still be in the hundreds not thousands of millions. All scientists and all educated people now face several sobering questions: Can we support well the present population and the future projected population of about 10 billion ($10^{10}$) without destruction of our environment and key life-support systems? Can we modify our technologies to produce sustainable systems? These are difficult questions. The ability of a space to support a given population of any animal depends on technologies and fluctuations in the natural systems—carrying capacity is difficult to quantify and depends on the time constants considered.

## II. CHANGE IN THE ATMOSPHERE

The air we breathe today is not the same air breathed by our great-grandparents. It is still dominated by photosynthetically produced oxygen and nitrogen, but for significant minor components like carbon dioxide, carbon monoxide, and methane there have been very considerable changes. Twenty thousand years ago, during the last ice age, atmospheric $CO_2$ concentrations were close to 200 parts per million (ppm) by volume. As the ice age terminated, this level rose to about 270 ppm and was moderately steady for thousands of years. With the Industrial Revolution and the burning of fossil carbon fuels (coal, oil, gas), the level is now over 350 ppm; between 1960 and 1985, it rose from 315 to 345 ppm. The rate of increase is still rising. In 1972, 16 billion tons of $CO_2$ were injected into the atmosphere, and in 1992 about 23 billion tons. Coal, oil, and gas are still our dominant sources of energy. How will the biosphere respond to a doubling of $CO_2$, a critical component involved in all photosynthetic processes? We are not sure. Similar proportional increases in the trace gas methane have been observed. [*See* GLOBAL CARBON CYCLE.]

All researchers agree that certain gases in the atmospheric—carbon dioxide, methane, nitrogen oxides, CFCs, and ozone (in that order)—and water vapor contribute to the "greenhouse effect" and provide the thermal blanket on our planet. All of these gases are increasing because of human actions. Researchers agree that the earth should get warmer and that it is getting warmer. The mean temperature of earth has increased about 1°C in the past century. The longer record over the past 20,000 years leaves not doubt about the correlation between global temperature and the levels of greenhouse gases. [*See* GREENHOUSE GASES IN THE EARTH'S ATMOSPHERE.]

This global change in atmospheric chemistry and the associated climate influences are largely a result of our energy and chemical industries. The changes are massive and the long-term influences are not adequately understood. The environmental scientist Wallace Broecker of Columbia University has aptly characterized our situation: we are playing Russian roulette with the planet.

A host of other changes are occurring in our atmosphere. As we are now well aware, the gas ozone in the stratosphere absorbs and protects us from lethal solar ultraviolet radiation. Certain chemicals, in particular those derived from organochlorine, bromine, and fluorine compounds, lead to destruction of this ozone layer in the upper atmosphere, especially at high latitudes. Ozone depletion, and particularly the "ozone hole," has been growing annually. Without this protective layer, there would be no life as we know it on the surface of this planet. [*See* ATMOSPHERIC OZONE AND THE BIOLOGICAL IMPACT OF SOLAR ULTRAVIOLET RADIATION.]

Many fossil fuels contain significant quantities of sulfur. When burned sulfur oxides (and nitrogen oxides) are produced, which when mixed with other atmospheric components produce acids. Thus in many parts of the world acid rain has become a major problem. Acid rain threatens human and animal health, kills trees and other vegetation, corrodes buildings, and is responsible for other environmental changes. [*See* ACID RAIN.]

It is worth noting here the different types of interactions that occur when various constituents of the atmosphere are changed. For example, increased ultraviolet radiation will reduce photosynthesis, which fixes carbon dioxide. Thus, one impact of ozone reduction can be an even faster rate of carbon dioxide increase in the atmosphere. The same is true for the impact of acid rain in reducing plant bioproductivity. Warming the oceans must lead to more surface evaporation and a higher water content (also a greenhouse gas) in the atmosphere and more rain somewhere. The great floods of 1993 in the United States and India could reflect such an influence and such fluctuations may become more common. Hot systems are more turbulent and less predictable than cool systems, as every cook knows.

## III. FRAGILE EARTH?

People often speak of the fragility of our planet. But earth is not fragile. As Broecker has stressed, for 4 billion years of earth history recorded in rocks, it has never boiled or totally frozen. There have been large fluctuations but the overall system has survived. And for the 4.5 billion years of earth history, and the geological record for 4 billion years, there is clear evidence of a prolific biomass of microorganisms that are still the major part of the biomass. Earth and parts of its biosphere are able to survive ice ages, continental drift, large meteorite impacts, large volcanic events, and so on. We have now discovered that microorganisms are prolific in the subsurface, even to depths of 4 km. But what we are slowly beginning to appreciate is that the human species, with its complex support systems, is fragile.

## IV. SOIL

Our supply of food and natural wood fibers depends on three main components: the temperature of a given region, the water supply, and soil. Although modern biotechnology may improve the yields of the species we grow, these three factors place the ultimate limits. Across the entire globe, our soil resources are deteriorating.

Soil is formed by the reactions of rock with air, water, and living organisms. Soil science is perhaps one of the most underdeveloped subjects of modern earth science. Although science strives toward the exact description of a system, necessary descriptions of soil on the microscale have only been possible in rare cases. The reactive components of soil consist of very small particles, and modern tools are required for these descriptions: geochemical analysis, structure analysis (e.g., high-resolution electron microscopy), and, in particular, surface analysis and particulate analysis. These tools were not available for normal descriptive procedures until very recently. [*See* SOIL ECOSYSTEMS.]

Soil is created by numerous processes and using many components. Debris from the parent rock sets the stage, in large part, for the future chemistry, mineralogy, and biology of the soil. Airborne dust may also be a significant component. New minerals and amorphous materials are formed by weathering; these minerals are normally hydrated, and frequently oxidized. A multitude of fine-grained particulate material is present as well. Soil is the home for a host of microorganisms, macroorganisms, and their debris, which are involved in key processes like nitrogen fixation, carbon fixation, and mineralization. Soil bacteria are excellent ion-exchange phases, and it provides the physical support system for the roots of plants, which are often finely adapted to local conditions. Soil contains gas and liquid phases that are present in its pore spaces and as surface films on inorganic and biological materials. Soil physics (thickness, porosity, permeability, rigidity, viscosity, strength, cohesion, compressibility, albedo) reflects all of these processes and materials.

Soil is a dynamic system whose biology varies with seasonal changes; its liquid and gas phases vary on many time scales. Wetting and drying are reflected in transient solution and precipitation processes, mostly on surfaces, of and in thin films of fine-grained materials. It is dynamic on all microscales, as it becomes wet or dry or warm or cold, and the biology changes in response to all changes in the physics and chemistry. The *in situ* interactions of all the components, living and dead, make this system very difficult to describe.

Although soil is complex, one overwhelming property is significant for the biosphere (and humankind) and is based on the question: Given a certain climate (temperature, rainfall, light, wind speed), what is the capacity of a soil to support photosynthetic organisms, carbon fixation, and the production of food and fiber on a truly sustainable basis? This same quality will determine the soil and plant modulation of the atmosphere by means of reactions of the soil with oxygen, carbon dioxide, and moisture, for example. For humankind, and for its economists, there are other key questions: Is the soil changing? How is it changing? How rapidly? Is it becoming more, or less, productive?

Today, one of the most dramatic changes occurring on a global scale is related to soil thickness. Erosion is the primary cause of change, but there are other causes, such as irrigation and salting, nutrient reduction because of overcropping, compaction, and human-induced anaerobic conditions. The relation between soil thickness and food and fiber productivity can be highly nonlinear. The introduction of many herbicides, pesticides, new plant species, and industrial pollutants like lead must surely change the total biologic–geologic interactions. Chemical-intensive farming methods, which appeared to be so successful over past decades, are now showing massive fatigue, even failure. And many people are worried about genetic engineering and the increase of monocultures, with their almost certain instability and vulnerability to fluctuations within the biosphere and the lithosphere.

How do we estimate the local and regional health of soil? I would suggest that there are simple ways to obtain a useful estimate of the health of soils on a local and regional basis. All organisms, from bacteria to trees, require a wide range of macronutrients and micronutrients. Table I shows the level of three key macronutrients in four great rivers of the world. Most of the water in these rivers passes through surface soil and rock. The levels of the macronutrients and the total dissolved inorganic material speak eloquently about the state of the soils from which the rivers drain. The Mississippi and Ganges river systems pass through younger soils that are loaded with rock-forming minerals; the Amazon and Negro systems flow through old, deeply weathered terrain. The water in the latter rivers indicates a low capacity to support intense bioproductivity, probably why this region has not been heavily populated by humans.

***TABLE I***
**Nutrient Species in Some Major Rivers (Part Per Million)**

| River | Dissolved calcium | Dissolved magnesium | Dissolved potassium | Total dissolved solids |
|---|---|---|---|---|
| Mississippi | 39 | 10.7 | 2.8 | 265 |
| Lower Amazon | 5.2 | 1.0 | 0.8 | 38 |
| Lower Negro | 0.2 | 0.1 | 0.3 | 6 |
| Ganges | 24.5 | 5.0 | 3.1 | 167 |

If we think of the great fertile regions of the world, such as Hawaii, southern Ontario, and the Nile Valley, all of these contain young soils with massive mineral debris. Plate boundaries are fertile, volcanic terrains are fertile, and regions that have been covered by sediments from diverse source rocks, as a result of river, ice, or wind erosion, are fertile. In contrast, the declining agricultural regions have not received new minerals or been subjected to a volcanic or mountain-building event for millions of years. The macronutrients and micronutrients in these regions have passed to the oceans. In fact, in some of these regions, the main input today may be from the airborne particulates in dust and rain.

According to the Worldwatch Institute, topsoil is being lost at a global rate of 0.7% per year, a rate that must be accelerating. If this is true, we are heading for a global food catastrophe. How long does it take to form a 10-cm-thick layer of fertile soil? (Note that I define fertile in terms of the capacity of soil to produce food and fiber.) The answer to this question is found in nature itself. In a place like Hawaii, we can watch well-dated volcanic flows evolve into a cattle ranch or a forest. In northern Canada, where clean rock was exposed after the ice retreated 10,000 years ago, there is often still no soil. A range of 1000–10,000 years seems to be needed to produce a farmland from hard primary, crustal rocks. For humans, this is essentially a norenewable resouce!

Conversely, rocks can be formed as a result of the erosion of soil (e.g., deltaic sedimentary rocks), and the formation of soil from these rocks may be much faster. What takes time to turn these rocks into fertile soil is the creation of porosity and other physical properties that allow the development of diverse biological support systems, and that will create water-holding systems. How fast these processes occur will depend critically on the history of sediment compaction.

For sustainable carbon fixation (i.e., agriculture), simple laws must apply. The rate of soil loss must not exceed the rate of soil formation. The nutrients taken out by crops must be balanced by input for the entire spectrum of essential elements. The ability of the soil to hold water must not be diminished because of agricultural practices. How often do we have the data to quantify such laws for large or small regions?

As mentioned earlier, the Worldwatch Institute has reported that global loss of topsoil is currently running at 0.7% per year. Although this figure may be challenged, it is cause for alarm, even if it is close to the actual rate. Further, the consequences of topsoil loss go far beyond the problems of the world food supply. Today, carbon dioxide in the atmosphere is rising faster than ever. If soil is failing, its carbon dioxide fixation processes will decline as well, and this coupling of problems has the potential to accelerate greenhouse warming.

Perhaps the most spectacular cases of the nonlinear results of soil erosion can be seen in regions of former tropical rain forest. In these areas, where evapotranspiration dominates the prehuman water cycle, the cutting of protective forests on steep topography can lead to catastrophic soil loss within a short time. In fact, reasonable productivity may drop to zero productivity within a period of years, or even days.

## V. ENERGY TECHNOLOGIES

There is no question that, to a large extent, our standard of living is related to the availability of energy. Energy resources are fundamental to our systems of education, health, nutrition, pleasure, communication, and so on. At the present time, most of the world's energy is derived from oil (40%), and oil consumption rose 3.1% globally in 1988. Next come coal at 30% and natural gas at 20%. Thus, fossil carbon supplies about 90% of the world's energy. Once great hydropower now supplies only 7% of global energy. $CO_2$, the byproduct waste gas from burning fossil carbon, is one of the greatest threats to changing world climate, temperature, and rainfall distribution. Present trends in the use of fossil carbon fuels are not sustainable, and the potential atmospheric pollution will almost certainly have intolerable consequences.

What are the options? First, without doubt, humankind must strive for population control and population reduction. This is already happening in the industrialized countries, even if the fact is not politically acknowledged. Nonetheless, in all parts of the world the demand for energy is rising, even where populations are stable or declining. But at this time in many of the developing and poorer countries, energy is needed, and in vast quanitities. If this needed energy were supplied, even transiently, from fossil carbon, particularly coal, the results could be a global disaster. And yet if cost-effective technology could be developed to eliminate the $CO_2$ released into the atmosphere (and acid components, etc.), the fuels might be usable. Although this problem is being discussed by some researchers, no such technology is in sight at this time.

Historically, it was not so long ago (100+ years) that, in much of the world, biomass (wood) was the dominant source of energy for a population of 1 billion. Population growth and the associated deforestation have removed this as a global option for the present. But for a few nations with low population density, excess farmland, excess food production, sound soil conservation, and robust soils, biomass fuels (wood, ethanol, methanol) are viable options. Biomass fuels basically use solar energy and recycle $CO_2$. However, let there be no illusions. The total production of biomass (or carbon fixation) annually from all cultivated land contains about $0.4 \times 10^{10}$ tons of carbon. If totally burned, the energy production would be $1.6 \times 10^{20}$ joules (J). Given that humans use about $4 \times 10^{20}$ J per year, the products from all cultivated land, if dedicated to fuel alone, would last us about 5 months. The carbon stored on land (90% in trees) is about $8 \times 10^{11}$ tons of carbon, equivalent to $20 \times 10^{21}$ J. This would last 5 years. It is obvious that the natural products of photosynthesis cannot satisfy the needs of 10 billion humans.

The future potential of hydropower is limited to a few areas. North America and Europe have already developed about half this potential, but for parts of Asia, South America, and Africa, 10% or less is currently used. It has been estimated that, in these countries, potential capacity exceeds 100,000 megawatts (equivalent to 100 typical nuclear power plants).

Yet water and the environment of rivers is increasingly needed for food production and irrigation. At the present time, almost half the reliable runoff from the continents is manipulated by humans. For a population of over 10 billion, the use of water will require planning with a wisdom never seen before. And often the source region of the water is far from the regions of energy needs (e.g., Brazil), or even in neighboring nations (e.g., Canada and the United States). The ecological impact of hydropower development (downstream and upstream) and the potential for geohazards (induced seismic activity), salinization, and other effects require the most careful analysis. There are also little-understood influences when major river systems no longer flow freely to the oceans. Change in light (fresh) water fluxes to the oceans can alter ocean current patterns and, hence, influence local and global climate regimes. And when a resource dependent on precipitation is developed, the risk of changing climate patterns must be clearly understood. When one considers the future needs for water for agriculture, industry, and urban centers as the human population doubles, one must contemplate that many or most of the world's great rivers will be diverted. Runoff from the continents will be replaced by evaporation. What will be the impacts, local and global, on climate? This change must be considered by those who model future climates.

There is simply no doubt that the ultimate, sustainable energy resource for *Homo sapiens* is the sun. Today, humans use about $4 \times 10^{20}$ J per year. Every day, the solar energy that arrives at earth is $1.5 \times 10^{22}$ J. Thus, if we could collect and use a few hours of solar energy per year, we could provide the energy now used by the entire population! There would be no change in the overall radiation balance of the earth. Given intelligence, the will, and the technology, this clearly must be our future energy resource, except for a few special geographic situations (near the North or South poles). To provide all our energy needs, we would need to collect and store all this energy on only about 0.01% of the earth's surface (much of it in deserts!).

In general, temperature increases with depth in the earth's crust, at a rate of 20°–30°C per kilometer. In volcanic regions, large masses of hot rock (1000°C) are present near the surface. A few countries, such as New Zealand and Iceland, have made major use of this energy source at competitive prices. There are problems, however, when very hot water is extracted, for it can be rich in components like hydrogen sulfide, arsenic, and heavy metals. These products can be largely eliminated by recycling the hot water to depth, thus avoiding direct discharge to rivers.

The energy in hot rocks, ultimately derived from the radioactive decay of isotopes of elements like uranium, potassium, and thorium in the deep earth, is impressive. One cubic kilometer of rock of mass $3 \times 10^{15}$ g, at 1000°C, will provide about $4 \times 10^{18}$ J as it cools to surface temperatures. Thus, 100 $km^3$ of hot rock, a 6-$km^3$ block, could supply world needs for a year. Current production of 1200°C magmas is about 15 $km^3$ per year, but the volume of associated hot rock is much larger. There are many well-known regions of the world, for example, New Zealand, Japan, Chile, Iceland, California, and East Africa, where such situations are common. Most sites occur near the great plate boundaries or rift systems of earth. But even under "normal" situations, geothermal energy could have valuable applications (e.g., for heating urban communities). The normal thermal gradient is about 30°C per kilometer of depth. A 1-$km^3$ block of rock at 3 km depth and at 90°C, in cooling to 30°C, would provide $2 \times 10^{17}$ J. A 1000-$km^3$ block of normal crust ($10 \times 10 \times 10$ km) could provide the world's annual energy needs.

In the 1940s and 1950s, nuclear fission energy appeared to have the potential to solve the world's energy needs using conventional uranium fuels. Estimates for uranium resources in the United States alone show that this energy resource could provide up to $6.6 \times 10^{25}$ J of electricity, a supply for 100,000 years! Nuclear energy cannot be dismissed from consideration as a future energy source for humankind. However, since the power applications commenced (nuclear energy now provides 15.6% of world electricity), problems, both real and conceptual, have arisen, including:

- the safety of nuclear power plants;
- the reliability of the operators;
- waste disposal;
- nuclear weapons as a potential by-product; and
- global social responsibility.

With the one great exception of Chernobyl, the record of nuclear power plant safety is in fact quite remarkable. The leading nuclear nations (except the U.S.S.R.), including the United States (98,000 megawatts), France (53,000 megawatts), and Japan (29,000 megawatts), have had no serious accidents, but there have been a host of "minor" leaks. But the problem is that an accident like Chernobyl can remove a huge area of land from safe human habitation for decades or centuries. We now have 428 reactors operating in 31 nations. But we should note—we are using the Model-Ts of nuclear reactors. New developments in fusion and fission technologies could change the present situation.

In developed nations, great care is taken in the training of all who work in nuclear power plants. Training of a licensed operator may take 10 years or more and requires a very special type of personality. The Chernobyl accident was basically caused by operator error with an early reactor design. Many nations who work with the International Atomic Energy Authority maintain adherence to strict international standards of inspection. A high degree of technological sophistication is required to safely enter the nuclear power game.

The wastes from nuclear reactors are extremely dangerous for at least 10,000 years. Even then, a few bioactive radionuclides remain. Until the development of nuclear power, humankind had never contemplated the need for 10,000-year garbage containment. In terms of disposal and monitoring, however, nuclear debris has a unique feature: it is easy to detect at enormous levels of sensitivity.

Can waste be stored with reasonable safety, so that there will be no significant leakage to the biosphere for a million years? To show if this is possible, the only recourse is to geological analogues. Salt deposits, oil, gas, sensitive materials like native copper, and sulfides have been stored in nature for hundreds even thousands of millions of years. In many cases, the leakage has been small. Can we duplicate nature? Many who have been deeply involved in the problem believe it is possible. However, we still have much to learn, and there is no hurry. Wastes should not be buried before undergoing long near-surface storage to remove excess heat production. Secure interim storage is feasible, with excellent demonstration provided by Sweden.

## VI. WATER

If the world human population reaches 10 billion, and if these 10 billion are to have adequate nutrition and supplies of clean water, there will be the necessity to manage the global water supply with great care and attention. When we consider the water resouce question, we see the potential for real limits or potential environmental disasters.

Global precipitation totals about 525,100 $km^3$ (1 $km^3 = 10^{15}$ g $= 10^9$ tons). However, much of this falls on the oceans (78%); and on land more than half evaporates. The reliable runoff available to humans is about 14,000 $km^3$ (much more occurs but only during floods). The reliable usable water is estimated to be about 9000 $km^3$. At this time, humans manipulate about 3500 $km^3$, almost 40%.

And after humans use water, its chemistry and biology are often drastically changed. It is interesting to note that North Americans use 5000 liters (about 1325 gallons) of water each day, for all their systems. If this were the rate for 10 billion humans, the amount would be $5 \times 10^3 \times 10^{10} \times 3.65 \times 10^2$ liters per year (or 18,000 $km^3$), twice the reliable supply!

For any given region, the water supply can be quantified. Perhaps it is best to think of the "island" model (e.g., a river may flow from one region to another, but who can guarantee that this inflow—through flow—will be maintained?). In an island model, the water inventory involves: the rainfall, the groundwater penetration, the evaporation or evapotranspiration, and the runoff to the oceans. A most important parameter is the reliability, or variation, of the rainfall. Over what time period does one average the availability of the total supply? The numbers used will, in part, reflect the technologies that may be available, or possible, to store water over long periods of time.

Certain realities must be addressed in preparing a management plan for the island water supply:

- There must be sufficient runoff to prevent salinization—runoff removes salts and other wastes from the land surface.
- The evaporation processes depend on the nature of the island's surface cover. This will vary depending on the vegetation, urban developments, transport systems, and so on.
- The potentially useful groundwater resources depend on the geology, the deep porosity, permeability, and the chemical nature of the rocks, which will influence the chemistry of the water and its suitability for various purposes, (e.g., agricultural and industrial).
- Surface storage removes land area and can increase evaporation.
- Subsurface storage is of sustainable use only if recharge exceeds withdrawal, and if the cycling does not reduce long-term porosity and permeability.

It is remarkable that, in many regions today, such basic data are not considered in the use of water. For example, groundwater is being mined in many regions.

A problem with potential to cause major global change involves the competing agricultural and energy uses of major systems of runoff. For example, when a major river is dammed for hydropower, the runoff to the oceans begins to fluctuate over large values, depending on fluctuations in rainfall. Downstream, the aquatic biodiversity and biomass are seriously perturbed. At sites of ocean discharge, the nutritional supply to the ocean biomass is seriously altered, and satellite pictures show us clearly that the marine biomass is concentrated near continental margins.

But perhaps most seriously of all, ocean current systems may be disrupted. About 10,000 years ago, the Gulf Stream current that transports energy to the Atlantic Arctic regions was perturbed. Northern areas went into a little ice age, called the Younger Dryas event. One explanation is that, for a period, the Mississippi River system was diverted into the St. Lawrence system. This placed a vast flow of light continental water onto the surface of the North Atlantic, which perturbed the Gulf Stream and the northern energy transport. Very rapidly the North Arctic froze over. This model warns us that, if a large outflow is changed, the patterns of ocean currents and ocean mixing may change and in turn affect local and global climate. Such phenomena must be considered when there are plans to modify major components of continental runoff!

## VII. WASTES

Over 90% of our energy resources, resources that fundamentally influence our living standards, come from coal, oil, and gas. All are nonrenewable and nonsustainable. Our energy industries live on the earth's capital. A recent report of the World Resources Institute shows that, for North America, if use of oil and natural gas continue at present rates, the reserves would last 10 and 14 years, respectively. The only significant reserves of conventional oil are in the Middle East. Large reserves of heavy oils and tar sands occur in other places. Of fossil carbon fuels, only coal appears as a longer-term resource, and more and more coal is being used across the world. But if all the world, all 5 billion or the future 10 billion people, used fossil carbon fuels at the same rate as the United States and Canada, and even Western Europe, the future would be disastrous.

The production of energy from fossil fuels generates enormous amounts of waste. We have already discussed the high levels of carbon dioxide and other greenhouse gases, as well as acid rain, that result from current coal technology. Some use terms like "clean" coal technology, but unless the carbon dioxide is controlled, a not impossible consideration as many geologists and chemists know, coal combustion is either very dirty, or just dirty. The economic and social consequences of even a small climate change or rise in sea level are simply mind boggling.

However, though gaseous emissions from fuel burning ($CO_2$, CO, $SO_x$, $NO_x$, etc.) are known, another mounting problem is ash disposal. In India, where high-ash coal (10–50%) is used, there is a vast problem with ash disposal. Coal is a material of highly variable chemistry and can contain significant amounts of elements like chromium, arsenic, lead, and uranium. The ash, being in a reactive, often glassy state, can be environmentally toxic. At present, river dumping is often the disposal method, creating severe problems such as flooding and high levels of toxicity. Across the globe, ash disposal is an increasing problem, as at least 1 $km^3$ is produced annually. Large-scale coal mining, often associated with oxidation of sulfides in coal and trace-metal leaching, can also perturb surface water and groundwater resources.

At a recent conference there was discussion on the impact of heavy metals and "xenobiotic" organics that flood modern soil–water–atmosphere systems on ever-increasing scales. At least 50,000 organics, which may have no natural analogues and whose environmental behavior is poorly known, are used in industry today. Increasingly, the trends are to eliminate dispersion by containment and, for organics, destruction by incineration. In the developed world, regulations and controls are much improved (e.g., the Rhine River system), but in much of the developing world there is still little control. The impacts of strange chemicals on the local and global ecology are little understood, but recent data from Eastern Europe are alarming. The global influence of a group of simple chemicals—CFCs—on ozone destruction illustrates the impact of what were once thought to be harmless chemicals.

The impact on soil, water, air, and ecological balance of the overuse of chemical fertilizers, herbicides, and pesticides is now increasingly recognized. On the other hand, the limits of "organic" farming are also recognized. It has been shown that, if organic farming alone were used, a much greater land area would be needed to feed the present world population and present reservations that preserve biodiversity would be threatened. As stated earlier, there has been a failure to quantify the chemical mass balance of agriculture. A wheat crop removes a quantifiable mass of K, Mg, Ca, and P per unit yield. Either by natural means or by fertilizers, a balance must be maintained. Such holistic studies are rare. Furthermore, irrigation is being increasingly used to meet the water requirements of present agriculture. The areas of the world where this has led to salinization of soils are widespread, and the affected acreage is enormous. Often, particularily in China and India, water is mined from aquifers, and the resulting productivity is totally nonsustainable.

The average citizen of the United States produces 762 kg of waste per year; the average for a German citizen is about 317 kg. The variety of modern waste materials is enormous. Though paper, plastics, glass, and food wastes dominate, we must not forget the modern chemicals and materials we use (batteries, film, metals, etc.). It is perhaps no surprise that in the city of Chicago, the incinerator fly ash particles were found to contain 69,000 ppm lead and 1500 ppm cadmium, more than a 1000 times the abundance in the earth's crust.

Given present waste production rates of about 1 kg per person per day worldwide, a population of 10 billion would produce about 3.6 $km^3$ of garbage per year, a volume exceeding all annual continental volcanism. Clearly something must be done, particularly because some of the plastics have a very long life. All of these problems have become more intense as more and more people move to cities. Urbanization has removed any hope for the low-level dispersion of wastes. Cities of 20 million will be common in the next century. Where will the garbage go? The impacts of urban waste disposal (e.g., untreated sewage) on aquatic systems are simply enormous.

Slowly we are moving to a system of recycling and reuse. All animal and vegetable waste, with care, makes good fertilizer, or in some systems can be used to generate methane and hydrogen. The last thing we should be doing is to take a product like paper, made from $CO_2$ by plants, and convert it to $CH_4$, a more potent greenhouse gas, via anaerobic bacteria. For environmental reasons, it would be better to burn it—$CO_2$ to $CO_2$.

## VIII. MINERAL RESOURCES

If we look around us, we notice that we are surrounded by modified materials, mostly derived from the top kilometer or so of the earth's crust. We live with concrete, glass, bitumen, ceramics, steel, copper, aluminum, stone, zinc, and so on. Our transport machines and computers contain components made from half the elements in the periodic table. Developed societies use about 20 tons of rock-derived materials per person per year. For a population of 10 billion with an affluent lifestyle, this means $2 \times 10^{14}$ kg of rock per year, or almost 100 $km^3$ per year. This quantity exceeds the volume of all the volcanism on the planet, under land and ocean, by an order of magnitude. Human actions are now the major component of the processes that modify the planet's surface.

Can we supply these materials for all humankind? If we have the necessary energy to drive the machines and transport the materials, the answer is probably yes. But it is also clear that, for the less common materials like copper and zinc, with rigorous attention to recycling, the modifications of the environment can be reduced from the scale of today. The new industrial trend to design a product for recycling must become the rule for the future. The savings in total use of materials, use of energy, and production of wastes makes recycling an economic necessity of the future.

A major component of the materials we use today is derived from wood. It is quite clear that the present consumption of wood in products for housing, paper, and packaging cannot continue. The environmental impact of the removal of forests is too serious to allow present careless harvesting to continue. Wood will not be a major material for the 10 billion humans of the next century. It will not be needed, given the potential of modern systems of building and communication.

## IX. BIODIVERSITY

On our planet, millions of species occupy almost every possible environment where liquid water exists, even if only transiently. There is life in Antarctic lakes and hot springs, biofilms on rocks in the high Arctic, and red algae in snow. In my garden, in London, Ontario, there are 100 plant species that I can see with the naked eye. The number increases by orders of magnitude if I use a hand lens, and increases again with an electron microscope. [*See* BIODIVERSITY.]

Why are there so many species? Is the vast number critical to our stability? This question is fundamental in understanding the consequences of the present dramatic species reduction of perhaps 100 species per day. The earth has survived massive extinction events. Yet the ecologist E. O. Wilson gives compelling examples of how the destruction of a single species can destabilize an environment. [*See* BIODIVERSITY, VALVES AND USES; EVOLUTION AND EXTINCTION.]

We have all observed certain influences of biodiversity and our environmental security. When there is a significant climate fluctuation, we notice that some species do a little better while others seem to vanish. One food crop does better—another fails. If we plant diverse crops, they—and we—survive. However, a fundamental influence of biodiversity is that the continental albedo does not experience wide fluctuations. Given diversity in ground cover species, the earth's color and reflectivity change little during a climate fluctuation, and $CO_2$ continues to be fixed and oxygen produced.

In the animal world, the importance of biodiversity is perhaps more obvious. We have all seen the interactions between insects and birds, or between mosquitoes and amphibians. All maintain a balance and, when it is disturbed, we see deleterious effects. The influences of human intrusion are often obvious, for example, soil erosion and forest damage in New Zealand when deer were introduced, or in Australia with rabbits. These examples compel us to contemplate the potential impacts of widespread introduction of genetically modified species.

## X. SUMMARY

Today, the major causes of change in our planetary environment are related to human activity. We are changing the composition of the atmosphere we breathe and the water we drink. We are changing the earth's temperature by greenhouse gas emissions. In many nations there is a critical shortage of water for agriculture and industry. Vast land areas are being degraded by soil erosion and salinization. We are reducing biodiversity. Given the increasing human population, if we continue as in the past, few rivers will flow regularly to the oceans and few forests will remain. Researchers do not quantitatively understand the impact on global environmental systems of such changes. However, wth the predicted doubling of the human population in the next century, there will be, must be, vast development of resources. But unless we shift to new sustainable tecnologies, there is potential for global disaster. When we change a natural system, we must ask a question: Is the change reversible?

## Glossary

**Acid rain** Rainwater that contains more acidic components, such as nitric acid and sulfuric acid, than rain showing no anthropagenic influences.

**Anthropogenic (change)** Change related to the actions of humans.

**Deltaic sedimentary rocks** Fine materials that accumulate where rivers meet the oceans, mainly made from the erosion of the surface of the continents (e.g., the Mississippi delta and the Ganges delta).

**Fertile soil** Soil with all parameters (chemistry, physics, thickness, etc.) such that it will be highly productive of food and fiber over a long time period; a soil of sustainable bioproductivity.

**Lithosphere** Outer shell of the solid earth, a layer, in general, 100–200 km thick.

**Monoculture** System (forest, agriculture, etc.) where one species dominates.

**Xenobiotic chemicals** Chemicals that do not occur in nature but are produced by human activities (e.g., plutonium and DDT).

## Bibliography

Baccini, P. (1989). "The Landfill." Berlin: Springer-Verlag.

Broecker, W. S. (1985). "How to Build a Habitable Planet." New York: Eldigio Press.

Broecker, W. S. (1987). Unpleasant surprises in the greenhouse. *Nature* **328,** 123–126.

Broecker, W. S., Kennett, J. P., and Flower, B. P. (1989). Routing of meltwater from the Laurentide ice sheet during the Younger Dryas cold episode. *Nature* **41,** 318–321.

Climate Alert (1992). Lagging food production. *The Climate Institute* **5,** 244–254.

Culotta, E. (1993). Is the geological past a key to the (near) future? *Science* **259,** 906–908.

Flavin, C., and Leussen, N. (1990). "Beyond the Petroleum Age: designing for a Solar Economy," Paper No. 100. Washington, D.C.: Worldwatch Institute.

Fyfe, W. S. (1981). The environmental crisis: Quantifying geosphere interactions. *Science* **213,** 105–110.

Fyfe, W. S. (1989). Soil and global change. *Episodes* **12,** 249–254.

Fyfe, W. S. (1992). Global change: Anthropogenic forcing—The moving target. *Terra Nova* **4,** 284–287.

Fyfe, W. S., Babuska, V., Price, N. J., Schmid, E., Tsang, C. F., Uyeda, S., and Velde, B. (1984). The geology of nuclear waste disposal. *Nature* **318,** 537–540.

Hatta, K., and Mori, Y. (1992). "Global Environmental Protection Strategy through Thermal Engineering." New York: Hemisphere Publishing.

Koshland, D. E. (1993). Clean thoughts on clean air. *Science* **261,** 1371.

Krauskopf, K. B. (1988). "Radioactive Waste Disposal and Geology." London: Chapman & Hall.

McLaren, D. J., and Skinner, B. J. (1987). "Resources and World Development," Dahlem Conference Reports. New York: John Wiley & Sons.

Postal, S. (1992). "Last Oasis—Facing Water Scarcity." New York: W. W. Norton.

Seifritz, W. (1990). $CO_2$ disposal by means of silicates. *Nature* **354,** 486.
Wilson, E. O. (1992). "The Diversity of Life." Cambridge, Mass.: Belknap Press of Harvard University.
World Resources Institute (1990–91). "World Resources." New York: Oxford University Press.
Worldwatch Institute (1992). "State of the World." New York: W. W. Norton.

# Global Carbon Cycle

**Berrien Moore III**
*University of New Hampshire*



The global carbon cycle is the flux of carbon from inorganic forms to organic compounds and back to inorganic chemical states. The global carbon cycle is being altered by human activities. From direct atmospheric measurements of carbon dioxide ($CO_2$) and indirect measurements from air bubbles trapped in ice layers, we know that the concentration of $CO_2$ has increased by roughly 25% since 1700. Similarly, the atmospheric concentration of methane ($CH_4$) has increased by over 100% since the early part of the 18th century. The primary human activities contributing to these changes are fossil fuel combustion and modifications of terrestrial ecosystems through land use. The increase in the atmospheric concentration of $CO_2$ and $CH_4$, as well as other greenhouse gases, because of human activity has produced serious concerns regarding the heat balance of the global atmosphere. Specifically, the increasing concentrations of these gases will lead to an intensification of the earth's natural greenhouse effect. Shifting this balance will force the global climate system in ways that are not well understood, given the complex interactions and feedbacks involved, but there is a general consensus that global patterns of temperature and precipitation will change, although the magnitude, distribution, and timing of these changes are far from certain. The uncertainty of future climate change, however, does not rest solely on issues of physical-climate system dynamics. The global carbon cycle is still not adequately understood or quantified. Using the best available estimates of sources and sinks of $CO_2$ does not lead to a balanced global budget. Using averages for the 1980s, sources of atmospheric $CO_2$ exceed identified sinks by 1.8–1.2 billion metric tons of carbon as $CO_2$ per year. This presents significant problems for projecting future atmospheric $CO_2$ concentrations.

## I. INTRODUCTION

Since ancient times humans have modified natural systems for food and fiber. In doing so they have altered the distribution of the basic chemical constituents that support life: carbon, nitrogen, oxygen, phosphorus, and sulfur. With the onset of the Industrial Age, this alteration has begun to appear at global scales. For example, the pool of carbon in the atmosphere in the form of carbon dioxide ($CO_2$) has increased in concentration from about 275 parts per million [ppm(v)] to almost 355 ppm(v) during the 225 years between 1765 and 1991 as a result of fossil fuel burning and forest clearing. The annual rate of increase is currently about 1.5 ppm year$^{-1}$. A direct record of this increase has been made since 1958 and there are a number of indirect records (from ice cores) of the increase over the past two centuries. Moreover, from the ice core records it is known that the concentration of carbon dioxide was relatively constant from the beginning of the present interglacial period (ca. 10,000 B.P.) to the onset of increases in the 18$^{th}$ century. Data for the concentration of atmospheric $CO_2$ from ice cores and from direct measurements are shown in Fig. 1. [*See* ASPECTS OF THE ENVIRONMENTAL CHEMISTRY OF ELEMENTS.]

The primary human activities contributing to this change are fossil fuel combustion and modifications of terrestrial ecosystems through land use (e.g., biomass burning and conversion of agriculture). For the period of 1980–1989, an average of 5.4 Pg C (1 Pg C = $1 \times 10^{15}$ g C = 1 billion metric tons carbon) per year as $CO_2$ was released to the atmosphere from the burning of fossil fuels, and it is estimated that an average of about 0.6–2.6 Pg C per year was emitted because of deforestation and land-use change during the same interval. Figure 2 shows an estimate of these two fluxes of anthropogenic $CO_2$ from the mid-18$^{th}$ century to the present. [*See* GLOBAL ANTHROPOGENIC INFLUENCES.]

Similarly, methane ($CH_4$) has been increasing in the atmosphere; however, its rate of increase has

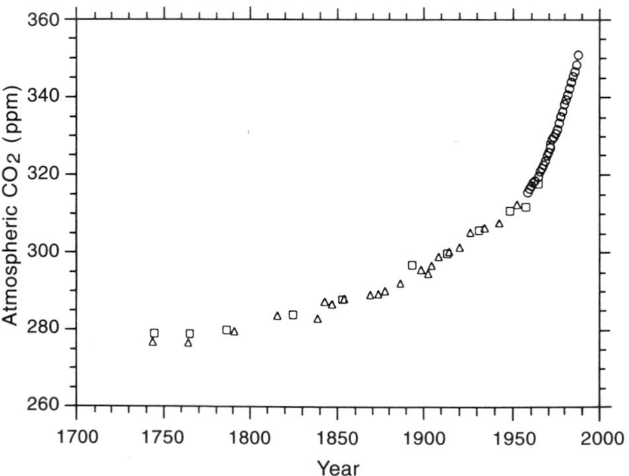


***FIGURE 1*** Historical atmospheric $CO_2$ concentrations ($pCO_2$) from ice cores. The open squares and triangles represent data derived from ice core measurements from Siple Station, Antarctica. The open circles are the annual averaged atmospheric measurements from the Mauna Loa Observatory (the Keeling record).

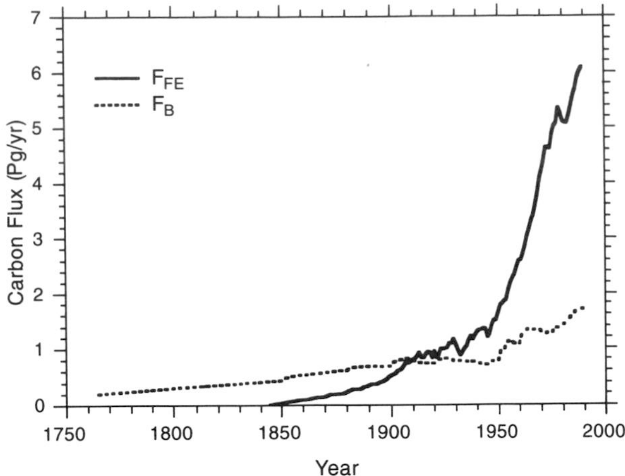


***FIGURE 2*** Estimates of $CO_2$ flux from fossil fuels and deforestation. The fossil fuel emissions record (e.g., Andres *et al.*, 1994) is accurate to within about 10%, but the estimate of biotic flux due to land-use change (Houghton, 1994) is less certain, especially for the less recent numbers, because of a lack of a detailed historical accounting of global land-use patterns. This figure shows that until about 1920, deforestation contributed more to the atmospheric flux than fossil fuel burning, which did not become significant until the start of the current industrial period in the 1860s.

slowed. Neither the causes of the increase nor the change in rate are understood. The picture for methane is complicated by its diverse and dispersed sources, by the complexity of its photochemical sink, and by its poorly understood biological sink in upland soils. Changes in the atmospheric concentration can result from changes in either sources or sinks or a combination of sources and sinks whose contributions vary over time.

The increase in the atmospheric concentration of $CO_2$ and $CH_4$, as well as other greenhouse gases, because of human activity has produced serious concerns regarding the heat balance of the global atmosphere. Specifically, the increasing concentrations of these gases will lead to an intensification of the earth's natural greenhouse effect. Shifting this balance will force the global climate system in ways that are not well understood, given the complex interactions and feedbacks involved, but there is a general consensus that global patterns of temperature and precipitation will change, although the magnitude, distribution, and timing of these changes are far from certain. General circulation model results indicate that globally averaged surface temperatures could increase by as much as

1.5–4.5°C in a world with an atmospheric concentration of $CO_2$ twice that of the preindustrial period [i.e., a world where the concentration would be roughly 550–580 ppm(v)]. [*See* GREENHOUSE GASES IN THE EARTH'S ATMOSPHERE.]

## II. THE CARBON CYCLE AND THE CARBON DIOXIDE PROBLEM

The present average concentration of carbon dioxide in the atmosphere is, as mentioned, 355 ppm(v), which is equivalent to a total carbon mass in the atmosphere of about 755 Pg C as $CO_2$ (Fig. 3). In these units of carbon mass, the rate of atmospheric increase is approximately 3 Pg C per year. Estimates of the amount of carbon in living organic matter on land vary between 450 and 600 Pg C so there is more carbon in the atmosphere than there is in all the world's forests—a rather unbelievable fact (Fig. 3).

There are, of course, other "pools" of terrestrial carbon. For instance, there is significantly more terrestrial carbon stored in soils. The organic carbon pool in the upper 1 m of the world's soils contains roughly 1200 Pg C as organic carbon. In those areas of the tropics where deep soils occur, the carbon stored below 1 m may add another 50 Pg C. The much less dynamic carbonate–carbon pool contains approximately 720 Pg C.

The net exchange of carbon between terrestrial vegetation and the atmosphere may be considered to be the sum of three fluxes: gross photosynthesis, autotrophic plant respiration, and heterotrophic (soil microbial) respiration. The two respiration fluxes are approximately equal and, in the absence of disturbance, their sum roughly balances the photosynthetic fixation flux of carbon. Interestingly, the gross fluxes between the terrestrial system and the atmosphere are similar to those for the oceans (approximately 100 Pg C per year; Fig. 3), but the considerably smaller terrestrial pool size leads to a much faster turnover time than for the oceans.

This approximate balance between carbon fixation and respiration has been upset by land-use practices and other anthropogenic changes to the global system. There is significant uncertainty on the exact quantitative changes of the role of terrestrial ecosystems in the global carbon cycle; however, there is little question that the quasi-equilibrium in terrestrial carbon storage no longer exists.

At least two factors govern the level of carbon storage in terrestrial ecosystems. First and most

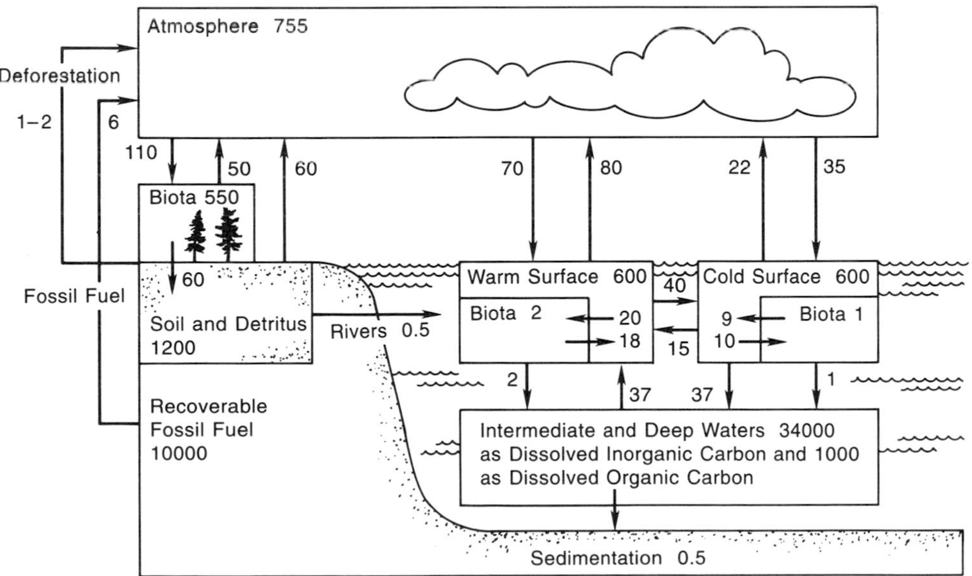


**FIGURE 3** The global carbon cycle [adapted from Moore (1985)]. The values within compartments are in units of Pg C, and the fluxes (arrows) are in units of Pg C $year^{-1}$.

obvious is the anthropogenic alteration of the earth's surface, such as through the conversion of forest to agriculture, which results in a net change in the carbon stored in those systems. This change generally reflects a net exchange of carbon between the terrestrial biosphere and the atmosphere. In general, there appears to have been a release of $CO_2$ to the atmosphere as a result of land use and land-use change. For instance, the clearing of forests for agriculture often reduces organic soil carbon stocks by 20 to 50%.

The second factor is more subtle; there are possible changes in net ecosystem production (carbon storage) that result from changes in atmospheric $CO_2$, other global biogeochemical cycles, and/or the physical-climate system. Ultimately, to adequately address such changes requires a much clearer understanding of the nitrogen and phosphorus cycles since they are the limiting nutrients in most terrestrial ecosystems, but our knowledge of the way these biogeochemical cycles relate to the carbon cycle compares poorly with our general understanding of the individual cycles themselves. The relationships between climate and terrestrial ecosystems must also be understood better.

The amount of carbon released globally from vegetation and soils as a result of deforestation and other land-use changes remains uncertain, although much progress has been made in recent years. Calculations of the net carbon loss from the global biotic inventory, which attempt to take into account the uncertainties concerning disturbance rates and biotic response, indicate that from 1700 to 1900 the total loss was approximately 120 Pg C, which is roughly one-half the amount of carbon dioxide that the combustion of fossil fuels has contributed to the atmosphere since 1860. Currently, approximately 5.5 Pg C are emitted annually in the burning of fossil fuels and approximately 1.5 Pg C from land deforestation (Figs. 2 and 3).

The estimate of an integral release of about 120 Pg C from deforestation for the period 1700 to the present is consistent with other calculations that use ratios of carbon isotopes, as found in ice cores and in tree rings, to infer how much $CO_2$ has come from deforestation; however, the temporal pattern from these two types of calculations do differ somewhat.

The other principal component in the global carbon cycle is the world's oceans. The oceans are by far the largest active reservoir of carbon (Fig. 3). Recent estimates of the total amount of dissolved inorganic carbon establish an amount of about 38,000 Pg C. Only a small fraction is carbon dioxide (0.5%); the bicarbonate ion, which makes up to 90%, and the carbonate ion, at slightly less than 10%, are the major forms of dissolved inorganic carbon. There is far less dissolved organic carbon, about 1000 Pg C (though this estimate remains in question), and even less particulate organic carbon. What role the oceans play in carbon dioxide exchange is, however, still unknown; they are without question a sink for anthropogenic $CO_2$, but the strength of this sink is somewhat unclear.

Although the oceans are the largest active reservoir of carbon and cover almost 70% of the globe, the total marine biomass contains only about 3 Pg C of carbon or just about 0.5% of the carbon stored in terrestrial vegetation. On the other hand, the total primary production of marine organisms is 30 to 40 Pg C per year, corresponding to 30 to 40% of the total primary production of terrestrial vegetation. However, only a relatively small portion of this production results in particulate organic carbon, which sinks and decomposes in deeper layers or is incorporated into sediments.

There is much uncertainty about the current environmental state and the future states of the planet. For instance, climate is sensitive to cloud extent and type. If greenhouse gases continue to increase in the atmosphere, will future cloud patterns change as a result of a change in the heat balance? If they change will there be more high, white clouds or more low, dark clouds? The uncertainty of future climate changes, however, does not rest solely on issues of physical-climate system dynamics and their representation in general circulation models. Understanding the carbon cycle is a key to comprehending the changing terrestrial biosphere and to developing a reasonable range of future concentrations of carbon dioxide and other greenhouse gases. Conversely, our predictions about the physical-climate system and climate change are confounded by the fact that the carbon cycle is still not adequately understood or quantified globally. Using the best available estimates of anthropogenic sources and sinks of $CO_2$ does not lead to a balanced global budget. Using averages for the 1980s, sources of atmospheric $CO_2$ exceed identified sinks by 1.8–1.2 Pg C per year, presenting significant problems for projecting future atmospheric $CO_2$ concentrations. In sum, these uncertainties are reflected in our uncertainty about the atmospheric lifetime of $CO_2$.

Without knowing the processes responsible for taking up current emissions, it is not possible to predict how sinks for anthropogenic $CO_2$ will evolve with time as climate, land use, the atmospheric $CO_2$ concentration, nitrogen loading, and other factors change.

The imbalance is, obviously, diminished by reductions in the estimate of the rate of deforestation or increases in the regrowth, if the oceanic uptake is underestimated, or if there are significant shorter-term natural variations in the concentration of carbon dioxide in the atmosphere that override the imbalances. The question of which combination of these possibilities is more likely is both important and fascinating; the answer is not obvious.

The next two sections look more closely at the two key subsystems that determine atmospheric $CO_2$: the terrestrial system and the oceanic system.

## III. ROLE OF THE TERRESTRIAL SYSTEMS IN THE GLOBAL CARBON CYCLE

The current net flux of carbon between the biota and the atmosphere due to land-use change is uncertain. Estimates of carbon flux, primarily from deforestation in the tropics, range from 0.4 to 2.5 Pg C year$^{-1}$. Three factors contribute to this range of uncertainty: (1) rates of deforestation, (2) the fate of deforested land (i.e., the amount of secondary forest regrowth and reclearing), and (3) the stock of biomass and soil organic matter and their response to disturbance, including anthropogenic reductions of carbon stocks within forests because of thinning or degradation. Models developed with improved geographic and temporal data on deforestation rates, better parameterization of the dynamic nature of deforestation and reforestation, and improved data on above- and belowground carbon response characteristics are under development. [*See* DEFORESTATION.]

The current rate of deforestation is unknown as there are only a few estimates of tropical deforestation available and they may be in error by as much as 50%. Moreover, most published sources of data on tropical deforestation come in nonspatial, tabular form. Therefore, both the rate and the geographic distribution of this critical forcing parameter need to be developed in a way that is objective, quantitatively reproducible, and useful as an input data set for numerical carbon models.

Two reports provide superficial information on the rate of tropical deforestation in the late 1970s. The first was a study carried out by the National Research Council. The second was a compilation provided by FAO and UNEP. There were only isolated studies done in particular countries for most of the 1980s. Several new attempts have been made to assess the current rate. The FAO is in the process of updating its 1981 assessment, but the results have not yet been released into literature.

Approximately 90% of the current annual net release of approximately 1.5 Pg C appears to be from the tropics, with 10 countries comprising two-thirds of the net release (Brazil, Columbia, Indonesia, Ivory Coast, Laos, Malaysia, Mexico, Peru, Thailand, and Zaire). At 0.6 Pg C year$^{-1}$, the net flux from Brazil is the largest single biotic source of biogenic carbon in 1988. Although the tropics provide a net source of carbon to the atmosphere, recent research points to the mid to high latitudes as an important sink. This observation places new importance on the role of temperate zone ecosystems. However, it is important to note that the magnitude of this estimated temperate zone sink is to a large degree determined by the value estimated for the tropical source term; there is also significant disagreement about its existence.

Satellite remote sensing is the only means for resolving discrepancies or quantifying temporal and spatial variations in deforestation rates. Ground-breaking research established that satellite remote sensing can quantify tropical deforestation and that there was no reason why satellite-based techniques could not be applied to a large area like the tropical forest belt to resolve the aforementioned controversies and uncertainty and thereby provide vastly improved forcing functions for global carbon models. This is now being realized in significant areas like the Amazon using high resolution land remote sensing imagery such as from the United State's LandSat and the French SPOT.

Turning to the second factor(s) that controls the amount of carbon stored in terrestrial ecosystems, there has been considerable work on other determents of terrestrial carbon storage motivated, in part, by the issue of a missing carbon sink. As mentioned earlier, it has been suggested that terrestrial vegetation may be "fertilized" by the increasing concentration of $CO_2$. Although this issue is controversial, it is possible that terrestrial ecosystems are responding to the rapid increase in atmospheric $CO_2$ concentration by producing more biomass and/or storing more soil carbon, thereby balancing the global carbon budget. For instance, the increasing concentration of $CO_2$ in the atmosphere may raise C/N ratios by either making plants more water efficient (more carbon fixed per $H_2O$ transpired) or through other indirect biochemical mechanisms (e.g., enhancing carbohydrate formation, which is generally not balanced by increased nitrogen uptake). Consequently, there is the possibility that higher $CO_2$ levels may lead to an increase in net primary production and perhaps net ecosystem production (carbon storage).

The issue of changes in carbon storage that are not directly linked to land use is important and challenging. For instance, if the terrestrial biosphere is currently a major sink for anthropogenic $CO_2$ because of changes in atmospheric $CO_2$ or past patterns of climate, this carbon could be quickly returned to the atmosphere in the future as a result of rapid climate change or other disturbances.

The change in atmospheric $CO_2$ concentration resulting from fossil fuel combustion, land management activities, and other human-induced disturbances of the global carbon cycle is strongly governed by the $CO_2$ exchange between the atmosphere and ocean. The ocean is believed to be the largest sink for atmospheric $CO_2$. In order to better appreciate the ocean's role in all of this, it is necessary to look a bit deeper into the particular biological, chemical, and physical processes that govern the largest sink for $CO_2$: the ocean.

## IV. ROLE OF THE OCEANS IN THE GLOBAL CARBON CYCLE

The rate of carbon uptake by the oceans is controlled by seawater temperature, by surface chemistry and biology, and by the various patterns of mixing and circulating which determine the amount of carbon transported from surface waters to the deep ocean.

The actual exchange of $CO_2$ between the sea surface and the atmosphere is by diffusion at the air–sea interface and hence is governed by the $CO_2$ partial pressure difference as the atmosphere, the sea surface wind velocity, and state of the ocean surface. The essential uncertainties are the distribution of the partial pressure of $CO_2$ in the surface of the ocean and the factors controlling this distribution.

From the perspective of the global carbon cycle, one could consider the partial pressure of $CO_2$ in seawater as a quasi-linear function, depending on alkalinity and on the concentration of $CO_2$ in seawater. Therefore, it is necessary to only keep track of the $CO_2$ and alkalinity in the sea surface.

This is far easier said than done. First, $CO_2$ dissociates in seawater, breaking into bicarbonate and carbonate ions. Further complicating this chemical phenomenon are the dynamics of biological and physical processes. Primary production consumes $CO_2$; respiration and decay processes produce $CO_2$. Each process affects the chemical equilibrium. Similarly, carbonate formation (shells) and dissolution alter alkalinity which, as mentioned earlier, also affect the partial pressure of $CO_2$ in seawater. The physical processes of circulation and mixing continually adjust the total inorganic carbon concentration throughout the ocean, and thereby continually change the $CO_2$ partial pressure distribution in the surface waters. It is really quite a myriad of activity, and it is remarkable that oceanographers have pinned down the role of the ocean in the carbon cycle as well as they have.

Five features about this role though do stand out: (1) the consumption of $CO_2$ in primary production in biologically active surface waters; (2) the enrichment of the deep water in $CO_2$ because of the decomposition and dissolution of detrital matter that originates from biological processes in the surface waters; (3) the sinking of water in polar regions, particularly in the North Atlantic, taking $CO_2$ with it followed by a general bottom water flow toward the equator; (4) the upwelling of this water in equatorial regions with a corresponding outgassing of $CO_2$ to the atmosphere and a general poleward flow of surface waters; and (5) accompanying the meridional circulation is the general turbulent mixing processes whereby the carbon-rich water at intermediate depths is continuously being exchanged (mixed) with water of less carbon content in the surface layers. All of these processes govern the exchange of carbon dioxide between the sea surface and the atmosphere, and all have been represented in models to varying extents; however, all do not play an essential or the same role in the perturbation problem posed by the increase in carbon dioxide.

The first two features taken together are often referred to as the "biological pump"; the biology pumps carbon to the bottom (Fig. 3). The most obvious pumping is the incorporation, in tissue or as carbonate in shells, into living organism of $CO_2$ that is dissolved in surface waters, lowering the partial pressure of $CO_2$, followed by the "shipping" of some of this $CO_2$ to the bottom "packed" in the remains of dead marine organisms.

As a consequence of the "biological pump," the concentration of dissolved inorganic carbon is not uniform with depth: the concentration of surface waters is 10–15% less than deeper waters. There is a corresponding depletion of phosphorus (and nitrogen) in surface water, even in areas of intense upwelling, because of biological uptake and the loss of the detrital material which also contains phosphorus (and nitrogen) as well as carbon.

The fate of the carbon that falls from the surface waters depends, in part, on its characteristics. If it is organic material then it is oxidized at intermediate depths, which results in an oxygen minimum and a carbon and phosphorus maximum. If the material is carbonate it dissolves, raising both alkalinity and the concentration of carbon, primarily at great depths where the high pressure increases the solubility of calcium carbonate.

Thus, where active, the biological pump lowers the partial press of $CO_2$ in surface waters and enhances the partial pressure in deep water not in contact with the atmosphere. It is as if the "biological pump" moves the partial pressure around in a way that allows carbon dioxide to work its way into the ocean. In areas of low production the partial pressure is often greater than that of the atmosphere, and $CO_2$ is released from the sea surface. In the natural preindustrial steady state the overall pattern, however, is in balance; the net exchange is zero. Furthermore, there is little reason to believe that the biological pump has changed over the last 300 years; therefore, its direct role in the perturbation problem, the "$CO_2$ problem," that has been induced by human activities is minimal.

An important aspect of biological activity that cannot be denied is its role in modifying the distribution of biogeochemical tracers (as just discussed). These distributions are uniquely valuable, providing a basis for inferring the rates of physical processes associated with ocean circulation or for validating other estimates of these physical processes. However, to use these tracer distributions for physical processes one must "remove" the biological signal since each change in alkalinity or in the concentration of phosphorus, total inorganic carbon, and oxygen is mediated by biological processes as well as by the motion of the water masses. By quantifying the biological processes and then exploiting oceanic chemical profiles, marine geochemists have been able to aid in determining the rates of circulation and mixing.

The role of the oceans in the carbon cycle is very dependent on its rate of overturning (the meridional circulation) and its mixing. In polar regions, ice formation leaves much of the salt "behind" still in solution. The result is an increase in salinity in these already cold waters and hence an increase in density. In certain high regions, evaporation exceeds precipitation, and this further increases the sea water density. As a consequence, these cold, dense surface waters can sink, and as such, they also have the potential to form, in effect, a pipeline or conveyor belt for transferring atmospheric $CO_2$ to the large reservoirs of abyssal waters that have long residence times.

This downward convection of surface waters in polar regions during "bottom water formation" creates a sink for carbon dioxide in high latitudes, but the balancing upwelling of carbon-rich waters in low latitudes creates a source. In other words, what goes down, i.e., cold polar water with $CO_2$, must come up, i.e., warm equatorial water with excess $CO_2$.

In addition to the bottom water formation in polar regions, there is water exchange between surface waters and intermediate waters due to vertical exchanges, in other words a form of turbulent mixing or diffusion, in association with the surface ocean currents, like the Gulf Stream.

These different exchange processes, bottom water formation and turbulent mixing, maintained by water motions renew the abyssal part of the oceans in the matter of a few hundred years in the Atlantic ocean, whereas the age of Pacific deep water is up to about 1500 years. Intuitively one realizes that this rather slow rate of ocean turnover limits the oceans as a sink for carbon dioxide.

This pattern of a rather slow oceanic turnover and mixing would appear to define a rather ponderous role for the oceans in the global carbon cycle. One might say that the sea appears to be only slowly responding to the marked perturbations of the atmospheric "pool" that humans are causing.

But does this imply that we can neglect understanding further the ocean's role in global carbon cycle? Quite the opposite. The implication is that even if we omit or misinterpret even rather minor process in the ocean, this may have a significant influence on our view of the likely future partitioning of excess carbon dioxide. The ocean is somewhat akin to a national economy. It might be rather slow to respond but misunderstanding what appear to be minor parameters can have costly effects.

## V. CONCLUSION

The realization of the critical role of living systems in all of the earth's (bio)geochemical cycles is a relatively recent discovery. The recognition of biotic factors as potential homeostatic controls of biogeochemical cycles has allowed for significant advances in our understanding of the natural metabolism responsible for the compositions of the atmosphere, oceans, and sediments on the surface of our planet. Since such a planetary metabolism is now, and has been for some time in the past, interactive, wherein physical, chemical, and biological processes are inextricably linked, quantification of the contribution of the biota is essential for a better understanding of global processes.

The carbon cycle is central to a better understanding of the globe. At the heart of this cycle are the roles of terrestrial ecosystems and the world's oceans, yet we are faced with obtaining an understanding that goes beyond terrestrial ecology or oceanography for we must now understand how the "coupled" atmosphere–land–ocean system is responding to human activities: Global change. The sum is, indeed, more than the collection of all of its parts.

## Glossary

**Greenhouse effect and greenhouse gases** Short-wave solar radiation can pass through the clear atmosphere relatively unimpeded, but long-wave terrestrial radiation emitted by the warm surface of the earth is partially absorbed and reemitted by a number of trace gases in the atmosphere known as greenhouse gases: the greenhouse effect. The main natural greenhouse gases are water vapor, carbon dioxide, methane, nitrous oxide, and ozone.

**Carbon cycle** Carbon in the form of carbon dioxide, carbonates, organic carbon, and other forms is cycled among various reservoirs, atmosphere, oceans, land and marine biota, litter and detritus, and on geological time scales, also sediments and rocks; generally designated as the global carbon cycle.

**Gross and net primary production** Rate of photosynthesis essentially equals gross primary production (GPP). Net primary production (NPP) equals GPP minus autotrophic respiration, which is the carbon stored by plants.

**Net ecosystem production (NEP)** NEP is net primary production minus heterotrophic respiration, which is the respiration associated primarily with microbial soil and litter respiration involved in converting organic material to carbon dioxide. In an unperturbed state, NEP is near zero and where positive is usually a reflection of peat and/or soil formation.

## Bibliography


Aber, J. D., and Melillo, J. M. (1991). "Terrestrial Ecosystems." Philadelphia: Saunders College Publishing.

Andres, R. J., Marland, G., Boden, T., and Bischoff, S. (1994). Carbon dioxide emissions from fossil fuel combustion and cement manufacture 1751-1991 and an estimate of their isotopic composition and latitudinal distribution. *In* "Proceedings from the 1993 Global Change Institute: The Carbon Cycle" (in press).

Bazzaz, F. A. (1990). The response of natural ecosystems to the rising global $CO_2$ levels. *Annu. Rev. Ecol. Syst.* **21,** 167–196.

Bazzaz, F. A., and Fajer, E. D. (1992). Plant life in a $CO_2$-rich world. *Sci. Am.* **266,**1, 68–74.

Bolin, B., ed. (1981). "Carbon Cycle Modeling: SCOPE 16." Chichester, England: John Wiley & Sons.

Bolin, B., Degens, E. T., Kempe, S., and Ketner, P., eds. (1979). "The Global Carbon Cycle: SCOPE 13." Chichester, England: John Wiley & Sons.

Bradley, R. S., ed. (1989). "Global Changes in the Past." Boulder, CO: UCAR/Office for Interdisciplinary Earth Studies.

Broecker, W. S., and Peng, T.-H. (1982). "Tracers in the Sea." Lamont-Doherty Geological Observatory, Columbia University, Palisades, NY.

Houghton, J. T., Callander, B. A., and Varney, S. K., eds. (1992). "Climate Change 1992: The Supplementary Report to the IPCC Scientific Assessment." Cambridge, England: Cambridge Univ. Press.

Houghton, J. T., Jenkins, G. J., and Ephraums, J. J., eds. (1990). "Climate Change: The IPCC Scientific Assessment." Cambridge, England: Cambridge Univ. Press.

Houghton, R. A. (1994). Emissions of carbon from land-use change. *In* "Proceedings from The 1993 Global Change Institute: The Carbon Cycle" (in press).

Houghton, R. A., Hobbie, J. E., Melillo, J. M., Moore, B., Peterson, B. J., Shaver, G. R., and Woodwell, G. M. (1983). Changes in the carbon content of terrestrial biota and soils between 1860 and 1980: A net release of $CO_2$ to the atmosphere. *Ecol. Monogr.* **53,** 235–262.

Houghton, R. A., and Skole, D. L. (1990). Changes in the global carbon cycle between 1700 and 1985. *In* "The Earth Transformed by Human Action." New York: Cambridge Univ. Press.

Keeling, C. D. (1982). The oceans and the terrestrial biosphere as future sinks for fossil fuel $CO_2$. *In* "Interpretation of Climate and Photochemical Models, Ozone and Temperature Measurements." New York: American Institute of Physics.

Keeling, C. D. (1986). Atmospheric $CO_2$ concentrations: Mauna Loa Observatory, Hawaii 1958–1986. NDP-001/R1, Carbon Dioxide Information Analysis Center, Oak Ridge National Laboratory, Oak Ridge, TN.

Keeling, C. D., Piper, S. C., and Heimmann, M. (1989). A three-dimensional model of atmospheric $CO_2$ transport based on observed winds. 4. Mean annual gradients and interannual variability. *In* "Aspects of Climate Variability in the Pacific and Western Americas." Washington, D.C: Geophysical Monograph 55, AGU.

Maier-Reimer, E., and Hasselmann, K. (1987). Transport and storage of $CO_2$ in the ocean: An inorganic ocean-circulation carbon cycle model. *Climate Dynamics* **2,** 63–90.

Moore, B. (1985). The oceanic sink for excess atmospheric carbon dioxide. *In* "Wastes in the Ocean," Vol. 4. Chichester, England: John Wiley & Sons.

Sarmiento, J. (1991). Oceanic uptake of $CO_2$: The major uncertainties. *Global Biogeochem. Cycles* **5,**4, 309–314.

Schlesinger, W. H. (1991). "Biogeochemistry: An Analysis of Global Change." San Diego: Academic Press.

Siegenthaler, U., and Sarmiento, J. L. (1993). Atmospheric carbon dioxide and the ocean. *Nature* **365,** 119–125.

Skole, D., and Tucker, C. (1993). Tropical deforestation and habitat fragmentation in the Amazon: Satellite data from 1978 to 1988. *Science* **260,** 1905–1910.

Tans, P. P., Fung, I. Y., and Takahashi, T. (1990). Observational constraints on the global atmospheric $CO_2$ budget. *Science* **247,** 1431–1438.

Wigley, T. M. L., and Schimel, D., eds., (1994). "The Carbon Cycle." Cambridge, England: Cambridge Univ. Press (in press).

# Grassland Ecology

**I. R. Sanders**
*Universität Basel*



The definition of grassland covers an extremely diverse range of ecosystems with vegetation dominated by grasses (Gramineae) or grasslike species, such as the sedges and rushes (Cyperaceae and Juncaceae). Although grasslands are found in the temperate, tropical, arctic, and montane regions of the world, they are almost all characterized by experiencing a seasonal period of low rainfall. Most grasslands cover terrain of relatively uniform relief and display little of the diversification in vegetation strata that can be found in woodlands or forests, however, they do maintain a remarkably diverse flora that can only coexist owing to the interacting environmental factors of climate, soils, grazing, and fire. At the small scale, below 10 m$^2$, grasslands are probably the most species-rich plant communities in the world. Knowledge of the structure and functioning of the aboveground and belowground floral and faunal communities is essential for the correct management for maintenance and preservation of these globally important ecosystems.

## I. INTRODUCTION

The world's grasslands are so diverse and the ecological processes that contribute to the functioning of these ecosystems so varied that this article can only cover important overall characteristics and processes of grassland ecosystems. For more precise ecological information pertaining to specific grassland habitats or general information on species competition, coexistence, or dynamic processes in vegetation, the reader should consult the texts listed in the bibliography.

### A. Natural, Seminatural, and Man-made Grasslands

The majority of the world's grasslands occupy the continental interiors lying between true forest and deserts. In some cases, grasslands are interdispersed with trees, for example, some tropical savannas ($\leq$15 trees ha$^{-1}$), although grasses or grasslike species comprise the dominant vegetation. A distinction should immediately be made between *natural grasslands,* which have formed as a result of climatic, edaphic, and biological factors, *seminatural grasslands,* which have been modified by human activities, and *man-made grasslands,* which have been planted and managed by humans (Fig. 1). These distinctions are critical in understanding how the grasslands have been formed, how they are maintained, and to what extent they are of biological value.

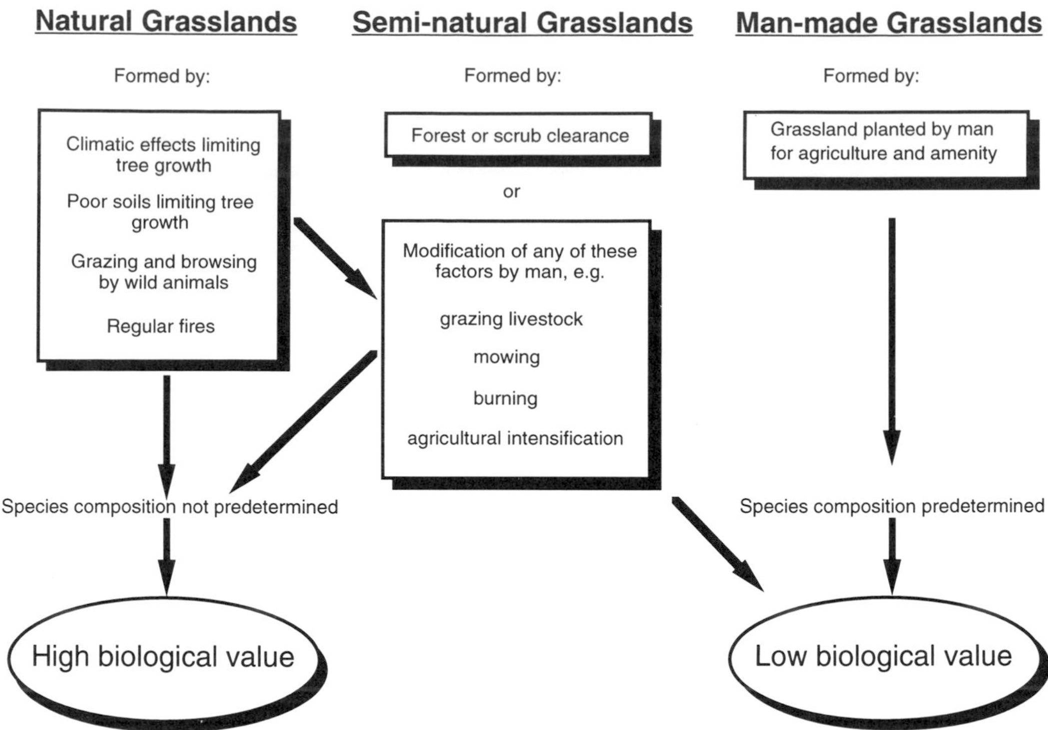


***FIGURE 1*** Classification of grassland types and their formation.

Natural grasslands have usually formed as a result of either low rainfall (between 50 and 130 cm year$^{-1}$) or low-nutrient soils which exclude the growth of trees, and/or grazing by animals. These two factors prevent succession from grassland to climax forest vegetation. Only in natural grasslands are plants thought to exist as a climax community that has reached some degree of equilibrium, although this is still an unresolved discussion among ecologists. The grassland community is, however, a dynamic environment of microsuccessions where deaths, replacements, and shifts in the distribution and dominance of species are continuously occurring.

Seminatural grasslands occupy much of the land surface and vary greatly in the extent to which they are influenced by humans and in the naturalness of their species composition. However, in general these are grasslands that have once been natural and have subsequently been modified by human actions (e.g., fire, domestic livestock grazing, mowing, and agricultural intensification). Seminatural grasslands can also include ecosystems that have been destroyed by human activities, for example, the woodland clearances that took place in Britain and Western Europe (probably beginning during the period of Neolithic man) produced semi-natural grasslands that were subsequently managed for the grazing of animals. In both instances, the existing plant communities are not predetermined by humans and the resulting vegetation has therefore developed over hundreds or thousands of years into a unique interacting community in which many species have undergone specialization and adaptation and are dependent on the grassland ecosystem for their continued survival. Many of the world's seminatural grasslands still support a high floristic diversity and many are of considerable biological value.

In contrast, man-made grasslands are often of little biological diversity, usually being composed of a few high-productivity grasses, with a few beneficial perennial plants, for example, nitrogen-fixing legumes. There are indeed man-made grasslands that have been planted and managed to re-create natural grassland types, for example, the restoration of the North American prairies. [*See* RESTORATION ECOLOGY.]

### B. Structure and Complexity

Although grasslands may appear to be fairly uniform and simple in structure, there are important features of composition and complexity that are common to most grasslands. The aboveground strata are fairly simple, constituting a main grass/herb layer with an underlying ground layer of subdominant species with prostrate or rosette growth habits and a carpet of mosses and lichens. Although the maximum height of vegetation in the tropical grasslands can reach around 4 m, it is usually much more restricted, owing to both climate and grazing activity. Grassland plant communities are, in fact, extremely diverse in terms of species composition and in the number of species that they can support [*See* SPECIES DIVERSITY.] In addition, the spatial arrangement of plants within the community is also complex and many studies in recent years have investigated how spatial patterns of plant species distribution in grasslands occur and how dynamic they are.

The belowground structure is extremely complex and poorly understood. Because of obvious practical reasons, ecological studies of grassland community structure have concentrated largely on the aboveground vegetation processes. However, the belowground structure of a grassland community exhibits considerable heterogeneity and can spatially occupy a greater depth than the corresponding aboveground vegetation height (depending on limitations imposed by the bedrock). For example, in the midwestern prairies of North America there may be several meters of soil that can be exploited by the roots, although the vegetation does not grow much above 1.5 m. In this case the root layer spatially occupies a much greater area than the aboveground vegetation. Rooting depth in other grasslands may be severely limited by bedrock, for example, in chalk grasslands with thin rendzina soils there may only be about 15–20 cm of soil depth that the plant roots can exploit before reaching solid bedrock. [*See* RHIZOSPHERE ECOPHYSIOLOGY.]

The belowground vegetation also exhibits a high degree of spatial and temporal structure among the different component species in rooting depth and different volumes of soil exploited by individual species. At the community level, the greatest density of roots occurs in the top layers of the soil. In West African savannas, the top 50 cm of soil contains 80% of the root biomass. In temperate species-rich calcareous and lowland meadows in Britain, the majority of roots (up to 80%) are contained within the top 10 cm of the soil. The most important feature, however, of the belowground component of all grasslands is the high biomass and productivity in comparison to that of the aboveground components (Table 1). For most grasslands it is difficult to provide comparisons of aboveground and belowground productivity because seasonal maximums of productivity in the two components are temporally separated; for example, in many temperate grasslands of Europe, maximum root productivity occurs in spring (late April and May) and maximum shoot productivity occurs in the following months (June and July). However, by looking at either seasonal maximums of the two components or by considering yearly averages of aboveground and belowground productivity, it is apparent that most of the plant productivity occurs belowground and that we can, therefore, expect a high degree of complexity in the belowground vegetation structure.

Yet measurements of root biomass are not a true representation of belowground plant productivity. Almost all individual plants comprising grassland ecosystems are occupied by mutualistic fungi

**TABLE I**

**Productivity of Aboveground and Belowground Components of Several Different Grassland Types (Values Expressed as Tons ha$^{-1}$)**

| Grassland type | Aboveground | Belowground | Percentage represented belowground[a] |
|---|---|---|---|
| Arid grassland (central Europe) | 1.33 | 25.90[b] | 95.1 |
| *Arrhenatherum*/ *Aloperurus* meadow (central Europe) | 3.09 | 10.50[b] | 77.3 |
| Prairie (North America) | 0.46 | 4.82 | 91.3 |
| Savanna (West Africa) | 5.60 | 19.00 | 77.2 |

[a] Percentage of total biomass represented by root production.
[b] Maximum seasonal belowground productivity.

known as *arbuscular mycorrhiza*. In nutrient-rich grasslands where the fungi are likely to provide little nutritional benefit to the plant, at least 40% of the root length of the community is occupied by these fungi. In soils where phosphorus is limited, mycorrhizal fungi can occupy an average of 70–80% of the root length in the plant community and this certainly represents a large amount of the belowground biomass. The high primary belowground productivity is a key feature of the grassland ecosystem. This high belowground biomass represents a large exploitable resource and this is reflected in both soil faunal and microbial biomass and activity, and in the high linkage in belowground food chains (up to 10 links). Though much progress has been achieved in understanding how aboveground vegetation processes affect the structure and functioning of the community below ground, processes are still poorly understood.

### C. Species Diversity

Plant diversity in natural grasslands (both tropical and temperate) can be extremely high. On the basis of area, at the level of 10 $m^2$ or less, grasslands are probably the most species-rich plant communities in the world. Many ecological studies of grassland ecology have attempted to explain how so many species are able to coexist. Estimates of grassland biodiversity have concentrated largely on counts of higher plant species and in some cases only the grass species. These estimates are extremely difficult to compare because of the scale on which they were conducted and the different criteria used to define grassland communities in each of the studies. In some investigations, many small areas of different habitats contained within the grassland were incorporated into the studies, resulting in artificially high species numbers. The highest estimates of grassland species diversity come from the tropical savannas of western and eastern Africa, where average areal species richness (average number of plant species per 10,000-$km^2$ area) attained more than 1750 species. Other tropical savannas appear to be less rich. The cerrados of Brazil support around 300 plant species per hectare. In temperate zones, grasslands can also attain high species richness. Typical floristic diversity in chalk and other limestone grasslands in middle and western Europe is around 40 species $m^{-2}$ and in Eurasia up to 75–80 species $m^{-2}$. In extreme cases, some of the temperate grasslands of Argentina (pampas) are extremely rich, supporting over 400 species of grasses.

In contrast to floristic diversity, faunal diversity of grasslands is generally low. Recent estimates reveal that only 5 and 6% of the world's birds and mammals, respectively, are adapted to grasslands, although grasslands cover approximately 23.1% of the world's vegetated land surface. This is partly due to the lack of diversity in aboveground structure of the vegetation and in the often extreme environmental conditions to which grassland-living fauna are subjected. Most estimates of grassland faunal diversity only account for vertebrates. Even in this case the majority (70%) of the vertebrates that inhabit grasslands live and breed below ground. Most of these are small rodents. An enormous nonvertebrate faunal biomass is contained both above and particularly below ground. This is not surprising when we consider the large plant and fungal biomass that occurs below the ground. As with the belowground vegetation structure, the belowground faunal biomass and function in the community have not been fully investigated. However, it is known that earthworms, nematodes, collembola, and mites constitute a large amount of the belowground fauna in both diversity and total numbers.

## II. MAJOR GRASSLAND TYPES AND THEIR DISTRIBUTION

The major grassland types are classified here on the basis of the climatic zones that they occupy (Fig. 2). Further subdivisions of grassland types have traditionally been made on the basis of climate (usually rainfall) and on the edaphic factors that are equally important in determining the composition of the plant communities. An attempt is made here to characterize the main features of temperate and tropical grasslands, although to gain access to information on specific grasslands the reader should consult the bibliography.

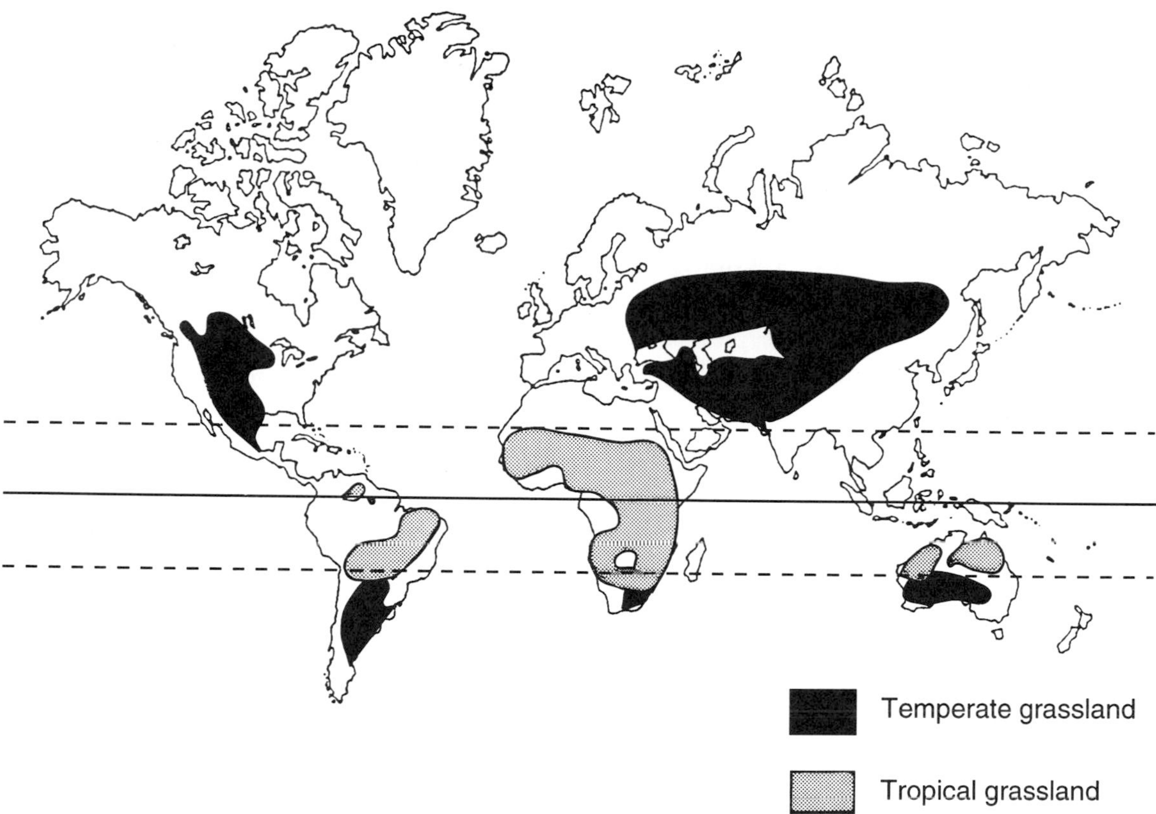


***FIGURE 2*** World distribution of tropical and temperate grasslands.

### A. Temperate Grasslands

Temperate grasslands once occupied large tracts of North America (*prairies*), Eurasia (*steppes*), Australia (*rangelands*), South America (*pampas* of Argentina and Uruguay), and South Africa (*veld*) (Fig. 2). The extent of these areas has now been greatly diminished by conversion to arable lands or have been greatly modified by intensive grazing regimes of domestic livestock and the use of fire. Most recent estimates of existing temperate grassland suggest that they cover between 6.7 and 12.5 million km$^2$ (or between 4.4 and 8.2% of the earth's surface).

With the exception of the South African grasslands, the world's natural temperate grasslands were heavily influenced by the grazing of large herbivores: in North America by bison and pronghorn antelope; in Eurasia by wild horse and ass, bison, saiga antelope, camels, and goats; in Australia by kangaroos; and in South America by guanaco. Whether grazing can explain the origin of temperate grasslands has been the subject of much debate. A combination of climatic change (reduced rainfall) and the evolution of specialized grazing animals is likely to have resulted in the formation of the great plains grasslands. However, others suggest that no grassland can naturally maintain itself without human activities, in particular fire followed by grazing.

Most temperate grasslands are characterized by having a soil with a rich humus layer that is subjected to periodic drying. The floral composition of the dominant species is related to climatic and edaphic factors of the area. Rainfall is probably the most important factor affecting whether or not grassland develops in a particular area. The component species and the physiological adaptations of those species are largely reflected in the climate and soil types. Temperature is particularly important in determining the species that are physiologically adapted to the area. Temperate grasslands can be composed of both $C_3$ and $C_4$ plants. *$C_3$ plants* fix carbon dioxide into three carbon compounds and *$C_4$ plants* initially fix carbon dioxide into four-carbon dicarboxylic acids. $C_4$ plants have a physiological advantage over $C_3$ plants at high light intensity and high temperature and exhibit a higher water-use efficiency, whereas $C_3$ plants have an advantage at moderate light intensity and temperature. Many temperate grasslands of Eurasia and North America exist in the latitudes that support both $C_3$ and $C_4$ plants and experience typical continental climates with hot, dry summers. In the prairies of North America, approximately 50% of $C_3$ and $C_4$ plants coexist at the 40°N latitude. North of this latitude many $C_4$ plants [e.g., big bluestem (*Andropogon gerardii*), little bluestem (*Andropogon scoparius*), and tall gramma grass (*Bouteloua curtipendula*)] are still able to coexist with $C_3$ species because they may be able to achieve maximum growth rates during the hottest part of summer. The coexisting $C_3$ species [e.g., prairie june grass (*Koeleria pyramidata*), basin wild rye (*Elymus cinereus*), and awnless brome-grass (*Bromus inermis*)] attain their maximum growth rate earlier in the year. The $C_4$ species appear to be restricted not so much by competition with $C_3$ species but by their sensitivity to chilling.

Edaphic factors are also extremely important in determining the species composition of the temperate grasslands. The ecology of seminatural chalk and other calcareous grasslands has come under much study as an example of how plants are able to coexist under edaphic conditions that exert nutritional constraints on the plants. In this respect the chalk downlands of southern Britain have been of particular interest as their history and management have been well recorded in local records back to the Domesday Book of A.D. 1086 and preceding that in the archaeological records dating back to Neolithic man. They are communities of plants whose management has remained unchanged for centuries and are, therefore, thought to be in a type of equilibrium, that is, the plant community will not change in species composition as long as this regime is maintained. The chalk grasslands of southern Britain are not natural and are not in their climax vegetation state. These downlands would have comprised either English elm (*Ulmus procera*), field maple (*Acer campestre*), and lime (*Tilia cordata*) or ash (*Fraxinus excelsior*), beech (*Fagus sylvatica*), and yew (*Taxus baccata*). From the period of Neolithic man onward the woodlands were cleared and then maintained as unimproved grassland for sheep grazing. Both sheep and rabbits (introduced in the twelfth century) play an important role in maintaining the characteristic short-turf grasslands in which a high number of species are present (over 40 higher plant species per square meter). All the species can tolerate the alkaline, calcareous soil, which imposes limitations on the availability of certain nutrients (e.g., $Fe^{3+}$, $Mn^{2+}$, and $K^+$) to the plants. Such plants are termed *calcicoles,* meaning that they are adapted to calcareous soils. Although the chalk grasslands of southern Britain can be divided into different grassland types, there are several species that are characteristic of these communities, for example, quaking grass (*Briza media*), tor grass (*Brachypodium pinnatum*), red fescue (*Festuca rubra*), sheep's fescue (*Festuca ovina*), upright brome (*Bromus erectus*), spring sedge (*Carex caryophyllea*), and glaucous sedge (*Carex flacca*).

Despite this high diversity there are a number of factors that are important in maintaining the species richness. First, the community could not be maintained as grassland or in its present diversity unless grazing was maintained by animals that crop the vegetation to a low level. Second, the species that comprise the communities could probably not exist if the nutrient availability were changed in some way, thereby allowing more competitive nutrient-loving species to become established within the community. Such nutrient amendments can occur in the form of manure from other animals (e.g., cattle) or by fertilizer additions of nitrogen and phosphorus. Changes in community structure that occur to fertilized grasslands are discussed later in this article.

### B. Tropical Grasslands

Tropical grasslands lie between the Tropics of Cancer and Capricorn (Fig. 2). *Savanna* is the term used to describe many tropical grasslands, although there are other specific terms such as *rangelands* (Australia), *cerrados,* and *llanos* (South America). Tropical grasslands occupy an estimated 24.6 million km$^2$ (16.3% of the earth's surface), although a large proportion of this is likely to be seminatural grassland that has been the result of tropical forest clearance. Unlike temperate grasslands, savannas are communities comprising mostly tussock-forming, xerophytic $C_4$ grasses of up to 3.5 m in height with a subcanopy consisting of some $C_3$ species. The tree cover varies greatly owing to both climatic and edaphic factors.

The climate of most savannas is characterized by high temperatures throughout the year with a seasonal period of drought. However, rainfall can vary from 500 to 3500 mm per year. In some cases the savanna may be subject to many months of drought, whereas other savannas may experience almost no drought. Classification and comparison of the ecology of savannas are difficult because of the great differences in climate and vegetation that exist among the major savanna ecosystems. However, it is now recognized that there is a continuum from arid, nonleached savannas through a moisture gradient to moist dystrophic savannas that merge into forests. The arid savannas are characterized by low-growing vegetation, large numbers of annuals, and low C:N ratios. The wetter savannas support more perennial species that are able to convert more carbohydrate into structural woody material, therefore giving rise to higher C:N ratios than found in the drier areas.

Interesting *physiological adaptations* are common to plants of savanna regions and allow them to cope with extremes of temperature and water loss. In the brazilian cerrados, it appears that plants have an adequate water supply from the soil and can transpire at a rate that allows them to cope with the high temperatures. Two strategies emerge, those plants that transpire all day and those that close stomata at midday, thereby reducing water loss during the period of maximum temperature. In both cases the stomata are closed at night to replenish water. Strangely, during the dry season the plants do not greatly change their transpiration rates and, therefore, must come under considerable stress from leaf heating when water content becomes low. However, many other plants reduce leaf heating by changing the amount of absorption of light and by leaf inclination. In the drier environments, such as the grasslands of the northern Transvaal (southern Africa), plants must be able to maintain transpiration to cool the leaves and be able to maintain an adequate rate of photosynthesis. The plant species of the African savanna differ greatly in their ability to do this. Some species can tolerate extremely low leaf water potentials without seriously affecting the rate of photosynthesis. Many of these plants are able to rapidly replenish lost water by having shallow roots that can obtain water from light dews. In addition, many of these species will have low rates of transpiration, thereby reducing water loss even at times when water is plentiful.

The savanna soils are generally poor in nutrient content (particularly phosphorus and nitrogen) and are also poorly drained owing to the presence of impermeable *lateritic crusts,* which can occur at a shallow depth. Many of these soils are acidic and contain high levels of aluminum, which can be toxic to many plants, and this has been suggested as a possible reason why trees have not become the dominant vegetation in some savannas. However, many plants do have ways of dealing with the high levels of aluminum and some are able to accumulate the element with no apparent harmful effects to plant growth. In addition, the activity of decomposers is reported to be extremely high, resulting in an extremely rapid turnover of organic matter. In contrast, other reports suggest that organic nitrogen pools may be relatively high, suggesting that in fact mineralization may actually be limited. Nitrogen in savanna soils is highly variable in form and spatial and temporal variation, which therefore makes it extremely difficult to understand how and the rates at which nitrogen is cycled in these ecosystems.

Evidence for the origin of savanna vegetation can be found in these features of the climate, soils, and the existing plant and animal species. It is clear that rainfall is not the only determinant of the grassland vegetation dominating the savannas. There are many trees, for example, the acacias, that are able to grow in the xerophytic grasslands of East Africa. Certainly, soils with lateritic crusts would be likely to inhibit the growth of deep-rooted trees and favor shallow-rooting grass and sedge species.

It has been suggested that where ancient flora of the Tertiary era exist in savannas, for example, in southern Africa, that these are in fact relicts of flora from a drier climatic era. The soils of the tropics can rapidly become degraded into the characteristic savanna soils with poor nutrient availability, which may have inhibited rapid forest development in these grasslands.

Many savannas are of secondary origin, having been formed from a combination of forest clearance followed by intensive grazing and burning regimes. Whether they are secondary in origin or not, many of these savannas are likely to have been subjected to burning for thousands of years and this is reflected in the ecological adaptations of many of the plants that occur. Substantial numbers of grasses are particularly suited to areas subject to fire (pyrophytic). In addition, many of the trees and shrubs of the savannas have adapted to seasonal fire and produce fruit either before the fires are likely to occur or immediately afterward. In both cases, seedlings that germinate after the fires are provided with a good source of available nutrients from the ashed plant material and this may also improve chances of seedling establishment owing to the absence of a vegetation canopy.

On both natural and seminatural savannas, grazing is also extremely important in maintaining the grassland vegetation, as is the case with the large herds of ungulates (antelope, wildebeest, and zebra) in Africa. Like the vegetation that has become adapted to these particular habitats, the grazing animals are also highly specialized to this vegetation type. The loss of many areas of natural savanna and overgrazing by domestic livestock have resulted in severe reductions in the numbers of specialized savanna animals, for example, the African elephant. But not only are the large vertebrates important in the grazing of savannas. Ants and termites play an extremely important function in removal of both live and dead vegetation from some savanna systems. In fact, in Australian savannas the biomass of termites has been reported to exceed that of the vertebrate grazers. In Brazilian savannas, leaf-cutting ants can remove approximately 50 kg of grass per hectare per day. The ants also have an important function in soil turnover and are responsible for changing the pH in the upper layers of the soil and reducing the amount of organic matter on the soil surface. In some savannas, termites play an important role in removing and digesting wood and other dead standing matter. Other termite species digest humic substances that occur in the soil and, therefore, play a similar role to that of earthworms. [*See* Soil Ecosystems].

## III. FACTORS AFFECTING MAINTENANCE OF GRASSLANDS AND THEIR SPECIES DIVERSITY

Many of the following factors are responsible, and in some cases essential, for maintaining grasslands and preserving their species richness. Overuse or misuse of any of these management practices can lead to degradation of the ecosystem by the influx of other species and the lowering of species diversity.

### A. Fire

Fire, whether natural or anthropogenic, is responsible for maintenance of many grasslands and prevents them from turning into scrub and subsequently, where climate permits, into forest. It is likely that fire was not responsible for the origin of most grasslands. However, in grasslands that have seen a decrease in fire, the result has in some cases been a succession into woodland. Apart from preventing succession, fire is recognized to play an important role in the ecology of many grassland species. [*See* Fire Ecology.]

Even taking into account the removal of plant material by grazers and decomposers, up to 50–60% of the yearly productivity can be burnt in a savanna fire. The ashing of this aboveground standing vegetation and litter layer provides plants with readily available nutrients. In addition, fire is also responsible as a stimulus to break seed dormancy in many plants. Reports of increased productivity in savannas following fire are between 30 and 90%. Fire also results in nutrient loss from a community; carbon is lost as carbon dioxide and nitrogen is lost in its gaseous form.

In African, South American, and Australian savannas, burning has been carried out by humans for thousands of years (up to 50,000 years in East Africa and 40,000 years in Australia). Consequently, the plants living in these savannas are well adapted to regular burning. Burning of grasslands by humans in areas where fires were infrequent, for example, some parts of the prairies, has caused changes in species composition. Introduction of readily burning fire tolerants [e.g., drooping brome (*Bromus tectorum*)] has led to more frequent fires and an increase in the abundance of such species and a reduction in natural species abundance. Burning causes little damage to belowground matter and is, therefore, often advantageous to rhizomatous plants or species with underground storage organs. Experiments on the effects of burning on calcareous grassland species composition resulted in rhizomatous plants (e.g., *Brachypodium pinnatum*) becoming dominant and a reduction in species diversity.

Apart from breaking seed dormancy, other adaptations enable plants to survive fires. Many shrubs are not adversely affected by fire and although leaves and shoots are burnt, buds and roots regenerate quickly. Other adaptations are mostly concerned with ensuring that seed matures early, that is, before the likelihood of fire, production of many, long-life seeds, or seed types that can become buried before fire arrives. Bunch spear grass (*Heteropogon contortus*) of the Australian savanna has adopted the latter of these strategies. It produces large amounts of seed that possess a hygroscopic awn and a barbed basal callus. The seed becomes rapidly buried in the top 1 cm of soil. Fire rapidly passes over the top litter layer, containing seeds of competitors, which are burnt away. The soil reaches only about 80°C, which is not high enough to kill the seeds of this species.

Fire tolerance among many shrubs is age dependent, with seedlings being particularly vulnerable during the first 12 months. Consequently, species composition of grasslands can be greatly affected by the frequency of fire. Furthermore, grazing interacts with fire to affect tree/shrub/herb ratios. Selective grazing is particularly important in changing the burning quality of vegetation, which will in turn affect both the frequency and intensity of fire. In addition, grazing can reduce the competitive ability of some fire-tolerant dominants directly following burning.

### B. Grazing and Mowing

Both grazing and mowing have a great impact on grassland communities by preventing succession. Removal or cropping of the vegetation can either increase or decrease species diversity. Unselective removal, that is, mowing, will generally have greater effects on plants with a tall growth habit, allowing more light to reach species with a low growth habit. The outcome of mowing or grazing on species diversity is dependent on which species are adversely affected. If the competitive dominants are most affected by mowing, then competition with the subdominants decreases thereby increasing species diversity and vice versa. This promotion of species coexistence is known as exploiter-mediated coexistence.

The effects of animal grazing are more complicated than those of mowing because animals differ greatly in their food preferences and therefore in selectivity of grazing. The importance of grazing became apparent on the downlands of southern Britain following the introduction of the myxoma virus into the rabbit population. Following the decline in rabbit numbers, a few grass species became dominant and species diversity was reduced.

Overgrazing often results in a decrease in species diversity, either by removal of all the vegetation or by selection for nonpalatable species, which ultimately become dominant. In poor alpine pastures in Europe, overgrazing by cattle results in an almost complete dominance by the nonpalatable monk's rhubarb (*Rumex alpinus*). This species produces large numbers of particularly tough seeds that are not destroyed in the cattle gut.

In addition, grazing of grasslands by domestic livestock has been responsible for the spread of many parasitic diseases beyond their natural distribution. Those parasites that have had the greatest effects on humans are ones that exhibit zoonosis, that is, they can be passed from one species of animal and parasitize a different animal species.

Rinderpest spread from northern Africa to southern Africa through domestic cattle, which resulted in the death of cattle and wild ungulates. As a further consequence, the trypanosomes causing trypanosomiasis (sleeping sickness) in humans, cattle, and wild ungulates increased in number owing to the development of the woodland following the rinderpest outbreak. The woodland is the habitat favored by the tsetse fly, which acts as a vector for trypanosomes.

Grazing animals also affect plant community structure by increasing habitat heterogeneity. Dung from herbivores results in nutrient-rich patches. Herding animals of the Serengeti (East Africa) concentrate where nutrient-rich patches increase vegetation quality. In turn this probably results in much dung being returned to these areas.

### C. Fertilizer Application and Agricultural Improvement

Addition of organic or inorganic fertilizers, particularly nitrates and phosphates, has been used to increase grassland production, especially on soils where nutrients are limiting. Experiments designed to investigate the effects of fertilizer additions on species diversity have been the subject of much study by plant ecologists and have been used for testing theories of species coexistence in plant communities.

The fact that many nutrient-poor grasslands are species-rich suggests ecosystems of low nutrient status promote species diversity. Indeed, in numerous experiments where nutrients have been applied to chalk grasslands, the species richness has declined as more competitive nutrient-prefering species became dominant. Two current theories as to how this occurs are the subject of much debate. It has been proposed that greater aboveground production also increases rates of competitive exclusion and consequently a decrease in species richness. In addition, ratios of resource availability have been put forward as determining productivity, which in turn affects species richness. Experimental manipulations of fertilizer addition to chalk grassland have resulted in changes in species richness that were not always related to increases in productivity but that were dependent on the ratio of nitrogen to phosphorus. It is clear that this type of agricultural improvement can have significant, detrimental effects on grassland community structure and consequently lower their biological value.

## IV. FRAGMENTATION, GLOBAL CHANGE, AND GRASSLAND SPECIES PRESERVATION

Although grasslands still occupy a large proportion of the world's surface, many of them are rapidly being degraded by human activities, especially by increased use as pasture and by agricultural intensification. Areas of true natural grassland are now extremely small and are diminishing. It is anticipated that in the next few years many grassland species will become extinct and many grasslands will be altered further and in some cases completely destroyed.

It is not just modification and destruction that threaten the world's grasslands, but also the degree of habitat fragmentation that can determine the fate of populations in grassland communities. Fragmentation may cause removal of resources or change birth, death, and species extinction rates. Rare and specifically adapted grassland species may potentially be some of the most threatened by extinction owing to their dependence on the grassland ecosystem and their loss of adaptive potential (i.e., reduction in genetic variability). It is essential that we understand the mechanisms by which fragmentation reduces the viability of populations and communities at the local level and also on regional and larger scales. The understanding of environmental effects of fragmentation on grassland floral and faunal populations is only in its infancy and correct management recommendations and strategies for ecosystem and species conservation cannot be made without this knowledge.

An understanding of how the components of grasslands will respond to global change must also be achieved. Elevation in levels of greenhouse gases (e.g., carbon dioxide) has great effects on the growth and functioning of plant species and is likely to affect how nutrient cycling takes place.

The ability of highly specialized grassland species to survive such changes must also be assessed. In addition, increased deposition of nitrates in rainwater, which act as fertilizers, may have significant short-term effects on community structure of grasslands. The likely effects of many other atmospheric pollutants are, as yet, unknown.

In some cases, attempts have been made to reintroduce extirpated species into grasslands and some of these have been successful. Reintroduction of the North American bison and pronghorn antelope into the prairies was successful in increasing the numbers of these species. However, the complexity and specialization of life histories of grassland species have resulted in great difficulties in their successful reinstatement. For example, respeated attempts to reintroduce the large blue butterfly (*Lycaena avion*) into British chalk grassland failed even though the butterfly's food plant, wild thyme (*Thymus drucei*), was present. Studies since have revealed that the butterfly requires a mutualistic partnership with ants (*Myrmica scabrinoides* and *M. laevonoides*) for the larvae to survive. The ants feed from honey glands of the larva and, in return, protect it from predators and parasites. Hence the butterfly cannot survive in chalk grasslands that do not support the ant hosts. The complexity of such interactions in nature makes the task of restoring habitats, whose locally extirpated species still occur elsewhere, almost impossible. [*See* INTRODUCED SPECIES.]

This article has demonstrated how grasslands contribute to the world's biodiversity, the complex ecology of these ecosystems, and how sensitive these environments are to degradation by human activities. To establish an effective strategy for grassland conservation, the ecology, functioning, biological, and sociological value of grasslands must be realized.

## Glossary

**Arbuscular mycorrhiza** Mutualistic fungi that occupy plant roots and provide the plant with improved phosphorus uptake.

**C3 plants** Plants that convert carbon dioxide into three-carbon compounds during photosynthesis.

**C4 plants** Plants that convert carbon dioxide into four-carbon compounds during photosynthesis.

**Calcicole plants** Plants that can tolerate soils with a high content of calcium carbonate.

**Cerrados** Tropical grassland of Brazil.

**Lateritic crusts** Mineral pans or crusts in the soil that are often impermeable to water and impenetrable to roots.

**Llanos** Tropical grassland of Venezuela.

**Pampas** Temperate grassland of South America.

**Prairie** Temperate grassland of North America.

**Rangeland** Grassland of Australia.

**Savanna** Tropical grassland of Africa.

**Steppe** Temperate grassland of Eurasia.

**Veld** Temperate grassland of South Africa.

## Bibliography


Begon, M., Harper, J. L., and Townsend, C. R. (1990). "Ecology: Individuals, Populations, and Communities," 2nd ed. Cambridge, Mass.: Blackwell Scientific.

Ellenberg, H. (1988). "Vegetation Ecology of Central Europe," 4th ed. Cambridge, England: Cambridge University Press.

Hornby, R. J. (1992). *In* "Global Biodiversity: Status of the Earth's Living Resources" (B. Groombridge, ed.). London: Chapman & Hall.

Hillier, S. H., Walton, D. W. H., and Wells, D. A., eds. (1990). "Calcareous Grasslands: Ecology and Management." Bluntisham, England: Bluntisham Books.

Huenneke, L. F., and Mooney, H. A., eds. (1989). "Grassland Structure and Function: California Annual Grassland." Dordrecht, Netherlands: Kluwer.

Huntley, B. J., and Walker, B. H., eds. (1982). "Ecology of Tropical Savannas." Heidelberg/Berlin: Springer-Verlag.

Scholes, R. J., and Walker, B. H. (1993). "An African Savanna: Synthesis of the Nylsvley Study." Cambridge, England: Cambridge University Press.

# Great Plains, Climate Variability

**W. K. Lauenroth and I. C. Burke**
*Colorado State University*



The Great Plains of North America is a vast treeless area in the center of the continent that stretches from Canada to southern Texas. It is an important agricultural region, producing both grain crops and beef cattle. Temporal and spatial climatic variability is one of the most notable features of the region. Precipitation and temperature extremes, as well as substantial interannual fluctuations, are characteristic of the temporal dynamics of the Great Plains climate. The region spans from the subhumid zone in the east to the arid zone in the southwest. Understanding current climatic variability is a necessary step in understanding the current distribution and dynamics of Great Plains ecosystems, as well as an important step in the process of predicting the potential effects of human-induced climatic change on the region.

## I. INTRODUCTION

The Great Plains of the United States occupies the majority of the center of the country. At the time of settlement by Europeans it was the home of largely nomadic tribes of indigenous peoples and large populations of bison, prairie dogs, elk, pronghorn, bighorn sheep, and deer. Since settlement in the 1800s, the forces of climate, economics, and social pressure have profoundly changed the Great Plains. In the 1800s settlers began reducing the herds of bison and replacing them with domestic livestock. The plains livestock enterprise began in Texas and by 1876 it had spread over the entire region. No sooner had the range livestock industry expanded to include the entire plains area than it began to decline. In 1862, Congress passed the Homestead Act and in 1874 the first barbed wire was sold in the United States. In the late 1800s, severe winter conditions devastated livestock herds in many areas. By the turn of the century, row crop production had replaced livestock grazing as the dominant land use in many parts of the plains. The first 30 years of the twentieth century experienced climatic conditions that were largely favorable to agriculture. During the 1930s an extended drought had severe impacts on agricultural production, causing widespread abandonment of farms in many areas of the plains and alteration of the distribution patterns of cropland. Recently, socioeconomic factors such as price supports and soil conservation programs have been more influential than climate in further alteration of the spatial distribution of croplands. [*See* AGROECOLOGY.]

Human activities at the end of the twentieth century may soon add a new dimension to the climate of the earth and the region. The release of greenhouse gases into the atmosphere is threatening to have large effects on both the temperature and precipitation regime of the Great Plains, which may in turn influence human use of Great Plains ecosystems. The objective of this article is to describe current levels of temporal and spatial variability in the Great Plains climate as a key initial step in understanding the climatic and ecosystem consequences of potential climate change. [*See* GREENHOUSE GASES IN THE EARTH'S ATMOSPHERE.]

Our analysis of climatic variability will focus on temporal and spatial fluctuations of mean annual precipitation (MAP), mean annual temperature (MAT), mean annual potential evapotranspiration (MAPET), and the mean annual number of days receiving measurable precipitation (MAPD). The reasons for choosing these variables instead of other climatic characteristics is that they are very important in determining the structure and function of Great Plains ecosystems as well as in determining the temporal and spatial patterns of human use. Furthermore, they are variables that will be changed by human-induced global climate change. Our interests in temporal climatic variability will concentrate on interannual scales. Interannual variability will be characterized using 20-year data sets for 296 stations throughout the Great Plains. These same stations will be used to evaluate spatial climatic variability. Our presentation of the temporal and spatial aspects of climatic variability will be interwoven because the importance of temporal variability must be understood in terms of its spatial context and vice versa for spatial variability. To provide an ecological context for our discussions of climate variability we will begin with an overview of the Great Plains that includes a description of the current distributions of ecosystems over the region.

## II. OVERVIEW OF THE GREAT PLAINS

### A. Geography and Physiography

The North American Great Plains has three geographic definitions. First, it is a physiographic region defined by landform that extends from southern Canada to central and southwestern Texas and from approximately the 98th meridian to the base of the Rocky Mountains. Second, within the United States it is an agricultural region that encompasses a 10-state area that includes the physiographic Great Plains (Fig. 1). Finally, there is a popular definition of the Great Plains that refers to the central portion of the North American continent that was largely grassland at the time of settlement. We use the term Great Plains to refer to the 10-state agricultural region that includes North and South Dakota, Nebraska, Kansas, most of Oklahoma, the north central portion of Texas, and the plains region of Montana, Wyoming, Colorado, and New Mexico.

The major surface features of the Great Plains are the result of complex interactions between wind- and water-driven processes as well as glaciation in the northeastern portion. Material eroded from the Rocky Mountains and carried eastward by major river systems has been deposited on the plains over the past several million years. After deposition, this material has been reworked by both wind and water erosion to produce the current surface. As it exists today, the Great Plains is a relatively flat surface that slopes from the base of the Rocky Mountains to approximately the 98th meridian. Elevation at the base of the mountains is approximately 1500 m, and 150 m where it meets the Mississippi River. The flatness of the surface is only relative and local topography results in elevation differences of up to several hundred meters over short distances.

### B. Climate

The distinctive features of the Great Plains climate are the result of two attributes: its location in the center of the continent and the presence of the Rocky Mountains as its western boundary. The first feature means that the Great Plains is located far from the moderating influence of oceans and therefore experiences much larger seasonal temperature fluctuations than are found in maritime regions. Because of the predominantly westerly flow of the atmosphere over the Great Plains, the Rocky Mountains produce a rainshadow effect, especially

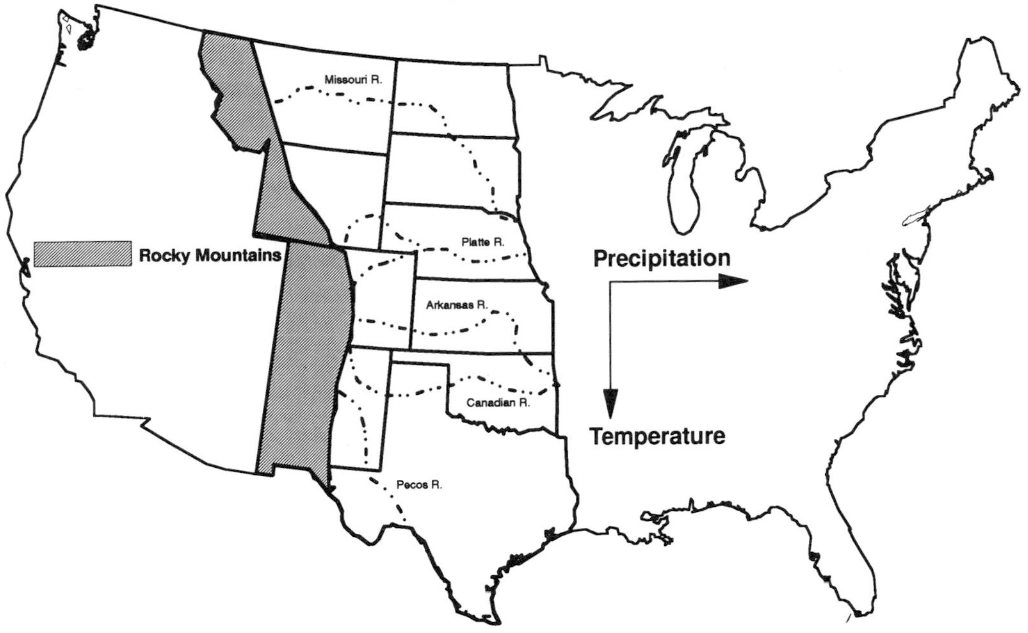

***FIGURE 1*** Geographic location of the Great Plains of the United States showing major rivers and climatic gradients.

for the western portion of the region. The general features of the regional climate are a north to south temperature gradient, a west to east precipitation gradient, and a very high atmospheric demand for water (Figs. 2 and 3).

Mean annual temperatures are less than 5°C at the northern extension of the Great Plains and greater than 20°C in the south (Fig. 2a). Forty percent of the region has MAT less than or equal to 8°C, 60% has MAT less than or equal to 11°C, and approximately 13% has MAT greater than 14°C. The climate is classified as temperate in the north and subtropical in the south. Mean annual precipitation is 30–40 cm in the semiarid western portion and up to 120 cm in the subhumid east (Fig. 2b). Approximately 25% of the western portion of the region has MAP less than or equal to 40 cm, 50% of the region has MAP less than or equal to 50 cm, and 80% has MAP less than or equal to 70 cm. Associated with this gradient in precipitation is a related gradient in the MAPD (Fig. 4). Less than 2% of the area in the Great Plains experiences more than 80 days per year with measurable precipitation and approximately half of the area has fewer than 60 such days each year.

Atmospheric demand for water (MAPET) is high in the Great Plains and spatial patterns of MAPET follow the temperature gradient, with southern sites having the greatest values and northern sites the smallest (Fig. 3). The majority of the area in the region has MAPET greater than 150 cm and 25% has values greater than 170 cm. The minimum value of MAPET for the Great Plains is 120 cm, which is the same as the maximum value of MAP. Less than 2% of the area in the region has MAP between 110 and 120 cm, whereas the entire region has MAPET equal to or greater than 120 cm. A common characteristic of grasslands worldwide and also of Great Plains grasslands is a ratio of MAP/MAPET that is less than 1. This ratio provides an indication of the adequacy of the water supply to meet the atmospheric demand. Most sites in the Great Plains have a value of MAP/MAPET considerably less than 1 (Fig. 5). Twenty percent of our 296 Great Plains sites had MAP/MAPET less than 0.3 and 60% had values less than or equal to 0.4. The gradients in MAT, MAP, MAPD, and MAPET result in important gradients in the structure and dynamics of ecosystems across the region, influencing species composition, net primary production, soil organic matter, and land use.

The climate of the Great Plains is influenced by three air masses, each with different temperature, moisture, and temporal-dynamic characteristics.

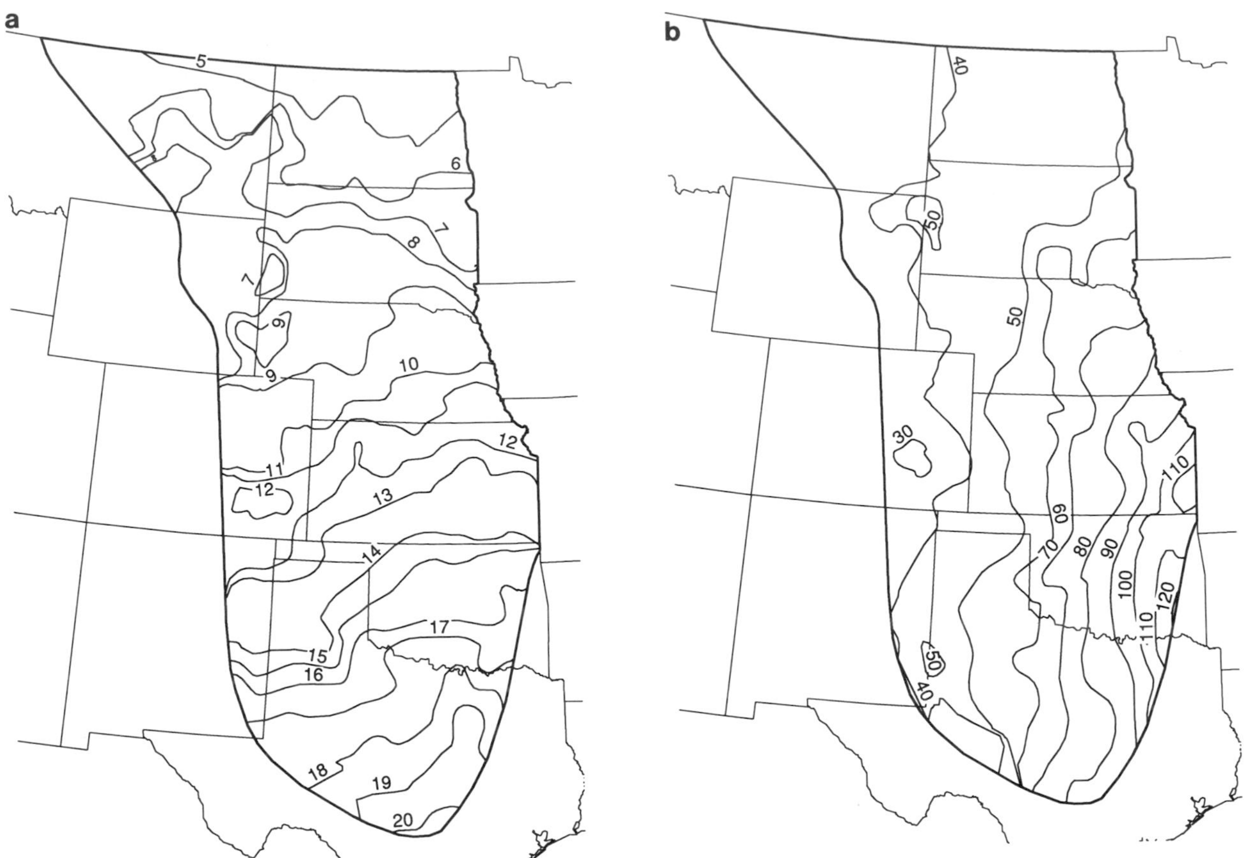


***FIGURE 2*** (a) Isotherms for mean annual temperature (MAT, °C) and (b) isohyets for mean annual precipitation (MAP, cm) for the Great Plains of the United States.

Winters are dry and dominated by strong westerly flow and the influence of a temperate maritime air mass that originates in the northern Pacific Ocean. To reach the Great Plains this air must cross several mountain ranges, and consequently it is most often a source of cool, relatively dry air. There is a decreasing winter snow gradient in the Great Plains from north to south. The position of the jet stream determines how often and how much of the region is subjected to the influence of the second important air mass, very cold dry air that originates over the North Pole. The presence of this air mass accounts for the extreme cold conditions in some winters. The frequency of these conditions decreases from north to south, resulting in northern areas experiencing episodes of polar air every winter and southern areas being exposed only infrequently.

Spring and summer are the wet seasons across the Great Plains. Peak precipitation occurs in April, May, June, and July throughout most of the region. In the extreme southwest the peak is shifted to late summer and in the southeast the peak tends to be bimodal with a second peak in fall and early winter. The strong westerly flow of winter begins weakening in the spring, when a warm, wet subtropical air mass with origins over the Gulf of Mexico begins to influence the region. Because this air mass enters the region from the southeast, its influence diminishes westward and northwestward, which explains a large part of the MAP gradient over the Great Plains.

## C. Ecosystems and Land Use

### *1. Grasslands*

The Great Plains comprises four grassland types (Fig. 6). Three of the types are contained entirely within the Great Plains; approximately one-half of the fourth type, the tallgrass prairie, lies outside

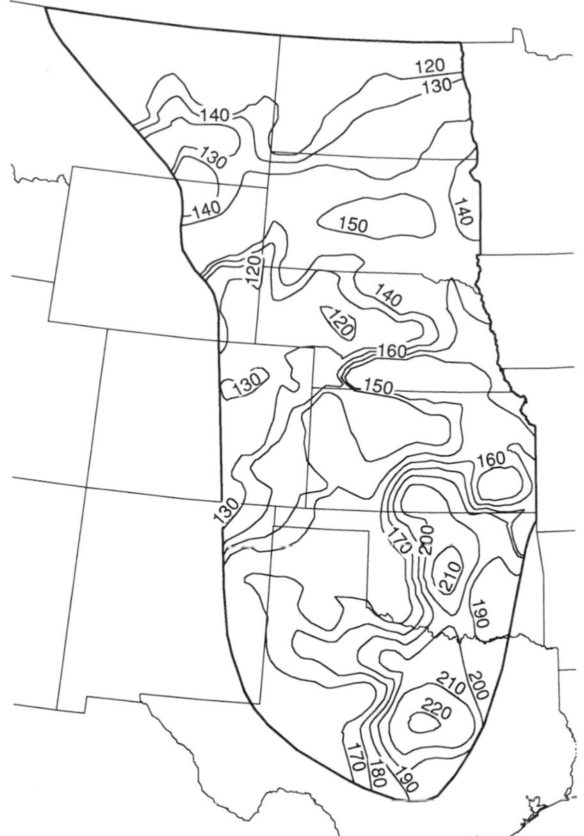


**FIGURE 3** Isopleths of mean annual potential evapotranspiration (MAPET, cm) for the Great Plains of the United States.

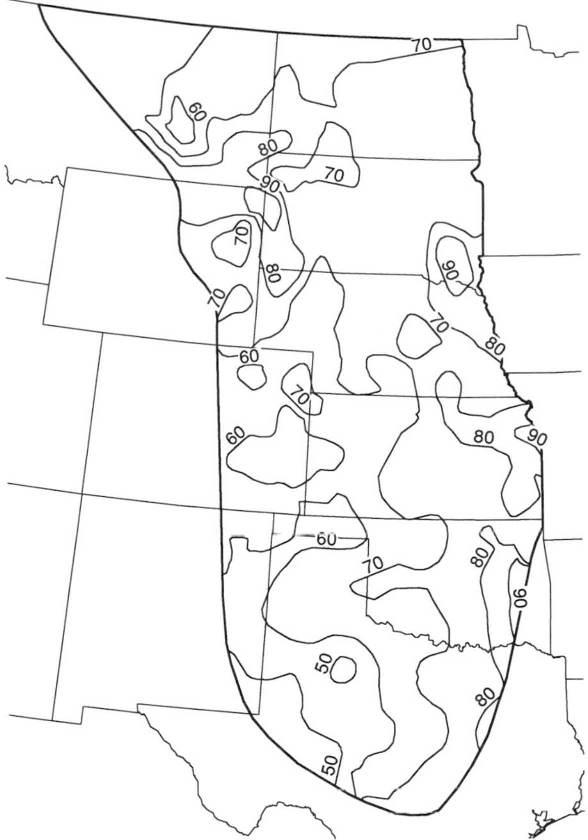


**FIGURE 4** Isopleths of mean annual number of days with measurable precipitation (MAPD, days) for the Great Plains of the United States.

the region. The distribution of the grassland types is almost entirely determined by the climatic gradients. The tallgrass prairie occupies the wettest, easternmost portion of the Great Plains. The dominant species in this type are all $C_4$ grasses in the central and southern parts and $C_4$ and $C_3$ grasses in the north. The most common $C_4$ species are *Andropogon gerardii, Schizachyrium scoparium, Panicum virgatum,* and *Sorghastrum nutans*. The most common $C_3$ species is *Stipa spartea*. In addition to occupying the wettest portion of the region, the tallgrass prairie is the most productive of the grassland types (Fig. 7). By contrast, the shortgrass steppe occupies the driest and one of the least productive parts of the region. The dominant species, *Bouteloua gracilis* and *Buchloë dactyloides,* are both $C_4$ shortgrasses.

The northern-mixed prairie occupies the area north of the shortgrass steppe and west of the tallgrass prairie. The boundary between the northern-mixed prairie and the tallgrass prairie occurs at

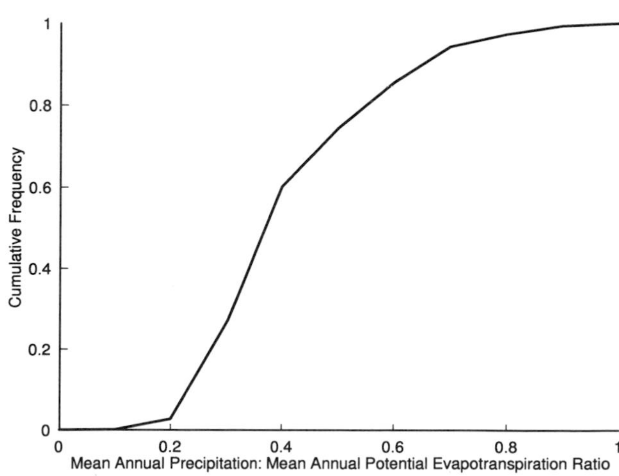


**FIGURE 5** Cumulative frequency of the ratio of MAP/MAPET for 296 weather stations in the Great Plains of the United States.

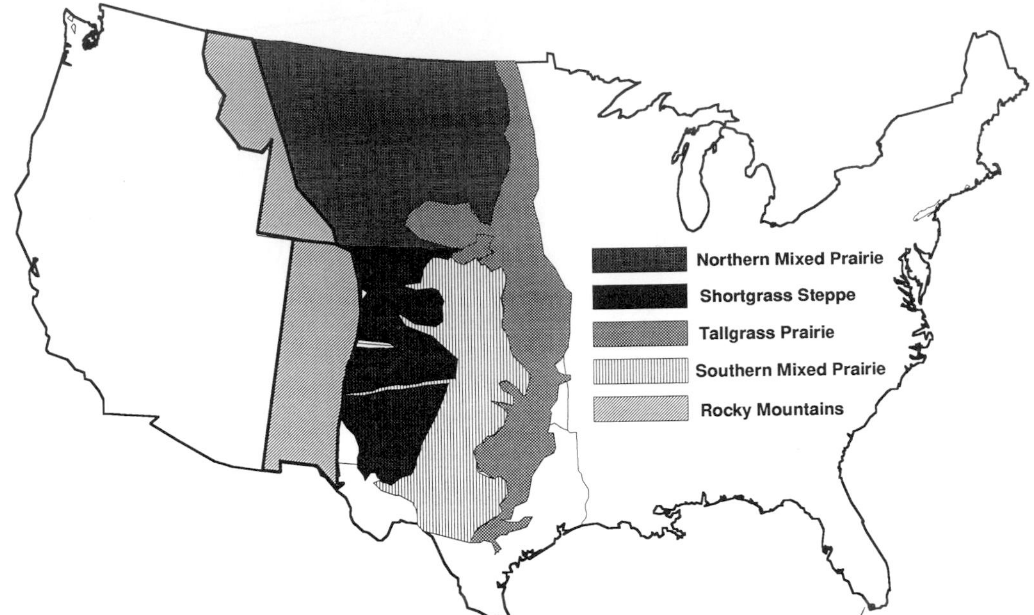


***FIGURE 6*** Geographic distribution of grassland types in the Great Plains of the United States. [Adapted from J. L. Dodd (1979). *In* "Perspectives in Grassland Ecology" (N. R. French, ed.), Springer-Verlag, New York.]

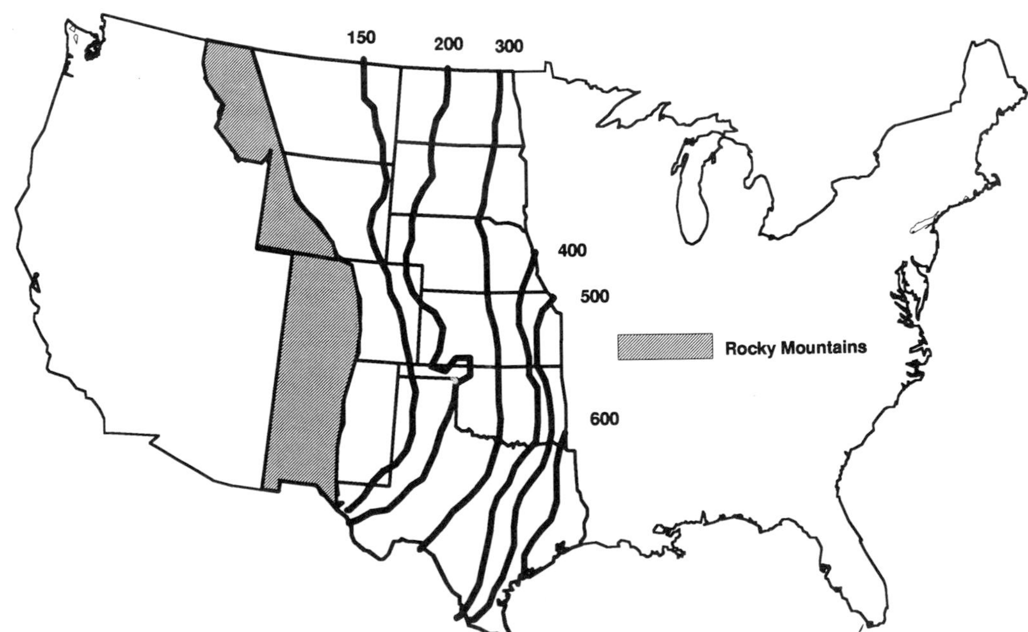


***FIGURE 7*** Isopleths of average annual aboveground net primary production (g/m$^2$) for the Great Plains of the United States. [Adapted from O. E. Sala *et al.* (1988). *Ecology* **69,** 40–45.]

approximately the 500 mm/year isohyet (Figs. 2b and 7). Annual aboveground net primary production is less than 300 g/m$^2$ throughout the northern-mixed prairie. The reason for describing it as a "mixed" prairie is that the dominants are a mixture of short and midheight grasses and also a mixture of $C_3$ and $C_4$ species. The dominant species are the $C_3$ grasses *Agropyron smithii* and *Stipa comata* as well as the dominant $C_4$ grasses from the shortgrass steppe. The southern-mixed prairie is located between the shortgrass steppe and tallgrass prairies in the southern portion of the region. This is a "mixed" prairie because the dominant species are a mixture of midheight and short $C_4$ grasses. The midheight grasses are *Bouteloua curtipendula* and *Schizachyrium scoparium*. The shortgrasses are the dominants from the shortgrass steppe. Annual aboveground production of grasslands in the Great Plains ranges from 200 to 400 g/m$^2$ year$^{-1}$ (Fig. 7). [*See* GRASSLAND ECOLOGY.]

### 2. Croplands

The current pattern of landcover in the Great Plains includes large areas of cropland in addition to native grassland (Fig. 8). The proportion of each of the grassland types that have been converted to cropland depends on both soil and climatic factors. The wettest areas have been exploited most heavily. The tallgrass prairie has been largely converted to cropland. The two driest types, the shortgrass steppe and the northern-mixed prairie, have the smallest proportion of cropland of the four grassland types.

The region is divided into four crop zones and a zone that is a combination of grazing and irrigated cropland. The principal crops are cotton in the southeast, winter wheat in the center, corn in the east central, and spring wheat in the north. The western edge of the region is grazing land except for local areas in which irrigation water is available.

Approximately 58% of the area of the Great Plains is grazing land supporting mostly native vegetation (Table I). The states with the largest proportions of grazing land are those in the western part of the Great Plains adjacent to the eastern edge of the Rocky Mountains. Thirty-one percent of the region is in crops and the eastern tier of states has the highest proportion of croplands corresponding to the highest precipitation.

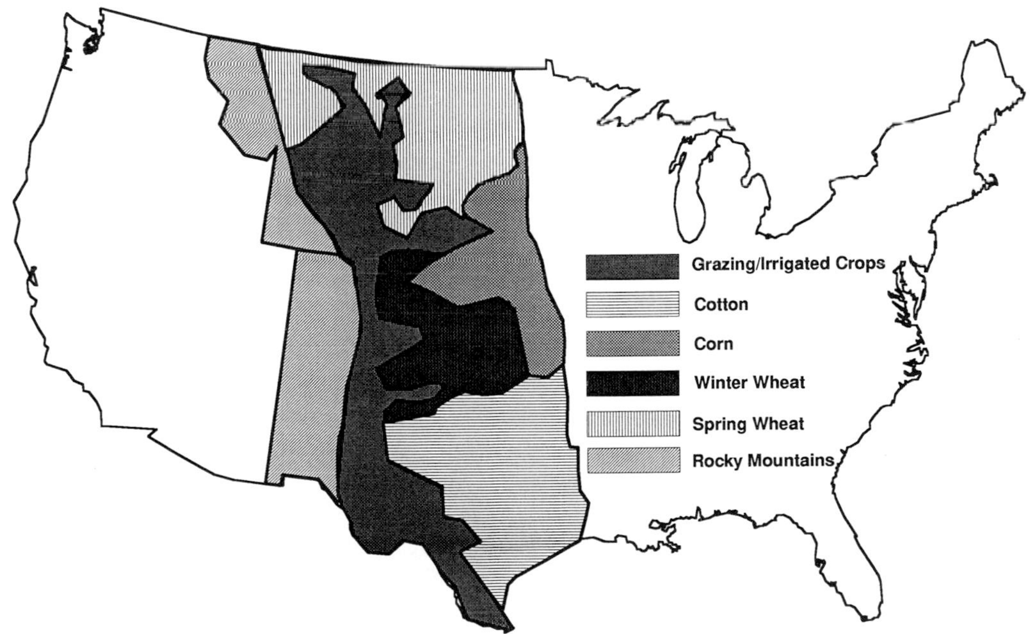


***FIGURE 8*** Geographic distribution of crop regions in the Great Plains of the United States. [Adapted from H. E. Dregne and W. O. Willis (1983). "Dryland Agriculture." American Society of Agronomy, Madison, Wisc.]

***TABLE I***

Agricultural Land Use (in Hectares and %) for the Ten Great Plains States[a]

| | Pasture | Grazing land | Forest | Cropland | Total |
|---|---|---|---|---|---|
| Colorado | $5.26 \times 10^5$<br>3.2 | $9.80 \times 10^6$<br>60.3 | $1.62 \times 10^6$<br>10.0 | $4.29 \times 10^6$<br>26.4 | $1.62 \times 10^7$<br>100.0 |
| Kansas | $8.91 \times 10^5$<br>4.5 | $6.84 \times 10^6$<br>34.6 | $2.43 \times 10^5$<br>1.2 | $1.18 \times 10^7$<br>59.6 | $1.98 \times 10^7$<br>100.0 |
| Montana | $1.21 \times 10^6$<br>4.7 | $1.53 \times 10^7$<br>59.8 | $2.11 \times 10^6$<br>8.2 | $6.96 \times 10^6$<br>27.2 | $2.56 \times 10^7$<br>100.0 |
| Nebraska | $8.50 \times 10^5$<br>4.5 | $9.35 \times 10^6$<br>50.0 | $2.83 \times 10^5$<br>1.5 | $8.22 \times 10^6$<br>43.9 | $1.87 \times 10^7$<br>100.0 |
| New Mexico | $8.10 \times 10^4$<br>0.4 | $1.66 \times 10^7$<br>84.9 | $1.90 \times 10^6$<br>9.7 | $9.72 \times 10^5$<br>5.0 | $1.96 \times 10^7$<br>100.0 |
| North Dakota | $5.26 \times 10^5$<br>3.3 | $4.41 \times 10^6$<br>27.5 | $1.62 \times 10^5$<br>1.0 | $1.09 \times 10^7$<br>68.2 | $1.60 \times 10^7$<br>100.0 |
| Oklahoma | $2.87 \times 10^6$<br>17.6 | $6.11 \times 10^6$<br>37.5 | $2.63 \times 10^6$<br>16.1 | $4.70 \times 10^6$<br>28.8 | $1.63 \times 10^7$<br>100.0 |
| South Dakota | $1.09 \times 10^6$<br>6.3 | $9.23 \times 10^6$<br>53.0 | $2.43 \times 10^5$<br>1.4 | $6.84 \times 10^6$<br>39.3 | $1.74 \times 10^7$<br>100.0 |
| Texas | $2.83 \times 10^6$<br>4.8 | $3.86 \times 10^7$<br>65.8 | $3.77 \times 10^6$<br>6.4 | $1.35 \times 10^7$<br>23.0 | $5.87 \times 10^7$<br>100.0 |
| Wyoming | $2.83 \times 10^5$<br>2.2 | $1.09 \times 10^7$<br>86.2 | $4.05 \times 10^5$<br>3.2 | $1.05 \times 10^6$<br>8.3 | $1.26 \times 10^7$<br>100.0 |
| Great Plains | $1.12 \times 10^7$<br>5.1 | $1.27 \times 10^8$<br>57.6 | $1.34 \times 10^7$<br>6.0 | $6.92 \times 10^7$<br>31.3 | $2.21 \times 10^8$<br>100.0 |

[a] Adapted from J. B. Newman (1988). *In* "Impacts of the Conservation Reserve Program in the Great Plains" (J. E. Mitchell, ed.), USDA Forest Service Gen. Tech. Rep. RM-158, pp. 55–59. USDA, Washington, D.C.

## III. CLIMATIC VARIABILITY

The connection between temporal and spatial climatic variability is a key idea for characterizing and understanding regional climate. An important connection between temporal and spatial climatic variability for the Great Plains is illustrated by the relationships between MAP and MAT and their interannual variability as represented by either the standard deviation (SD) or the coefficient of variation (CV). The SD is an index of absolute variability and the CV is an index of relative variability. The reason these relationships have both temporal and spatial significance for the Great Plains is because of the clear geographic patterns in precipitation and temperature (Fig. 2). The relationship between MAP and its SD is positive, with the SD increasing as the mean increases (Fig. 9a).

By contrast, the general feature of the relationship between MAP and its CV for sites in the Great Plains as well as world wide is that variability decreases as the mean increases (Fig. 9b). This means that the driest sites have the highest relative interannual variability in precipitation and the wettest sites the lowest variability. Relative variability in MAP is highest in the west, particularly in the southwest, and lowest in the east (Fig. 10). The low and variable nature of precipitation in the western portion of the Great Plains makes water a relatively more important variable as a control on ecosystem structure and function than it is in the east. Relationships between net primary production and annual precip-

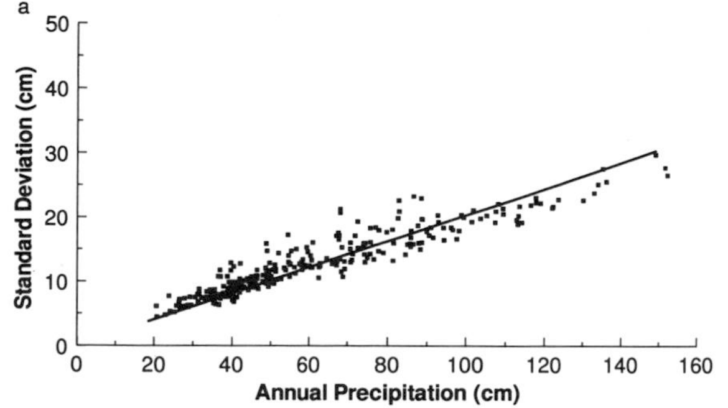


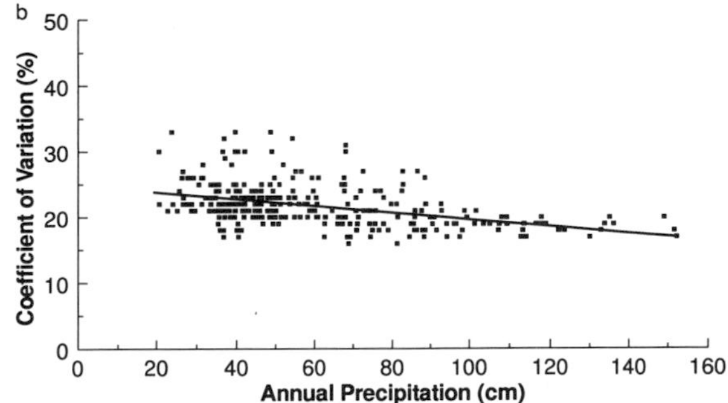


***FIGURE 9*** Relationship between (a) the standard deviation and the mean ($r^2 = 0.85$) and (b) the coefficient of variation and the mean ($r^2 = 0.20$) for annual precipitation for 296 weather stations in the Great Plains of the United States.

itation are commonly important for explaining interannual variability in production in the western Great Plains but not in the east.

In contrast to precipitation, the SD of MAT does not have a relationship to the mean (Fig. 11a). The size of the standard deviation does not vary with the size of the mean; cold sites have the same absolute variability as hot sites. The CV of MAT has a negative and nonlinear relationship to the mean (Fig. 11b). Sites in the north with low MAT have high relative variability, whereas the hot southern sites have low relative variability (Fig. 12). At least part of the explanation for this pattern is related to the frequency of the occurrence of episodes of polar air during the winter and very low temperatures. These low winter temperatures can contribute substantially to MAT. Northern sites have a higher frequency of episodes of polar air than do southern sites. The slight west–east component of the temperature gradient across the plains (Fig. 2a) also shows up as a slight increase in the relative variability, especially in the southern half of the region.

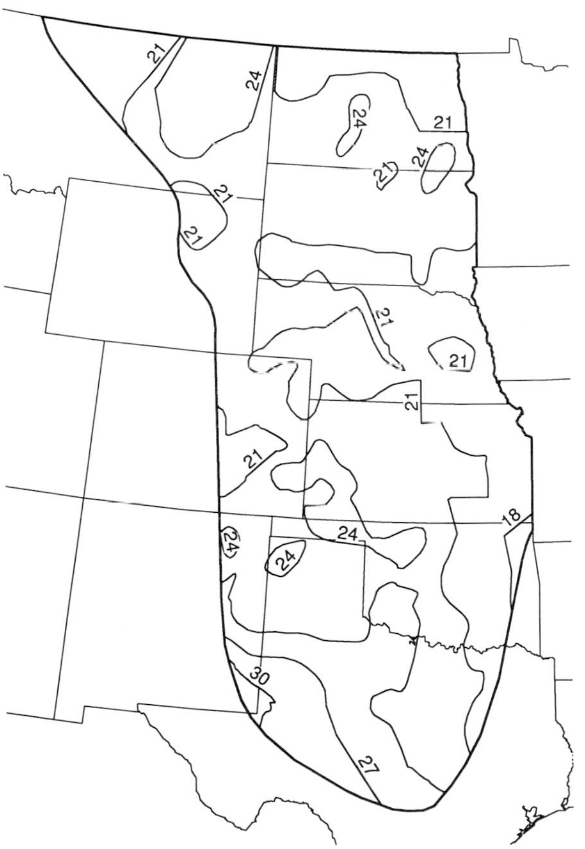

***FIGURE 10*** Geographic distribution of the coefficient of variation of mean annual precipitation (%) in the Great Plains of the United States.

The mean and the SD of MAPD are positively related (Fig. 13a). Absolute variability in MAPD is lowest when MAPD is low and highest when MAPD is high. The relative variability (CV) in MAPD is negatively related to the mean, indicating again that on a standardized basis, dry sites have greater variability in MAPD than do wet sites (Fig. 13b). Spatially, this translates into a predominantly east–west gradient in the CV (Fig. 14). Sites along the eastern edge of the region have the lowest relative variability in MAPD and sites in the west, and especially the southwest, have the highest variability.

Although absolute variability in water input variables (MAP and MAPD) is high and positively related to amounts, variability in the atmospheric

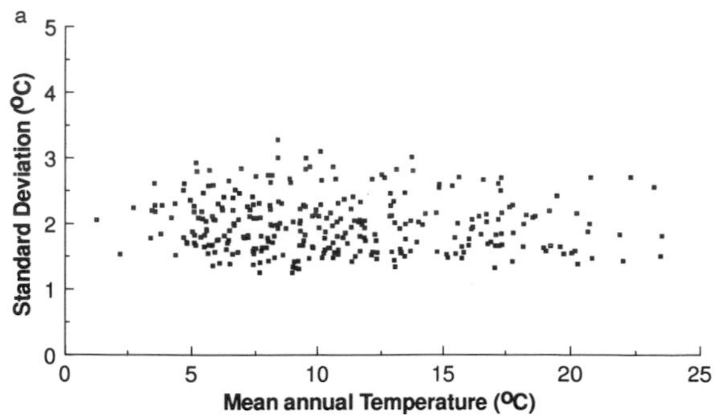


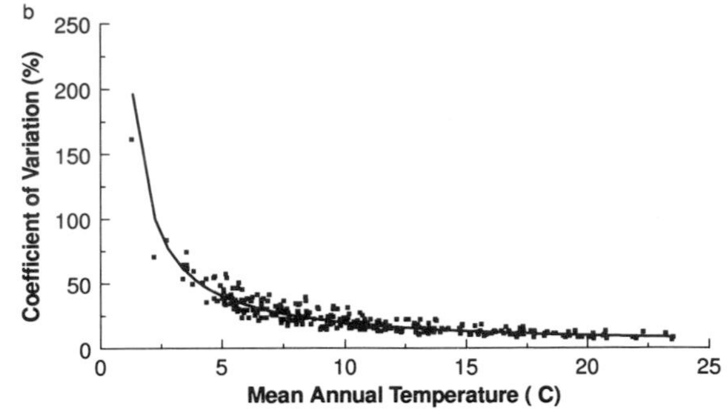


***FIGURE 11*** Relationship between (a) the standard deviation and the mean and (b) the coefficient of variation and the mean ($r^2$ = 0.81) for annual temperature for 296 weather stations in the Great Plains of the United States.

demand for water, as indexed by MAPET, is low and invariant across the range of amounts of MAPET (Fig. 15a). Absolute variation (SD) for MAPET ranges from slightly greater than 1 cm to more than 3 cm, whereas MAPET ranges from approximately 100 cm/year to more than 200 cm/year. Relative variability (CV) in MAPET is negatively related to the amount of MAPET, indicating that sites with very large atmospheric demands for water experience less relative year-to-year variability in MAPET than do sites with low mean MAPET (Fig. 15b). The CV of MAPET is small throughout the region, ranging from near 1 in the southwest to greater than 2 in the northeast (Fig. 16).

***FIGURE 12*** Geographic distribution of the coefficient of variation of mean annual temperature (%) in the Great Plains of the United States.

## IV. POTENTIAL EFFECTS OF CLIMATE CHANGE

What, if anything, can we say about the potential effects of climate change based on our analysis of the current climate of the Great Plains? The answer to this question relies on knowledge of how climate is expected to change. Predictions of doubled $CO_2$ scenarios using global circulation models suggest that central North America will experience an increase in temperature and either a slight increase or decrease in precipitation. The increase in temperature will occur throughout the year, but the largest increases will be in the winter. By contrast, precipitation is predicted to not change or to increase slightly in the winter and to decrease slightly in the summer. [*See* GLOBAL ANTHROPOGENIC INFLUENCES.]

How might these changes influence Great Plains climates? Under this scenario, increases in MAT will be accompanied by increases in MAPET because both are affected by the surface energy balance. The relationship between MAPET and MAT for our

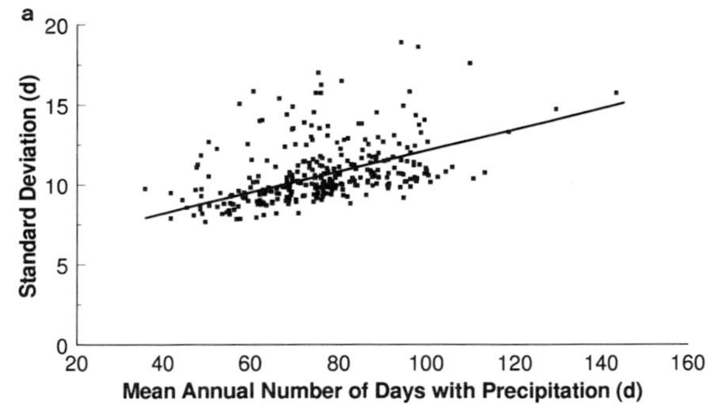


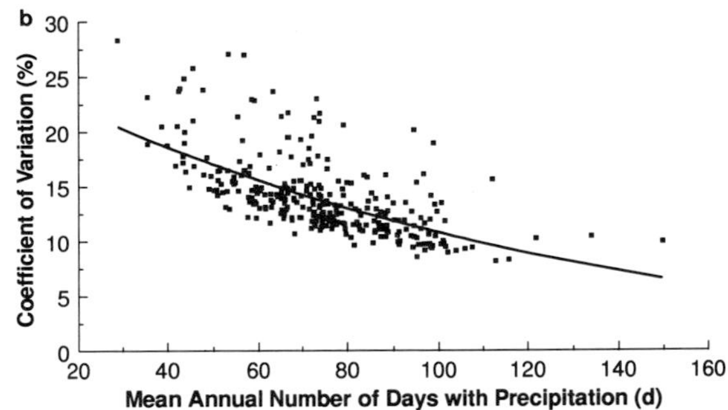


***FIGURE 13*** Relationship between (a) the standard deviation and the mean ($r^2 = 0.17$) and (b) the coefficient of variation and the mean ($r^2 = 0.44$) for annual number of days with measurable precipitation in the Great Plains of the United States.

Great Plains data set can be described by the equation MAPET = 94 + 5* MAT ($r^2 = 0.71$), suggesting that MAPET will increase by 5 cm for each 1°C increase in temperature. Current consensus predictions of increases in MAT for central North America range from 1.5 to 3.5°C, corresponding to increases in MAPET of 7.5 to 17.5 cm. The larger amount represents an 8% increase in MAPET for the sites that currently have the highest values of MAPET and 20% for sites with the smallest current values. Increases in MAPET without increases in MAP will shift the distribution of the ratios of MAP/MAPET to the left (Fig. 5), reflecting an increase in water deficits. The response of Great Plains ecosystems, including crop types, to increased MAT and decreased MAP/MAPET will likely be a geographic shift eastward. Increases in both MAT and MAPET may also result in decreases in their relative variabilities (Fig. 11b and 15b).

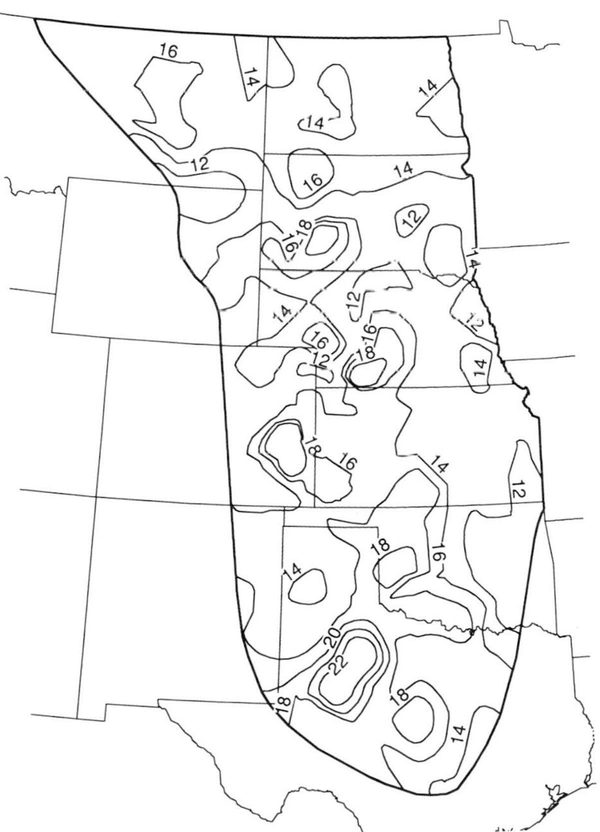

***FIGURE 14*** Geographic distribution of the coefficient of variation of mean annual number of days with measurable precipitation (%) in the Great Plains of the United States.

If increases in MAT are accompanied by increases in MAP, the effect on water deficits will be ameliorated. The magnitude of the amelioration will depend on the size of the increase in MAP and its seasonal distribution. Increases in the winter will do less to relieve water deficits than increases in the summer. If increases in MAT are accompanied by decreases in MAP or even decreases in summer rainfall, the effects on water deficits will be magnified. Although there is still a great deal of uncertainty about exactly how Great Plains climates will change as a result of global change, we have enough information about climates and ecosystems to know that certain combinations of changes in MAT and MAP will have negative effects on Great Plains ecosystems.

## V. SUMMARY

Spatial climatic patterns in the Great Plains of the United States can be defined by three important

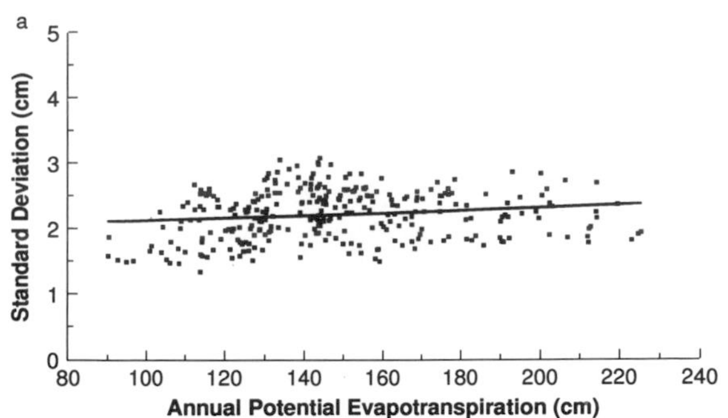


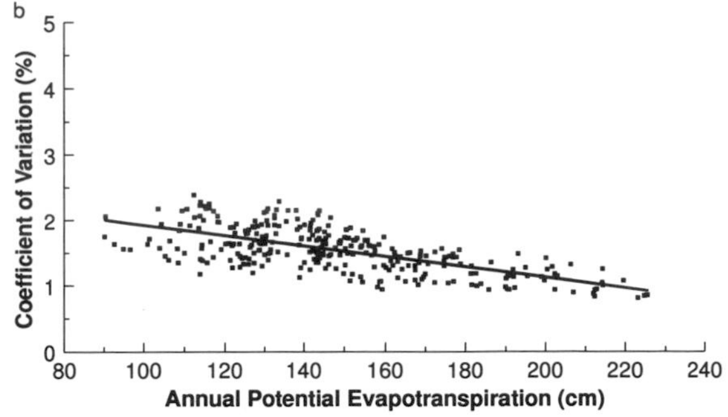


***FIGURE 15*** Relationship between (a) the standard deviation and the mean and (b) the coefficient of variation and the mean ($r^2 = 0.26$) for annual potential evapotranspiration in the Great Plains of the United States.

features: a predominantly north–south gradient in MAT from less than 5°C in the north to greater than 20°C in the south; a predominantly west–east gradient in MAP from 30–40 cm in the west to 120 cm in the east; and a north–south gradient of MAPET that ranges from less than 120 cm in the north to more than 220 cm in the south.

***FIGURE 16*** Geographic distribution of the coefficient of variation of mean annual potential evapotranspiration (%) in the Great Plains of the United States.

In addition to the high spatial variability across the region, the interannual variability is also high and this temporal variability is highest in the driest and coldest portions of the region. The CVs of MAT range from 10% in the south to greater than 45% in the north. The CVs for MAP are less variable than for temperature and range from less than 20% in the east to more than 30% in the southwest. The relative variability in MAPD parallels that of MAP, ranging from less than 2% in the east to more than 20% in the southwest. The atmospheric demand for water as indexed by MAPET is the least spatially and temporally dynamic characteristic of the Great Plains climate. CVs for MAPET range from near 1% in the southwest to greater than 2% in the northeast. This high and relatively constant (on an interannual basis) atmospheric demand for water is responsible for keeping the ratios of MAP/MAPET less than 1 throughout the region. The ecological interpretation of this is that ecosystems throughout the Great Plains are always in a water deficit at some time in every year. This is very likely an important part of the explanation why the area was originally occupied by native grasslands rather than forests.

Human-induced climate change will almost certainly result in increased MAT and increased MAPET. Under such conditions, the geographic distribution of current ecosystems, including crops, will likely shift eastward to maintain adequate water balance. Current global circulation models vary in their predictions of precipitation changes for the

Great Plains. Because precipitation is a major control over the structure and function of ecosystems in the Great Plains, there is considerable uncertainty about the influence that climate change will have on the region.

## Acknowledgments

The authors acknowledge the assistance of of Martha Coleman, Tammy Bearly, and Scott Warner for data analysis, Howie Epstein for map production, and Jose Paruelo for manuscript review.

## Glossary

**Air mass** A portion of the atmosphere with a uniform horizontal distribution of temperature and water content.

**Annual aboveground net primary production** The aboveground portion of the amount of energy (in terms of dry weight) that plants capture in a year in excess of their respiratory needs.

**$C_3$ grasses** A species of grass whose initial products of photosynthesis are 3-carbon acids. These plants tend to make the majority of their annual growth during the coolest parts of the growing season.

**$C_4$ grasses** A species of grass whose initial products of photosynthesis are 4-carbon acids. These plants tend to make the majority of their annual growth during the warmest parts of the growing season.

**Coefficient of variation** An index (expressed as %) of the relative variability of a collection measured as the standard deviation divided by the average.

**Ecosystem** An interacting system that includes all of the plants and animals of an area along with their physical environment.

**Greenhouse gas** Gas that is radiatively active such that it traps infrared radiation reradiated from the earth's surface, leading to increased temperatures.

**Potential evapotranspiration** Maximum amount of water that could be lost from an ecosystem, assuming that all moisture requirements can be met.

**Standard deviation** An index of the variability of a collection of items measured as the difference of each from the average item in the collection.

**Subtropical air mass** Air mass that originates in the subtropical zone, having a latitude of 25° to 35°N or S.

**Temperate climate** Climate with regular winter seasons of freezing weather alternating with summer seasons that are either hot or only warm but of long duration.

## Bibliography


Bryson, R. A., and Hare, F. K. (1974). "Climates of North America, Volume 11." Amsterdam and New York: World Survey of Climatology.

Burke, I. C., Kittel, T. G. F., Lauenroth, W. K., Snook, P., and Yonker, C. M. (1991). Regional analysis of the central Great Plains: Sensitivity to climate variability. *BioScience* **41,** 685–692.

Dodd, J. L. (1979). North American grassland map. Frontispiece. *In* "Perspectives in Grassland Ecology" (N. R. French, ed.). New York: Springer-Verlag.

Dregne, H. E., and Willis, W. O. (1983). "Dryland Agriculture." Madison, Wisc.: American Society of Agronomy.

Houghton, J. T., Jenkins, G. J., and Ephraums, J. J. eds. (1990). "Climate Change: The IPCC Scientific Assessment." Cambridge, England: Cambridge University Press.

Lauenroth, W. K., and Milchunas, D. G. (1991). Shortgrass steppe. *In* "Natural Grasslands: Introduction and Western Hemisphere. Vol. 8A. Ecosystems of the World" (R. T. Coupland, ed.), pp. 183–226. New York: Elsevier Press.

Lauenroth, W. K., Milchunas, D. G., Dodd, J. L., Hart, R. H., Heitschmidt, R. K., and Rittenhouse, L. R. (1994). Effects of grazing on ecosystems of the Great Plains. *In* "Ecological Implications of Livestock Herbivory in the West" (M. Vavra and W. A. Laycock, eds.), pp. 69–100. Society for Range Management.

Newman, J. B. (1988). Overview of the present land-use situation and the anticipated ecological impacts of program implementation. *In* "Impacts of the Conservation Reserve Program in the Great Plains" (J. E. Mitchell, ed.), USDA Forest Service Gen. Tech. Rep. RM-158, pp. 55–59. Washington, D.C.: USDA.

Sala, O. E., Parton, W. J., Joyce, L. A., and Lauenroth, W. K. (1988). Primary production of the central grassland region of the United States. *Ecology* **69,** 40–45.

Shantz, H. L. (1923). The natural vegetation of the Great Plains. *Ann. Assoc. Amer. Geographers* **13,** 81–107.

Weaver, J. E., and Albertson, F. W. (1956). "Grasslands of the Great Plains." Lincoln, Neb.: Johnsen Publishing Co.

Webb, W. P. (1931). "The Great Plains." Boston: Ginn and Co.

# Greenhouse Gases in the Earth's Atmosphere

*M. A. K. Khalil*
*Oregon Graduate Institute*



## I. THE GREENHOUSE EFFECT

The interactions among life, the oceans, and the atmosphere control the greenhouse effect on Earth, creating a warm and stable climate. This article is about the human role in the cycles of atmospheric gases that, by increasing the greenhouse effect, may lead to a warmer world in the future.

Let us consider first how the greenhouse effect comes about. The sun, the Earth, and the atmosphere radiate a spectrum of energies characteristic of their temperatures. The amount of energy in each band of wavelengths is determined, more or less, by Planck's blackbody radiation law. The peak energy, therefore, is radiated at wavelengths that are inversely proportional to the temperature according to Wein's displacement law, and the total energy radiated is proportional to the fourth power of the temperature according to the Stefan–Boltzman law. Consequently, the hotter the blackbody, the more energy it radiates, and the spectrum shifts toward visible light. Stars like the sun radiate most of their energy as visible light; the Earth by comparison is cold and radiates mostly in the infrared part of the electromagnetic spectrum.

The greenhouse effect, so called because of its analogy to common greenhouses used to grow vegetables and heat homes, is a natural phenomenon that keeps the surface of the Earth warm and suitable for life. The atmospheric greenhouse effect does not work quite like normal glass greenhouses, which trap heat mostly by preventing convection.

Simply put, the atmospheric greenhouse effect happens because the solar radiation arriving at the top of the atmosphere is of such high energy, being mostly visible light from the sun, that it is not readily absorbed by most of the atmospheric gases, except by ozone in the stratosphere and some by $O_2$, $H_2O$, and aerosols (total of 20%). About a third is reflected, mostly by the clouds and some by the Earth's surface. The rest of the light that shines on the Earth is absorbed. For the Earth to have a constant temperature, it must radiate about the same amount of energy as it receives, otherwise, it would continuously get colder or warmer. The energy radiated by the Earth back toward outer space is at much lower frequencies than the sunlight it receives. There are many atmospheric constituents that can absorb portions of this low-energy radiation; in recent times these have been called the "greenhouse gases." Water vapor and $CO_2$ are the most important such gases, but $O_3$, $CH_4$, and $N_2O$ also contribute substantially to the greenhouse effect. The man-made chlorofluorocarbons $CCl_3F$

(F-11) and $CCl_2F_2$ (F-12) have also been found to have the potential for causing global warming. All the greenhouse gases are present at very low concentrations relative to the major constituents of the atmosphere, namely, nitrogen and oxygen. After a layer of the atmosphere absorbs this heat from the Earth, it radiates it back, half outward and half back to the Earth's surface. So at the Earth's surface we receive this added energy that makes it warmer. [*See* ATMOSPHERIC OZONE AND THE BIOLOGICAL IMPACT OF SOLAR ULTRAVIOLET RADIATION.]

The greenhouse gases absorb the Earth's radiation in specific energy bands determined by their quantum mechanical properties. How effective a gas is in causing global warming is determined by which portion of the Earth's radiation spectrum it absorbs, by its concentration, and also by the concentrations of other gases that compete to absorb energy from the same radiation bands. As the concentration of the gas becomes high, it absorbs most of the radiation in its bands and thus further increases in the concentration have a diminishing effect on global warming; this is called the *band saturation* effect. At low concentrations, the increase of a gas has a proportional effect on global warming. If, however, the gas absorbs at energies that other gases already absorb, then the effects of increasing concentrations are also reduced; this is the *band overlap* effect. Curiously enough, atmospheric gases that are naturally present in the atmosphere do not trap much of the energy near the maximum wavelength of the Earth's radiation, which is in the 8- to 12-$\mu$m region (except for one ozone absorption band). This is called window region because energy radiated at these wavelengths passes through the atmosphere to outer space. The few gases that absorb in the window region are unaffected by band overlaps and can be very effective in causing global warming. It happens that the main chlorofluorocarbons, $CCl_3F$ and $CCl_2F_2$, absorb in this region, giving them importance in the global warming issue that is far beyond what their extremely low concentrations would suggest.

A heuristic model of the Earth's temperature can be derived by equating the energy that arrives at the Earth's surface per unit time with the energy that leaves during the same unit of time. Such a model is a tool that allows readers to think quantitatively about the greenhouse effect. The Earth receives $S_0$ (0.136 watts/cm$^2$) at the top of the atmosphere or a total of $\pi r^2 S_0$ where $r$ is the radius of the earth (watts). Of this, a fraction $A_a$, the atmospheric albedo, determined mostly by clouds, is reflected back to space, and a fraction $f$ is absorbed by the atmosphere, for instance, by the stratospheric ozone layer. The remaining amount, $\pi r^2 S_0 (1\text{-}f\text{-}A_a)$, reaches the Earth's surface. Of this, the surface reflects a fraction $A_0$, the surface albedo, back into space. The remaining amount $\pi r^2 S_0 (1\text{-}f\text{-}A_a)(1\text{-}A_0)$ is absorbed by the Earth's oceans, plants, and soils. The Earth radiates energy over its entire surface in the amount of $4\pi r^2\sigma T^4$, where $T$ is the average temperature of the Earth. The atmosphere absorbs a fraction $F$ of the Earth's infrared or longwave radiation, or a total of $4\pi r^2 F\sigma T^4$ ($F \leq 1$). Assuming that half the energy absorbed by the atmosphere is radiated back to the Earth and the other half goes outward to space, the Earth receives energy directly from the sun equal to $(1\text{-}A_a\text{-}f)(1\text{-}A_0)\pi S_0 r^2$ and from the atmosphere equal to $(4\pi r^2 F\sigma T^4)/2 + \pi r^2 S_0 f/2$, and this is balanced by the radiation from the surface in the amount of $4\pi r^2\sigma T^4$. This gives the equilibrium surface temperature ($T$) of the Earth as in Equation (1), and similar considerations give the temperature of the atmosphere ($T_a$) as in Equation (2):

$$T = \{(S_0/4\sigma)[(1\text{-}A_a\text{-}f)(1\text{-}A_0) + {}^1/_2 f]/(1 - {}^1/_2 F\}^{1/4} \quad (1)$$

$$T_a = \{(S_0/4\sigma)\ {}^1/_2[(1\text{-}A_o)(1\text{-}A_a\text{-}f) + f/F]/(1 - {}^1/_2 F\}^{1/4} \quad (2)$$

Heating of the surface unavoidably leads to convective transport of energy to achieve thermodynamic equilibrium and transport by latent heat fluxes. In a more complete model, these fluxes are taken into account. [*See* ATMOSPHERE–TERRESTRIAL ECOSYSTEM MODELING.]

Increasing concentrations of trace gases increase $F$, causing the Earth's temperature to rise according to the following formula derived from Equation (1):

$$\delta T = T\left(\frac{1}{8(1\text{-}F/2)}\right)\delta F \quad (3)$$

In Equation (2), 10% change in $F$ ($\delta F = 0.08$)

would lead to a 5°C change in average temperature. Moreover, according to these equations, if the atmosphere did not absorb the Earth's infrared radiation, the surface temperature would be about 255°K ($A_a$ = 0.24, $A_o$ = 0.07, $f = F = 0$, $\sigma = 5.67 \times 10^{-12}$ watts/cm$^2$/K$^4$). With atmospheric absorption the temperature at the surface is 285°K, and the atmosphere is at 255°K ($f$ = 0.18, $F$ = 0.85). The absorption of infrared radiation by $H_2O$, $CO_2$, $CH_4$, $O_3$, and $N_2O$, represented by the parameter F, causes the Earth's surface to be about 30°C warmer than if these gases were not present and all other factors were the same.

Calculations based on the Global Change Research Center's one-dimensional climate model show that about half the natural greenhouse effect is due to water vapor. Without $CO_2$ the average temperature would be cooler by 12°C; without $CH_4$, $N_2O$, or tropospheric $O_3$ it would be 0.7, 0.6, and 0.8°C cooler, respectively. The stratospheric ozone layer causes the Earth's surface to be cooler by about 12°C because it absorbs incoming solar radiation, preventing it from reaching the surface. The $f$- factor in Equations (1) and (2) reflect this effect.

It is hard to imagine, however, that if the greenhouse gases did not exist, all other factors could be the same. The existence of the oceans necessarily leads to water vapor in the atmosphere and the formation of clouds. Clouds then increase the Earth's albedo. The surface of the Earth has a low albedo because of the high absorption by the oceans and the vegetation on land that compensates for the high reflectivity of the clouds. If there were no oceans or atmosphere, the Earth would perhaps be more like the moon. It would be cooler than the present Earth because of the lack of greenhouse gases and the higher reflectivity of its surface, but this effect would be partly compensated by the lack of clouds and the absorption and scattering of sunlight by the atmosphere. The net effect of these opposing processes would be an Earth that is 15°C cooler rather than the often quoted 30°C "greenhouse effect." The opposing forces of the Earth's albedo, determined by vegetation, oceans, and the clouds on the one side and causing cooling, and the greenhouse gases and surface absorption, also affected by life and causing warming, provide the possibility for the Earth's environment to have a stable temperature needed to sustain life as we think of it by various feedbacks that can balance these opposing forces. The existence of the greenhouse effect and the ozone layer is therefore closely tied to more fundamental processes, some of which are directly related to life on Earth. These processes may control the Earth's temperature not only by causing the greenhouse effect but also by changing cloud distributions and the albedo of the Earth by feedbacks and buffering processes.

Feedbacks can either amplify global warming and climate change or stabilize it in spite of changing atmospheric composition. One of the well-known feedbacks is the ice–albedo feedback. As the Earth warms, it loses ice in the polar regions. The highly reflective ice is replaced by a nonreflective ocean, causing more of the sun's energy to be absorbed and thus increasing the warming of the Earth—a positive feedback. There are buffers, such as the oceans, for instance, that, by absorbing heat, slow down global warming from increased greenhouse gases or any other cause of increased radiative forcing. Similarly, when the causes of the increasing radiative forcing are eliminated, oceans allow the Earth to cool slowly back to its original state.

## II. THE ROLE OF HUMAN ACTIVITIES IN GLOBAL CHANGE

Next we look at the sources, sinks, and trends of the "greenhouse gases." Most atmospheric gases, other than diatomic molecules, can absorb the infrared radiation from the earth. As mentioned earlier, the most important are $H_2O$ and $CO_2$, with additional contributions from $CH_4$, $O_3$, $N_2O$, $CCl_3F$, and $CCl_2F_2$. Except for the chlorofluorocarbons, these gases occur naturally in the earth's atmosphere. Human activities related to the production of food and energy are causing increases in the concentrations of $CO_2$, $CH_4$, and $N_2O$. The man-made chlorofluorocarbons are used in various consumer and industrial applications (See Section III, D).

Ozone ($O_3$) is formed by chemical reactions in the atmosphere and is at the core of global atmospheric chemistry. There is evidence that human activities are affecting ozone, possibly causing increases in the troposphere of the northern hemisphere and decreases in the stratosphere. These changes have complex effects on the Earth's temperature, atmospheric chemistry, and the nature of the radiation reaching the Earth's surface.

The global content of water vapor also changes with climate. When the Earth is cool, the atmosphere holds less water than when it is warm. As greenhouse gases, both ozone and water vapor change with changing climate and may thus affect the climate by causing feedbacks.

In the rest of this article we will focus attention on the long-lived "source gases" that are emitted directly into the atmosphere by agricultural and industrial activities, some in addition to natural processes, others entirely from human activities. Long-lived gases, with atmosphere lifetimes of several years at least, are of particular interest in the greenhouse effect because these gases have the potential to accumulate in the atmosphere and reach concentrations that are high enough to affect the global temperature. In addition to the five main stable greenhouse gases, $CO_2$, $CH_4$, $N_2O$, $CCl_3F$, and $CCl_2F_2$, there are many others, some with natural sources and some entirely man-made, that also behave similarly (Table I). Individually these gases have small effects on the environment because of low concentrations and slow increases; but together, they may be important also.

Two ideas are useful in explaining the role of human activities on the greenhouse effect. The quantitative evaluation of human activities on the cycles of trace gases is stated by a global mass balance model. The simplest form is to balance the global rate of change ($dC/dt$) with the total emissions into the atmosphere ($S$) and the effective annual removal of the gas by various processes specific to the gases, such as chemical or depositional removal ($C/\tau$):

$$dC/dt = S(t) - C/\tau(t) \qquad (4)$$

Trends ($dC/dt \neq 0$) arise only if the sources are increasing or if the effective removal processes are decreasing. There are many ways, not always obvious, by which one or both of these processes can occur. $\tau$ is called the "lifetime"; it is rarely constant and can change for various reasons. Later we will see that the "greenhouse gases" mentioned earlier are all increasing because of human activities.

The second idea is to devise a formula by which to rank man-made gases according to their effectiveness for causing global warming. This is called the "global warming potential" (GWP) and is stated as:

$$\mathrm{GWP}(t) = \frac{\int_0^t Q\chi\,dt}{\int_0^t Q\chi_0\,dt} \approx [Q\tau/Q_0\tau_0][(1 - e^{-t/\tau})/(1 - e^{-t/\tau_0})] \qquad (5)$$

This equation ties together the heat-trapping efficiencies of molecules expressed in Equations (1) and (2) (under the $F$ parameter, which is the composite effect of all gases, including band saturation and overall effects) and the persistence of the trace gas as expressed in its lifetime [Equation (4)]. The GWP expresses the warming potential of a mole (or a molecule) of a trace gas released to the atmosphere, relative to a standard gas (in the denominator), usually taken to be $CO_2$. Here $\chi$ (t) is the fraction of the mole left after $t$ years, during which time the gas is consumed by atmospheric or surface sinks. Q is the radiative forcing or the trapping of the Earth's energy leading to global warming [Equations (1) and (2)], and $Q_0$ is the radiative forcing for a mole of $CO_2$. So, $Q/Q_0$ is the effectiveness of a molecule of a trace gas in trapping heat relative to a molecule of $CO_2$. For methane this ratio is about 20, making methane about 20 times as effective as $CO_2$ in trapping the Earth's radiation. For $N_2O$, F-11, and F-12, at their present atmospheric concentrations, this ratio is 200, 12,000, and 15,000, respectively (see Table I). The remaining term accounts for the fact that a mole of a longer-lived gas persists in the atmosphere for many more years than a mole of a

**TABLE I**
Lifetimes, Trends, and Global Warming Potentials of Selected Greenhouse Gases from Human Activities

| Gas | Lifetime (yr) | Concentrations[a] (ppxv) Present | 200 years B.P. | Trends[b] (ppxv/yr-decadal) | Global warming potentials[c] Integration times (yr) 0 | 20 | 500 |
|---|---|---|---|---|---|---|---|
| Main anthropogenic greenhouse gases | | | | | | | |
| $CO_2$ | 120[d] | 350 | 275 | 1.6 | 1 | 1 | 1 |
| $CH_4$ | 10 | 1,686 | 700 | 16 | 20 | 35 | 4 |
| $N_2O$ | 150 | 302 | 285 | 1 | 200 | 270 | 190 |
| F-11 | 60 | 198 | 0 | 9 | 10,000 | 4,500 | 1,500 |
| F-12 | 130 | 329 | 0 | 16 | 20,000 | 7,100 | 4,500 |
| Other gases | | | | | | | |
| F-22 | 15 | 120 | 0 | 7 | — | 4,100 | 510 |
| F-113 | 90 | 60 | 0 | 5 | — | 4,500 | 2,100 |
| F-114 | 200 | 15 | 0 | — | — | 6,000 | 5,500 |
| F-115 | 400 | 5 | 0 | — | — | 5,500 | 7,400 |
| $CCl_4$ | 50 | 150 | — | 2 | — | 1,900 | 460 |
| $CH_3CCl_3$ | 6 | 146 | — | 5 | — | 350 | 34 |
| $CF_3Br$ | 110 | 2 | 0 | 0.3 | — | 5,800 | 3,200 |

[a] The units ppxv mean parts per x by volume, where x = m = million, x = b = billion, and x = t = trillion. Units are $CO_2$ in ppmv, $CH_4$ and $N_2O$ in ppbv, and the rest in pptv. The concentrations are also reported for preindustrial times as 200 years before present.

[b] The trends are variable and are reported here as averages for 1980–1990. In recent years the trends of $CH_4$, F-11, F-12, $CH_3CCl_3$, and $CCl_4$ are declining rapidly. Trends of $CCL_4$ and $CH_3CCL_3$ have already become zero and the concentrations are falling.

[c] The GWPs are adopted from International Panel on Climate Change (1990). "Climate Change." Cambridge University Press, New York. They are corrected for $CH_4$ as in World Meteorological Organization (1991). "Global Ozone Research and Monitoring Project," Report No. 25. WMO, Geneva.

[d] The lifetime of $CO_2$ is complicated and listed here for reference. It is not used in the GWPs reported.

shorter-lived gas. This allows us to attach a greater global warming potential to gases that, once put into the atmosphere, will continue to cause global warming for many years, or decades, compared to gases that disappear quickly.

The global warming potential is an artificial construct with many deficiencies, but it provides a means to compare widely differing gases using more than just their radiative properties. One problem associated with the GWP is that the index is biased against long-lived gases even if their low concentrations and no potential for substantial future increases make them innocuous for global warming. Second, gases transform into each other and cause changes in their own lifetimes; this further complicates an accurate calculation of the GWP; and finally, if the simplified version of Equation (5) is used, the lifetime of the reference gas, $CO_2$, is not a simple exponential decay, which leads to further complications in obtaining an accurate index. For completeness, the global warming potentials for various man-made greenhouse gases are reported in Table I for integrations over different lengths of time ($t$) in Equation (5).

## III. TRENDS AND GLOBAL BALANCES OF THE GREENHOUSE GASES

Atmospheric concentrations of $CO_2$ and the trace gases have been measured only during the last 30 years. For F-11 and F-12, the methods for measurements at the low concentrations in the atmosphere were invented only 20 years ago. We can, however, determine the concentrations of many gases by analyzing air trapped in polar ice cores. Snowfall on the polar ice caps is a mixture of ice crystals and air between snowflakes. In time, as snow falls, it compacts the previous snow, turning it into ice, and air gets trapped in bubbles. The deeper the ice, the older the air in the bubbles. The trapped air, because of the low temperatures and no source of light, is preserved in a deep freeze and is unaffected by chemical and biological processes. Cores drilled in the polar ice caps contain a record of atmospheric composition, including $CO_2$ and $CH_4$, that extends back 160,000 years; fewer data exist for $N_2O$. The ice core record shows that, during the last several hundred years, the concentrations of $CO_2$, $CH_4$, and $N_2O$ have undergone dramatic increases. The concentrations of the chlorofluorocarbons are found to be below detection limits in uncontaminated ice core samples, indicating their recent and entirely anthropogenic origin. [*See* GLOBAL ANTHROPOGENIC INFLUENCES.]

### A. Carbon Dioxide

The carbon dioxide concentrations during the last 250 years are shown in Fig. 1. It is apparent that concentrations rose from about 280 ppmv (parts per million by volume) to the present levels of about 350 ppmv, and they continue to rise (Fig. 1B). The causes for this increase are twofold. First, the burning of fossil fuels, which is a most important element of the transition from preindustrial to modern times, is a major cause of the increased $CO_2$. In recent years, it is estimated to emit some 6 petagrams of carbon per year as $CO_2$ (1 pG = $10^{15}$ g). The other cause is deforestation and land-use change, which may contribute about 2 pG carbon/yr in recent times. As land is cleared or as forests are burned, the standing carbon is released as $CO_2$, but in comparison, crops or whatever else replaces forests recovers very little carbon from the atmosphere and cycles it over short times (of a year or so for crops), thus leading to a net increase in atmospheric $CO_2$. How the biosphere responds to increases of $CO_2$ is a most complex question and currently a subject of active debate and scientific investigation.

Annually, a small amount of $CO_2$, 0.2–0.4 pG C, comes from the atmospheric oxidation of methane.

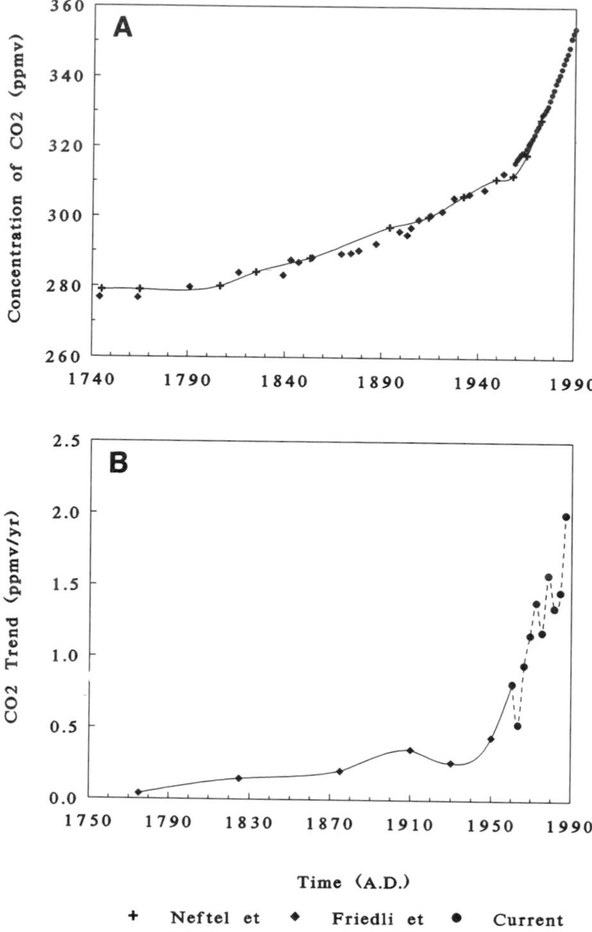


***FIGURE 1*** (A) The concentration of $CO_2$ during the last 250 years. (B) Trends of $CO_2$ derived from the data in (A). [From M. A. K. Khalil and R. A. Rasmussen (1995). *In* "Chemistry, Composition and Climate of the Atmosphere" (H. B. Singh, ed.), Chap. 3. Von Nostrand Reinhold, New York.]

Another 0.6 pG/yr comes from the oxidation of CO from nonmethane sources. Only some of this is "fossil" carbon that has a small effect on the buildup of $CO_2$; the rest is from carbon cycling over decadal time scales with methane, other hydrocarbons, or CO as intermediaries.

The atmospheric lifetime of $CO_2$ is often quoted as 50–200 years. It is not a constant, however, because the carbon in $CO_2$ cycles through several reservoirs, most notably the biota, the ocean mixed layer, and the deep oceans. The flows and semipermanent storage in reservoirs cause the apparent lifetime of a mole of $CO_2$ put into the atmosphere to change over time, starting at a relatively short

5–10 years after it is released into the atmosphere and becoming about 200 years after a few decades and even longer afterward.

### B. Methane

The concentrations of methane during the last 1000 years are shown in Fig. 2, and the trends during the recent decade are shown in Fig. 3. Methane has many sources; almost all are influenced by human activities. The major sources are natural wetlands, domestic cattle, other animals both wild and domestic, and rice agriculture. Next there are many smaller sources such as biomass burning, landfills, leakages from natural gas production and use, coal mining and waste management, lakes, oceans, and termites. Finally, there are some very small sources that probably do not greatly affect atmospheric methane. For details of the global methane cycle, see the Bibliography (Table II). According to records compiled by the United Nations Food and Agriculture Organization (FAO), the sources such as rice agriculture and cattle have increased by factors of two to three during the last century. Other sources, such as biomass burning, coal mining,

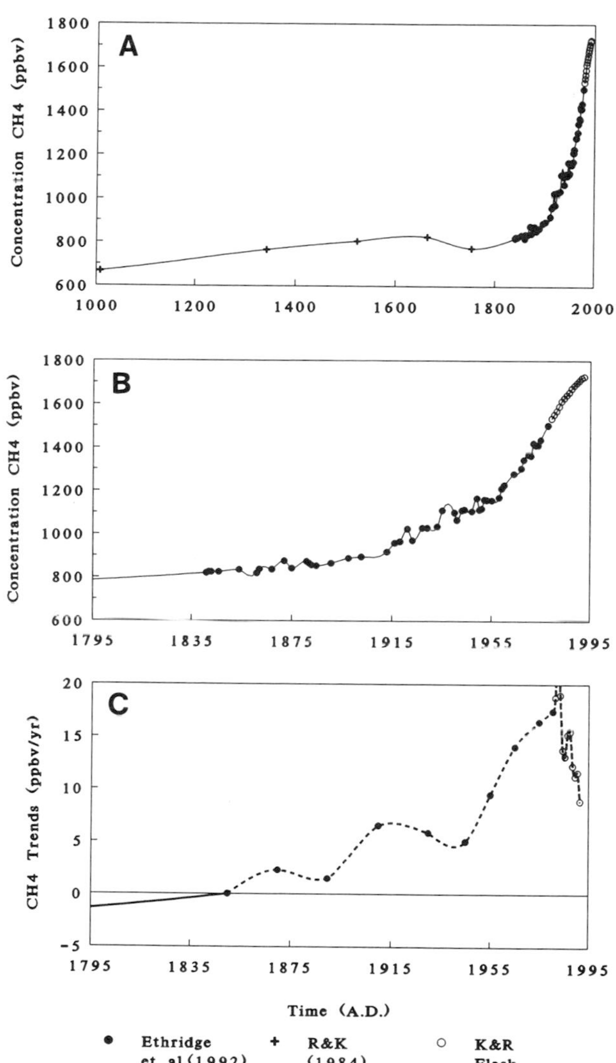


***FIGURE 2*** (A) The concentrations of $CH_4$ during the last 1000 years and (B) during the last 300 years. (C) The trends of methane during the last 300 years derived from (B). Methane decreased during the little ice age. More recently, the rates of increase have peaked and are declining now. [From M. A. K. Khalil and R. A. Rasmussen (1995). *In* "Chemistry, Composition and Climate of the Atmosphere" (H. B. Singh, ed.), Chap. 3. Von Nostrand Reinhold, New York.]

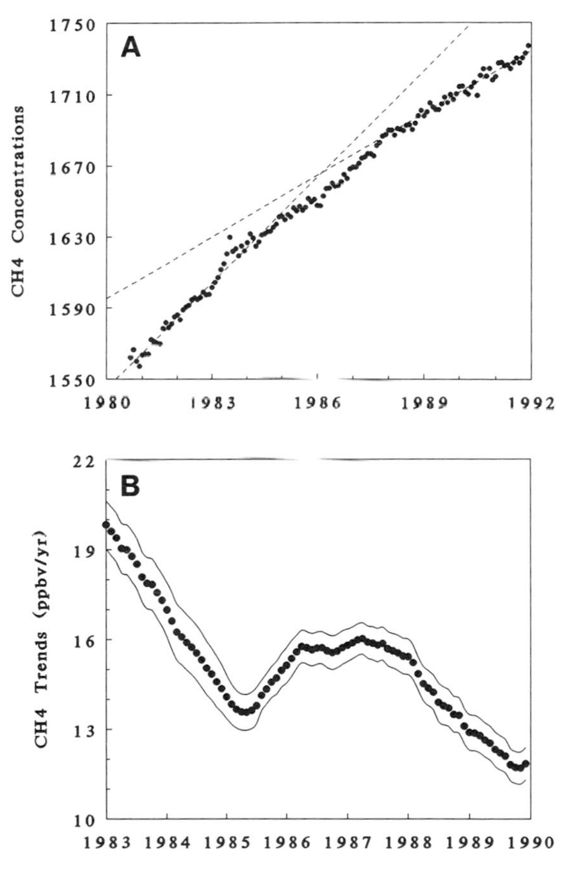


***FIGURE 3*** (A) The concentrations, as global monthly averages, and (B) trends of methane with 90% confidence limits (solid lines), based on direct atmospheric measurements during the last 15 years. This is an expanded view of the later part of Fig. 2C.

***TABLE II***
Estimates of Methane Emissions from Various Sources[a]

| Source | Methane (Tg) | Range (if available) |
|---|---|---|
| Natural sources | | |
| Wetlands[1] | 110 | |
| Termites [2] | 20 | 15–35 |
| Open ocean[2] | 4 | |
| Geological[2] | 10 | 8–65 |
| Wild fire[3] | 2 | 1–13 |
| Total[b] | 150 | 2–5 |
| Anthropogenic sources | | |
| Rice[4] | 65 | 55–90 |
| Animals[5] | 79 | |
| Manure[6] | 15 | |
| Landfills[6] | 22 | |
| Wastewater treatment[6] | 25 | 27–80 |
| Biomass burning[3] | 50 | |
| Coal mining[7] | 46 | 25–50 |
| Natural gas[c, 7] | 30 | |
| Other anthropogenic[2, 7] | 13 | 7–30 |
| Low-temperature fuels[2] | 17 | |
| Total[b] | 360 | |
| $^{14}$C-depleted sources[b, d] | 120 | |
| All sources[b] | 510 | |

[a] The values in this table are based on an evaluation of emissions from each source by various authors. References are cited in full in M. A. K. Khalil, ed. (1993). "Atmospheric Methane: Sources, Sinks and Role in Global Change." Springer-Verlag, Berlin, Sources: [1] Matthews; [2] Judd *et al.*; [3] Levine *et al.*; [4] Shearer and Khalil; [5] Johnson *et al.*; [6] Thorneloe *et al.*; [7] Beck *et al.*

[b] Total number rounded to nearest ten.

[c] These numbers were estimated from data in Beck *et al.*

[d] $^{14}$C-depleted sources were considered to be the "geological" source, plus emissions from coal mining, natural gas, other anthropogenic (mainly industrial and transportation fossil fuel combustion), and low-temperature fuel combustion.

landfills, and waste management, have no doubt also increased. On the other hand, sources such as wetlands may have decreased because many wetlands have been drained in the process of expanding human habitation and changing land use.

Methane is removed from the atmosphere mostly by reacting with OH. It is also removed by dry soils and possibly other chemical processes in the troposphere and stratosphere. To understand the trends of methane it is necessary to know the trends of hydroxyl radicals (OH). The present global lifetime of methane is about 10 years. If OH decreases, methane lifetime would increase and concentrations would rise even if the emissions remained constant. Similarly, if OH concentrations rise, the trends of methane are reduced. [*See* GLOBAL CARBON CYCLE.]

The cycle of OH is more important than its effect on methane alone. In most of the Earth's atmosphere, far from the relatively polluted air of cities, OH is the major oxidant that removes dozens and perhaps hundreds of gases from the atmosphere. Changes in OH, particularly decreases, may cause many gases to build up in the atmosphere that together may increase global warming and change the distribution of ozone in the atmosphere.

The primary mechanism for forming OH is the photolysis of $O_3$ ($O_3 + h\nu \rightarrow O(^1D) + O_2$), which forms $O(^1D)$—an excited oxygen atom. It reacts with water vapor to form OH ($H_2O + O(^1D) \rightarrow 2OH$). The OH radicals so formed are very reactive and persist in the lower atmosphere only for a second or so. Most of them react with methane and carbon monoxide. The oxidation of methane ($CH_4 + OH \rightarrow CH_3 + H_2O$) continues until the fragments are removed or end up as CO and $H_2$. CO that comes from methane oxidation and from many other sources itself ends up as $CO_2$ ($CO + OH \rightarrow CO_2 + H$). When NO and $NO_2$ ($NO_x$) are present in sufficient quantities, the oxidation cycle of methane produces $O_3$, which is a feedback for further production of OH. Similarly, the H atoms formed from the reaction of OH with CO, and OH with $H_2$, cycle back to OH by first forming $HO_2$ and then OH after the $HO_2$ reacts with $O_3$ or NO. What does all this tell us about methane?

First, the cycles of methane, OH, CO, and $O_3$ are closely tied together. If methane and CO increase, OH can decrease, causing a feedback that can accelerate the trend of methane beyond what direct emissions may suggest. So the relationship between industrial and agricultural emissions of methane and its atmospheric concentrations is not straightforward; if we double the emissions of methane, the atmospheric concentration becomes more than double. Second, we see that although ozone is not the primary oxidant for removing methane and other trace gases, it is necessary to produce OH, and therefore indirectly affects trace gas cycles in the atmosphere. Without $O_3$ in the lower atmosphere (troposphere), it would be very difficult to remove many gases and pollutants from the atmosphere. Third, under several conditions, such as with high $NO_x$, OH can be recovered so that the loss of OH after it reacts with $CH_4$ or CO is not necessarily the end of the OH cycle.

Experiments have shown that during ice ages, methane concentrations are only about 350 ppbv (parts per billion by volume), compared to interglacial periods when the concentrations are about 650–750 ppbv. During an ice age we also expect CO concentrations to be very low compared to present or interglacial conditions. Under these conditions we might think that OH concentrations should be very high during an ice age, because there is so little methane and carbon monoxide to remove OH. This, however, is not likely. At the same time, in an ice age, tropospheric ozone, $NO_x$, and water vapor should also be much lower than under warm interglacial conditions. Reduced water vapor and ozone would reduce OH production and hence bring the ice-age concentrations of OH back to levels not much higher than at present. The stability of OH suggests that the major reduction of atmospheric methane observed during ice ages may well be dominated by reduction of emissions. This seems likely also from looking at the major natural sources of methane, which are wetlands, wild animals, and a few small sources. All of these would be reduced during an ice age, particularly the wetlands, which, at middle and higher latitudes, would freeze.

The same ideas are applicable to the transition from the old atmosphere, sometimes referred to as preindustrial, to the present. As shown in Fig. 2, atmospheric concentrations have risen from 750 ppbv about 200 years ago to about 1700 ppbv now. By considering the CO budget, in which about half the present emissions are from human activities, one can conclude that CO must also have increased during the last 200 years. This should have caused a decrease of OH, which should be accelerating the buildup of $CH_4$, CO, and many other gases. This effect is compensated by increasing ozone in the troposphere. For the most recent decade, it has been suggested that OH may even be increasing. This could happen because, in addition to increase of tropospheric $O_3$, the solar uv radiation necessary to dissociate $O_3$ may be increasing if the chlorofluorocarbons have reduced stratospheric ozone, allowing more uv to reach the troposphere.

There is yet another connection between chlorofluorocarbons and methane in the environment. In the stratosphere, chlorofluorocarbons release Cl atoms that catalytically destroy $O_3$. The reaction of methane with Cl atoms is an important mechanism for stopping the Cl atom from continuing to destroy stratospheric ozone ($CH_4 + Cl \rightarrow HCl + CH_3$). The HCl is a relatively stable reservoir of Cl and eventually is transported to the troposphere, where it is removed from the atmosphere. Methane affects stratospheric ozone in other ways as well. The reaction of methane with OH in the stratosphere produces water vapor that can destroy ozone. Yet on balance it can be argued that reducing methane emissions too much, say to the levels of 200 years ago, may not be good for the ozone layer. This is because methane has an atmospheric lifetime of about 10 years and its concentrations would therefore fall to preindustrial levels within a couple of decades. The chlorofluorocarbons, on the other hand, have lifetimes of 50–100 years and would persist in the environment for centuries. With less methane to remove the chlorine from the stratosphere and the longer persistence of the chlorofluorocarbons, stratospheric ozone may be reduced to low levels for a long time.

From Figure 3 it is also apparent that methane trends are falling dramatically in recent years. The exact causes are not known but could be related to the slowdown in most of the anthropogenic sources and the hypothesized increase of OH. Recent data show that the acreage of rice fields and world cattle populations are no longer increasing. Other sources, such as biomass burning, landfills, and natural gas leakages, particularly from the former Soviet Union, may also have slowed down or been reduced compared to in previous years. The case of rapid slowdown of methane trends is interesting because it is happening without overt intervention from international agreements such as the "Montreal Protocol," which is responsible for the dramatic slowdown in the rates at which the chlorofluorocarbons are now increasing in the atmosphere.

### C. Chlorofluorocarbons: F-11 and F-12

The chlorofluorocarbons $CCl_3F$ (F-11) and $CCl_2F_2$ (F-12) are best known by their duPont trade name as Freons®. There are in fact many Freons® (chlorofluorocarbons) that have various industrial and consumer applications (see Table III). The Freons® are designated as F-xyz, where $x = C - 1$, $y = H + 1$, and $z = F$ (where C, H, and F are the numbers of carbon, hydrogen, and fluorine atoms in the molecule—for example, $C_2H_2F_2Cl_2$ is F-132).

F-11 and F-12 were among the first chlorofluorocarbons to be manufactured and used in consumer and industrial applications, mostly as refrigerants, propellants for spray cans, and, in recent years, for blowing insulating foams for various applications such as home insulation and packaging for fast foods. For these and several other chlorofluorocarbons there are practically no destruction processes (sinks) at the Earth's surface or in the lower atmosphere. Higher up, in the stratosphere, there is enough high-energy radiation to photodissociate these molecules (and most others). The process releases chlorine atoms (Cl), which can destroy stratospheric ozone by the reaction $Cl + O_3 \rightarrow ClO + O_2$, but the reaction $ClO + O \rightarrow Cl + O_2$ recovers the chlorine atom (Cl) that can go on to destroy another ozone molecule, and so on. The net result of these two reactions is to rapidly take $O_3 + O$ into two $O_2$ molecules. Something has to stop this process, otherwise we would have no stratospheric ozone left. As mentioned earlier, methane can remove the chlorine atom from this cycle by the reaction $CH_4 + Cl \rightarrow HCl + CH_3$. The atmospheric lifetimes of F-11 and F-12 are about 50 and 100 years, respectively, due mostly to removal in the stratosphere.

It is the effect of the chlorofluorocarbons on the stratospheric ozone layer, as discussed earlier, that was the main cause for signing the Montreal Protocol, which calls for a complete ban on the production of these and many other "analogous" chemicals before the turn of the century. But the added concern that these chlorocarbons could cause global warming was also used to justify the current bans. As greenhouse gases, F-11 and F-12 are singularly effective. They are, molecule for molecule, some 10,000–20,000 times more effective than $CO_2$. This is because they absorb the Earth's infrared radiation in the window region discussed earlier.

Figures 4 and 5 show the measured concentrations and trends of F-11 and F-12. It is apparent that the concentrations have undergone major changes during the last decade, but the most dramatic is the recent rapid decline in the trends. It is likely that atmospheric concentrations will soon stop increasing and will then start to decline slowly, characteristic of their 50 to 100-year lifetimes.

The global budget of F-11 is given in Table III and Fig. 6. The table shows that most of the F-11 produced by the chemical industry is still in the environment; it also shows where the rest has gone. Figure 6 shows the past, present, and future trends calculated from the knowledge of industrial production, use, and emissions. The calculated con-

**TABLE III**
Reservoirs and Emissions of F-11 in 1991[a]

| Category | Gg | Percentage of total produced |
|---|---|---|
| Refrigeration | 89 | 1.0 |
| Foams | 1031 | 11.3 |
| Aerosol + other | 24 | 0.3 |
| Unreported | 166 | 1.8 |
| Stratosphere | 741 | 8.1 |
| Troposphere | 5360 | 58.6 |
| Ocean, surface | 27 | 0.3 |
| Ocean, deep | 6 | 0.1 |
| Soils | 1 | 0.0 |
| Removed, stratosphere | 1709 | 18.7 |
| Total | 9152 | 100 |

[a] From M. A. K. Khalil and R. A. Rasmussen (1993). *J. Geophys. Res.* **98**(D12), 23091–23106.

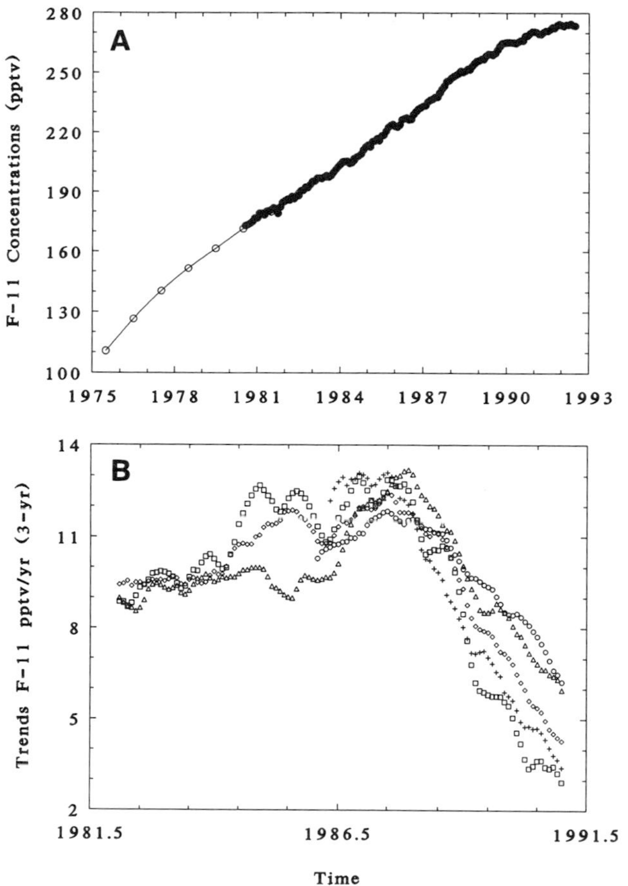


**FIGURE 4** (A) The concentrations of $CCl_3F$ (F-11) during most of the last two decades as monthly and global averages. (B) The trends of F-11 at various locations during 3-year overlapping periods of time. The rates of increase are rapidly falling and may soon lead to a peak in the F-11 concentration, after which time atmospheric concentrations will begin to fall.

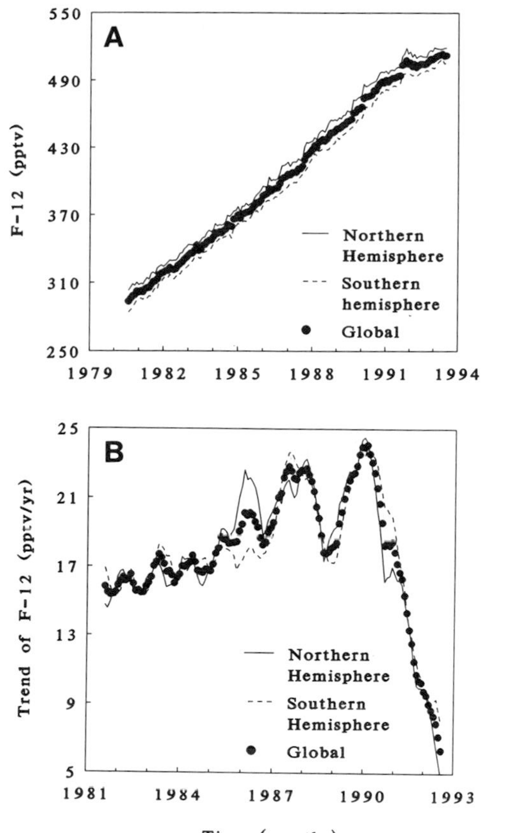


**FIGURE 5** (A) The concentrations and (B) trends of $CCl_2F_2$ (F-12) during the last decade as monthly and global averages. The trends are similar to those of F-11 (see Fig. 4 legend).

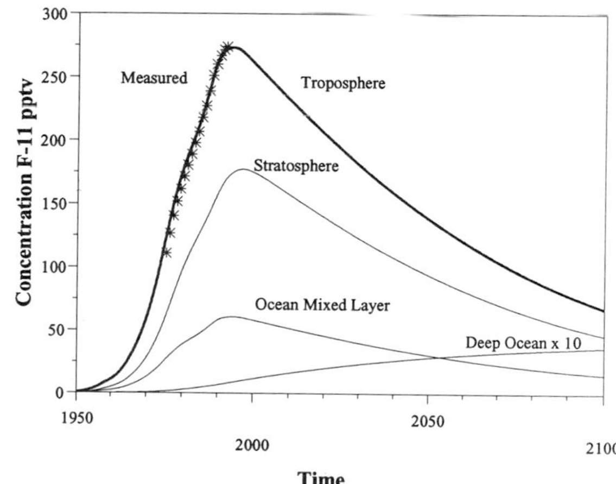


**FIGURE 6** The distribution of F-11 in various reservoirs as annual average concentrations from the time it was first put into use to the present and beyond. [For details see M. A. K. Khalil and R. A. Rasmussen (1993). *J. Geophys. Res.* **98**(D12), 23091–23106.]

centrations are compared with atmospheric measurements in Fig. 7. There are not many gases for which one can achieve such exact agreement between theoretical and measured global atmospheric concentrations.

## D. Nitrous Oxide

The concentrations of $N_2O$ during the last 1000 years are shown in Fig. 8, and the trends during the recent decade are shown in Fig. 9. The understanding of the interactions of nitrous oxide in the atmosphere is not as complete as that for $CO_2$, $CH_4$, and the CFCs. It is known that $N_2O$ is, molecule for molecule, about 200 times as effective as

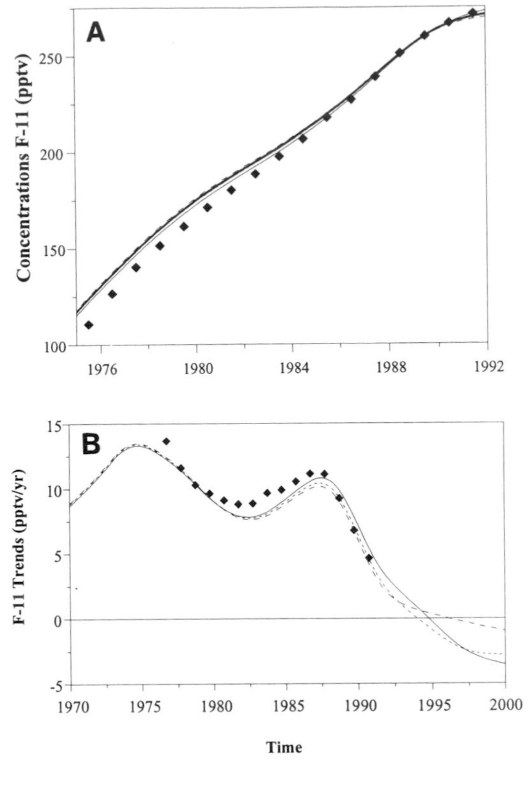


***FIGURE 7*** The recent record of (A) concentrations and (B) trends of F-11 (an expanded view from the tropospheric concentration reported in Fig. 6). The lines show results of mass balance calculations [more detailed versions of Equation (4)]. In these calculations, industry estimates (represented by AFEAS—Alternative Fluorocarbons Environmental Acceptability Study) and data on how long F-11 remains in various uses such as refrigerators and polyurethane foams are used to calculate the expected concentrations. The match between theory and experiment, for concentrations, trends, and particularly the timing of trend changes, is better than for most other gases of interest in global change science. [For details see M. A. K. Khalil and R. A. Rasmussen (1993). *J. Geophys. Res.* **98**(D12), 23091–23106.]

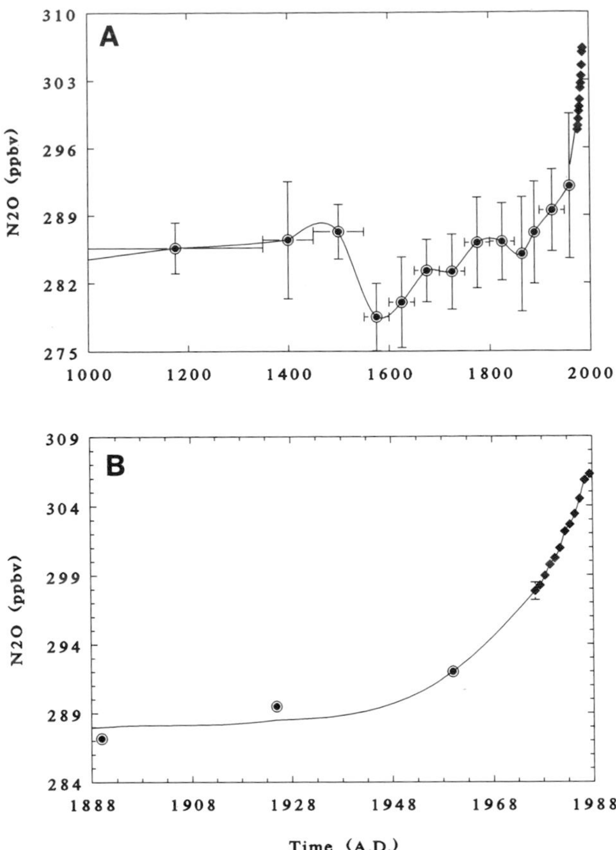


***FIGURE 8*** The concentrations of $N_2$ during the last (A) 1000 years and (B) 100 years. There is a variability of the ice core data, probably due to analytical techniques. The concentrations of $N_2$ have risen by about 7% and were lower during the little ice age than during warmer recent periods.

$CO_2$ in causing global warming. It is also a source of NO in the stratosphere that catalytically destroys stratospheric ozone. The major process for removing $N_2O$ from the atmosphere is in the stratosphere by reactions with $O(^1D)$, giving it an effective atmospheric lifetime of about 150 + 50 years. [*See* NITROUS OXIDE BUDGET.]

For a long time it was thought that the major anthropogenic source of $N_2O$ was fossil fuel burning. The rate of increase of $N_2O$ in the atmosphere was found to be small at about 0.3% per year. This trend fitted the expected increase from fossil fuel use. There have been relatively few systematic studies of $N_2O$ compared to the efforts expended on understanding $CO_2$, the chlorofluorocarbons, and methane. Perhaps this happened because the increase was so slow and $N_2O$ was regarded as a problem related to fossil fuel use, which was already identified as a source of global warming from $CO_2$. Recently, however, there has been a complete change in our understanding of atmospheric $N_2O$, and now we find ourselves knowing very little about the origins of the trend and the role of human activities in its cycle.

It was shown recently that the previous experiments to determine how much $N_2O$ came from coal-fired power plants were affected by an artifact in the sampling and analysis procedures. It turns

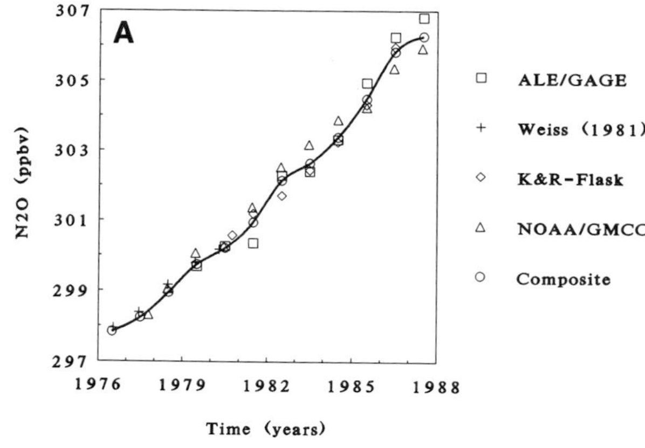


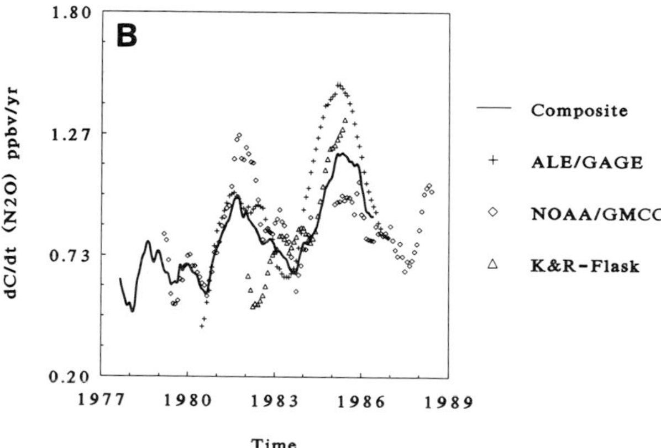


***FIGURE 9*** The recent (A) concentrations (annual and global averages) and (B) trends of $N_2$ based on direct atmospheric measurements (an expanded view of the later part of Fig. 8B). The trends have varied considerably and may be cyclical. [For details see M. A. K. Khalil and R. A. Rasmussen (1992). *J. Geophys. Res.* **97**(D13), 14651–14660.]

out now that practically no detectable amount of $N_2O$ comes from the power plants. There are conditions, however, in which small amounts of $N_2O$ may be produced either in the combustion process or shortly after the effluent leaves the stack of a power plant.

Where, then, does anthropogenic $N_2O$ come from to account for the unequivocally observed atmospheric increase? Many of the sources previously identified, but mostly ignored because they seemed so small compared to the power plants, suddenly become very important. In fact, it seems now that there is no one large source; instead there are many small ones (Table IV). This circumstance makes it very difficult to precisely determine the global emission rates from all the sources. The emission rates (or fluxes) become so small as to be on the fringes of detectability. This happens not only because each source is small to begin with, but also because most sources are spread out over large surface areas. This is quite the opposite of having most of the $N_2O$ come from a concentrated "point" source such as the chimney of a power plant.

An important recent observation is that the rate of increase is getting faster, but because we do not have a good understanding of what is causing the present trends, no one knows whether this observation will be sustained. This is in contrast to other gases, such as methane and the chlorofluorocar-

***TABLE IV***
**Natural and Identified Anthropogenic Sources of $N_2O$ (in Tg/yr)[a, b]**

| | Emissions | Range | Uncertainty factor |
|---|---|---|---|
| Anthropogenic sources | | | |
| Biomass burning | 1.6 | 0.2–3 | 15[1] |
| Power plants | 0.0 | 0.0–0.2 | 20[2] |
| Nylon manufacture | 0.7 | NK | NK[3] |
| Nitrogen fertilizer | 1.0 | 0.4–3 | 8[4] |
| Sewage | 1.5 | 0.3–3 | 10[5] |
| Cattle, agriculture | 0.5 | 0.3–1 | NK[6] |
| Aquifers, irrigation | 0.8 | 0.8–2 | NK[7] |
| Automobiles | 0.8 | 0.1–2 | 20[6, 8] |
| Global warming | 0.3 | 0.0–1 | NK[9] |
| Land use change | 0.7 | NK | NK[10] |
| Atmospheric formation | NK | NK | NK |
| Total | 8 | 5–10 | |
| Natural sources | | | |
| Soils | 12 | — | — |
| Oceans | 3 | — | — |
| Total | 15 | | |

[a] This table is taken from M. A. K. Khalil and R. A. Rasmussen (1992). *J. Geophys. Res.* **97**(D13), 14651–14660; this contains full references to the following sources: [1] Coffer *et al.* (1991), lower limits from Crutzen and Andrea (1990); [2] Linak *et al.* (1990), Sloan and Laird (1990), Yokoyama *et al.* (1991), and Khalil and Rasmussen (1991); [3] Thiemens and Trogler (1991); [4] Eichner (1990) and primary references therein; Conrad *et al.* (1983); [5] Kaplan *et al.* (1978); [6] Khalil and Rasmussen (this work); [7] Ronen *et al.* (1988); [8] EPA (1986); [9] Khalil and Rasmussen (1989b); [10] Matson and Vitousek (1990) and Luizao *et al.* (1989); McElroy and Wofsy (1987) and WMO (1985).

[b] NK = not known; uncertainty factor = max/min of the range. Some numbers, especially in the ranges, are rounded.

bons, which are slowing down in their rates of accumulation. It means that in time $N_2O$ is likely to become relatively more important to global change science than it is now or has been in the past.

There are some distant linkages between the environmental role and cycle of $N_2O$ and the other gases discussed earlier. First, the increasing use of nitrogen-based fertilizers may be causing some, or possibly much, of the increase of $N_2O$. In recent decades particularly, nitrogen fertilizers are used increasingly in rice agriculture instead of the usual organic fertilizers. This shift in agriculture may simultaneously lead to a decrease of methane emissions and an increase of $N_2O$.

## IV. PERSPECTIVES AND CONCLUSIONS

The surface temperature as determined by the greenhouse effect is a crucial component of the Earth's climate. Indeed, the issue of global climate change is often narrowed to global warming and surface temperature change for a prescribed increase in the concentrations of greenhouse gases, such as a doubling of $CO_2$.

We do not know enough about the Earth's climate to be able to control the temperature at will. The past studies of the Earth's environment have, of necessity, looked separately at each element of the system and even at the cycle of each gas, but in this article we have seen that there are many connections between the cycles of gases, climate, atmospheric chemistry, and all the other pieces that constitute the global environment. There are many connections, but which ones are strong enough to make the climate stable? Or to make it unstable? We need to know which connections are so fundamental that without knowing them we cannot hope to understand the Earth's environment, create a complete theory of the climate, or take steps to avert undesirable global environmental change without creating new and perhaps more insidious problems.

We do know, however, that during the last 200 years the concentrations of trace gases that are involved in the greenhouse effect have increased considerably. Carbon dioxide has increased by 25%, methane has increased by a factor of about 2.5 (or 250%), and $N_2O$ has increased by 7%. New greenhouse gases, such as the chlorofluorocarbons F-11 and F-12, as well as many others in less amounts, have been added to the atmosphere. Yet in spite of these sizable changes, observations show that the global surface temperature of the atmosphere has increased by about 0.5°C, and even this increase can be attributed to a number of other possible causes unrelated to the buildup of greenhouse gases.

## Acknowledgments

I have benefited from numerous discussions with Professor R. A. Rasmussen. I have benefited also from discussions with my current students, R. M. MacKay, F. Moraes, M. J. Shearer, Y. Lu, Z. Ye, and T. Marshall. I thank Dr. J. Pinto of the United States Environmental Protection Agency for his constructive comments on the manuscript. Financial support for this work was provided in part by a grant from the Department of Energy (DE-FG06-85ER60313) and from the resources of Andarz Co.

## Glossary

**Feedbacks** Process whereby a change in one variable, such as the earth's temperature, may set into motion processes that either cause more increases in the temperature (positive feedback), thus amplifying the original disturbance, or reduce the effect of the perturbation (negative feedback). An example is the ice–albedo feedback, in which global warming, if it causes melting of polar ice, would also cause highly reflective ice to be replaced by absorbing water, which reduces the net albedo of the earth and causes more global warming. Feedbacks are expected to be very important in affecting global warming.

**Global mass balance** Fundamental tool for understanding the behavior of trace gases in the environment. It relates the concentrations, often observed, with the emissions from natural and anthropogenic sources and the processes that remove the gas from the environment. In its simplest manifestation it says that the rate of change of the atmospheric burden of a gas, or its trend (grams/yr), is equal to the current emissions minus the current losses by all processes. If the losses can be taken to be first order, then the loss term is the atmospheric burden divided by the lifetime. In this case there are three basic terms in the mass balance: the burden (total grams of the gas in the atmosphere), the emissions (grams/yr) and the lifetime (yr). If any two are measured, the third can be deduced.

**Global warming potential** Index that is used to compare the effectiveness of a molecule in causing global warming. It incorporates both the efficiency with which the molecule traps the earth's infrared radiation and the persistence of the molecule in the atmosphere. The idea is that all else being the same, a longer-lived molecule is more significant because once put into the atmosphere, it can cause global warming for a long time compared to a molecule that would disappear sooner.

**Greenhouse effect** A natural phenomenon in which the earth's surface warms up because of the gases in the Earth's atmosphere. The earth's atmosphere is mostly transparent to sunlight. Solar radiation that is not reflected by clouds or the surface is absorbed, causing the ground to heat up. The surface then radiates energy outward in the infrared (at much lower energies than incoming sunlight). This energy is readily absorbed by the atmosphere, causing it to heat up. The atmosphere reradiates the energy it receives back to the surface and skyward. By receiving the energy reradiated by the atmosphere, the Earth's surface gets even warmer. This additional warming is the greenhouse effect.

**Greenhouse gases** Term commonly used for gases that have a strong potential for causing the greenhouse effect or global warming. These gases, in order of importance, are $H_2O$, $CO_2$, $CH_4$ $N_2O$, and $O_3$ in the natural atmosphere, and now also for the global warming caused by human activities. Adding to the effect from human activities are the chlorofluorocarbons $CCl_2F_2$ and $CCl_3F$ and a number of other gases with lesser effects.

**Lifetime** Measure of the persistence of a molecule in the atmosphere. If a gas is in steady state in the atmosphere, and we know that the total emissions into the atmosphere are $S$ grams/yr while the total amount of the gas in the atmosphere is $C$ grams, then the lifetime is $C/S$ yr. The concept gets more complicated if the emissions and removal of the gas change in space and time, if a change in the gas concentration changes its own removal rate, or if the gas cycles between reservoirs that have widely differing exchange times and destruction times.

**Longwave and shortwave radiation** Usages of these terms vary. Shortwave radiation refers to solar radiation that arrives at the earth. It has a range of wavelengths from about 0.15 to 5 $\mu$m (peaking at about 0.6 $\mu$m). Visible light is included within this band. Longwave refers to the radiation by the earth, which varies over wavelengths from about 5 to 100 $\mu$m (peaking at about 15 $\mu$m).

**Trace gas concentrations** Concentrations of atmosphere gases are often expressed as parts per million by volume (ppmv), parts ber billion (ppbv), or parts per trillion (pptr). One part per million means that there is one molecule of the gas for every million molecules of air; one part per billion (ppbv) is one molecule of the gas per billion molecule of air; and similarly for a part per trillion.

## Bibliography


Fleagle, R. G., and Businger, J. A. (1980). "An Introduction to Atmospheric Physics." New York: Academic Press.

Intergovernmental Panel on Climate Change (IPCC) (1990). "Climate Change." New York: Cambridge University Press.

Khalil, M. A. K., ed. (1993). "Atmospheric Methane: Sources, Sinks and Role in Global Change." Berlin: Springer-Verlag.

Khalil, M. A. K., and Shearer, M. J., eds. (1973) "Chemosphere. Vol. 26. Atmospheric Methane."

Khalil, M. A. K., and Rasmussen, R. A. (1992). The global sources of nitrous oxide. *J. Geophys. Res.* **97**(D13), 14651–14660.

Khalil, M. A. K., and Rasmussen, R. A. (1995). The changing composition of the earth's atmosphere. *In* "Chemistry, Composition and Climate of the Atmosphere" (H. B. Singh, ed.), Chap. 3. New York: Van Nostrand Reinhold.

Khalil, M. A. K., and Rasmussen, R. A. (1993). The environmental history and probable future of fluorocarbon-11. *J. Geophys. Res.* **98**(D12), 23091–23106.

National Academy of Sciences (1992). "Policy Implications of Greenhouse Warming." Washington, D.C.: National Academy Press.

Warneck, P. (1988). "Chemistry of the Natural Atmosphere." New York: Academic Press.

World Meterological Organization (WMO) (1985–1991). "Global Ozone Research and Monitoring Project." Reports No. 16 (3 volumes), 1985; No. 18 (2 volumes), 1988; No. 20 (2 volumes, 1989); No. 25 (1 volume), 1991. Geneva, Switzerland.

Wuebbles, D. J., and Edmonds, J. (1991). "Primer of Greenhouse Gases." Chelsea, Michigan, Lewis Publishers.

# Hydrologic Needs of Wetland Animals

**James P. Gibbs**
*Yale University*



Hydrology influences the distribution of wetland animals through a complex set of ecological interactions. First, the hydrologic regime of a wetland determines the chemical and physical characteristics of a wetland's substrate, such as the load and spatial distribution of oxygen, nutrients, and toxins in a wetland's sediments. Second, substrate characteristics, in concert with patterns of inundation, control the composition of wetland plant communities and rates of primary and secondary productivity in a wetland. Third, these factors shape communities of aquatic macroinvertebrates and forage fish, which in turn influence the distribution of large predatory fish and semiaquatic reptiles, mammals, and birds. Many effects of wetland hydrology on wetland animals are mediated through the trophic relationships among wetland animals. For example, amphibians thrive in wetlands that are too temporary to support predatory fish. Large-scale hydrologic processes affect the size, isolation, and connectivity of wetlands and thereby influence the distributions of many wetland birds and fishes. Some wetland animals modify wetland habitats and hydrologic processes. Humans readily manipulate wetland hydrology for the benefit and, more often, the detriment of the wetland fauna.

## I. OVERVIEW

High biological productivity and strong selection pressures in wetlands have produced many wetland-associated animals that are not found in other habitats. The distribution of these animals is influenced by wetland hydrology through a complex set of ecological interactions. These begin with the effect that a wetland's hydrology, defined as the source, velocity, renewal rate, and timing of water in a wetland, exerts on the chemical and physical properties of a wetland's substrates. Nutrient levels, oxygen availability, and toxin loads in wetland sediments are determined mainly by the source of the water flowing into a wetland. Turbulence and the amount of suspended particulate matter carried by the water are determined by water velocity. Renewal rates determine the turnover rates of wetland waters and how rapidly nutrients and toxins are flushed and replenished. Finally, the potential for wetland ecosystems to become established and to undergo succession is determined by the frequency of inundation. These hydrologic factors interact over time to determine the composition, richness, and spatial organization of wetland plant communities. Hydrophytes directly control the composition of macroinvertebrate and forage fish communities, which in turn influence what amphibians, predatory fish, mammals, and birds occur at a wetland. Six generic groups of wetland animals, and their relationship to patterns of wetland inundation, hydrophyte communities, water chemistry, substrate composition, wetland geometry, and human influences, are discussed in the following sections. The focus is primarily on freshwater wetlands of North America. [*See* LIMNOLOGY, INLAND AQUATIC ECOSYSTEMS.]

## II. WETLAND ANIMALS

### A. Aquatic Macroinvertebrates

The wetland-dependent macroinvertebrates include an enormous number of species, but four groups comprise the majority: the insects, molluscs, crustaceans, and annelid worms. The most diverse of these groups is the aquatic insects, which occur throughout all habitat zones (the bottom, deep waters, and shallow, vegetated areas) in most wetlands. The midges (Chironomidae, Diptera) comprise perhaps the most widespread group of aquatic insects in North America, and larval forms often represent the most abundant macroinvertebrates in wetlands. Most larval midges are filter feeders that build tubes upon plant stems or in bottom sediments. Other ecologically important, widespread groups include the mosquitos (e.g., *Culex*), dragonflies and damselflies (Odonata), mayflies (Ephemeroptera), caddis flies (Trichoptera), stone flies (Plecoptera), crane flies (Tipulidae), and water beetles (e.g., Corixidae, Belostomatidae, and Notonectidae). Among these groups, only the water beetles are entirely aquatic; members of other groups typically have aquatic larval stages that undergo a synchronized transformation ("hatch") into adult forms, which may live for periods ranging from 1–2 days (some mayflies) to weeks or months (many dragonflies). [*See* INVERTEBRATES, FRESHWATER.]

The molluscs include the many species of filter-feeding clams and herbivorous snails. The most widespread and common are fingernail clams (e.g., *Muscalium*). Larger clams and mussels are equally widespread but less common, although they are sometimes abundant along large rivers and lake fringes. Most larger clams dwell on the bottom of wetlands, whereas the fingernail clams often occur in submergent vegetation. Planorbid, lymnaeid, and physid snails occupy many wetlands throughout North America, live upon stems, stalks, and leaves of aquatic vegetation, and graze on epiphytes. [*See* AQUATIC WEEDS.]

The crustaceans, like the insects, are commonly encountered throughout most wetlands. Especially common in open water and upon aquatic vegetation are isopods (e.g., *Asellus*) and amphipods (e.g., *Hyallela*). In protected open-water areas, small zooplankters, including cladocerans (e.g., *Daphnia*) and copepods (e.g., *Cyclops*), are abundant. Crayfish (e.g., *Cambarus* and *Procambarus*) are common bottom-dwelling molluscs in many wetland ecosystems.

Annelids, mainly oligochaetes, can be extremely abundant in some wetlands (e.g., *Tubifex* worms) and burrow in the wetland bottom or attach to aquatic vegetation.

### B. Amphibians

Amphibians comprise two main groups, the frogs and toads (Anurans or tailless amphibians) and the salamanders (Urodeles or tailed amphibians). Most species have complex life cycles that include an aquatic, usually herbivorous, larval stage adapted for rapid growth, and a terrestrial, carnivorous adult stage adapted for dispersal among wetlands. Nearly all North American frogs are dependent on wetlands for larval habitat. The major groups include the Ranidae, whose members may be mostly terrestrial but return to wetlands to breed and sometimes to overwinter (e.g., members of the leopard frog complex, such as *Rana pipiens*). Other ranids, such as the green shore frogs (e.g., the bullfrog *Rana catesbeiana*), are primarily aquatic and typically leave wetlands only to disperse to new areas. Other large families of anurans in North America include the tree frogs (Hylidae), which are mostly arboreal but migrate to wetlands to breed, and the toads (mainly Bufonidae and Scaphiopedae), which often live as adults in xeric, terrestrial habitats but require pools as larvae.

Among North American salamanders, all are wetland-dependent except for members of the Plethodontidae, although even this family includes the Desmognathinae, a large group of stream-associated species. Wetland dependency ranges from strictly aquatic groups [e.g., mudpuppies (*Necturus*), hellbenders (*Cryptobranchus*), conger eels (*Amphiuma*), and sirens (*Siren*)], to semiaquatic groups that spend much of the growing season in wetlands but overwinter on land [e.g., newts (Salamandridae)], to primarily terrestrial species that return to wetlands only to breed [e.g, mole salamanders (Ambystomidae)].

### C. Fishes

The majority of wetland fishes in North America, in terms of numbers, species richness, and biomass, are small forage fishes, such as killifishes (*Fundulus*), shiners (*Notropis*), sunfishes (*Lepomis*), and mosquito fishes (*Gambusia*). In deeper-water wetlands or wetlands connected to permanent water bodies, larger, bottom-feeding species occur, such as ictalurid catfish (e.g., bullheads) and carp (*Cyprinus carpio*), as well as carnivorous species such as pickerel (*Esox*), perch (*Perca*), and bass (*Micropterus* and *Morone*). [*See* FISH ECOLOGY.]

### D. Reptiles

Turtles are the most diverse group of wetland reptiles. Most turtles are highly aquatic and leave water only to lay their eggs and to disperse to new habitats. Commonly encountered, highly aquatic turtles in North America include the snapping turtle (*Chelydra serpentina*), mud and musk turtles (Kinosternidae) and softshells (Trionychidae). The family Emydidae includes many other highly aquatic turtles, such as cooters, sliders (*Pseudemys*), painted turtles (*Chrysemys*), map turtles and sawbacks (*Graptemys*), and terrapins (*Malaclemys*), as well as other partially aquatic species that travel regularly among disjunct wetlands (e.g., the spotted turtle, *Clemmys guttata*) or that use wetlands mainly for hibernation (e.g., the wood turtle, *Clemmys insculpta*).

The most common wetland-dependent snakes are the water snakes (*Nerodia*). These use wetlands primarily for feeding on fish, frogs, and crayfish, and rest above water on hydrophytes or on land at other times. Most wetland-dependent snakes are viviparous, that is, they bear live young, and therefore do not undertake migrations to terrestrial habitats to lay eggs, as do turtles. Many other snakes, however, and some lizards, live primarily in uplands but feed opportunistically in wetlands [e.g., king snakes (*Lampropeltis*), rat snakes (*Elaphe*), and garter snakes (*Thamnophis*)].

### E. Birds

Several large groups of birds occur nearly exclusively in wetlands. These include many carnivorous (mainly fish-eating) species, for example, the loons (Gaviidae), herons and bitterns (Ardeidae), several waterfowl species (Anatidae), the kingfishers (Alcedinidae), terns (Sternidae), and several raptors (Accipitridae, e.g., the Osprey, *Pandion haliaetus*). Omnivorous groups include the grebes (Podicipedidae), many species of diving and dabbling ducks (Anatidae), cranes (Grudiae), rails (Rallidae), and gulls (Laridae). Primarily insectivorous species include the shorebirds (Charadriidae) and many songbirds. Unlike other wetland inhabitants, birds are highly mobile and often use disjunct wetlands seasonally or even daily. [*See* BIRD COMMUNITIES.]

### F. Mammals

There are relatively few, wholly wetland-dependent mammals in North America. The most prominent are two rodents, the muskrat (*Ondatra zibethicus*) and beaver (*Castor canadensis*). Other obligate wetland mammals include several primitive, carnivorous shrews ("water shrews" in the genus *Sorex*), two lagomorphs, the swamp rabbit (*Sylvilagus aquaticus*) and marsh rabbit (*S. palustris*), and one mustelid, and river otter (*Lutra canadensis*). Wetlands are a key resource, however, for many otherwise terrestrial species that use wetlands for feeding and cover, for example, moose (*Alces alces*), raccoon (*Procyon lotor*), and mink (*Mustela vison*).

### G. Trophic Relationships among Wetland Animals

The effects of wetland hydrology on wetland animals are typically mediated through trophic relationships among species. Understanding the hydrologic needs of wetland animals is therefore enhanced by identifying key links in wetland food chains that involve wetland animals (Fig. 1) and the habitat associations of each major faunal group (Fig. 2). Macrophytes are colonized by epiphytes, which are grazed by many invertebrate animals, such as snails. Dead macrophytes and their associated bacteria are the main food source for detritivorous insects. Many filter-feeding organisms, such as the zooplankters and clams, sieve fine particulate matter from the the water column. Macroinvertebrates are, in turn, the main food source for forage fish, waterfowl, shorebirds, and songbirds. Forage fishes are the primary prey of larger, predatory fish, snakes, turtles, and birds. Larger predatory fish are eaten by wading birds, raptoral birds, large reptiles, and mammals. Amphibians, which are herbivorous as larvae and eat aquatic insects as adults, are eaten by fish and snakes, wading birds, and mammals. Other wetland animals are predominantly herbivorous, such as muskrats and geese. Occasionally trophic relationships shift among species, as is the case with frogs, which as larvae are prey for aquatic insects, but as adults prey on aquatic insects. Finally, competitive relationships often occur among morphologically similar species, for example, different species of tadpoles, but competing forms are sometimes morphologically quite dissimilar, as is the case of ducks that compete directly with fish for aquatic insect prey.

## III. HYDROPERIOD AND WETLAND ANIMALS

The duration, frequency, and periodicity of flooding vary considerably among wetlands and

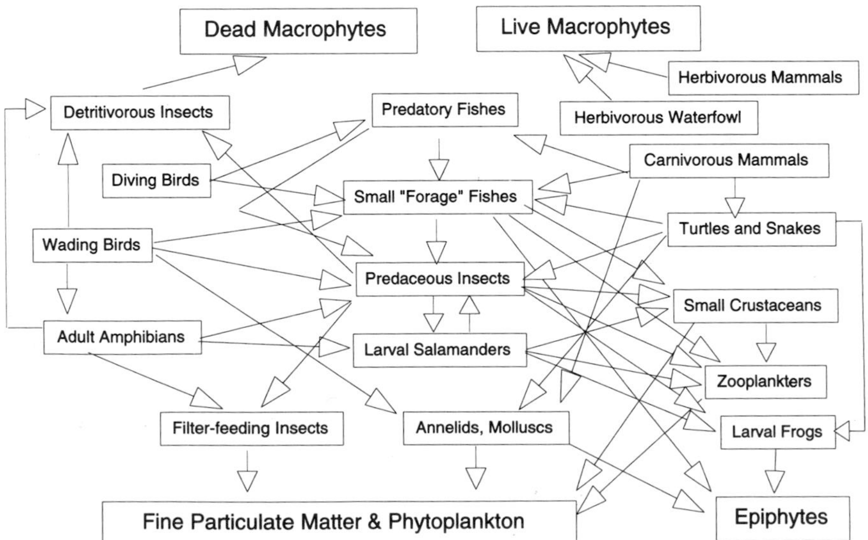


***FIGURE 1*** Hypothetical trophic relationships among major groups of freshwater wetland animals. Arrows originate at consumer groups and illustrate the complexity of food webs and interdependencies among faunal groups.

| | Habitat Type | | | | | | | |
|---|---|---|---|---|---|---|---|---|
| Faunal Group | Sediment Interior | Sediment Surface | Macrophyte Surface | Water Column, Vegetated | Water Column, Open | Water Surface, Vegetated | Water Surface, Open | Adjacent Upland |
| Filter-feeding Insects | * | * | * | | | | | |
| Detritivorous Insects | | * | * | | | | | |
| Predaceous Insects | | * | * | * | * | * | * | |
| Zooplankters | | | * | * | * | | | |
| Small Crustaceans | | * | * | * | * | | | |
| Annelids | * | * | * | | | | | |
| Molluscs | * | * | * | | | | | |
| Larval Frogs | | | * | * | | | | |
| Larval Salamanders | | | | * | * | | | |
| Adult Amphibians | | | | | | * | * | * |
| Small "Forage" Fishes | | * | * | * | * | | | |
| Predatory Fishes | | | | * | * | | | |
| Turtles | | | | * | * | * | | * |
| Snakes | | | | * | * | * | * | * |
| Diving Birds | | | | | * | | * | |
| Wading Birds | | | | | | * | | |
| Herbivorous Waterfowl | | | | | | * | * | * |
| Herbivorous Mammals | | | | | | * | * | * |
| Carnivorous Mammals | | | | | | * | | * |

***FIGURE 2*** Habitat associations of wetland animals. Asterisks indicate significant association between a given habitat type and members of a particular wetland faunal group.

strongly influence which animals occupy a wetland. At one end of the flooding spectrum are small (<1 hectare), temporary pools that remain dry and largely unvegetated throughout most of the year but are inundated for perhaps 1–2 months. At the other extreme are bogs and fens and large bodies of open water with fringing vegetation that remain flooded for centuries. Intermediate along this gradient are wetlands that flood seasonally and for irregular periods and that develop extensive, spatially heterogeneous hydrophyte communities, and those that remain flooded except during drought years and that develop well-defined and relatively stable zones of different hydrophyte taxa.

Biotic productivity in wetlands is generally highest in wetlands with pulsing hydroperiods, intermediate in slowly flowing systems, and lowest in permanently flooded, stagnant situations. Patterns of diversity and productivity in wetland animals follow this general pattern, with some notable exceptions. The productivity and diversity of wetland invertebrates and fish, and consequently many waterbirds and mammals, is highest in semipermanent wetlands, for example, those flooded during most of the growing season but that remain dry otherwise. Most amphibians, however, occur in transient pools. Trends in reptile diversity and productivity along this inundation gradient are not clear.

Semipermanent and permanent water bodies are the primary habitats of fish, which generally lack the ability of many other wetland inhabitants to disperse overland or enter dormancy when wetlands dry. Thus, areas of sufficiently oxygenated water must occur in wetlands throughout the year to support fish populations, especially those of large-bodied "game" species. Large, seasonally flooded wetlands with pools or channels that remain permanently flooded, and lakes with well-developed littoral vegetation, tend to support the most diverse fish communities. Similarly, fish-eating birds and mammals frequent the habitats occupied by the fish species upon which they prey.

In contrast, temporary, unvegetated pools are the primary breeding habitats of amphibians, and offer two main advantages. First, amphibian larvae are able to exploit a pulse of primary productivity that occurs shortly after pools fill with water. Second, pools that are transient enough will not support fish, which are major predators on amphibian eggs and larvae. Use of such pools, however, is associated with three liabilities: predation, desiccation, and competition. Aquatic insects and other amphibian larvae, especially salamander larvae, occur in temporary pools, are important tadpole predators, and can prevent successful metamorphosis. Pools that dry often enough to preclude large predators, however, may also be of insufficient duration to permit metamorphosis of amphibian larvae. Finally, pond productivity declines as the initial flush of nutrients associated with pool filling is sequestered by various organisms as the breeding season proceeds.

Many other wetland animals exploit intermittently flooded wetlands. Waterfowl frequently travel among shallow pools, ditches, and puddles, particularly egg-laying females, to feed on the pulse of invertebrate production that occurs following flooding. These feeding pools often are quite remote from nesting areas on islands and hummocks at more permanent wetlands, where deep waters and dense vegetation can protect nesting birds and their eggs from predators. Similarly, snakes and turtles can often be found in temporarily flooded wetlands, and seek refuge from subsequent low-water periods by either migrating or entering dormancy in a wetland's sediments.

Several species of wetland animals have life histories closely adapted to periodic inundation and drying of their wetland habitats. For example, the snail kite (*Rostrhamus sociabilis*) often skips breeding during drought periods and is nomadic across large regions, but breeds prolifically during wet years. Key to the persistence of this species is flooding periodicity; populations persist as long as the inter-drought interval is sufficient to enable populations to rebuild themselves prior to the next drought. Changes in the natural periodicity of flooding events potentially threaten the viability of snail kite populations.

## IV. HYDROPHYTES AND WETLAND ANIMALS

Wetland hydrology directly influences the distribution of wetland macrophytes, which provide the primary habitat for wetland animals. Much of the variation in structure, cover, and food for the wetland fauna is associated with variation in the species diversity, density, and structural characteristics of wetland plants. Many wetland animals show affinities for particular life-forms of hydrophytes (e.g., submergent plants, floating-leaved plants, or emergents), and some are associated with specific plant taxa.

Most aquatic invertebrates seek protection from predation and safe oviposition sites in macrophytes. In general, plants with dissected leaves support more invertebrates than plants with broad leaves. Macroinvertebrate diversity and density generally are greater in vegetated than in open-water areas and, within vegetated areas, are greatest in areas of mixed emergent and submergent vegetation. Shading by erect hydrophytes also influences macroinvertebrate abundance; unshaded areas of submergent vegetation often support dense populations owing, in part, to elevated water temperatures in unshaded areas.

Similarly, shallow, vegetated wetland habitats fringing deep-water areas are critical nursery habitats for fishes, many species of which use stands of particular wetland plant species as spawning habitat. These areas provide protective cover for eggs and larval fishes and food resources for young fishes. Like aquatic insects, many fishes prefer unshaded, shallow areas where warm water temperatures hasten the development of eggs and larvae.

Many wetland animals incorporate several dissimilar hydrophyte taxa or life-forms into their life cycles. Waterbirds, such as rails, dwell within dense stands of emergent vegetation, whereas others, such as bitterns, herons, and shorebirds, forage mainly along the interface of wetland vegetation and open water. Still others, such as grebes and many dabbling ducks, are associated with floating and submergent vegetation, whereas loons and cormorants forage in deep, open waters. Most of these birds, however, build their nests within dense stands of emergents, on small islands, or in trees or shrubs; thus, the proximity of these diverse habitat types within a wetland is critical to many breeding waterbirds.

Fish communities provide another example of the importance to wetland animals of spatial heterogeneity in wetland habitats. Most larger species of bottom-feeding and predatory species undergo small-scale migrations between feeding and spawning areas in shallow wetlands and resting areas in deeper, open-water habitats. These migrations often occur nightly, with fishes returning to deep waters to rest during the day. Critical to these migratory movements are the links established between shallow, vegetated areas and deep-water zones by seasonal regimes of wetland flooding.

Hydrophytes also vary considerably in their quality as forage for various wetland herbivores. For example, the dominant hydrophyte in a particular wetland strongly affects the local population density of muskrats. Marshes dominated by cattail (*Typha*) often support two to five times the densities of muskrats supported by marshes dominated by bulrush (*Scirpus*). Similarly, beavers occur at low densities where their preferred food items, cottonwoods (*Populus*) and willow (*Salix*), are uncommon.

## V. WATER CHEMISTRY AND WETLAND ANIMALS

Three significant water chemistry parameters that influence distributions of wetland animals are acidity, salinity, and oxygen levels. The hydrological components that most influence these parameters are the source and renewal rates of a wetland's waters. Highly acidic waters are characteristic of open-water wetlands that receive low-pH inputs and that have substrates with low buffering capacity; acidity directly influences the invertebrate, amphibian, and fish communities in these wetlands, and thereby indirectly shapes bird and mammal communities. In contrast, wetlands with extensive emergent plant communities are naturally buffered and tend to be circumneutral to alkaline; water acidity plays a limited role in the distribution of wetland animals in such wetlands. Salinity generally reaches high levels only in wetlands that receive saline source waters, have low rates of water renewal, and that dry periodically. Oxygen levels are lowest in wetlands with shallow, stagnant, warm waters and highest in wetlands with deep, cool, running source waters.

The relationship between water acidity and the distribution of wetland animals can be complex. For example, many aquatic, fish-eating birds frequent wetlands with moderately acidic waters (pH 5.5–6.0). Although fish populations often are reduced in wetlands with waters of pH <6.0, suspended material may precipitate from the water column at low pH, resulting in increased water transparency. This may improve encounter rates with prey and capture success for birds that pursue their prey underwater, and thereby compensate for decreased fish density. The reduction or loss of fish from moderately and strongly acidified waters may also reduce competition between fish and insectivorous birds for macroinvertebrate prey. This may explain why dabbling ducks frequent waters of relatively high acidity that fish-eating ducks avoid. Highly acid waters, that is, pH <5.0, generally support few fish or amphibians and low-diversity communities of aquatic invertebrates. Waterbird and mammal use of such wetlands is limited.

Salinity determines what animals occupy a wetland primarily through its effects on the composition of plant and invertebrate communities, and through its direct, physiological effects on wetland animals. Difficulty in maintaining water balance leads many dabbling ducks, particularly those brooding young, to avoid wetlands with highly saline waters. Owing to their highly permeable skin, amphibians are intolerant of even moderately saline conditions. For similar reasons, highly saline conditions often limit the hatching success and survival of fish eggs and larval fish. Many mammals, such as muskrats, are tolerant of highly saline conditions, but populations often are limited by the low quality of forage plants that grow in saline wetlands.

Oxygen levels strongly influence the distributions of wholly aquatic, nonmigratory wetland animals, that is, the macroinvertebrates and fishes. Hypoxic conditions are common in wetland waters; particularly in the warm, residual waters that occur following dry-downs of semipermanent and ephemeral wetlands. Wetland animals have acquired an array of physical adaptations to cope with anoxia, which include development of gills and other organs specialized for gaseous exchange, mechanisms for increasing water movement across gas-exchange membranes, and modification of respiratory pigments to improve oxygen-carrying capacity. Behavioral adaptations also are critical, and include dormancy or low locomotor activity during periods of oxygen stress and migration from hypoxic to oxygen-rich environments. For example, sticklebacks (*Culaea* spp.), small fishes common to freshwater marshes in North America, can face severe oxygen limitation during winter months. To cope, sticklebacks migrate vertically in the water column to the ice–water interface, where, aided by their pointed snouts, they use microlayers of dissolved oxygen and trapped air bubbles that occur directly beneath the ice layer. Sticklebacks also undergo horizontal movements to find specific microhabitats where oxygen may be more available, such as spring seeps in the wetland substrate and the underwater openings to muskrat lodges. Finally, sticklebacks curb their activity levels during winter, thereby lowering their metabolic rates and reducing oxygen demands.

## VI. SUBSTRATE EFFECTS ON WETLAND ANIMALS

The quantity, velocity, and sediment loads of source waters initially determine the characteristics of a wetland's substrate; the substrate is then modified over time by the hydrophytes that become established. The substrate characteristics most important to wetland faunas are dissolved oxygen levels, nutrient and toxin loads, sediment composition, and sediment stability. In addition to affecting hydrophyte distributions and, secondarily, distributions of wetland animals, substrate features directly determine what benthic invertebrates can inhabit a wetland's bottom. Poorly oxygenated and frequently shifting sediments support relatively fewer benthic organisms than well-oxygenated stable sediments. Because benthic invertebrates anchor food chains within the wetland animal community, substrate effects on the benthos indirectly affect the many fish, amphibians, and reptiles that rely on benthic invertebrates as a food resource. Fish communities also are directly affected by wetland substrates because the quality of spawning, hatching, and larval habitats is closely associated with the characteristics of wetland substrates.

## VII. WETLAND GEOMETRY AND WETLAND ANIMALS

In many regions, large-scale hydrologic processes originally created a mosaic of variably sized and isolated wetlands within a matrix of upland habitats. Wetland size, connectivity, and isolation together represent three aspects of wetland geometry that influence the distribution of many wetland animals, particularly birds. Small wetlands often support fewer species of birds than large wetlands, and isolated wetlands support fewer species than wetlands in complexes. Aquatic, fish-eating birds are particularly sensitive to wetland area, and often shun small wetlands (<10 hectares). Small wetlands may not provide a sufficient quantity of habitat or prey for many large-bodied species, and short interwetland distances may provide certain species with alternate foraging sites while minimizing their time in flight. Larger wetlands generally support a wider diversity of vegetative life-forms, and therefore may provide a greater variety of habitats for wetland species than smaller wetlands. This may account for the frequently observed pattern of increased fish species diversity in larger compared to smaller wetlands.

Connectivity among wetland environments is critical to wetland animals, particularly amphibians. Aquatic breeding and terrestrial nonbreeding habitats must be contiguous for amphibian populations to persist. Local breeding populations undergo frequent extinction and recolonization events, and individuals often are exchanged among populations. The importance of migration in amphibian population ecology is implicated by the presence in many species of a juvenile stage adapted strictly for dispersal. Destruction of larval or adult habitats, or the connections between them, for example, riparian corridors, often drives local amphibian populations to extinction. Finally, many fishes undergo seasonal migrations over hundreds of kilometers into river floodplains, and return to the river channels as floodplain waters recede.

## VIII. HABITAT MODIFICATION BY WETLAND ANIMALS

Whereas the distributions of many wetland animals represent a response to hydrologically derived, physical and chemical characteristics of wetland habitats, the activities of a few wetland animals can significantly alter wetland habitats and hydrologic processes. Dam-building activity of beavers is a prominent example. Beavers alter stream channels and transform riparian forests into freshwater marshes and ponds with stands of dead timber. Beaver activities also permanently modify soils beneath dam sites and even change local topographies; dams reduce current velocities and cause massive silt depositions, which remain long after the beavers' dams disintegrate.

Another example of habitat modification is provided by muskrats. High levels of herbivory by muskrats can result in "eat outs," whereby essentially all emergent plant material in a marsh is consumed over a few years. Similarly, geese and macroinvertebrates have also been observed to consume extensive amounts of the standing crop of hydrophytes over short periods in some wetlands. Such drastic, animal-caused modification of the wetland environment not only affects habitat availability for other wetland animals, but can alter patterns of energy flow and biotic productivity within wetland ecosystems. Other, less drastic examples of habitat modification include the activities of alligators, which construct small basins that serve as critical refuges for many fishes, and important feeding areas for fish-eating birds, during low-water periods. Crayfish burrows serve in many areas as moist refuges for water snakes, salamanders, and frogs.

## IX. HUMAN INFLUENCES ON WETLAND HYDROLOGY AND WETLAND ANIMALS

Because the structure and composition of wetland environments are so closely linked with wetland hydrology, manipulation of wetland hydrology provides humans with a ready means to alter communities of wetland plants and animals. Humans accomplish this primarily by building dams at wetland outlets to regulate the depth and duration of inundation, and by constructing channels and ditches to dewater (drain) wetlands and prevent prolonged flooding. Artificially regulated water levels often result in impoverished animal communities because few species of wetland animals have life histories adapted to the stable water regimes that result. This occurs even where water control structures were originally built with the intention of benefiting wetland animals, for example, at wildlife refuges. Management strategies for many water impoundments now seek to emulate the "natural" flooding regimes that were present before impoundments were built. Wetland drainage has obvious and drastic effects on wetland animal communities and has resulted over the last 150 years in a loss of about half of the original habitat available to the wetland fauna of the United States. Among the most imperiled wetland habitats are the easily drained, intermittently flooded basins and pools that are critical to such a large proportion of the wetland fauna.

Other human activities impoverish the wetland fauna in more subtle ways, even where wetlands receive nominal legal protection. Destruction of the uplands adjacent to wetlands destroys the habitat interface that is critical for amphibians and reptiles that migrate between wetland and upland habitats. Reductions in the connectivity among wetlands, for example, by construction of roads and levees, blocks the migration routes of amphibians, fish, and birds, and can lead to local population extinctions. In contrast, dredging of canals that link major river systems causes sudden mixing of fish and reptile faunas that may have evolved in isolation for thousands of years. Anthropogenic inputs of nutrients and toxins drastically alter the chemical environment of wetlands and render many sites unsuitable for wetland animals. Finally, introduction of exotic species of hydrophytes, for example, purple loosestrife (*Lythrum salicaria*) and common reed (*Phragmites australis*), and exotic wetland animals, for example, carp, nutria (*Myocaster coypu*), and bullfrogs, outside their native range, has greatly altered wetland environments throughout much of North America, mostly to the detriment of the native wetland fauna.

## Glossary

**Benthic** Located on the bottom of a water body, or pertaining to bottom-dwelling organisms.

**Buffering capacity** Refers to the ability of a solution to resist changes in pH when acid or alkaline substances are added.

**Detritivores** Organisms that consume dead plant or animal material.

**Emergent vegetation** Rooted plants that tolerate flooded soil but not extended periods of submersion.

**Floodplain** Broad, flat areas adjacent to rivers and streams that become inundated during floods.

**Hydrology** Source, velocity, renewal rate, and timing of water in a wetland.

**Hydrophyte** Plants thriving on moist or flooded soils.

**Littoral** Shoreline zone where sunlight penetrates to a wetlands substrate and supports rooted plant growth.

**Macroinvertebrate** Organisms lacking backbones and generally visible to the human eye.

**Submergent vegetation** Plants that grow and reproduce while completely submerged.

**Wetland** Transitional zones between dry land and deep water where saturation with water is the dominant factor determining the types of soils, plants, and animals present.

## Bibliography

Good, R. E., Whigham, D. F., and Simpson, R. L., eds. (1978). "Freshwater Wetlands." New York: Academic Press.

Greeson, P. E., Clark, J. R., and Clark, J. E., eds. (1979). "Wetland Functions and Values: The State of Our Understanding." Minneapolis: American Water Resources Association.

van der Valk, A., ed. (1989). "Northern Prairie Wetlands." Ames: Iowa State University Press.

# Industrial Development, Arctic

**L. C. Bliss**
*University of Washington*



## I. INTRODUCTION

The Arctic has been the major world biome to be explored with only a few areas developed for their nonrenewable resources. Petroleum exploration in northern Alaska began in the 1940s, followed by providing natural gas to Point Barrow in 1950. Oil was discovered in the Canadian subarctic in 1920 and a refinery built at Norman Wells, NWT, followed in 1936. Coal has been mined by the Norwegians on Svalbard since the 1890s. Greenland has deposits of coal, cryolite, graphite, lead, uranium, and zinc, but most deposits are of poor quality and are not economic to mine. Russia has vast deposits of coal and heavy metals within the Arctic that have been utilized since the 1940s and has huge gas and oil fields, some in operation since the 1960s and early 1970s. [*See* NORTHERN POLAR ECOSYSTEMS.]

Exploration and development of resources have been slow, due to severity of climate, remoteness of these lands from markets, costs of developing resources, and the need for having environmental regulations. Ecological research has been closely associated with resource development in Canada and the United States but is much more limited in Russia. In the United States the passage of the National Environmental Policy Act (NEPA) in 1970 required that an Environmental Impact Statement (EIS) be prepared for each project. This has been followed by enforceable state and federal regulations for development and production. The government of Canada established guidelines for northern pipelines in 1970 and more recently for mining, but laws comparable to NEPA have not been passed.

In the late 1980s Russian scientists and groups of northern people became more concerned with the massive development of petroleum fields, pipelines, and large mine operations. In 1988 the Central Committee of the Communist Party and the USSR Council of Ministers adopted a major resolution on the environment. The new State Committee for Environmental Protection (Goskompriroda) has the power to regulate land use, protect surface and groundwater, regulate atmospheric contamination, and protect wildlife, vegetation, fish, and marine animals. A major problem of enforcement remains. However, in 1989 General Secretary Gorbachev and the Presidium of the Council of Ministries placed a moritorium of 5–7 years on further development of the huge gas fields on the Yamal Peninsula in western Siberia (Fig. 1) until more detailed environmental studies were conducted and development plans modified to lessen environmental impacts. In December 1989 the Supreme Soviet passed a resolution on ecology by

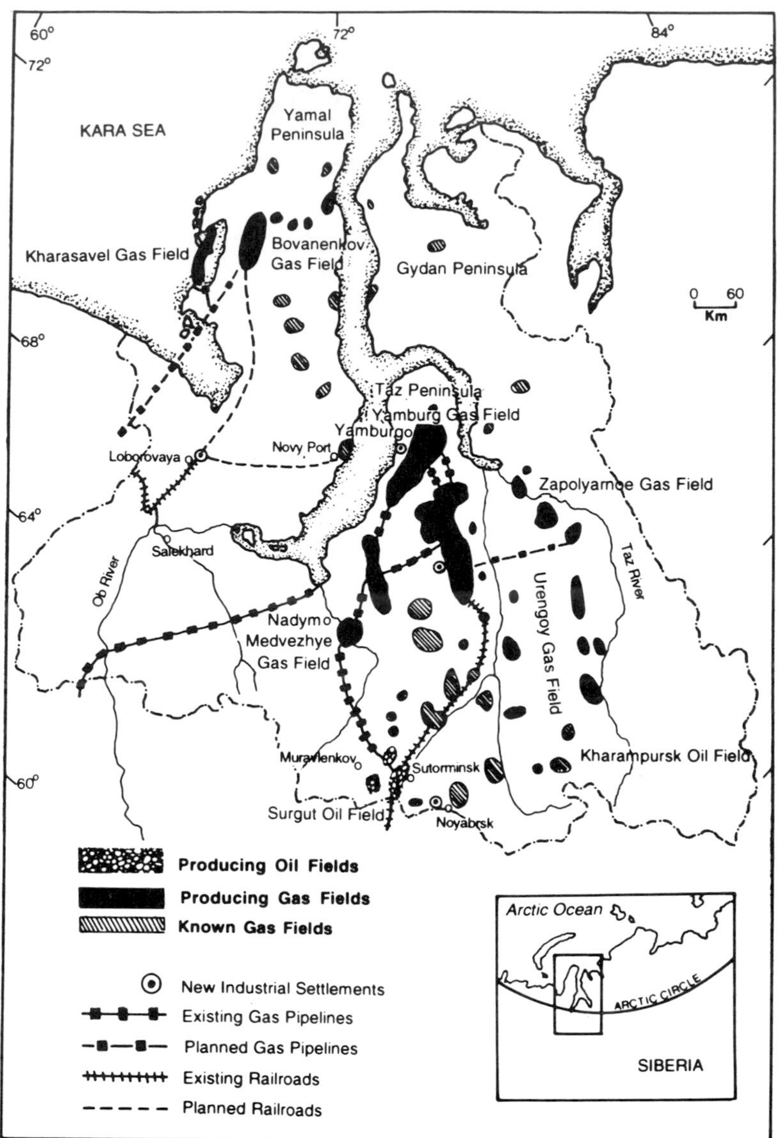


***FIGURE 1*** Location of major gas and oil fields, cities, and railways within western Siberia.

urging the Soviet Council of Ministers, Councils of Ministers of the Autonomous Republics, and the Soviet Ministries and Soviet Departments to define measures for granting assistance to ecological disaster areas, such as the Aral Sea, the Chernobyl nuclear power station, nuclear testing on Novaya Zemlya, and petroleum development in the Arctic and subarctic. It is well established that very major environmental problems have occurred in the Russian Arctic, and only time will tell whether these impacts will be lessened and future environmental problems reduced.

With the establishment of the Home Rule Act for Greenland in 1979 the control of mineral policy became a central issue. Prior to 1988 all revenue from mining went to Denmark, but future revenue will be shared equally between the two countries.

## II. MINING

### A. North America

Prior to the 1970s, mining in northern North America was largely confined to the boreal forest.

Uranium was discovered on the shore of Great Slave Lake in the 1930s. Gold mining has been important at Yellowknife, NWT, from the 1940s and at various locations in the Yukon and Alaska from 1896 to 1902. Lead, zinc, copper, and silver are mined in Alaska and the Yukon Territory. Low-grade coal (bituminous shale) occurs in the Smoking Hills, near Coppermine, NWT, and has been burning since at least the first European people reached this area (1826). The large release of $SO_2$ and soluble Al and Mn has resulted in very acid soils (pH 2.7–3.2) and pond waters (1.8–2.4) in the vicinity of these deposits. Consequently, within 40 m of the burning sea cliffs, there is no vegetation. From 80 to 320 m there are a few scattered plants of *Artemisia tilesii* and some *Arctagrostis latifolia* beyond 320 m. This illustrates the limited number of vascular plants that can tolerate such a severe environment.

Important lead and zinc mines are located in the Canadian Arctic on Strathcona Sound, northern Baffin Island, and a second mine at Arvik on Little Cornwallis Island. Both mines produce high grade ores that are stockpiled in winter and shipped in summer by ship. These mines employ some Inuit (Eskimos) from nearby settlements. The Mary River iron ore deposit in northeastern Baffin Island is of high grade, but mining operations have not begun.

There are huge deposits of coal in arctic Alaska as well as large deposits of copper and other minerals. The coal deposits are estimated to be about 3 trillion metric tonnes, but development awaits new transportation systems and environmental technology. The Red Dog Mine near Kotzebue is the largest zinc mine in the United States. Alaska is still an important gold-mining state, although only a small amount is placer mined from alluvial deposits in the Arctic, near Nome.

Mining in the Arctic is similar to that within temperate regions with the exception of problems related to temperature in winter and to permafrost. If the permafrost contains large amounts of ice, special precautions are necessary in the construction and maintenance of roads, buildings, and tailing dumps. New diking techniques have been developed to prevent problems of leaching and thermal erosion of massive ice wedges and lenses. Mines removing heavy metals in the North American Arctic have not yet faced abandonment, thus disposal of mining equipment and any rehabilitation of tailings, camp, and supply sites have yet to be faced. The lack of smelters in the North American Arctic has prevented all of the environmental problems discussed next relative to the Russian Arctic and subarctic.

### B. Greenland

The cryolite mine at Igvitut, now closed, began operation in the 1860s. The large lead–zinc mine at Maarmorilik in west central Greenland and the smaller mine at Mestersvig in east central Greenland are also closed. There is hope that future mineral discoveries will be economically feasible and that the New Mining Act for Greenland (1991) will enable the country to receive much needed revenue.

### C. Svalbard

The Svalbard Archipelago has been under Norwegian sovereignty since 1925 which enables them to exploit natural resources. The island of Spitsbergen contains considerable deposits of coal that have been mined by various companies since the 1890s. Swedish mines were established in 1911 but due to economic conditions they sold some mines to Norwegians and other mines were purchased by Russians. In recent years the Norwegian and Russian mines have produced about 500,000 tonnes of coal per year. There has been limited exploration for petroleum, but to date no commercial reserves have been reported.

### D. Russia

Northwest Russia is one of the most polluted regions on earth. The combination of open air nuclear tests on Novaya Zemlya, radioactive contamination of the Kara and Barents seas from effluent dumped into rivers and bays, the huge coal mines at Vorkuta, the nickel mines and smelter at Norilsk, and the massive mining and smelting operations on the Kola Peninsula have resulted in large areas of air, water, and soil contamination.

There are 13 coal mines near Vorkuta, a city of about 220,000 people. Like most Russian arctic cities, it was founded in 1933 as a gulag. The largest mine, Vorgashorskaya, is capable of producing 18,000 tonnes of coal per day. Miners in this region are very militant over their difficult working and living conditions. Their wages are high by Russian standards, yet food and basic commodity prices have soared. Coal sold to the government brings $1.50 to $3.00 per tonne, yet was sold in western Europe for up to $50 per tonne in 1992.

The world's largest nickel mining and smelting operations are in the region surrounding Norilsk. Until 1991 it was a closed city to western people. Norilsk was built by prison labor in the 1930s. A nickel smelter under construction was moved from Monchegorsk on the Kola Peninsula to Norilsk at the start of World War II. The former Soviet Union was the world's largest producer of nickel (30%), and in 1992 the Norilsky Nickel Company accounted for 87% of the total nickel and 67% of the copper production in Russia. Although the residents of Norilsk have favorable material living conditions, environmental degradation is severe. The smokestacks of the smelters emit 2.5 million tonnes of pollutants, mostly $SO_2$, per year. The Norilsk region is probably the world's greatest source of atmospheric sulfur. The air is also heavy laden with phenols, heavy metals, chlorine, and formaldehyde. Heavy metals accumulate in winter snow and at spring melt, enter lakes and rivers that are now highly polluted. Pollutants have accumulated in soils and vegetation, entering the food chain and threatening reindeer and fish that are consumed by people. Air pollution drifts over the tundra and northern forests depositing sulfur as acid precipitation and as a solid that has killed stands of *Betula, Larix,* and *Picea* over a 3800-$km^2$ area. Arctic haze has become common at high latitudes in recent years and Norlisk is believed to be the major source. In the late 1980s, medical records indicated a 60% increase in cancer cases and a 270% increase in birth defects in this region.

While Moscow makes demands for environmental cleanup, no monies are provided. Environmental protection efforts are locally funded; they include environmental monitoring and research, investment in pollution control technologies, and promotion of recycling and resource recovery. With the current state of the economy and the lack of funds for equipment maintenance, the Hope smelter, built using Finnish technology near Norilsk in 1981, has pollution control equipment designed to scrub 85% of sulfur emissions, but has fallen into disrepair and operated at only 40 to 60% capacity in 1992.

## III. OIL AND GAS RESOURCES

### A. North America

#### *1. Exploration and Development*

Petroleum exploration, conducted in northern Alaska in the 1940s, resulted in the discovery of natural gas, which has been supplied to Point Barrow since 1950. Oil and gas exploration increased along the arctic coast of Alaska and in the Mackenzie River Delta region of Canada in the mid 1960s. Exploration led to the discovery of oil and gas at Prudhoe Bay, Alaska, in early 1968 and a year later in the Mackenzie River Delta region. New oil fields were discovered in Alaska in the 1970s and 1980s. Large gas fields and limited oil fields have been discovered in the northern Canadian arctic islands in the 1970s and 1980s (Fig. 2).

Most of the early terrain problems associated with summer road construction, summer seismic activity, and uncontrolled off-road vehicles have been minimized through changed seismic operation, use of winter snow and ice–snow roads, and government regulations. Some winter roads (Fig. 3) have been used for more than 3 years with resultant limited damage to vegetation, other than broken shrubs, and limited areas of permafrost melt. Recent winter seismic lines run in the Alaskan Arctic Wildlife Range indicate that multiple passes through wet sedge–shrub tundra result in a deeper active layer and that surface water remains in depressed track surfaces. The potential for permafrost melt is often greater within the forest–tundra or the northern portion of the boreal forest due to the higher levels of net radiation compared with the shrub tundra to the north. When trees are removed within the forest–tundra, a higher percentage of the net radiation is dissipated as soil heat flux and

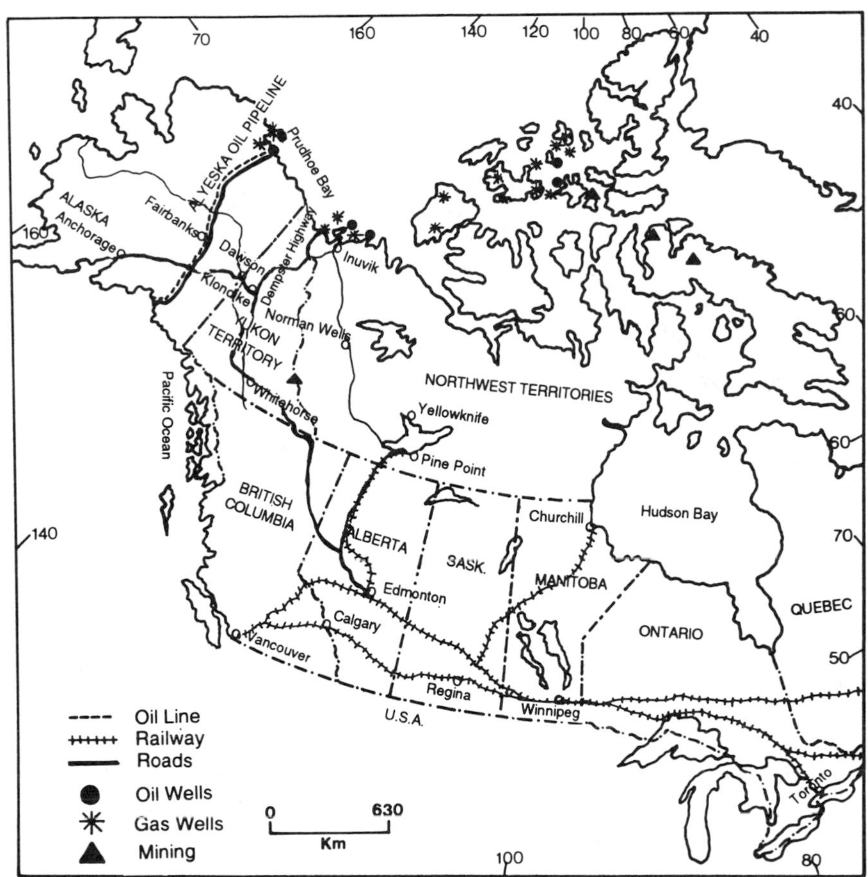


**FIGURE 2** Location of oil and gas fields and transportation systems within northern Alaska and Canada.

**FIGURE 3** Winter road constructed from snow and ice (fourth winter) near Parsons Lake, NWT, Canada.

is effective in melting permafrost. In both northern Alaska and mainland Canada, gravel pads are 2–3 m thick for exploration (Fig. 4) and production wells (Fig. 5), associated roads (Fig. 6), pump stations (Fig. 7), equipment storage areas, and camps, providing a sufficient thermal barrier from the ice-rich permafrost below. However, there have been problems with the seepage of drilling fluids from sumps (ponds) within the thick gravel pads associated with oil wells at Prudhoe Bay. It had been assumed that the permafrost table would rise into the gravels, preventing seepage. This demonstrates the need for plastic liners to retain fluids until they can be removed by tanker trucks and reinjected into wells.

Road dust from the Dalton Highway (Alyeska Pipeline Haul Road) is high in Ca, which over time may reduce or eliminate acidophylic species

**FIGURE 4** Gas drilling rig and camp on a gravel pad near Parsons Lake northeast of Inuvik, NWT, Canada.

**FIGURE 5** Large gravel pad with 24 oil wells and a small pump facility. Note the drill rig that is reworking one of the wells (Prudhoe Bay, Alaska).

**FIGURE 6** Dalton highway and the Alyeska pipeline near Toolik Lake in northern Alaska.

**FIGURE 7** Pump station No. 4 on the Alyeska oil pipeline, the northern part of the Brooks Range.

of moss and speed the leafing of shrubs adjacent to these roads. This is not a serious environmental problem compared with early problems of cross-road drainage with spring runoff, because of a lack of culverts or where culverts filled with gravel or ice at freeze up. Water moving parallel to roads and increased flow from culverts can result in combined thermal and hydraulic erosion that can lead to thermokarst topography. Fortunately, with road maintenance this is a minor problem in the Prudhoe-Kuparuk oil fields and the associated highway to the south.

A detailed mapping study of three areas within the Prudhoe Bay oil field from 1968 to 1983 showed that the cumulative effects of roads and gravel pads for wells, supply camps, and storage yards have resulted in some blocked drainages in this flat landscape, resulting in flooding in 9% of the mapped area and 3% of the total field (500 km$^2$). Thermokarst occupied 15% of the mapped area. These findings, although significant, are quite small compared with the potential levels of flooding and thermokarst that might have occurred had not stringent regulations been developed and enforced. A major review of the pipeline, the Alyeska project, and the petroleum exploration in the Mackenzie River Delta region indicated that these massive projects have been constructed and maintained in an environmentally acceptable manner. This would not have been the case if strict environmental regulations had not been enforced initially.

Prior to construction of the pipeline, when the project was under litigation, it was determined that 689 km of the pipe needed to be elevated because of the high ice content of the permafrost compared with the original plan to elevate only 220 km. The 120-cm-diameter pipe was redesigned to be placed on highly engineered steel support members with ammonia heat exchangers to prevent heat transfer to the supports set into the permafrost within the forested portion of the line (Fig. 8). The original design called for wooden supports. Culverts, bridges, and low-water crossings (stream fords) permitted free passage of fish. The original 280 designated fish streams that the pipeline crossed were eventually increased to 450 fish streams. The significant increase in culverts and bridges to accommodate fish resulted from limited data from fish surveys prior to construction, and thus the enforcing agency biologists could demand more structures than actually needed. This points to the need for adequate biological surveys prior to developing detailed construction plans. The same holds true for the elevated and buried sections (Fig. 9) of the pipeline to accommodate the passage of moose and caribou. Survey studies since construction indicate that moose and male caribou cross freely under the pipeline, but that caribou females and calves tend to avoid the highway, pipeline, and adjacent rangeland. Biologists who have studied wildlife along the pipeline and road believe that frequent road traffic and work crews along the road are a greater deterrent to wildlife than the passive pipeline.

**FIGURE 8** Alyeska oil pipeline illustrating the elevated pipe, the ammonia heat exchanges on each support member, and the adjacent road used in construction and now maintenance of the pipeline, near the Yukon River, south of the Brooks Range (Alaska).

**FIGURE 9** Section of buried pipeline to permit caribou crossing. A revegetated work area is shown in the center (northern Alaska).

A major component of the pipeline project was related to land restoration. Much of the disturbed 12,000 ha has been revegetated using various seed mixes and fertilizer mixes for each of the four designated climatic sections (arctic tundra, alpine tundra, Brooks Range, interior boreal). Seeding was done by fixed-wing planes and hydroseeding by trucks. Mulching was applied in special sites. Revegetation via seeding has proved successful in most regions, though more difficult in the wet coastal tundra. The native grasses *Calamagrostis canadensis* and *Arctagrostis latifolia* have been used along with selected strains of arctared creeping red fescue, boreal creeping red fescue, nugget Kentucky bluegrass, meadow foxtail, and engmo timothy as northern agronomic species. Over time (2–4 years) native species have begun to take over.

Offshore wells have been drilled within the Beaufort Sea, north of Alaska and the Mackenzie River Delta in Canada. Small islands, spits, and artificial islands have been used north of the Kuparuk oil field in Alaska. These artificial islands were built using gravels pumped from the shallow continental shelf. In Canada additional wells have been drilled in summer from ships anchored in relatively shallow waters. In the 1970s Canmar had three drill ships operating north of the Delta.

In the High Arctic of Canada, wells were drilled in winter from ice platforms by pumping sea water onto the ice to freeze that resulted in platforms 6–7 m thick. Drilling rigs and camps were brought to the platform by trucks over ice roads and by cargo planes. Drilling was limited to the 3 to 4 coldest months of winter. New gas fields have been found, but no production has resulted. A major concern in drilling wells from either a ship in summer or on an ice island in winter is the need to prevent a blow-out should a zone of high pressure be entered. Thus blow-out preventers, often 5 to 9, were "stacked" on a typical well. When drilling from a ship, well heads in shallow waters were placed 12 to 15 m below the ocean floor in a "glory hole" to prevent scowering from potential icebergs.

Many experiments have been conducted near Barrow, Alaska and in the Mackenzie Delta to determine ways of putting out oil fires in both summer and winter. The coastal waters of the Beaufort and Chukchi seas contain large populations of marine mammals, fish, and sea birds. Thus there is a great need to highly regulate drilling and production operations offshore so as to greatly reduce polluting these waters that would have a severe impact on native peoples that depend on marine organisms for part of their food.

### *2. Petroleum Spills and Fire*

Small oil and diesel spills occurred during construction of the Alyeska pipeline, resulting in trucks turning over on slippery winter roads. Winter spills are less damaging than summer spills, provided they can be cleaned up, the surface burned lightly, and fertilizer added to stimulate plant growth. Several studies in Alaska and Canada have shown that wet sedge vegetation shows considerable recovery in 1–3 years but that upland tussock and shrub tundra recovers more slowly (5–10+ years). Spills within black spruce forest kill the trees but with a 1- to 3-year lag. This is believed to result from the secondary effects of toxic sulfur released by sulfur-reducing bacteria. Both oil and diesel fuel kill vegetation on contact but regrowth often occurs from shoot bases and latent buds of shrubs.

Burning oil spills should only be done in winter after removal of most of the oil. Repeated burning results in destroying much of the peat and soil microflora, and produces a tar-covered surface that greatly inhibits reestablishment of vegetation. With summer spills, unless most of the oil can be pumped into trucks with minimal trampling and no burning, more damage is done in the cleanup operation than if the sites are left alone, at least until winter when light burning is attempted.

Forest-tundra, tussock tundra, and shrub tundra fires can occur in unusually dry summers. Fire destroys most aboveground plants, including part of the peat, but regrowth is often more rapid than one might predict. Cottongrass and shrub species recover quite rapidly and where native grasses are present, they often reseed an area within 1–3 years. In upland tundra near Inuvik, NWT, aboveground plant production can reach 50–90% of control sites within 2–3 years.

Where fire is common in the forest–tundra regions of Canada and Russia, fires often result in the dominance of shrub tundra because the shrub species are able to resprout and the establishment of tree seedlings is difficult due to lack of a seed source and a series of cool summers.

Exploration and potential development methods need to be different in the more environmentally severe High Arctic. In those lands there is much less vegetation, almost no peat layer except in wet sedge–moss meadows, and almost no species of grass that can serve as succession species to revegetate disturbed soils. In addition, these soils frequently erode severely, thus increasing problems relative to road construction, well pads, and support camps. Gravels are abundant along the rivers in northern Alaska and have been successfully mined there and in deep borrow pits for construction. There are few areas in the High Arctic that contain comparable gravel deposits. Surface gravels and rocks often have to be scraped from stone nets and stripes to provide adequate material for runways, roads, and pads for camp construction. Consequently, most of the exploration drilling has been done in winter (Fig. 10) with sites cleaned up prior to spring melt, and the rig moved the next fall after freeze up.

## B. Russia

The West Siberian sedimentary basin (>3,000,000 km$^2$) in the Tyumen Oblast is the largest structural

**FIGURE 10** Exploration gas well on the Fosheim Peninsula, Ellesmere Island, northern Canada.

basin in the world. Hydrocarbon deposits, mostly oil, are found within the southern, boreal forest portion and large gas fields occur to the north within the Arctic. This huge basin, east of the Ural Mountains and Ob River, was an uninhabited wilderness except for hunters and trappers prior to discovering oil and gas in the early 1960s. The immense Samotlor oil field was discovered in 1965 and placed into production in 1969. Production at the Fedorovo oil field began in 1973 and by 1982 this field accounted for 6% of the Soviet national output.

Strict environmental laws and regulations have not been enforced with regard to construction and operation of these oil fields, pipelines, supply bases, towns, and cities. In 1985 it was reported there were 343 breaks in pipelines in one field, resulting in 1000 tonnes of oil, 600 tonnes of which reached rivers. Each year it is estimated that 1,000,000 tonnes of oil are released from broken pipelines in the Tyumen and Tomsk Oblasts; the result of poor welding and caustic erosion of pipes. The released oil has destroyed large areas of the winter grazing range of reindeer herds and contaminated many lakes and rivers that have severely impacted fish and water fowl. There is little indication that these oil spills have been cleaned up.

Towns and cities, ranging in size from 10,000 to 188,000 (1984 data), have been developed in only 20 to 30 years. These towns and cities are located on rivers or along the coast where water transportation permits the movement of drilling rigs, construction materials, and food and supplies for people. The larger towns and cities have airports and there is a limited railway system (Fig. 1) that serves this huge oil and gas region. A railway connects the cities of Surgut and Nizhnevartovsk within the oil fields and Novy Urengoy, Nadym, and Yamburg to the north in the gas fields. Large portions of this single-track line are built on sands with resultant problems of shifting road bed and tracks. A railway is presently under construction to the huge gas fields at Bovanenkov on the Yamal Peninsula. There are few paved or gravel roads; many of them are constructed of concrete slabs welded together and placed on a sand base. Winter road travel is largely via ice or snow-packed roads over the vast areas of swamps and lakes.

Nadym (Fig. 11) and Novy Urengoy, each with a population of more than 50,000, are the major support cities for the large gas fields of the Gydan Peninsula. Yamburg, further north, was built to house only workers who stay up to 6 months. This work camp is operated in a similar manner to those in arctic Alaska and Canada. The Urengoi gas field is the largest in Russia, producing about 50% of total production in Russia; about 13 trillion $ft^3$ per year.

Gas pipelines were buried, but in places have "floated" to the surface or have become exposed due to severe erosion with spring runoff waters. Surface disturbance for the construction of a single pipeline varies from 0.05 $km^2\ km^{-1}$ in well-drained lands to 0.25 $km^2\ km^{-1}$ in swamps. Some of the transmission routes contain three to eight parallel

**FIGURE 11** Massive new apartment building under construction (right) and one occupied for 1 year on the left in Nadym, Russia.

**FIGURE 12** Six parallel gas pipelines, each 1.4 m in diameter, on the Gydan Peninsula north of Nadym, Russia.

**FIGURE 14** Hydrate plant for removal of water from the gas. Note the collecting pipes from wells within the Yamburg gas field (Russia).

pipelines with large areas of bulldozed land between each line (Fig. 12). The lack of enforced regulations has permitted the use of track vehicles during summer and winter, with resultant surface disturbances of large areas [supply camps, support facilities, pump stations, drill sites (Figs. 13 and 14), and supply roads]. A study of ecological problems in the Yamal-Nenets Autonomous District estimated that over 25% of the area has received catastrophic disturbance, 15 to 25% with critical disturbance, and up to 15% with acute disturbance. The acute and critical levels of disturbance refer to land-use modifications that disrupt traditional uses by native peoples. The Yamal-Nenets Autonomous District covers 6200 km$^2$, of which 450 km$^2$ has completely destroyed vegetation within the oil and gas fields and 1800 km$^2$ along pipelines and roads.

**FIGURE 13** Drilling rig in operation within the Yamburg gas field. Note the amount of surface disturbance and ponded water in the vehicle tracks around the well and the supply road in the upper left (Taz Peninsula, Russia).

Ice-rich soils and lakes are present on the Taz and the Gydan peninsulas, but are less frequent than on the Yamal Peninsula with its lower relief and larger areas of wetlands and lakes. Considering the amount of surface disturbance and the pollution of lakes and rivers that has resulted from the existing oil and gas fields in western Siberia, it is fortunate that delays were imposed on developing the huge gas fields and transportation systems on the Yamal Peninsula in order to produce more environmentally acceptable procedures for development and production. Only time will tell whether new technology will be successfully employed.

Recognizing that environmental regulations are necessary to significantly reduce thermal and water erosion, experimental plots have been established at Nadym, Yamburg, and Bovanenkov to determine the best plant species to use in a revegetation program.

## IV. TRANSPORTATION SYSTEMS

### A. North America

A transportation network is essential for northern development. Roads, railway beds, and airport runways must be elevated with gravel to provide adequate insulation from permafrost and to enable wind to remove snow. Road construction in arctic Alaska follows guidelines established by the Environmental Protection Agency. These guidelines include placing fill over undisturbed vegetation and soil to maximize the insulative layer of peat. The placement of extra large culverts is essential to accommodate spring runoff, often over relatively flat lands that in summer discharge little water. Controlling permafrost degredation and culvert icings to prevent spring flooding upslope of roads are essential to reduce road maintenance and to preserve vegetation. These and other stipulations and their enforcement have resulted in relatively few environmental problems in terms of impeded drainage and permafrost melt relative to the successful operation of the Dalton Highway that parallels the Alyeska oil pipeline (Fig. 6) and the Dempster Highway in the Yukon Territory and the Northwest Territories in Canada. The effect of vehicle and train traffic on the movement or large mammals has been studied in Scandinavia and North America. The results show that animals are turned back by frequent traffic and that when snows are deep, moose will move along railway lines only to be killed by oncoming trains.

As stated earlier, the lack of gravels in the High Arctic greatly limits the development of roads and pads for drilling permanent wells and supply camps. Sands have been substituted successfully in some areas and will continue to be used.

### B. Russia

Much of the supplies used in exploration and development of wells in the Russian arctic are moved by ship and barges along the arctic coast and on the major rivers. The oil and gas fields in western Siberia occur below huge areas of surface sands and thus sand-based roads are surfaced with concrete slabs welded together. Supply camps, production wells, and other support facilities built on a sand base are often surrounded by wetlands, heavily scarred with vehicle tracks (Fig. 15).

A limited network of railroads extends to several of the newer northern towns and petroleum fields.

**FIGURE 15** Sand road leading to a sand pad with nine gas-producing wells within the Yamburg gas field (Taz Peninsula, Russia).

Building and maintaining these railways over ice-rich permafrost and where sands are the major source of building material result in difficult engineering and environmental problems.

## V. CONCLUSIONS

The Arctic, which covers vast areas of northern Alaska, Canada, and Russia, contains rich deposits of heavy metals, coal, and petroleum. The 1970s and 1980s have seen major development of some of these resources, but this development is always tempered with the difficulties induced by severe cold and long winters, the presence of permafrost, massive runoff of melt water in spring, and huge expenses associated with bringing these resources to market.

Only strict environmental regulations can prevent serious problems of pollution and maintenance of intact pipelines, roads, and railways in lands where the threat of permafrost melt is an ever-present problem. These problems can be surmounted as they have been in Alaska and Canada. The present challenge is to help the Russians learn how to lessen the severe environmental impacts to their north, impacts that include concerns for the Nenets who depend upon their reindeer herds for food, clothing, and transportation. Much has been learned in 25 years on how to construct and maintain viable facilities in the north. Challenges remain on how best to proceed in the more hostile lands of northern Russia and the High Arctic of Canada.

## Bibliography


Alexander, V., and van Cleve, K. (1983). The Alaska pipeline: A success story. *Annu. Rev. Ecol. Syst.* **14,** 443–463.

Bliss, L. C. (1983). Modern human impact in the Arctic. *In* "Man's Impact on Vegetation" (W. Holzner, M. J. A. Werger, and I. Ikusima, eds.), pp. 213–225. London: Dr. W. Junk Pub.

Bliss, L. C. (1990). Arctic ecosystems: Patterns of change in response to disturbance. *In* "The Earth in Transition: Patterns and Processes of Biotic Impoverishment" (G. M. Woodwell, ed.) pp. 347–366. Cambridge: Cambridge Univ. Press.

Dyson, J. (1979). "The Hot Arctic." Boston: Little, Brown, and Co.

Luzin, G. P., Pretes, M., and Vasiliev, V. V. (1994). The Kola Peninsula: Geography, history and resources. *Arctic* **47,** 1–15.

Osherenko, G., and Young, O. R. (1989). "The Age of the Arctic: Hot Conflicts and Cold Realities." Cambridge: Cambridge Univ. Press.

Vilchek, G. E., and Bykova, O. Y. (1992). The origin of regional ecological problems within the northern Tyumen Oblast, Russia. *Arctic Alp. Res.* **24,** 99–107.

Walker, D. A., Cate, D., Brown, J., and Racine, C. (1987). Disturbance and recovery of arctic Alaskan tundra terrain: A review of recent investigations. U.S. Army Corps Engineers. CRREL Rept. 87-11.

Walker, D. A., and Walker, M. D. (1991). History and pattern of disturbance in Alaskan arctic terrestrial ecosystems: A hierarchical approach to analyzing landscape change. *J. Appl. Ecol.* **28,** 244–276.

# Insect Demography

***James R. Carey***
*University of California*



Insect demography is the study of insect populations and the processes that shape them. The field is organized around three main areas. The first is the life table, which is a detailed description of the mortality of a cohort giving the probability of dying and other statistics. The second is analysis of reproduction, which summarizes the lifetime and age-specific reproduction of cohorts. The third area is population, which is concerned with the consequences of birth and death rates on population growth rate, age distribution, and other population properties. The main tool used in the study of population process is stable population theory, which states that in the absence of migration, populations consisting of a single sex (usually female) subjected to fixed birth and death rates will eventually attain constant growth rates and age structures that are independent of initial conditions. This property—the tendency of models with fixed rates to converge to identical states that are independent of initial conditions—is referred to as ergodicity and is fundamental to a wide range of demographic population models.

## I. INTRODUCTION

Insect demography is concerned with four aspects of insect populations: (1) *size*—the number of individuals in the population; (2) *distribution*—the arrangement of the population in space at a given time; (3) *structure*—the relative frequency of age and sex groupings; and (4) *change*—the growth or decline of the total population or one of its structural units. The first three—size, distribution, and structure—are referred to as *population statics,* whereas the last—change—is referred to as the *population dynamics*.

The starting point for demography is a simple schematic referred to as the *life course,* the end points of which represent the birth and death of an individual. It is divided by age into phases or stages. The passage from one stage to another is termed an *event*. The sequence of events and the duration of intervening stages thoughout the life of the organism constitute the life course. Age is an important dimension of life and grouping individuals into age classes or at least distinguishing between young and old is useful. The age of egg hatch is denoted $\eta$ (eta), the age of pupation as $\pi$ (pi), the age of eclosion (first day of adulthood) as $\varepsilon$ (epsilon), the age of first reproduction as $\alpha$ (alpha), the age of last reproduction as $\beta$ (beta), and the oldest possible age as $\omega$ (omega) (see Fig. 1). The notation for adult traits, particularly $\alpha$, $\beta$, and $\omega$, are convention in demography, whereas the notation for the preadult are not. In

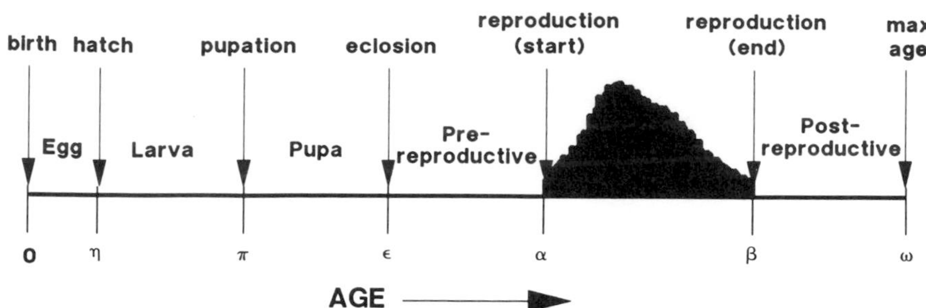


***FIGURE 1*** Schematic diagram of an insect life course. [Redrawn from J. R. Carey (1993). "Applied Demography for Biologists." Oxford University Press, New York.]

general, age is the characteristic, central variable in almost all demographic analysis and serves as a surrogate for more fundamental measures (e.g., physiological state and duration of exposure to risk). [*See* LIFE HISTORIES.]

Two levels above the individual are distinguished in demography. The first level is the *cohort,* defined as "a group of same-aged individuals" or, more generally, as "a group who experience the same significant event in a particular time period, and who can thus be identified as a group for subsequent analysis." For example, all individuals born on the same day are considered a cohort. The second level is the *population,* defined as a group of individuals coexisting at a given moment.

## II. INSECT LIFE TABLES

A life table is a detailed description of the mortality of a population that gives the probability of dying and various other statistics at each age. There are two general forms of the life table. The first is the *cohort life table,* which provides a longitudinal perspective in that it includes the mortality experience of a particular cohort from the moment of birth through consecutive ages until none remains in the original cohort. The second basic form is the *current life table,* which is cross-sectional. This table assumes a hypothetical cohort subject throughout its lifetime to the age-specific mortality rates prevailing for the actual population over a specified period.

Both cohort and current life tables may be either *complete* or *abridged.* In a complete life table the functions are computed for each day of life. An abridged life table deals with age intervals greater than one day such as over a complete stage (e.g., larval period), where precise determination of daily survival is difficult. The distinction between complete and abridged has to do with the length of the age interval considered. Both forms of the life table may be either *single decrement* or *multiple decrement.* The first of these lumps all forms of death into one and the second disaggregates death by cause.

### A. Complete, Cohort Life Table

One of the most common life tables used for the study of adult insects in the laboratory is the complete, cohort life table, an example of which is given in Table I for 1.2 million medflies. Data are gathered for this table by monitoring survival and mortality in a cohort from emergence through the death of the last individual. The data are organized by age $x$ (col. 1) and by the number surviving to age $x$, $N_x$ (col. 2). Five main life table parameters are computed using these data. The first function is $l_x$(col. 3), defined as the fraction of the original cohort surviving to age $x$. The formula for computing $l_x$ from the numbers alive at age $x$, $N_x$, is

$$l_x = \frac{N_x}{N_0},$$

where $N_0$ denotes the initial number in the cohort (1.2 million). For example (from Table 1), 0.98320

***TABLE I***
Complete Cohort Life Table for 1.2 Million Medflies

| Age (days) $x$ (1) | Number living $N_x$ (2) | Fraction surviving $l_x$ (3) | Period survival $p_x$ (4) | Period mortality $q_x$ (5) | Frequency of deaths $d_x$ (6) | Expectation of life $e_x$ (7) |
|---|---|---|---|---|---|---|
| 0 | 1,203,646 | 1.00000 | 1.00000 | 0.00000 | 0.0000 | 20.84 |
| 1 | 1,203,646 | 1.00000 | 0.99856 | 0.00144 | 0.0014 | 19.84 |
| 2 | 1,201,913 | 0.99856 | 0.99599 | 0.00401 | 0.0040 | 18.87 |
| 3 | 1,197,098 | 0.99456 | 0.99492 | 0.00508 | 0.0050 | 17.95 |
| 4 | 1,191,020 | 0.98951 | 0.99362 | 0.00638 | 0.0063 | 17.03 |
| 5 | 1,183,419 | 0.98320 | 0.99247 | 0.00753 | 0.0074 | 16.14 |
| — | — | — | — | — | — | — |
| 20 | 575,420 | 0.47806 | 0.90772 | 0.09228 | 0.0441 | 8.26 |
| — | — | — | — | — | — | — |
| 50 | 10,782 | 0.00896 | 0.84854 | 0.15146 | 0.0014 | 6.66 |
| — | — | — | — | — | — | — |
| 80 | 181 | 0.00015 | 0.93370 | 0.06630 | 0.0000 | 20.56 |
| — | — | — | — | — | — | — |
| 170 | 2 | 0.00000 | 1.00000 | 0.00000 | 0.0000 | 1.50 |
| 171 | 2 | 0.00000 | 0.00000 | 1.00000 | 0.0000 | 0.50 |
| 172 | 0 | 0.00000 | — | — | — | — |

of the original 1.2 million medflies survived to age 5 days, 0.47802 survived to age 50, and 0.00015 survived to age 80. This survival function is a monotonically decreasing function with age (Fig. 2).

The second and third main life table functions are $p_x$(col. 4) and $q_x$(col. 5), referred to as period survival and age-specific mortality, respectively. These functions are complements of each other as given in the equations

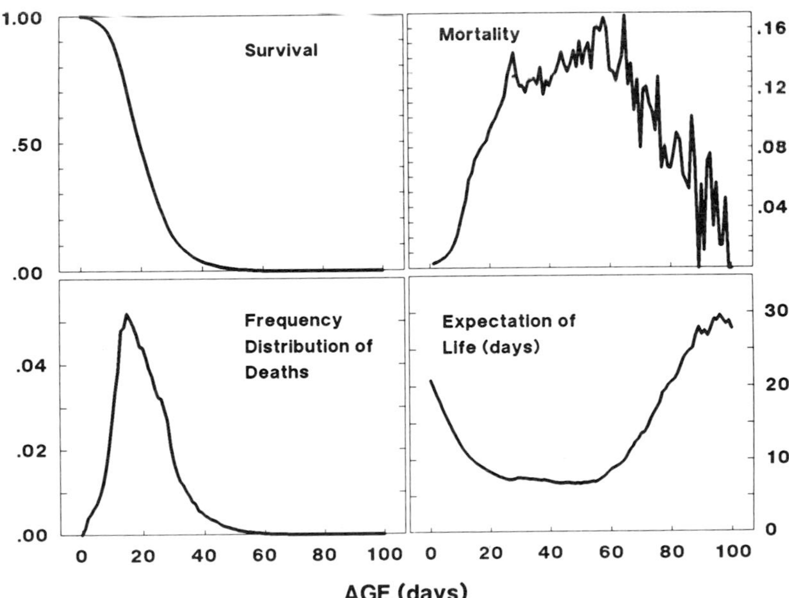


***FIGURE 2*** Four main life table functions for 1.2 million medflies. [Redrawn from J. R. Carey (1993). "Applied Demography for Biologists." Oxford University Press, New York.]

$$p_x = \frac{l_{x+1}}{l_x}$$

$$q_x = 1 - \frac{l_{x+1}}{l_x}$$

or

$$q_x = l - p_x.$$

As examples from Table 1 (col. 4 and 5), the fraction of medflies alive at age 5 that survived to age 6 was 0.99247 ($p_5$). Conversely, the fraction of medflies alive at age 5 that died prior to age 6 was 0.00753 ($q_5$). At Day 50, the fraction 0.84854 of those alive survived to Day 51 ($p_{50}$) and the fraction 0.15146 of those alive died prior to Day 51 ($q_{50}$). As shown in Fig, 2 the $q_x$ function increased from age 0 to age 20, then leveled off at around 0.15 to age 60, at which time it began decreasing. The $q_x$ function is fundamental to all other life table functions because its value at a specific age is independent of values at other ages.

A fourth important life table function is $d_x$ (col. 6), defined as the fraction of the original cohort that dies in the interval $x$ to $x + 1$. This function represents the frequency distribution of deaths in the cohort and, though usually of a different shape, is the analog of the "bell curve" in traditional statistics. The formula for this function is

$$d_x = l_x - l_{x+1}.$$

As an example from Table 1, 0.0074 of the original cohort died in the interval 5 to 6 days ($d_5$) and 0.0014 died in the interval 50 to 51 days ($d_{50}$). The frequency distribution of deaths is often skewed toward the younger ages (left) with a long right tail at the older ages. The reason for this long right tail is that few of the original individuals are alive at the most advanced ages. For example, nearly 50% of all deaths in the medfly cohort occurred prior to age 20 days and over 99% of all deaths occurred prior to age 50 days.

The last main life table function is life expectancy at age $x$, $e_x$(col. 7), defined as the number of days remaining to the average individual age $x$. The computational formula for this function is

$$e_x = \frac{1}{2} + \frac{l_{x+1} + l_{x+2} + l_{x+3} + \ldots + l_\omega}{l_x}.$$

For example, from Table 1 the life expectancy of a medfly at emergence, $e_0$, was 20.84 days, at age 50 ($e_{50}$) it was 6.66 days, and at age 80 ($e_{80}$) it was 20.56 days. The increase in remaining life expectancy at age $x$ at older ages as shown in Fig. 2 is due to the leveling off and decrease of the mortality at advanced ages.

Understanding the complete cohort life table is important because the notation and organizational framework used in constructing this table are fundamental to all other types of life tables and because understanding survival and mortality is basic to a wide range of biological subdisciplines, including applied and basic ecology, evolution and environmental biology.

## B. Cohort Mortality: Selected Topics

### *1. Force of Mortality*

The force of mortality at age $x$, denoted $\mu(x)$, is the instantaneous mortality rate representing the limiting value of the age-specific mortality rate when the age interval to which the rate refers becomes infinitessimally short. It is given as

$$\mu(x) = - \frac{dl(x)}{l(x)dx}.$$

Also known as the instantaneous death rate and hazard rate, $\mu(x)$ is used extensively in gerontology, actuarial mathematics, demography, and the analysis of life spans because it has many useful properties, the most important of which is that its value does not depend on the length of the age interval. In contrast, the magnitude of age-specific mortality depends on the length of the age interval. A formula used to estimate the force of mortality at age $x$ is given as

$$\mu_x = \frac{1}{2n} ln_e \left( \frac{l_{x-n}}{l_{x+n}} \right),$$

where $n$ denotes the age band width. For example, if $n = 1$ then the force of mortality is computed using survival values one time unit before and one time unit after $x$.

### 2. Gompertz Formula

In 1825, the English actuary Benjamin Gompertz developed a model in an attempt to describe the observed age-specific pattern of mortality. Gompertz reasoned that the inability to withstand destruction increased exponentially with age. The Gompertz formula is given as

$$\mu_x = ae^{bx},$$

where $\mu_x$ is the force of mortality at age $x$, $a$ denotes the initial mortality rate, and $b$ is the exponential mortality rate coefficient. The Gompertz model approximates the medfly mortality data shown in Fig. 2 for the first 2 weeks using parameter values of $a = 0.01$ and $b = 0.09$. This model does not describe the data at the remaining ages because of the leveling off and decrease in mortality. However, the Gompertz formula is widely used in gerontology and demography to model mortality in species whose mortality patterns increase exponentially with age.

### 3. Heterogeneity

The concept of heterogeneity stems from the notion that not all individuals are equally endowed at birth (or emergence) for lifetime potential—some individuals are "frail" and others are "robust." Cohort heterogeneity is invoked as an explanation for various patterns of mortality using the following reasoning: As the cohort ages, it becomes more and more selected because individuals with higher death rates will die out in greater numbers than those with lower death rates, thereby transforming the population into one consisting mostly of individuals with low death rates. Thus mortality patterns may level off at advanced ages because the only individuals remaining alive are those with low death rates. This is one possible explanation for the leveling off and decrease in medfly mortality rate at advanced ages shown in Fig. 2.

## III. REPRODUCTION

Reproduction can be characterized in two ways: the process that results in offspring and the per capita *rate* of offspring production in a given period of time. The first is physiological and the second is demographic. *Reproductive rate, renewal rate, recruitment rate,* and *natality rate* all denote the same thing and are frequently used synonymously.

### A. Per Capita Reproductive Rates

The per capita age-specific reproductive rate over specified age intervals or over a cohort's lifetime is referred to as the *reproductive age schedule.* A *gross schedule* is one in which mortality is not taken into account and a *net schedule* weights reproduction by the proportion or the number in the cohort that survive to each age class.

An example of the age-specific reproductive schedule for female medflies is given in Table II. The data are organized in five columns starting with age $x$ (col. 1). Midpoint survival (col. 2) gives the average of exact survival to age $x$ ($l_x$) and exact survival to age $x + 1$, ($l_{x+1}$). That is, $L_x = (l_x + l_{x+1})/2$. The sum of this column gives the expectation of life at emergence for the average female medfly. In the example, the average female medfly lived 22.6 days.

The number of offspring produced by the average female age $x$, denoted $M_x$, is given in column

**TABLE II**
Age-Specific Reproduction in 1000 Female Medflies

| Age $x$ (1) | Midpoint survival $L_x$ (2) | Eggs/day $M_x$ (3) | Net maternity $L_xM_x$ (4) | Age-weighted maternity $xL_xM_x$ (5) |
|---|---|---|---|---|
| 0 | 1.000 | 0.0 | 0.000 | 0.00 |
| 1 | 0.998 | 0.0 | 0.000 | 0.00 |
| ⋮ | ⋮ | ⋮ | ⋮ | ⋮ |
| 10 | 0.988 | 28.4 | 28.06 | 280.59 |
| 11 | 0.932 | 27.8 | 25.91 | 285.01 |
| 12 | 0.922 | 29.1 | 26.83 | 321.96 |
| ⋮ | ⋮ | ⋮ | ⋮ | ⋮ |
| 104 | 0.001 | 2.1 | 0.00 | 0.00 |
| 105 | 0.000 | 0.00 | 0.00 | 0.00 |
| Totals | 22.631 | 697.5 | 439.47 | 5216.51 |

3. The lower case, $m_x$, is used to denote the number of female offspring produced by the average female age $x$. The sum of this column gives the total fertility rate, denoted TFR and defined as the total number of offspring (both female and male) produced by a female in a hypothetical cohort that lived to the last day of possible life. In the example, this hypothetical female produced 697.5 offspring.

The product of column 2 (midpoint survival) and column 3 (offspring produced at age $x$) is denoted $L_xM_x$ and is referred to as the net maternity schedule. The sum of this column is the net fertility rate, NFR. This parameter is defined as the number of offspring that the average female will produce in her lifetime. In the example, the average female produced nearly 440 offspring in her lifetime. The product of columns 1 and 4 is contained in column 5 and weights the net fertility schedule by age. The sum of this column is used to compute the mean age of parenthood in the cohort:

$$\text{Mean age of parenthood} = \frac{1}{\text{NFR}}\sum_{x=0}^{\omega} xL_xM_x.$$

In the example, the mean age of parenthood is (sum col. 5)/(sum col. 4) = 5,216.5/439.5 = 11.9 days.

## B. Reproductive Heterogeneity

Not all individuals in a cohort produce the same number of offspring at a given age. Parameters that express various aspects of this heterogeneity include *reproductive interval,* which gives the daily frequency of offspring production among individuals, *daily parity,* which denotes the fraction of the cohort that produce specified numbers of offspring at a given age, *cumulative partiy,* which expresses the fraction of living individuals within a cohort that have laid specified sum total offspring at given ages, and the *concentration of reproduction,* which gives the frequency distribution of lifetime reproduction ranked by individual. The offspring produced by insect females are often deposited in clutches and thus *clutch size* designates the offspring groupings and can be expressed as an age schedule. Most insects produce more than one *offspring type,* the most common of which is sex.

An example of daily parity for the medfly is presented in Table III. The numbers in each column represent the percentage of females alive on a given day that produce 0, 1 to 25, 26 to 50, over 50 eggs. This abridged table reveals that no females laid any eggs until age 3, at which time slightly less than 20% produced from 1 to 25 eggs. Anywhere from 10 to 20% of mature females produced no eggs on a given day and only a small percentage ever produced over 50 eggs on any one day. In general, examination of within-cohort heterogeneity in reproduction provides additional insight into the extent of reproductive variation among individuals.

# IV. POPULATION

Population has a variety of meanings, including "the collection of entities about which information is sought" used in statistics, "an interbreeding group of organisms" used in ecology and population biology, and "a group of individuals coexisting at a given moment" used in demography. From the perspective of population growth, the meaning extends beyond population merely as a collection of individuals. Rather population implies individuals that produce offspring and eventually die. Indeed, the study of population must be grounded in understanding the fundamental and simultaneous processes of birth and death.

**TABLE III**
Percentage of Living Medfly Females on Selected Days Whose Egg Production Was 0 or Ranged from Either 1 to 25, 26 to 50, or Was Greater Than 50

| Age (days) | Range of offspring production | | | | Total |
|---|---|---|---|---|---|
| | 0 | 1–25 | 26–50 | >50 | |
| 0 | 100.0 | 0.0 | 0.0 | 0.0 | 100.0 |
| 1 | 100.0 | 0.0 | 0.0 | 0.0 | 100.0 |
| 2 | 100.0 | 0.0 | 0.0 | 0.0 | 100.0 |
| 3 | 81.2 | 18.8 | 0.0 | 0.0 | 100.0 |
| ⋮ | ⋮ | ⋮ | ⋮ | ⋮ | ⋮ |
| 10 | 8.7 | 37.0 | 54.3 | 0.0 | 100.0 |
| ⋮ | ⋮ | ⋮ | ⋮ | ⋮ | |
| 20 | 18.2 | 31.8 | 50.0 | 0.0 | 100.0 |
| ⋮ | ⋮ | ⋮ | ⋮ | ⋮ | ⋮ |
| 29 | 17.6 | 20.7 | 58.8 | 2.9 | 100.0 |

### A. The Balancing Equation

The simplest of all population models is the crude rate model, which is based on the *balancing equation*. This equation relates the total population one day to the total population the preceding day. Let $\text{Pop}_0$ denote the initial number in the population, then the balancing equation is given as

$$\begin{aligned}\text{Pop}_1 = \ &\text{Pop}_0\\ &+ \text{births}\\ &- \text{deaths}\\ &+ \text{in-migrants}\\ &- \text{out-migrants}.\end{aligned}$$

This model partials out the relative contribution of birth, death, and migration. For simplicity, migration is typically neglected in demographic analysis. Because births and deaths represent totals, these terms can be reexpressed as

$$\begin{aligned}\text{births} &= \text{Pop}_0 \times b\\ \text{deaths} &= \text{Pop}_0 \times d,\end{aligned}$$

where $b$ and $d$ denote per capita birth and death rates, respectively. Substituting these terms into the equation yields (excluding migration terms)

$$\begin{aligned}\text{Pop}_1 &= \text{Pop}_0 + (\text{Pop}_0 \times b) - (\text{Pop}_0 \times d)\\ &= \text{Pop}_0(1 + b - d).\end{aligned}$$

Note that if $b - d = 0$ the population at time $t$ will equal the population at time $t + 1$, if $b > d$ the population will increase, and if $b < d$ the population will decrease. The population at time $t = 2$ (i.e., $\text{Pop}_2$) is determined as

$$\begin{aligned}\text{Pop}_2 &= \text{Pop}_1(1 + b - d)\\ &= \text{Pop}_0(1 + b - d)(1 + b - d)\\ &= \text{Pop}_0(1 + b - d)^2.\end{aligned}$$

The relationship for $t$ number of time units is

$$\text{Pop}_t = \text{Pop}_0(1 + b - d)^t.$$

Substituting $N_t$ for $\text{Pop}_t$ and $\lambda$ for $(1 + b - d)$ yields

$$N_t = N_0\lambda^t,$$

which represents the discrete (geometric) version of the basic model of population growth. This equation is used to model insect populations without overlapping generations. Thus $\lambda$ in this context denotes the per generation growth rate. This model illustrates the most fundamental property of population growth in which the number in the population changes geometrically. The distinction between an arithmetic versus a geometric series of numbers is that in the former the numbers change by a common difference, whereas numbers in the latter change each time step by a common ratio. For example, the series 1, 4, 7, 10, and 13 is arithmetic because the numbers differ from the preceding one by a common difference of 3 ($4 - 1 = 3$, $13 - 10 = 3$, etc.). In contrast, the series 1, 3, 9, 27, and 81 is geometric because each number differs from the preceding one by a common ratio of 3 ($3/1 = 3$, $81/27 = 3$, etc.). The basic population model is geometric because $\lambda$ is the common ratio of the numbers at time $t + 1$ and time $t$.

The crude rate model has three assumptions: (1) the population is not structured by age (homogeneity assumption); (2) birth and death rates remain fixed; and (3) closed population (i.e, no migration). The only major conceptual difference between the crude rate model and the stable population model (Lotka's equation) covered later is that of homogeneity. Age structure adds a more realistic and interesting dimension but does not change the manner of growth (geometric). Although this simple equation is unrealistic in the sense that no population can change at a constant rate forever, it is fundamental to demography by defining population change as a compounding process, by establishing a foundation for examining population growth patterns over short periods, and by providing the initial framework from which to build more complicated models such as those with feebuck terms (logistic equation).

### B. Stable Population Theory

The fundamental basis of stable population theory was formulated by the Swiss mathematician

L. Euler and brought to prominence by the actuary A. J. Lotka, who proposed the equation

$$1 = \int_0^\omega e^{-rx} l(x) m(x) dx,$$

where $e$ is the exponential function, $r$ is the intrinsic rate of increase, $l(x)$ is survival to age $x$, and $m(x)$ is the number of female offspring to a female age $x$. The objective of the stable model is to trace the dynamic characteristics of a population that starts off with an arbitrary age structure and is submitted from that moment on to a specified demographic regime. The assumptions of the stable model are basically the same as those of the crude rate model except age structure is considered. Like the crude rate model, the stable model assumes fixed birth and death rates, closed population, and only one sex (female). The stable model has two conclusions with regard to populations with fixed birth and death rates. First, that the age distribution of a population is completely determined by the history of fertility and mortality rates. Second, that particular schedules of birth and death rates set in motion forces that make the age structure and the rate of increase point toward an inherent steady state that is independent of initial conditions. [*See* POPULATION REGULATION.]

The stable population model is useful in four respects: (1) it provides a simple starting point for addressing fundamental questions about the process of population renewal; (2) it identifies the two most important population parameters—age distribution and growth rate—their interdependence, and their relationship with the cohort parameters of birth and death; (3) its output suggests a direction for population growth and a population age pattern. The sign of growth rate and the general skew of the age distribution are often more important than are the actual numbers; and (4) it is capable of providing actual numbers that may serve as a frame of reference for the rate of growth, the number in each of the age classes, and the overall population at a specified time.

An organizational and computational framework for determining the intrinsic rate of increase, $r$, and the stable age distribution, SAD, is given in Table IV using a hypothetical population of six age classes. Data are organized by age $x$ (col. 1) and include midpoint survival, $L_x$ (col. 2), female offspring at age $x$, $m_x$ (col. 3), and net maternity, $L_x m_x$ (col. 4). An estimate for $r$ can be made using the equation

$$r \approx \frac{\ln_e R_0}{T},$$

where $R_0$ is the sum of the net maternity function referred to as the net reproductive rate and $T$ is the mean age of parenthood (=3.51). The $r$ estimate using this formula is 0.499, which is quite close to its exact value of $r = 0.512$ that is obtained using numerical methods.

The fraction age $x$ in the stable population, denoted $c_x$, is determined using the formula given in Table V for $c_x$. For example, the fraction of the total population in age class 3 in Table IV is computed using the entry corresponding to age 3 in column 6 divided by the sum of column 6. That is, $c_3 = (0.121/2.073) = 0.058$. The interpretation of this specific result is that 0.058 of the total population will be contained in age class 3 at stability.

A summary of the main parameters and the associated formulae that are derived from the Lotka equation is presented in Table V. The *net reproductive rate* is defined as the number of female offspring produced by the average newborn female. An alternative definition is the factor by which a population changes each generation. The *intrinsic rate of increase* is the instantaneous rate of increase in the stable population. The *finite rate of increase* is the factor by which a stable population increases each time step. This parameter is the geometric rate of increase. For example, a stable medfly population that possesses a finite rate of increase of 1.115, as in the example, will increase its numbers each day by this factor or 11.5%. The *intrinsic birth rate* is the instantaneous rate of death. The doubling time of a population expresses the time required for the population to increase by two-fold. The *mean generation time* ($T$) of a population is defined as the time required

**TABLE IV**
Computation of Stable Population Parameters for a Hypothetical Population (Intrinsic Rate of Increase, $r = 0.512$).

| Age $x$ (1) | Midpoint survival $L_x$ (2) | Female offspring $m_x$ (3) | Net maternity $L_x m_x$ (4) | Stable net maternity function $e^{-rx}L_x m_x$ (5) | Exponential survival product $e^{-rx}L_x$ (6) | Stable age distribution $c_x$ (7) |
|---|---|---|---|---|---|---|
| 0 | 1.000 | 0.0 | 0.000 | 0.000 | 1.000 | 0.482 |
| 1 | 0.987 | 0.0 | 0.000 | 0.000 | 0.592 | 0.285 |
| 2 | 0.856 | 0.0 | 0.000 | 0.000 | 0.307 | 0.148 |
| 3 | 0.562 | 5.6 | 3.147 | 0.677 | 0.121 | 0.058 |
| 4 | 0.322 | 7.2 | 2.318 | 0.299 | 0.042 | 0.020 |
| 5 | 0.145 | 2.1 | 0.305 | 0.024 | 0.011 | 0.005 |
| 6 | 0.000 | 0.0 | 0.000 | 0.000 | 0.000 | 0.000 |
| Totals | | 14.9 | 5.770 | 1.000 | 2.073 | 1.000 |

**TABLE V**
Symbols, Parameters, Formulae, and Example Values for the Mediterranean Fruit Fly for Stable Population Parameters Derived from the Lotka Equation

| Symbol | Parameter | Formula | Example value |
|---|---|---|---|
| Growth rates | | | |
| $R_0$ | Net reproductive rate | $\sum_{x=\alpha}^{\beta} l_x m_x$ | 414.2 female offspring |
| $r$ | Intrinsic rate of increase | $-\sum_{x=\alpha}^{\beta} e^{-rx} l_x m_x$ | 0.109 |
| $b$ | Intrinsic birth rate | $1/\sum_{x=\alpha}^{\beta} e^{-rx} l_x$ | 0.193 |
| $d$ | Intrinsic death rate | $b - r$ | 0.084 |
| $\lambda$ | Finite rate of increase | $e^r$ | 1.115 per day |
| Growth times | | | |
| DT | Doubling time | $\frac{\ln 2}{r}$ | 6.4 days |
| $T$ | Mean generation time | $\frac{\ln R_0}{r}$ | 40.8 days |
| | (alternative formula)[a] | $\frac{\sum_{x=\alpha}^{\beta} x l_x m_x}{\sum_{x=\alpha}^{\beta} l_x m_x}$ | 41.1 days |
| Age distribution | | | |
| $c_x$ | Stable age distribution | $\frac{e^{-rx} l_x}{\sum_{x=0}^{\omega} e^{-rx} l_x}$ | — |
| — | Egg stage | $100 \times \sum_{x=0}^{3} c_x$ | 29.8% |
| — | Larval stage | $100 \times \sum_{x=4}^{12} c_x$ | 49.7% |
| — | Pupal stage | $100 \times \sum_{x=13}^{23} c_x$ | 15.6% |
| — | Adult stage | $100 \times \sum_{x=24}^{\omega} c_x$ | 4.9% |

[a] Also referred to as the mean age of parenthood.

for the population to change by a factor of the net reproductive rate. An alternative definition of mean generation time is the mean age of parenthood, which characterizes $T$ as the mean interval separating births of one generation from those of the next. The *stable age distribution* (SAD) is the fraction of the total stable population that is age $x$. The stable age distribution and the intrinsic rate of increase are both long-run consequences of closed populations subject to constant birth and death rates.

## C. Social Insect Populations: A Simple Model

Eusocial insect colonies such as honeybee colonies constitute a special kind of population in that virtually all births are directly attributable to a single individual, the queen, whereas all deaths are attributable to the group. They are subject to the "balancing equation" like any population. But unlike populations in which each female has the potential to reproduce, the contribution toward growth through births due to the *individual* (the queen) is offset by the sum of deaths in the *group*. Consequently, colony size cannot exceed the point where the number of deaths per day in the colony is greater than the maximum daily number of offspring that a queen is capable of producing. Colony growth will only occur after all individuals that die are first replaced.

A simple model of this relationship is derived as follows. Let $e_0$ denote the worker expectation of life at birth. Then $(1/e_0) = d_w$ denotes the per capita number of worker deaths in the stationary population. For example, if an individual in a colony of 1 million lives an average of 45 days, then a total of 22,222 deaths will occur each day (1 million divided by 45). Because there are 1440 minutes in one day, a queen must produce around 15 eggs per minute (22,222/1440) or one egg every 4 seconds just to replace the number that die in this hypothetical colony.

The level of egg production required for queens to maintain the colony is given as

$$\text{(Per capita deaths)} \times \text{(Number in colony)}$$

Let $b_q$ denote the maximum number of eggs that a queen can produce in one day. Because the expression $d_w$ gives the per capita deaths, the product of this term and the maximum number possible in a colony, denoted $N^*$, will give the number of eggs that a queen must produce. This relationship is expressed as

$$d_w = N^* b_q$$
$$N^* = \frac{b_q}{d_w}.$$

In words this expression states that the upper colony size limit is equal to the product of the expectation of life of workers (days) and the maximum daily egg production rate of the queen.

Two implications emerge from this model. First, an upper limit for colony size must exist owing to demographic constraints. Even if physiological reproductive limits are not considered, a finite amount of time is needed for workers to pick up the eggs for placement in the brood chamber, as with ants and termites, or for queens to move between cells and oviposit, as is the case for bees and wasps. Second, once the maximum colony size is attained, the only way for growth to continue is by the addition of more reproductives (queens). Multiple queen colonies (polygyny) are common in some species of termites and ants. This is the functional equivalent of budding. Queen addition in honey-bees results in (or is the cause of) swarming. A major point here is that the only way that they can increase in number is by colony fission rather than by colony growth.

## D. Population Projection

A matrix formulation of the Lotka model is known as the Leslie matrix. It is useful for illustrating and studying the transient properties of populations as they converge to the stable state and provides a technique of cohort-component projection. The Leslie matrix is given as

$$\begin{bmatrix} F_0 & F_1 & F_2 & \cdots & F_{\omega-1} & F_\omega \\ p_0 & 0 & 0 & \cdots & 0 & 0 \\ 0 & p_1 & 0 & \cdots & 0 & 0 \\ 0 & 0 & p_2 & \cdots & 0 & 0 \\ \vdots & \vdots & \vdots & \cdots & \vdots & \vdots \\ 0 & 0 & 0 & \cdots & p_{\omega-1} & 0 \end{bmatrix} \begin{bmatrix} N_{0,t} \\ N_{1,t} \\ N_{2,t} \\ N_{3,t} \\ \vdots \\ N_{\omega,t} \end{bmatrix} = \begin{bmatrix} N_{0,t+1} \\ N_{1,t+1} \\ N_{2,t+1} \\ N_{3,t+1} \\ \vdots \\ N_{\omega,t+1} \end{bmatrix},$$

where the top row gives the birth elements $F_x = (m_x + p_x m_{x+1})/2$, the subdiagonals $p_x = L_{x+1}/L_x$ are the period survival elements, and the vectors $N_{x,t}$ and $N_{x,t+1}$ denote the numbers at age $x$ at times $t$ and $t + 1$, respectively. A population is projected through time by first entering an initial number of individuals into one or more age classes and multiplying the Leslie matrix by the age vector through a process of one-step iteration and resubstitution. As an example, consider a population with three age classes starting with $N_{0,t} = N_{1,t} = N_{2,t} = 1$ and Leslie matrix elements of $F_0 = 0$, $F_1 = 3$, $F_2 = 2$, $p_0 = 0.7$, and $p_1 = 0.4$. Two iterations may be computed as follows:

$$\begin{bmatrix} F_0 & F_1 & F_2 \\ p_0 & 0 & 0 \\ 0 & p_1 & 0 \end{bmatrix} \begin{bmatrix} N_{0,t} \\ N_{1,t} \\ N_{2,t} \end{bmatrix} = \begin{bmatrix} N_{0,t+1} \\ N_{1,t+1} \\ N_{2,t+1} \end{bmatrix} \quad \text{3-age-class Leslie matrix}$$

$$\begin{bmatrix} 0 & 3 & 2 \\ 0.7 & 0 & 0 \\ 0 & 0.4 & 0 \end{bmatrix} \begin{bmatrix} 1.0 \\ 1.0 \\ 1.0 \end{bmatrix} = \begin{bmatrix} 5.0 \\ 0.7 \\ 0.4 \end{bmatrix} \quad \text{Iteration \#1}$$

$$\begin{bmatrix} 0 & 3 & 2 \\ 0.7 & 0 & 0 \\ 0 & 0.4 & 0 \end{bmatrix} \begin{bmatrix} 5.0 \\ 0.7 \\ 0.4 \end{bmatrix} = \begin{bmatrix} 2.9 \\ 3.5 \\ 0.3 \end{bmatrix}. \quad \text{Iteration \#2}$$

The results for selected time periods between 0 and 20 days are presented in Table VI, which shows the approach to stability (constancy) of growth rate and age distribution. The growth rate during the first five days ranged from 1.1-fold to nearly 2.4-fold, whereas by Days 10 and 11 its fluctuations ranged only from 1.48 to 1.64. By Days 18 to 20, growth rate differed by only 0.04 on successive days. Approximately 63% of the total population was contained in age class 0 at 18 to 20 days. In contrast, the percentage in this age class of 1 to 5 days ranged from 43 to 82%. This damping of growth rate illustrates convergence to the stable age distribution (SAD) and the intrinsic rate of increase ($r$).

**TABLE VI**

Iterations of Projection Matrix for 20 Time Steps Showing the Number within Each Age Class, the Total Number at Each Time Period, and the Factor by Which the Population Changes, $N_{t+1}/N_t$ ($= \lambda$)

| Time | Number at age $x$: $N_{0,t}$ | $N_{1,t}$ | $N_{2,t}$ | Total $N_t$ | $\lambda$ $N_{t+1}/N_t$ |
|---|---|---|---|---|---|
| 0 | 1.0 | 1.0 | 1.0 | 3.0 | 2.03 |
| 1 | 5.0 | 0.7 | 0.4 | 6.1 | 1.10 |
| 2 | 2.9 | 3.5 | 0.3 | 6.7 | 2.17 |
| 3 | 11.1 | 2.0 | 1.4 | 14.5 | 1.20 |
| 4 | 8.9 | 7.7 | 0.8 | 17.4 | 1.96 |
| 5 | 24.9 | 6.2 | 3.1 | 34.2 | 2.39 |
| ⋮ | | | | | ⋮ |
| 10 | 318.4 | 119.6 | 37.5 | 475.5 | 1.8 |
| 11 | 433.9 | 222.9 | 47.8 | 704.6 | 1.64 |
| ⋮ | | | | | ⋮ |
| 18 | 6,737.8 | 3,143.4 | 760.4 | 10,641.7 | 1.59 |
| 19 | 10,951.1 | 4,716.5 | 1,257.4 | 16,925.0 | 1.55 |
| 20 | 16,664.2 | 7,665.8 | 1,886.6 | 26,216.6 | 1.58 |

A more realistic example of projection is given in Fig. 3 for a medfly population starting with preovipositional adults. Two aspects of this projection merit comment. First, the change in the proportion of the total population in each stage was

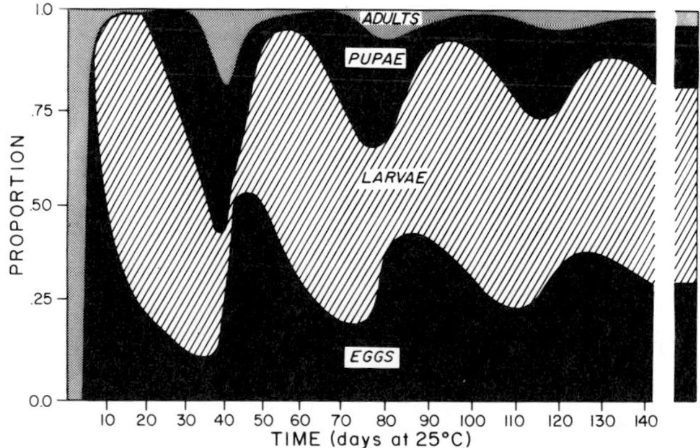


**FIGURE 3** Convergence to the stable age distribution in the medfly. [With permission from J. R. Carey, (1982). *Ecol. Modelling* **16,** 125–150.]

wavelike and variable, particularly at the early stages. For example, the proportion in the larval stage varied from over 75% at around 3 weeks to less than 10% at 6 weeks. Similarly, the proportion in the adult stage varied from less than 1% to over 20%. This variability illustrates the effects of initial age structure on population waves—the boom and bust cycles of a population when its age structure departs from the stable distribution. Second, the proportion of the total population in each stage damped with each generation. This damping occurred because of generation overlap—daughters, mothers, and even grandmothers were reproducing simultaneously and, consequently, individuals were present in all preadult age classes. The total number of offspring produced by all reproducing females approached a fixed number relative to the total in the population. And because death rates were constant at each age, this fixed proportion of newborn "matured" into fixed proportions at each subsequent age. The proportions at the far right in Fig. 3 correspond to the medfly stable age distribution given in Table V.

## E. Fundamental Properties of Populations

### *1. Age Structure Transience*

If two populations, one with fixed birth and death rates and the other with these rates changing, were to begin with a narrow age distribution, say a single female, each would undergo a sequence of demographic changes. As current births reached maturity, a new surge of births would "echo" the initial female's production of offspring, and this process would continue. In populations with fixed schedules, these echoes would be periodic and distinct initially but they would eventually become damped. In populations with unfixed schedules the surges may or may not be distinct or periodic depending on the specific features of the changing schedules and, by definition, would never become stable. The main point is this—variation in age structure and growth rate will exist in both populations. A population growing with fixed schedules not yet converged to a stable state may produce patterns that are virtually indistinguishable over a short period from those produced by a population subject to changing schedules.

### *2. Independence of Initial Conditions*

As a past age distribution becomes more remote, its form makes less and less difference to the shape of the current age distribution. The same factors that cause the transient effects of an initial age distribution to disappear from the stable population will operate for any time path of fecundity and mortality rate. After a suitably long period, the effect of an initial age distribution is swamped by the cumulative effect of the time pattern of vital rates. In short, the age distribution of any closed population is entirely determined by the mortality and fecundity rates of recent history. Hence given any observed age structure, it is impossible to determine either the initial population's size or the age structure.

### *3. Mortality Declines and Life Expectancy*

Generally speaking, mortality decline saves many young lives when mortality is higher, which contributes more to gains in life expectancy. This is often referred to as "squaring" the survival curve. When mortality is lower, mortality declines save mostly older lives, which contribute less to gains in life expectancy.

### *4. Fertility and Mortality*

Fertility differences usually have a far greater impact on current age distributions than do mortality differences. This is because the role of fertility in shaping age distributions is simpler than that of mortality by the fact that the differences in fertility operate in a single direction. In contrast, mortality differences have only second-order effects on age distributions. That is, a change in mortality tends to change all cohorts, implying a small effect of mortality on the immediate age distribution. When mortality changes in a gradual and monotonic fashion, the age distribution tends to adapt continuously and closely to current mortality conditions. Thus the age structure tends to differ very little from that which would have resulted from the existence of the indefinite past of current mortality conditions.

### *5. Speed of Convergence*

Speed of convergence refers to the process of convergence of a population from arbitrary initial conditions to the stable form. Three factors determine the rate of convergence: (1) for a fixed value of net reproductive rate, the speed of convergence increases as the mean of the net maternity function decreases; (2) for a fixed shape of the net maternity function ($l_x m_x$), the speed of convergence increases as the value of net reproductive rate increases. In other words, the higher the rate of increase, given the same pattern of reproduction, the faster the population will converge to stability. (3) The speed of convergence to stability depends not on the shape of the net maternity function but on the shape of the stable net maternity function (i.e., net maternity weighted by the exponential term, i.e., $e^{-rx} l_x m_x$). The smaller the mean of the stable net maternity function, the faster the speed of convergence.

### *6. Population Momentum*

Like physical objects, which have a tendency to continue moving once in motion, increasing populations have a tendency to continue growing. This is referred to in demography as population momentum. It is the extent to which a population continues to change in size after it adopts replacement-level rates of mortality and fertility. The momentum of a population can be regarded as the opposite of the intrinsic rate of increase, which indicates the growth rate implicit in a set of vital schedules and independent of initial age structure. In contrast, population momentum describes the growth potential due to age structure alone.

## F. Ergodicity: The Unifying Thread of Population Models

A large and significant body of theory, methods, and applications in demography is concerned with-the transitions that individuals experience during their lifetime as they pass from one state of existence to another, for example, from being alive to being dead, from being age 20 to age 21, or from living in one region to living in another region. These refer to the "states of existence" whereas the transition rates refer to the "rules of transition." By subjecting the "states of existence" to specified "rules of transition," a set of properties emerge that apply to a wide range of demographic population models. This common thread is ergodicity—the property of having the present state of a population independent of its makeup in the remote past. The unifying ergodic threads for each of a number of different demographic population models are described as follows.

### *1. The Leslie Matrix*

The *age structure* (SAD) and *growth rate* ($r$) of two populations beginning with different initial conditions will be fixed and identical when projected into the distant future if they are both subject to the same regime of constant age-specific birth and death rates. This property of the deterministic case is referred to as *strong ergodicity*.

### *2. Stochastically and Deterministically Varying Vital Rates*

The *age structure* and *long term growth rate* of two populations structured by age and beginnning with different initial conditions will be similar when projected into the distant future if they are both subject to the same *set* of age-specific birth and death *probabilities*. This property of stochastic models is referred to as *weak stochastic ergodicity*. The age structure and long-term growth rate of two populations structured by age and beginning with different initial conditions will follow identical trajectories when projected into the distant future if they are both subject to the same *deterministic sequence* of age-specific birth and death rates. In this case an external cycle is superimposed on an intrinsic or internal demographic one (the generation cycle). This property is referred to as *weak deterministic ergodicity*.

### *3. Two-Sex Model*

The *age structure, growth rate,* and *sex composition* (intrinsic sex ratio) of two populations structured by age and sex beginning with different initial conditions and subject to the same regime of constant sex-specific birth and death rates will be constant and identical when projected into the distant future.

### 4. Multiregional Models

The *region-specific age structure, regional shares* of the total population, and *overall population growth rate* of two populations identically divided into regions that begin with different initial conditions will be constant and identical when projected into the future if they are subject to the same fixed regime of age-specific birth, death, and migration rates.

### 5. Structured by Age and Genotype

The *age composition, growth rate,* and *genotype frequency* of two populations structured by age and allelic traits beginning with different initial conditions will be identical when projected into the distant future if they are both subject to the same selection regime and fixed schedules of age-specific birth and death rates. Delays in attainment of genotypic frequencies in populations are caused by age structure effects and reproductive lags due to the necessary inclusion of developmental times from newborn to mature adult.

### 6. Social Insects (Honeybees)

Social insect populations, such as the honeybee populations, structured by individual age within a colony as well as by age of colony that are subject to a fixed regime of worker survival rates, queen reproduction, swarming thresholds, swarming properties, and colony survival rates will eventually attain a *stable individual age-by-colony distribution,* a *stable distribution of colonies by age,* and a *constant rate of colony increase.* The resulting population will consist of a fixed overall age distribution of individual worker bees within a fixed distribution of different-aged colonies.

There are several reasons why an understanding of these generic properties of demographic models is important. First, only by understanding the general properties of models is it possible to begin to build general theories of population. For example, analogs of the intrinsic rate of increase and the stable age distribution are properties of all demographic models that can be framed in matrix form and with ergodic properties. In addition, it is possible to address questions of the common characterisitics of the modeled population, including growth rates, structure, momentum, convergence rates, sensitivity to parameter changes, and so forth. Second, a powerful set of mathematical and statistical tools can be brought to bear on all models. These tools do not have to be rediscovered for each new model. Third, understanding the interconnections among the demographic models fosters new perspectives and applications. For example, knowledge of stochastic models may suggest ways to incorporate stochasticity into multiregional models and, in turn, anticipate the "hybrid" properties of stochastic and multiregional models. Fourth, data available for one model might be directly applicable to another, thus reducing the need for completely new sets of expensive data.

## Glossary

**Cohort** Group of same-aged individuals or, more generally, a group of individuals who experience the same significant event in a particular time period.

**Demography** Study of populations and the processes that shape them.

**Ergodicity** Tendency of population models with fixed rates to converge to identical states that are independent of initial conditions.

**Force of mortality** The instantaneous death rate in a population.

**Intrinsic rate of increase** Growth rate of a closed population that has been subject to constant age-specific birth and death rates over a sustained period.

**Life table** Device for organizing data on death and for providing detailed descriptions of mortality, survival, and expectation of life by age class for a population.

**Mean generation time** Time required for a stable population to change by a factor equal to the net reproductive rate, or the mean age of parenthood.

**Net reproduction rate** Number of female offspring produced by the average newborn female in her lifetime.

**Stable age distribution** Age structure of a closed population that has been subject to constant age-specific birth and death rates over a sustained period.

## Bibliography

Carey, J. R. (1993). "Applied Demography for Biologists with Special Emphasis on Insects." New York: Oxford University Press.

Carey, J. R., Liedo, P., Orozco, D., and Vaupel, J. W. (1992). Slowing of mortality rates at older ages in large medfly cohorts. *Science* **258,** 457–461.

Caswell, H. (1989). "Matrix Population Models. Sunderland, Mass.: Sinauer Associates.

Chiang, C. C. (1984). "The Life Table and Its Applications." Malabar, Fla.: Robert E. Krieger Publ. Co.

Keyfitz, N., and Beekman, J. A. (1984). "Demography through Problems." New York: Springer-Verlag.

Manton, K. G., and Stallard, E. (1984). "Recent Trends in Mortality Analysis." Fla.: Academic Press.

McDonald, L., Manly, B., Lockwood, J., and Logan, J., eds. (1989). "Estimation and Analysis of Insect Populations" Berlin: Springer-Verlag.

Pressat, R. (1985). "Dictionary of Demography." Oxford, England: Basil Blackwell.

Tuljapurkar, S. (1990). "Population Dynamics in Variable Environments: Lecture Notes in Biomathematics." New York: Springer-Verlag.

# Insect Interactions with Trees

***Jeffry B. Mitton***
*University of Colorado*



Herbivorous insects use trees as food sources, mating sites, and sites for oviposition. A single tree species may serve as the host for hundreds of species of insects that rely on it partially or fully as a food source, mating site, and oviposition site. The activities of herbivores cost trees photosynthetic tissue, nutrients, and fluids, and these losses may be reflected as negative impacts on growth, size, viability, and reproduction. Because insects can disperse to seek hosts, and because their generations are much shorter than those of their hosts, insects appear to have evolutionary advantages over their hosts. However, persistence of trees in the face of long-term pressure from herbivorous insects indicates that trees are not defenseless stocks of food waiting to be found. Trees have evolved a variety of defenses against the myriad of insects that use them as food and sites for reproducing. Some adaptations appear to be generalized defenses against herbivores, whereas other traits may have evolved in a coevolutionary arms race between a host species and one or several herbivores. The evolution of trees and their associated insects continues today; herbivorous insects often display genetic variation influencing host choice, and host species often exhibit genetic variation influencing resistance to herbivores.

Insects and trees are also bound in mutualisms, or interactions that benefit both parties. In pollination mutualisms, the plants benefit by having their pollen dispersed and ovules fertilized, while insects benefit from food presented in the form of pollen, nectar, and nutrients from extrafloral nectaries and, in some cases, from nest and brood sites on the trees. In defense mutualisms, insects defend the plant from herbivores in return for food and nest sites on the tree. Intense interactions between insects and plants, such as those in obligate mutualisms, can link evolutionary changes together so tightly that the phylogenetic relationships among a group of related plants is reflected in their associated group of insects.

## I. GRAND DISPARITIES IN MOBILITIES AND GENERATION TIMES

The grand disparities between the generation times and mobilities of trees and the insects that use them would seem, at first glance, to favor insects. Trees are large, conspicuous, sedentary organisms with life spans that are typically measured in centuries. Many insect species have one or more generations per year, are highly mobile, and use trees as food, mating sites, and sites to rear their offspring. Insects can search for their host trees, and when they find a susceptible host, their short generation times allow them to fully utilize the resource. Consider the plight of the tree. An immense and immobile cornucopia of cellulose, resin, and leaves, incapable of any evasive or defensive action, stands for several centuries as a beacon for burrowing, crawling, and flying insects searching for food. Furthermore, consider the odds in this battle; more than 200 species of insects use ponderosa pine (*Pinus ponderosa*) as a source of food, and several hundred insects are found on a representative tree in a tropical rain forest. Although the disparities in generation time and mobility seem to provide the insects with the demographic and evolutionary advantages, insects continue to feed and live on trees—the insects have not eaten themselves out of their homes, or pushed their hosts to extinction.

## II. EVOLUTION AND COEVOLUTION

### A. Herbivores Decrease the Fitnesses of Trees

An evolutionary perspective is critical to understand the balance of forces in the relationships between herbivorous insects and their host trees. First of all, herbivory has a direct impact on trees. Certainly, the impact is most apparent when insect activity directly kills the trees. For example, hundreds of thousands of mountain pine beetles, *Dendroctonus ponderosae,* will converge on a ponderosa pine, *P. ponderosa,* chew through the bark, and kill the tree by tunneling through the phloem, cutting and clogging the tubes that transport water and nutrients. But less drastic impact on trees can also be very important. A caterpillar slowly chewing the leaf of a sugar maple may appear innocuous, but this consumption is a direct loss of photosynthetic tissue. Chronic herbivory damages 15–20% of *Eucalyptus* leaves in Australia and 3–10% of deciduous leaves in North America. Comparisons of *Eucalyptus* trees treated with insecticides and unprotected trees reveal that chronic herbivory decreases the growth rates by approximately 50%. Herbivory that denudes trees, as during epidemic outbreaks of the silkworm caterpillar on maples and beeches, or during epidemics of the western budworm on Douglas fir, not only stops growth but also depletes stored energy. By reducing the growth of the tree, herbivory may also compromise the ability of the tree to compete for light, and by reducing the size of the tree, herbivory may reduce viability and fecundity.

### B. Plant Defenses

In the 1950s, Vincent Dethier and Gottfried Fraenkel stirred the imagination of evolutionary biologists by suggesting that plant secondary compounds, such as flavonoids, tannins, phenols, and monoterpenes, were not simply metabolic byproducts, but were deterrents against herbivores. Trees may be immobile and conspicuous, but they are not defenseless. Paul Ehrlich and Peter Raven proposed that the presence of chemical deterrents both limited the access of insects to plants and mediated the ecological release that allowed adaptive radiations of both insects and plants. If a population of plants evolved a deterrent that protected it from its traditional herbivores, it might thrive and expand in its defended niche, proliferating new forms in a protected radiation. Similarly, a population of insects that evolves a mechanism to handle previously impervious defenses gains access to an uncontested resource, or a clade of resources. [*See* Forest Insect Control.]

In response to numerous and varied herbivores, plants have evolved a wide variety of defenses. Thorns and spines deter many of the larger herbivores. Soft, fuzzy pubescence can deny access by many insects to leaves and stems, and oil glands can exude compounds such as coumarins, which irritate or burn. Trees have evolved a series of defenses. Thick bark and tough cuticles on needles and leaves deter some insects. A variety of digestibility-reducing compounds defend against a broad array of herbivores; the tannins of oaks and the phenols of pines are released from intracellular compartments when herbivores chew on leaves. These compounds may simply dilute the nutritious components of the leaves or may bind to proteins, precipitating them and rendering them inaccessible to the digestive systems of herbivores. Thus, a deer or caterpillar can chew on the needles of ponderosa pine but will get virtually no protein. In this way, trees discourage herbivores from eating them.

### C. Ancient Interactions

Many of the associations between insects and their host trees are truly ancient. A 50-million-year-old fossil leaf of the genus *Cedrela* contains the characteristic mining pattern of the lepidopteran genus *Phyllocnistis*. Today, larvae of the genus *Phyllocnistis* are mining the leaves of *Cedrela* in Costa Rica. Similarly, 25-million-year-old fossil oak leaves bear the characteristic mines of the lepidopterans associated with oaks today. Closely related species of leaf beetles of the genus *Blephardia* feed on closely related species of sumac of the genus *Rhus* in northern Africa and the deserts of the Middle East. Studies of flavonoids of the sumacs, the feeding preferences of the beetles, and the current distributions of species support the hypothesis that the association between these beetles and sumacs predates the evolution of the contemporary species. Geographical, life-history, and paleobotanical evidence indicates that aphids of the subtribe Melaphidina have been associated with sumac genus *Rhus* from before the time that sumac crossed the Bering land bridge from Asia to North America 48 million years ago.

### D. Genetic Variation

The impacts of herbivory are not uniform among populations, or among the individuals within a population. If genetic variation confers differential susceptibility to herbivory, then natural selection can cause an evolutionary change in the population. Thus, herbivorous insects may evoke evolution in their hosts of adaptations that reduce susceptibility to herbivory. In turn, some intermediate degree of resistance to herbivory will evoke an evolutionary response in a population of herbivores with genetic variation for their ability to circumvent or overcome the plant defenses. Individuals with the capability to overcome plant defenses have a greater number of hosts to choose from, allowing them to utilize trees not available to other members of the population. If this advantage increases their reproductive success, natural selection will increase the proportion of insects able to overcome the plant defenses. Genetic variation for resistance to herbivores is common in plant populations, and genetic variation enabling insects to overcome plant defenses is common in herbivorous insects.

### E. The Red Queen

Evolutionary biologists generally concede that plants and animals live in environments that vary both in time and in space. The metaphor of the Red Queen, the character in Lewis Carroll's "Alice in Wonderland" who had to run constantly just to stay in the same place, is used by evolutionary biologists to describe the plight of plants besieged by a shifting set of herbivores. In pine forests, an epidemic of bark beetles may be followed by an epidemic of western budworms, and by endemic populations of scale insects, aphids, and cone beetles. Rising pressure from any group may evoke an evolutionary response that leaves the host less susceptible to that particular group of insects, but no less susceptible, and conceivably more susceptible, to the remainder of the herbivores. For example, a comparison of ponderosa pines parasitized by dwarf-mistletoe, *Arcethobium vaginatum,* and the mountain pine beetle, *D. ponderosae,* revealed that the parasitic plant and the beetle favored trees with distinctly different nutritional and defensive characteristics. As climates change, and the geographic ranges of trees and insects shift, a host population can be exposed to an unfamiliar herbivore that imposes novel evolutionary responses. Host populations may constantly evolve, as the Red Queen runs, from its contemporary herbivores, without ever gaining an evolutionary advantage. [*See* PLANT–ANIMAL INTERACTIONS.]

### F. Coevolution and Concordant Phylogenies

If an association between a group of herbivores and their hosts is persistent in time, the evolution in a host species of an adaptive novelty that conveyed a defensive advantage could evoke an evolutionary response in the herbivores. Coevolution is a pattern of reciprocal evolutionary responses, and it has suggested the metaphor of the evolutionary arms race between predator and prey, parasite and host, and herbivore and host. If coevolution continues over evolutionary spans of time, then speciation by a host may cause diversification and speciation in the herbivore, building concordant phylogenies in groups of trees and their associated insects. Many groups of plants and herbivores do not exhibit concordant phylogenies—interactions between herbivores and plants are diverse, so any single model of evolution will only apply to some groups. However, concordant phylogenies are clearly evident for chrysomelid beetles of the genus *Phyllobrotica* and their host plants of the genus *Scutellaria* (Fig. 1).

The mutualism of fig trees of the genus *Ficus* and their pollinators, fig wasps of the family Aponinae, has created concordant phylogenies in the trees and insects. Figs are pollinated only by fig wasps, and fig wasps mate and develop inside the fruits of figs. The life cycles of figs and fig wasps are intricately and obligately intertwined; the female has evolved specialized pockets to carry pollen, and larvae consume the seeds produced by the pollination activity of their parents. The seeds lost to hungry larvae are just part of the cost invested by figs to guarantee pollination. With just a few minor exceptions, each of the hundreds of fig trees is associated with a single species of fig wasp, and closely related species of figs are occupied by closely related fig wasps. [*See* Ecology of Mutualism.]

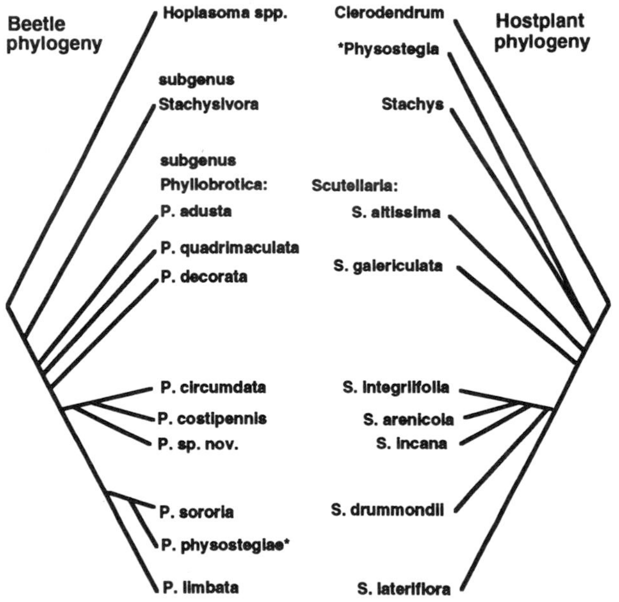


***FIGURE 1*** Comparison of the phylogenies of beetles of the genus *Phyllobrotica* and their host plants in the genus *Scutellaria*. Each species of *Phyllobrotica* is limited to a single species of *Scutellaria*. The similar phylogenetic relationships in beetles and their hosts suggest that speciation in the host plants has been followed by speciation in the beetles. [From B. Farrell and C. Mitter (1990). *Evolution* **44,** 1389–1403.]

## III. CASE STUDIES

### A. The Mountain Pine Beetle and Ponderosa Pine

Though it is difficult to unambiguously demonstrate coevolution between long-lived trees and herbivorous insects, the present interactions between the mountain pine beetle, *Dendroctonus ponderosae,* and ponderosa pine, *Pinus ponderosa,* suggest that these species and their progenitors have evolved morphological, physiological, and behavioral adaptations in reciprocal responses.

Bark beetles fly moderate distances, on the order of a few hundred meters, in search of a host tree. The mountain pine beetle is capable of using several host species, but it tends to favor ponderosa pine at the lower elevations, lodgepole pine (*Pinus contorta*) at higher elevations, and it will settle for limber pine (*Pinus flexilis*) at the highest elevations. Beetles land on the bole or trunk of a tree, and taste the tree by chewing the bark. At this point, the beetle may break off the encounter by flying off to find a more suitable host, or it may begin to bore into the bark.

The resin of ponderosa pine contains many kinds of monoterpenes, and the relative concentrations of the monoterpenes can vary dramatically among trees. These compounds impart a fragrance to individual trees, so that some smell like vanilla, others like cinnamon. Herbivores undoubtedly make greater distinctions among the bouquets than do humans. Choice experiments with both squirrels and bark beetles reveal that some monoterpenes, such as myrcene, are avoided by herbivores, whereas others, such as $\alpha$-pinene, are favored.

As the beetle bores into the tree, it severs radial ray ducts in the sapwood and phloem, and resin floods the wound. The beetle pushes the resin out of its tunnel, creating a cone of pitch that dribbles from the bore hole. If the resin pressure is sufficiently high, the beetle can be "pitched out" of the tree, and become trapped in the drying resin. Bark beetles have evolved the ability to detoxify inhaled and ingested monoterpenes to less toxic terpene alcohols.

The resin pressure of a tree varies dramatically from time to time, and thus susceptibility to mountain pine beetles varies over time. Trees are especially susceptible to beetle attack during a sustained drought, when injured by lightning or wind, or if they are compromised in some other way, such as by fungal infection of the roots. A healthy tree with moderate or high resin pressure can pitch out a few errant beetles, but virtually all trees will succumb to a mass attack by hundreds or thousands of beetles. Beetles martial a mass attack by attracting nearby beetles with an aggregation pheromone, which is synthesized from components of the resin of the host. Female *D. ponderosae* initiate the attacks on ponderosa pine, and shortly after the vapors of monoterpenes are inhaled through the tracheal system, they are detoxified to terpene alcohols that appear in the hemolymph and are secreted into the hindgut as the aggregation pheromone *trans*-verbenol. Some experimental evidence suggests that bacteria in the gut of bark beetles play a role in the synthesis of aggregation pheromones. When a sufficient number of beetles have bored through the bark, these leaks in the resin system will reduce resin pressure to nearly zero, providing easy access for more beetles to become established in the tree.

Mountain pine beetles bore through the bark and establish feeding tunnels and egg galleries in the phloem. This narrow strip of tissue in the inner bark is essentially two dimensional, so that the space available for feeding of adults and larvae is potentially limiting. Once beetles are established in a tree, they stridulate and release *exo*-brevicomin, an antiaggregation pheromone, to prevent local densities from becoming too great. The antiaggregation pheromone signals that a host tree is fully occupied and causes the mass attack to be diverted to a nearby tree. There is some evidence that the antiaggregation pheromone is produced by the fungal symbiont of the beetle from substrates in the resin. Thus, the antiaggregation pheromone reliably signals the inoculation into the tree of the fungal symbionts.

Mountain pine beetles have mycangia, shallow pits on the surface of the head that have evolved to carry their fungal symbionts of the genus *Ceratocystis*. The fungi are introduced into the phloem and sapwood by the beetle, and the fungi soon clog the tubes that conduct nutrients and water. As mycelia, the fungi penetrate extensively into the phloem and sapwood. When a tree is girdled by fungus, the crown soon withers and the tree dies. The death of the tree changes the amount of oxygen and water in the galleries, allowing the fungi to produce abundant mycelia and spores. The adults and their offspring feed on the fungi growing in the feeding tunnels and the egg galleries. Thus, the symbiotic fungi carried and introduced by the beetle assist in the killing of the tree and provide food for the adults and their offspring.

Several inferences can be taken from the interactions of mountain pine beetles and ponderosa pine. It is likely that monoterpenes were originally metabolic intermediates of terpene metabolism, and that the concentration of monoterpenes was increased by natural selection when monoterpenes deterred insects from attacking the trees. Today, all the conifers produce large amounts of monoterpenes. Perhaps an early *Dendroctonus* progenitor evolved the ability to detoxify monoterpenes, gaining access to the conifers. It is ironic that the originally toxic, repellent monoterpenes are now used by beetles as cues to find preferred trees and to produce a pheromone that triggers mass attacks. Unhindered by competition from other herbivores, *Dendroctonus* flourished and diversified, so that today virtually every species of conifer is attacked by one or more species of *Dendroctonus*. The mycangia and fungal symbionts appear to be adaptations specific to bark beetles for killing trees and feeding offspring.

### B. Herbivory on Pinyon Pine

Plant resistance to herbivores is variable within populations and over time. The genotype of a tree contributes to resistance, but the environmental component is often important, sometimes overriding. A stand of trees may be uniformly resistant under optimal conditions and uniformly susceptible during a severe drought. Genetically determined variation among individuals is often greatest under a moderate degree of environmental stress. This point is illustrated with an example of herbivory on pinyon pine in a stressful environment.

In A.D. 1065, Sunset Crater, near the present town of Flagstaff, Arizona, began a 200-year eruption that buried 2000 $km^2$ under a deep layer of pea-sized cinders. Pinyon pine, *Pinus edulis,* invaded the cinder fields from surrounding sites with sandy-loam soils in the last 800 years. Trees growing in the cinders are water and nutrient stressed and herbivory is sustained at levels not seen in surrounding environments with typical soils. The stem moth *Dioryctria albovittella* feeds on the emerging needles in spring, and the infestation on some trees is so severe that most of the growing branch tips are killed, forcing the proliferation of internal branches. Chronic herbivory produces rounded, densely branched, small trees that contrast sharply with the typical growth forms of uninfested trees. Resistant and susceptible trees are interspersed across the homogeneous cinder fields, suggesting that resistance is primarily a genetic trait.

An electrophoretic survey of protein polymorphisms revealed genetic differences between trees that were resistant and susceptible to the stem moth. Resistant trees are more heterozygous at genes coding for enzymes than are the susceptible trees (Fig. 2). Furthermore, in both the resistant and susceptible groups, older trees are more heterozygous than young trees, indicating that viability increases with heterozygosity. In the adjacent environments with sandy-loam soils, herbivory is much less severe, and heterozygosity is not associated with either herbivory or viability.

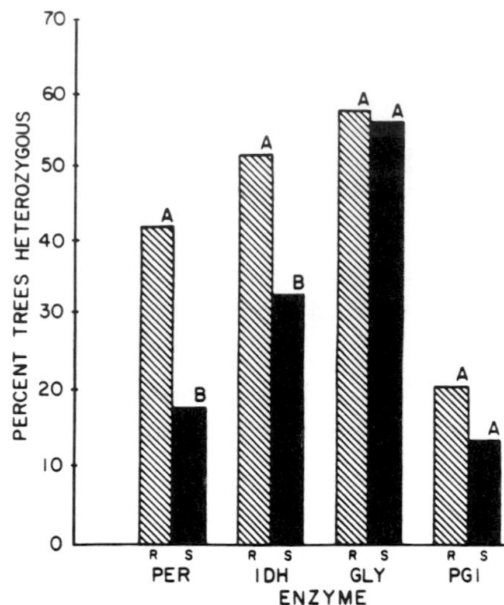


***FIGURE 2*** The enzyme heterozygosities of pinyon pine, *Pinus edulis,* resistant and susceptible to attack by the stem moth *Dioryctria albovittella,* on the cinder fields at Sunset Crater, Arizona, U.S.A. Resistant trees are significantly more heterozygous at peroxidase (PER) and isocitrate dehydrogenase (IDH) than susceptible trees, whereas these groups are not significantly different at glycerate dehydrogenase (GLY) and phosphoglucose isomerase (PGI). [From S. Mopper *et al.* (1991). *Evolution* **45,** 989–999.]

### C. Latex and Resin Tubes

Tube or canal systems that transport latex and resin have evolved at least 40 times. These tubes carry adhesive compounds that trap insects as the fluid dries, or toxic compounds such as coumarins, alkaloids, catechols, monoterpenes, and resin acids. Both latex and resin defend plants against herbivorous insects. Brian Farrell and colleagues proposed that latex and resin tubes, although structurally different, comprised a recognizable syndrome of defense against herbivores, and they tested the hypothesis that plant groups with tubes would be more abundant and diverse than related groups lacking tubes. In 13 of 16 comparisons between sister groups of similar ages, the group with tubes had more species, sometimes by two orders of magnitude, than the group without tubes. Surveys of lowland habitats in Manu National Park, Peru, yielded data suggesting a contemporary advantage for species with either latex or resin tubes. Seven species with tubes were compared with sympatric, closely related species lacking tubes. The local abundances and habitat breadths were consistently greater in the plant species protected by tubes. These observations on diversity, abundance, and niche breadth are consistent with the hypothesis that the acquisition of defenses against herbivores enhances the spread and diversification of plant species.

## D. Hybrid Zones as Safe Sites for Herbivores

Hybrid zones are natural laboratories for the study of interactions between plants and herbivores. If hybridizing species have evolved resistance to insect herbivores, and the adaptations differ between the hybridizing species, then resistance to herbivores may break down with the scrambling of genes in hybrid zones. This reasoning applies to any mode of resistance, whether it is produced by the accumulation of toxic compounds or by phenological differences in the timing of flushing, maturation, and senescence of leaves. Surveys of insect densities in hybrid zones have revealed that insects reach far higher abundances on hybrid and backcrossed genotypes than on pure species. Fremont cottonwoods (*Populus fremontii*) hybridize with narrowleaf cottonwoods (*P. angustifolia*) along a 13-km section of the Weber River in Utah. The gall-forming aphid *Pemphigus betae* exhibits a 75-fold difference in survival on individual trees in the hybrid zone and is most abundant on backcrossed genotypes, which have unbalanced mixtures of genes from the parental species. Surveys conducted in six consecutive years revealed that the mean density of galls was two orders of magnitude higher in the hybrid zone than in the areas occupied by pure species. Experimental transfers of aphids onto trees in the hybrid zone and in the zones of pure species demonstrated that trees in the hybrid zone were more susceptible to aphid attack. The leaf-feeding beetle *Chrysomela confluens* has the same pattern of abundance in this hybrid zone; 94% of the *C. confluens* population was restricted to the hybrid zone.

Hybrids between species of eucalypts support greater abundances and diversities of insects than do pure species. Thirty-eight species of insects were surveyed in a hybrid zone between *Eucalyptus amygdalina* and *E. risdonii* near Hobart, Tasmania. The majority of the insects surveyed were significantly more abundant in the hybrid zone than in zones of pure species, and several rare species were nearly restricted to the hybrid zone. Most of the insects were not on the $F_1$ hybrids (first-generation hybrids) but on backcrosses between hybrids and either of the pure species.

The studies of cottonwood and *Eucalyptus* hybrid zones suggest that hybrid zones are critically important for herbivorous insects. The genetic recombination among plant species probably disrupts the physiological and morphological traits conveying resistance to the pure species, providing insects with susceptible hosts.

## Glossary

**Backcross** Individual or genotype produced by crossing an $F_1$ hybrid with one of its parental species.

**Clade** Group of species descended from a common ancestral species, or a group of closely related species.

**Electrophoresis** Separation of molecules in a supporting medium with a direct electric current. The migration rates of the molecules vary with their size and their net charge.

**Evolution** Change in the genotypic constitution of a population.

**Herbivore** Plant-eater.

**Heterozygous** Bearing two different copies of a gene. Contrasts with homozygous, meaning bearing two copies of the same gene.

**Monoterpene** One of a group of low-molecular-weight, toxic, volatile hydrocarbons produced by plants and known to affect the physiology or behavior of animals.

**Mycangium (plural, mycangia)** Cup-shaped repository bearing fungi, typically paired, on the head of many species of bark beetles.

**Natural selection** Differential reproduction of genotypes.

**Phenol** Class of secondary plant substances having six to ten carbons arranged in one or two benzene rings, often involved in plant defenses against herbivores.

**Pheromone** Chemical emitted by an organism that induces a behavioral or physiological response in another organism of the same species.

**Phloem** Inner bark of trees, containing the sieve tubes that conduct food and water.

**Phylogeny** Genealogy or evolutionary relationships among a group of species or higher taxa.

**Resin** Complex mixture of terpenoid compounds found in coniferous trees and believed to play a role in their defense.

**Sapwood** Newly formed outer wood, just inside the cambium of a tree.

**Tannin** Group of toxic, stringent, nonnitrogenous compounds containing phenols, glycosides, and hydroxy acids.

## Bibliography

Berg, C. C., and Wiebes, J. T. (1992). "African Fig Trees and Fig Wasps." Amsterdam: North-Holland.

Ehrlich, P. R., and Raven, P. H. (1964). Butterflies and plants: A study in coevolution. *Evolution* **18,** 586–608.

Farrell, B. D., Doussourd, D. E., and Mitter, C. (1991). Escalation of plant defense: Do latex and resin canals spur plant diversification? *Amer. Nat.* **138,** 881–900.

Farrell, B., and Mitter, C. (1990). Phylogeny of insect/plant interactions: Have *Phyllobrotica* leaf beetles (Chrysomelidae) and the lamiales diversified in parallel? *Evolution* **44,** 1389–1403.

Fritz, R. S., and Simms, W. L. (1992). "Plant Resistance to Herbivores and Pathogens." Chicago: University of Chicago Press.

Furth, D. G., and Young, D. A. (1988). Relationships of herbivore feeding and plant flavonoids (Coleoptera: Chrysomelidae and Anacardiaceae: *Rhus*). *Oecologia* **74,** 496–500.

Futuyma, D. J., and Peterson, S. C. (1985). Genetic variation in the use of resources by insects. *Annu. Rev. Entomol.* **30,** 217–238.

Hickey, L. J., and Hodges, R. W. (1975). Lepidopteran leaf mine from the early Eocene Wind River Formation of northwestern Wyoming. *Science* **189,** 718–720.

Linhart, Y. B., Snyder, M. A., and Gibson, J. P. (1994). Differential host utilization by two parasites in a population of ponderosa pine. *Oecologia* **98,** 117–120.

Mitton, J. B., and Sturgeon, K. B. (1982). "Bark Beetles in North American Conifers." Austin: University of Texas Press.

Mopper, S., Mitton, J. B., Whitham, T. G., Cobb, N. S., and Christensen, K. M. (1991). Genetic differentiation and heterozygosity in pinyon pine associated with resistance to herbivory and environmental stress. *Evolution* **45,** 989–999.

Moran, N. A. (1989). A 48-million-year-old aphid–host plant association and complex life cycle: Biogeographic evidence. *Science* **245,** 173–175.

Morrow, P. A., and Fox, L. R. (1989). Estimates of presettlement insect damage in Australian and North American forests. *Ecology* **70,** 1055–1060.

Whitham, T. G. (1989). Plant hybrid zones as sinks for pests. *Science* **244,** 1490–1493.

Whitham, T. G., Morrow, P. A., and Potts, B. M. (1994). Plant hybrid zones as centers of biodiversity: The herbivore community of two endemic Tasmanian eucalypts. *Oecologia* **97,** 481–490.

# Intertidal Ecology

**Clinton J. Dawes**
*University of South Florida*



Intertidal ecology is the study of the interaction of marine organisms and their environment in communities found between lowest low tide and highest high tide. This will include biological associations found below the maritime communities, areas wetted by wave activity (spray or supralittoral fringe), the intertidal or eulittoral region, and the subtidal fringe (eulittoral fringe). Below this level is the subtidal (sublittoral) zone, which is never exposed to the air.

## I. INTRODUCTION

Part of the beauty of a rocky shoreline is the vertical zonation of benthic marine plants and animals (Fig. 1). This zonation is not random and, as in terrestrial systems, is primarily caused by a combination of abiotic and biotic factors resulting in highly distinctive communities of marine organisms in the intertidal region. Water temperature is the major physical factor governing the geographical distribution of marine organisms; tides are the primary controlling factors in the distribution of organisms across habitats; and space and light are the limiting resources in an intertidal community. There is an extensive literature dealing with intertidal ecology because the topic has been of interest since the beginning of the nineteenth century. [*See* OCEAN ECOLOGY.]

The abiotic factors of temperature, tides, exposure (wave action, topography), substrate, and climate will be considered first along with the biotic factors of competition, predation, and grazing. Examples of intertidal communities on rocky (hard substrate) and unconsolidated coasts (sand, gravel, and boulder beaches) and adaptations of the inhabitats to the abiotic and biotic factors will follow.

## II. ABIOTIC FACTORS

### A. Temperature

Temperature is the most fundamental factor for all organisms because of its effects on molecular activities; at higher temperatures, molecules have more energy and reactions (enzyme activity) will be faster. Thus, temperature is the major factor controlling geographical distribution of marine organisms. The mean annual latitudinal thermal gradient of seawater ranges from a surface maximum of 26 to 30°C in the tropics to a minimum of −1.9°C in lower latitudes (near the poles). The unusual

***FIGURE 1*** Zonation on a rocky cliff in southern California. 1 = spray; 2 = high intertidal barnacle zone; 3 = middle intertidal mussel zone; 4 = low intertidal region, an area of large brown seaweeds (kelps).

steep temperature gradient that presently exists, compared to the more narrow gradients of past geological history, has resulted in well-defined biogeographical regions, especially for organisms found in the intertidal region. Thus, seven latitudinal groups of organisms can be distinguished that are associated with biogeographical regions: in the northern hemisphere, Arctic (one region), cold-temperate (three regions), warm-temperate (four regions), and tropical (four regions), and in the Southern Hemisphere, warm-temperate (five regions), cold-temperate (five regions), and Antarctic (one region).

Because of the regular exposure of intertidal habitat to air, temperature regimes are much more complex than that experienced by subtidal organisms. The main source of heat in the intertidal zone is direct solar radiation. Temperature fluctuations can be severe for intertidal organisms at higher levels in temperate climates. Intertidal organisms can be exposed to temperatures ranging from 5 to 33°C over an 8-hr period during submersion and exposure on a summer day. In general, few larger animals or plants are found in the intertidal zone of subtropical and tropical areas when compared to temperate climates due to the intense solar radiation, the effect of desiccation, and the wide ranges in temperature. For example, the dominant plants in the intertidal zone of the Florida Keys are microscopic cyanobacteria, organisms able to withstand a range of 40°C, in contrast to the large, fleshy, brown rockweeds of temperate zones.

Tide pools are of particular interest because they support organisms that may not otherwise survive in the intertidal zone. The water in tide pools initially has the same conditions as that of the ebbing seawater. Depending on the position of the tide pool with relation to the intertidal zone and its size and depth, the air temperature will warm or cool (even to freezing) it. Comparative studies on these pools have demonstrated tolerances of organisms not only to temperature but also to changes in salinity (due to rain or evaporation).

### B. Tides

Tides are the rhythmic rise and fall of sea levels caused primarily by the gravitational attraction of the sun and moon on the earth. Based on the size of the basin and latitude of the site, an intertidal community may have semidiurnal (two highs and two lows every lunar day, which lasts 24 hr and 50 min), diurnal (daily with a single high and low during a lunar day), or mixed tides (semidaily ranging from a single high or low to irregular cycles and with unequal highs and lows). The tidal cycles will increase to maximum and minimum amplitudes (spring tides) every 14.5 days. The frequency and length of tidal exposure will control the distribution of intertidal organisms on a vertical and seasonal scale.

Thus, intertidal organisms usually experience varying periods of exposure to the air because of vertical position and tidal amplitude of that region. The concept of "critical tide levels" has been used to explain zonation. Critical tide levels are levels in the intertidal zone where there are marked increases in the duration of exposure or submergence. These occur at crests and troughs on daily, monthly, or annual intervals, being most evident in areas with mixed tides.

### C. Substrate

As shown in Section IV, rocky coasts support highly distinctive communities when compared to unconsolidated beaches. The topography and features of the substrate (solidarity, texture, porosity, solubility, color) all influence the type of habitat in the intertidal zone. In the tropics, soluble, porous limestone supports a variety of boring (urchins, cyanobacteria, and green algae) organisms in an otherwise hostile (intense radiation, high rates of desiccation) environment. This can be contrasted with dark, nonporous, and nonsoluble volcanic rock, which increases in temperature and solar radiation, and limits organisms to its surface.

### D. Exposure

Exposure results from a combination of wave action (coastal energy) and shore topography (steepness and ruggedness). Intertidal plants growing on a jutting headland will encounter larger waves than a protected shoreline. Shorelines with lower gradients will reduce the effect of breaking waves compared with a cliff. Organisms found on high-energy coasts are typically tougher and have more highly adapted attachment organs than those found in areas of low energy. In general, there is an upward shift (expansion) of vertical zones with an increase in wave energy and usually a more rapid removal of competitive organisms resulting in greater species diversity.

### E. Climate

Climatic factors include air temperature, humidity, prevailing winds, winter storms, solar radiation, cloud cover, and rain or snow. Organisms in the intertidal zone will be exposed to all of these factors. On temperate coasts, there may be extensive cloud cover and rain or snow, resulting in exposure to diluted seawater and low irradiance and temperatures. In contrast, intertidal zones of tropical coasts frequently experience high air temperatures (35°C) and irradiances (2500 $\mu$m of photons $m^{-2}$ $sec^{-1}$) and severe desiccation (water loss of 20 to 30%). Winter storms and distinct seasonal patterns of sand deposition can dramatically influence intertidal community composition. It is apparent that the structural (morphology, anatomy) and functional (reproduction, physiology) adaptations of intertidal organisms vary greatly depending on where they inhabit a rocky shore or sandy beach.

## III. BIOTIC FACTORS

### A. Zonation

The study of the distribution of organisms in an intertidal habitat has resulted in a variety of approaches, ranging from simple line transects and quadrats of smaller (meters) defined communities to analyses of large (hectares, kilometers) areas using numerical analysis and phytosociological procedures. It is difficult to randomly sample an irregular, patchy intertidal community. The primary tool is usually a type of line transect where 25-$cm^2$ to 1-$m^2$ quadrates are sampled at fixed intervals from the spray to subtidal zones. Abundance of each species can be assessed by percent cover (for dense, uniform vegetation) or number of individuals counted (if distinct). A turf community might require percent cover measurements and then separation and weighing to determine the contribution of each species. All the techniques are intended to define the intertidal communities in terms of species presence, dominance, and distribution. In general, zones are defined by "important" or "dominant" species and are linked to abiotic characteristics, particularly position with relation to tide. [*See* CONTINENTAL SHELF ECOSYSTEMS.]

The principal zones, as defined earlier, are shown in Fig. 1 in relation to high and low wave activity and are of importance to both hard and unconsolidated coasts. The effect of the topography, wave energy, and substrate will greatly influence the distribution of organisms of the intertidal community. Thus, in high-energy regions the zonal patterns will be widely separated with opportunities for intermediate communities to develop, whereas in protected areas of low wave energy the communities will be compressed.

### B. Competition

The lower limits of intertidal species can be controlled by interference competition, whereas the upper limits are usually controlled by abiotic factors such as desiccation. As an example, the red algae (Rhodophyta) *Gigartina papillata* and *Gastroclonium coulteri* occur in the upper and middle intertidal zone, respectively, on the central California coast. If the higher intertidal species, *G. papillata,* is removed, *G. coulteri* will not expand, as it is sensitive to increased level of desiccation. In contrast, if *G. coulteri* is removed, the upper species, *G. papillata* will expand into the lower zone. The reason why *G. papillata* cannot expand normally into the lower zone is because it is a slower growing species and *G. coulteri* can grow faster. In turn, *G. coulteri* is prevented from extending in the intertidal fringe and subtidal zone by species of *Phyllospadix,* a sea grass whose true roots and rhizomes can outcompete the red alga.

Another example of interference competition are turf-forming algae and crustose species that are competing for space. Turf-forming algae are faster growing than crustose species and can occupy large areas. If the turf species are grazed out, then grazing-resistant crustose species will dominate. Interference competition occurs when carnivores, by removing sessile herbivores, permit the establishment of less competitive species. Also, the removal of the large blades of the brown (Phaeophyta) alga *Laminaria hyperborea* that form a canopy has resulted in a significant increase in biomass of algae that grow epiphytically on the stipes, again a competition for space and light.

Exploitative competition can be seen where plants in dense stands tend to be small, whereas those in more dispersed populations are larger. For example, sporelings of the red seaweed *Chondrus crispus* (Irish moss) germinated in dense patches on glass slides were two to three times smaller than those grown at 50% that density.

### C. Grazing and Predation

Episodic and long-term variations in intertidal communities occur owing to predation. A number of well-documented studies have shown that sea urchin predation on intertidal communities resulted in the removal of all sea grasses and seaweeds. In contrast, the range increase of the sea otter *Enhydra lutris* in the past 20 years on the California coast appears to have reduced the intertidal mussel populations as well as other prey and opened areas for seaweed colonization. In this last example, the mussels were competing for the same space as some of the larger seaweeds, such as the brown sea palm *Postelsia palmaeformis.*

A proven adage is that if there is vegetation, there will be herbivores. Studies of intertidal plants have shown that reduction of herbivory by seaweeds is possible through temporal, spatial, structural, and chemical "escapes." Temporal escapes include life histories in which the dominant morphology is crustose during periods when grazers are present. Also some seaweeds have life histories in which only spores are present when grazers are in the intertidal zone. Spatial escapes might be where plants grow between rocks or in the high-energy intertidal fringe so that sea urchins cannot reach them. Structural escapes include the tough stipes of brown seaweeds, calcification of a number of red and green seaweeds, and crustose morphologies. Finally, chemical escapes include production of secondary products such as phenols (brown algae), sulfuric or malic acid in the cell vacuole (*Desmarestia* spp.), or other terpinoids such as udoteal in the green (Chlorophyta) alga *Udotea flabellum*.

## IV. ROCKY COASTS

### A. Types of Rocky Coasts

Rocky coasts can be divided into their geographical landforms—coastal cliffs, shore platforms, and limestone coasts. The different types of coast will support characteristic communities that, in turn, are influenced by the degree of their exposure (topography and wave action). These coasts are subjected to mechanical erosion by waves, freezing, chemical and salt weathering (and solution for limestone), and bioerosion through microflora, grazing, boring, and burrowing.

Seacliffs occur on all of the world's oceanic coasts, with the largest gradients in regions of vigorous waves that are common to temperate latitudes (Fig. 2). Shore platforms are horizontal or gently sloping surfaces produced along a shore by wave erosion (Fig. 3). Limestone coasts are usually associated with warm climates and young calcareous rocks (Fig. 4), where weaker waves, higher temperatures, and highly varied marine life result in chemical and biological activity.

### B. A Temperate Intertidal Community

The rocky coasts of the northeast Pacific stretch from the tip of Baja California, Mexico, to Alaska,

***FIGURE 2*** Seacliffs along the coast of South Australia. (Photograph by Dr. R. A. Davis, University of South Florida.)

***FIGURE 3*** A sea platform on Kangaroo Island, Victoria, Australia. (Photograph by Dr. R. A. Davis, University of South Florida.)

about 20,000 km. The rocks are mainly sedimentary and the intertidal substrate ranges from massive rock benches to sandy beaches. Surface water temperatures change seasonally with means of 5 to 20°C annually.

Intertidal zonation of an open rocky coast is controlled by the extent of tidal exposure, as demonstrated on high-energy rocky coasts such as those of southern California. Usually four well-defined horizontal zones can be recognized because of the large (2–3 m) tidal amplitude (see Fig. 1). Zone 1, only rarely wetted by storm waves and spray, is usually covered by small, filamentous green algae and cyanobacteria or bare rock. Fauna common to this zone are few and include the limpet *Collisella digitalis*, gastropods (*Littorina keenae*), and isopods (*Ligia* spp.). Zone 2 is the high intertidal "barnacle zone," where dense populations of *Balanus glandula* are exposed to air for long periods each day. Seaweeds common to this zone include the turf-forming red algae (*Endocladia muricata, Gigartina papillata*) and the brown rockweed *Pelvetia fastigiata*. Snails are also common here (*Littorina scutulata, Tegula funebralis*). Zone 3 (mussel zone) is the middle intertidal and is exposed to air for short periods twice a day, with the mussel *Mytilus californianus* and the gooseneck barnacle *Pollicipes polymerus* dominating. On more protected sites, chitons (*Katharina tunicata, Nuttallina californica*) and the red alga *Iridaea flaccida* will dominate. Zone 4 is the low intertidal region and is uncovered only during spring (lowest) tides. This region can have carpets of the surfgrass *Phyllospadix* spp. and a variety of brown algae (kelps), particularly *Laminaria setchellii*.

### C. A Tropical Intertidal Community

The Caribbean Sea lies between latitudes 23 and 10° N and has a hard substrate ranging from black igneous rock to light-colored limestone. Surface water temperatures range from 20 (winter) to 33°C

***FIGURE 4*** Limestone intertidal region of West Summerland Key in the Florida Keys, United States. 1 = spray (white to black) zone; 2 = upper, yellow intertidal zone; 3 = lower golden intertidal zone; 4 = subtidal fringe.

(summer), although some areas experience lower temperatures due to upwelling (Isla Margarita, Venezuela). Because of the small tidal amplitude (29 to 110 cm), three general zones can be identified based on the color of rocky coasts throughout the Caribbean (supralittoral, and upper and lower intertidal; Fig. 4). In contrast to a temperate intertidal community, larger size species of seaweeds are limited to the lower intertidal region owing to desiccation and solar radiation.

Zonation of a limestone coast is evident in the lower islands of the Florida Keys (Fig. 4). The spray zone can be identified by color, ranging from white (uppermost) to gray to black. This region is not well developed except on coasts with wave action (Barbados, Florida Keys). The white (supralittoral fringe) region is where the sea meets the land and may support mangroves and succulent composites. The gray to black portion of the limestone platform is colored by endo- and epilithic cyanobacteria (*Callothrix* spp., *Lyngbya* spp.). Filamentous species of red seaweeds (*Catenella repens* and *Polysiphonia* spp.) and littorinoid gastropods (*Littorina* spp., *Nerita* spp.) and isopods (*Ligia baudiniana*) can occur in the spray zone.

The intertidal region is yellow to gold in color and can be divided into an upper, more desiccated region and a lower area that is frequently wetted by moderate tides. The upper intertidal area is characterized by small barnacles (*Chthamalus* spp., *Tetraclita* spp.) and vermnetids (*Petaloconchus* spp., *Spiroglyphus* spp.) and usually an algal turf consisting of filamentous red and green seaweeds. The lower intertidal region is dominated by a mixture of filamentous red (*Centroceras calvulatum, Bostrychia tenella, Ceramium* spp., *Polysiphonia howeii*) and brown (*Giffordia* spp., *Sphacelaria* spp.) algae along with fleshy forms (*Acanthophora* spp., *Laurencia* spp., *Sargassum* spp.). Barnacles and mussels (*Mytilus exustus*) are also common in this zone, with an algal turf up to 1 cm in height. The algal turf is particularly well developed at the lower edge of the intertidal zone, the subtidal fringe (Fig. 4, zone 4).

## V. UNCONSOLIDATED BEACHES

### A. Types of Beaches

A beach is an accumulation of loose (unconsolidated) sediment extending from low tide to the base of sand dunes or bedrock (Fig. 5). Waves are the most fundamental force operating on a beach. Thus, beaches are one of the most variable of coastal landforms and exist on both primary (youthful) and secondary (mature) coasts wherever unconsolidated sediments occur. Barrier islands, long sandy islands found along secondary coasts, have some of the longest beaches extending parallel to the coast and form almost 20% of the world's coastline.

The beach, whether 1 or 100 m wide, can usually be divided into three zones: backshore, foreshore, and nearshore (Fig. 5). The backshore extends landward from the highest tide mark to the dunes or cliffs, the foreshore is the intertidal zone of the beach, and the nearshore is the zone from lowest tide level extending seaward. The most dynamic parts of a beach are usually the foreshore and nearshore, where waves break and longshore currents develop.

The foreshore or intertidal portion of a beach usually has a steeper gradient than the backshore in beaches that are adding sediment (accreting). In contrast, beaches subjected to high wave energy tend to lack a backshore and have a steep gradient. The slope or gradient along with particle size of beaches will indicate the wave energy of that coast; larger particle sizes and steeper slopes are associated with high levels of wave energy. The foreshore will show ridges and runnels, which are intertidal sand bars formed due to breaking waves. In the nearshore region, longshore bars can form in the breaker zone that parallel the shore.

### B. Zonation of a Beach

The lack of hard substrate and constantly shifting sediment limit the type of organisms to those able to bore or burrow or move over the surface, as

***FIGURE 5*** Zonation of a beach on the Virginia coast of the United States. 1 = maritime zone; 2 = backshore; 3 = foreshore; 4 = nearshore. (Photograph by Dr. R. A. Davis, University of South Florida.)

shown for a moderate-energy beach on the coast of Virginia (Fig. 5). Compared with an intertidal community on hard substrate, beach communities have low diversity and the only plants are microscopic (diatoms, cyanobacteria, unicellular green algae).

A study of the extensive (322 km) barrier islands on the coast of Texas demonstrated a well-defined macrobenthic zonation. The ghost crab (*Ocypode quadrata*) community (six species) occupied the backshore and upper foreshore, having a low species diversity, low density of individuals, and being mainly infaunal (burrowing) forms, all of which were detritus feeders and scavangers. The burrows of the ghost crab were the dominant biogenic sedimentary structures. Other members of the backshore and upper foreshore community include the tube polychaete (*Scolelepis squamata*) and amphipods.

Communities with greater diversity occupied the lower foreshore (21 species) and bar-trough system of the nearshore (19 species). Haustorid ampiphods were the single most important group in both communities, followed by *Scolelepis squamata*. In the lower foreshore, another polychaete, *Lumbrineris* sp., and the surf clam *Donax variabilis* were abundant. In the bar-trough region, a community dominated by *Callianassa islagrande* (ghost shrimp) and the amphipod *D. variabilis* had a relatively high diversity of infaunal species. Other members of this community were the sand dollar *Mellita quinquiesperforata* and *Lumbrineris* sp. The bar-trough community had a mixture of trophic levels, ranging from deposit feeders to detritus feeders.

The ecotone or boundary between the backshore and the bar-trough system of the nearshore is the foreshore. The highest diversity and density of infaunal organisms was in this ecotone. Species found in both adjacent communities were abundant but restricted in the foreshore ecotone and thus there is a mixture of trophic forms.

## Glossary

**Algae** General term for photosynthetic, unicellular to massive plants that lack vascular (conducting) tissue and usually are of simple morphology. (See also *seaweed*.)

**Biogeographical** Referring to distribution of organisms on a latitudinal and longitudinal basis.

**Chlorophyta** Algae having the same photosynthetic pigments (chlorophyll a and b, carotenes, xanthophylls) as vascular plants.

**Cyanobacteria** Formally called "blue-green algae" and prokaryotic in structure (lacking membrane organelles), and thus related to bacteria but having true chlorophyll a.

**Phaeophyta** "Brown" algae containing chlorophyll a and a special xanthophyll, fucoxanthin, which can result in a brown coloration.

**Rhodophyta** "Red" algae owing to the masking of chlorophyll a by special water-soluble pigments (phycoblins), especially the red-reflecting pigment, phycoerythrin.

**Sea grass** Plants more closely connected to lilies, and thus vascular plants that have grasslike leaves but are not true grasses. These plants have evolved back into the ocean from land.

**Seaweed** Common term for the larger forms of algae, particularly the larger brown algae called kelps and red algae that form larger plant bodies.

## Bibliography

Dawes, C. J. (1981). "Marine Botany." New York: John Wiley & Sons.

Lewis, J. R. (1964). "The Ecology of Rocky Shores." London: English University Press.

Mathieson, A. C., and Nienhus, P. H. (1992). "Ecosystems of the World. Vol 24. Intertidal and Littoral Ecosystems." Amsterdam: Elsevier.

Stephenson, T. A., and Stephenson, A. (1972). "Life between Tidemarks on Rocky Shores." San Francisco: Freeman.

Trenhaile, A. S. (1987). "The Geomorphology of Rocky Coasts." Oxford, England: Clarendon Press.

# Introduced Species

**Daniel Simberloff**
*Florida State University*



Introduced species are those that, deliberately or inadvertently placed in a new habitat by humans, colonize regions isolated from their existing geographic ranges.

## I. INTRODUCTION

Introduced species have caused enormous damage to agriculture, waterway fouling, erosion, and spread of plant and animal diseases, including those of humans. These impacts are usually on more or less human-made communities. For example, the typical agricultural community comprises crop plants or animals, themselves introduced, subsidized by inputs of fertilizer, food, and weed and disease control. From an ecological and conservation standpoint, however, although damage to such anthropogenous communities may be economically devastating, the most important effects of introduced species are on communities not otherwise substantially disturbed by human enterprise. Nowadays no community is truly "pristine." The burgeoning human population affects most communities directly, and indirect effects of various pollutants and global warming reach every corner of the earth, but one can still recognize some communities as almost pristine in that they are not directly harvested and we intend them to serve as undamaged refuges for their inhabitants. Yet introduced species have often altered even these communities.

This fact is surprising, because one of the two articles of conventional wisdom about introduced species is that they are more important in disturbed than in pristine habitats; the other is that island communities are particularly likely to be damaged. Charles Elton, who brought introduced species to the attention of ecologists in 1958, cited numerous examples of devastating invasions of disturbed continental communities. He resurrected the idea of "biotic resistance" as an explanation: in undisturbed areas, a battery of competitors, predators, parasites, and diseases thwarts most invaders. In disturbed areas, the numbers of such defending species are greatly reduced, hence resistance is believed to be lowered (see Section V,C).

Detecting patterns in the establishment and effects of introduced species is complicated by two problems with the data. First, surviving introduced species are much more likely to be noticed than those that disappear, and those producing big ecological and/or economic effects are much more likely to be studied and reported. Second, there are few thorough studies of the ecological effects of introduced species. Rather, the literature abounds with untested hypotheses. For example, the American mink (*Mustela vison*), imported to fur farms in Britain and Sweden around 1930, escaped and began to spread rapidly in both countries in the 1950s. This was about the same time that the otter (*Lutra lutra*) began to decline precipitously in both countries, and for years many game biologists saw a causal nexus. Subsequent research established that, in the otter's favored habitat, it outcompetes the mink, and that pollution, particularly with organochlorine pesticides, has had far greater impact on the otter. Just as introduced species can be blamed for things they have not done, so they can fail to be implicated in tragedies when they are, in fact, the culprits. Forest bird species have declined drastically on Guam, and several species are probably extinct. This collapse is now known to be due to the Australian brown tree snake (*Boiga irregularis*), accidently introduced around 1950. Yet even within the last ten years, the disappearance of the birds has been confidently ascribed to organochlorine pesticides.

## II. ARRIVAL AND SURVIVAL

The vast majority of propagules die without issue or perhaps reproduce for a few generations before the population disappears, but there is no way to estimate the fraction of propagules that gives rise to ongoing populations because it is very unlikely that the arrival and disappearance of a few individuals would be witnessed. In Britain, of *known* introductions, 240 of 409 insect species and 166 of 488 vascular plant species failed. But who knows the great number of propagules of different species that arrived and quickly went extinct? It must be at least hundreds for both taxa.

Exactly why most introductions fail to produce ongoing populations is unknown, but several factors are surely important. First, a propagule must arrive in a suitable habitat or be able to move to a suitable habitat. Probably most propagules end up in unsuitable habitats and die there—terrestrial seeds fall in water, parasites of both plants and animals fail to find hosts, microbes land in environments of intolerable pH, and so on. Further, even if adult individuals can survive at a site, they may be unable to reproduce or the offspring unable to live. For example, the geographic ranges of many plant species are limited not by the ability of the adults to survive but by their inability to produce seed or failure of seeds to germinate at the margins of the range. Thus, individual plants may appear healthy at the range margins, but they are all progeny of more central populations.

Biotic as well as physical factors might prevent survival and/or reproduction of the initial propagule. A disease organism or high densities of a voracious predator, for example, could be critical. In one of very few experimental introductions, R. Levins and H. Heatwole introduced 90 individuals of the lizard *Anolis pulchellus* from Puerto Rico onto the nearby tiny island of Palominitos. Though the habitat is almost certainly suitable for survival, and reproduction occurred, the population disappeared within a year, probably because of predation by the resident *A. cristatellus*. In a few instances, either deliberately introduced or naturally invading phytophagous insects have failed to become established because of predation or parasitism. For example, the sphingid moth *Erinnyis ello* invaded an isolated cassava plot in Brazil, but no larvae reached first instar because eggs and parasites were too heavily attacked by parasitic wasps. Crucial biotic determinants of initial success and reproduction need not all be enemies of the invading species. Seeds of a plant species that survive and germinate cannot produce an ongoing population if the species must be pollinated by a missing animal species.

If the abiotic and biotic environments at the site of introduction do not preclude an invader, propagule size probably plays a role. Problems of breeding and increase at low population size, collectively called the "Allee effect," can become so severe in a small propagule that establishment is highly improbable. Difficulty finding mates is one such possibility. Very small populations are also at high risk of extinction for many other reasons—collectively these lead to the concept of the "minimum viable population size." For example, various demographic and genetic problems are exacerbated in very small populations. Thus, one would expect propagule size to contribute to probability of initial survival and increase, a pattern seen among game birds introduced throughout the world. Some insect introduction data show a similar trend. [*See* Evolution and Extinction.]

This pattern is not universal, however. For the many flies, beetles, and wasps introduced as biological control agents, there are no differences in success rate among propagules of size 1–20, 21–100, and 101–1000, respectively. It is important to realize that survival and initial reproduction are stochastic processes, and one cannot predict initial success very accurately from propagule size alone. Consider the European house sparrow (*Passer domesticus*), of which 8 pairs were introduced in Brooklyn, New York, in 1851, never to be seen again. The next year, another 50 birds were unsuccessfully introduced near the same site. In 1853, the same number were again released nearby, and this propagule established a rapidly growing and spreading population. Subsequently supplemented by a few other releases, the sparrow dispersed throughout most of North America. Similarly, in the biological control literature are numerous examples of "replicate" releases with very different trajectories. Of course, because the same individuals are not used in different introductions of the same species, no two introductions are true "replicates." Individuals might be genetically different in two replicates, for example, in ways that might account for different results, but it is unlikely that such detailed information will ever be available for most propagules of potential introduced species.

Numerous ecologically important introductions have resulted from surprisingly small propagules. Several insects successfully used in biological control started as propagules of 20 or fewer individuals. The North American muskrat, *Ondatra zibethica*, spread throughout much of Europe from an initial propagule of five individuals released in Czechoslovakia in 1905. In 1938, one pair of the Himalayan tahr (*Hemitragus jemlahicus*) escaped into a South African nature reserve, and a damaging population of 300 animals was established by 1972. The first propagule of many surviving introduced species of plants and animals was unrecorded, but the circumstances of some (escape from culture or cultivation, transoceanic dispersal) suggest that not many individuals could have been involved. In 1869, a scientist studying silkworms brought several eggs of the gypsy moth (*Lymantria dispar*) from Europe to Massachusetts. Just one cage of caterpillars broke in a windstorm, but these established a population that spread throughout the Northeast and established beachheads elsewhere.

## III. SPREAD

Animals, plants, and microorganisms disperse by many means—passive transport by wind, water, or animals, and active movement by swimming, flying, or on the ground. Most species have stylized behaviors that enhance dispersal and characteristic life-cycle stages during which dispersal is particularly likely. Even passive movement by animals, plants, and microorganisms often starts with a process that places the propagule in the vicinity of a transport agent. Plants explosively release their seeds to wind and water currents, and spiderlings assume postures and spin silk to facilitate ballooning. For many animals and some plants, dispersal is particularly likely to terminate in a site suitable for survival and reproduction. Planktonic larvae of various marine invertebrate species, for example, settle in response to chemical and/or tactile cues that signal suitable habitat.

Human activities, particularly a growing volume of increasingly faster transportation of both goods and people plus increased trade in animals and plants, have been most responsible for a recent flood of species introductions. For example, many aquatic, marine, and terrestrial organisms have moved great distances in ballast water or soil. It is no surprise that ports were the major sites of plant introduction after European colonization of other continents. Air travel and the increase in both commercial shipping and tourism have accelerated the flood of exotic propagules in the twentieth century.

Faced with this array of dispersal mechanisms, E. C. Pielou distinguished between two rather distinct forms of spread. In the first, diffusion, a species' spread more or less closely approximates increasing concentric circles of which the circumferences become progressively more warped. "Diffusion" here does not mean the Brownian movement of particles but simply gradual and more or less regular spread. The Colorado potato beetle (*Leptinotarsa decemlineata*) in Europe and the Japanese beetle (*Popillia japonica*) in the United States are good examples. The rate at which such species' ranges expand is a function of the dispersal and behavioral biology of the species. The warping of the range circumference as diffusion proceeds is probably caused by heterogeneities in the physical environment. For example, the Atlantic Ocean and Chesapeake Bay have prevented the Japanese beetle from spreading evenly in all directions from its point of introduction in New Jersey (Fig. 1). The unsuitability of ocean as a habitat for a terrestrial beetle is easy to deduce, but the influence of other habitat gradients may be subtler.

In contrast to quite regular diffusion, some introduced species have spread irregularly from the outset or after a short period of circular range expansion. Often several foci arise simultaneously by long-distance "jumps," each subsequently serving as a base for slower circular growth or yet another long-distance jump. The gypsy moth spread more or less gradually and regularly in New England and the Middle Atlantic states, but it has been recorded thousands of kilometers away from the eastern population in the Midwest and West. These jumps were probably achieved by eggs attached to undersides of motor vehicles traveling from the Northeast. Other means of jump-dispersal might include the rare transport of a plant, animal, or microbe propagule by a bird or in a swift upper air current.

Jump-dispersal and diffusion are not completely dichotomous categories. Each move in a gradual diffusion constitutes a jump, and the length of jump that characterizes jump-dispersal is arbitrary. Gypsy moths regularly "balloon" up to a kilometer

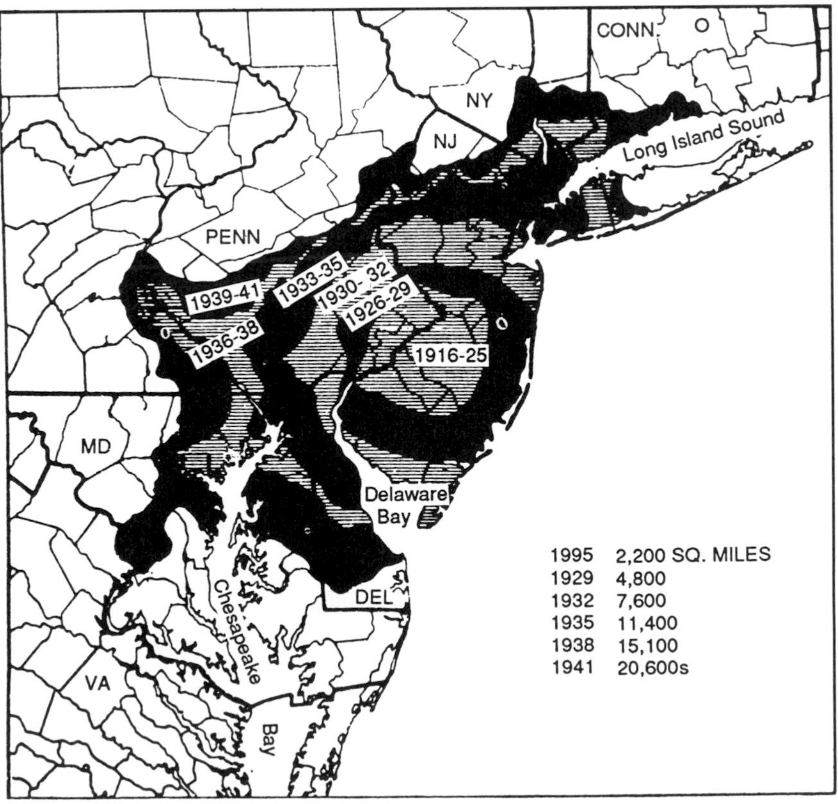


***FIGURE 1*** Spread of the Japanese beetle, *Popillia japonica,* in the United States, 1916–1941. It has since extended much farther. [From United States Bureau of Entomology and Plant Quarantine, 1941.]

on threads spun by small caterpillars, but strong winds occasionally carry them still farther, perhaps even as far as a car or truck might. Transport farther than a kilometer is probably rare, but the distribution of larval movement distances is a continuum. Generally, a jump entails a propagule that lands far outside the existing range of a species.

The variety of dispersal means might induce pessimism about the ability to predict the rate and direction of spread. Particularly for jump-dispersal, which often rests on such idiosyncratic events as adhesion of a piece of mud to a bird's foot, prediction would seem difficult. Yet there has been much modeling of the spread of introduced species, both post facto curve-fitting from observed range changes and prior prediction from assumptions about biological traits and the characteristics of various dispersal mechanisms. The most extensively developed models are epidemiological ones concerned with microbial disease agents. Many ecological models are based on the tenuous assumption of diffusion in the strict sense (that individuals move at random throughout their lives) and, though successful in specific instances, have not allowed the accurate prediction of the changing size and shape of the geographic ranges of most introduced species.

The reason is probably that the spread of an introduced species usually occurs by several means at once, often a combination of short-distance diffusion and long-distance jumps. For example, many mapped invasions suggest diffusive spread from each focus, and new foci continually established outside the existing range. Then diffusion from all foci fills in unoccupied area until all ranges are continuous. For a wide range of combinations of dispersal mechanisms, the plot of the square root of occupied area versus time since introduction is expected to be sigmoidal. That is, there is an initial period of slow spread, then an increase in the rate of spread, and finally deceleration and cessation of range growth as inhabitable area is filled. The central period, after the initial slow growth, may appear linear over a long time span. The period of slow spread may be so short as to be invisible at most plotted time scales, whereas the ultimate deceleration may not yet have been reached. For several well-studied invasions, parts of the plot, predicted from demographic and dispersal characteristics, are well supported by observations (e.g., Fig. 2).

The sudden spread of an introduced species can be tied not to its inherent biological properties but to a habitat change. A species that had reached its maximum extent by whatever combination of dispersal means can then begin to expand again. For example, the aphid *Hydaphis tatarica* was restricted to a small area of southern Russia, apparently by the limited range of its host, Tatarian honeysuckle (*Lonicera tatarica*). It was in the process of spreading gradually westward and had just reached the Moscow region when it was scientifically described in 1935. However, as the honeysuckle was planted as an ornamental throughout much of central and southern Europe, the aphid's range increased greatly and irregularly, often to areas not contiguous with the original range.

## IV. EFFECTS OF INTRODUCED SPECIES

One must bear in mind that many hypotheses on effects are unsupported by adequate data, and often alternative hypotheses have not even been considered. Nevertheless, one can recognize numerous effects of introduced species on one another or on

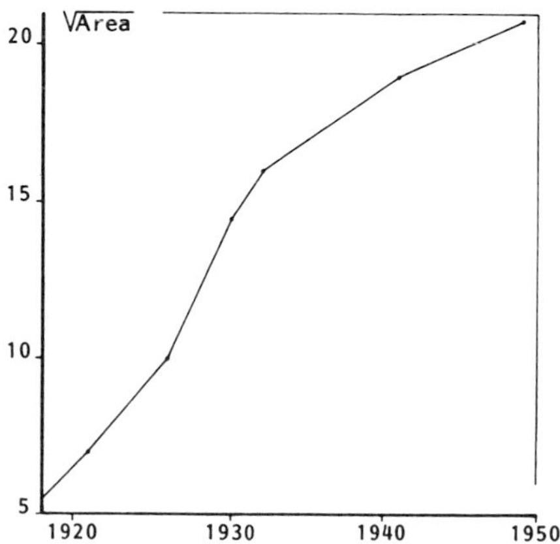


***FIGURE 2*** Expansion rate of the European starling in North America. [After R. Hengeveld (1989). "Dynamics of Biological Invasions." Chapman & Hall, London.]

native species. A species may kill, eat, or displace a second species directly; these are direct effects. On the other hand, a species can change the habitat of another species or reduce its prey or host. These are indirect effects, which are defined as effects of one species on the relationship between two others. Indirect effects are often subtle but can be very important.

### A. Direct Effects

An introduced species can directly affect another as a pathogen or parasite. Myxoma virus deliberately introduced from South America into Australia in 1951 killed over 99% of the huge European rabbit population, though subsequent evolution of both virus and rabbit has allowed some recovery. The impact of a pathogen can range from minor to enormous, depending on the particular way in which it weakens infected individuals and the fraction of the population that is infected.

An introduced species can prey on a native animal or be an herbivore of a native plant. Bird species have been eliminated all over the world by introduced rats, mustelids, and feral cats, dogs, and pigs. The most famous case is that of the lighthouse keeper's cat on Stephen Island (New Zealand). The cat arrived in 1894 and eliminated the entire population of the Stephen Island wren (*Xenicus lyalli*) within a year. Caterpillars of the introduced gypsy moth have virtually defoliated many native trees of the Northeast, especially common oak species, and have probably effected a permanent change in forest composition. Introduced predators and herbivores can also attack other introduced organisms, an ability used in attempts to control introductions (see Section VI).

It has proven more difficult to demonstrate competition by an introduced species against a native, but observational evidence is often quite convincing. For example, introduction of the barn owl (*Tyto alba*) to the Seychelles Islands coincided with the decline of the endemic Seychelles kestrel (*Falco araea*), probably because of nest-site competition. The limiting resource does not even have to be used in the same way for an introduced species to harm a native one. In Bermuda, nest boxes on poles for eastern bluebirds (*Sialia sialis*) are used as perches by the introduced great kiskadee (*Pitangus sulpuratus*), which prevents the bluebirds from nesting.

Introduced plants can affect native plants by allelopathy. For example, the African ice plant (*Mesembryanthemum crystallinum*) has devastated native vegetation in coastal California. It is an annual and accumulates salt throughout its life. When it dies, rain and fog leach the salt into the soil, where it suppresses growth and germination of native species.

The most important ecological effect of introduced species on native communities is modification of the habitat, because such modification can affect entire ecosystems. In the eighteenth and nineteenth centuries, much of the northeastern North American coast consisted of mudflats and salt marshes, not the current rocky beach. This change was wrought by the European periwinkle (*Littorina littorea*), which eats algae on rocks and rhizomes of marsh grasses. This snail slowly spread southward after introduction in Nova Scotia about 1840. Because the physical structure of the entire intertidal zone of a large region changed, virtually the whole biotic community dependent on that structure changed also. Similarly, introduced feral pigs (*Sus scrofa*) have modified entire ecosystems in the Hawaiian archipelago and the Great Smoky Mountains National Park of Tennessee. They selectively root out and feed on plant species with starchy bulbs, tubers, and rhizomes, and they modify soil characteristics by thinning the forest litter, mixing organic and mineral soil layers, and increasing leaching of many minerals. In some areas, they have greatly aided invasion by exotic plants. Some effects of both periwinkles and pigs are indirect (see Section IV,B), but many are direct, in that native species are harmed because their habitat is destroyed.

Introduced plants can modify an entire plant community, and therefore the animal community, by changing the fire regime. Around 1900, the cajeput tree, *Melaleuca quinquenervia,* was introduced from Australia into south Florida, where it has displaced less fire-adapted species such as cypress over thousands of hectares (Fig. 3). Several intro

***FIGURE 3*** *Melaleuca quinquenervia* invading the eastern part of the Everglades National Park, Florida.

duced plants act similarly as fire-enhancers in Hawaii.

Introduced plants can modify the habitat by constituting substantial forests where none had existed. Examples include riverbanks in the arid southwestern United States, where Eurasian salt-cedar (*Tamarix* spp.) has had far-reaching effects. Salt-cedars have deep roots that allow them to thrive in situations where other trees cannot, such as parts of the Colorado River floodplain. Mangroves were unknown in Hawaii, where soft substrates in sheltered bays and estuaries were unforested. In 1902, seedlings of red mangrove (*Rhizophora mangle*) were planted on Molokai. Now, by natural dispersal and possibly deliberate plantings, this mangrove has spread to other islands and forms forests up to 20 m high. The consequences of this invasion are not well studied but must be enormous. For example, healthy mangrove swamps drop about 4000 kg of leaves annually per hectare, and the roots form critical habitat for fishes and crustaceans and accumulate sediment. South American water hyacinth (*Eichhornia crassipes*) was introduced into Florida in the late nineteeth century and spread to clog over 50,000 acres of waterways by the 1950s (Fig. 4). The complete cover led to lower dissolved-oxygen levels in infested waters, increased temperatures, smothered beds of native plants, and decreased populations of some animals.

Finally, introduced species can hybridize with native ones, potentially modifying the native species or even changing it so much that it is not regarded as the same species. The Seychelles Islands subspecies of the Madagascar turtle dove (*Streptopelia picturata rostrata*) has been destroyed by hybridization with the introduced Madagascar subspecies (*S. p. picturata*). Introduced species can also hybridize with other introduced species. Introgression from cultivated sorghum (*Sorghum bicolor*) has rendered shattercane (*Sorghum drummondi*) and Johnsongrass (*Sorghum halepense*) more serious pests. Finally, hybridization between an introduced species and a native one can produce a new pest. North American cordgrass (*Spartina alterniflora*) was introduced in shipping ballast to southern England. A subsequent hybrid with the noninvasive native *S. maritima* was sterile, but when one of these hybrid individuals underwent a doubling of chromosome number, a fertile, invasive form (*S. anglica*) was produced.

***FIGURE 4*** Water hyacinth (*Eichhornia crassipes*) choking the Oretaga River (Florida) in the 1940s.

### B. Indirect Effects

Indirect effects of introduced species, in which one species affects the interaction between two others, can be idiosyncratic and complex. Consider the mite *Pediculoides ventricosus,* accidentally introduced into Fiji. It attacked larvae and pupae, but not eggs and adults, of the coconut leaf-mining beetle during the dry season. Adult beetles then oviposited and died, converting the beetle population to one with synchronous, nonoverlapping generations. The consequent absence of larvae and pupae during certain periods caused the mite population to plummet, as did those of two native parasitoids that had previously controlled the beetle. The mite and parasitoids did not live long enough to survive the intervals between occurrences of the host stages they required for oviposition, and the beetle population exploded.

This example is probably not more complicated than many that arise in nature, but detection of such a phenomenon is difficult. Because species can interact through shared prey or hosts and through shared predators, parasites, and pathogens, as well as through many types of habitat modification, the range of possible indirect effects is enormous.

The chestnut blight fungus (*Cryphonectria parasitica*) came from Asia to New York on nursery stock in the late nineteenth century. In less than 50 years it spread over 91 million ha of the eastern United States. The American chestnut (*Castanea dentata*) had been a dominant tree in many regions, comprising up to 25% of canopy individuals. It was virtually eliminated. However, the fungus affected far more than the chestnut. For example, several insect species host-specific to the chestnut are now endangered or extinct. The oak wilt disease (*Ceratocystis fagacearum*) has increased on many native oak species because the especially susceptible red oak (*Quercus rubra*) increased greatly in the absence of the chestnut.

Introduced species can be vectors or reservoirs of disease to which they are resistant but to which native species are susceptible. The major reason many native Hawaiian bird species have gone extinct and others are threatened is destruction of upland forest habitat. However, a key contributing factor is avian pox and malaria vectored by Asian songbirds introduced in the late nineteenth and early twentieth centuries. An introduced species can even serve as a reservoir for a disease that was not introduced with it. For example, on Puerto Rico the introduced small Indian mongoose (*Herpestes auropunctatus*) carries rabies but did not introduce it. That human diseases brought by Europeans to North and South America, Australia, New Zealand, and various small islands throughout the world were devastating to many native peoples is well known. There is every reason to think that other animals were equally affected by diseases introduced by Europeans, and the same may also be true of plants, though scientific proof is often lacking.

Any introduced species that greatly modifies the habitat can easily change a balance between two native competitors, between a predator and its prey species, between a parasite and its host, and so on. For example, rooting and herbivory by pigs (Section IV, A) could favor one existing plant species to the detriment of another by modifying the nutrient regime. Similarly, grazing by periwinkles must have favored marine invertebrates whose larvae can settle and grow on rocks over those that establish themselves in mud. So both the pig and the periwinkle, in addition to direct interactions with other species, almost certainly generated indirect effects by changing how other species interact with one another.

### C. Economic Effects

It is impossible to estimate the total economic costs of all introduced species, but they are staggering. The chestnut blight was estimated to have eliminated $25 million worth of trees by 1911, independently of ecological effects (Sections IV,A, and IV,B). The entrance of the sea lamprey (*Petromyzon marinus*) into the Great Lakes after construction of the Welland Canal has virtually destroyed the large trout fishery. Within 25 years it had fallen from about 2 million kg to less than one-tenth that amount in Lake Huron alone. In addition to this loss, the United States Fish and Wildlife Service spends approximately $5 million annually poisoning lamprey larvae. Many millions of dollars have been spent to control European starlings (*Sturnus vulgaris*) in the United States; their roosts can number a million individuals, and they eat enormous quantities of agricultural produce (Fig. 2). In 1960 a starling flock flew into a commercial airline flight taking off from Boston, causing a crash and 62 fatalities. The total cost of this bird in North America has not been calculated. The successful campaign to eradicate coypus (*Myocastor coypus*) in England (see Section VI) cost $4 million. No estimate is available for control expenses before that or for ecological or crop damage. An Asian mite (*Varroa jacobsoni*) that kills honeybees has rapidly spread from Wisconsin to many other states, and prospects for control are not yet clear. Without control, the United States Department of Agriculture estimates that the honey and beeswax industries alone would lose $109 million annually.

Two notable recent invaders of the United States are the European zebra mussel (*Dreissena polymorpha*) and the Africanized honeybee or "killer bee" (the African subspecies of the honeybee, *Apis mellifera adansonii*, or hybrids of this subspecies with others). The bee, accidentally released in 1957 in Brazil, has stung to death hundreds of people and many domestic animals in Latin America. It has now reached Texas. The main concern, in addition to the honey industry, is that domestic honeybees in the United States pollinate about $10 billion worth of crops each year. If Africanized bees were to destroy the domestic beekeeping industry, costs to growers for domesticated colonies could rise greatly. The United States and Mexico have already spent millions of dollars to control the bee and slow down its rate of advance. (On the other hand, Africanized bees are at least somewhat resistant to the mite *V. jacobsoni*.) [*See* INTRODUCED SPECIES, ZEBRA MUSSELS IN NORTH AMERICA.]

The zebra mussel probably arrived in 1988 in the Great Lakes from Europe in ballast water. Their main damage so far is fouling of water and power plant intakes, but decomposing mussels pollute water, and their sharp shells are a beach hazard. They may also damage fisheries by competing for food plants. One estimate of potential damage is $4 billion by the year 2000 in the Great Lakes alone, and they have already spread to the Hudson River.

It is apparent from these examples that the sorts of costs that have been estimated entail health or existing commercial activities—timber, fisheries, agriculture—as well as maintenance costs to support these. Aesthetic and other amenities are generally assigned an economic value through their market costs. Ecological costs are not easily estimated in this way. Is the cost of an extinction of a native species that had not had a market value only the lost potential of its genes to genetic engineers? Did its existence have nonmarket value for humans? Indeed, is the value of a species only to be calculated in terms of realized or potential human use? These questions demonstrate why the costs of introduced species cannot be estimated, even if scientific knowledge increases (Section I).

Introduced species also generate incalculable economic benefits. The majority of North American crop plants are of Old World origin, as is the honeybee (*Apis mellifera*) that pollinates so many of them. So are most domestic animals. Many introduced insect species and some introduced pathogens successfully control introduced agricultural and other pest species (see Section VI). For each dollar invested in research and transport of such biological control agents, agricultural benefits have averaged about $40. As in the attempt to estimate ecological costs, attempts to estimate benefits are problematic and often rest on incalculable values. For example, introduced game animals and ornamental plants produce numerous benefits, but only some of them can be estimated economically (e.g., money spent on hunting). They also have staggering ecological costs, including many of the inimical effects on native species discussed in Sections IV,A, and IV,B. Yet attempts to extirpate introduced game animals and ornamental plants have occasionally engendered enormous controversies, at least partly because the antagonists have different noneconomic values.

## V. THE GEOGRAPHY OF INVASION

### A. The Special Vulnerability of Islands

It is widely believed that islands are far more easily invaded than mainland and that effects of introduced species are more drastic on islands (Section I). The hypothesized reason for this enhanced invasibility is that island species have been subjected to less rigorous natural selection, largely because islands harbor fewer species. Thus they evolve to become less competitive. The poor data on failed invasions make it difficult to assess this view, because one cannot determine whether a higher fraction of propagules of introduced species survive on islands than on mainland. Available data are almost exclusively about survivors. About 80 introduced mammal species have survived on islands, and only 50 on continents. Similarly, about 300 bird species have persisted on islands against about 100 on continents. The pattern is not always so dramatic, but it seems generally true that there have been more surviving introductions on islands than on mainland. For all taxa, the survivors are almost all continental species. [*See* Island Biogeography, Theory and Applications.]

It is not apparent that these species have survived more often on islands than on continents because they have outcompeted island species. In fact, it is difficult to find any incontrovertible examples of a mainland invader eliminating an island species by outcompeting it for a limited resource. This is not to say that such events have not occurred; resource competition is notoriously difficult to document in field situations.

By contrast, mainland invaders have demonstrably affected island species in several important ways, often to the point of extinction. The best known is predation. Many island biotas lack species that fill particular ecological roles on continents, so it is not surprising that certain types of mainland species have prospered on islands. Small oceanic islands, and even some large ones like New Zealand, generally lack mammalian predators other than bats. Thus indigenous island species have evolved in ways that make them particularly vulnerable to such species—fearless behavior and ground-nesting in birds are two examples. Thus it is small wonder that 90% of the land and freshwater birds extinguished since 1600 are island species. Many were preyed on by introduced rats, mustelids, feral dogs and cats, and the small Indian mongoose (*Herpestes auropunctata*)—or by the most voracious mammalian predator, *Homo sapiens*. The Mauritius dodo, rapidly eliminated by the first European colonists, is the most widely discussed extinct bird.

Habitat destruction by grazing by introduced pigs, rabbits, sheep, cattle, horses, and other species has probably extinguished more island species than has predation. Such grazing can eliminate a plant species or can reduce the quantity and quality of a habitat so much that various animal and plant species are prone to extinction from the forces that conspire to place small populations at risk. Indirect effects of such activities (Section IV,B) can produce further extinctions. Although such habitat destruction is often depicted as far greater on islands than on mainland, there has been no systematic study. Certainly introduced species have caused similar habitat destruction on continents. The effects of pigs in the Great Smoky Mountains National Park were described earlier. Even greater regions have been affected: the native plant communities of the North American intermountain West have been largely eliminated by Eurasian grazing animals.

Disease vectored by introduced species is also believed to devastate island communities more than mainland ones. Avian malaria and pox vectored by introduced Asian songbirds have facilitated the decline of native Hawaiian birds (Section IV,B). Similarly, the native bulldog rat (*Rattus nativitatis*) of Christmas Island was quickly eliminated by disease carried by introduced ship rats (*R. rattus*). Introduced diseases have also devastated continental communities; Asian rinderpest devastated many African ruminants in the late nineteenth century. No comprehensive study has been performed, but it is likely that the probability of such an epidemic is greater on islands.

Whatever the relative probabilities of survival and impact of introduced species on islands and continents, subsequent extinction of native species is more likely on islands. Island populations are more geographically restricted, so they are less likely than mainland populations to have a refuge unexposed to predators, grazers, or diseases. On continents, huge ecological damage has often been accompanied by rather few species extinctions. Over 98% of the original forest of the eastern United States was destroyed after European colonization, but only three landbird species went extinct. By contrast, over 150 landbird species and subspecies have been eliminated on islands during the same period.

### B. Ecological Imperialism: The Effects of Eurasian Invaders

Many authors have noted that Eurasian species seem more likely to invade and disrupt other biotic communities than vice versa. For example, almost all great human contagious diseases but syphilis are of Old World origin, and there are numerous historical accounts of epidemic devastation of American, Australian, and oceanic peoples upon contact with European explorers. Similarly, most heralded plant crises have been caused by introduced Eurasian pathogens. In addition to the deficiencies in data on introduction failures (Section I), absence of information on opportunities for invasion stymies an assessment of this apparent trend. For example, about half the insect species deliberately introduced into the United States for biological control through 1950 came from Europe, and almost none came from Africa. Because such introductions comprise a large fraction of surviving introduced insect species in the United States, it would not be surprising if the list of survivors had a European accent even if nothing predisposed European species to be more likely to survive. Likewise, several bird-introduction societies flourished in the Hawaiian islands and introduced at least 69 songbirds and doves from Europe, Asia, Australia, and North and South America in an attempt to augment what they saw as a boring avifauna. To my knowledge, no one has released a native Hawaiian species on a continent.

Nevertheless, several authors posit the intrinsic invasion superiority of Eurasian species. Alfred Crosby describes an "ecological imperialism" of Eurasian, and particularly European, species that have changed the face of the earth as they have spread during the past millennium. He attributes their success to a synergism among many species—plants, pathogens, and animals (especially humans)—that combined to make a juggernaut that destroyed native peoples and their ecosystems. In his view, Eurasian plant species evolved with the very grazers that devastated the Americas and Australia, so they were adapted to them, and the grazers underwent thousands of years of evolutionary adaptation to humans as well as to their cultivated plants. Others point out that such grazers need not have been adapted to particular cultivated plant species to have aided Eurasian plants (including uncultivated ones) to spread.

There is no doubt that Eurasian introduced species are disproportionately represented among surviving introduced species throughout much of the world. The reason for this trend cannot be given without much more information on patterns of movement and of persistence and disappearance. It need not be a consequence of inherently greater competitive ability of the average Eurasian species than of those of other regions. For example, Eurasian introductions may have been attempted more often.

### C. Disturbed Habitats and Biotic Resistance

One dogma of the literature on introduced species is that they are particularly likely to invade disturbed habitats, for the same reason that they are thought to invade islands disproportionately: fewer resident species (Section I). The idea is that fewer resident competitors, predators, parasites, and pathogens will present a biotic resistance to the invader in disturbed habitats than in undisturbed ones. Lack of information on failed invasions and opportunity for invasion (Sections I and V,B) besets a quantitative analysis of this idea. However, some kinds of disturbed habitats surely contain large numbers of surviving introduced species. Agricultural fields are heavily and routinely disturbed by site preparation, chemical inputs, and harvest. Their invasion by exotic insects and weeds is legendary. Similarly, freshwater reservoirs, especially in the American West, harbor large numbers of introduced fishes and usually undergo a management regime that entails major disturbance, and exotic plants typically invade forests through trails or disturbances wrought by animals, such as rooting by pigs.

However, some evidence on invasibility of disturbed habitats is ambiguous. Agricultural habitats, for example, are disturbed but are also new habitats created by humans. Agricultural plants provide new habitats for phytophagous insects, which, in turn, constitute new resources for parasitoids and predators. Also, because of their economic importance, these habitats have been particularly well studied, so invasions and their consequences are especially well known. This bias in research effort also casts doubt on claims that pristine habitats are very rarely invaded by biological-control agents introduced for agricultural pests. Western freshwater reservoirs have been the targets of strenuous efforts to ensure that introduced species survive. There are often repeated deliberate releases and various procedures to eliminate "undesirable" native species.

Further, disturbed habitats that seem especially easily invaded are those disturbed by humans. Habitats that routinely undergo natural disturbance may not be particularly prone to invasion by introduced species. For example, old-growth forests of the American South dominated by longleaf pine (*Pinus palustris*) and wire grass (*Aristida* spp.) are maintained by a frequent disturbance—fires. Without fires every few years, they are replaced by hardwoods, yet they do not appear to be easily invaded. In a well-studied longleaf stand, all introduced plants were within 2 m of access roads. The introduced South American fire ant (*Solenopsis invicta*) has heavily colonized all agricultural and residential habitats in this region and has largely replaced the native *S. geminata,* but in old-growth longleaf pine forests, it is found only along roads.

Some observations are consistent with the more general form of the biotic resistance hypothesis—that invasion success is more likely where resident species are fewer. Islands are typified by low species number and seem to be more easily invaded (Section V,A). Similarly, many freshwater habitats of the United States that have been heavily invaded by fishes—reservoirs, coldwater lakes and streams, desert springs—had relatively few native species. However, the exact mechanisms that tend to preclude invaders from species-rich sites and allow them into the depauperate ones are not well understood (Sections II and V,A).

## VI. ERADICATION AND CONTROL

Once an introduced species is believed to be an economic or aesthetic problem, thoughts turn automatically to control measures. Reproduction and dispersal by living organisms greatly complicate control. In the earliest stage of an introduction, when geographic spread is still proceeding slowly (Section III), eradication may be possible. During the first 30 years after the gypsy moth invasion, citizens of Massachusetts attempted to eradicate it by a combination of strenuous procedures and actually succeeded in greatly depressing populations. However, the campaign was then halted because damage was so low. Now that this invasion covers such a huge range, complete eradication seems impossible, except perhaps at isolated new foci generated by jump-dispersal.

If an animal is not very dispersive, a sufficiently sustained and comprehensive effort may eliminate it. The coypu escaped from English fur farms in 1932 and soon established a population in East Anglia that undermined riverbanks, greatly damaged marsh vegetation, and ate crop plants. A 10-year, sophisticated trapping campaign eradicated them by 1990. This success rested on the determination to see the project through to completion and the fact that individual coypus do not move far. A similar campaign to control the introduced American mink was unsuccessful. Though the mink has a lower rate of population increase than the coypu, individuals can move much farther. Even dispersive introduced species can be eradicated if their movements are greatly restricted, as on a small island. Rabbits have been exterminated on Round Island (151 ha) near Mauritius and on Philip Island (190 ha) of the Norfolk group between Australia and New Zealand.

Most eradication campaigns, however, are not likely to succeed. The United States Department of Agriculture has coordinated numerous campaigns to eradicate introduced pest insects with few successes. For example, the Mediterranean fruit fly (*Ceratitis capitata*) has been the target of repeated expensive eradication campaigns in California; the effort in 1981–1982 alone cost $150 million. Sophisticated techniques are combined, such as blanket aerial spraying of chemical insecticide and mass rearing, sterilization, and release of males to induce females to mate fruitlessly. Yet the frequency and spatial proximity of appearances of new individuals suggest that the effort has failed and that an ongoing population is probably established.

Control rather than elimination is therefore the usual goal. Myriad methods are used. For plants, mechanical methods of control are often possible, though these may be very labor-intensive. Native vegetation has been restored to one area on the Colorado River after salt-cedars were largely removed by bulldozing and their roots destroyed by a large device hauled behind the bulldozer. In northern California, the introduced bush lupine (*Lupinus arboreus*) is partially controlled at a Nature Conservancy reserve by periodic volunteer "bush bashes."

For pest insects and plants, chemical pesticides continue to be heavily used. Broad-spectrum chemicals such as DDT led to numerous ecological and health problems, including biological magnification and pollution. Thus they have been replaced by chemicals with more focused effects, such as hormones. Two general problems with chemical control are cost and the evolution of resistance in target species, which exacerbates the cost problem. American farmers spend at least $2 billion annually on about 450 million kg of insecticides and herbicides.

A popular alternative to chemical control is biological control, the introduction of nonnative antagonists of the pest—predators, parasites, or pathogens. Biological control is not new; over a thousand years ago the Chinese used green weaver ants (*Oecophylla smaragdina*) to control caterpillars and beetles on citrus. Nowadays biological control is a much more systematic and sophisticated effort, and about 20% of all projects lead to economically significant control of the target pest with few other evident direct costs. Biological control is often touted as an environmentally safe alternative to pesticides, but the evidence is equivocal. Early biological-control introductions have often been problematic. The small Indian mongoose was introduced into the West Indies, Hawaiian Islands, Mauritius, and Fiji to control introduced rats. Its record on rat control is disappointing, but it has extinguished endemic island snakes and lizards and contributed to the decline of native birds. Similarly, the predatory snail *Euglandina rosea* from Florida and Central America was introduced onto islands throughout the world to control the giant African snail (*Achatina fulica*), introduced for food. It has never been effective in this role but has been implicated in extinctions of endemic Hawaiian and Tahitian snails.

Despite recently heightened attention to such unintended side effects, the ability of living organisms to disperse well beyond their area of release and the absence of intensive monitoring in many habitats are worrisome. For example, the Argentine moth *Cactoblastis cactorum* was introduced on Nevis in the Leeward Islands, West Indies, to control various species of prickly pear (*Opuntia*). In 30 years it island-hopped by its own flight all the way to Florida, where it has already eliminated from the wild a candidate endangered species, the semaphore cactus (*Opuntia spinosissima*).

## VII. CONCLUSION

The thread that unites the various parts of the introduced-species tapestry is unexpected effects. Many introductions have been deliberate—for biological control, sport hunting and fishing, ornamental plants and birds, crops or timber, "humane" disposal of unwanted pets. Even those conducted by scientists have led to numerous surprising ecological and economic problems. Unintended introductions, through unregulated or loosely controlled commerce and travel, have similarly caused incalculable damage. The particular forms of damage are often complex and could only have been predicted, if at all, by lengthy, detailed scientific study. Further, eradication has proven extremely difficult, and control is often expensive and incomplete. The clear conclusion is therefore that any proposed introduction for any purpose, even to remedy the effects of a prior introduction, should be viewed with great caution. Further, costs of prevention are likely to be far less than those of a cure after introduction has occurred.

## Glossary

**Allelopathy** Negative chemical effects of plants on one another.

**Biological control** Use of natural enemies (predators, parasites, and pathogens) to control pest populations.

**Biological magnification** Increasing concentration of a chemical substance in increasingly higher levels of a food web.

**Competition** Interaction in which organisms of the same or different species use a common resource that is in short supply or interfere with one another's access to such a resource.

**Propagule** Group of individuals capable of initial population increase in a site. For many species, a single fertilized female or an adult female and male constitute a propagule.

## Bibliography


Crosby, A. W. (1986). "Ecological Imperialism. The Biological Expansion of Europe, 900–1900." Cambridge, England: Cambridge University Press.

Drake, J. A., Mooney, H. A., diCastri, F., Groves, R. H., Kruger, F. J., Rejmanek, M., and Williamson, M., eds. (1989). "Biological Invasions: A Global Perspective." Chichester, England: John Wiley & Sons.

Ebenhard, T. (1988). Introduced birds and mammals and their ecological effects. *Viltrevy* (*Swedish Wildlife Research*) **13**(4), 1–106.

Elton, C. S. (1958). "The Ecology of Invasions by Plants and Animals." London: Chapman & Hall.

Groves, R. H., and Burdon, J. J., eds. (1986). "Ecology of Biological Invasions." Cambridge, England: Cambridge University Press.

Hengeveld, R. (1989). "Dynamics of Biological Invasions." London: Chapman & Hall.

Kauffman, W. C., and Nechols, J. E., eds. (1992). "Selection Criteria and Ecological Consequences of Importing Natural Enemies." Lanham, Md.: Entomological Society of America.

Long, J. L. (1980). "Introduced Birds of the World." New York: Universe.

Mooney, H. A., and Drake, J. A., eds. (1986). "Ecology of Biological Invasions of North America and Hawaii." New York: Springer-Verlag.

Spongberg, S. A. (1990). "A Reunion of Trees." Cambridge, Mass.: Harvard University Press.

# Introduced Species, Zebra Mussels in North America*

**Don W. Schloesser**
*National Biological Service*



For thousands of years, ranges of aquatic organisms have been changing in response to natural dispersal mechanisms and environmental restrictions. This natural process almost always takes decades to occur and rarely has been perceived as introducing new organisms into new geographically distant areas. However, human activities have accelerated the process of species dispersal several hundredfold during the past two centuries. For example, of 1354 introductions of fishes (237 species) into 140 countries, only 7 introductions have been classified as ancient, occurring before 1800, when major marine transportation systems were initiated. It has been suggested that the rate of new fish introductions has declined on a global scale because the majority of fishes have been introduced into most suitable recipient areas, thus creating a globally common fish fauna.

During the past 100 years, many aquatic exotic species (i.e., species introduced from a foreign country) have been transported from other continents and become established in North America. Within the Laurentian Great Lakes, there are at least 139 established exotic species, and the existing fauna and flora no longer resemble the biota present before Europeans settled the area. Although few detailed analyses have been performed, one such analysis indicated that the rate of establishment of exotic species has accelerated in the last 30 years. For example, in the mid-1980s, several exotic species of economic and ecological importance were introduced into the Laurentian Great Lakes, including fish (ruffe, *Gymnocephalus cernuus;* round goby, *Neogobius melanostomus;* tubenose goby, *Proterorhinus marmoratus*), zooplankton (spiny water flea, *Bythotrephes cederstroemi*), and mollusks (zebra mussels, *Dreissena* spp.). Reasons for the increased rate and survival of these recent introductions are unknown. However, the advent of cleaner waters, which allow increased species diversity in harbors where ballast water is obtained, and faster transoceanic shipping, which moves species between continents, probably has contributed to the recent increase in exotic species in North America. These

* Contribution 847 of the National Biological Service, Great Lakes Science Center, 1451 Green Road, Ann Arbor, Michigan 48105.

introductions would likely have continued without serious attention if the zebra mussel, *Dreissena polymorpha,* had not become an important aquatic nuisance that threatens to disrupt human use of water.

The discovery of zebra mussels in North America in 1988 raised concern for water users because the species can become abundant enough to obstruct the flow of water in human-made structures such as pipes and screens. This work reviews the biology, distribution, and impacts of zebra mussels in the context of its discovery in the Laurentian Great Lakes and its impending spread to most surface waters of North America.

## I. INTRODUCTION

Zebra mussels (*Dreissena* van Beneden 1835) are small bivalve mollusks that occur throughout most of Europe and western portions of the former Soviet Union (Fig. 1). In 1988, zebra mussels were discovered in Lake St. Clair, located between Lakes Huron and Erie in the Laurentian Great Lakes. The potential for introduction of zebra mussels into North America by human activities was recognized in 1921 by Johnson: "The possibility of . . . the zebra mussel being introduced [into the United States] is very great." The most probable dispersal mechanism for zebra mussel introduction into North America was recognized (Bio-Environmental Services) in 1981: "The occurrence of its veliger larvae in the plankton . . . greatly enhances its potential for introduction to the [Laurentian] Great Lakes in ship ballast water."

These anecdotal predictions and later discovery of zebra mussels in a major world-wide shipping corridor indicate that mussels were introduced into North America with the discharge of ballast water obtained from a freshwater harbor in Europe or the former Soviet Union, where mussels were present. Mussels are believed to have been discharged as free-swimming larvae (i.e., veligers) in 1986, and they quickly became established in Lake St. Clair and western Lake Erie. Zebra mussels can live in a broad range of environmental conditions and will probably spread to most surface waters throughout North America. [*See* INTRODUCED SPECIES.]

***FIGURE 1*** Zebra mussels (*Dreissena polymorpha*) found throughout most of Europe and portions of the former Soviet Union. [From T. F. Nalepa and D. W. Schloesser, eds. (1993). "Zebra Mussels: Biology, Impacts, and Control." CRC Press, Boca Raton, Fla.]

## II. TAXONOMY

Zebra mussels are separated from other closely related taxa by the classification scheme shown in Table I. Zebra mussels are in the phylum Mollusca, class Bivalvia, subclass Lamellibranchia, superorder Eulamelliobranchia, order Veneroida, and family Dreissenidae. Although zebra mussels are not closely related to marine mytilid mussels (e.g., *Mytilus* spp.), there are many similarities between the two groups, including shell shape, heteromyarian anatomical muscles, larval veliger stage, and byssate attachment. Both species are also important biofouling organisms. Some native North American taxa, such as the false dark mussel (*Mytilopsis*

**TABLE I**
Classification of Zebra Mussels and Other Close Evolutionarily Related Taxa

Phylum Mollusca
- Class Bivalvia ( =Pelecypoda)
  - Subclass Lamellibranchia
    - Superorder Filibranchia (characteristics: usually attach by byssal threads and are primarily found in marine habitats)
      - Order Mytiloida
        - Mytilidae (mussels)
        - Ostreidae (oysters)
        - Arcidae (arc shells)
        - Anamidae (jingle shells)
    - Superorder Eulamelliobranchia (characteristics: rarely attach by byssal threads and live primarily in freshwater habitats)
      - Order Veneroida
        - Sphaeriidae (fingernail clams)
        - Unionidae (freshwater mussels)
        - Corbiculidae (Asiatic clams)
        - Dreissenidae (zebra mussels)

*leucophaeta*), may be mistaken for zebra mussels. Most taxa that look similar to zebra mussels in North America are close phylogenetic taxa that can be separated by examination of shell morphology.

The shell of the zebra mussel is a distinctive feature of the genus, but high variability exists between individuals within the genus and species (Fig. 2). The common name, zebra mussel, is derived from the pattern of stripes on the shell, which may resemble a mammalian zebra. The commonly applied species name, *polymorpha* (Pallas), refers to many different forms of shell color and pattern. In Europe and the former Soviet Union, shell patterns have been used to differentiate between geographical races of mussels.

Species nomenclature within *Dreissena* is not well agreed upon. Currently, there seem to be three or four recognized taxonomic authorities listing between three and seven existing species of *Dreissena*. However, well over 100 subjective species have been named in the European and former Soviet Union literature. Up to 20 species of existing and fossilized forms of *Dreissena* have been scientifically described. Presently, several commonly found species listed in the European literature include *D. polymorpha, D. distincta,* and *D. bugensis*. In

***FIGURE 2*** Six shell types of zebra mussels (*Dreissena polymorpha*) demonstrating the high variability in shell color and pattern found in Europe and the former Soviet Union. [From T. F. Nalepa and D. W. Schloesser, eds. (1992). "Zebra Mussels: Biology, Impacts, and Control." CRC Press, Boca Raton, Fla.]

addition to shell color and pattern, some variability occurs in length, width, and height of shells of mussels of the same age within a population. Morphological and genetic analysis has identified two species of zebra mussels in North America: *D. polymorpha* and *D.* (*rostriformis*) *bugensis*. The external morphological distinction between the *D. polymorpha* and *D. (r.) bugensis* is primarily in shell shape. *D. (r.) bugensis* has a rounded edge and the zebra mussel has an angular edge separating the bottom surface of the shell from the top surface. The taxonomy of the Dreissenidae in Europe and North America is presently being evaluated using morphological and genetic characteristics.

## III. SPECIES CHARACTERISTICS

Zebra mussels successfully occupy a habitat selected by few other freshwater species. Zebra mussels have unique characteristics in life cycle, feeding, benthic existence, habitat selection, reproductive potential, and population structure that allow them to be one of the most serious aquatic pest species in the world.

### A. Life Cycle

The life cycle of zebra mussels is similar to that of marine bivalves, but it is unique among freshwater bivalves (Fig. 3). Development of eggs and sperm (i.e., gametogenesis) occurs during the cold season, when a vast majority of the population exists as juveniles and adults. Gametogenesis is usually complete in early to late spring. Spawning begins in spring when water temperatures increase to 10–15°C. Eggs and sperm are released into the water, where external fertilization occurs. The sex ratio of mature zebra mussels is generally one to one. Release of eggs and sperm in temperate climates begins in spring, peaks (one or more pulses) during summer (optimum 20–22°C), and decreases in fall as water temperatures decrease. Spawning usually lasts 4 to 5 months in temperate climates but may be extended up to 8 months when environmental requirements are optimal. Eggs released from mature mussels need to be fertilized within 5 hr and remain viable up to 22 hr. For successful fertilization, the water temperature must exceed 10°C. Within 1 day after fertilization, planktonic larvae usually develop and begin to swim.

Planktonic life stages of zebra mussels (generally termed veligers) are usually completed in 1 to 2 weeks but may last as long as 4 weeks. There are five planktonic and pseudoplanktonic forms of mussels: egg, trochophore, straight-hinged, umbonal, and pediveliger. Each of the five forms has distinct morphological characteristics: eggs are fertilized in the water and are embryological in form; trochophore forms have cilia (i.e., velum) but lack shell structures; straight-hinged forms have straight hinge shells and tufts of velum; umbonal forms have rounded hinges and reduced velum; and pediveliger forms lose velum and begin to search substrates for attachment. Planktonic forms grow 4 to 7 $\mu$m/day. Early settled mussels may be easily resuspended in turbulent water. Once they become benthic, individuals attach to substrates, and the juvenile stage begins.

Most juvenile mussels become mature when they reach 6–10 mm in length, and a few may mature as small as 5 mm. Because larvae are primarily produced in middle to late summer, most juvenile mussels do not attain reproductive size before the onset of colder weather; therefore, they do not reproduce until the following year. However, some juvenile mussels spawned in early summer may attain sexual maturity and reproduce in the year in which they are spawned. Mussels usually grow to a minimum of 10 mm long in their first year of life, with most individuals being sexually mature. Juvenile and adult mussels may live 2 or 3 years in the Laurentian Great Lakes. In Europe and the former Soviet Union, some individual mussels are believed to live for up to 7 years, although the typical age is between 4 and 5 years.

### B. Feeding

Zebra mussels feed by filtering food from the water. Filtering begins in the larval stage about 24 hr after fertilization. In the absence of acceptable food, larvae remain active for 1 to 2 weeks at optimal temperatures and then die. Veligers ingest small food particles (i.e., $<4$ $\mu$m in diameter) and are highly sensitive to the type of food needed for development. Feeding stops for 1–2 days after larvae settle out of the water. Settled mussels quickly metamorphose into juvenile mussels and active filter feeding resumes. In general, mussels continually filter water for food throughout their juvenile and adult states. However, in the absence of food in the water, juvenile and adult mussels may survive for extended periods (i.e., 6–12 months) without feeding.

Mussels effectively remove all suspended materials from the water they filter. Suspended materials are either excreted as feces or encapsulated in mucus and discarded as pseudofeces. Feces is excreted from mussels after available nutrients have been ingested and utilized for metabolic needs such as growth, respiration, and reproduction. Pseudofeces includes discarded materials that are filtered out of the water but not used for metabolism. Both feces and pseudofeces may remain separated from the water as sediment or they may be resuspended by biological and physical activities.

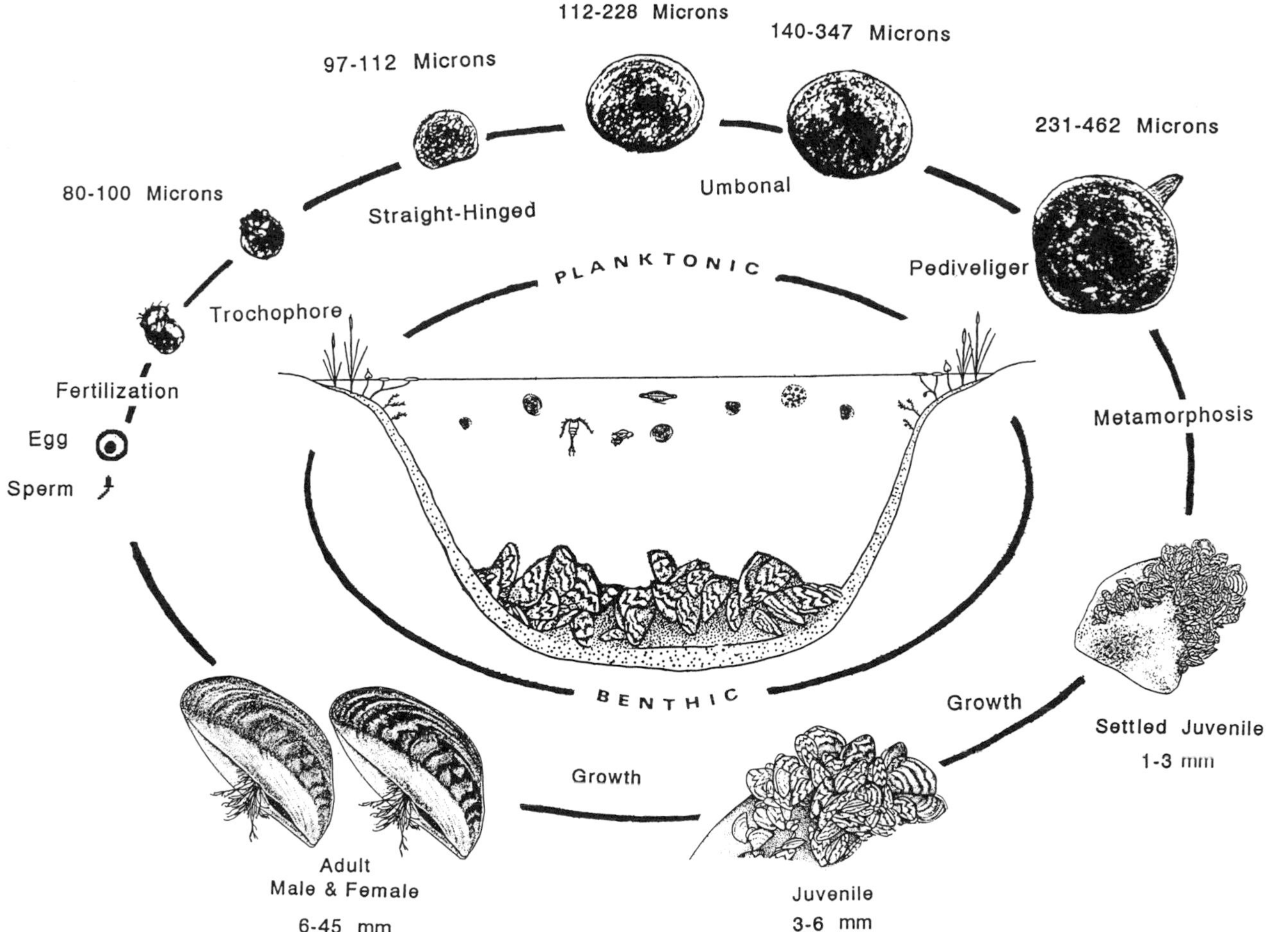


***FIGURE 3*** Idealized life cycle of zebra mussels (*Dreissena* spp.).

Populations of zebra mussels can affect the suspended particulate concentration in the water by their filtering activity. An individual mussel of average length (ca. 20 mm long) filters about 1 liter/day. In the Netherlands, a relatively small population density of 600/m$^2$ was found to filter an entire lake in 12 days, resulting in a substantial decrease in suspended materials in the water column and an increase in water clarity. In western Lake Erie, an increase in water clarity of 85% and a decrease in chlorophyll-a of 43% between 1988 and 1989 are attributed to increased numbers of zebra mussels. This filtering by zebra mussels has been calculated to exceed that of all other forms of filter feeders, including herbivorous zooplankton. A theoretical population density of 7400 mussels/m$^2$ in western Lake Erie (far below the observed maximum mean range of 112,000 to 342,000/m$^2$ observed) could filter the basin in 1 day. In Lake St. Clair of the Great Lakes, relatively low densities of adult mussels (i.e., <1000/m$^2$) have been estimated to consume 4 to 18% of the annual phytoplankton production. Predictive models of filtering by zebra mussels have determined that they have the potential to reduce the chlorophyll of large bays of the Great Lakes by 80%.

Zebra mussels are dependent on food, and it is believed that their success as biofoulers is partially related to feeding. Water movements around zebra mussels result in relatively high populations of mussels. Water flowing by mussels delivers food and other environmental requirements, such as oxygen and calcium, and removes toxic by-products of feces and pseudofeces from the immediate area. Faster growth rates and higher densities of mussels have been found in relatively fast-moving waters of lakes, rivers, and pipes, compared with slower-moving waters in the same water body. In addition, unidirectional flow of water through pipes seems to be more beneficial to mussels than the turbulent water flow typical of natural lake systems and rivers with impediments to laminar flow.

### C. Benthic Existence

Juvenile and adult zebra mussels attach to any firm substrate by the use of byssal threads (Fig. 4). Mussels will attach to rocks, wood, debris, crayfish, native clams, breakwalls, docks, navigation structures, and pipes. The byssate form of existence enables zebra mussels to attach to hard surfaces and contributes to fouling of human-made structures.

The evolution of byssate structures and the triangular heteromyarian shell of ancestral clams have resulted in mussel-type organisms, such as zebra mussels, that are very successful at filling a niche

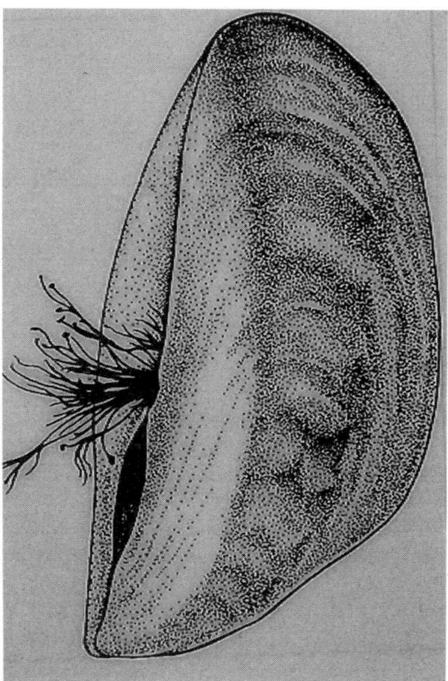

***FIGURE 4*** Zebra mussel (*Dreissena polymorpha*) showing ventrally located byssal threads used to hold the organism to firm surfaces.

not occupied by other mollusks. The retention of the larval byssus (used in most bivalves to maintain stability during metamorphosis) into adult life enabled dreissenids to escape the benthic-burrowing niche of early ancestors. Bysally attached bivalves are thought to have given rise to cemented (e.g., oysters), swimming (e.g., scallops), and boring (e.g., tube worms) bivalves. In effect, byssal structures are a modified extension of the foot used by clams to burrow into sediments. The evolution of the triangular heteromyarian shell has been important to substrate utilization because it allows mussels to reduce the amount of energy required to remain byssate in water currents. Mussels may orient shells to reduce resistance (e.g., anterior end toward currents), whereas homomyarian shell forms create substantial current resistance. In addition, the ventral widening of mussel shells results in a downward push on shells caused by water currents. Evolution of zebra mussels into organisms living on substrates has resulted in the inability of these organisms to occupy the niche from which they evolved; in effect, zebra mussels cannot live in sediments as did their early ancestors. However, zebra mussels have been observed to live on soft mud in unique environmental conditions by forming interconnecting mats covering the bottom and as very small individuals that do not sink in mud.

As many as 200 threads may hold a single mussel to a substrate. There seem to be two distinct types of threads, temporary and permanent, that differ in length, thickness, number, and attachment arrangement. Temporary threads are longer (about 50%), thinner (about 40%), and fewer in number (less than 10 versus several hundred) than permanent threads. Temporary threads are usually laid down in a tripod manner, attaching within hours after a mussel contacts a firm surface. These threads can easily be detached by a mussel, thereby aiding the mussel in search of adequate substrate type. Permanent threads are shorter, thicker, and greater in number than temporary threads and are laid down directly below the shell at a rate of about 5 to 12 per day. Permanent threads are not usually detached. However, mussels as large as 2 cm may detach and search substrates at a rate of about

25 cm/hr. In general, older mussels attach more quickly than young mussels, and most thread production is complete within 1 week after temporary attachment. Threads can be added at any time in response to varying water velocities and resulting stresses caused by attempts to remove mussels from substrates. Mussels may remain attached to substrates at water velocities up to 2.5 m/sec, although settling is hindered at velocities exceeding 2.0 m/sec.

### D. Habitat Selection

Although it is the unique byssus structure that allows individual zebra mussels to adhere to firm substrates, it is their preference to aggregate together in limited microhabitats that creates the collective biofouling characteristics of the organism. In early settlement and juvenile stages of development, mussels actively search and select microhabitats that are similar to pipes in many ways. For example, colonization of cement blocks indicates that 72% of mussels occur in block holes, 24% on sides, and 4% on tops. Mussels search for areas of consistent water flow and minimal pressure changes (e.g., nonwave zone). In laboratory experiments, up to 80% of early settlement mussels chose to become byssate within tubes versus 20% outside tubes. In the natural environment, newly settled mussels are often found concentrated in areas at the periphery of adult mussel beds and later move to areas occupied by adult mussels. This implies that mussels aggregate in pipes as a response to favorable habitat conditions, as well as being deposited by settlement. Whether mussels chose pipe-type environments as a response to hydrodynamic or chemical factors, genetic tendencies to aggregate, or a combination of these is unknown.

### E. Reproductive Potential

Zebra mussels release enormous quantities of eggs and sperm into the water for external fertilization. This type of reproduction is a very successful strategy that increases the genetic diversity, survival, and dispersal of the species. Zebra mussels have high fecundities, ranging from 10,000 eggs per female in their first year of life to 1,700,000 per female in their third and fourth years. Greater numbers of male than female zebra mussels may also increase fertilization through the random chance of many sperm contacting one egg. High sperm concentrations of 1,000,000 per milliliter of water may occur in highly colonized areas. Continual release of gametes with one or more peak spawning events over many months results in some seasonal reproductive success in most water bodies. The lack of adult zebra mussels protecting developing young from environmental factors is a major cause for extreme seasonal and annual variation in the numbers of juvenile and adult mussels in the environment.

The survival of enormous numbers of fertilized eggs, planktonic veligers, and newly settled veligers is highly variable and difficult to relate to specific environmental parameters. Potential wide dispersal of relatively abundant planktonic larvae increases survival if suitable habitat conditions, such as pipes, are found. However, young planktonic and early settled mussels are most vulnerable to mortality and typically account for 99% of the total life cycle mortality when suitable habitat conditions do not exist. About 10 to 40% mortality may occur in the planktonic stages and 25 to 99% mortality in early settlement. When environmental conditions are optimal, early settled veligers may attain densities as high as $1,000,000/m^2$. However, densities of juveniles and adults rarely exceed $30,000/m^2$. In general, environmental tolerances of larvae increase with increasing size. Generally, water temperature must be between 10° and 24°C and pH must be between 7.4 and 9.4. Calcium ions must be above 40 mg calcium per liter and oxygen levels must be above 20% saturation. Early stages of larval development are very sensitive to turbulent water, availability of food, and other environmental conditions. Other factors affecting larval mortality may include predation by fish larvae, predation by zooplankton, ingestion by filter-feeding zebra mussels, and bacterial attack.

### F. Population Structure

The population structure of juvenile and adult zebra mussels, as determined by length–frequency analysis, indicates that growth rates of mussels in different populations are highly variable, and the abundance of mussels may change dramatically within a short period. In Europe, growth rates may vary by twofold within the same lake. However, two basic types of mussel populations can be distinguished based on shell size and growth. One type is long-lived (some individuals living up to 7 years), slow-growing (i.e., maximum growth of young less than 1 cm/year), and relatively small (maximum shell size <35 mm). The other type is short-lived (few individuals living up to 5 years), fast-growing (i.e., maximum growth greater than 1 cm/year), and relatively large (maximum shell size >40 mm).

In the Great Lakes, mussel populations resemble the short-lived, fast-growing, relatively large-shelled population type. In Lake St. Clair, growth rates of mussels have been shown to be exceedingly fast, as much as 0.5 mm/day, compared with 0.07 mm/day in Dutch lakes. The maximum length of mussels in western Lake Erie (i.e., 4.2 cm) is similar to the maximum shell length of fast-growing populations in British lakes (4.0 cm). However, between 1988 and 1992, the mean and maximum lengths of mussels in western lake Erie decreased, possibly due to environmental conditions and decreased food supplies caused by the filtering of high numbers of mussels.

Changes in the population structure of mussels in western Lake Erie near Monroe, Michigan, indicate that the abundance of mussels may change rapidly and dramatically (Fig. 5). In February 1989, two distinct age groups were present in relatively low numbers: 1-year-old mussels accounted for 22% (250 mussles) and 2-year-old mussels made up 78% (900 mussels) of the sampled population. This relatively small population accounted for the reproductively viable sampled population between February and July 1989. Then, in July, zebra mussel reproduction entered logarithmic reproduction and survival. Veligers increased by 733%, and juvenile and adult mussels increased by 300% in the summer (compared to February). By August, numbers of newly spawned young-of-the-year mussels accounted for 99% (67,106 mussels) of the sampled population.

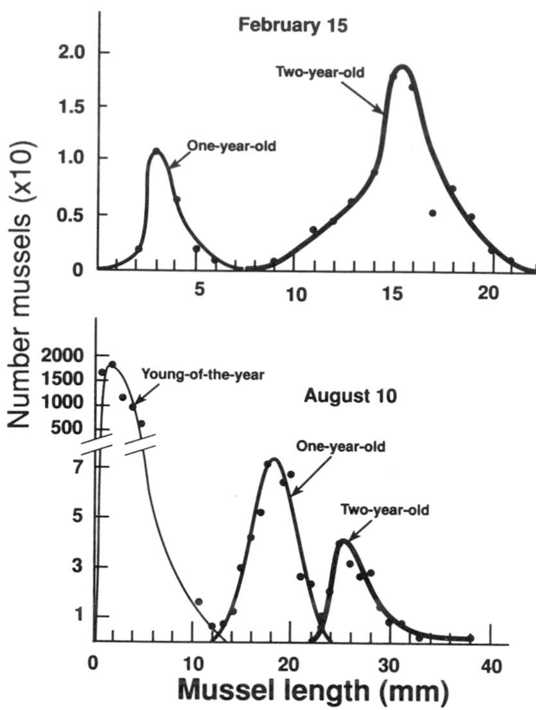


***FIGURE 5*** Length–frequency distributions of zebra mussels (*Dreissena polymorpha*) in western Lake Erie February 15 and August 10, 1989. [Modified from D. W. Schloesser and W. Kovalak (1991). Infestation of unionids by *Dreissena polymorpha* in a power plant canal in Lake Erie. *Journal of Shellfish Research* **10,** 355–359.]

Together, 1- and 2-year-old mussels accounted for less than 1% (600 mussels) of the sampled mussels. Observations of mussel settlement in the summer of 1989 indicate that most of the increase in mussel numbers occurred during a 6-week period between June and August, when conditions were optimal for mussel survival. The maximum mean density of mussels observed on rock substrates in open waters of western Lake Erie was 340,000/m$^2$.

## IV. DISTRIBUTION

### A. Dispersal Mechanisms

The number of natural and human-mediated dispersal mechanisms of zebra mussels is high for a potential invader species (Table II). Most of the natural dispersal of mussels is mainly attributable to downstream movements of water, which carries planktonic mussels and sedentary and attached mussels

**TABLE II**
**Natural and Human-Mediated Dispersal Mechanism and Direction of Dispersal of Zebra Mussels (*Dreissena polymorpha*)**

| Dispersal mechanism | Direction | | | |
|---|---|---|---|---|
| | Upstream | Downstream | Overland | Intercontinental |
| Natural | | | | |
| Currents | | X | | |
| Animals | X | X | X | |
| Human | | | | |
| Canals | X | X | | |
| Ballast water | X | X | | X |
| Ship exteriors | X | X | | |
| Ship interiors | X | X | | X |
| Fishing vessel wells | X | X | | |
| Navigation buoys | X | X | | |
| Marine equipment | X | X | X | |
| Fishing equipment | X | X | X | |
| Fish culture | X | X | X | X |
| Bait release | X | X | X | X |
| Aquarium release | X | X | X | |
| Firetruck water | X | X | X | |
| Amphibious planes | X | X | X | |
| Scientific research | X | X | X | |
| Intentional release | X | X | X | X |

colonizing drifting debris. In addition, juvenile and adult mussels can possibly be dispersed by extending mucous threads (up to 15 mm long) from the surface of the water by capillary action and drifting with surface water currents. Dispersal caused by turtles, crayfish, and birds may occur but is not well documented for zebra mussels. By far, the greatest dispersal vector of zebra mussels is humans (Table II). In Europe, most mussels spread in the eighteenth and nineteenth centuries through human activities, especially the development of interconnecting waterways associated with the Industrial Revolution. For example, zebra mussels invaded the lower Rhine River in 1835 but did not invade upstream waters of the Rhine until the late 1960s, when mussels were probably transported via motorboats. The high number of possible overland dispersal mechanisms (Table II) will increase the speed and extent of mussel invasion in North America. Intercontinental dispersal mechanisms of zebra mussels may result in North America being the source of introduction to other continents.

### B. Europe

The distribution of zebra mussels before the nineteenth century was restricted to very small areas in and adjacent to the Black, Caspian, and Azov seas. The greatest influences on the geographical range of zebra mussels were glacial epochs, which led to the restriction and slow redispersal of mussels as glaciers advanced and retreated over periods of tens of thousands of years. Glacial geographical separation of mollusks in small isolated areas and natural selection resulted in the evolution of many species of *Dreissena*. The range of zebra mussels was determined primarily by natural dispersal mechanisms and local environmental restrictions. The limited distribution of zebra mussels in 1800 was a result of thousands of years of natural processes (Fig. 6A).

Between 1800 and 1900, zebra mussels more than doubled their range in Europe (Fig. 6B). Mussels spread mainly in rivers and canals interconnected by humans for trade. The doubling of the range of mussels in only 100 years exceeded that which had occurred during the previous several thousand

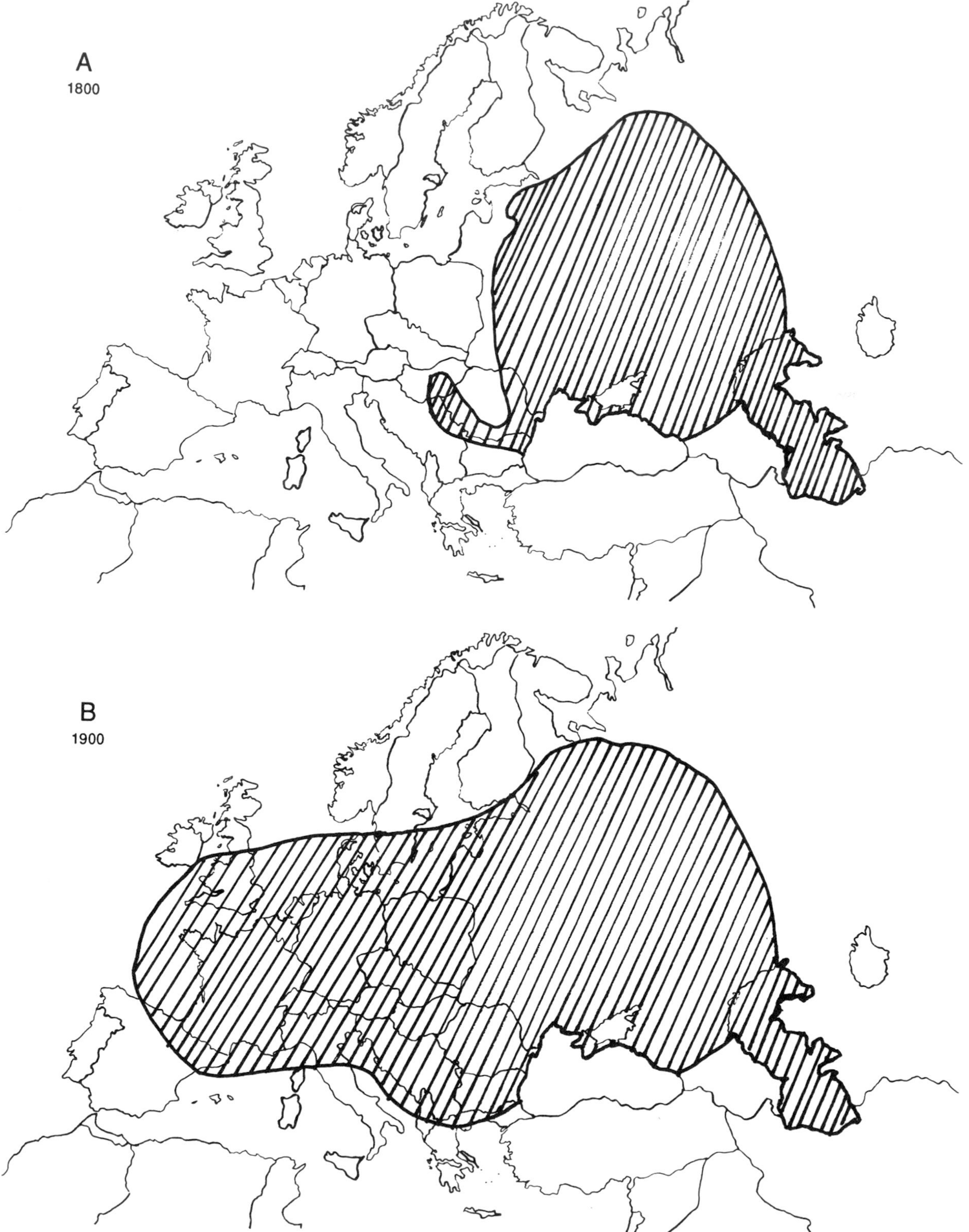


***FIGURE 6*** Distribution of zebra mussels in Europe and the former Soviet Union in (A) 1800, (B) 1900, and (C) the present.

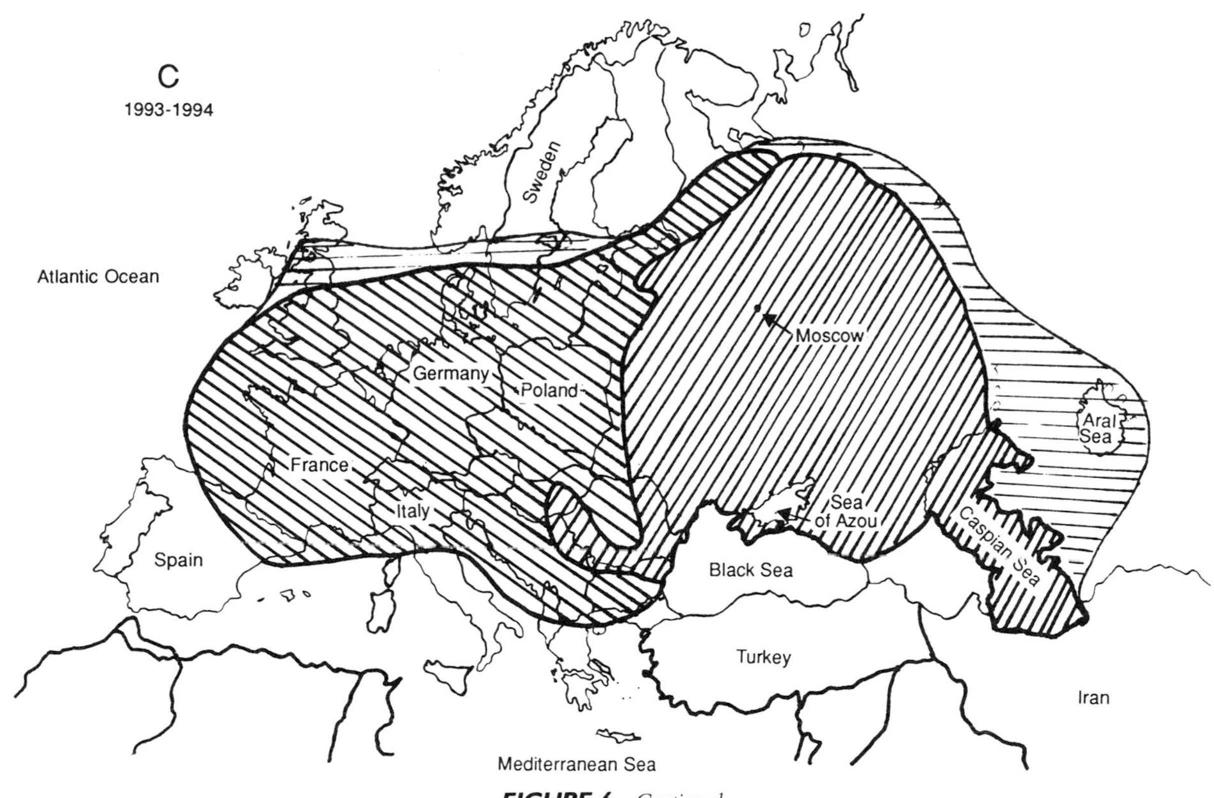


**FIGURE 6** *Continued*

years. Human-mediated dispersal mechanisms (Table II) provided transporation opportunities to zebra mussels that resulted in its common name of wandering mussel ("Wundermussel").

Mussels may still be expanding their range and changing local distribution in response to human-induced changes in water characteristics, such as trophic status. The absence of mussels from some areas of Europe, such as the Iberian Peninsula and Italy, is attributed to mussels not yet being dispered to those areas (Fig. 6C). Waters in Scandinavia are believed to be free of mussels because most of the waters in this region are too soft to support shell formation. In Poland, mussels have been extirpated in some polluted waters and have recolonized some waters where pollution has decreased.

### C. North America

The zebra mussel was discovered in June 1988 in Lake St. Clair of the Laurentian Great Lakes (Fig. 7B). Length–frequency analysis of known populations in 1988 and review of studies performed before 1988 indicate that mussels became established in Lake St. Clair and the western most portion of Lake Erie downstream from Lake St. Clair in 1986 (Fig. 7A). This initial founding population occurred in a range about 75 km north to south (N–S) by 25 km east to west (E–W). The relatively large initial range is supported by genetic studies of mussels that indicate the introduced population contained a high level of genotypic diversity. Apparently, the population was established by a substantial number of individuals and did not undergo a bottleneck of genetic reduction after introduction, or several introductions occurred within a short period.

In 1988, zebra mussels occurred downstream of the founding population throughout two-thirds of Lake Erie—a range of 110 km N–S by 225 km E–W. The range of mussels expanded between 1986 and 1988, primarily through natural mechanisms (e.g., veliger drift). In 1988, juvenile and adult mussels were seen on beaches, drifting debris,

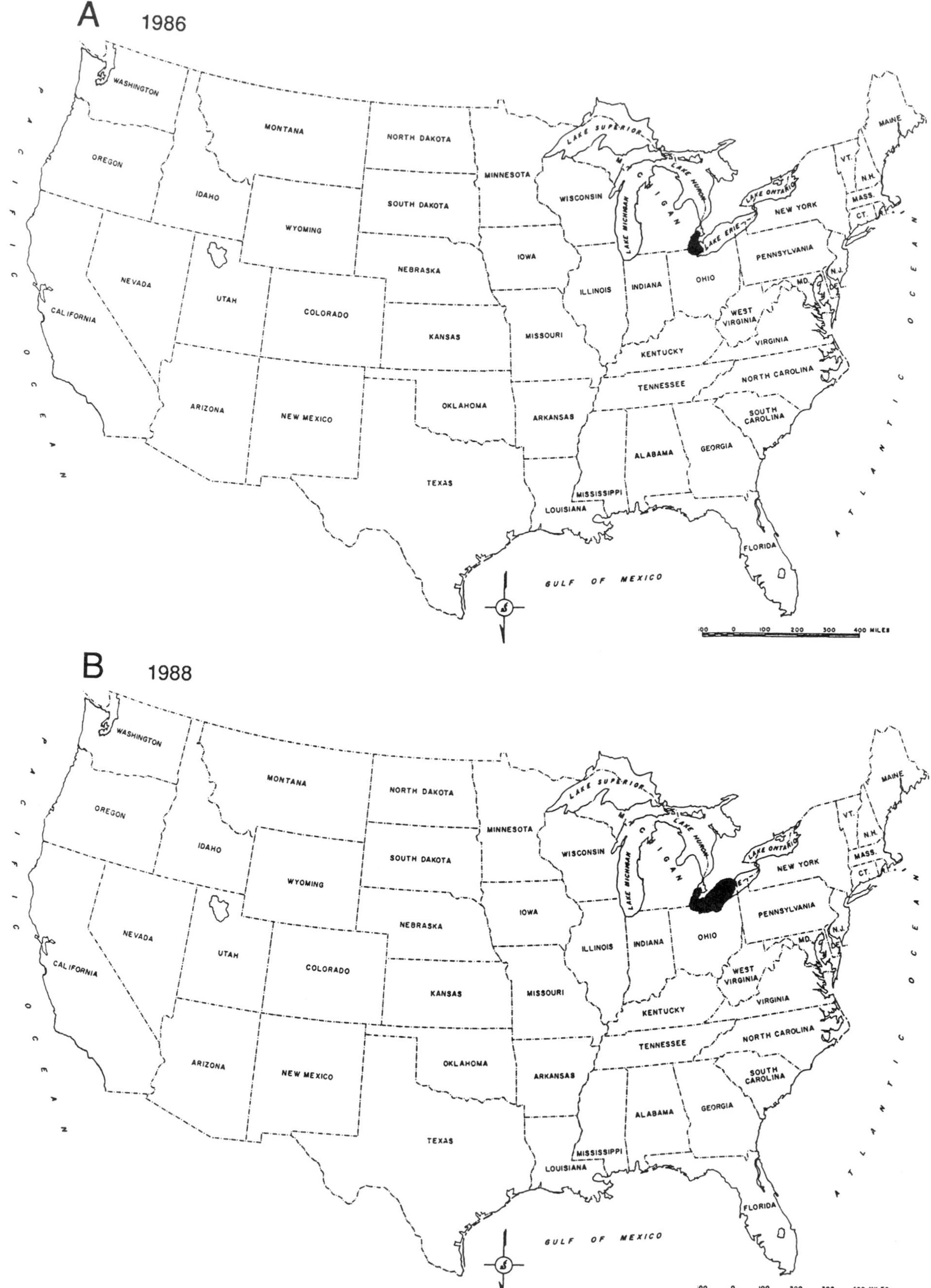


***FIGURE 7*** Known distribution of zebra mussels (*Dreissena polymorpha*) in North America in 1986, 1988, 1990, 1992, and 1994 (A–E).

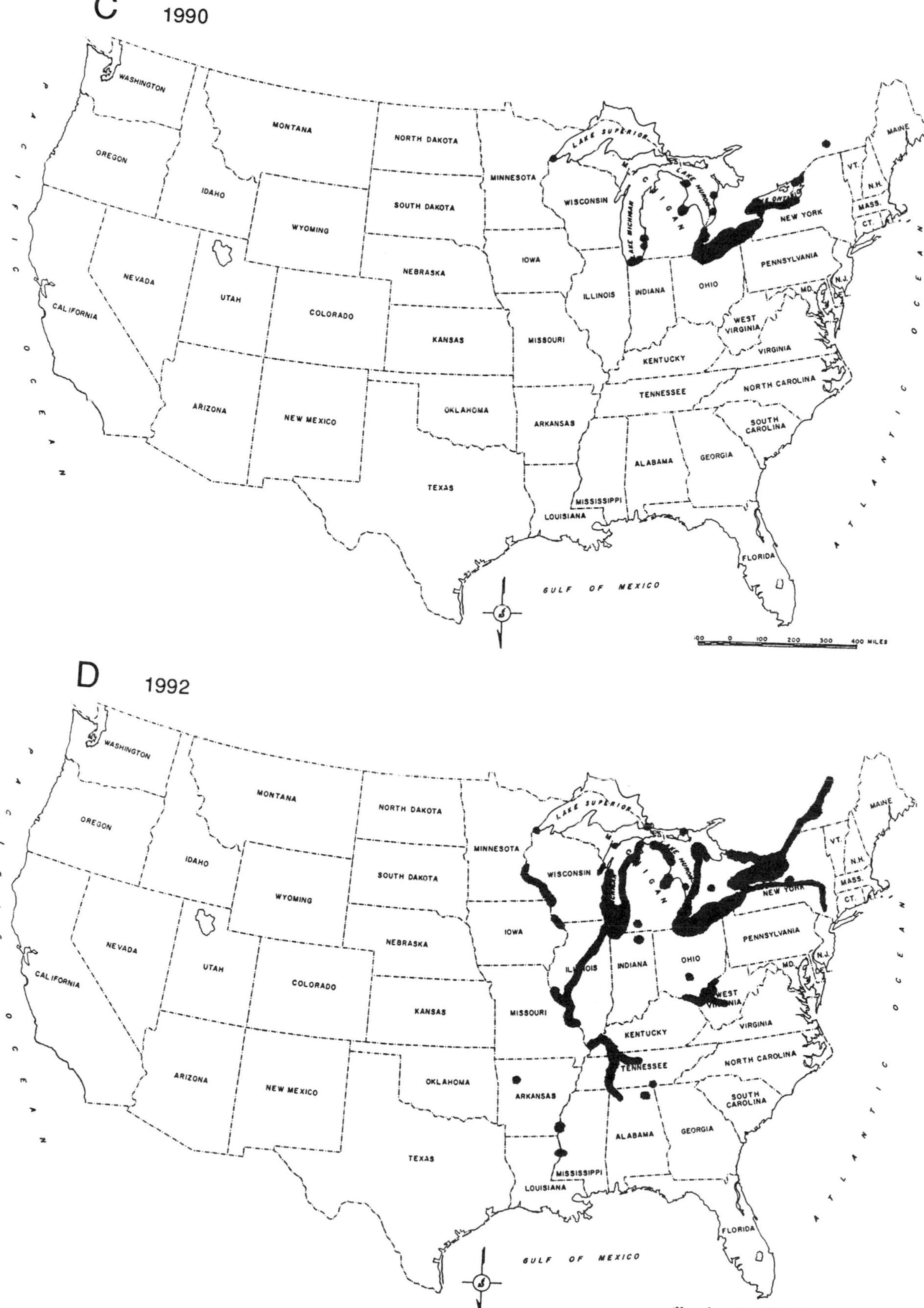


**FIGURE 7** *Continued*

**FIGURE 7** *Continued*

dock materials, navigation aids, and boat hulls. In 1990, mussels were found primarily downstream of their 1988 range, along western, southern, and eastern shores of Lake Ontario (Fig. 7C). In addition, mussels were found upstream from Lake St. Clair in several harbors and bays of Lakes Huron, Michigan, and Superior. The range of mussels extended to the extreme eastern and western ends of the Great Lakes—600 km N–S by 1400 km E–W. This rapid rate of dispersal is attributed primarily to human-mediated mechanisms such as veligers being transported in water carried by boats and juvenile and adult mussels attached to outside surfaces of boats originating in and moving upstream from Lake St. Clair (Table II).

In 1992, mussels were found outside the Great Lakes water basin as far south as Vicksburg, Mississippi, west to Minneapolis, Minnesota, and north and east to Québec, Québec (Fig. 7D). In 1992, the range extended 1900 km N–S by 1700 km E–W. In spring of 1994, mussels occurred primarily within the range recorded in 1992 but were found more continually along stretches in infested rivers (Fig. 7E). The rapid spread south of the Great Lakes basin is primarily attributed to natural dispersal of mussel veligers in water passing through the Chicago diversion (a human-created waterway) into the Illinois River and then into the Mississippi River. Mussels were found in the upper and lower Mississippi, Cumberland, Tennessee, and Niagara rivers. Distributions north of the intersection of the Illinois and Mississippi rivers and east of the Mississippi River are primarily attributed to boats traveling throughout the Mississippi River and its connecting tributaries, such as the Ohio River. However, transport and separate introductions of mussels to these rivers by trailored boats is also believed to have occurred. One commercial barge with zebra mussels attached to the outside hull traversed a distance of 15,884 km through the Mississippi and Illinois rivers in 1992.

In addition, mussels moved out of the Great Lakes across New York State, a distance of about 400 km, by diversion of water into the Erie Canal.

### D. Potential Range

The rate of future range expansion of mussels is difficult to determine. The majority of mussel dispersal to date is attributed to the downstream movement of water in interconnecting waterways (e.g., the Mississippi River) and the upstream movement of boats. Downstream dispersal by passive transport is inevitable and will occur at a relatively fast rate. However, dispersal by human activities will depend on efforts to prevent the success of human-mediated dispersal mechanisms. Because trailered boats can disperse mussels, isolated noninfested water bodies are at risk of invasion by zebra mussels.

In Europe, few environmental parameters are closely associated with the occurrence of zebra mussels. Thirty years of studies of mussels in Polish lakes indicate weak correlations between specific environmental parameters and the presence or absence of mussels. However, the occurrence of mussels is weakly related to the overall trophic status of the water. Mussels are usually not found (or are found in low densities) in eutrophic, shallow lakes, which are circulated by wind. Mussels are most commonly found in lakes that are moderately mesotropic (i.e., not rich or poor in nutrients). In general, mussels are generally most often found in large (>30 ha), deep (mean >3 m), hardwater (>1–2 meq/liter), clear (transparency >2 m) lakes. In addition, mussels occur in most streams more than 30 m wide but rarely in streams smaller than 30 m.

One of the most widely accepted predictive parameters for the future range of zebra mussels is water temperature. However, this limiting factor may vary among geographical populations. In Europe, generally accepted threshold temperatures are 11° to 12°C for growth and 15° to 17°C for reproduction. However, threshold temperatures of 6°C for growth and 12°C for reproduction have been reported for zebra mussels in waters in the Netherlands. In Lake St. Clair, threshold temperatures for mussels seem to be 10° to 12°C for growth and 14° to 16°C for reproduction. European data indicate that the long-term lethal temperature for zebra mussels is about 33°C. In some North American populations, mussels seem to have a higher long-term lethal temperature (up to 36°C) when acclimated to relatively high temperatures.

The potential range of zebra mussels in North America has been estimated based on the existing range and corresponding environmental parameters in Europe (Fig. 8). Environmental parameters, such as water temperature, pH, and dissolved calcium concentration, indicate that zebra mussels will probably spread over a large portion of the United States and southern Canada. They will probably be relatively abundant in large, hardwater streams, rivers, and lakes, and relatively uncommon in small, shallow, and highly productive and polluted environments. Mussels are unlikely to occur in calcium-poor regions such as the Canadian Shield, New England, the Adirondacks, and much of the Pacific Northwest.

To date, the common zebra mussel seems to have higher tolerance limits of environmental parameters than does the quagga mussel. In general, zebra mussels can tolerate lower pH, higher temperature, higher salinity, and higher concentrations of specific bacterial endotoxins than the quagga. Therefore, the postinvasion distributions of these two taxa in North America are likely to be different.

### E. Abundance

At present, there are no specific methods to determine the potential maximum densities of mussel populations in specific waters of North America. This information would be extremely important for water users because it would aid in planning for control and, therefore, the amount of money needed to minimize impacts. Maximum densities of mussels observed in parts of the Great Lakes (700,000/m$^2$) far exceed those observed in Europe (100,000/m$^2$) and could not have been predicted based on the European literature. Similar to the occurrence of mussels, densities of mussel populations seem to be related to the overall trophic status of the water. Mussels are found in low densities in eutrophic, shallow lakes, which are circulated by

***FIGURE 8*** Potential range of zebra mussels (*Dreissena polymorpha*) in North America. [Courtesy of D. Strayer, The New York Botanical Garden, Millbrook, New York].

wind, and in higher densities in lakes that are moderately mesotrophic.

Extensive review of the European and former Soviet Union literature indicates that populations of mussels typically fluctuate in density over time. Population densities can undergo three orders of magnitude increase or decrease in less than 1 to 10 years. The dramatic degree of mussel invasion and infestation in the Great Lakes can be compared to only one other lake—Lake Balaton, Hungary. Documentation of the mussel's invasion into Lake Balaton parallels, visually and descriptively, the invasion into western Lake Erie. Unfortunately, little quantative information exists concerning the densities and impacts of zebra mussels in Lake Balaton.

## V. SOURCE OF NORTH AMERICAN POPULATIONS

The founding population of zebra mussels in North America, appears to have originated from the southern portion of its range in Europe. However, six phenotypes of mussels, characteristic of different regions in Europe and the former Soviet Union, have been found in North America (Fig. 2). Proportions and similarity indices among mussel phenotypes in European and Lake Erie populations indicate that mussels in Lake Erie most closely resemble a population in the southern Black Sea–Caspian Sea region or one in the southeastern Caspian Sea–Aral Sea region. Supporting evidence for the founding population originating in the general area of the Black and Caspian seas is that two benthic fishes (round nose goby, *Neogobius melanostomus,* and the tubenose goby, *Proterorhinus marmoratus*) found in southern Russia were found in the Laurentian Great Lakes shortly after zebra mussels were introduced. This indicates multiple introductions of several species originating from one geographical area.

The source of the second North American Dreissana species (*D. rostriformis bugensis*) is unknown. However, it primarily occurs in river basins of the Black and Caspian seas of the former Soviet Union. This also indicates that this region is the founding source of several exotic species introduced in the 1980s. Regardless of the source of mussels now in North America, impacts resulting from existing and possible new introductions of zebra mussel species are likely to be similar because many of the species have similar characteristics.

## VI. ECONOMIC IMPORTANCE

Concern about the economic impacts caused by zebra mussels is mainly related to the mussel's ability to foul pipes. Biofouling reduces pipe diameter, which may reduce and possibly stop water flow. A few mussels do not, for the most part, create fouling problems, but when massive encrustations of mussels occur, the impacts are substantial. Fouling occurs at densities far below the maximum of $114{,}000/m^2$ in European waters and $700{,}000/m^2$ in the Great Lakes.

The potential loss of water to cities and agriculture facilities caused by zebra mussels can be costly. In the Great Lakes water basin, the economic disruption to businesses and communities has been estimated at $400 million per year. The total impact may reach several billion dollars per year throughout North America. These costs are new for water withdrawal facilities because formerly there was no organism like the zebra mussel in North America that necessitated extensive and specialized controls.

Three primary areas of concern are associated with zebra mussels and most water withdrawal facilities: biofouling of surfaces exposed to water, internal blockage of pipes, and increased potential for corrosion of structures. Fouling of external surfaces by mussels occurs on most human-made materials used in and near water-pumping facilities, such as docks, structural supports, screens, filters, booms, gears, pumps, and canal walls. Surface biofouling may prevent proper operation of secondary water-supply structures. Internal blockage of pipes may be caused by massive encrustations of layered mussels that effectively reduce or occlude pipes, large clumps of mussels that break off (i.e., druses) and are transported to pipe orifices, and individual mussels that obstruct small-diameter pipes. Corrosion caused by zebra mussels occurs because the organisms effectively isolate a layer of detritus and water between substrates and mussels. This layer becomes an ion-reducing environment (i.e., anaerobic), and the pH is lowered, becoming acidic and increasing corrosive activity.

In Europe and the former Soviet Union, zebra mussels have not been extensively documented as a major biofouler. This is attributed to the lack of industrialization when mussels slowly expanded their range in the early 1800s. For example, zebra mussels became of great interest in the mid-1820s when they were discovered in London and several places throughout Europe before major modern industrialization began. As industrialization expanded in the middle to late 1800s, mussels were observed to biofoul pipes in many places in Europe, so water facilities being built in this period were designed with control measures to mitigate impacts of mussels on water supplies. Generally, *Dreissena* is known as a biofouling organism in Europe that can attain maximum densities of $114{,}000/m^2$ and form layers up to 1.5 m thick, thereby disrupting operations of water treatment and electrical generating facilities, but the history of mussels as potential biofoulers has not been extensively documented because problems were often mitigated before major fouling occurred. Similarly, major fouling problems caused by zebra mussels in many areas of North America will be minimized because of experiences in western Lake Erie in the late 1980s.

### A. Water Treatment Plants

In Europe, zebra mussels fouled waterworks mainly between the turn of the century and the middle 1900s. Mussels interrupted water supplies in Hamburg in 1886 and in Rotterdam in 1887. In 1895, water supplies in Berlin were putrified by dying mussels when they died en masse within confined spaces of pipes, thereby imparting a putrid taste and odor to the water. The waterworks were shut down for 27 days as a result of decaying mussels. Problems with water supplies occurred in 1925, 1944, and 1947 at the Great Yarmouth Waterworks in Great Britian. These early problems resulted in engineering designs for waterworks and power plants that now substantially reduce the impacts of mussels in Europe.

In North America, the most dramatic fouling occurred at a waterworks located on the shore of western Lake Erie at Monroe, Michigan. This facility was invaded by zebra mussels in 1989 before the potential fouling capabilities of the organism were widely known in North America. A few mussels were discovered in the waterworks in January, 1989. No reproduction of mussels occurred between January and late May, and low reproduction (<50 veligers/liter) occurred between late May and early July. In midsummer of 1989, however, zebra mussels entered exponential reproduction with high survival rates, resulting in dramatic colonization of the waterworks by the fall of 1989. Reproduction of mussels increased 8-fold in early July (400 veligers/liter), decreased substantially in mid-August (25 veligers/liter), and increased by 10-fold in late August (500 veligers/liter). In September, delivery of water to the waterworks was interrupted because of zebra mussels. This waterworks had not lost a day of service between 1924 and 1989, so this interruption of service was a highly unusual and unexpected event! After September 1989, reproduction decreased to low levels (<10 veligers/liter) and subsequently stopped.

In early fall, large numbers of relatively small mussels (<1 cm long), probably resulting from successful reproduction in July and August, were washing up on beaches and colonizing substrates in western Lake Erie. Disruptions of service occurred again in December 1989 and February 1990 before control measures could be implemented at the Monroe waterworks. Similarly, several municipal waterworks withdrawing water from Lake Erie also reported reduced intake capacity attributed to zebra mussels. Mussels were believed to have increased the roughness and decreased the effective diameter of intake pipes at many facilities. In one waterworks, drinking water withdrawal was reduced 45% because of mussels, but none experienced the quick and massive invasion seen at the Monroe waterworks. Uninterrupted service at other facilities is partly attributed to effective information exchange about the biofouling experience at Monroe and subsequent development of appropriate control methods.

At present, mussels have affected most drinking water supplies with which they have come in contact in North America. Waterworks along Lakes Michigan, Erie, Huron, and Ontario have also been affected by zebra mussels, but to a lesser degree than that experienced in western Lake Erie. Reduced water-intake efficiences up to 50% have been reported over most of the mussel's range in the Great Lakes. Frazzle ice has also become a major problem in pipes withdrawing water from shallow water where zebra mussels occur. The increased turbulence caused by zebra mussels in pipes results in the formation of ice that normally would not freeze in the absence of turbulence. The ice builds within pipes and decreases water pumping efficiencies. In addition, water turbidity levels have dropped so low in the presence of zebra mussels that waterworks have had to add clay to assist with the flocculation process necessary for effective water treatment.

### B. Power Plants

In Europe, early problems of mussel fouling in waterworks helped eliminate fouling of many power plants because major power facilities were built in the middle to late 1900s, after control methods for mussels had been developed. However, in 1921 a major hydroelectric plant in northern Germany was shut down because of a loss of head pressure in the water-pumping system as a result of mussel fouling. Mussels in layers up to 1.5 m thick covered tunnels and open channels used to deliver water to power plant systems.

The first power plant to be dramatically affected by zebra mussels in North America is located on the shore of western Lake Erie at Monroe, Michigan. In fall of 1988, densities of mussels were low (less than $200/m^2$), and the population was composed of mostly young-of-the-year up to 10 mm long and a few individuals 15–20 mm long. Between December 1988 and February 1989, higher densities of 200 to $3000/m^2$ were found. In the summer of 1989, exponential reproduction occurred, along with high survival rates, resulting in dramatic colonization by fall of 1989. Mussel densities during the period increased from a maximum of $3000/m^2$ to $700,000/m^2$. In fall of 1989, mussels formed continuous layers 2.5 to 12.0 cm thick on all firm substrates leading into the power plant, including canals, structural supports, trash racks, and traveling screens. Clusters of mussels (i.e., druses) sloughed from canal surfaces formed piles more than 2 m deep at the base of canal walls (about 500 $m^3$ in 1989). Trash bars were 75% occluded, and traveling screens prevented up to 2 $m^3$ of mussels/day from entering the water system. However, veligers and small mussels (<9 mm long) passed through traveling screens and became attached to surfaces within screenhouses that supply water to the power plant. Druses sloughed from screenhouse walls were transported into tubes of main system condensers (used to cool steam from power generators) and service water systems (used in fire protection and ash handling). Tubes (19 mm in diameter) of the steam condensers became up to 35% occluded with druses, reducing power-generating efficiency. Service water lines were found to contain mussels on pipe interiors, elbows, spigots, valves, meters, and access covers. Blockage within service water systems could prevent proper operation of fire and environmental cleaning systems.

### C. Socioeconomic Problems

The public response to the zebra mussel far exceeded any response to previous exotic species. This is attributed to the large number of people that use or receive water withdrawn from surface waters affected by mussels and the ease with which the general public can visualize zebra mussels as the cause for their concern. For example, clogged pipes that threatened water supplies potentially affected everyone within a city regardless of their interests. In addition, mussels are highly visible through the news media or by a walk on the beach. A zebra mussel toll-free telephone service in Ontario allowed active participation between the general public and direct sources of information. Documentation of this service indicated that interest by the public and popular news media was cyclical and corresponded to the warmer months of the year, when the public and mussels were most active (Fig. 9). In 1992 alone, the province of Ontario distributed 15,000 pieces of literature at the request of the general public. In one community located on the shore of the Great Lakes, 70% of the general public was aware of the presence of zebra mussels in nearby waters and the possible fouling problems they could cause. About 10% of the public believed they were economically affected, and 70% believed the ecosystem was affected by zebra mussels.

The zebra mussel infestation problem in the Laurentian Great Lakes was the primary reason for the creation and passage of the federal Nonindigenous Aquatic Nuisance Prevention and Control Act of 1990. The legislation has four purposes: to coordinate research, control, and information exchange; to develop and carry out enviromentally

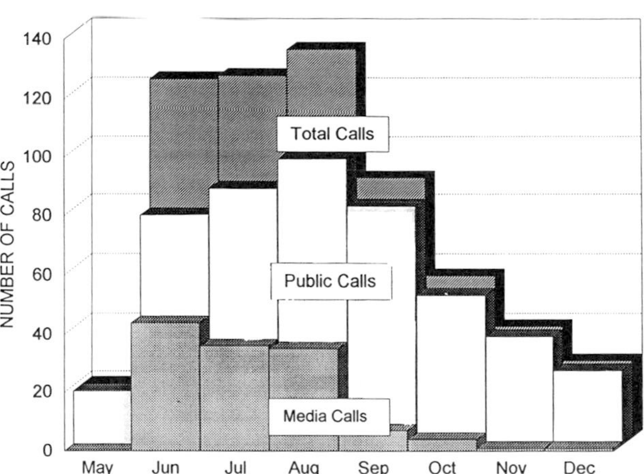


***FIGURE 9*** Number of telephones calls received from the public and the news media concerning zebra mussels (*Dreissena* spp.), May to December 1992. [Courtesy of Beth McKay, Ontario Federation of Angles and Hunters, Peterborough, Ontario.].

sound control methods; to minimize economic and ecological impacts; and to establish a research and technology program to benefit state governments.

## D. Preparing for Zebra Mussels

The important steps to take in preparing to minimize potential economic impacts of zebra mussels include the following: (1) evaluate the potential for invasion and colonization of the water body from which water is used; (2) evaluate the uses of the water; (3) determine the vulnerability of the water system to biofouling by mussels; (4) initiate a monitoring program to determine when mussels infest a water supply; (5) develop a proactive, potential plan to control and minimize impacts of mussels on the use of water; and (6) reserve some resources to modify the plan control plan in case the proactive plan fails. Potential for invasion and colonization is relatively easy to determine but difficult to quantify in terms of the degree of colonization. Factors such as limiting environmental parameters, potential dispersal mechanisms that may introduce mussels, and proximity of mussels to the water supply may all affect how, when, and to what degree mussels may invade and colonize water bodies. There may be hundreds of vulnerable points in a water system to consider when evaluating uses and how mussels may affect the use of water. Control plans are often individually designed for each vulnerable point of a water system and usually include several types of control-specific points and safeguards to ensure uninterrupted water service. Control measures may include one or all of the three basic control types: biological, physical, and chemical.

## Glossary

**Byssus** Tuft of strong filaments secreted by a mollusk gland located in the foot and used for attachment.

**Ecological niche** Position or status of an organism within its community, including where it lives and what it does.

**Exotic species** Organism introduced into a country from another country.

**Gametogenesis** Formation of reproductive cells.

**Heteromyarian** Bivalve mollusks with two adductor muscles of unequal size.

**Homomyarian** Bivalve mollusks with two adductor muscles of about the same size.

**Taxonomy** Science of classification as applied to living organisms, including the study of species formation.

**Veliger** Water-carried larval stage of some mollusks.

## Bibliography

Mills, E. L., Leach, J. H., Carlton, J. T., and Secor, C. L. (1993). Exotic species in the Great Lakes: A history of biotic crises and anthropogenic introductions. *J. Great Lakes Res.* **19**(1), 1–54.

Nalepa, T. F., and Schloesser, D. W. eds. (1993). "Zebra Mussels: Biology Impacts, and Control." Boca Raton, Fla.: CRC Press.

Neumann, D., and Jenner, H. A., eds. (1992). "Zebra Mussel *Dreissena polymorpha;* Ecology, Biological Monitoring, and First Applications in the Water Quality Management." Deerfield Beach, Fla.: VCH Publishing.

Rosenfield, A., and Mann, R. eds. (1992). "Dispersal of Living Organisms into Aquatic Ecosystems." College Park: University of Maryland Press.

Shtegman, B. K., ed. (1964). Biology and control of *Dreissena:* A collection of papers. *Acad. Sci. USSR, Inst. Biology and Inland Waters, Leningrad* **7,** 1–145. (Translated from the Russian)

Stanczykowska, A. (1977). Ecology of *Dreissena polymorpha* (Pallas) (Bivalvia) in lakes. *Polskie Arch. Hydrobiol.* **24**(4), 461–530.

Strayer, D. L. (1991). Projected distribution of the zebra mussel, *Dreissena polymorpha,* in North America. *Canad. J. Fisheries and Aquatic Sci.* **48,** 1389–1395.

Welcomme, R. L. (1988). "International Introductions of Inland Aquatic Species," Fisheries Technical Paper. Rome: Food and Agriculture Organization of the United Nations.

# Invertebrates, Freshwater

**James H. Thorp**
*University of Louisville*



Freshwater invertebrates consistently play decisive roles in the functioning of all aquatic ecosystems through their diverse participation in energy flow, nutrient cycling, and population regulation. Despite their importance in stream and lake ecosystems, only about 5% of all invertebrate species live in fresh waters and approximately half of the 40+ phyla of metazoa and heterotrophic protozoa have freshwater representatives. This relatively low faunal diversity and phyletic representation is indicative of the strenuous environmental conditions found in fresh water, including low osmotic state, thermal extremes and variability, and both ecological and evolutionary instability. Environmental instability, especially the tendency of habitats to become parched, has required special life-history adaptations for survival in ephemeral aquatic ecosystems, such as dormant propagules in bryozoans and encysted life stages in tardigrades. The nature of an invertebrate's feeding, respiratory, osmotic, and reproductive systems reflects both the unique habitat conditions to which it has adapted and its evolutionary history.

## I. DIVERSITY PATTERNS OF FRESHWATER INVERTEBRATES AND THEIR POSSIBLE CAUSES

About half of the approximately 40 phyla of either unicellular, heterotrophic protozoa, or multicellular animals (metazoa) have freshwater representatives (Table I). The freshwater members of these phyla probably constitute less than 5% of all invertebrates. Fewer than one million living species of invertebrates have been described from terrestrial, marine, and freshwater habitats, but this may represent less than 15% of the actual number of protozoa and metazoa. North America contains about 15,000–20,000 species of freshwater invertebrates in temperate and boreal zones, and even greater numbers of classified and undescribed species certainly exist in the tropics. Some representatives of this fauna are shown in Figs. 1 and 2.

Of phyla having freshwater representatives, almost all are less diverse in inland waters than in either marine or terrestrial environments. For example, most species in the phylum Anthro-

**TABLE I**
Composition and Diversity of Freshwater Invertebrate Fauna

| Taxa | Representative common names | Diversity in North American fresh waters[a] | Diversity in all habitats[b] |
|---|---|---|---|
| Protozoa[c] | Amoebae; ciliates; flagellates | 3,000 | 65,000 |
| Phylum Porifera | Sponges | 25 | 9,000 |
| Phylum Cnidaria | Hydra; freshwater jellyfish | 8 | 9,000 |
| Phylum Platyhelminthes[d] | Planaria | 300+ | 20,000 |
| Phyum Nemertea | Ribbon worms | 3 | 900 |
| Phylum Gastrotricha | Gastrotrichs | 100+ | 500+ |
| Phylum Rotifera | Rotifers | 1,200 | 2,000 |
| Phylum Nematoda | Roundworms | 500 | 100,000+ |
| Phylum Nematomorpha | Horsehair worms | 12 | 230 |
| Phylum Mollusca | Snails; mussels; clams | 600 | 100,000 |
| Phylum Annelida | Earthworms; leeches; polychaetes | 500 | 14,000 |
| Phylum Bryozoa | Moss animals; bryozoans | 22 | 4,500 |
| Phylum Entoprocta | Entoprocts | 1 | 150 |
| Phylum Tardigrada | Water bears | 180 | 600 |
| Phylum Arthropoda | Crayfish; water mites; insects | 11,000 | 1,000,000+ |

[a] Estimates based only on the fauna of the United States and Canada.
[b] Approximate number of terrestrial, marine, and freshwater species.
[c] Protozoa is a nontaxonomic name for the six to seven phyla from the kingdom Protista with unicellular, heterotrophic species; estimates for free-living taxa only.
[d] Only free-living flatworms included in estimate (no trematode flukes).

poda—the most diverse freshwater taxon—are either terrestrial (e.g., insects and mites) or marine (mostly crustaceans). All members of the small phylum Tardigrada (water bears) and the huge phylum Nematoda (roundworms) require moist environments, but only a minority live in conventional lotic (flowing water) or lentic (e.g., lakes and wetlands) habitats. Instead, they primarily inhabit the film of water collecting on plant surfaces (tardigrades), live in moist soil (free-living nematodes), or infest tissues of plants and animals (endoparasitic nematodes). Many other taxa listed in Table I comprise principally marine species, such as the phyla Porifera (sponges), Cnidaria (e.g., jellyfish and sea anemones), and Mollusca (including marine snails and squids). Two phyla most closely identified with freshwater environments are free-living Rotifera (wheel animals) (Fig. 1) and parasitic Nematomorpha (horsehair worms) (Fig. 2). [*See* Hydrologic Needs of Wetland Animals.]

What accounts for the relatively low diversity of taxa[1] in lakes and streams and the scarcity of phyla dominated by freshwater members? Although a definitive answer to this question is presently unavailable, certain prominent chemical, physical, biological, and evolutionary factors undoubtedly contribute to the dearth of freshwater invertebrate taxa.

Perhaps the foremost chemical impediment is the low osmotic state of almost all inland waters except salt lakes. This factor probably explains sufficiently the total lack of echinoderms (e.g., sea urchins) and virtual absence of cnidarians from inland waters. A related problem that can be severe in headwater streams and ephemeral ponds is drought; desiccation engenders both osmotic and physical stresses, either of which can cause death unless appropriate

[1] Compared to freshwater ecosystems, terrestrial habitats contain fewer phyla but many more species, whereas oceans are more diverse in both phyla and total species.

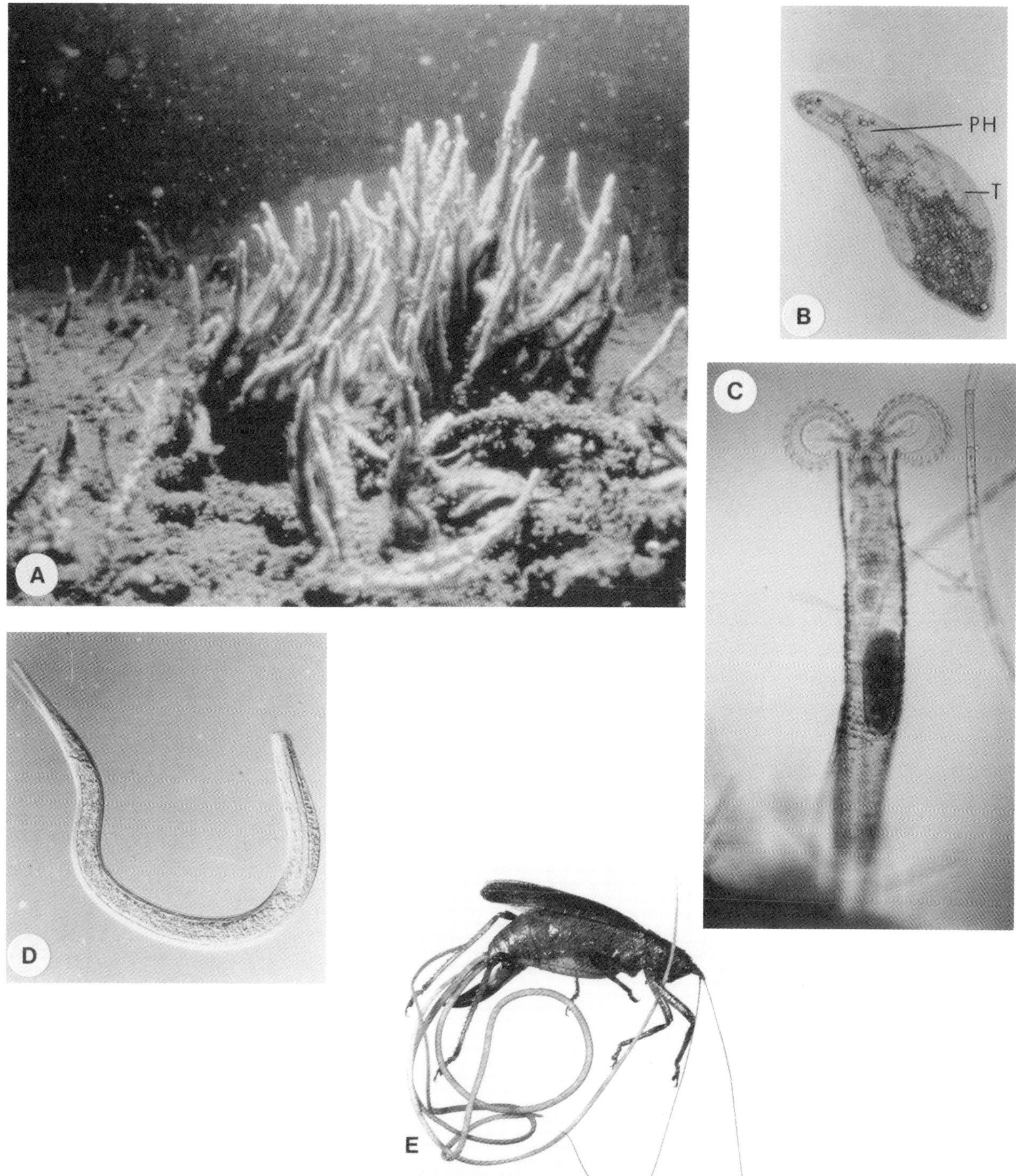


**FIGURE 1** Photographs of representative invertebrates: (A) sponge *Spongilla* (Porifera) (from Frost); (B) turbellarian flatworm *Strongylostoma* (Platyhelminthes) (from Kolasa); (C) sessile *Limnias* (Rotifera)—note the paired ciliated "wheel" organs at the upper portion of the animal (from Wallace); (D) free-living roundworm (Nematoda); (E) horsehair worm (Nematomorpha) emerging from its insect host (D–E from Poinar); (F) portion of a *Lophomorpha* colony (Bryozoa) (from Wood); (G) water bear *Echiniscus* (Tardigrada) (from Nelson). [From J. H. Thorp and A. P. Covich, eds. (1991). "Ecology and Classification of North American Freshwater Invertebrates." Academic Press, New York.]

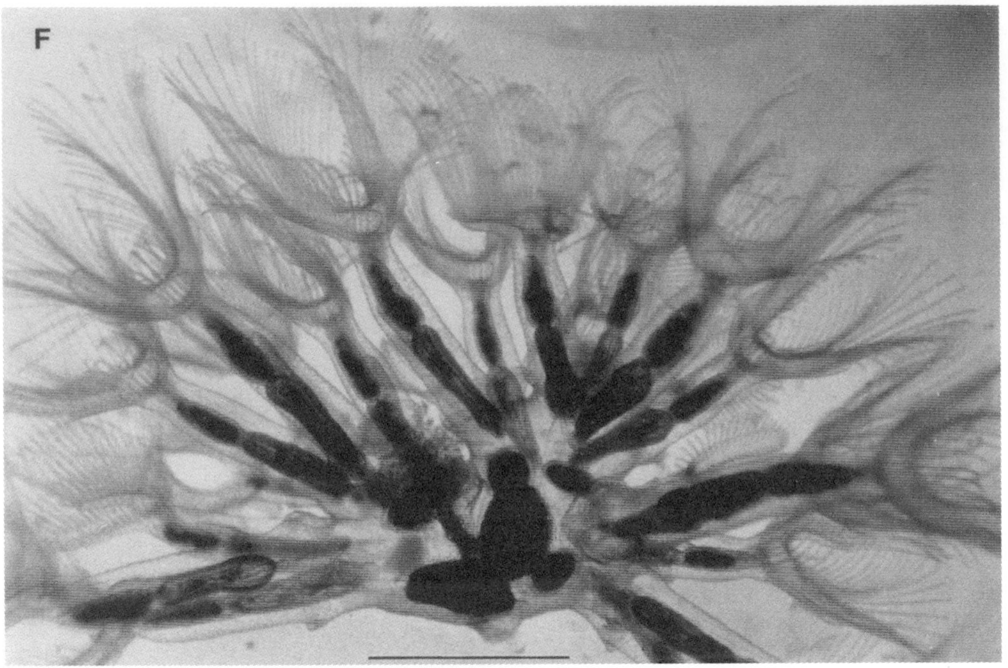


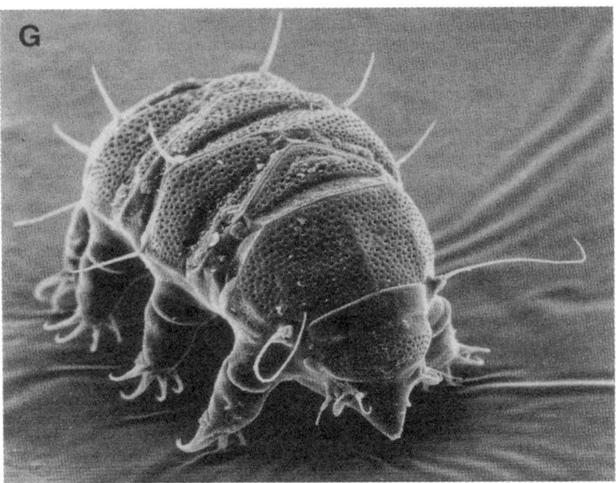


**FIGURE 1** *Continued*

biological or ecological responses (such as encystment or migration to deeper ponds) have evolved. Other chemical stresses sometimes encountered in freshwater habitats include low oxygen concentrations and extreme pH values. [*See* LIMNOLOGY, INLAND AQUATIC ECOSYSTEMS.]

Physical barriers that may have eliminated some potential immigrants include absolute temperature extremes, thermal variability, high current velocity, silt burial (especially in floodplain rivers), and occasional scarcity of hard surfaces. Aside from terrestrial habitats, where temperatures are most extreme and variable, the thermal nature of rivers and lakes is more stressful than all but a few marine habitats. Water currents can be both beneficial (bringing food and oxygen and reducing local thermal stratification) and detrimental to life (e.g., shear stress, silt burial, and forced emigration to less hospitable environments). Invertebrates able to cling to hard surfaces or seek shelter under or

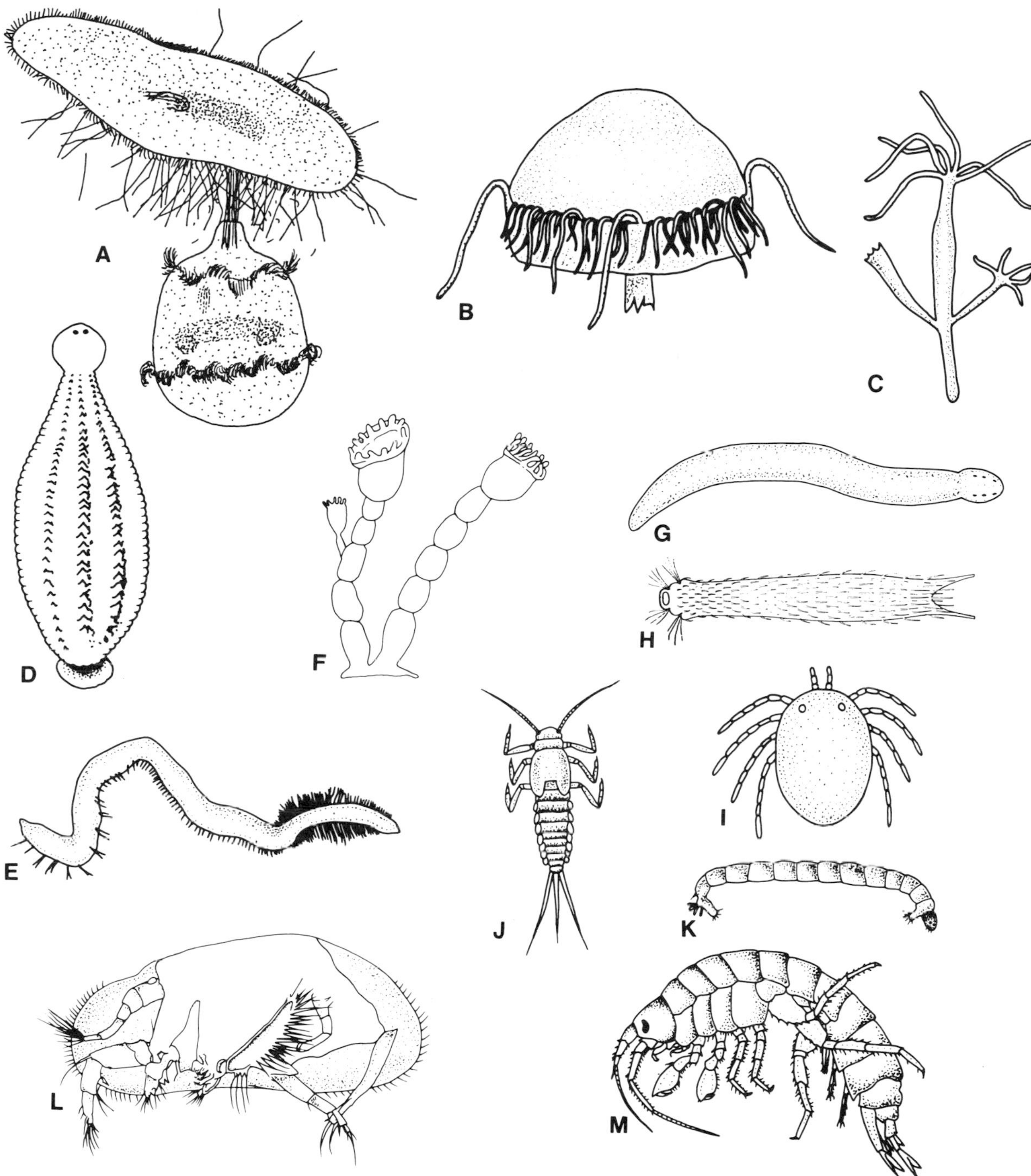


***FIGURE 2*** Drawings of various freshwater invertebrates: (A) capture of *Paramecium* by *Didinium* (protozoa: Ciliata) (from Taylor and Sanders); (B) medusa *Craspedacusta* (Cnidaria); (C) polyp *Hydra* (Cnidaria); (D) leech *Placobdella* (Annelida); (E) aquatic earthworm *Branchiura* (Annelida); (F) *Urnatella* (Entoprocta); (G) ribbon worm *Prosotoma* (Nemertea); (H) *Chaetonotus* (Gastrotricha) (from Strayer and Hummon); (I) adult water mite (Arthropoda: Acari); (J) *Baetis* mayfly numph (Arthropoda: Insecta); (K) midge larva *Chironomus* (Insecta); (L) ostracod *Candona* (Arthropoda: Crustacea); (M) amphipod scud *Gammarus* (Crustacea). [From J. H. Thorp and A. P. Covich, eds. (1991). "Ecology and Classification of North American Freshwater Invertebrates." Academic Press, New York.]

within substrates often tolerate high current velocities, but bouyant or weak-swimming, pelagic organisms find rivers difficult. Consequently, benthic cnidarian polyps (mostly hydra) are common in rivers and lakes, whereas pelagic medusae (like *Craspedacusta sowerbii,* the single jellyfish in North America fresh waters) are extremely rare. Furthermore, invertebrates with pelagic larvae encounter potentially severe problems in flowing waters. Firm surfaces are important in freshwater habitats for many species, such as net-spinning caddisflies, and their absence influences species diversity. Likewise, many sessile marine species require hard surfaces, and community diversity is typically the lowest on soft bottoms. Thus, marine invertebrates requiring firm substrates, such as barnacles, tunicates, and corals, would have difficulty moving up, and surviving in, sediment-laden coastal rivers even without salinity barriers.

Although biological constraints imposed by differences in competitors, predators, parasites, and food resources may have slowed adaptive radiation of marine invertebrates into freshwater systems, chemical and physical obstacles were probably much more formidabe; however, a significant biological hindrance to successful species immigration relates to life-history phenomena. As discussed in a later section, species with free-swimming larvae are extremely common in oceans but rare in fresh waters, the principal exceptions being planktonic copepods (microcrustacean arthropods) and aquatic insects.[2] To alter this major life-history feature would have been evolutionarily difficult for species in many phyla. Furthermore, conditions in habitats such as headwater streams, ephemeral ponds, or shallow Arctic lakes are extremely rigorous relative to those of most other aquatic habitats because of periodic drying or freezing of the entire water body. Survival there often demands life-history stages that are immune to those physical and chemical stresses (as described in Section IV,B,4); these adaptations are lacking in most invertebrates. [*See* Speciation and Adaptive Radiation.]

Finally, the relatively low diversity of freshwater habitats is probably associated in part with their evolutionary and ecological instability. Ecosystems generally increase in species diversity as they grow older in evolutionary time. Consequently, endemic species are more likely to occur, for example, in an ancient rift lake, such as Lake Baikal in Siberia, than in a relatively new lentic system such as Lake Superior. This phenomenon contributes to the greater overall diversity of animals in river systems than in lakes, because the latter are usually evolutionarily young whereas rivers are more continuous in time despite their meandering channels.

## II. ECOLOGICAL ROLES OF FRESHWATER INVERTEBRATES

Although the foregoing discussion may have unintentionally demeaned the importance of freshwater invertebrates by emphasizing their minor part in the total diversity of the kingdoms Animalia and Protista, ecological contributions of these diminutive creatures are hardly small. Indeed, they consistently play decisive roles in the functioning of all aquatic ecosystems through diverse participation in energy flow, nutrient cycling, and population regulation.

Freshwater invertebrates occupy all heterotrophic functional feeding groups, such as algal grazers, filter feeders, shredders, carnivores, and detritivores. As important components of aquatic food webs, invertebrates bridge the gap between primary producers and fish or other high trophic level consumers. As examples, snails and some mayflies graze benthic algae (primarily diatoms, green algae, and cyanobacteria) from surfaces of vascular plants, rocks, woody debris, and soft sediments. Filter-feeding sponges, mussels, caddisflies, and copepods eat phytoplankton, bacteria, and/or dead organic matter in suspension within the water column. Carnivorous dragonfly nymphs, true bugs, copepods, nematodes, and hydra prey on various protozoa and metazoa; cannibalism among aquatic predators is not exceptional. Invertebrates

[2] Aquatic insects, which spend most of their lives immersed in water, either have aquatic larvae and terrestrial adults or both stages are primarily aquatic.

relying on dead particulate organic matter (POM) for energy and nutrients are common in both lakes and rivers; they include oligochaete worms, planaria, and midges. The original source of POM for these detritivores may have been dead animals, leaves fallen from riparian trees, algae, or decaying aquatic vascular plants (little of this last group can be eaten while alive). Before detritivores ingest organic matter, however, the food may have required processing by shredders (e.g., some mayflies and beetles) and thorough colonization by surface bacteria and fungi. Often ingested POM is not completely assimilated; instead, the microbial flora attached to food particles serve as the primary carbon source that is assimilated into consumer tissues. Few invertebrate species occupy only one functional feeding group in all ecosystems, seasons, and life stages. Also, it is usually unrealistic to assign many invertebrates (e.g., crayfish) to a single feeding guild because these omnivores eat a great variety of living and dead animals and plants. All feeding guilds of freshwater invertebrates fall victim in turn to predators like benthic and pelagic fish, salamanders, turtles, a few snakes, and some waterfowl.

By recycling carbon and other nutrients, invertebrates reduce loss of energy and vital elements to sediments or downstream and shorten the time to recycle material through community food webs. This recycle process is termed nutrient cycling in lakes and nutrient spiraling in rivers. Without invertebrate contributions to recycling processes, aquatic ecosystems would support far fewer species owing to lower available productivity and many shallow ponds would soon fill with nondecomposed plants.

In addition to roles in nutrient cycling, parasitic and predaceous invertebrates may function as density-dependent regulators of population sizes of lower trophic level species. Furthermore, by preying selectively on dominant species, a carnivore or herbivore may influence relative abundance and species diversity within an entire community far out of proportion to the predator's abundance; such species are sometimes called "keystone predators." The important role of predators has been demonstrated by *in situ* experiments examining the potential regulatory effects of (a) predaceous phantom midge larvae feeding on zooplankton in northern lakes, (b) stonefly nymphs consuming mayflies in Rocky Mountain streams, and (c) dragonflies preying on midges and other benthic invertebrates in a southeastern thermal reservoir.

## III. THE INVERTEBRATE–*HOMO SAPIENS* LINKAGE

Few people appreciate the importance of invertebrates to humans and many seem to dislike them as a group, perhaps because of a misplaced dislike of all things "buglike." Although it is easy to demonstrate the many benefits of terrestrial and marine invertebrates (e.g., crop pollination, animal waste decomposition, and food), the positive linkage between freshwater invertebrates and humans is weaker and negative relationships are occasionally quite strong.

The worst aspects of freshwater invertebrates concern parasitism and transmission of diseases. For instance, malaria is caused by a sporozoan parasite (kingdom Protista: phylum Microspora) that is transmitted to humans through the bite of an adult anopheline mosquito. As most people realize, larval mosquitoes are aquatic and primarily reside in temporary pools lacking fish. Schistosomiasis, or bilharzia, is one of the world's most widespread and serious parasitic infections. *Schistosoma* spp. (phlym Platyhelminthes), the tropical trematode parasite involved, penetrates human skin and migrates through blood vessels to various target organs. Control of this fluke species has focused on improving sanitary conditions and eradicating the intermediate snail hosts. Secondary problems of freshwater invertebrates are linked to (a) terrestrial adults of a few biting insects with aquatic larvae, such as mosquitoes, blackflies, and some midges; (b) a minority of leech species that such blood from unwary waders; and (c) unusual economic impacts, such as clogging of pipes in power plants and water utilities by zebra mussels (*Dreissena polymorpha*) and Asiatic clams (*Corbicula fluminea*). [*See* PARASITISM, ECOLOGY.]

On the positive side, freshwater invertebrates provide food (e.g., southern crayfish and some clams), freshwater pearls, and "seed pearls" (for use in producing cultured marine pearls), and they were a source material for buttons before the advent of plastic. They have served as simple test models for research having human applications; examples include neurobiological research on the crayfish *Procambarus clarkii* and cancer-related, developmental studies with common hydra. Primitive human cultures sometimes used shells of freshwater mussels as currency and decorations. From a human-related ecological viewpoint, freshwater damselflies and other invertebrate predators help control larvae of noxious biting insects. And of course, fishing would be a waste of time without invertebrates as source of food for our finned quarry!

## IV. FUNCTIONAL ANALYSIS OF FRESHWATER INVERTEBRATES

### A. Pertinent Aspects of Freshwater Habitats and Invertebrate Life

The morphological, physiological, and ecological characteristics necessary to live in inland waters differ widely according to the ecosystem's physical, chemical, and biological features. The most widely studied inland aquatic ecosystems are epigean (=surface waters) wetlands, lakes, and streams. Freshwater invertebrates are generally more diverse within flowing waters than in lentic ecosystems, as a partial consequence of the greater heterogeneity and permanence of lotic ecosystems. Compared to lakes, streams are usually more turbulent, less thermally stratified, more oxygenated throughout the water column, less variable in surface temperatures, and more diverse in substrates and microhabitats. The depth and thermal stratification of lakes create dramatically different habitats that in turn lead to distinctive fauna and flora (Fig. 3). Equivalent horizontal and vertical differences in stream fauna are rarely as evident, but zonal differences begin to appear with increases in depth beyond limits of light penetration; these patterns are most apparent in large rivers.

Less research has focused on hypogean habitats, such as underground caves and streams (hyporheic and phreatic zones). Cave faunas are often composed of multiple species having obligate (troglobitic) or facultative (troglophilic) relationships with those habitats. Troglobites, such as the blind, white cave crayfish *Orconectes inermis,* are often characterized by reduction in eye structure and lack of body pigments. In association with relatively constant, low temperatures and minimal energy levels in these detrital-based ecosystems (all or most energy is derived from sources external to the cave), troglobitic invertebrates are typically long-lived and iteroparous with low fecundity—all features of a "K" life-history pattern.[3]

Freshwater habitats are defined as having salinities of less than 1 ppt (parts per thousand), or 1 gram of total dissolved solids per liter of water; most are usually considerably less than 1 ppt. This compares to an average salinity of 35 ppt for full-strength seawater and intermediate values for estuarine waters. Inland salt and "soda" lakes are common to prairies and dry, western regions of North America. These lentic ecosystems are typically studied by "freshwater ecologists" despite the fact that their salinities are greater than 1 ppt and sometimes exceed the concentration of seawater by several times! One such inland habitat, Mono Lake (near the eastern slope of the Sierra Nevada range), is so saline that its only metazoan residents are brine flies, the brine shrimp *Artemia salina,* and migratory birds like the eared grebe. Some waterfowl molt their flight feathers while at Mono Lake and depend entirely on brine shrimp for energy to grow new feathers and continue migrating.

Temperatures within the near-surface waters of epigean habitats normally reflect recent ambient temperatures; exceptions are deep, large-volume lakes, spring-fed brooks, and geothermal streams. Hot spring ecosystems present physiological challenges to which few invertebrates have adapted. Upwelling regions within thermal pools and streams are typically too hot for metazoan life; but

[3] Note: avoid damaging or collecting cave animals, as their populations are usually small and the community is thereby highly susceptible to destruction; isolated species could easily be driven to extinction!

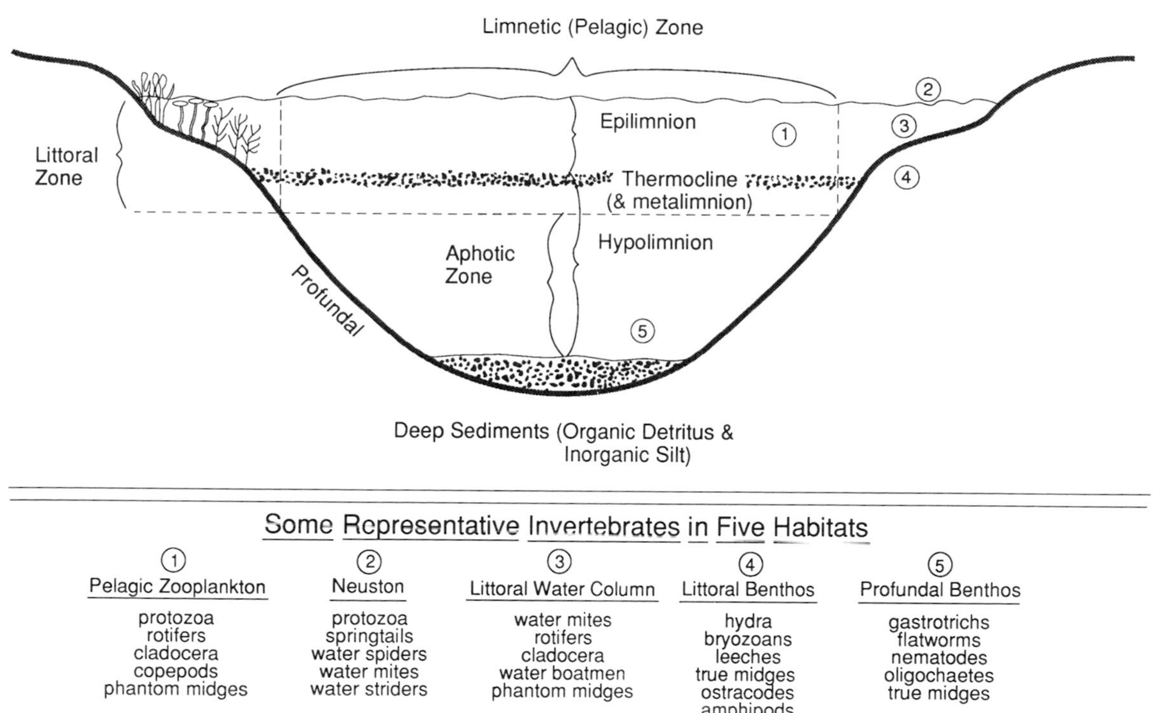


***FIGURE 3*** Biotic and abiotic zones within a lake, along with a list of some representative freshwater invertebrates found within those zones. [From J. H. Thorp and A. P. Covich, eds. (1991). "Ecology and Classification of North American Freshwater Invertebrates." Academic Press, New York.]

where temperatures have cooled to ≤40°C, conditions become suitable for a very few species of invertebrates, such as some nematodes, oligochaetes, water mites, and ostracods and a few other crustaceans. High levels of dissolved sulfur compounds in many geothermal waters also produce stressful conditions.

## B. Physiological Adaptations to Freshwater Environments

### *1. Aspects of Feeding by Freshwater Invertebrates*

#### a. Carnivory

Freshwater ecologists have conducted extensive research on feeding biology and the relative importance of food limitation versus predation (carnivory and herbivory) in regulating aquatic communities. The importance of carnivory seems to depend on a variety of factors, including food web complexity and linkage strength, food selectivity of predators (generalist versus specialist), prey susceptibility (e.g., motility, size, reproductive rate, and density of prey), and the environment's equilibrium state. Although carnivores occur among many invertebrate orders in over half the freshwater phyla, apparently only a few species exert enough influence to be considered keystone predators.

Carnivores can be divided into "engulfers" and "piercers." The former consume solid tissue (whole animals or parts), whereas the latter suck liquid tissue (blood or liquified solid tissue). The more common "engulfing" technique is employed by members of all phyla containing freshwater carnivores (Fig. 4). Predators must usually be larger and strong enough to hold their victims with ripping body appendages; dragonflies and some cyclopoid copepods are good examples of this type of engulfer. Predaceous rotifers and nematodes lack grappling appendages and must engulf their microscopic prey whole. An exception to the predator/prey size relationship is the common hydra, which can capture relatively massive prey by stunning its victim with neurotoxin injected with dartlike nematocysts. "Piercers" include true bugs (hemip-

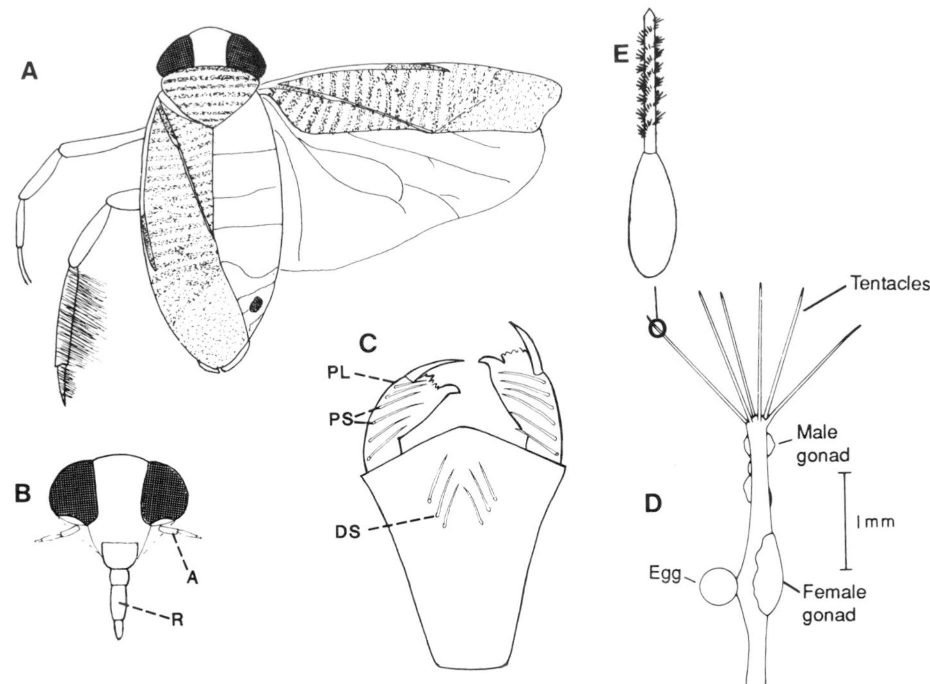


***FIGURE 4*** Predaceous freshwater invertebrates: (A) corixid bug *Sigara* (dorsal view with right wing extended), a piercer; (B) head of another piercer bug, *Notonecta* (A = antenna, R = rostrum); (C) prementum of the damselfly *Enallagma* (see also Fig. 7E) (DS = dorsal setae, PL = palpal lobe, PS = palpal setae) (A–C from Hilsenhoff); (D) hydra showing gonads and tentacles; (E) one of several kinds of nematocysts in cnidarian tentacles (D–E from Slobodkin and Bossert after Hyman). [From J. H. Thorp and A. P. Covish, eds. (1991). "Ecology and Classification of North American Freshwater Invertebrates." Academic Press, New York.]

terans), aquatic spiders, and some leeches. Such predators usually overpower their quarry, but leeches frequently attach to animals larger than themselves. Depending on the victim's size and nature, a piercer may consume blood or inject enzymes into the prey's body and then suck up liquefied, formerly solid tissue (Fig. 4).

### b. Benthic Grazers

Aquatic herbivores that feed at bottoms of lakes and streams graze upon a carpet of energy-rich aufwuchs, the assemblage of microinvertebrates, algae, fungi, and bacteria living attached to the outer surfaces of rocks, woody debris, and macrophytes. Because the aufwuchs is often difficult to detach, these microherbivores must either pick and tear living material from substrates with appendages—as do many beetles, mayflies, and caddisflies—or rasp the attached film from benthic surfaces with toothlike radulas—a technique employed by most snails. Although the role of grazers is less evident at a macrocommunity level than the effects of large carnivores, ecologists have amply demonstrated that benthic herbivores can have impressive influences on the nature of microhabitat assemblages.

### c. Shredders and Detritivores

Invertebrates using benthic particulate organic matter (POM) as a primary energy source may operate in opposing portions of the food web by processing fallen leaves as they enter the system and by consuming dead aquatic plants and animals after the carbon has passed through living components of the aquatic ecosystem. In forested headwater streams, where leaves falling or washing into the water are the primary source of carbon, the community could not function without leaf shredders, such as many larval craneflies and beetles, which tear leaves into pieces small enough for consumption. Recently abscised leaves are more easily digested and nutritive, but all leaves colonized by microbes having energetic value. Shredders occur in most aquatic systems but play their greatest role in small, erosional streams. In contrast, detritivores, which also participate in all aquatic food webs, are especially important in lakes and depositional areas of streams. These infaunal or epifaunal invertebrates eat small particulate detritus in a somewhat selective (e.g., some mayflies) or relatively nonselective fashion (e.g., aquatic earthworms). Both shredders and detritivores compensate for the low energy value of the basic material consumed (dead and decomposing plant and animal POM) by eating massive quantities of detritus and assimilating the microbial flora that has colonized it.

#### d. Suspension Feeding

Freshwater and marine invertebrates in many phyla obtain energy and nutrients by capturing dead organic matter and living organisms (bacteria, algae, protozoa, and metazoa) that live in, or are swept up into, the water column. Suspension feeding is the primary feeding mode in sponges, many rotifers, bryozoans, mussels, numerous microcrustaceans, and insect larvae in several orders.

Suspension feeding can be divided into four subcategories: (a) true filter feeders; (b) setal-net feeders and scan-and-trap consumers; (c) contact suspension feeders; and (d) tentacular predatory-suspension feeders. The true filter feeders, which include net-spinning caddisflies, employ sievelike structures to retain particles carried by a water current. Setal-net feeders and the more selective scan-and-trap consumers have setose appendages on the head or thorax that are modified for extracting POM from water and/or for generating feeding currents; copepods and some mayfly nymphs fit within this subcategory. Contact suspension feeders move POM along mucus-covered ciliary tracts on a feeding structure, such as the molluscan gill, bryozoan lophophore, or blackfly cephalic fan appendages. Freshwater mussels are considered contact suspension feeders by some scientists but true filter feeders or deposit feeders by other ecologists. Tentacular predatory-suspension feeders are rare in fresh water, being represented only by hydra and freshwater jellyfish (both could equally well be classified as true carnivores). Some suspension feeders may also procure food by other means. For example, net-spinning caddisflies normally eat the dead POM that collects on their nets but will quickly become carnivorous if a small invertebrate strays near them; in this way they obtain more easily assimilated and energetically beneficial food.

### 2. *Organismal Respiration in Aquatic Species*

A relatively few species of freshwater invertebrates consistently (e.g., aquatic spiders and some bettles), frequently (pulmonate snails), or occasionally (e.g., crayfish) acquire oxygen directly from the atmosphere. Spiders and some insects, such as larval mosquitoes and dytiscid beetles, penetrate the air–water interface and either form an air bubble, which is then taken below the surface, or absorb atmospheric oxygen directly through internal, tubelike tracheae while at the surface. Pulmonate snails have a mantle cavity that is richly lined with blood vessels; on a frequent basis they rise to the surface and absorb air into a lunglike cavity. This ability also allows them to migrate short distances between pools if the original habitat dries. Crayfish normally use gills enclosed within a branchial chamber to absorb oxygen from water; however, under low-oxygen conditions or when migrating over land in damp weather, they will emerge partially or completely from the water and respire atmospheric oxygen. A unique method of obtaining nondissolved oxygen is found in some insects with specialized stylet-siphons that can pierce the outer surface of aquatic weeds to obtain oxygen bubbles produced by the plant.

A more common respiratory technique is to procure dissolved oxygen entirely from the surrounding water. This can be accomplished using either cutaneous (integumentary) respiration (i.e., across surface membranes) or specialized respiratory structures. In both cases the animal relies on a high surface-to-volume ratio to acquire oxygen by diffusion. This ratio can be attained through body shape (e.g., flat, or long and tubular) or the proliferation of respiratory structures (e.g., branching tracheal tubes or thin external gills).

Cutaneous respiration is employed by many small and a few large organisms. Very small, sessile or motile organisms, such as protozoa, rotifers, and gastrotrichs, require no special respiratory structure and transpire gases across surface membranes. Although sessile sponges can be thick and massive, a proliferation of internal, water vascular canals places the choanocytes and most other cells in immediate contact with oxygenated water. Some larger but thin ($\leq$2 mm thick) invertebrates, such as planaria, also respire through the body wall, but these animals are almost always sluggish. It is also possible for larger, thick eumetazoa (e.g., oligochaete worms and larval dipteran insects) to rely on cutaneous respiration if they have a high surface-to-volume ratio (usually they are tubular) and a relatively efficient circulatory system.

Three basic types of specialized respiratory structures have evolved: gills, lungs (e.g., pulmonate mantle cavity), and tracheae. The first two required concomitant development of circulatory systems to transport oxygen to inner cells. Gills are simple to complex, ectodermal evaginations that are thinner than normal epidermis. Among freshwater invertebrates, true gills are found only in Mollusca (prosobranch snails and bivalve mussels and clams), Annelida (mostly oligochaete worms and polychaetes), and Arthropodia (e.g., decapod crustaceans and many aquatic insects). Bivalve ctenidia, or gills, have a dual respiratory-feeding function. An analogous structure to true gills is a lophophore, the respiratory-feeding organ of bryozoans. All gills must be kept wet to function and prevent collapse of gill lamellae; they are most efficient when held in an externally or internally generated current (such as those produced by a crayfish's gill bailers). Tracheae, the air-filled, tubular invaginations that ramify throughout the body of an insect or aquatic spider, evolved in terrestrial arthropods but are still useful to some aquatic species because oxygen diffuses to inner cells faster through air-filled canals than through solid tissue. Some insects that rise to the surface to obtain oxygen (as described previously) capture air bubbles under wings or the ventral surface and gradually absorb oxygen through internal tracheae. As oxygen is depleted in the bubble, it diffuses inward from the surrounding water. Because diffusion rates are slow and the bubble volume continues to diminish, the insect must occasionally rise to the surface to replenish its air. More permanent oxygen storage is achieved by a process called "plastron respiration." Certain beetles and true bugs trap air within a semirigid structure or vertical struts or a thick pile of hydrophobic hairs. The plastron design reduces loss of air volume over time, enabling the insect to survive longer on the air diffusing into the air-filled plastron. Plastron respiration occurs only in highly oxygenated environments and is usually restricted to habitats with frequent water fluctuations that cause short, intermittent exposure of the insect to the atmosphere. Other insects, like larval stoneflies, caddisflies, and mayflies, have evolved tracheal gills that extend outward from the body in thin gill-like lamellae.

### *3. Osmotic and Excretory Concerns*

Like other organisms, freshwater invertebrates must maintain proper osmotic balance, control ion flux, and regulate nitrogenous products; however, their osmotic and excretory concerns are often considerably different from those faced by terrestrial and marine invertebrates. Lacking a terrestrial invertebrate's need to avoid water loss and subsequent damaging ionic concentrations, freshwater and marine invertebrates can excrete waste nitrogen as highly toxic ammonia rather than investing energy to package concentrated urea, guanine, or uric acid. Unlike marine invertebrates, which face only slight osmotic and ionic gradients between external and internal fluids, freshwater species are constantly confronted with an influx of water and a loss of ions to a more dilute surrounding milieu, upsetting the delicate osmotic and ionic balances.

To maintain their intracellular fluids hyperosmotic to the environment, freshwater invertebrates must continually expend energy to transport ions. All freshwater invertebrates regulate ions to a small or large degree, but fewer species are adept at osmotic regulation; fortunately, the ionic and osmotic concentration of freshwater ecosystems rarely fluctuates significantly (unlike in estuaries). Most species, especially freshwater mussels, tolerate more dilute internal fluids than those that characterize marine species. For instance, internal fluids in the unionid mussel *Anodonta cygnea* are only 4–5% as concentrated as seawater, whereas osmotic concentrations in other freshwater invertebrates are 10–40% of seawater.

Life in hypersaline environments (>35 ppt) is impossible for most inland invertebrates, as the salinity sometimes exceeds 200 ppt (200,000 mg/liter). However, the brine shrimp and certain fly species survive by maintaining internal fluids at about 10% of external osmolarity. To accomplish this remarkable feat they drink large quantities of water (to combat diffusion loss) and excrete enormous amounts of salt through gill tissue at a high energetic cost.

Salinities of inland waters can influence species distribution and faunal composition. Aside from the harshness of hypersaline lakes, very soft waters (approximately 0.01 ppt or 10 mg/liter) seem to eliminate some shelled species that have difficulty actively absorbing enough calcium carbonate ions to maintain adequate shell thickness. Shells and exoskeletons of freshwater molluscs and crustaceans typically have less calcium than occurs in related marine species, making them relatively more susceptible to predators.

### 4. Reproductive and Life-History Patterns

Reproductive and life-history patterns of freshwater invertebrates are quite different from those of their marine relatives but often comparable to those of closely related terrestrial species. Aquatic insects, which have terrestrial ancestors, possess similar developmental patterns (e.g., ametabolous, hemimetabolous, and holometabolous development) and may have one, two, or multiple generations per year (i.e., univoltine, bivoltine, and multivoltine development). However, larval periods of freshwater insects are almost always longer than life spans of the often terrestrial adults, whereas the reverse can be true in completely terrestrial species. Most species live less than a year, except in colder climates where larval periods may extend for two or more years. Other phyla with significant numbers of terrestrial and freshwater species also show similar reproductive and life-history patterns in these disparate environments.

Two of the more striking contrasts in life-history phenomena between marine and freshwater invertebrates involve colonial life and larval development. The only true colonial, freshwater taxa are bryozoans (unless you consider sponges to be colonies). In contrast, many species of marine tunicates, cnidarians, and bryozoans are truly colonial, and other species, such as some polychaetes and bivalve molluscs, benefit from living in cemented clumps without tissue connections. As discussed previously, species with free-swimming larvae are extremely common in the ocean but rare in fresh waters, the principal exceptions being planktonic copepods and aquatic insects. Marine and freshwater species of decapod and amphipod crustaceans commonly carry abdominal eggs; however, larvae emerge from eggs in a very early stage (nauplius) in marine species but in a postlarval stage in freshwater species. Consequently, crayfish and freshwater shrimp can provide some parental care for their progeny, but brooding is much less common in marine decapods. Causes for this dissimilarity in life-history characteristics are not entirely clear, although they may reflect interecosystem differences in habitat permanency, dispersal needs, biotic interactions, and physical constraints.

Given differences in temporal variability, it is not surprising that reproductive techniques of their constituent fauna often differ between marine and freshwater communities. Sexual reproduction in freshwater invertebrates usually involves internal fertilization, but numerous marine taxa broadcast gametes. Budding, fission, and fragmentation are typical asexual processes in marine invertebrates. Among freshwater species, budding is mostly limited to hydras and bryozoans, and fission/fragmentation occasionally transpires in flatworms. Multiple fission (polyembryony) characterizes some life stages of some trematode flukes. Metagenesis (alternation of generation between polyps and medusae) occurs in many marine cnidarians but is found in freshwater species only in the rare hydrozoan jellyfish *Craspedacusta*.

Three other interesting reproductive responses to inhospitable conditions are present in fresh waters: formation of asexual propagules, encystment, and parthenogenesis. To survive adverse changes in the environment, some freshwater taxa produce resistant, seedlike propagules by asexual means; examples are gemmules in sponges (much less common in marine species) and statocysts in freshwater bryozoans. Encystment, a process found in tardigrades and some other freshwater taxa, involves production of a resting, or dormant, life stage, usually protected by a special capsule. If the environment dries completely, the encysted organism can survive long periods until rains return. Ofter this dormant stage is transported great distances by wind or vertebrates, thus serving as a dispersal mechanism. Parthenogenesis, the most widespread form of asexual reproduction, is especially common in freshwater species, particularly rotifers. Although many forms of parthenogenesis exist, all involve production of eggs that can develop into adults without fertilization by spermatozoa. This phenomenon is so distinctive that rotifers can often be separated into classes on the sole basis of reproductive system and, in some species, males have never been collected by ecologists!

Although freshwater invertebrates are often ignored or belittled when compared to their larger and usually more brightly colored marine relatives, it should be clear from this short discourse that they are fascinating creatures in their own right. A great many scientific questions involving freshwater invertebrates need to be investigated and merely await the attention of the next generation of aquatic biologists.

## Glossary

**Aufwuchs** Group of microinvertebrates, algae, fungi, and bacteria living attached to outer surfaces of rocks, woody debris, and macrophytes; this film of organisms is consumed by various invertebrate grazers.

**Epigean** Surface water habitats, such as streams, lakes, and wetlands (as opposed to underground, or hypogean, habitats).

**Functional feeding group (or guild)** Group of organisms with similar feeding habits, as defined by the investigator; these often relate to either the type of food ingested or the means used to procure food; examples are filter feeders, algal grazers, carnivores, and detritivores.

**Heterotroph** Organism that cannot synthesize its own organic molecules from inorganic substances and must instead consume other organisms; includes all multicellular animals, fungi, and protozoa.

**Hyperosmotic** Solution whose solute osmotic concentration is greater than that of a reference solution; thus, internal fluids in an organism may be more concentrated (hyperosmotic) than the surrounding environment.

**Hypogean** Underground aquatic habitats (as opposed to epigean, or surface, habitats).

**Iteroparous** Reproducing at multiple times during the life of an organism (opposite of semelparous).

**Lentic** Standing water habitats (e.g., wetlands, lakes, and ephemeral pools).

**Lotic** Flowing water habitats (e.g., creeks and rivers).

**Nematocyst** Usually dartlike, threaded organelle with venom or sticky substance that is injected into or on a victim; used on predation, defense, and sometimes locomotion; found in all Cnidaria and one species of Ctenophora.

**Parthenogenesis** Development of an ovum into an adult without fertilization by a male.

**Tracheae** Chitinized, often branching, respiratory tubes that extend inward from external spiracles to the hemocoel of many terrestrial and aquatic arthropods (principally insects and spiders).

**Troglobite** Obligate cave-dwelling organism (found only in hypogean habitats).

**Troglophile** Faculative cave-dwelling organism (found in both hypogean and epigean habitats).

## Bibliography

Balcer, M. D., Korda, N. L., and Dodson, S. I. (1984). "Zooplankton of the Great Lakes." Madison: University of Wisconsin Press.

Barnes, R. S. K., and Mann, K. H. (1991). "Fundamentals of Aquatic Ecology." London: Blackwell Scientific.

Peckarsky, B. L., Fraissinet, P., Penton, M. A., and Conklin, D. J., Jr. (1990). "Freshwater Macroinvertebrates of Northeastern North America." Ithaca, N.Y.: Cornell University Press.

Thorp, J. H. (1986). Two distinct roles for predators in freshwater assemblages. *Oikos* **47**(1), 75–82.

Thorp, J. H., and Covich, A. P., eds. (1991). "Ecology and Classification of North American Freshwater Invertebrates." New York: Academic Press.

Ward, J. V. (1992). "Aquatic Insect Ecology. 1. Biology and Habitat." New York: Wiley.

# Island Biogeography, Theory and Applications

*Jianguo Wu*
*Desert Research Institute*

*John L. Vankat*
*Miami University, Ohio*



Island biogeography is the study of pattern in the distribution of species on islands as influenced by ecological and evolutionary processes related to island characteristics such as isolation and area. The MacArthur–Wilson theory of island biogeography asserts that two processes, immigration and extinction, determine the species diversity of an island's biota. As the number of species on the island increases, the immigration rate decreases and the extinction rate increases. The immigration rate decreases with isolation (distance effect), whereas the extinction rate decreases with island area (area effect). When the immigration and extinction rates are equal, an equilibrium of species diversity is reached. The extinction of species on an island and their replacement by the immigration of new species results in species turnover. The equilibrium theory of island biogeography has been one of the more influential concepts in modern biogeography, ecology, and evolutionary biology. It also has had a major influence on conservation biology, particularly with attempts to develop a theoretical basis for the design of nature reserves. However, several aspects of the theory remain unsubstantiated, and therefore uncritical application of the equilibrium theory of island biogeography to nature conservation is unwarranted.

## I. INTRODUCTION

Islands as ecological systems have such salient features as simple biotas and variability in isolation, shape, and size. These characteristics and their large numbers facilitate both intensive and extensive studies with the repeatability necessary for statistical validity. Since Darwin, islands have provided particularly important and fruitful natural experimental laboratories for developing and testing hypotheses on evolution, biogeography, and ecology. The theory of island biogeography has been one of the more important products of island studies. [*See* GALÁPAGOS ISLANDS.]

Eugene G. Munroe (1948, 1953) first developed the concept of an island having an equilibrium species number when he examined species-area relationships in his study of the distribution of butterflies in the West Indies. Unfortunately, Munroe's ideas were unrecognized by biogeographers and ecologists until the 1980s, partially because they appeared only as a small portion of his dissertation at Cornell University and as an abstract in the proceedings of a regional conference. Later, Frank W. Preston (1962) made a significant contribution to the early development of island biogeography theory. However, it was Robert H. MacArthur and Edward O. Wilson (1963, 1967), working independently of Munroe and Preston, who provided a coherent, comprehensive theory with elegant mathematical models in their seminal, landmark monograph "The Theory of Island Biogeography." MacArthur and Wilson's work triggered explosive growth in the scientific literature on insular habitat studies and essentially transformed research on island biogeography from largely descriptive works to a more quantitative and predictive new stage. Modified and extended by many others, the MacArthur–Wilson theory continues to occupy a central position in basic and applied biogeography and ecology.

Many studies inspired by the theory have been conducted not only on oceanic and continental islands but also on a variety of other insular habitats, involving numerous different taxonomic groups of plants, animals, and microbes. The theory has also had a profound impact on conservation biology in both theory and practice. Although its inspirational role has been tremendous, the validity and applicability of the theory have been hotly disputed.

## II. THE EQUILIBRIUM THEORY OF ISLAND BIOGEOGRAPHY

### A. Species–Area Relationship

A monotonic increase in the number of species with increasing island area (or sampling area of terrestrial communities) has long been recognized. Among various mathematical expressions for the species–area relationship, a power function of the following form has been shown to best fit observations in many cases:

$$S = cA^z \tag{1}$$

or

$$\log S = z \log A + \log c, \tag{2}$$

where $S$ is species diversity (the number of species, i.e., species richness), $A$ is area, and $c$ and $z$ are positive constants. $c$ usually reflects the effect of geographical variation on species diversity, and $z$ has a theoretical value of 0.263 and usually varies between 0.18 and 0.35. Linear regression typically has been used to estimate values for these constants.

Although Equation (1) has been widely used, it provides little explanatory or predictive power. Thus, three hypotheses have been proposed to account for the species–area relationship: (1) the habitat diversity hypothesis by C. B. Williams (1964); (2) the passive sampling hypothesis by E. F. Connor and E. D. McCoy (1979); and (3) the dynamic equilibrium hypothesis independently by Munroe (1948, 1953), Preston (1962), and MacArthur and Wilson (1963, 1967).

According to the habitat diversity hypothesis, the species–area relationship results from a positive correlation existing between area and habitat diversity and between habitat diversity and species diversity. In the passive sampling hypothesis, insular taxonomic groups are simply viewed as subsets or samples from larger communities. Therefore, species diversity is a function of both sampling size and intensity; B. D. Coleman *et al.* (1982) proposed a similar explanation that is referred to as the random placement hypothesis. The dynamic equilibrium hypothesis, the topic of this article, has been by far the most influential in island biogeography and ecology. [*See* Species Diversity.]

### B. Basic Tenets and Mathematical Models of the Equilibrium Theory

MacArthur and Wilson theorized that species diversity on an island is primarily determined by two processes: immigration and extinction. If there are a limited number of empty niches or habitats available on an island, the more species already there, the less likely a successful immigration of a new species and the more likely an extinction of a species already present. For a given area and degree of isolation (usually represented by distance to the island's colonization source), immigration and extinction rates are expected to monotonically decrease and increase, respectively, as the number of established species increases. For a given number of species, the immigration rate decreases with distance from the source of colonizing species because species have different dispersal abilities—this is called the "distance effect." On the other hand, the extinction rate will decrease with island area because the larger the area, the larger the species' populations and thus the smaller the probability of extinction—this is referred to as the "area effect." [*See* EQUILIBRIUM AND NONEQUILIBRIUM CONCEPTS IN ECOLOGICAL MODELING.]

When the rates of immigration and extinction are equal, the island's biota is at a state of dynamic equilibrium, that is, species diversity remains relatively constant while a compositional change of species continues. These changes in species composition result from concurrent extinction of existing species and replacement by immigration of new species and are called "species turnover." The rate of change in species composition is called turnover rate. In theory, the turnover rate at equilibrium is equal to the immigration or extinction rate (Figs. 1 and 2). If different groups of organisms are treated in parallel instead of collectively, it has been asserted that the turnover rate generally increases from higher to lower organisms.

The mathematical description of the MacArthur–Wilson theory takes the general form

$$\frac{dS(t)}{dt} = I - E, \tag{3}$$

where $I$ is the immigration rate and $E$ is the extinction rate. Assuming homogeneity, temporal consistency, and additivity of immigration and extinction rates among species, a linear relationship between the rates and species diversity would be expected (Fig. 1). In other words, if all the colonizing species have similar per-species rates of immigration and extinction, if these rates are constant

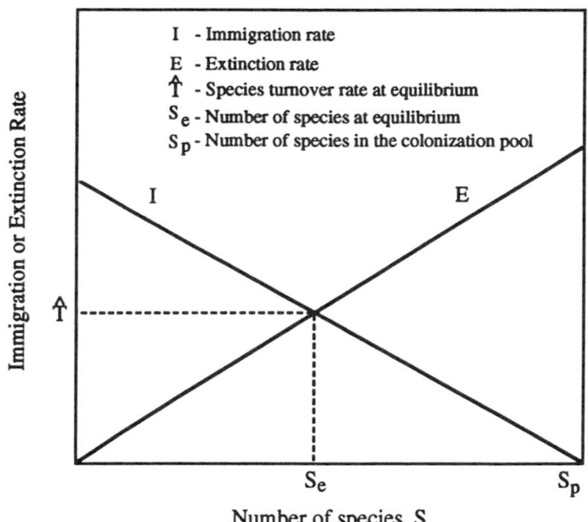


**FIGURE 1** The simplest version of the equilibrium model of island biogeography with linear relationships between the rates of immigration and extinction ($I$ and $E$, respectively) and the number of species present ($S$) on an island. These relationships assume homogeneity, temporal consistency, and additivity of immigration and extinction rates among all species in the species pool. An equilibrium value of the number of species ($S_e$) is reached when the rate of species immigration from a colonizing pool of size $S_p$ is equal to the rate of species extinction on the island. Species turnover rate at equilibrium ($\hat{T}$) is equal to the immigration or extinction rate.

with changing species diversity, and if species interactions are insignificant, it holds that

$$I(s) = I_0(S_p - S(t)) \tag{4}$$

and

$$E(s) = E_0 S(t), \tag{5}$$

where $I_0$ is the per-species immigration rate or immigration coefficient, $E_0$ is the per-species extinction rate or extinction coefficient, and $S_p$ is the total number of potential immigrants in the colonization pool. Substituting Equations (4) and (5) into (3) yields

$$\frac{dS(t)}{dt} = I_0 S_P - (I_0 + E_0)S(t). \tag{6}$$

This equation gives rise to the rate of change of species diversity with respect to time. The number

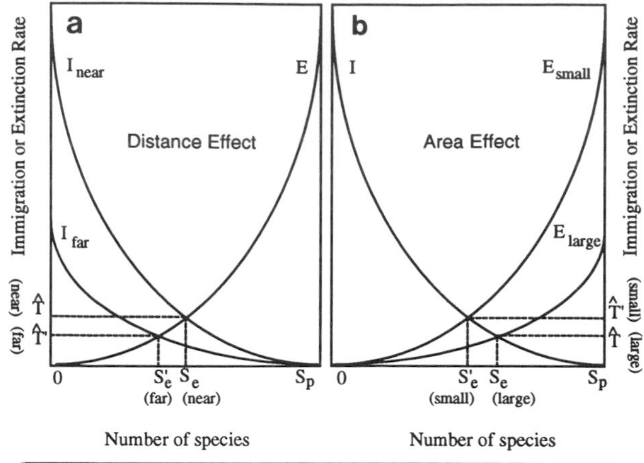


***FIGURE 2*** The main tenets of the MacArthur–Wilson theory of island biogeography. The number of species on an island is determined by two processes: immigration and extinction. With an increase in the number of species on an island, the immigration rate ($I$) decreases monotonically and the extinction rate ($E$) increases monotonically. For a given number of species, the immigration rate is smaller for more distant islands (distance effect, part a), whereas the extinction rate is larger for smaller islands (area effect, part b). As a result, the number of species at equilibrium ($S_e$) is greater on larger and less distant islands than on smaller and more remote ones. Species turnover rate at equilibrium ($\hat{T}$) is greater on less distant and smaller islands. [Redrawn from D. S. Simberloff (1974). "Equilibrium theory of island biogeography and ecology." *Annu. Rev. Ecol. Syst.* **5**, 161–182.]

of species on an island at any time $t$ can be calculated by integrating the differential equation, that is,

$$S(t) = \frac{I_0}{I_0 + E_0} S_p - \left( \frac{I_0}{I_0 + E_0} S_p - S(0) \right) \exp(-(I_0 + E_0)t), \quad (7)$$

where $S(0)$ is the initial number of species on an island.

The equilibrium value of species diversity, $S_e$, can readily be obtained from Equation (6) by setting the rates of immigration and extinction equal (i.e., $I = E$):

$$S_e = \left( \frac{I_0}{I_0 + E_0} \right) S_p. \quad (8)$$

Equation (8) states that the equilibrium species diversity is only determined by per-species immigration and extinction rates and the size of the species pool. The rate of change in species number can also be related to the equilibrium value as

$$\frac{dS(t)}{dt} = (I_0 + E_0)(S_e - S(t)). \quad (9)$$

That is, the change rate of species diversity is proportional to the difference between the equilibrium and nonequilibrium values of species diversity. Furthermore, the change rate becomes negative for $S(t) > S_e$ and positive for $S(t) < S_e$. Substitution of Equation (8) into (7) yields

$$S(t) = S_e - (S_e - S(0)) \exp(-(I_0 + E_0)t), \quad (10)$$

which indicates the relationship among the values of species diversity at initial, nonequilibrium, and equilibrium states. From this equation, one can predict the time required for an island in disequilibrium to approach or return to the equilibrium state. This can be done by solving Equation (10) for $t$, resulting in

$$t = \frac{\ln R}{I_0 + E_0} \quad (11)$$

where

$$R = \frac{S(t) - S_e}{S(0) - S_e}. \quad (12)$$

$R$ is the ratio of the departure of $S(t)$ from equilibrium at time $t$ to its initial deviation at $t = 0$. The (exponential) relaxation time, $T_r$, is usually defined as the time required for a departure of species diversity from equilibrium to decrease to $1/e$ (or 36.8%) of the initial departure, that is,

$$T_r = -\frac{\ln(1/e)}{I_0 + E_0} = \frac{1}{I_0 + E_0}. \quad (13)$$

From this equation, 90 and 95% complete relaxations take 2.303 and 3 relaxation times, respectively.

It is important to notice again that the foregoing quantitative description is necessarily based on the assumption that $I_0$ and $E_0$ are the same for all species, are constant in time, and are unaffected by species interactions. However, in nature immigration and extinction capabilities are species specific, temporal variability exists, and interactions occur among species. MacArthur and Wilson pointed out that nonlinear concave curves may represent the rate–species diversity relationship more realistically than do the straight lines. They argued that species with the best colonizing abilities were most likely to establish first and that the weakest competitors would be expected to go extinct first. A gradient in species colonizing abilities would cause rapid initial drop in the overall immigration rate and the combined effects of diminishing population size and increasing species interference could yield an extinction curve that was approximately exponential (Fig. 2).

There also is empirical evidence that the per-species rates of immigration and extinction tend to decrease and increase, respectively, with increasing species diversity. In particular, some studies on birds have suggested that both curves are very concave, especially the immigration curve. Some studies also show that the rates of immigration and extinction have a roughly log-normal distribution among species, with differences ranging over many orders of magnitude. It has also been suggested that rate curves incorporating stochastic variations would be more appropriate than deterministic lines. MacArthur and Wilson commented that the modifications in shape of the two curves would not be critical, as long as the curves are monotonic.

### C. Predictions of the MacArthur–Wilson Equilibrium Theory

Many controversies have developed in the literature over what the original equilibrium theory of island biogeography does and does not predict. However, it is generally accepted that the core of the MacArthur–Wilson theory consists of the concept of species equilibrium and the postulation that the immigration and extinction curves are monotonic. Major predictions based on the equilibrium theory include the following: (1) there is an equilibrium for a given island biota that is achieved when the extinction and immigration rates are equal; (2) the immigration rate is primarily affected by the distance between the island and its continental colonizing source, and the extinction rate varies primarily with island area; (3) for a given island, the immigration rate decreases and the extinction rate increases with increasing number of species already on the island; (4) the number of species at equilibrium increases with island area and this increase should be faster on more remote islands; (5) the number of species at equilibrium decreases with island–continent distance and this decrease should be faster on smaller islands; and (6) the species turnover rate at equilibrium is greater on less distant and smaller islands (Fig. 2).

## III. MODIFICATIONS AND EXTENSIONS OF THE ORIGINAL EQUILIBRIUM THEORY

The curves of immigration and extinction rates are central to the equilibrium theory of island biogeography. Studies have suggested that (1) the immigration rate, at least for plants, may initially increase with increasing species diversity as the original unfavorable environmental conditions of the island are ameliorated by the first colonists, and (2) slight rate heterogeneity and weak species interactions may alter both concave and convex curvatures, whereas great rate heterogeneity, temporal variability, and strong species interactions can produce significant changes in the rate curves such as discontinuities and nonmonotonic features. J. Wu and J. L. Vankat examined the effects of several different sets of immigration/extinction rate–species diversity curves on the predictions of the MacArthur–Wilson theory through computer simulations. They found that different monotonic rate–species diversity curves do not affect the basic predictions of the theory of island biogeography; however, the level of equilibrium species diversity can be substantially affected. On the other hand, nonmonotonic rate–diversity curves may result in potential multiple equilibria of species diversity.

Questions about the assumption of independence between immigration and extinction rates also have produced modifications in the original MacArthur–Wilson model. For example, researchers have claimed that the extinction rate of insular populations may be reduced when immigration of conspecific individuals provides demographic reinforcement and augments genetic variability of existing populations. This phenomenon has been referred to as the "rescue effect." It suggests that distance directly affects not only the immigration rate but the extinction rate as well. More specifically, with greater distance the extinction rate may increase as immigration of conspecifics decreases. This may initially increase the turnover rate before its expected decrease. The rescue effect is most likely when the immigration rate is very high or at least approaches local recruitment rate; therefore, it may not occur with truly isolated islands. Another example of the lack of independence between immigration and extinction rates is that island area may affect both, at least to the degree that larger islands are more likely to receive immigrants than smaller ones. The phenomenon of increasing probability of immigration with island area has been called the "target effect."

Questions also have been raised about the assumption of the original MacArthur–Wilson theory that evolution rates are always much slower than colonization rates and therefore can be safely ignored. This assumption leads to the conclusion that the emergence of new species on an island results only from immigrations. Some studies on the isolated Hawaiian Archipelago have shown that the autochthonous speciation rate has exceeded the immigration rate and, with adaptive radiation, has produced a species–area equilibrium involving a high degree of endemism. Other studies have also shown that island forms can evolve rather rapidly and differentiation to the subspecific level may occur in a single generation. By restricting gene flow between islands and their species pools, isolation could facilitate interisland evolutionary divergence and increase endemism. A long-term equilibrium of species diversity involving speciation, immigration, and extinction has been called the "taxon cycle"; it describes the process of ecological and evolutionary diversification of species populations on islands.

The original simplistic equilibrium concept has been expanded to include four phases of an insular species equilibrium: the noninteractive equilibrium, the interactive equilibrium, the assortative equilibrium, and the evolutionary equilibrium. The first, the noninteractive equilibrium, is a temporary or short-term phase that involves a balance between immigrations and extinctions occurring when species populations are thought to be too small for interspecific interactions to be significant. The second phase, the interactive equilibrium, is reached as population sizes grow and some of the less fit original colonizers are lost owing to competitive exclusion. Following this, the third phase of assortative equilibrium involves a long time period of colonizations by better-adapted species and extinctions of lesser-adapted species. This continual sorting of species composition, with or without change in species diversity, results in a nonrandom, coadapted set of species. The fourth phase, the evolutionary equilibrium, reflects the impact of natural selection on species diversity, as it represents the balance between the addition of species via increased species coadaptation and adaptation to physical environment and the deletion of species through extinction as niches evolve to be narrower. Over geological time, evolution tends to gradually increase the equilibrial number of species—a number that is steady over ecological time.

Another development was Wu and Vankat's quantitative synthesis of the theory of island biogeography. They produced a comprehensive system dynamics simulation model incorporating a variety of modifications and extensions, including area, distance, competition, habitat diversity, target, and rescue effects. A main purpose of their work was to provide a user-friendly simulation tool to enhance understanding of the internal structure and predictions of the theory when the modifications are taken into account.

## IV. APPLICATIONS OF THE THEORY OF ISLAND BIOGEOGRAPHY

MacArthur and Wilson conjectured that the theory of island biogeography should be applicable to insular continental habitats and habitat patches at various spatial scales. Since then, the theory has been applied to a variety of "habitat islands" such as individual plants, caves, lakes or ponds, mountaintops, microcosms, and patches of terrestrial ecosystems. In fact, it was the theory of island biogeography that most noticeably brought scientists' attention to spatial patchiness and the effects of habitat size and isolation on ecological and evolutionary processes.

The theory of island biogeography, therefore, has served as a conceptual framework for studies of the impacts of habitat fragmentation on biological diversity and for research in conservation biology in general. It also has inspired theoretical investigations of population dynamics in heterogeneous environments and, more recently, provided an impetus to the development of the field of landscape ecology, particularly in North America. For example, consideration of the effects of patch area and interpatch distance has been central to both empirical and theoretical studies of the dynamics of populations and flows across landscape mosaics. The equilibrium theory also has been adapted to help explain mass extinctions in geological time, as diversity at different taxonomic levels (species, genera, or family) is treated as the product of a dynamic equilibrium between origination and extinction. For example, the MacArthur–Wilson model was employed to explain changes in diversity of North American land mammal genera during the past 12 million years. However, such paleobiological models have received much criticism. [*See* MASS EXTINCTION, BIOTIC AND ABIOTIC.]

Almost from its inception, the theory of island biogeography has had enormous impacts on both the theory and the practice of nature conservation. Indeed, among its numerous applications, the theory finds its widest, most conspicuous, and yet most controversial usage in the design of nature reserves to maximize species diversity—with nature reserves being perceived as islands in a sea of human-transformed habitats. Such applications gained tremendous momentum during the early 1970s when general design principles based on the species–area relationship and the equilibrium theory were proposed. These principles became widely publicized in prestigious journals and books, including the "World Conservation Strategy" published by the International Union for the Conservation of Nature and Natural Resources in 1980. The so-called general design principles included: (1) a large reserve is superior to a small one; (2) a single large reserve is better than several small reserves with the same total area; (3) when two or more reserves are inevitable for some specific habitat or species, interreserve distance should be as short as possible; (4) corridors between reserves are recommended to increase interreserve immigration; and (5) a circular shape is optimal because it minimizes dispersal distances within the reserve. [*See* NATURE PRESERVES.]

The relevance of these design principles/recommendations has been heavily criticized. For example, several authors have pointed out that there is little evidence of a dynamic equilibrium in continental habitat islands, yet the assumption of equilibrium is essential to the theory of island biogeography. Also, the theory itself does not directly address questions of shape, including what is optimum (if there is an optimal shape). Several researchers have reached the conclusion that the matter of shape is trivial in the design of nature reserves, if the mechanisms controlling species diversity dynamics for reserves and for islands are comparable. Another criticism is that while corridors between reserves certainly may increase immigration, facilitate gene flow, and reduce local extinctions through the rescue effect, the effectiveness and significance of corridors may greatly depend on their content and dimensions, interreserve distance, and specific species involved. Moreover, corridors may increase the spread of disease, disturbance, and exotic species. Hence, decisions about corridors ought to be case specific rather than blindly follow general principles.

A particularly controversial aspect of applying the MacArthur–Wilson theory to nature conservation has been whether a single large or several small reserves (SLOSS), with the same total area, would better protect species diversity. The answer to the SLOSS question depends on the slope of the species–area curve, the proportion of common species in the small reserves, and the gradient of colonizing abilities among species in the available pool. Indeed, both theoretical analysis and empirical evidence have suggested that in some circumstances several small reserves may have more species than a single large one. Several small reserves may have compensating advantages such as greater overall habitat heterogeneity, lower intra- and interspecific competition, reduced spread of some diseases, disturbances, and exotic species, and more habitat for edge species. Moreover, the debate over SLOSS has overlooked the complexity of species diversity dynamics. Factors such as minimum viable population (MVP), minimum area to sustain MVP, and minimum dynamic area to maintain the ecosystem integrity must be considered in questions concerning nature conservation. In conclusion, it is generally accepted that the species–area relationship and the theory of island biogeography are equivocal with respect to the SLOSS issue. [*See* LANDSCAPE ECOLOGY.]

## V. CRITIQUES OF THE THEORY OF ISLAND BIOGEOGRAPHY

The equilibrium theory of island biogeography, like many other ecological models, is difficult to test. For the purpose of validation, the following necessary but not necessarily sufficient conditions have been suggested: (1) a strong species–area relationship; (2) an equilibrium state of species diversity; (3) an appreciable species turnover rate; and (4) detectable distance and area effects. Therefore, a strong species–area relationship alone neither suffices to validate the equilibrium theory itself nor to warrant its application to a particular set of insular habitats. MacArthur and Wilson clearly pointed out in their original work that a species–area curve does not prove the existence of an equilibrium.

Many attempts have been made to verify or falsify the MacArthur–Wilson model; however, direct experimental examinations, which would be most convincing, have been uncommon. D. S. Simberloff and E. O. Wilson designed an experimental test involving removing the fauna from a group of six small red mangrove islands in the Florida Bay while leaving two similar ones as controls. They concluded that their results supported the theory, although the turnover rates were substantially overestimated. Another experimental test demonstrated the dynamic equilibrium and the area effect but did not clearly detect the distance effect. Other work purporting to validate the equilibrium theory has been controversial. Studies often have been insufficient or erroneous because of misunderstanding the theory or misinterpreting the data collected. Recent work shows that the testing processes may also be confounded by artifactual ratio correlation when relative rates of immigration, extinction, and turnover are used.

Since the late 1970s, the validity of the equilibrium theory has been increasingly questioned and criticized on several grounds. In spite of an extraordinary number of studies regarding this theory, it has been frequently criticized for lack of convincing evidence to support the existence of a species equilibrium, species turnover, and area and distance effects. The theory also has been criticized for overaggregating many affecting factors into essentially two variables, area and distance, and leaving out species-specific population demographic and genetic information that may provide the mechanism for species diversity dynamics. Several recent reviews concluded that the equilibrium theory of island biogeography remains insufficiently validated and its application to nature conservation is premature. These criticisms have led to efforts to develop nonequilibrium theories of island biogeography that emphasize the importance of characteristics of island biotas such as habitat heterogeneity and transient (nonequilibrium) dynamics.

## VI. CONCLUDING REMARKS

MacArthur and Wilson's dynamic equilibrium theory has had revolutionary impact on island biogeography in particular and on biogeography in general by providing a conceptual framework and insight into the dynamics of geographical patterns of species diversity. It has been one of the more influential concepts in biogeography, ecology, and evolutionary biology. However, several aspects of the equilibrium theory remain unsubstantiated, and although it appears to hold in some specific cases, it does not in many others. Uncritical application of the theory to nature conservation is unwarranted and may lead to misleading conclusions.

In general, as with other equilibrium paradigms in ecology, the equilibrium theory of island biogeography is being treated with increasing skepticism. Nevertheless, the criticism does not completely negate the value of the theory of island biogeography. Qualitative use of the theory can result in valuable contributions to studies even when its quantitative application may be invalid. More specifically, the theory was and still is important in guiding scientists in constructing conceptual frameworks for addressing relevant questions in which patchiness and isolation otherwise may not have been sufficiently emphasized or even identified. In fact, as indicated in the preface of their book, even MacArthur and Wilson did not believe that the equilibrium model would exactly fit all field observations; instead, they hoped that the theory would provide stimulus and impetus for advancing "new forms of theoretical and empirical studies." In this regard, the equilibrium theory of island biogeography is one of the most successful theoretical developments in the history of biogeography and ecological science.

## Glossary

**Area effect** Phenomenon in which species extinction rate is reduced with increasing island area.

**Distance effect** Phenomenon in which species immigration rate is reduced with increasing distance between an island and its continental source of colonizing species.

**Extinction curve** Curve relating species extinction rate to the number of species already present on an island.

**Immigration curve** Curve relating species immigration rate to the number of species already present on an island.

**Island biogeography** Study of pattern in the distribution of species on islands as influenced by ecological and evolutionary processes related to island characteristics such as isolation and area.

**Rescue effect** Phenomenon in which species extinction rate is reduced when immigration of conspecific individuals provides demographic reinforcement and augments genetic variability of existing populations.

**Species–area curve** Curve relating number of species to island area.

**Species diversity** Either the absolute number of species (species richness) or a measure that incorporates both the number of species and their relative abundance.

**Species turnover** Changes in species composition resulting from concurrent extinction of existing species and replacement by immigration of new species.

**Target effect** Phenomenon in which species immigration rate increases with increasing area of islands, when distance is constant.

## Bibliography


Begon, M., Harper, J. L., and Townsend, C. R. (1990). "Ecology: Individuals, Populations, and Communities," 2nd ed. Oxford, England: Blackwell Scientific.

Brown, J. H., and Lomolino, M. V. (1989). Independent discovery of the equilibrium theory of island biogeography. *Ecology* **70,** 1954–1959.

Heaney, L. R., and Patterson, B. D., eds. (1986). "Island Biogeography of Mammals." London: Academic Press.

MacArthur, R. H., and Wilson, E. O. (1967). "The Theory of Island Biogeography." Princeton, N.J.: Princeton University Press.

Myers, A. A., and Giller, P. S., eds. (1988). "Analytical Biogeography: An Integrated Approach to the Study of Animal and Plant Distributions." London: Chapman & Hall.

Schoener, T. W. (1988). On testing the MacArthur–Wilson model with data on rates. *Am. Nat.* **131,** 847–864.

Shafer, C. L. (1990). "Nature Reserves: Island Theory and Conservation Practice." Washington, D.C.: Smithsonian Institution Press.

Wu, J. (1989). The theory of island biogeography: Models and applications. *Chin. J. Ecol.* **8**(6), 34–39.

Wu, J., and Vankat, J. L. (1991). A system dynamics model of island biogeography. *Bull. Math. Biol.* **53**(6), 911–940.

# Keystone Species

**L. Scott Mills**
*University of Idaho*



The keystone species concept refers to species whose removal would cause a drastic change in the composition or abundance of plants and animals in a community. The original connotation referred to predators whose removal would cause the subsequent loss of additional species. Since then, however, the term has included keystone prey, plants, links, and modifiers, whose loss could either increase or decrease species diversity. Implicit in the use of the term is the assumption that keystones comprise only a minority of species in a community. The term has played a vital role in demonstrating that species differ in their strengths of interactions with other members of their community. The lack of an operational definition limits the application of the "keystone" concept to ecological or environmental research and management, although the concept of differing interaction strengths should be embraced and incorporated into the decision-making process.

## I. INTRODUCTION

Current conservation efforts have encountered a difficult impasse: on one hand is the desire to protect entire communities or ecosystems, and on the other are the legal and practical limitations that place primary focus on the management of single species. In an attempt to bridge the gap between ecosystem and single-species management, biologists and planners have advocated that keystone species be special targets for efforts to protect biodiversity. Such an approach has been justified on the grounds that keystone species are those whose removal (or loss) will have particularly large impacts on the rest of the community, causing the loss of a myriad of other species. Before giving priority protection to certain "keystone" species, however, we must first examine the biological basis of the term and the implications of using the term in policy decisions. [*See* BIODIVERSITY.]

## II. HISTORY OF THE KEYSTONE SPECIES CONCEPT AND RELATED TERMS

The term "keystone species" was originally coined by Robert T. Paine in 1969. Paine observed that certain carnivores in the intertidal zone (such as the starfish *Pisaster*) controlled the density of mussel species that were capable of competitively excluding other species from the community. When the starfish was experimentally removed, the dominant mussels increased in abundance and overall species diversity declined.

Following Paine's pioneering work, the term has been applied to a variety of species at different trophic levels, with three common elements. First, the keystone interacts strongly in direct and indirect ways with a variety of species; consequently, perturbations to the keystone propagate throughout the community and affect species that may seem far removed, both taxonomically and ecologically, from the keystone. Second, the presence of these keystone species is deemed to be crucial to maintaining the organization and diversity of their ecological communities. Third, it is implicit that a given community has only a few keystones, so that these species are exceptional, relative to the rest of the community, in their significance.

Several other terms are related to keystone species, or extend the original connotation. For example, Steven Carpenter and co-workers refer to species causing "trophic cascades," where the removal of an upper-level predator causes a cascade of productivity changes across trophic levels. In particular, a substantial decrease in predatory fish such as bass or salmon resulted in increased numbers of their plankton-eating prey, which decreased herbivorous zooplankton, ultimately increasing phytoplankton biomass.

Laurence E. Gilbert also emphasized indirect effects and expanded on the perception of keystones. In studying tropical food webs, he observed that an assortment of pollinators (such as bees, moths, and hummingbirds) and seed dispersal agents (including birds and bats) played vital role by visiting scattered plants of many taxonomic groups. Because their foraging was critical to a variety of plants in different food webs, Gilbert called these animals "mobile links."

In all cases, keystone species are assumed to have effects on community or ecosystem processes that are disproportionate to population abundance. How dramatic the effect must be to earn the keystone label has not been quantified, but a recent review by L. Scott Mills and others indicates that ecologists have typically defined as keystone those populations whose removal is expected to result in the disappearance of at least half of the assemblage analyzed. However, we shall see that the keystone term has been used in other contexts as well, prompting the conclusion that it is not currently possible to apply any single operational definition—such as a "50% loss" criterion—to the term's myriad of uses.

Importantly, the history of the keystone species concept has included a flurry of critics and negative evidence. In several studies, experimental investigation of presumed "keystone" species revealed that neither total number nor diversity of other species were changed upon removal of the keystone. On the basis of these studies, several authors have argued that the term is misleading because it implies that "keystone" is a particular property of the species, analogous to species characteristics such as "predatory" or "high growth rate." In contrast, the critics note, the "keystone" role is often particular to and dependent on the environmental setting, current species associations, and responses of other species. Clearly, there remains a pressing need for studies addressing how strongly different species interact with each other, and how general those interactions are across time and space.

Despite the critiques, the term "keystone species" has become firmly entrenched in the scientific literature and continues to be used by ecologists. More recently, the term has been embraced by policy advocates. The assumed importance of keystone species has prompted the position that species interactions be considered an important attribute in biodiversity protection efforts.

Furthermore, management to protect keystone species has been recommended to resolve general policy and land-use dilemmas. For example, it has been suggested that management for individual keystone species should be a focus for the management of whole communities, and that restoration of keystone species is necessary for reestablishment of ecosystem structure and stability.

Because historical use of the term "keystone species" is key to how the term is applied for conservation purposes, I will further consider historical connotations by describing five main types of keystone species. This categorization is not meant to imply mutually exclusive groups or an exhaustive review of the term's application, but rather to show the diversity of keystone effects that have been described.

## III. VARIOUS MEANINGS OF THE CONCEPT

### A. Keystone Predator

The original "keystone species" was an intertidal carnivore, a starfish, whose removal resulted in greatly decreased species diversity. Subsequent to this early work, a great number of predators have been described as keystones because their removal or loss results in reduction in species diversity, or change in species composition. For example, sea otters were regarded as keystone predators because they limit the density of sea urchins that in turn eat kelp, which forms the basis of a unique community type. In other cases, large predators such as lions or coyotes have been called "keystone" because their loss could initiate numerical cascades whereby their prey increases, so that prey abundance at the next level decreases, and so on across several trophic levels. [*See* SPECIES DIVERSITY.]

Although loss of keystones originally implied a decrease in species richness, others have labeled predators "keystone" simply because the species in question has a major effect on community structure. For example, fire ants have been referred to as keystone predators because they ate several species harmful to agriculture, so that in their absence their prey increased in numbers of individuals and species.

### B. Keystone Prey

Prey species have a variety of effects on species diversity. If a prey species has a high population growth rate or other characteristics allowing it to be relatively invulnerable to predation, then its presence may increase a predators' numbers; increased predator numbers could then, in turn, decrease the abundance of other prey species that are more susceptible to predation. In this case, the relatively invulnerable prey has been considered a keystone prey whose loss would increase species diversity. On the other hand, if the predator switches to the predation-tolerant prey when the numbers of other prey species are low, then sensitive prey that otherwise may have been driven to extinction may coexist in the presence of the predation-tolerant keystone prey. Counter to the first case, in this example the loss of the keystone would decrease species diversity.

### C. Keystone Mutualists

Some species have been considered to be keystone because they are essential to mutualistic relationships. "Mobile link" pollinators and seed dispersers have been cited as keystones because their activities are critical to the persistence of plant species that in turn support several disparate food webs. Other examples of this type of keystone species include mammalian dispersers of certain mycorrhizal fungi. The numerous fungi species whose fruiting bodies are underground cannot be dispersed by wind; instead they rely on mammals such as flying squirrels and red-backed voles, which eat the fruiting bodies and then disperse them as they roam.

### D. Keystone Hosts

If mobile links, or keystone mutualists, depend on ecologically important host plants, then it follows that these hosts also receive the label "keystone." In tropical forests, palm nuts, figs, and nectar have been considered keystone resources because during seasonal periods of food scarcity they are critical to primates, squirrels, rodents, and birds. In this case, this small group of plants provides essential support to a large percentage of the vertebrate biomass. Animals have often been called keystone hosts for similar reasons. For example, a diverse array of species eat termites or other insects that occupy their burrows, this securing the keystone label for termites.

### E. Keystone Modifiers

The activities of animals can also affect living conditions of other species without directly eating them or being eaten. For example, the North American beaver has been described as a keystone species because its dams alter hydrology, biogeochemistry, and productivity on a wide scale. Likewise, gopher tortoise burrows in the coastal plains of the southeastern United States provide a unique habitat upon which a range of species depend.

In many cases foraging behavior causes direct habitat modification. Feeding sapsuckers create sap wells in tree bark that provide resources for other herbivores, and pocket gophers maintain mountain meadow communities because their diet of roots slows down aspen invasion of the meadows. In an interesting historical application, a "keystone herbivore hypothesis" has been used to explain the late Pleistocene extinction of approximately half of the mammalian genera with body masses of 5–1000 kg. This theory posits that elimination of large herbivores initiated vegetational changes that negatively impacted mammalian fauna.

Finally, keystone foraging effects have been extended to include multiple species. In the southwestern United States, several species of kangaroo rats have been referred to collectively as a "keystone guild" because their removal led to substantial changes in vegetation type and accompanying changes in the rodent community.

## IV. USEFUL CONTRIBUTIONS OF THE KEYSTONE SPECIES CONCEPT

There are two fundamentally important contributions of the keystone species concept to ecology and conservation. First, it has focused attention on the fact that all species are not equal in terms of interactions within their community. Instead of envisioning food webs or communities to be connected by multiple interactions of equal strength (Fig. 1A), we now realize that many species may have only weak relationships with each other while a few have particularly strong interactions; these might be drawn as bold lines between the "keystone" and other members of its interaction web or food web (see species "W" in Figs. 1B and 1C).

The second important contribution of the keystone paradigm is its implication that only a small minority of species have interactions that strongly affect community composition. In other words, reference to a particular species as keystone implies that it is unusual, standing out from the majority of the other species in its effects on community structure or function. To illustrate the consequences of this assumption, Fig. 2 shows stylized distributions, under several different sce-

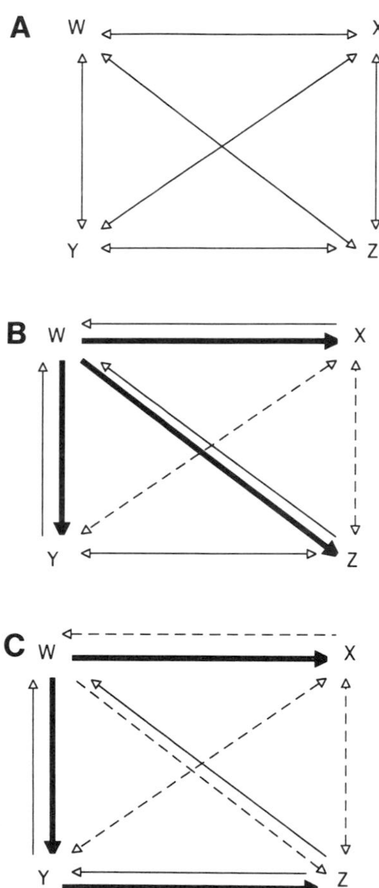


***FIGURE 1*** A schematic of a simple hypothetical community of four species with contrasting food web dynamics. In (A), the four species interact with nearly equal interaction strengths, as expected in "traditional" food webs. In (B) and (C), one species (species "W") is a strong interactor, or "keystone." In (B), these strong interactions (represendt by bold arrows) operate directly on the other species. In (C), there are also indirect effects, such as the strong effect on "Z" being mediated by "Y." Note that the strengths of interactions for other species vary.

narios, of "community importance" values. These values go one step beyond interaction strength to refer to a change in species richness upon loss of a given species. If a community has "keystone species," we would expect a distribution of community importance values as in Fig. 2A, with only a few (keystone) species having large effects on the composition or structure of the community. In contrast, food web theory has generally assumed that species-by-species community importance values are drawn from symmetrical distributions (Fig. 2B) or else are equal across species (Fig. 2C). Thus, the keystone concept assumes a unique distribution of interaction strengths.

What do the data say about this distribution in interaction strengths? To date, only one study has addressed the distributions of interaction strengths in part or all of an ecological community. In brief, it found that only two of seven species of intertidal grazers strongly affect brown algae, which are their major food and also a profound modifier of the local environment. These results support the assumption of skewed interaction strengths (Fig. 2A), although the generality of the result awaits confirmation across a variety of communities. Indeed, this important study illuminates the need for examination of interaction strengths for an assortment of species in different communities and under diverse conditions.

The apparent dichotomy between food web theory and the keystone species concept is worth exploring for its significance to conservation strategies. If many or most species are of similar importance (Figs. 2B and 2C), any efforts to save only a few species will inevitably fail to protect the rest. Conversely, if there are only a few "keystone" species having strong interactions/community effects, then detailed understanding and protection

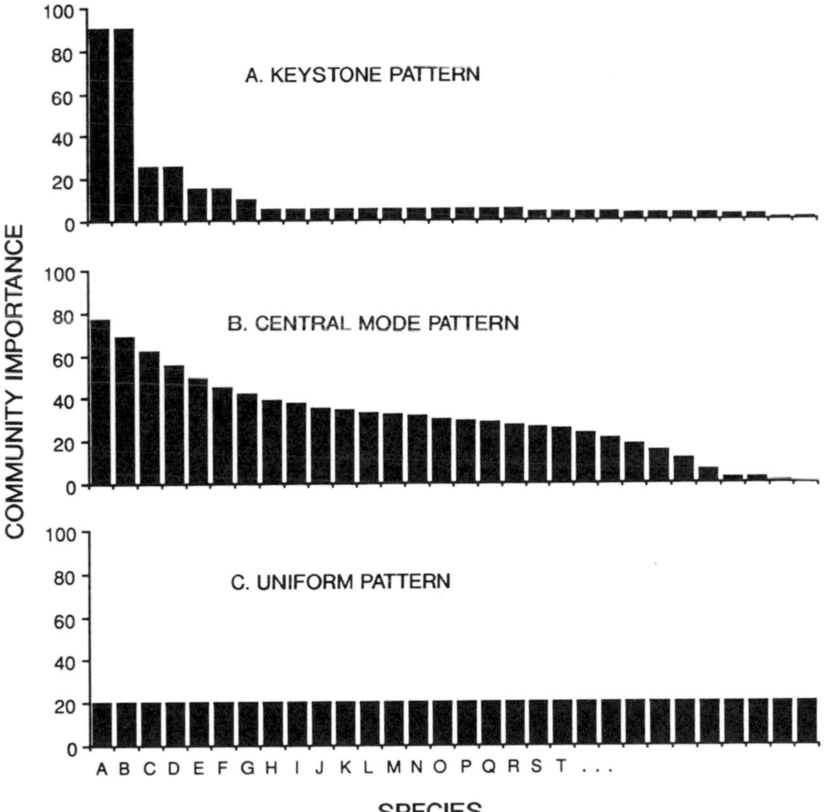


***FIGURE 2*** Expected distributions of community importance values (percentage of species lost from a community upon removal of a given species) for a hypothetical community based on (A) the keystone species model and (B and C) food web theory. [Adapted from L. S. Mills *et al.* (1993). *Bioscience* **43,** 219–224.]

of these few taxa would be pivotal to the well-being of the overall community.

## V. THE KEYSTONE SPECIES CONCEPT AND CONSERVATION POLICY

What role should the keystone species concept play in conservation efforts? Currently, implementation of the Endangered Species Act often amounts to what has been called "emergency room conservation," whereby the bulk of conservation resources are spent on single species that are on the brink of extinction. In the absence of more comprehensive biodiversity legislation and/or increased funding and support for the Endangered Species Act, it has been suggested that keystone species could bridge the gap between single species and more comprehensive ecosystem-level approaches.

There are, however, several compelling liabilities with relying on keystone species in such a manner. First, there is no reason to expect that a keystone species will necessarily be responsive to human-caused perturbations. That is, keystone species have no direct relationship to "indicator species," which might be used to indicate the extent to which habitat loss or pollution, for example, are threatening an ecosystem. Although it is true that losing a keystone species may initiate further cascades of extinctions of other species, the keystone may or may not be particularly susceptible to being lost. Conversely, the loss of an indicator species will not necessarily cause the cascade of species changes expected with a presumed keystone species.

A related concern is that a conservation criterion that favors the maintenance of keystone species may fail to protect other species of interest to conservationists or the public at large. For example, spotted owls, wolverines, and California condors may have little role in the maintenance of species richness in their respective habitats, yet the protection of these charismatic species has been advanced because their fates are thought to indicate the integrity or health of their habitats, or because the viability of such species requires large areas. In addition, the simultaneous removal or loss of a collection of nonkeystone species could have effects as large as removal of a single keystone.

An even more fundamental challenge in reliance on keystone species is the lack of definition. Before keystone species become the centerpiece for biodiversity protection or habitat restoration, we must be able to say what is and is not a keystone species; otherwise, subjectively chosen subsets of species will be so labeled, whereas other species of similar importance could be ignored. Clearly, a first step is to acknowledge that impact due to biomass dominance needs to be distinguished from that due to inordinately large per capita effects. Community dominants, such as a dominant tree species in a forest, play a critical role, but their effects can be attributed to sheer biomass and not to the disproportionate individual effects that keystones are expected to have. An operational "dividing line" to quantify one species as keystone and another as nonkeystone is not yet possible. However, more formal development of how species vary in community importance values will certainly lead to more rigorous field work and protocols for identifying keystones.

Field approaches to identify keystone species will need to be based on an intimate knowledge of natural history of the system. The strongest, but most difficult, approaches involve perturbation experiments whereby the candidate keystone species are removed and the responses of a predefined assemblage of species are monitored. Such tests require adequate experimental replication and careful attention to defining the relevant assemblage, as well as contemplation of time scales over which responses should be measured. Careful comparative analyses have also shown great potential. These might include comparing different areas with and without local reintroductions or extinctions of particular species.

In conclusion, both the complexity of ecological interactions and a lack of knowledge militate against the uncritical application of the keystone species concept for applied purposes. The term has considerable heuristic value, however, so that integration of such information into management plans and policy decisions will provide important ecological insight into the particular role that certain species play in their native communities.

There is no doubt that the keystone species concept has been fundamentally important to both ecology and conservation. It has drawn attention to the fact that some species have particularly strong interactions with other members of the community, and that there may well be only a few of these strongly interacting species in a given community. Although measurement of interaction strengths (or community importance values) of particular species will be difficult, such as emphasis would incorporate much-needed information into ecological and environmental analysis.

## Glossary

**Community** Collection of species (or populations) that are connected by effects of one population on the life history or genetic constitution of the other.

**Food web** Schematic representation of direct and indirect feeding relationships among populations in a community. Food webs are often generalized to include nonfeeding interactions, such as effects on habitat structure.

**Interaction strength** Magnitude of one species' influence on other members of its community, including both direct and indirect effects.

**Mobile link** Species whose foraging movements are critical to certain plants that are fundamental to independent food webs.

**Trophic cascade** Changes in biomass of species at different trophic levels in response to an initial change at a higher trophic level.

## Bibliography

Carpenter, S. R., and Kitchell, J. F. (1988). Consumer control of lake productivity. *Bioscience* **38,** 764–769.

Gilbert, L. E. (1980). Food web organization and the conservation of neotropical diversity. *In* "Conservation Biology: An Evolutionary–Ecological Perspective" (M. E. Soulé and B. A. Wilcox, eds.), pp. 11–33. Sunderland, Mass.: Sinauer.

Mills, L. S., Soulé, M. E., and Doak, D. F. (1993). The keystone-species concept in ecology and conservation. *Bioscience* **43,** 219–224.

Paine, R. T. (1969). A note on trophic complexity and community stability. *Amer. Nat.* **103,** 91–93.

Paine, R. T. (1992). Food-web analysis through field measurement of per capita interaction strength. *Nature* **355,** 73–75.

Terborgh, J. (1986). Keystone plant resources in the tropical forest. *In* "Conservation Biology: The Science of Scarcity and Diversity" (M. Soulé, ed.), pp. 330–344. Sunderland, Mass.: Sinauer.

Wootton, J. T. (1994). Predicting direct and indirect effects: An integrated approach using experiments and path analysis. *Ecology* **75,** 151–165.

# Krill, Antarctic

**Stephen Nicol**
*Australian Antarctic Division*



Antarctic krill is one of the larger species of the order of pelagic crustaceans known as euphausiids. It is found solely in the waters of the Southern Ocean surrounding Antarctica, with its major concentrations near the continental shelf break, islands, and, at some times of year, the ice edge. Antarctic krill is the pivotal species in the food web of the Southern Ocean, forming a link between the microscopic algae of the water and the myriad larger animals such as fish, squid, seals, birds, and whales that depend on it for food. The huge abundance of this one species, which enabled the whale, seal, and penguin populations to flourish, has also attracted the attention of the world's fishing fleets. Once touted as a potential major source of protein for the developing world, krill is now being marketed mainly as an expensive additive in processed food or as a feed for growing aquaculture industries. The fishery throughout most of the 1980s was the largest crustacean fishery in the world and by far the largest fishery in the waters off Antarctica. The catch of krill has, however, declined markedly since the disintegration of the Soviet Union. Concerns over how potentially huge catches of krill might affect the other wildlife of the Antarctic led to the signing of an innovative treaty—the Convention on the Conservation of Antarctic Marine Living Resources (CCAMLR)—that took an ecosystem approach to fisheries management, recognizing the place of krill at the center of the Antarctic food web.

## I. INTRODUCTION

The word "krill" comes from the Norwegian *Kril,* meaning "young fish," but it is now used as the common term for the euphausiids, a family of pelagic marine shrimps found throughout the oceans of the world. There are 85 species of euphausiids, ranging in size from the smallest (some millimeters long) to the largest deep-sea species, which can reach 15 cm in length. Antarctic krill, scientifically named *Euphausia superba* by the American naturalist Dana in 1850, is one of the bigger species, growing to a maximum size of 6.5 cm and weighing over a gram (Fig. 1).

Several notable features set the euphausiids apart from other crustaceans. The gills are exposed below the carapace unlike those of most other advanced crustaceans, which have gills sheltered within it. There are luminous organs (photophores) at the

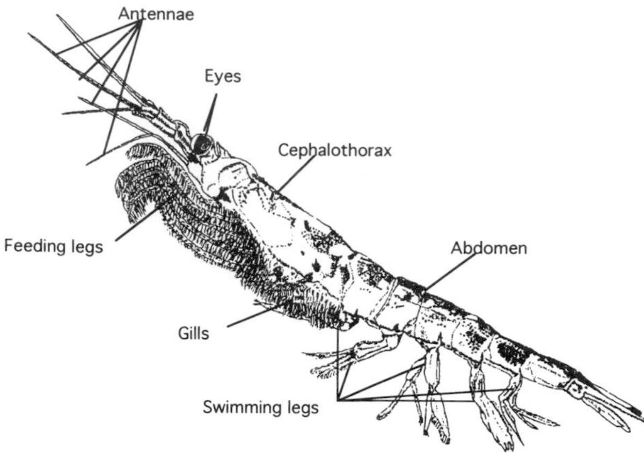


***FIGURE 1*** Adult Antarctic krill showing features referred to in text. (J. Cox, Australian Antarctic Division, Tasmania, Australia.)

base of the swimming legs, as well as pairs of photophores at the genital segment of the abdomen, near the mouthparts, and in the eyestalks. These organs produce electric blue light that can be directed by lenses that focus light produced by the luminescent cells behind them. However, the general body plan is similar to that of many familiar crustaceans. The fused head and trunk—the cephalothorax—contains most of the internal organs—the digestive gland, stomach, heart, gonads, and, externally, the sensory appendages: the two large eyes and two pairs of antennae. The limbs of the cephalothorax are modified into highly specialized feeding appendages; the 11 mouthparts are modified for handling and grinding the food and the six pairs of food-collecting limbs are capable of trapping food particles from the water and of moving them to the mouth. The muscular abdomen has five pairs of swimming legs (pleopods), which move in a smooth paddling rhythm. The pairs of swimming legs are joined at the base by tiny adhesive patches that resemble Velcro® fasteners so that the paired legs act in unison. Krill are heavier than water and stay afloat by swimming in bursts, which are interspersed with short resting bouts. When resting, krill spread the tips of their swimming legs to act like parachutes, slowing their rate of descent.

Adult Antarctic krill are referred to as micronekton, which means that they are more independently mobile than the plankton, which are drifting animals and plants at the mercy of water movements. The term nekton embraces a wide diversity of animals from krill to whales. Krill pass through a planktonic phase when young but, as they grow, they become more able to move through their watery environment and to maintain themselves in particular areas.

Despite *Euphausia superba* being referred to as the Antarctic krill, there are at least four other species of krill that inhabit the waters surrounding Antarctica. These other, smaller species of krill have been little studied but they are locally abundant and are important members of the pelagic community, even if they are somewhat overshadowed by their larger relative.

Despite their biological importance, and many years of study, little is known about Antarctic krill. This is partly a function of their difficult biology but also is a result of their remote and inaccessible habitat. The Southern Ocean is a troublesome place to work in the best of times; it is expensive to get to and requires specialized ships to withstand the pounding of the seas of the "roaring forties" and "furious fifties" and the rigors of working in the pack ice. The summer is short and, during the winter, the sea where krill is found in summer ices over and darkness falls for much of each day. Few attempts have been made to study the Southern Ocean in midwinter, so little is known of the biology of the organisms that have been enclosed beneath the icy blanket that covers some 20 million square kilometers. The mystery of where krill go in winter is an enduring one and is one of the many questions that remain concerning the biology of these enigmatic crustaceans.

Until the late 1970s, attempts to keep krill alive in laboratories for extended periods had always failed. Then aquaria were successfully designed to maintain krill, first on the Antarctic continent and then in temperate latitudes. The results of this accomplishment are still being felt, because for the first time scientists could work year-round with the living animal rather than with dead, preserved specimens, which yielded few of their secrets when examined in laboratories far away from Antarctica. The 1970s and 1980s saw a surge in discoveries about the biology of Antarctic krill; this understanding of their physiology, behavior, and ecology was made possible by the ability to work with living animals. Divers revealed the intimate structure of swarms of krill in the open ocean and new sonar technology began to reveal the details of their distribution, but, as is often the case, these investigations revealed krill to be a far more complex and confusing animal than was first thought.

## II. DISTRIBUTION

Although Antarctic krill are found throughout the Southern Ocean, they are not evenly distributed in the area between the Polar front and the Antarctic continent. The British *Discovery* investigations begun in the 1930s provided the first clues to unraveling the puzzle of where to find krill. Results from research and fishing vessels have begun to map the general distribution of krill in the Southern Ocean. Not surprisingly, the areas where krill is most abundant coincides well with the areas where the greatest harvest of baleen whales occurred (Fig. 2). Antarctic krill appears to be most abundant in areas where there are oceanographic or bathymetric features such as fronts or shelf breaks. Krill is scarcer in the open ocean than near the ice-edge or the sub-Antarctic islands, particularly in the South Atlantic. On a large scale during summer, krill concentrates in the ice-edge zone of the East Wind Drift, being distributed along the unusually deep (~500 m) continental shelf break. In the Antarctic Circumpolar Current, krill is concentrated in the gyres formed in the South Atlantic sector and around the major island groups such as South Georgia and the South Orkneys, as well as in the complex water masses that swirl around the Antarctic Peninsula.

The picture of seasonal krill distribution is complicated by the lack of information for times of the year other than summer. Krill may follow the retreating ice-edge in a wave during spring and remain restricted in its longitudinal distribution during summer. Similarly, the zone of maximum krill concentration may expand during autumn, remaining at the periphery of the pack ice. Alternatively, krill may remain where it is concentrated during summer or may disperse under the ice or

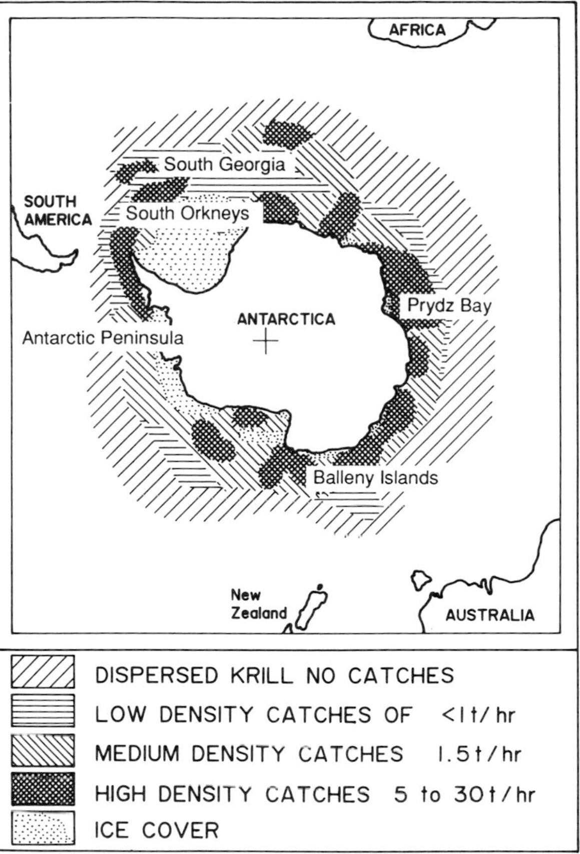


***FIGURE 2*** Map showing main areas of Antarctic krill distribution, derived from scientific and fisheries data. Shading indicates commercial catch rates. (J. Cox, Australian Division, Tasmania, Australia.)

retreat to the ocean floor during winter. Knowledge of the exact details of the seasonal distribution of krill will await the results of studies focusing on the relationship between the krill and the pack ice during the winter months.

Large-scale records of presence or absence of krill have helped to build a picture of its occurrence on a broad geographic scale, but major questions remain about the factors that are controlling its detailed local abundance. Krill is a schooling animal and vast swarms can cover many square kilometers and may contain millions of tons of krill. It is likely that such congregations of krill are in fact composed of a large number of discrete schools separate from each other and maintaining their integrity over time. Each school can be composed of animals of one size, or sex, or developmental stage, and these schools probably contain up to 50,000 krill per cubic meter. That we know so little about these schools is largely a function of the crude tools that have been available for their study.

Traditionally, large plankton nets have been used to sample pelagic animals such as krill. Such nets are towed behind ships for known periods and the relationship between the number of animals caught and the volume of water filtered is calculated. Because patches of krill can be separated by areas of open water that can be many thousands of times greater than those occupied by the densely packed krill, a net towed through an area where krill is aggregated is likely to miss most of the krill and so underestimate its importance. Even if the net encounters a dense school, the larger animals may be strong enough to evade the net and smaller animals may slip through the meshes. If each school in a particular area is made up of unique groupings of individuals, a net that is towed through many of these schools will get a jumbled picture of the biological integrity of these aggregations, and recent studies have shown that in such cases, up to 26 net tows are necessary to sample the whole population adequately.

Techniques for examining krill distribution have improved recently. Modern sampling systems, utilizing several electronically operated nets that can be opened and closed at discrete depths, can take multiple samples from different depths. When combined with the ability to "visualize" krill aggregations at depth with scientific echosounders, these multiple opening/closing nets can selectively sample parts of an observed swarm or can sample a number of individual swarms. Echosounders work by sending out pulses of high-frequency sound that are scattered by objects that have densities different from that of the surrounding seawater. The beam of reflected sound gives information about the animals that make up the reflecting layer and can also indicate the extent of the layers.

Unfortunately, even the most sophisticated modern hydroacoustic equipment is still unable to provide a complete picture of small-scale krill distribution. It is still difficult to determine whether the echo received back from a group of animals in the water is coming from krill or from other animals. Additional uncertainty is introduced when converting reflected sound to krill density. Krill can undergo quite extensive daily vertical migrations and is often found in the top few meters at night and even occasionally during the day. This is a portion of the krill population of unknown size that is not detected by echosounding gear, which usually looks downward from the hull of the research vessel.

The distribution patterns of adult, larval, and immature stages of krill are likely to be affected by forces acting on different space and time scales. Young krill are entirely planktonic, being carried with the water mass where the eggs were laid or to which they have swum after hatching. As they grow they become more powerful swimmers, are less at the mercy of the currents, and probably are more able to maintain themselves in a particular area. There are distinct areas around the coast of the Antarctic and sub-Antarctic islands where krill are perennially abundant. This enrichment may result from favorable alignments of currents or from the ability of krill to maintain themselves in or migrate to the areas of highest productivity. The ultimate reason for these persistent krill concentrations is likely to be a complex interaction between the behavior of krill and the oceanography.

Whether these semipermanent, large-scale concentrations of krill constitute separate biological stocks is a matter of great interest, particularly for fisheries management. A biological stock is one in which there is limited immigration or emigration and which can generally be distinguished genetically from other populations of the same species. So, fishing on one of these stocks is unlikely to affect other stocks in the short term. This makes if feasible to subdivide the population into manageable areas and for quotas to be set that reflect local conditions. Studies that have attempted to discriminate between stocks of krill from various areas of the Southern Ocean have suggested that there is one oceanwide stock, so subdivision of the population for management purposes may have to be based on less biologically realistic criteria.

## III. LIFE HISTORY

Where krill go during the long, dark, icy winter remains one of the greatest mysteries of its life cycle. Because it is difficult to penetrate the pack ice during winter, it is difficult to study what goes on during much of the year. Ships breaking pack ice in early spring frequently turn over floes with krill on the underside, and studies using divers and submersibles have reported that, at this time, the underside of the ice, which is rich in its own specialized algal flora, can be alive with feeding krill. Krill may spend the winter months in the complex folds and caverns under the pack ice feeding off the under-ice community. Krill may turn more carniverous in winter, feeding off the zooplankton in the water column. Other studies have shown that krill are sometimes found near the ocean floor, so some portion of the population may spend the winter in the deep water feeding off the remains of the summer blooms of algae that have sunk to the sediments.

The life history of krill has been pieced together using a combination of laboratory investigations and field studies. In late spring and during summer, female krill produce large lipid-rich ovaries. The reproductive process requires a considerable intake of food and can only take place in regions where food is highly abundant. Before spawning, the weight of a ripe ovary can account for up to a quarter of the body weight of a female. It is likely that females can produce a number of broods in one season, which generally lasts from late December to March. The body of the female becomes more and more distended as the ovary expands prior to spawning (Fig. 3), and after spawning a large cavity is left behind. The female lays up to 10,000 eggs at one time and they are fertilized as they pass out of the genital opening by sperm liberated from spermatophores that have been attached to the females by the males. Once free in the ocean, the eggs begin to sink and develop. Krill eggs are extremely fragile and are rarely found in any great numbers in plankton nets, so the exact vertical location and timing of spawning are difficult to pinpoint.

The egg takes 8 days (at 0°C) to hatch into the nauplius larva, which continues to sink as it develops into the second nauplius in another 5 days (Fig. 4). This is the first swimming stage and the larva then begins the long 2000-m ascent to the surface, still fueled by the rich lipid store laid down in the egg. The well-lit surface zone of the ocean is where the food of krill is found and it is likely that there is a "critical period" during which the larvae must arrive at the surface and find food to survive.

Once in the sunlit layers the larva begins to change into a form that is more recognizable as a euphausiid. The tail grows, the eyes develop, limbs and antennae sprout, and at this phase—the calyptopis—the animal begins to feed. The larvae are voracious feeders, consuming vast quantities of the smaller plants and protozoans in the surface layer of the ocean. The larvae develop rapidly, molting regularly through a bewildering array of stages during which they add more limbs and segments. The number of stages varies depending on factors such as food availability and temperature. In the first year of life the larvae also begin to exhibit many of the behavior patterns of the adults. They undergo the diurnal vertical migrations that are typical of many planktonic organisms, staying deeper in the water column during the day and rising to the surface to feed during the night.

Within 3 to 5 months of hatching, krill larvae face the second major challenge of their short lives; they must endure the first Antarctic winter. In March the pack ice begins to expand, covering the summer feeding ground of the krill, and as the sun begins to wane the food for the teeming masses of krill larvae begins to disappear. The primary food of larval and adult krill is the phytoplankton, microscopic plants of the ocean that are suspended in the upper water column where the light is sufficient to allow growth. As winter approaches, light begins to diminish, pack ice grows and further cuts out the sunlight, and the water column becomes unstable. During summer the water column stratifies as the sun warms the surface layer and the plants are prevented from sinking out of the illuminated zone by a density barrier called the thermocline. In autumn the thermocline breaks down and the plants can sink out of the surface layer and are lost to the surface-living animals. The krill larvae now face a darkening world in which food is scarce. Possible food sources for larvae in winter are ice algae, detritus, and other animals of the zooplankton—certainly they do not have the energy resources to overwinter without feeding.

First-year krill larvae emerge from the pack ice in spring as advanced furcilia and it is probably

***FIGURE 3*** Mature female (upper) and male (lower) Antarctic krill. Note distended cephalothorax of female krill just prior to spawning. (Photograph S. Nicol Australian Antarctic Division.)

at this point that they begin to exhibit schooling behavior. As they develop into juveniles over their second summer this behavior becomes more pronounced. The juveniles come more and more to resemble the adults in form and function, reaching some 25 mm in length before their second winter starts.

Euphausiids, unlike many crustaceans, continue to molt regularly throughout their life. Smaller, short-lived species may produce a new exoskeleton

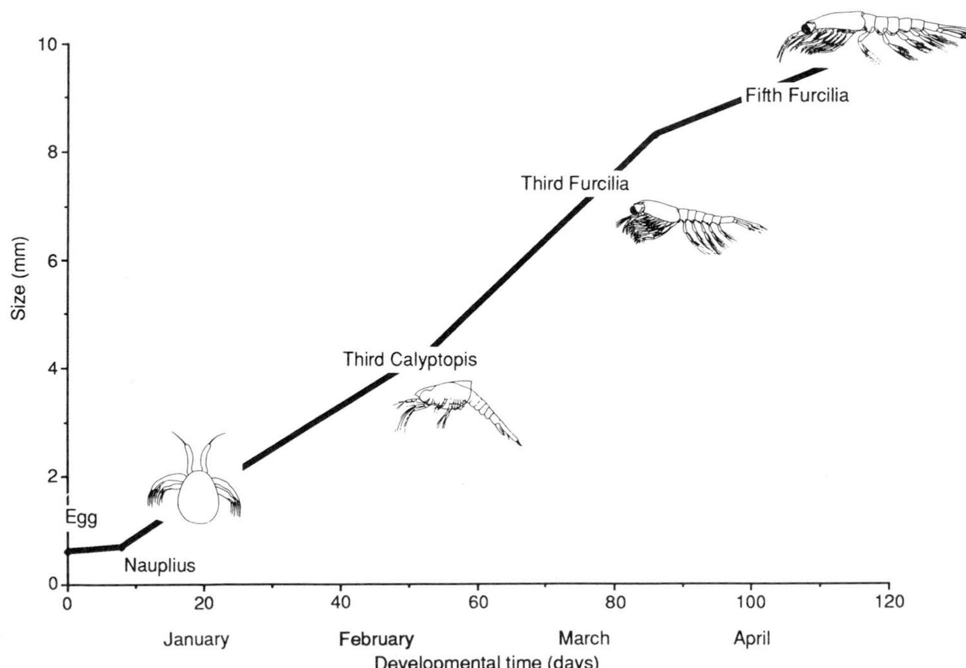


***FIGURE 4*** Growth and development of krill in its first summer. Spawning is indicated as occurring in early January but may occur throughout the summer into early March, with consequent delays in the development of the various larval stages. Timing is derived from laboratory studies at 0°C and development is likely to be slower in nature.

every week. For Antarctic krill, the molting rate seems to be largely dependent on temperature, though there are also size and seasonal effects. In the laboratory, at 0°C adult Antarctic krill molt approximately every 27 days, whereas at 3°C they molt approximately every 14 days. These molting rates, if maintained year-round in the ocean, would result in the production of 13 to 26 cast exoskeletons per krill per year, respectively, at the ambient temperatures close to the ice-edge. A medium-sized adult casts off approximately 6% of its dry weight as molt at ecdysis—a significant drain on the resources of the animal and possibly one of the major fluxes of carbon from the euphotic zone in Antarctic waters.

Krill continue to molt regularly even if deprived of food. Laboratory studies show that adult Antarctic krill can survive more than 200 days of starvation, and during this period the animals continue to molt—reducing their body size at each molt. The ability to survive long periods of starvation by shrinkage may be a partial answer to the question of what happens to the huge population of these animals during the long, dark Antarctic winter. Euphausiids are not rich in lipids and wax esters, which are often stored by crustaceans that experience long-term food shortages. Shrinking krill may be able to survive the winter by utilizing their own body protein as fuel. Such size reduction has an additional bonus as it reduces the energy expenditure needed to keep from sinking. Thus the mean size of individuals in a population at the end of winter could be considerably smaller than it was at the end of summer. Adults also begin to lose their secondary sexual characteristics, further complicating the picture. It thus becomes increasingly difficult to separate out size and age classes from the krill population and this hampers efforts to understand the population dynamics of this species.

The age structure of the population is often easier to examine if the number of year classes is known and this depends on knowing the longevity of krill. Early workers suggested a life span of 2 years, but once facilities became available for the successful rearing of krill in laboratories, the potential life span was extended to 11 years. Although it is now accepted that krill survive in nature for considerably longer than 2 years, the exact age structure of the krill population remains in considerable doubt.

Understanding the population dynamics of a species requires that a known population be sampled repeatedly. Because of its oceanic habit, it has not been possible to sample regularly the same local population of krill with any degree of certainty over short time scales let alone interannually. Thus it has not proved feasible to follow changes that occur within a population of krill as it grows and overwinters in the ocean. This lack of field data makes it hard to interpret laboratory-derived information on growth and molting and to integrate experimental results into investigations of the population dynamics of Antarctic krill.

## IV. KRILL IN THE ANTARCTIC ECOSYSTEM

Krill has often been referred to as the keystone species in the Antarctic ecosystem because of its great abundance, its conspicuousness, and its position in the food web—between the microscopic plants of the phytoplankton and the large vertebrate predators such as whales, seals, and penguins (Fig. 5).

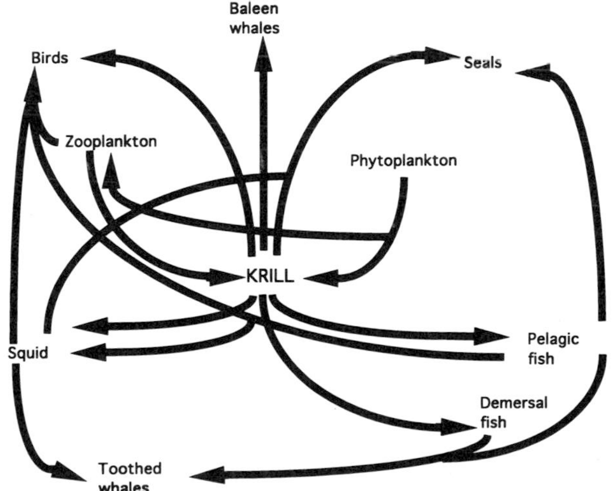


***FIGURE 5*** Antarctic food web showing the central role of krill, between the microscopic phytoplankton and zooplankton and the large vertebrate and invertebrate consumers. For clarity, the decomposition and regeneration cycles have been omitted.

The first feeding stage of krill is only ~2 mm long whereas the adults can reach 65 mm, and this 30-fold length increase is probably accompanied by changes in feeding behavior and preference. Little is known of the changes that occur in the trophic behavior between the truly planktonic larval stages and the nektonic adult stage. The original ideas about how krill feed were based on an examination of the limbs that form the "feeding basket" and by assuming that this complex of spines and hairlike appendages acted like a simple net or sieve that rather passively removed phytoplankton from the water. It is now thought that krill feed in a much more complex fashion: detecting food from a distance, orienting toward the food source, and selectively feeding on different elements of the plankton, both phytoplankton and zooplankton. Planktonic food sources for krill range from tiny flagellates 4 $\mu$m in diameter, to the largest diatoms several hundred microns across, to copepods several millimeters in length. Krill are voracious feeders and swarms can graze down phytoplankton blooms extremely rapidly, and this in turn fuels its rapid summer growth and reproductive rate. For much of the year, however, there is little food available in the pelagic realm and at this time they may turn to alternative food supplies such as ice algae, benthic detritus, and zooplankton.

It is perhaps not surprising that much more is known about what eats krill than what krill eats. Many vertebrates from Antarctic waters have been commercially harvested over long periods and many of the seals and birds have been targeted for detailed long-term ecological studies, so records of the presence of Antarctic krill in the stomachs of vertebrates are frequently available. Baleen whales were probably the major single predator on Antarctic krill before many of their species were hunted to the brink of extinction (Table I). Nowadays, the krill-feeding whales have been reduced to a fraction of their original population size and other elements of the Southern Ocean ecosystem are now more important as krill consumers. The extent of competition between the land-based vertebrates such as seals and seabirds and the pelagic predators such as squid and whales for the krill resource is unknown, but any such competition may, in the long-term, affect the ability of the great whales to recover. [*See* Antarctic Marine Food Webs.]

**TABLE I**
**Estimated Current Consumption of Krill by Major Predators**

| Predator | Krill consumption (range of estimates in millions of tons) |
|---|---|
| Whales | 34–43 |
| Seals | 63–130 |
| Birds | 15–20 |
| Squid | 30–100 |
| Fish | 10–20 |
| Total | 152–313 |

In the prewhaling days, the baleen whales of the Southern Ocean consumed approximately 190 million tons of krill each year. Currently their consumption is only on the order of 40 million tons, and the difference—~150 million tons—has been referred to as the "surplus krill" population. The suggested availability of this huge tonnage led to fisheries investigations into this last, large, unexploited marine stock.

## V. THE KRILL FISHERY

Exploitation of the marine resources of the Antarctic region prior to the 1960s concentrated on the mammals—seals and whales—and it was not until stocks of these species became severely depleted that attention focused on other living resources. The recent development of Antarctic fisheries can be linked to the overfishing of many of the traditional fishing grounds throughout the world and to the declaration of 200-mile fisheries exclusion zones in the 1970s. Initial interest in fishing on a commercial scale turned to stocks of Antarctic fish particularly around the Antarctic Peninsula and the sub-Antarctic islands. The major fishing nation in the Southern Ocean was the Soviet Union, and it began commercial operations for finfish in this region in 1967. The fishery commenced with a highest-ever recorded catch of nearly 400,000 tons, reached a second peak in the 1977/1978 season when over 350,000 tons were landed, and then declined because of overexploitation of these resources (Fig. 6). All commercially harvested species are now covered by some form of Conservation Measure under the Convention on the Conservation of Antarctic Marine Living Resources. Some areas are now completely closed to finfish harvesting and some species are wholly protected. It seems unlikely that there will ever again be large harvests of fish from Antarctic waters.

Ever since the huge abundance of krill became known, there has been speculation that it might form a suitable target for a fishery. There has been a fishery for a small species of krill—*Euphausia pacifica*—off Japan since the 1950s, but Antarctic krill has always attracted the greatest attention. Exploratory fishing expeditions first harvested Antarctic krill in 1961/1962 and throughout the 1960s ships from the Soviet Union continued to make sporadic, small catches of krill. During this period, technologists attempted to develop suitable catching gear and products from krill and the fisheries scientists determined where the best concentrations lay. Full-scale commercial krill fishing operations were under way by the mid-1970s.

The krill catch gradually increased during the late 1970s as the fishery moved out of its experimental phase, reaching a peak in 1982 when 528,201 tons were landed, 93% of which was taken by the Soviet Union. A drop in catches has accompanied the breakup of the Soviet Union. Currently (1994), fishing boats from four nations are involved in the fishery for Antarctic krill: Chile, Russia, Poland, and Japan, with Japanese boats taking 80% of the catch.

The initial view of the Antarctic krill resource was that it was a vast untapped source of protein for the "starving masses," but the economics of the fishery do not lend it to this purpose. Sending trawlers to the Southern Ocean is expensive and the fishing season is short. Traditionally, most commercial vessels have fished for fewer than 150 days per year in Antarctic waters, although there is now a trend toward a winter fishery off South

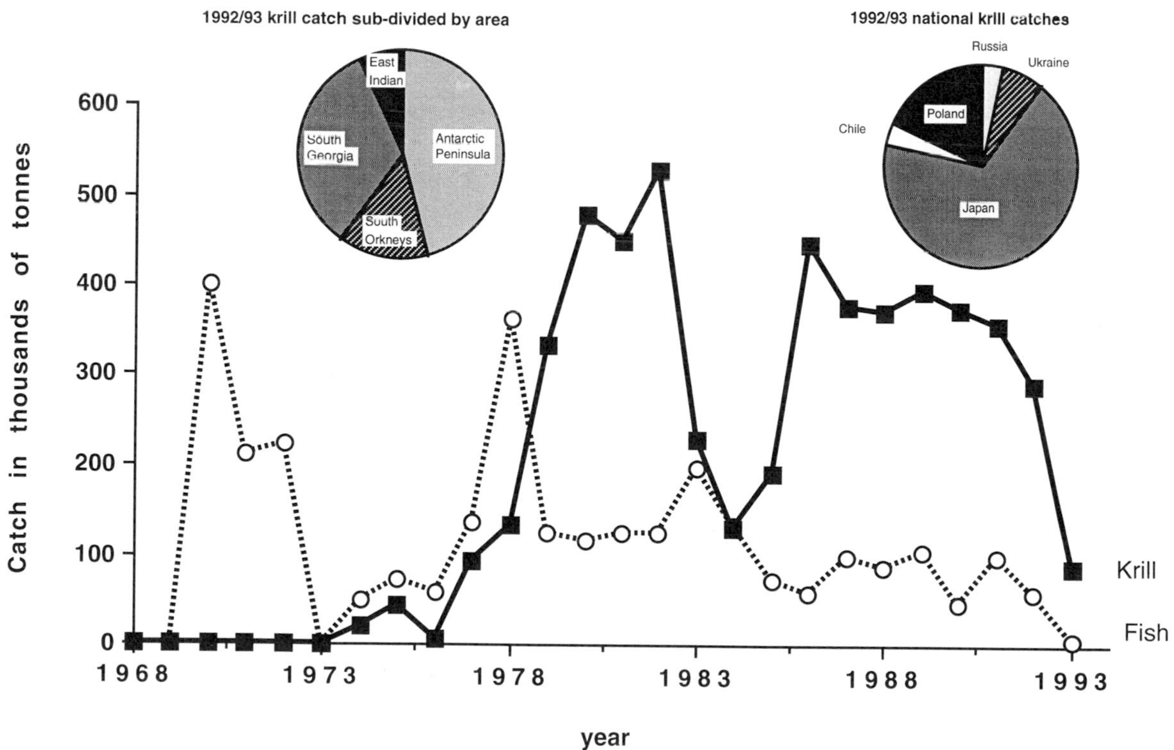


***FIGURE 6*** Fisheries of the Antarctic region, the catch of fish and krill from the Southern Ocean, the proportional national catches in 1992/1993, and the areas where krill were caught in 1992/1993. (From data submitted to CCAMLR.)

Georgia. In most cases, though, summer krill fishing operations must be combined with fisheries on other species in different areas during the rest of the year.

Krill fishing in the Antarctic is carried out by large (~100 m) trawlers throughout the year. The large Soviet fleet used to fish during all seasons, following the retreating ice-edge during summer in the South Atlantic, with most of the winter catch coming from the South Georgia area. Historically, catches have been highest in the vicinity of the South Orkneys, South Georgia, and the Antarctic Peninsula. The fishery has concentrated on the Atlantic Sector, although there have been substantial catches in the Indian Ocean Sector by both Japan and the Soviet Union. The Pacific Ocean Sector, with its lower krill biomass and great distances from ports, has been largely left alone by the krill fishing fleets.

Krill are caught in pelagic trawls towed at speeds of less than 3 knots. The swarms are located by the ship's echosounders and fishing boats aim to catch around 10 tons per haul. In areas of high krill abundance, a catch of this size can often be achieved in less than 15 minutes. Larger hauls (>10 tons) are avoided because krill get crushed in the net by the weight of the catch and, if landed, the processing capacity of the ship becomes swamped. If the catch falls below 3 tons per haul or 50 tons per day, the ships search for new fishing grounds. Fleets avoid areas where the krill are feeding, because the catch is tinted green, has a "grassy" taste, and tends to spoil rapidly owing to internal enzymatic activity. The Japanese fleet also makes a distinction between transparent or "white krill," which are firm-bodied and "look delicious," and "red krill," which are flaccid and easy to crush. There can also be problems with the by-catch of jellyfish, salps, or larval fish that are sometimes associated with the krill swarms.

Because of the expense of fishing, there has been a trend toward producing high-value krill products. Several other factors have influenced this decision. Krill contains high levels of very active enzymes that begin to break down the body tissues immediately after death. This means they must be processed within 3 hours of capture if they are to be used for human consumption and within 10 hours if they are going to be used for animal feed. Because krill must be processed rapidly, factory processing ships or shore-based facilities cannot be used and the catch must be dealt with on the trawlers. This further drives up the cost and increases the complexity of the fishery operation.

Krills also have a fluoride content in their shells that is 40 times higher than that permitted by the U.S. Food and Drug Administration in items for human consumption. Krill meat, however, has acceptably low fluoride concentrations. Unfortunately, the shell fluoride is highly mobile and leaches out into the flesh immediately following death. So, krill for human consumption must be peeled as quickly as possible following capture. Krill meal also contains four times the European Community's permissible fluoride levels for animal food. As a consequence, krill meal must be mixed with other fish meals for feeding to pigs and poultry and is not deemed safe at all for cattle or for breeding animals. However, krill meal has been used with success for feeding farmed salmonids, which seem to be able to cope with the high fluoride levels without significantly increasing their muscle fluoride, and for feeding fur-bearing animals such as mink, where the tissue fluoride levels are of little consequence. Animals whose natural diet includes krill, such as fish, penguins, and seals, seem to be able to handle the high fluoride levels found in their prey. The fluoride problem, coupled with the need for rapid processing of the catch and the expense of fishing in the Southern Ocean, has strongly influenced the types of products that emerge from the fishery.

A wide variety of end products have emerged from the krill fishery, partly because of attempts to overcome technical difficulties with the raw material, but also because of problems associated with producing a marketable product from a new and untested food source. Krill are smaller than most commercially exploited crustaceans and do not readily substitute for any other item, so they must compete with established species for a place on the table or at the farm.

The first commercially released foodstuff made from krill was a paste made from coagulated krill protein. This was sold under the brand name "Okean" in the USSR in the late 1970s, and subsequently attempts have been made to produce krill butter, minced krill, and krill protein precipitates, all of which can be used as food additives.

The most valuable products that come from the krill fishery at present are those based on the whole animal. Despite the fluoride problem there is still a market in the Far East for whole krill either frozen or boiled and then frozen. In Japan a premium is placed on ripe females as these are richest in fat and are deemed to have the best flavor. Whole krill are also used for aquaculture feed and for fishing bait, but more than half the Japanese catch is used for human consumption.

Peeled tail meats are probably the form in which krill will gain most acceptance and are the most valuable product of the fishery. Peeling processes have been modified from those developed for larger crustaceans and are now able to produce acceptable tail meats that can be used either as an additive in processed food or as a food in their own right. Roller peeling methods developed by Polish and Japanese technologists can now process 500 kg of krill per hour on board ship with a yield of 10–25%. The tails are light pink and have a good smell, a springy texture, and a sweet shellfish taste, and after blanching they shrink to the size of elongated rice grains. They store well and after defrosting can form gels and emulsify other substances and are thus useful in foodstuffs such as burgers, stews, sauces, and fish sticks. Canned tail meats are also manufactured and krill meat in brine appears to be quite acceptable as a shrimp substitute. This latter market holds the future of the krill fishery. If krill tails can be sold as a low-cost alternative to larger shrimp tails, then they are competing in an expanding market that pays a premium price for its raw material. A 1982 FAO study on the krill fishery estimated that krill could capture up to 10% of the edible shrimp market or roughly 60,000 tons of processed krill per year.

Other forms of krill have also been produced for a variety of purposes. Meal for animal feed has already been mentioned and, although processes for low-shell (hence low-fluoride) meals have been developed, it still remains uneconomic to gear up the fishery to produce solely low-grade materials. Krill meal is expensive and is produced only as a by-product of the higher-quality food manufacturing processes and as a method of utilizing low-grade krill or surplus catch. Krill meal also suffers from a few peculiarities that limit its usefulness. It absorbs water readily, the manufacturing process is relatively low yield (~15%), it has been observed to ignite spontaneously, and, when used in chicken feed, has the effect of turning egg yolks red.

Whole krill or processed krill waste has been shown to be a good food source for farmed fish, particularly salmonids. A krill diet seems to promote feeding and growth in some farmed fish and it also imparts a rosy color to the flesh of salmonids because of the carotenoid pigments that it contains. Antarctic krill as an aquaculture food supply in Japan currently faces stiff competition from the fishery for North Pacific krill (*Euphausia pacifica*), which caught 108,000 tons in 1991/1992, almost all of which was used to feed farmed fish.

It has been suggested that krill have various pharmaceutical and medicinal qualities. Polish and Russian literature promotes krill on the basis of its healthy qualities, and krill have been used for the treatment of atherosclerosis and for lowering serum cholesterol levels.

Various chemicals are found in krill and several of these have been investigated as possible by-products of commercial value. Krill are rich in proteolytic enzymes and in certain vitamins and their precursors, but markets for these substances are not sufficiently developed to warrant on-board extraction procedures at present. The shell of krill, like that of all crustaceans, contains chitin, which accounts for around 4% of the dry weight of the animal. Chitin is a valuable polysaccharide polymer similar in structure to celllulose and can be simply altered to produce another polymer—chitosan. Chitin and chitosan have a wide range of potential applications in water treatment, the medical field, and for a variety of agricultural, biochemical, and industrial purposes. One of the limitations on the industrial development of chitin is the lack of sufficient raw material and the krill fishery has been suggested as a potential source.

The future trend will be to use as much of the krill resource as is possible to repay the huge costs of fishing in Antarctic waters. Even now the Japanese fleet is simultaneously producing valuable tail meats, processing the waste material for aquaculture feed, and recovering protein from the processing water. Development of new products, more efficient processing of krill, and full utilization of the resource will be the most profitable strategy for the krill industry.

The krill fishery appears to have reached the end of its first phase of expansion and, indeed, has entered a decline. The Japanese fishery is probably economic but it has taken a long time to establish the market for krill products and there does not presently seem to be much room for rapid expansion. Additionally, competition from the North Pacific krill resource may limit any penetration into the Japanese aquaculture market. The catch by countries of the former Soviet Union is less predictable as there are many forces, economic and political, that might shape their attitudes to such a major operation as the krill fishery. Other nations may enter the krill fishery, but the large capital investment and uncertain market for krill products will probably delay any further expansion of the fishery for a while.

## VI. REGULATION OF THE KRILL FISHERY

The fishery is now only partially regulated, so the krill stocks remain vulnerable to rapid expansion should the economic situation improve. Concerns over the effects of an uncontrolled expansion of the krill fishery on the Antarctic ecosystem led to the establishment in 1980 of a regime for the management of the fisheries of the Southern Ocean—the Convention on the Conservation of Antarctic Marine Living Resources (CCAMLR).

For much of the 1970s and 1980s the Southern Ocean provided high annual total catches fluctuating between 100,000 and 700,000 tons, and since 1973 the mainstay of the Antarctic fisheries has been that for krill. In the late 1970s the catches of krill and finfish were increasing annually and it had become apparent that several species of fish had been overfished to the extent that they were in dire need of some form of protection. Fears began to be voiced concerning the effect of a large and uncontrolled fishery on the species that was the prey for most of the vertebrates of Antarctica and the surrounding seas. It was in this atmosphere that an international convention was formulated whose primary aim was the protection of the Antarctic ecosystem as a whole. This unique convention, the Convention on the Conservation of Antarctic Marine Living Resources, came into force in April, 1982, and it was the first, and so far only, fishing treaty that has tried to manage the whole ecosystem rather than the individual species that are of commercial interest. The Convention is administered by a Commission (also known as CCAMLR) that has met annually since 1981 and that has established a Scientific Committee and a number of Working Groups to aid it in its decision making.

CCAMLR has now been acceded to by 27 nations and the European Community, and its members include the 12 original Antarctic Treaty nations and all of those engaged in fishing activities in the Southern Ocean. There is no doubt that the possible rapid expansion of the krill fishery was the prime impetus for the formulation of CCAMLR, but it was more concerned with the effects that a large harvest of krill might have on the large populations of land-based vertebrates and on the potential for recovery of the great whales than because of an intrinsic desire to protect krill for its own sake.

Several unique features of the CCAMLR approach to managing the Southern Ocean ecosystem set it apart from any other such treaty. The Convention was established while the krill fishery was still in its infancy rather than once stocks had been dangerously depleted, as is the case with most fisheries treaties. It should be noted, however, that by the time the treaty was in force, the annual catch of krill was in excess of half a million tons per year, making it one of the world's largest fisheries. CCAMLR also broke new ground by enshrining the principle of "rational use" of the living resources of the Southern Ocean—with several provisos—while at the same time espousing an ecosystem approach to resource management.

The krill fishery has been the largest fishery in the Southern Ocean for the entire lifetime of CCAMLR. Despite this, it was not until its 10th meeting, in 1991, that CCAMLR took its first step toward managing the krill harvest when it set a precautionary catch limit of 1.5 million tons per year for the South Atlantic Sector. It may seem paradoxical that a "precautionary limit" would be set that is so much higher than the level of harvesting yet so much lower than the apparent potential of the fishery, but this is actually in keeping with the spirit of the Convention. The precautionary limit is seen as that point at which more complex management procedures must take over. It is based on the best estimates of the annual production of krill in the area and allowances have been made in the calculations for the demands of the myriad of krill-consuming vertebrates that the Convention requires be taken into account. CCAMLR was designed to act in advance of the fishery; to manage the resources so that they do not reach an undesirable level of depletion. This means that catch levels should be specified well in advance so that the industry has plenty of scope for planning its future and the scientists can refine their management models in advance of any increases of the fishery. That the precautionary limit is so much lower than the initial suggestions of the potential yield of krill is a reflection of the cautious approach that CCAMLR was designed to take. That the specified level is so much higher than the current catch is merely a function of the great abundance of krill in the area in question. In 1992, CCAMLR set a second precautionary limit for the krill fishery, this time for the Western Indian Ocean Sector, and again the limit specified—390,000 tons—was far in excess of the maximum tonnage ever landed from this area.

Scientific management of any fishery requires considerable knowledge of the species being harvested. In the case of krill, much of this knowledge does not exist. Abundance of krill in the open ocean is extremely difficult to measure partly because it has an extremely patchy distribution and also because it exists over a huge area, most of which is ice covered for half the year. Biological characteristics such as stock identity, yield, reproductive rate, longevity, mortality, and recruitment, which are standard measures used in fisheries management, are all poorly understood for krill. The biology of krill seems designed to make it difficult to understand its life-style and this has resulted in a lack of proposals for the management of this crucial species.

## VII. CONCLUSIONS

Antarctic krill is still a little understood species. Many critical aspects of its life history are poorly documented and it will be many years before a better understanding of its distributional biology is achieved. However, the central role that krill occupies in the high Antarctic ecosystem is well known, although the effect of large harvests of krill on the already altered Southern Ocean ecosystem is less certain. The fishery, which prompted so much interest in this species during the 1970s and 1980s, has gone into a decline, but the evidence suggests that this may be a lull before exploitation begins again in earnest. The krill fishery spawned CCAMLR, which is a pioneering attempt to apply the principles of sustainable development on a global ecosystem scale. Perhaps the hiatus in the krill fishery will allow the further development of the innovative management schemes being promoted through CCAMLR, which will provide an example for other harvesting regimes throughout the world.

## Glossary

**Antarctic Circumpolar Current** Eastward-flowing current that encircles the Antarctic continent carrying the largest amount of water of any current system on Earth.

**Antarctic Coastal Current** Counter-current that flows westwards along the coastal margins of Antarctica.

**Antarctic Divergence** Zone of upwelling at the boundary between the two counter-rotating current systems.

**Nekton** Aquatic animals that are capable of independent movement through the water. Their distribution is less dependent on the small-scale movements of the water.

**Pelagic** Animals of the open ocean.

**Plankton** Aquatic animals and plants that are suspended or floating in the water. Planktonic organisms are, by and large, at the mercy of water movements at all scales.

**Polar front** Oceanographic boundary that separates the Southern Ocean from the waters to the north. It is a distinct frontal system formed by the sinking of the denser, north-flowing Antarctic surface water beneath the warmer, less dense, sub-Antarctic surface water.

**Precautionary limit** Catch limit used in fisheries management that is designed to limit the increase in size of the fishery to a level that is estimated conservatively to be sustainable when more robust management methods are not available.

**Southern Ocean** Those portions of the Indian, Atlantic, and Pacific Oceans that lie south of the Polar front.

## Bibliography


Hamner, W. M. (1988). Biomechanics of filter feeding in the Antarctic krill *Euphausia superba:* Review of past work and new observations. *J. Crust. Biol.* **8**(2), 149–163.

Marr, J. W. S. (1962). The natural history and geography of Antarctic krill (*Euphausia superba* Dana). *Discovery Rep.* **32,** 33–464.

Miller, D. G. M., and Hampton, I. (1989). "Biology and Ecology of Antarctic Krill (*Euphausia superba* Dana): A review," BIOMASS Scientific Series, No. 9. Cambridge, England: SCAR/SCOR.

Nicol, S. (1990). The age old problem of krill longevity. *Bioscience* **40**(11), 833–836.

Nicol, S., and de la Mare, W. (1993). Ecosystem management and the Antarctic krill. *Amer. Sci.* **81**(1), 36–47.

Quetin, L. B., and Ross, R. M. (1991). Behavioural and physiological characteristics of the Antarctic krill *Euphausia superba. Amer. Zool.* **31,** 49–63.

Ross, R. M., and Quetin, L. B. (1988). *Euphausia superba:* A critical review of annual production. *Comp. Biochem. Physiol.* **90B,** 499–505.

# Land–Ocean Interactions in the Coastal Zone

**William A. Reiners**
*University of Wyoming*



The coastal zones of the world constitute the most complicated features of the earth system because of the three-way interface between atmosphere, land, and ocean. Geographic variation in the geological, hydrological, hydrodynamical, climatologic, chemical, and biological properties make it extremely difficult to generalize about coastal zones. Unifying characteristics are the high physical energies encountered there. Collectively, coastal zones play a role in earth system functioning well beyond the geographic area they represent. In addition to its complexity and global importance, the coastal zone is also the most intensely altered portion of the globe from human activity. As a result, human activities in the coastal zone may be the most far-reaching for global functioning as well as for human welfare.

## I. DEFINITION OF THE COASTAL ZONE

A general definition of the coastal zone is the region where factors of land affect sea and factors of sea affect land. This is a noncontroversial, but also not very useful definition. An international definition used for maritime legal matters is the 200 nautical mile limit from land over which coastal nations exert sovereignty. Unfortunately, this legal definition does not have much scientific utility. A working definition adopted by scientists who are directing their attentions to the global implications of coastal zone change is: the zone extending from upward limits of coastal plains on land to the outer edges of continental shelves of the sea (Fig. 1). This scientific definition is expansive and leads to considerable variation in the width of the coastal zone from place to place, but it has the advantage of embodying the full range of important interactions necessary for a reasonably complete analysis of these extraordinarily complicated and important ecosystems. [*See* CONTINENTAL SHELF ECOSYSTEMS.]

What is the extent of the global coastal zone? It is impossible to accurately measure the linear extent of the coastal zone worldwide because it presents a classic problem of fractal geometry—the number changes with the precision of measurement. Esti-

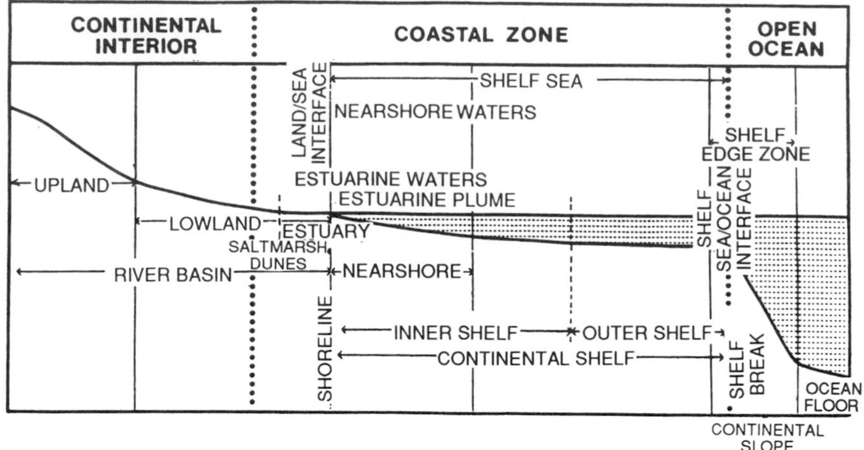


***FIGURE 1*** A diagrammatic cross section of the region where land meets the sea, including designations for continental interior, open ocean, and the object of interest—the coastal zone. [From P.M. Holligan and H. de Boois (1993). "Land–Ocean Interactions in the Coastal Zone. Science Plan." International Geosphere–Biosphere Program, Stockholm.]

mates range between 500,000 and 1,000,000 km worldwide. Uncertainty of the zone's length compounds, of course, estimation of its area. An estimate for coastal zone area is on the order of 510,000,000 km,$^2$ or 8% of the world's surface area (Fig. 2).

The distribution of coastal zone area worldwide is quite uneven. Widths of coastal plains on land and continental shelves of the sea range from zero to hundreds of kilometers. Wide coastal plains and continental shelves tend to be geographically coincident, typically occurring along passive continental margins such as those around the Arctic Ocean (Fig. 2). Extensive shelves can also exist in tectonically active areas such as those in the Austral-Asian region (Fig. 2).

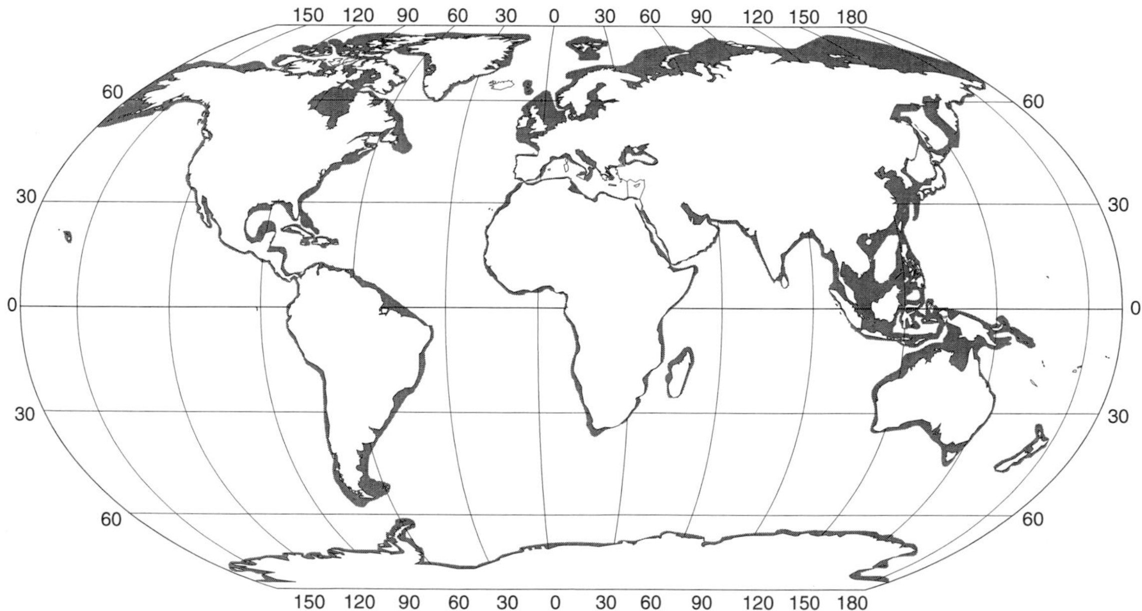


***FIGURE 2*** Global map highlighting the continental shelves of the world, which, together with coastal plains (not shown) of about equal extent, comprise the coastal zone as defined here. [Redrawn from H. Postma and J. J. Zijlstra, eds. (1988). "Ecosystems of the World 27. Continental Shelves." Elsevier, Amsterdam.]

## II. NATURE OF THE COASTAL ZONE

### A. Variation in Physiography

The geology of the land margin dictates the form, shape, and extent of the submerged shelf and associated coastal plain. In fact, coasts can be classified in terms of the prevailing geotectonic characteristics of the continental margin. In general terms, sea-floors spread from rifts located more or less in the centers of oceans toward the continents, where they typically subduct, causing orogenic and volcanic activity on the coastal margins. The degree of tectonic activity along a specific coastal margin determines the basic form of that margin. Within this context, margins can be classed as "collision coasts" or "active coasts," where collision, subduction, and obduction are active, or as "trailing edge" or "passive" coasts, where there is little such activity.

Trailing edge coasts tend to have flat coastal plains, wide continental shelves, gently dipping continental slopes, and extensive sediments, including beach–dune complexes enclosing coastal lagoons (Fig. 3). Collision coasts typically have abrupt, steep cliff lines, narrow continental shelves, steeply dipping coastal slopes, and thinner sediment beds, usually with little or no subaerial deposits (Fig. 3).

These features more or less parallel coastlines. Intersecting parallel features are rivers, which are associated with areas where fresh water meets salt water. These areas are called estuaries. They can be divided into three main types: coastal plain estuaries, including river deltas, drowned river valleys, and embayments of coastal lowlands; fjords, which are mouths of formerly glaciated valleys on mountainous coasts; and lagoons, bodies of water formed behind sediments paralleling coastlines. Estuaries are highly dynamic zones where less dense, fresh water mixes with salt water in complex ways, and where fluvial materials, including nutrients, organic carbon, and inorganic sediments, are delivered to the sea. Estuaries are areas of very high physical and biological activity because of the instabilities in adjacent freshwater and saltwater bodies, and because of the abundant resources supplied for both photosynthesis and heterotrophic metabolism. Estuaries tend to be more numerous and broad on trailing edge coasts, where their influences can dominate coastal properties.

Associated with river mouths are depositional zones called deltas, which are some of the most valuable and highly utilized areas of the earth's landscapes for mankind. Deltas are also complex in themselves, containing deltaic plains, distributary channels, river-mouth bars, open and closed interdistributary bays, tidal flats, tidal ridges, beaches, beach ridges, dunes and dune fields, and swamps (including tropical mangrove swamps, or mangels) and marshes. Lateral, coastal currents often carry river-borne material along coastlines so that the influences of estuaries can be felt far from the deltas themselves. Examples of deltas are shown in Fig. 4.

All of these geological features are found to varying extents around islands as well as continents. Special mention should be made of coral reefs, however, which are not only important kinds of islands for human habitation, but are very important as reservoirs of biological diversity. Coral reefs occur in warm, tropical waters and consist of cemented and unconsolidated calcareous materials built up through the cumulative action of coralline organisms. There are four kinds of reefs: fringing reefs bordering larger land features with no deep-water channel between them and land; barrier reefs bordering landmasses at greater distances and separated by wide and deep channels; atolls consisting of rings of coral unconnected with any other landmass; and patch reefs of various forms associated with larger reefs. [*See* CORAL REEF ECOSYSTEMS.]

### B. Coastal Zone Hydrodynamics

Imposed on all of these physiographic variants of coastlines is ocean circulation in the form of currents, tides, and storms. These arise from the global ocean currents modified by the Coriolis force, upwelling of currents against continental shelves, tides, and temperature and salinity gradients. An example of the complex forces directing coastal currents is shown in Fig. 5. The direction, strength, periodicity (as with tides), and storm variants asso-

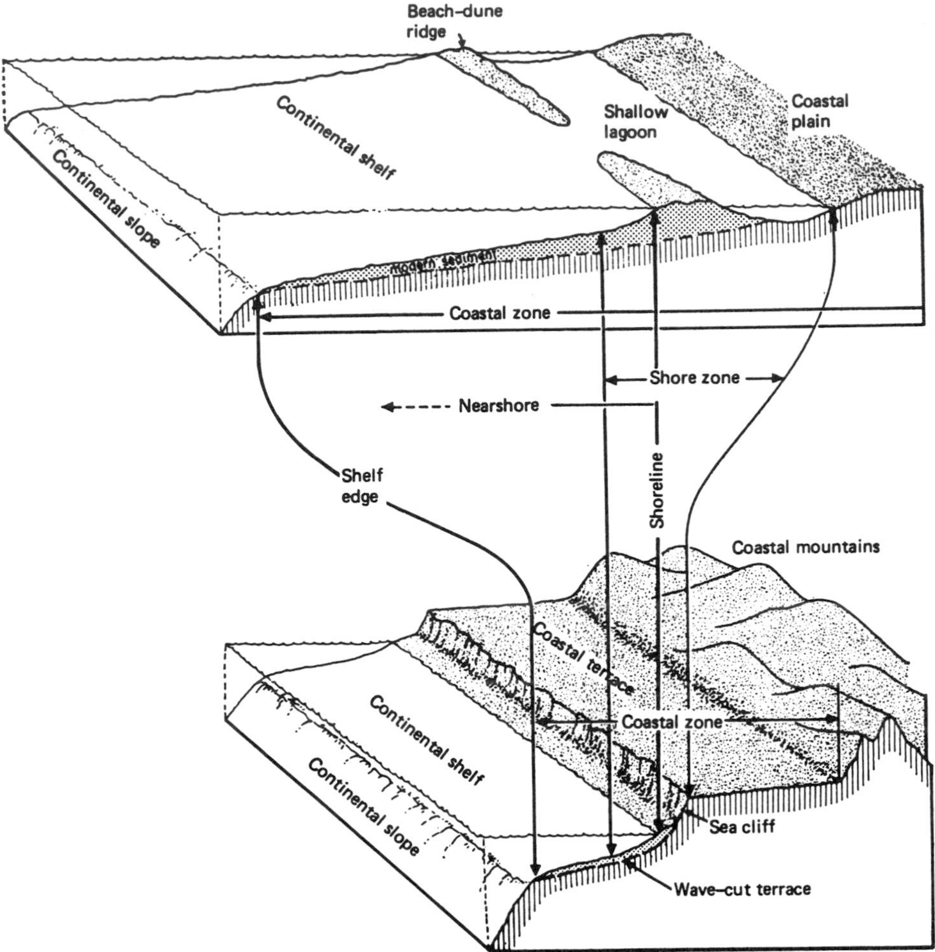


***FIGURE 3*** Block diagrams demonstrating coastal characteristics of trailing edge continental margins (above) and collision continental margins (bottom). [From J. P. Kennett (1982). "Marine geology." Prentice–Hall, Englewood Cliffs, N.J.]

ciated with these movements underlie the fundamental geomorphology, mixing patterns, formation of density–temperature barriers to water movement ("fronts"), and the resultant biological activity of stretches of coastline. Fronts are regions of intense motion and mixing of bodies of water having dissimilar temperatures and/or chemistries. The physics underlying all of these properties is complex but necessary for understanding the structure and function of coastal ecosystems.

### C. Coastal Climate

A third major variable is climate. Coastal climate is generally controlled by latitudinal position but is regionally modified by characteristics of adjacent water. Cold, upwelling water, such as off the coast of Peru, causes low vapor pressure and low temperature of the overlying air. As this air moves over land and is heated, the relative humidity drops, decreasing the probability of precipitation. Conversely, warm waters, such as those associated with the Gulf Stream, raise the temperature and humidity of overlying air leading to foggy, cloudy, high-precipitation regimes.

Water temperature itself is of profound geochemical and biological significance as chemical and biological processes are highly temperature dependent. For example, precipitation of calcium carbonate—a critical aspect of marine geochemistry—is partially controlled by temperature through both chemical and organic means. The kinds of

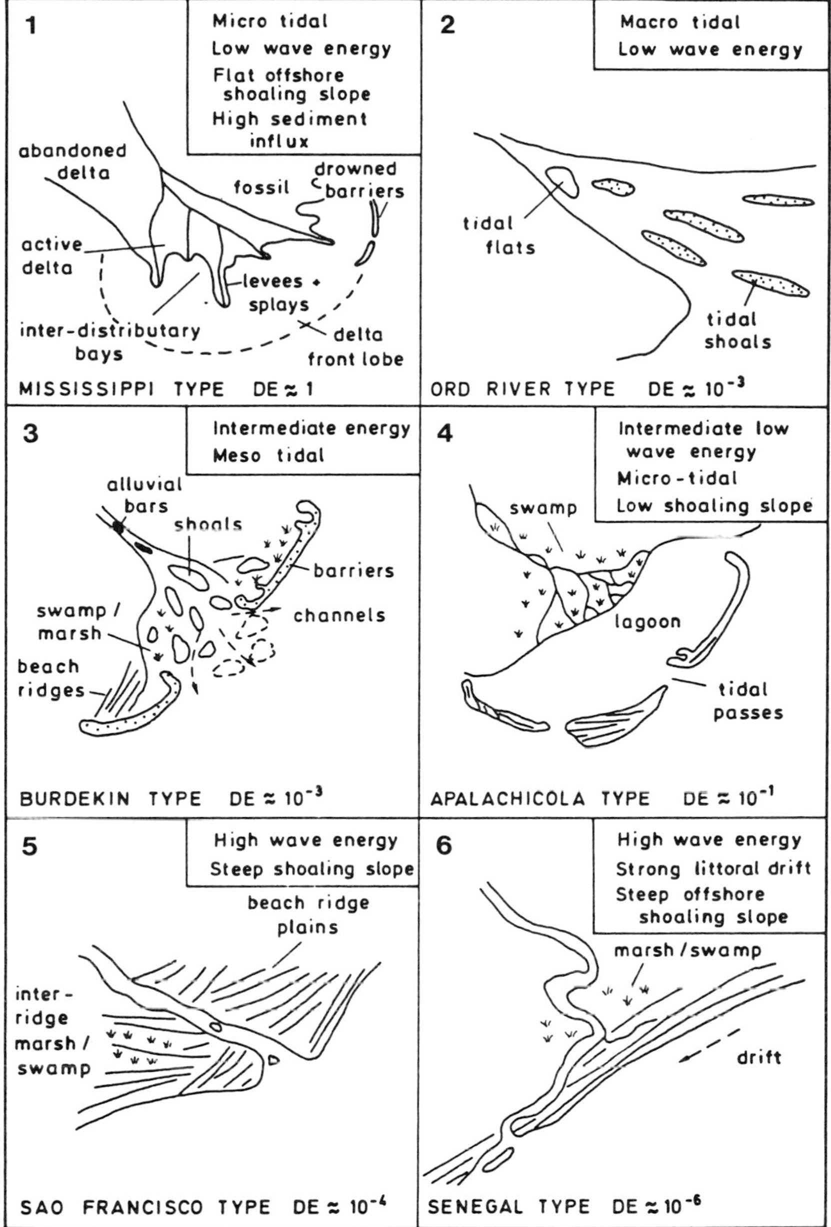


***FIGURE 4*** Types of deltas from around the world. Different morphologies result from different relationships between stream discharge and marine wave and tidal processes. [From R. W. G. Carter (1988). "Coastal Environments." Academic Press, New York.]

phytoplankton, macroalgae, or marsh and swamp vegetation are partially controlled by air and seawater temperature.

The enormous heat capacity of the sea and interchange of energy through evaporation and condensation confer a large moderating influence on air temperature and humidity. Regions near the sea typically, but not universally, feature moderate variability in temperature, extensive periods of fog or cloudiness, and diurnal sea breezes due to differential heating of air over land and sea. These are generally moderating influences on the coastal environment. Some of these environments are also susceptible to storms such as hurricanes and typhoons, which can have long-lasting influences on terrestrial vegetation through

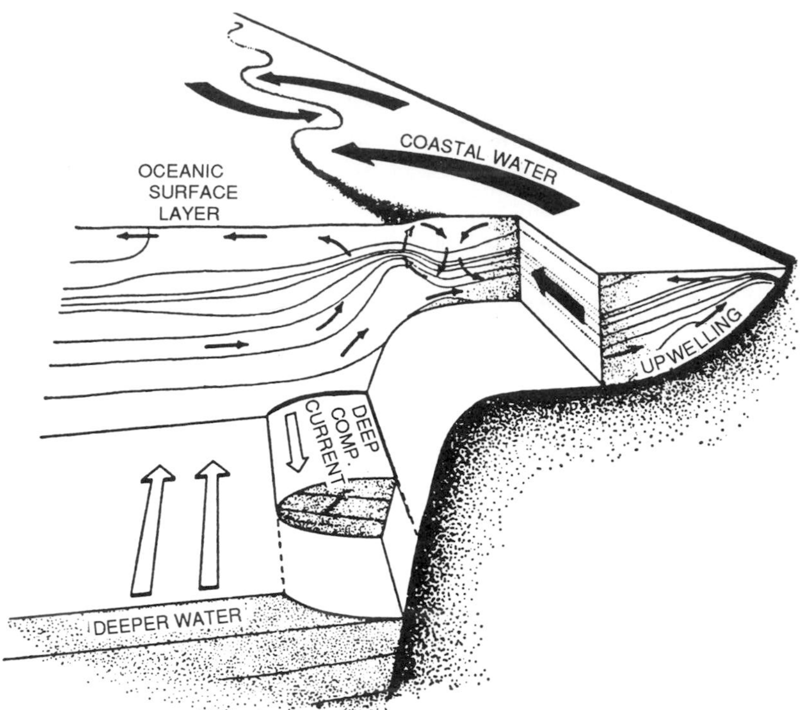


***FIGURE 5*** An illustration of the complex hydrodynamics along the West African Coast where northward winds driving the Benguela Current create a very productive upwelling zone. [From T. Beer (1983). "Environmental oceanography." Pergamon Press, Oxford, England.]

mechanical damage by wind and chemical damage by salt spray. Estuarine ecosystems and coral reefs can be altered through storm flow, freshwater inundation, or intensified sedimentation. Episodic events are as important as average conditions in shaping the physical and biological properties of coastal ecosystems.

## D. Biota and Ecology of the Coastal Environment

The physical influences of hydrodynamics and climate combine to create a zone of high physical energy in the forms of sensible heat (temperature times mass), or latent energy (water vapor), or momentum derived from large-scale pressure gradients. When moving masses of air and water come into contact with each other or with a solid object (like land), energy exchanges must take place. These often act as erosive forces on submarine bottoms and land surfaces and promote the mixing of waters, which influences the distribution of chemicals and, ultimately, the biotic structure of coastal ecosystems.

Mixing of top waters with bottom waters, together with upwellings from deep oceanic waters in some areas and river inputs in others, provides a relatively nutrient-rich environment. Equally important, this nutrient-rich water is brought near the surface, where there is ample sunlight to drive photosynthesis. This relatively high primary production is sometimes augmented in parts of the coastal zone with additional amounts of organic carbon imported from land. Imported, or allochthonous, organic carbon is derived from river flows in estuarine areas, from tidal outflows from coastal wetlands, from general shoreline erosion, and, to a small extent, from groundwater flows. The combination of high primary productivity plus allochthonous carbon often sustains relatively high levels of biological metabolism in the coastal zone. Thus, coastal zones are environments of high biological as well as high physical energy.

It is impossible to adequately describe the amazing diversity of biotic forms and ecosystems included in the coastal zone in a brief exposition, especially when coastal plains are included in the definition of the coastal zone. Physiognomically defined ecosystems range from terrestrial forests, fields, and swamps to riverine communities, salt marshes, mangrove swamps, coral reefs, intertidal rocky shores, intertidal mudflats, beaches, and deeper shelf waters. Within each of these types, specific forms of plants, an astonishingly wide range of animal phyla, and broadly diverse microbes locally aggregate to form unique and sometimes highly diverse biological assemblages. Primary producers range from terrestrial plants to submerged, sessile, higher plants like sea grasses or sessile macroalgae to both fixed and planktonic algae. Consumer organisms range from subterranean groups such as terrestrial earthworms to marine, bottom-dwelling (benthic) clams and bristle worms on and in mudflats and beaches, to attached animals such as corals and barnacles (at least in some stage of their life cycles), to drifting zooplankton and highly vagile fishes such as tuna and mackerel. [*See* INTERTIDAL ECOLOGY.]

The highest rates of primary productivity in marine environments are associated with the coastal zone, where rates are typically three times higher than in oceanic waters. High primary productivity in intertidal and subtidal systems, such as salt marshes, mangrove swamps, sea grass flats, and kelp beds, coupled with organic carbon supply from river flows, leads to high energy flow through food chains to heterotrophic organism such as zooplankton, benthic and pelagic invertebrates such as oysters and shrimp, and larger vertebrates such as fish and marine mammals. This high secondary productivity in the coastal zone extends its effects to deeper waters. Coastal zone zooplankton are believed to support about 95% of the world's fishery yield. This represents 87% of the global finfish yield and most of the shellfish catch. [*See* MARINE PRODUCTIVITY.]

## III. INTERACTIONS BETWEEN LAND AND SEA

One way to organize the many interactions between land and sea is to describe energy and material fluxes between them. Ignoring tectonics, these interactions are unsymmetrical, with more energy (including latent heat) being directed from sea to land, and more matter (excluding water) moving from land to sea.

### A. Energy Fluxes

Two kinds of energy are transported from land to sea. The first is the energy of momentum in river flow to the coastal zone. This energy is responsible for much of the higher-elevation features of deltas. River flow, together with the much larger energies of tidal currents and waves, helps to drive the mixing process that is so important for gas and nutrient supply to biota throughout the lower reaches of deltas and estuaries. This flow also contributes to the transport of sediment to positions where it becomes redistributed and deposited by marine processes (see Fig. 4). The second form of energy transport from land to sea is more indirect—the heating of air over land and transport as sensible heat to air over sea, and to seawater itself through conduction. This is a relatively unimportant flux in the global climate system but may be locally significant.

Transport of energy from sea to land takes several significant forms. One is the energy of momentum in wind itself, which is sometimes very important in destructive subtropical storms and more frequently in the transport of marine aerosols. A second is the transport of sensible heat and especially latent heat from air over sea to the land atmosphere. These are very large energy fluxes but are highly variable geographically.

A third energy flux from sea to land is the energy of momentum behind oceanic currents, tidal bores, and waves. These are large energies, and during storms they can be enormous. They are responsible for most of the transport and deposition of sediments across the coastal zone and out to the shelf slope, for nutrient mixing from the bottom sediments to the euphotic zone, and for much of shoreline morphodynamics. Wave action in particular acts on subaerial surfaces of the coastal zone, reducing cliffs, producing and destroying terraces, and forming and re-forming barrier islands and associated beaches and dunal complexes. Waves, currents, and tides act on submarine features such as mud-flats, channels, and submerged bars. The energy of momentum behind water movement is a primary factor shaping the physical elements of coastal zone complexes and driving biological and chemical processes.

### B. Fluxes of Matter from Land to Sea

All land surfaces are connected with the sea through the transport of materials via fluvial and aeolian transport. Land provides a flow of fresh water and dissolved materials, both organic and inorganic, as well as suspended particulates that provide organic energy, nutrients, and the sedimentary material that is ultimately reshaped into coastal deposits like sand dunes and shelf muds.

The amount of fresh water running from land to the coastal zone via river discharge is about 33–47 $\times$ $10^3$ $km^3$/yr and a small amount, perhaps 2–2.5 $\times$ $10^3$ $km^3$, is contributed by direct seepage of groundwater into the oceans (Table I). River water itself is derived from surface flow and groundwater. On a global average, approximately 30% of annual river discharge comes from ground-water, but this varies widely with individual rivers. River discharge is concentrated in a relatively small number of rivers; approximately 40% of surface flow comes from 16 of the major rivers of the world that collectively drain only 23% of the land area. These facts once more highlight the geographic variability of the coastal zone.

Water is also transported from the atmosphere above land to that above sea, although in lesser amounts than in the reverse direction. This flux amounts to about 10 $\times$ $10^3$ $km^3$/yr of water as vapor (Table I).

Immense amounts of sediment and dissolved and particulate materials are associated with these river flows. Sediment yield per unit of land catchment varies from 5–6 to 28,000 tons/ha/yr, with a global average of 150 tons/$km^2$/yr. This leads to a total delivery of about 13.7 $\times$ $10^9$ tons/yr. Flood discharge and bed load transport amount to about 10% of this total.

The extreme range in sediment yield is reflected in the geography of sediment delivery to different sectors of the global coastal zone. Not too long ago it was thought that most of the sediment was delivered by a small number of large rivers, mostly in Oceania, southern Asia, and South America. The situation is currently being reassessed in terms of a higher evaluation of the contributions of small rivers draining steep terrains of coastal mountain ranges.

Rivers carry dissolved materials and fine, particulate organic matter that is important for geochemical and biological functions in the coastal zone (Table I). Organic carbon and organic and inorganic nitrogen and phosphorus are particularly critical substances for biogeochemical processes as well as biological productivity in the coastal zone. Through most of the coastal zone, fluxes represented in Table I are immediately deposited in the zone for autotrophic and heterotrophic utilization.

The flux of chemicals via groundwater flow to the ocean is not known. It has not been measured or calculated, but it is probable that groundwater is richer in inorganic species and poorer in organic species than is river water.

Other materials carried from land to sea are wind-blown dust, pollutants, and miscellaneous other classes of materials. The amount of dust delivered is 530–851 $\times$ $10^6$ tons/yr, largely from the eolian plateau region of China to the North Pacific, and from the Sahara to the mid-Atlantic. The amount and composition of pollutants carried from land to sea vary widely by region.

As Table I shows, large amounts (comparable with fluvial delivery) of inorganic nitrogen are deposited from land-derived air to the ocean through rainout and dry deposition. In addition to this are deliveries of amino acid-N, some of which come from land to the ocean via atmospheric transport. On the other hand, the transport of phosphorus, mainly in dust, is relatively low compared with that of fluvial flux. Almost 50% of this total phosphorus flux comes from plumes arising from the Saharan region.

### C. Fluxes of Matter from Sea to Land

More water is transported from sea to land as vapor via the atmosphere than from land to sea as liquid

**TABLE I**
Material Fluxes between Land and Sea

| Substance | Units | Land to sea | Sea to land |
|---|---|---|---|
| Atmospheric Fluxes | | | |
| Water vapor | $10^3$ km$^3$/yr | 10 | 78 |
| Dust | $10^6$ tons/yr | 530–850 | — |
| Sea spray aerosols | $10^6$ tons/yr | — | 100–1000 |
| Inorganic nitrogen | $10^6$ tons/yr | 23 | ? |
| Amino acid nitrogen | $10^6$ tons/yr | 8–13 | ? |
| Phosphorus | $10^6$ tons/yr | 1 | ? |
| Dimethyl sulfur-S | $10^6$ tons/yr | — | <1 |
| Fluvial Fluxes | | | |
| Surface water | $10^3$ km$^3$/yr | 33–47 | — |
| Groundwater | $10^3$ km$^3$/yr | 2 | — |
| Sediment | $10^6$ tons/yr | 13,700 | — |
| Dissolved inorganic ions | $10^6$ tons/yr | | |
| $Cl^{1-}$ | | 308 | — |
| $Na^{1+}$ | | 269 | — |
| $SO_4^{2-}$ | | 143 | — |
| $Mg^{2+}$ | | 137 | — |
| $K^{1+}$ | | 52 | — |
| $Ca^{2+}$ | | 550 | — |
| $HCO_3^{1-}$ | | 1980 | — |
| $H_4SiO_4$-Si | | 180 | — |
| Dissolved inorganic-N | $10^6$ tons/yr | 24 | — |
| Dissolved organic-N | $10^6$ tons/yr | 6 | — |
| Dissolved and particulate P | $10^6$ tons/yr | 1 | — |
| Dissolved organic-C | $10^6$ tons/yr | 100–250 | — |
| Particulate organic-C | $10^6$ tons/yr | 420–570 | — |

because of evapotranspiration on land. Gross precipitation on land is estimated as 111 × $10^3$ km$^3$/yr, but some fraction of that precipitation, perhaps 30% on a worldwide basis, is recycled on land; that is, it represents some reprecipitation of water generated from evapotranspiration from land surfaces. Given that estimate, transport of water vapor from sea to land is about 78 × $10^3$ km$^3$/yr on a global basis.

A critical flux from sea to land is salt water into coastal aquifers. This is not a high flux, as groundwater hydrostatic levels are more or less in a steady state except where withdrawals for agriculture or residential or industrial use are increased. To the extent that climate change, land subsidence, sea level rise change, or loss of recharge area alters the hydraulic relationships between fresh water and groundwater in coastal aquifers, there will be a net saltwater flux in or out of those aquifers. Extraordinary high tides driven by storms can drive salt water as much as 50 km up estuarine rivers, contaminating riverbed aquifers far inland from the coastline. Although the long-term flux may not be great, a small flux of salt water can deplete or destroy freshwater resources in coastal aquifers, which is a critical aspect of global change owing to impacts on ecological, agricultural, industrial, and residential water requirements. Perched freshwater aquifers are very susceptible to contamination from pollutants as well.

The sea produces between 1000 and 10,000 × $10^6$ tons/yr of salt spray aerosols with radii less than about 20 $\mu$m. Approximately 10% of these aerosols are transported from sea to land for a total of 100–1000 × $10^6$ tons/yr. Salt is generally deleterious to land plants in large amounts, but in some places where sulfur and boron are limiting for geological reasons, sulfate and borate may be supplied chiefly by these aerosols.

An estimate of dimethyl sulfide (DMS) vapor emission to the atmosphere is 38 × $10^3$ tons/yr. The amount of DMS, or its derivative, sulfate, transported from sea to land is not known. The impacts of this on land have not been estimated but are presumably slight compared with those of salt-spray aerosols and terrestrially derived pollutants.

## IV. BIOGEOCHEMICAL PROCESSES IN THE COASTAL ZONE

An unusually broad range of physical-chemical conditions is found in the coastal zone. In marine environments alone, conditions range from brightly lighted, nutrient-rich, high-oxidation states of the surface waters to dark, organic carbon-rich, low-oxidation states of some benthic areas. The relatively close juxtaposition between utterly different chemical and biological environments in the coastal zone fosters fast rates of cycling by biogenic elements. Abrupt changes in dominating processes can occur in distances of centimeters. Elements such as nitrogen, phosphorus, sulfur, iron, and other metals can cycle rapidly between organically bound forms, dissolved forms such as free and complexed ions, and inorganically bound forms (Fig. 6). Under certain conditions, high rates of molecular nitrogen may be fixed by nitrogen-fixing bacteria and algae. In others, copious amounts of reduced sulfur compounds, such as hydrogen sulfides or methane or nitrogen oxides, may be emitted from soils, sediments, and waters to the atmosphere.

A biogeochemical phenomenon of profound importance in the context of global change is the extent of organic matter deposition in the coastal zone. Coastal zones have long been the loci of most petroleum formation, but in the present the importance of sedimentation, burial, and preservation of organic carbon lies in understanding the global carbon cycle. As humankind has accelerated the conversion of fossil fuel carbon to atmospheric $CO_2$, the question of the fate of this $CO_2$ is critical to understanding the outcome of this globally significant process. Nearshore areas, or the continental slope bounding the continental shelves, are important sites of study in this regard.

Among the interesting and important biogeochemical reactions that occur in the coastal zone are three cases involving transport between sea and land through the atmosphere. The first of these is the production of dimethyl sulfide. Many macroalgae and phytoplankton produce DMS as cellular solutes, which may be volatilized from elutions and cellular breakdown products of these organisms and lost to the atmosphere. DMS is oxidized in the atmosphere to sulfate, becoming a sulfate aerosol contributing to the sea–land transport of this sometimes limiting element on land, and perhaps slightly to acid deposition as well. DMS is also a nucleating agent for water droplets and thus cloud formation, so that it may indirectly influence climate by increasing atmospheric reflectance of incoming solar radiation. DMS is produced in the open sea as well as coastal zone, but DMS produced near land is more likely to be transported to land or to affect sunlight over land.

The second important land–sea transport interaction is the production of wind-blown dust on land and subsequent transport to sea. Dust plumes are important to the global radiation budget by increasing Earth's albedo. But dust deposited on the sea takes on special significance in its fertilizing effect on oligotrophic waters and enhancement of the sinking rate of organic detritus to the sediments.

The third case is the production of sea-spray aerosols, some of which are transported to land. These aerosols are generated by the eruptions of droplets from waves to the overlying air, evaporation of the water, and transport upward of hygrophilic salts through turbulent transport. The chemical composition of sea-spray aerosols departs slightly from seawater composition but generally is dominated by major seawater ions, including chloride, sodium, sulfate, potassium, and calcium.

## V. HUMAN INFLUENCES ON THE COASTAL ZONE

Humankind has exploited the coastal zone for many resources throughout history and this exploitation is accelerating. Worldwide it is estimated that 70% of the human population lives on coastal

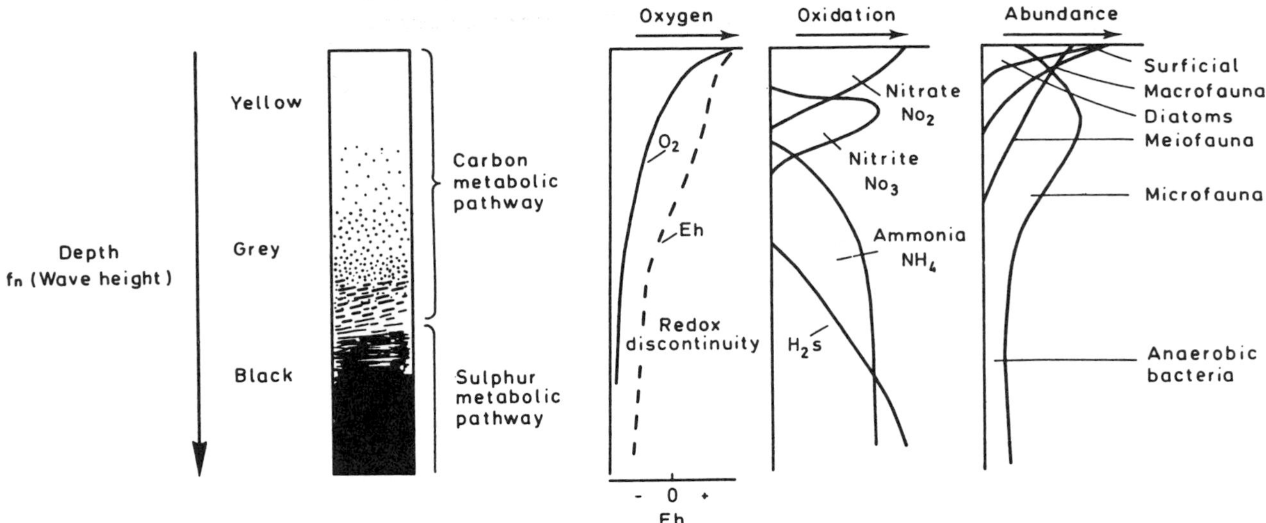


**FIGURE 6** Diagrammatic profile through a sandy beach soil at a particular depth below mean wave height. A similar profile could be found in mudflats, deep-water sediments, salt marshes, and even on land where the water table is near the surface. Starting at the left, the first column demonstrates increasingly dark colors with depth associated with declining oxygen availability and concomitant low redox conditions (second column). Most heterotrophic organisms in the upper part of this profile oxidize carbon using oxygen as an electron acceptor; most organisms in the lower part of the profile directly or indirectly use sulfate as an electron acceptor. Following mineralization, inorganic nitrogen tends to be oxidized to nitrate in the upper, aerobic part of the profile, but remains in the reduced ammonium form in the anaerobic, lower part of the profile (third column). Nitrite is of intermediate oxidative state and is most abundant in intermediate depths. Sulfate is reduced as a result of anaerobic catabolism in the lower part of the profile, producing sulfides. The fourth column from the left indicates the typical abundances of photosynthetic diatoms at the surface; macrofauna also near the surface; intermediate-sized meiofauna somewhat deeper where there is less oxygen; microfauna even deeper where aerobic conditions are marginal; and finally, anaerobic bacteria living in the lowest part of the profile where oxygen is absent. [From Carter R. W. G. (1988). "Coastal Environments." Academic Press, New York.]

plains. In the United States, 75% of the population lives within 80 km of oceanic or the Great Lakes coasts. Much of the world's arable lands and industrial investment lies on the coastal plains and in lower river valleys. Coastal plains, particularly deltas, have long been developed for agricultural, trade, residential, and industrial uses. Coastal waters have been used for transportation, fishing, recreation, and waste disposal. Coastal zones more generally have always been exceptionally important in terms of commerce and military operations. In fact, human utilization for multiple, sometimes conflicting, purposes in the coastal zone are so massive that human influences must be considered in a summary of interactions in this area as much as are nonhuman influences.

The most obvious, and perhaps historically most important, human influence has been the direct effect of overfishing on native fisheries, both finfish and shellfish. Today, however, there are also many forms of indirect impact on the coastal biota. Human impacts can start far up the watersheds of river systems by changing the amount, timing, and chemical nature of water delivery to the coastal zone. Changes in land use such as logging, conversion of forest to agricultural land, development of irrigation works, or construction of impoundments can have enormous impacts on the normal hydrologic and sedimentary balances of coastal zones. [*See* MARINE BIOLOGY, HUMAN IMPACTS.]

Humans have had extensive impact on the geomorphology of developed coastlines through engineering works that reorganized established patterns of deltaic or long-shore sediment deposition or that completely eliminated some critical coastal habitats. The balance of erosion and deposition in some barrier beach zones or sea cliff areas has been entirely altered so that the geography of the coastline has been changed. In other areas, vast marshes have been eliminated through diking and drainage. These are direct changes. Some activities, such as overconsumption of fresh water in delta sediments or diverting or decreasing the rate of sediment delivery, have led to alarming rates of land subsidence in delta regions.

Chemical impacts are, in a way, more perfidious and subtle. Increased fertilizer applications upstream or directly on the coastal plains ultimately leads to enrichment or eutrophication of marine coastal ecosystems. This can be manifested as an increase in primary production and, in some cases, an abrupt alteration in the species composition and food webs of these areas. Sometimes these changes are seriously deleterious, such as promotion of "red tides" with toxic effects on commercially valuable biota and humans themselves. Even more subtle and less well understood are the effects of toxic wastes and radioactive substances in the complicated nearshore environments of the coastal zone.

These kinds of human influences are generally proportional to the population densities along the coastline and the degree to which that population has a well-developed industrial infrastructure. Such damaging influences can be mitigated where governments have the will and capacity to regulate such activities. Where governments are weak in this regard, the damage can be so extensive and complete as to render useless the potential biological, recreational, and residential values of the coastal zone.

There is one more level of indirect effects that need to be considered in the coastal zone—the potential effects of human-induced climate change. Through changes in land use and combustion of fossil fuels, it is possible that the world climate system may become significantly altered within the next 50 years. Such an alteration would influence virtually all of the interactions already mentioned. One special impact, uniquely critical in coastal environments, would be a potential rise in sea level. Although that rise is not considered to be significant at the present time, scientific consensus on this varies over time so that sea level rise is a special consideration that planners must keep in mind.

In summary, the coastal zones of the world can be viewed as a highly variable reticulum wrapping around the continents and islands of the globe. This reticulum represents a zone of extraordinarily high interaction between the atmosphere, oceans, and land and incorporates a high population of humans together with their multiple impacts. The coastal zone is enormously important to global functioning and humankind, and our understanding of its structure, function, and variability is critical to the welfare of both.

## Glossary

**Allochthonous** Material, often organic, transported into an ecosystem from another source area.

**Autotroph** Organism that gains its energy from physical sources, either light through photosynthesis (photoautotroph) or by the oxidation of reduced inorganic compounds such as sulfide (chemoautotroph).

**Coastal plain** Low plain extending along a coast.

**Coastal zone** Zone extending from upward limits of coastal plains on land to the outer edges of continental shelves of the sea.

**Consumer** See heterotroph.

**Continental shelf** Seaward extension of the adjacent continent from the shoreline to the shelf break or outer edge.

**Coral reef** Shallow-water zone constructed primarily of calcium carbonate by coralline organisms (algae or coelenterates).

**Delta** Nearly flat plain composed of alluvial sediments deposited at its mouth by divergent branches of a river; often, but not always, triangular in shape.

**Estuary** Region where fresh water meets salt water.

**Heterotroph** Organism that gains its energy from the metabolism of organic carbon.

**Hydrodynamics** Movements of fluids, in this case the fresh and salt waters of the coastal zone.

**Latent heat** Energy represented by the heat of vaporization of water vapor in a parcel of air.

**Mangrove swamp** Low- to medium-height forest ecosystem composed of various species of woody plants that can tolerate high salinities of the intertidal and high tide zones of gently grading coast lines. Limited to frost-free conditions of the tropics and subtropics.

**Obduction** Crustal plate accretion caused by thrusting along collision margins of marine and continental crusts.

**Orogeny** Deformation and alteration processes, such as folding, faulting, metamorphosis, and plutonism, leading to mountain building.

**Perched freshwater aquifer** Lens of fresh water within an aquifer lying on top of more dense salt water.

**Phytoplankton** Small, often microscopic algae that drift in water.

**Primary producer** See autotroph.

**Sensible heat** Energy represented by the density and temperature of a parcel of air.

**Sessile** Firmly attached to a substrate such as soil, rock, or coarse sedimentary material.

**Subduction** Movement of the ocean floor downward into the earth's mantle at the margins of continents.

**Tectonics** Movement and deformation of the earth's crust on a large scale.

## Bibliography


Beer, T. (1983). "Environmental Oceanography." Oxford, England: Pergamon Press.

Berner, E. K., and Berner, R. A. (1987). "The Global Water Cycle. Geochemistry and Environment." Englewood Cliffs, N.J.: Prentice–Hall.

Carter, R. W. G. (1988). "Coastal Environments." London: Academic Press.

Davis, R. A., Jr. (1978). "Coastal Sedimentary Environments." New York: Springer-Verlag.

Dyer, K. R. (1986). "Coastal and Estuarine Sediment Environments." Chichester, England: John Wiley & Sons.

Hekstra, G. P. (1989). Global warming and rising sea levels: The policy implications. *The Ecologist* **19,** 4–13.

Holligan, P. M., and de Boois, H. (1993). "Land–Ocean Interactions in the Coastal Zone. Science Plan," International Geosphere–Biosphere Global Change Report No. 25. Stockholm: IGBP.

Holligan, P. M., and Reiners, W. A. (1992). Predicting the responses of the coastal zone to global change. *Adv. Ecol. Res.* **22,** 211–255.

Kennett, J. P. (1982). "Marine Geology." Englewood Cliffs, N.J.: Prentice–Hall.

Ketchum, B. H. (1983). "Ecosystems of the World 26. Estuaries and Enclosed Seas." Amsterdam: Elsevier.

Milliman, J. D. (1990). River discharge of water and sediments to the oceans: Variations in space and time. *In* "Facets of Modern Biogeochemistry. A Festschrift for Egon Degens" (V. Ittekkot, S. Kempe, W. Michaelis, and A. Spitzy, eds.), pp. 83–101. Berlin: Springer-Verlag.

Postma, H., and Zijlstra, J. J., eds. (1988). "Ecosystems of the World 27. Continental shelves." Amsterdam: Elsevier.

UNESCO (1978). "World Water Balance and Water Resources of the Earth," USSR Committee for the International Hydrological Decade, Studies and Reports in Hydrology 25. Paris: UNESCO Press.

Walsh, J. J. (1988). "On the Nature of Continental Shelves." San Diego, Calif.: Academic Press.

Wright, L. D. (1978). River deltas. *In* "Coastal Sedimentary Environments" (R. A. Davis, Jr., ed.), pp. 5–68. New York: Springer-Verlag.

# Landscape Ecology

**R. J. Hobbs**
*CSIRO, Australia*



Landscapes are heterogeneous areas of land, usually hectares or square kilometers in area, composed of interacting ecosystems or patches. Landscape ecology aims to study the patterns and processes operating at this scale, and focuses on landscape structure, function, and change. Heterogeneity is an important, but difficult to measure, feature of most landscapes, and the spatial distribution and interrelationships between landscape patches determine functions such as biotic movement and fluxes of water, energy, and nutrients. Landscapes are seen as an important scale for many land management and conservation problems, but as yet landscape ecology has few general principles or a sound theoretical framework that can be applied to these problems.

## I. INTRODUCTION

Landscape ecology is a relatively young science that aims to study patterns and processes at a larger scale than has traditionally been considered in ecology. Most ecological study takes place at scales small enough to be easily sampled and analyzed, that is, in marked plots or individual patches or stands of vegetation. Only recently has it been recognized that patterns and processes at larger scales are very important from a number of perspectives, including land management and conservation. Similarly, it is only recently that methods have become available to deal adequately with the sampling and analysis of landscapes. These include geographic information systems and remote sensing.

Landscape ecology is thus currently in a state of rapid development. It is moving away from its largely descriptive European roots and is developing more quantitative and analytical methods. Many methods are new and relatively untried, and as yet a strong theoretical framework has not developed. Large amounts of terminology are being generated, much of which may not have lasting value. This article provides an overview of important current concepts in landscape ecology and focuses on those that are liable to be relevant to other fields of environmental science.

## II. CONCEPTUAL FRAMEWORK

Ecological systems consist of structures and processes at a number of different scales. The simplest unit is that of the individual organism, and collections of organisms of the same species make up populations, while collections of populations of different species compose communities. Communities and their abiotic environment can be viewed as ecosystems. Each of these levels of organization has its own set of properties and ecologists have developed a wide array of descriptive and theoretical ways to consider the patterns and processes operating at each level. Here a further level in this hierarchy is considered, that of landscapes.

A landscape is defined as an area of land, at the scale of hectares to square kilometers, that consists of a collection of different but interacting patches (also called landscape elements). Patches may comprise different ecosystems (e.g., lakes, rivers, forest, grassland), different land uses (e.g., urban, agricultural, nature reserve), or different community types, successional stages, or alternative states within a particular ecosystem (e.g., postfire, polestage, and old-growth forest stands).

Landscape ecology considers three main aspects: structure (or pattern), function (or process), and change (Fig. 1). The characteristics of individual patches and their spatial relationship with other patches determine landscape structure, whereas landscape function is determined by physical, chemical, and biotic transfers between patches. Landscape change results from either changes in individual patches or changes in patch configurations and interrelations. The three aspects of landscapes are closely interlinked: structure strongly influences function, which can feed back into structure, and landscape change can affect both structure and function.

## III. LANDSCAPE STRUCTURE

### A. Spatial Patterns and Scales

Landscape mosaics are commonly complex entities consisting of numerous patches of varying types in a variety of configurations (Fig. 2). Spatial pattern can be described in numerous ways, and some of the commonly used descriptors of pattern are indicated in Fig. 2 and described in Table I. Our understanding of landscape patterns has increased greatly with the advent of remote sensing and geographic information system technologies. Remote sensing offers the possibility to acquire large amounts of data on the characteristics of the earth's surface in a relatively easy, repeatable, and analyzable way (Fig. 3). Geographic information systems allow us to develop spatially explicit data bases on a wide range of landscape features (Fig. 4), which in turn allow quantitative analysis of pattern.

An assessment of landscape pattern requires that the scale of investigation be defined, and this scale will be determined by the types of question being asked and the types of organism or process being studied. Many problems, misunderstandings, conflicting results, and misapplication of research findings arise from failures to determine the relevant scale of study or to ensure that studies are conducted at similar scales. Figure 5 indicates the

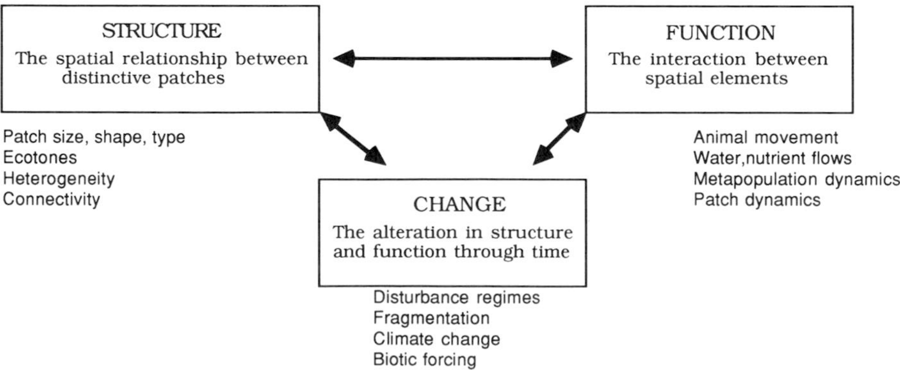


***FIGURE 1*** The principal characteristics of landscapes are structure, function, and change. Examples of phenomena that contribute to these characteristics are indicated.

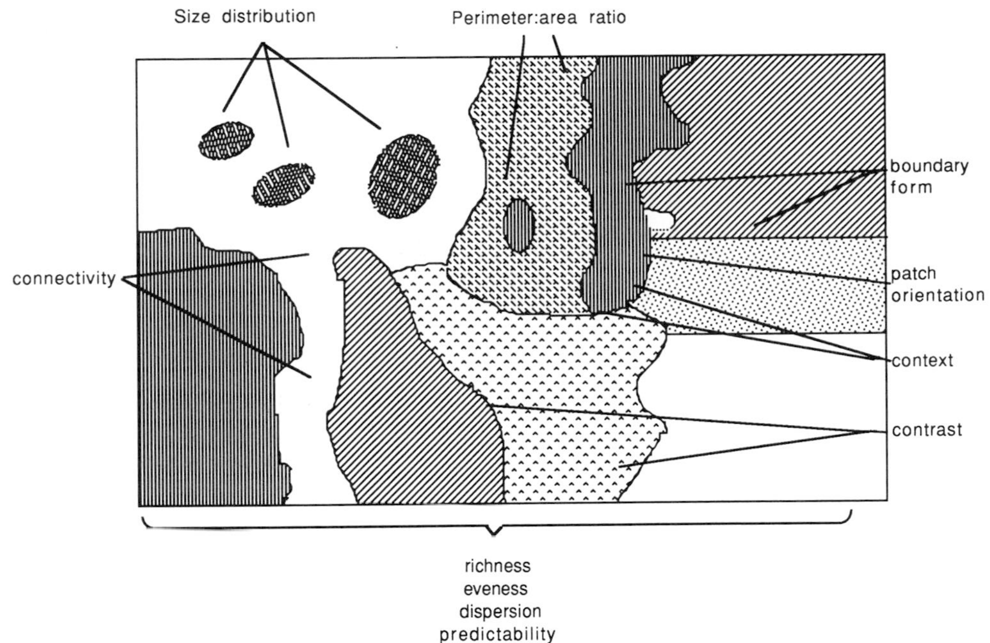


***FIGURE 2*** A hypothetical landscape made up of several different types of patch (indicated by different shadings), in different size and configurations. A range of the characteristics commonly measured is discussed in Table I. [Redrawn from J. A. Wiens, N. C. Stenseth, B. Van Horne, and R. A. Ims (1993). *Oikos* **66,** 369–380.]

three main levels of interest in landscape ecology. Though the landscape level is the primary focus here, it is important to recognize that individual landscapes occur in a regional setting. Indeed, boundaries between landscapes are frequently little more than convenient lines drawn by humans. A frequently used natural landscape unit is the catchment, as this has natural topographically determined boundaries.

Also of importance is the patch level, that is, the level of the units that make up landscapes. Individual patches have a set of characteristics (e.g., size,

***TABLE I***
**Measurable Features of Landscape Mosaics, as Indicated on Fig. 2**

| Feature | Description |
|---|---|
| Size distribution | Frequency distribution of sizes of patches of a given type |
| Boundary form | Boundary thickness, continuity, linearity (e.g., fractal dimension), length |
| Perimeter : area ratio | Relates patch area to boundary length; reflects patch shape |
| Patch orientation | Position relative to a directional process of interest (e.g., water flow, animal movement) |
| Context | Immediate mosaic-matrix in which a patch of a given type occurs |
| Contrast | Magnitude of difference in measures across a given boundary between patches |
| Connectivity | Degree to which patches of a given type area are joined by corridors into a lattice of nodes and links |
| Richness | Number of different patch types in a given area |
| Evenness | Equivalence in numbers (or areas) of different patch types in a mosaic (the inverse of the degree of dominance by one or a few patch types) |
| Dispersion | Distribution pattern of patch types over an area |
| Predictability | Spatial autocorrelation: the degree to which knowledge about features at a given location reduces uncertainty about variable values at other locations |
| Grain | Complexity of the landscape as determined by the size of individual patches; a fine-grained landscape is composed of many small patches, whereas a coarse-grained landscape has fewer, larger patches |

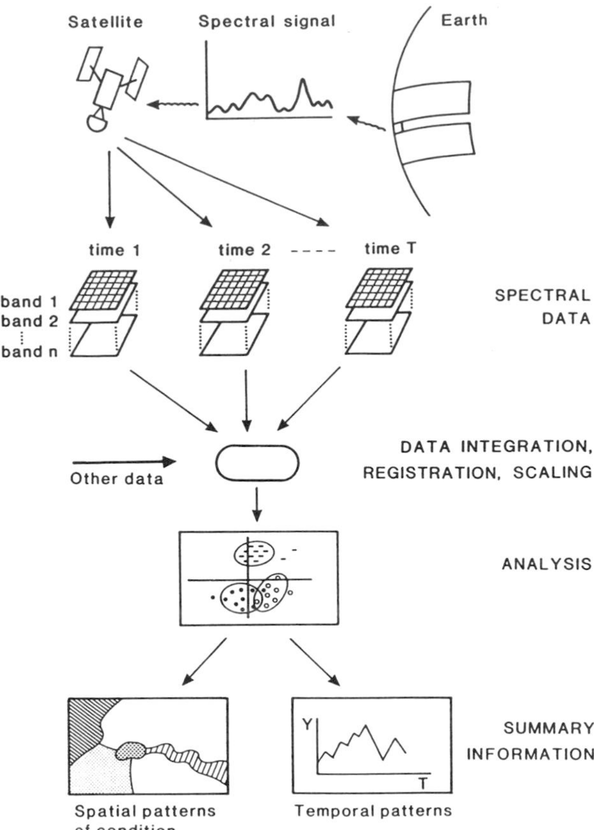


***FIGURE 3*** Remote sensing of the earth surface by satellites provides landscape-scale information on a wide array of surface characteristics. By sensing and recording different regions of the electromagnetic spectrum, the satellite provides numerous spatially explicit data layers that are repeated in time. These data can then be processed and analyzed to produce displays of spatial and temporal variation in surface characteristics. Modern images provide resolution on the order of meters and allow accurate assessment of many of the features needed to analyze landscape pattern. Because data are also collected over time, landscape dynamics can be analyzed directly from the remotely sensed data. [Reproduced from J. Wallace and N. Campbell (1990). *In* "Remote Sensing of Biosphere Functioning" (R. J. Hobbs and H. A. Mooney, eds.), pp. 291–304, Springer-Verlag, New York.]

composition, age) that can influence processes both within the patch and at the landscape level. The recognition that patterns and processes at one level can be influenced by patterns and processes at other (higher and lower) levels is essential if we are to come to grips with the complexity of large-scale ecological patterns and processes.

Pattern in the landscape is the result of the interaction of many influences. At a broad scale, patterns of vegetation composition and structure can be related to regional gradients in climatic variables such as temperature and rainfall, and to changes in soil and landform type and topography. Within any given region, climate, soil type, and landform generally determine the broad vegetation patterning. [*See* Soil Ecosystems.]

Within that broad patterning, however, numerous smaller scales of pattern may be present. These may be determined by finer-scale variations in soil characteristics or microclimate, but may also be the result of other factors (Fig. 6). Species turnover across the landscape and patchy distribution of populations of individual species can add to the complexity of vegetation patterning. This patchiness can result from chance dispersal events, response to localized disturbances or response to microenvironmental variation. [*See* Plant Ecophysiology.]

The history of disturbances determines the distribution of patches of different age and successional stage across the landscape, and may in some cases be responsible for the development of mosaics of patches in alternative semipermanent vegetation states. The components of the disturbance regime will determine the scale and pattern of variation observed, and each disturbance type may produce different vegetation responses. Individual disturbance types can also produce different responses depending on factors such as environmental variations within the disturbed areas, weather characteristics following the disturbance, and interactions with other disturbances. Important landscape-scale disturbances include fire, drought, infrequent frosts or periods of higher than normal temperatures, severe storms, localized soil disturbance by animals, treefalls, and insect outbreaks. In addition to the natural disturbance regime, human disturbance is now an important component of many ecosystems. Human activities such as vegetation removal, timber or soil extraction, and introduction of nonnative plants and animals produce a further overlay of variation on the landscape, as well as altering the natural regime.

### B. Landscape Heterogeneity

Landscape heterogeneity is a complex multiscale phenomenon, involving the size, shape, and com-

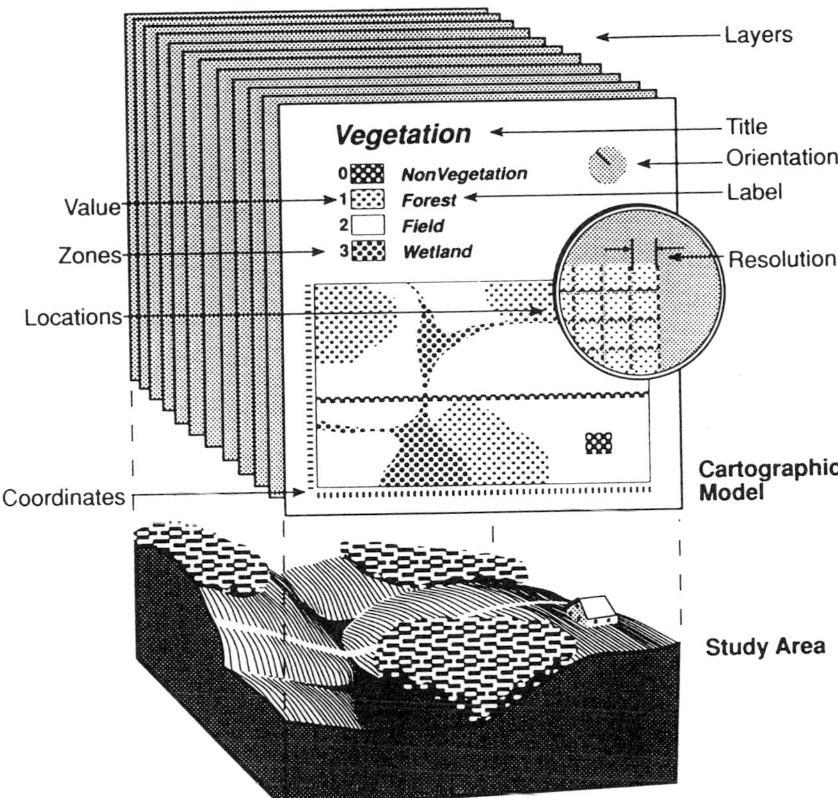


***FIGURE 4*** Geographic information systems provide a system for capturing, storing, checking, manipulating, analyzing, and displaying data that are spatially referenced to the earth. They generally consist of layers of spatially explicit data on various landscape features (e.g., topography, vegetation types, land use types, species distributions, etc.). They can be used as decision support tools and in predictive modeling; for instance, microclimate can be predicted from geographic position and topography. [Reproduced from C. D. Tomlin (1992). *In* "Geographical Information Systems. Principles and Applications." (D. J. Maguire, M. F. Goodchild, and D. W. Rhind, eds.), pp. 361–374. Longman Scientific and Technical, Harlow, England.]

position of different landscape patches and the spatial (and temporal) relations between them. Adequate methods for measuring and assessing its significance have yet to be developed. Landscape heterogeneity is frequently examined at two levels—differences in the heterogeneity of landscape units between landscapes (regional scale) and differences in the heterogeneity within a given landscape unit within landscapes (landscape scale) (Fig. 5). Various indices have been developed for use on both of these scales, some of which are indicated in Fig. 5, and include combinations of estimates of numbers of different landscape units present, areas occupied by different landscape units, and lengths of landscape unit perimeters. Other approaches consider the underlying patterns of plant species richness and variations in evenness. To date, most studies using these indices have investigated their ability to quantify landscape pattern in space and with time. In general, the indices have been considered useful if they have indicated differences between landscapes already known to differ substantially, and there are few instances of indices providing new insights or allowing extrapolation to other situations, or where the relationship between the index and functional aspects of heterogeneity have been explored.

Changing the scale of measurement affects the results obtained. Changing the grain (spatial resolution) and extent (total area of study) affects mea-

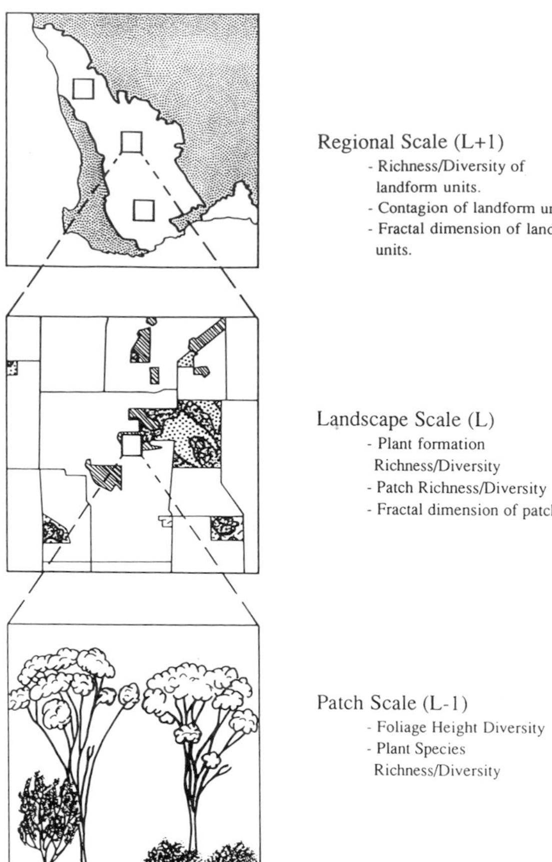


***FIGURE 5*** The major spatial scales important in landscape ecology. Measures of heterogeneity used at each scale are indicated. Processes at one level (L) may be affected by heterogeneity at higher (L + 1) or lower (L-1) levels. Hence when considering the landscape scale, larger scales (e.g., regional) and smaller scales (e.g., patch) have to be considered. [From P. Cale and R. J. Hobbs (1994). *Pac. Conserv. Biol.* **1,** 183–193.]

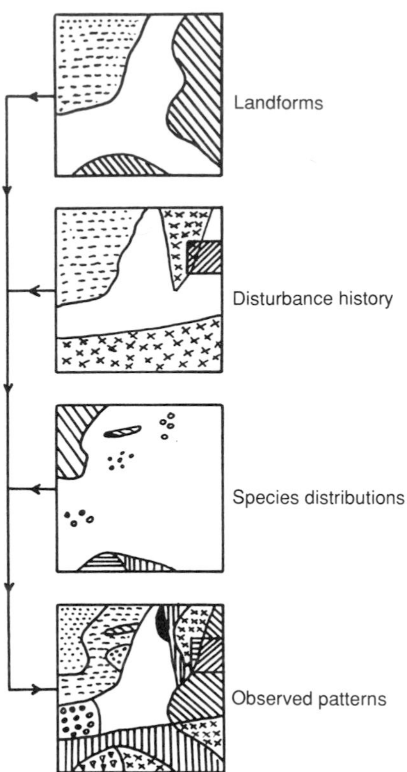


***FIGURE 6*** Observed landscape pattern, consisting of a mosaic of patches of different compositions and structure, results from a complex interplay of the effects of physical features such as landform, topography, and local climate, disturbance regime, and distributions of individual species. [From R. J. Groves and R. J. Hobbs, (1992). *In* "Biodiversity of Mediterranean Ecosystems in Australia" (R. J. Hobbs, ed.), pp. 47–60. Surrey Beatty and Sons, Chipping Norton, New South Wales, Australia.]

sures of heterogeneity because they are sensitive to the number of patches detected, which changes with changes in scale. The spatial configuration of patches influences the rate of change in their number with changing scale.

An adequate description of landscape heterogeneity must include not only a description of the number, sizes, and configurations of patches, but also some characterization of the structure and composition within them. Study of landscape heterogeneity has tended to focus on one of these scales (i.e., configuration of landscape units within the landscape or configuration within landscape units), but seldom both together.

The interpretation of differences in heterogeneity between landscapes is difficult. If one landscape has a higher index than another, this may or may not have any significance when particular processes or functions are considered. The significance of the measured difference in heterogeneity depends on how well measured heterogeneity corresponds to functional heterogeneity. Functional heterogeneity is process-dependent and hence can be defined only in the context of a particular process (e.g., animal movement or water flow). The measures of landscape heterogeneity currently available are frequently difficult to interpret and should be used with caution in both research and management.

### C. Connectivity

Connectivity refers to the degree to which patches of a given type are joined by corridors into a lattice of nodes and links. The term "corridor" has been used to describe a suite of different structures, with different modes of origin and different functions. A corridor can generally be considered to be a linear patch. Almost any strip of vegetation could be viewed as a corridor in some contexts. The important component of the corridor is that it allows movement *from* somewhere *to* somewhere else. Corridors can occur as natural environmental features, such as riparian strips, or can be created by human activities, such as clearing of adjacent vegetation (in which case a remnant corridor results) or modification of the vegetation in a linear strip (as in powerline corridors). Corridors can also be constructed by humans, as in the case of hedgerows and windbreaks. Corridors may have different properties depending on their characteristics. Width is especially important, and a distinction can be drawn between strip corridors, which are wide enough to have an interior that is not dominated by edge effects, and line corridors, in which edge effects permeate the entire structure. Other structures, such as highway underpasses and greenways, have also been called corridors.

In terms of function, corridors may act as important components of a regional conservation system by retaining important species or providing representative examples of native vegetation types that complement those in reserves. They may also serve as faunal habitat, alter landscape fluxes, provide shelter, reduce wind and water erosion, and enhance the aesthetic appeal of a landscape.

### D. Ecotones and Edges

A component of landscape structure that has received increasing attention recently is the ecotone, or edge between adjacent patches. Ecotones can be considered at a variety of scales, ranging from the biome down to the individual patch (Fig. 7). Ecotones are considered important because they represent the boundary between different patches

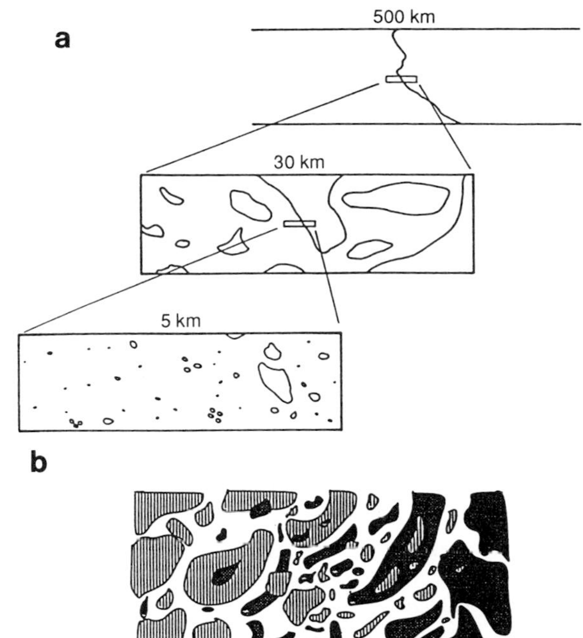


***FIGURE 7*** A hypothetical biome transition zone, showing (a) increased fragmentation of patches at decreasing scales of observation and (b) the mosaic pattern across the ecotone. The pattern changes from relatively large patches in the core area of each biome to small patches at the ecotone. [From J. R. Gosz (1993). *Ecol. Appl.* **3,** 369–376; and J. R. Gosz (1992). *In* "Landscape Boundaries. Consequences for Biotic Diversity and Ecological Flows (A. J. Hansen and F. di Castri, eds.), pp. 55–75. Springer-Verlag, New York.]

through which various landscape flows pass. The ecotone between different biome or vegetation types could be expected to show the first responses to global climatic changes.

Edges have become increasingly important as landscape fragmentation has increased (see Section V,B). Patch size and shape strongly affect the amount of edge that is present (Fig. 8). A set of phenomena known as "edge effects" are associated with edges. These result from physical, chemical, and biotic transfer into patches from adjacent patches. The edges of isolated fragments experience different microclimatic conditions, receive more nutrients transferred from adjacent patches, and may have a higher incidence of weed invasion or predation than the interior of the fragment. Such changes can result in changes in vegetation structure and floristic and faunal species compositions.

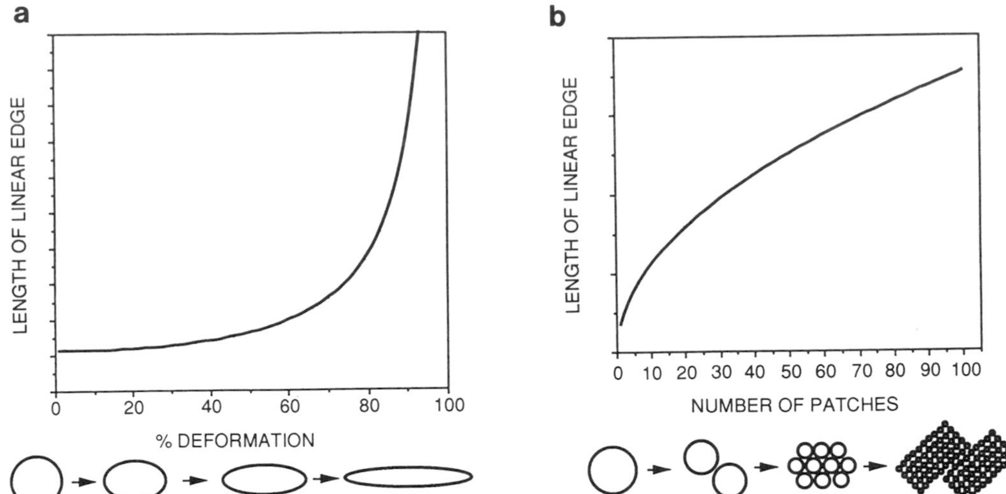


***FIGURE 8*** Graphical representation of the relative amount of edge (measured as a two-dimensional attribute of a habitat patch) in (a) circular versus oval patches with equal areas, where the x axis is the percentage reduction in the shorter axis of the remnant, and (b) an equal area divided into an increasing number of smaller circular patches. [From T. D. Sisk and C. R. Margules (1993). *In* "Nature Conservation 3: The Reconstruction of Fragmented Ecosystems: Global and Regional Perspectives" (D. A. Saunders, R. J. Hobbs, and P. R. Ehrlich, eds.), pp. 57–69. Surrey Beatty and Sons, Chipping Norton, New South Wales, Australia.]

The distance to which edge effects permeate a patch varies with the type of patch and the feature being considered: for instance, microclimatic changes may occur only over tens of meters, whereas changes in predation rates or weed invasion may extend much further.

## IV. LANDSCAPE PROCESSES

### A. Movement of Biota

It is generally accepted that fauna need to move across the landscape for a variety of reasons, including dispersal and resource acquisition, and that movement is required to counter the potential effects of fragmenting faunal populations into small, isolated units. Movement can be thought of as minimizing the impacts of demographic stochasticity and inbreeding depression. Some argue, however, that the requirement for faunal movement may have been overstated, and that corridors may not be required to foster it when it is necessary. Though movement along corridors is frequently assumed to occur, there have been relatively few studies that have shown that corridors are actually required for movement. Studies that have been frequently cited as illustrating corridor use for faunal movement do not, in fact, provide clear evidence. The types of study required to establish unequivocally that corridors are important for faunal movement are difficult and costly to design and implement and require intensive, long-term observations. Few studies provide good data on animal movement, although recent studies of marked or radio-tagged animals have indicated that some species do use corridors for movement in preference to moving across open ground. Other studies have shown that corridors can act in complex ways, enhancing movement of some species in some cases, while inhibiting that of others.

An important function of animal movement is the recolonization of patches that have suffered local extinction. The concept of metapopulations has received increasing attention, especially in the context of fragmented landscapes (see Section V,B). A metapopulation can be regarded as a spatially structured population consisting of distinct subunits, or subpopulations, occupying patches separated by space or barriers and connected by dispersal movements. Subpopulations characteristically undergo periodic extinctions and reestablish following recolonization by dispersers from other elements of the metapopulation. Metapopulation theory deals with the factors that determine the likelihood of local extinctions and subsequent recolonization. In general, subpopulations in small, isolated patches are considered more likely to go extinct, whereas large, well-connected patches are more likely to be recolonized, resulting in different probabilities of a species occurring on patches of different sizes and connectivity (Fig. 9). Clearly, the validity these relationships depends on the extent to which our perceptions of the influence of size and connectivity match the actual influence of these parameters on the species involved.

### B. Fluxes: Water, Nutrients, Material

Fluxes of water, nutrients, and material are often important determinants of landscape patterning and can also be strongly affected by that patterning. Surface and subsurface hydrology is determined by geology, landform, and surface characteristics. Patch characteristics and configurations can influence interception, infiltration, evapotranspiration, and runoff patterns. Thus, for instance, perennial vegetation in semiarid areas can intercept and utilize more water than annual cropland, with the result that runoff and water input to groundwater will be greater under an annual crop. Distribution of water across the landscape clearly influences the distribution of patch types, for instance, riparian forests and swamps can develop only where there are large amounts of water available close to the surface. [*See* ECOLOGICAL ENERGETICS OF ECOSYSTEMS.]

Nutrient and material fluxes across the landscape result primarily from the processes of erosion by wind and water. Redistribution of nutrients over long time periods has resulted in areas of the landscape accumulating nutrients that have been eroded from other areas. This in turn exerts a strong influence on the types of patch present in each area. Resource-rich patches may be particularly important from many points of view, in terms of both vegetation composition and faunal assemblages and also potential human utilization. Landscape patterning can influence erosional processes, with some patch types acting as interceptors of eroded material. For instance, in managed landscapes, windbreaks act both to reduce the degree of wind erosion and to intercept eroded material. Riparian strips also intercept eroded material and nutrients transported in runoff, and hence affect nutrient and material loadings in waterways. Riparian strips are frequently viewed as "buffer zones," and this concept can be extended to any type of patch that protects adjacent patches from nutrient or material inputs. [*See* NUTRIENT CYCLING IN FORESTS.]

Nutrients can be transferred across the landscape in a variety of other ways. Particularly important vectors are animals, which may feed in one patch and defecate in another, resulting in a transfer of nutrients between patches. In extreme cases, this can significantly increase the nutrient input into recipient patches. Fire also acts to redistribute nutrients in smoke and ash, although it is difficult to quantify this effect.

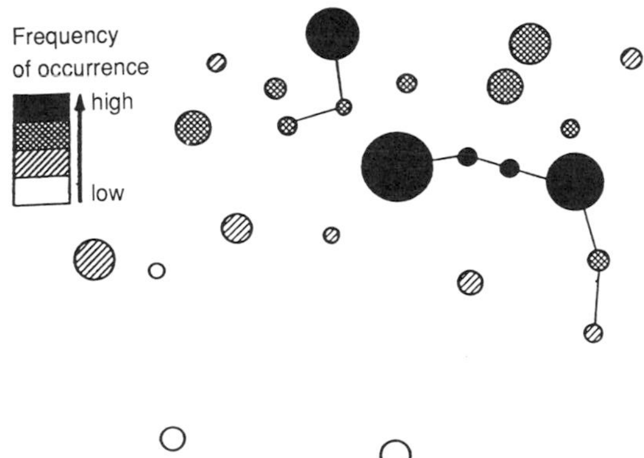


***FIGURE 9*** Expected pattern of occurrence in patches differing in size and isolation for a species with a recolonization rate too small to compensate for local extinction immediately. Large, well-connected patches have a higher probability of occurrence than small, isolated patches. [From P. Opdam (1991). *Landscape Ecol.* **5,** 93–106.]

## V. LANDSCAPE CHANGE

### A. Natural Change

Natural landscapes are in a constant state of flux. Patch composition and configuration change in response to a variety of processes (Fig. 10). Of pri-

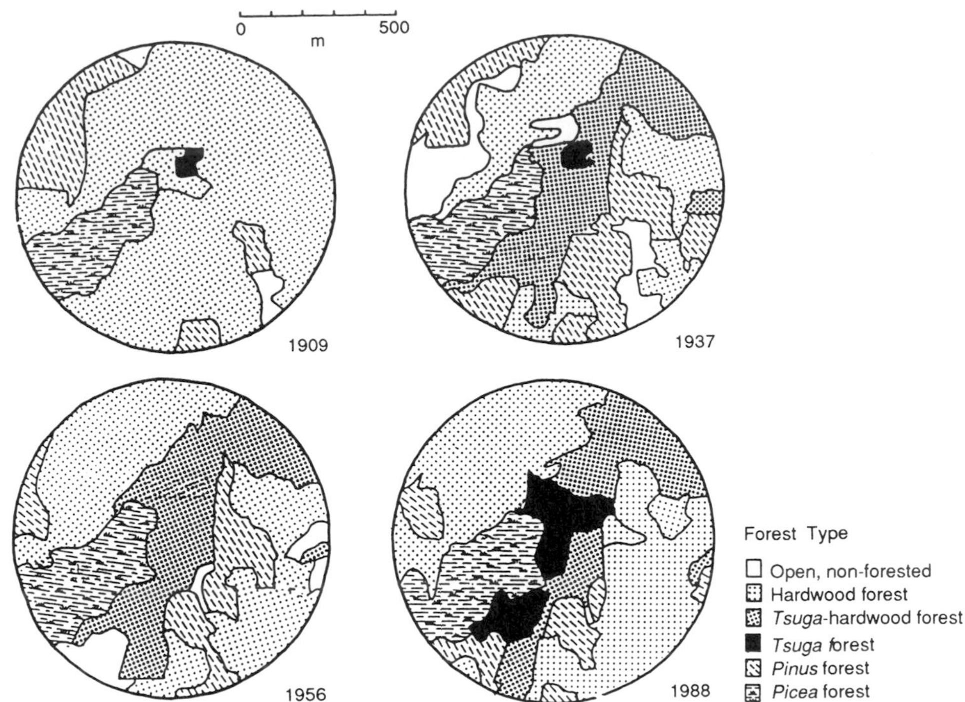


***FIGURE 10*** Maps depicting forest types within a 1-km-diameter circular area in Harvard Forest, Massachusetts, over the period 1909–1988, showing changes in both patch composition and landscape structure and heterogeneity. Changes in patch type result from succession following abandonment of agricultural land and from response to hurricane damage and other disturbances. [From D. L. Foster, T. Zebryk, P. Schoonmaker, and A. Lezberg (1993). *J. Ecol.* **80,** 773–786.]

mary importance are climatic changes and disturbance regimes. Climate varies greatly over a number of different time scales, ranging from the long-term changes that take place over thousands of years in response to glacial/interglacial cycles to relatively short-term phenomena such as drought cycles over periods of decades. A considerable body of evidence shows that natural vegetation responds to long-term climatic changes by migrating across the landscape. However, individual species migrate at different rates and, at any given time, the composition of the landscape depends on which species are already present, which species are migrating in, and how these two sets of species interact. Past landscapes in any particular area are liable to have been quite different from those present today.

Infrequent episodic climatic events are particularly important agents of landscape change. Events such as droughts, exceptionally high rainfalls, or windstorms are capable of switching patches from one type to another very quickly. For instance, a grassland patch can become a shrubland patch following a year of exceptionally high rainfall. In this respect, episodic events can be considered as particular types of disturbance. Any form of disturbance, such as fire, flood, storm, or landslide, can also act in this way. As discussed earlier, the distribution of patches across the landscape is determined by the overall disturbance regime.

Landscape change may also be biotically forced, as in plant species migrations, but the effects of animals and pathogens should also be noted. Changes in abundances of herbivores can lead to changes in patch types and configurations, and pathogen spread can significantly alter landscape composition and structure. [*See* BIOLOGICAL CONTROL.]

There has been considerable debate over the idea that, although individual landscape patches are in a constant state of flux, over a large enough area the landscape as a whole is in a state of equilibrium. In other words, over the entire landscape, the distri-

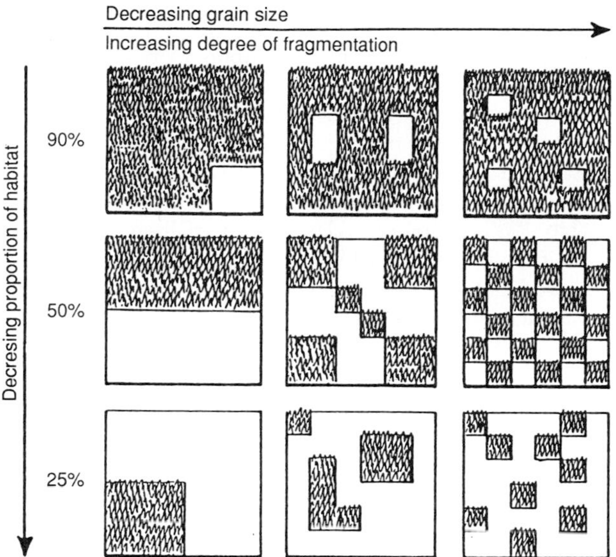


***FIGURE 11*** The effects of human land exploitation have two components in the spatial configuration of two patch types such as native vegetation (shaded) and agriculture (open), namely, habitat loss and habitat fragmentation. Thus, landscapes may have or be managed for a coarse- or fine-grained structure. [From P. Angelstam (1992). *In* "Ecological Principles of Nature Conservation. Applications in Temperate and Boreal Environments" (L. Hansson, ed.), pp. 9–70. Elsevier, London.]

case a certain minimum area is required, known as the "minimum dynamic area." The validity of the assumption of equilibrium landscapes depends greatly on the scale considered, and recent writers have questioned its generality. Examination of areas where large tracts of natural ecosystems persist indicates a distribution of patch types and ages that is far from equilibrium. [*See* EQUILIBRIUM AND NONEQUILIBRIUM CONCEPTS IN ECOLOGICAL MODELS.]

### B. Human-Induced Change

Human activities superimpose another set of factors that affect landscape structure and functioning. Clearance for agriculture, harvesting of timber and fauna, grazing of livestock, and introduction of exotic plants, animals, and pathogens all have dramatic effects. Many of these simply extend the kind of impacts already discussed. For instance, introduced livestock change patch compositions, alter nutrient flows, and alter disturbance regimes. Introduced plants can also alter patch composition and structure, and in some cases significantly alter landscape patterning. [*See* INTRODUCED SPECIES.]

An important effect of human activities is the increasing fragmentation of natural ecosystems, leading to the creation of small remnant patches with varying degrees of connectivity. Fragmentation results from agricultural development and forestry clear felling. As the degree of fragmentation increases, both the proportion of native vegetation in the landscape and the grain size decrease (Fig. 11). This results in reduced areas of natural ecosystems, loss of species, and significant changes in landscape fluxes. Species losses result first from loss of habitat and then, as more habitat disappears, from metapopulation effects, whereby subpopulations on increasingly small and isolated patches become increasingly susceptible to local extinction, as discussed earlier (Fig. 12). [*See* DEFORESTATION.]

## VI. RELEVANCE TO MANAGEMENT AND CONSERVATION

In the past, most land and conservation management has been directed at individual parcels of land,

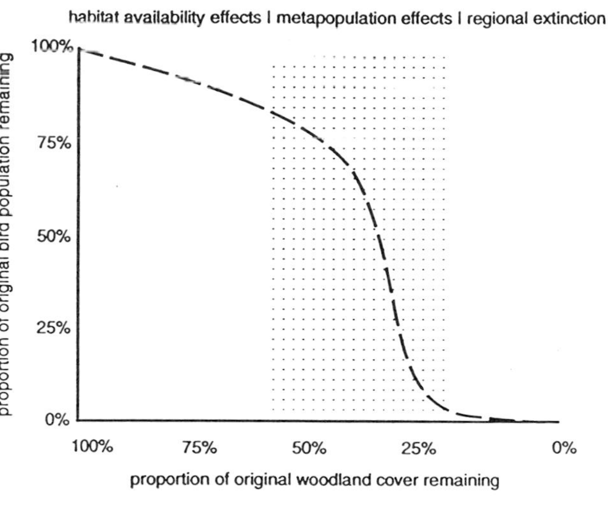


***FIGURE 12*** A theoretical model of the respective roles of habitat availability and metapopulation effects in regional declines of bird species. [From G. Fry and A. R. Main (1993). *In* "Nature Conservation 3: The Reconstruction of Fragmented Ecosystems: Global and Regional Perspectives" (D. A. Saunders, R. J. Hobbs, and P. R. Ehrlich, eds.), pp. 225–241. Surrey Beatty and Sons, Chipping Norton, New South Wales, Australia.]

that is, individual landscape patches. It is becoming increasingly recognized that the appropriate scale for many land use planning and management decisions is that of the landscape. Understanding of the interrelations between different parts of the landscape is growing, as is the realization that activities in one patch are liable to have impacts on adjacent patches. This has profound implications for conservation in particular, where the emphasis has been on the acquisition and management of individual protected areas. No reserve exists in isolation from the surrounding landscape, and in the long term the impacts of events in that landscape matrix are likely to be as great if not greater than those of events within the reserve itself. The recognition of the importance of metapopulation dynamics and the potential role of connecting corridors also has profound importance for conservation management. Reserves must now be viewed as part of a network representing more than the simple sum of its individual parts. Similarly, the potential importance of buffer zones is now being recognized and incorporated as part of the biosphere reserve concept.

Management of landscapes rather than individual landscape patches is not a simple matter, and adequate tools and approaches are still in the early stages of development. However, the use of such approaches is essential if sustainable production and long-term conservation of biodiversity are to be achieved.

## Glossary

**Connectivity** Degree to which patches of a given type are joined by corridors into a lattice of nodes and links.

**Corridor** Linear patch that differs from the surrounding vegetation (or land use) matrix and connects at least two nonlinear patches that were connected in historical time.

**Fragmentation** Process whereby natural vegetation is reduced in area and restricted to remnants or fragments of varying sizes and degrees of isolation from one another, usually as a result of clearance of land for agricultural or forestry production.

**Landscape** Area of land, at the scale of hectares to square kilometers, that consists of a collection of different but interacting patches.

**Metapopulation** Spatially structured population consisting of distinct subunits, or subpopulations, separated by space or barriers (usually in fragmented landscapes) and connected by dispersal movements. Subpopulations characteristically undergo periodic extinctions and reestablish following recolonization by dispersers from other elements of the metapopulation.

**Patch** Basic ecological, relatively internally homogeneous unit, collections of which make up a landscape. Patches may be distinct ecosystems, vegetation types, successional stages, or areas of different land uses.

## Bibliography


Forman, R. T. T., and Godron, M. (1986). "Landscape Ecology." Chichester, England: Wiley & Sons.

Hansen, A. J., and di Castri, F., eds. (1992). "Landscape Boundaries. Consequences for Biotic Diversity and Ecological Flows." New York: Springer-Verlag.

Hobbs, R. J., and Saunders, D. A., eds. (1993). "Reintegrating Fragmented Landscapes: Towards Sustainable Production and Nature Conservation." New York: Springer-Verlag.

Hudson, W., ed. (1991). "Landscape Linkages and Biodiversity." Washington, D.C.: Island Press.

Naiman, R. J., and Décamps, H., eds. (1990). "Ecology and management of Aquatic–terrestrial Ecotones." Paris: UNESCO, and Cornforth, UK.: Parthenon.

Naveh, Z., and Lieberman, A. S. (1990). "Landscape Ecology. Theory and Application," student edition. New York: Springer-Verlag.

Saunders, D. A., and Hobbs, R. J., eds. (1991). "Nature Conservation 2: The Role of Corridors." Chipping Norton, New South Wales, Australia: Surrey Beatty and Sons.

Saunders, D. A., Hobbs, R. J., and Ehrlich, P. R., eds. (1993). "Nature Conservation 3: The Reconstruction of Fragmented Ecosystems: Global and Regional Perspectives." Chipping Norton, New South Wales, Australia: Surrey Beatty and Sons.

Turner, M. G., ed. (1987). "Landscape Heterogeneity and Disturbance." New York: Springer-Verlag.

Turner, M. G., and Gardner, R. H., eds. (1991). "Quantitative Methods in Landscape Ecology." New York: Springer-Verlag.

Vos, C. C., and Opdam, P. eds. (1993). "Landscape Ecology of a Stressed Environment. IALE Studies in Landscape Ecology 1." London: Chapman and Hall.

Zonneveld, I. S., and Forman, R. T. T., eds. (1990). "Changing Landscapes. An Ecological Perspective." New York: Springer-Verlag.

# Life Histories

**Henry M. Wilbur**
*University of Virginia*



The life history of an organism is a detailed description of events in its life cycle and how the ecological relationships of the individual change as it passes through each stage in its life cycle. These events determine how the individual's expectation of survival and reproductive potential shift as it passes from one life history stage to another.

## I. INTRODUCTION

Life histories can range from the rather simple life cycle of a unicellular organism, such as a bacterium, alga, or protozoan that grows and then divides into two identical organisms, to the complex life cycles with several morphologically discrete and ecologically distinct stages of many parasites. Most organisms with complex life histories have one stage adapted to dispersal and another stage adapted to growth, development, and reproduction.

## II. SEEDS AND EGGS

Many eggs, spores, and seeds are designed for passive dispersal through space or time. Some of these propagules are able to survive long periods of dormancy. Many spores and seeds and a few types of eggs, for example, are dispersed long distances by wind or water currents. Mechanisms for dispersal through space and time are adaptive in changing environments in which the conditions favoring growth and reproduction are transient. Classic examples of changing environments are ephemeral pools, disturbed ground, light gaps in the forest, and hosts for parasites. Seeds of the common mullein (*Verbascum thapsus:* Scrophulariaceae), a weedy plant commonly seen along roadsides and railroad embankments, can remain dormant in the soil for a century until a disturbance to the soil brings them to the surface where they germinate.

Dispersal may be assisted by structures, such as wings on seeds, that enhance the effect of the wind. Many seeds are adapted for dispersal by animals. Examples are fruits that are eaten by birds and mammals, which carry the seeds away from the parent plant. Other seeds hitch rides on passing animals by means of hooks or barbs.

The mother provides the dispersal for eggs in many animals by locating habitats that hold a promise of opportunities for growth and survival. Female flies, for example, locate dung or rotting fruits and carcasses in which to oviposit. Many parasitic insects spend most of their adult life searching for hosts on which they can deposit eggs or larvae. Other examples are herbivorous insects that search for new plants on which they can oviposit and frogs that search for recently created pools where they can lay eggs free of competitors and predators.

## III. THE TIMING OF REPRODUCTION

Many organisms reproduce only once and then die. These *semelparous* organisms include annual and biennial plants as well as some long-lived yet still monocarpic plants, such as bamboos and the so-called century plant (*Agave*) of the American southwest. Not all "annual" plants have a 1-year life cycle because many have seeds that can persist in the soil for decades until conditions are propitious for germination and reproduction. Other *iteroparous* species reproduce more than once. Many turtles, for example, have reproductive lives spanning tens of years without evidence of senescence. Trees may have reproductive lives of more than a thousand years.

Individuals partition resources among growth, storage, survival, and reproduction differently throughout their life cycle. In the juvenile stage, all resources are committed to growth and maintenance. The timing of maturation is presumed to be determined by the optimal time to shift resources from these two purposes to reproduction as energy and nutrients devoted to current reproduction may detract from growth and future reproductive potential. Semelparous organisms tend to devote all available resources to their current reproduction, whereas iteroparous organisms devote resources to growth, storage, and maintenance as well as reproduction. The timing of maturity can be determined by strictly genetic mechanisms or it can be a sensitive response to environmental conditions. The effect of competition, unfavorable weather, or pollution may fall directly on growth rates, which indirectly affect the timing of reproduction.

Another aspect of allocation between growth, storage, and reproduction is the timing of maturity. In many metazoans there is a positive relationship between age and reproductive success. Larger females are able to produce more, and often larger, seeds or eggs. Older females of many vertebrates are generally more successful than younger mothers because of size, social status, or experience. Older males are also more successful parents because they are larger or more experienced. There is a tradeoff between the advantages of early reproduction and the advantages of delayed maturity. Early reproduction is advantageous because of the risk of mortality to the individual and the demographic benefit of reproducing early that is analogous to an early deposit of money in a account accruing interest. The advantage of delayed reproduction is a larger body size and hence fecundity and perhaps a higher social standing that ensures higher survival for parents and offspring.

The decision to begin reproduction incurs a cost to future growth and survival that is presumed to depend on the amount of resources committed to each bout of reproduction. This current reproductive investment can be allocated by a female to a few large seeds (or eggs) or many small ones. In plants, the balance between seed size and seed number seems to depend on the ecology of seedling establishment. Plants that live in competitive environments, such as forests where light and nutrients are limiting, tend to produce large seeds. Weedy plants that favor high dispersal with a sweepstakes to find open environments with low competition tend to produce many small seeds. An extreme case is the orchid that produces tens of thousands of very small seeds, which only germinate when they are in association with mycorhizal fungi on which they rely for water and nutrients.

The age of last reproduction is generally determined by survival rates rather than a breakdown of the reproductive machinery. That is, most individuals are killed before they wear out. Senescence occurs in a few organisms, especially those with prolonged periods of parental care, such as humans and elephants. Senescence as an adaptation is of great interest to evolutionary biologists, who suggest that older parents may be of greater value as grandparents or leaders in a social group than as parents because of their greater cultural wisdom and falling life expectancy. Senescence may also be an indirect result of selection on the timing of the expression of deleterious traits. If selection acts to postpone the expression of deleterious traits then they may accumulate and be expressed later in life, when the reproductive value of an individual is low.

Widely differing reproductive schedules exist in males and females in species with separate sexes depending, in part, on the breeding system and the differing demands of reproduction on males and females. Males may mature earlier and at a smaller size than females in species with a positive relationship between fecundity and body size in females, but not males. In highly polygynous breeding systems, such as in elephant seals, only a few males in a population dominate the mating system. Males have very little chance of mating successfully when small so they delay maturity and invest in increasing their body size. [*See* SEX RATIO.]

Plants with perfect flowers can invest in both male function (pollen production) and female function (seed production). These investments can be done sequentially (protandry when male function is first; protogeny when female function is first) or simultaneously. The consequence of different male and female investments to individual success and the genetic structure of populations is an area of much study in ecological genetics and population ecology.

## IV. LIFE CYCLE GRAPHS

The life history of an individual can be represented in several ways. The life cycle graph of the mole salamander, for example, is a diagrammatic representation of discrete life history stages in this pond-breeding amphibian. All individuals in some populations appear to have the potential to reproduce in either an aquatic paedomorphic form or a terrestrial metamorphic form. Females generally mature at age one if they take the paedomorphic route and at age two if they metamorphose. Males, on the other hand, nearly always mature at age one if they take either route. This diagram of the possible pathways can be developed into a quantitative model by determining the probabilities that an individual will take a particular path. The arrows leading from each stage plus the probability of death in that stage must all sum to unity. The life history model is made complete by adding arrows to indicate fecundity of each type of female.

## V. LIFE TABLES

The life cycle graph is a diagram that captures the major features of a life history. This device focuses on the identification of discrete life history stages and the transitions between them. These transitions may involve changes in morphology, behavior, physiology, and habitat use or they may merely reflect changing probabilities of survival and reproduction. More detailed models include descriptions of the probability of survival and the expected fecundity for every year—or other appropriate time units—in the organism's potential life span. These *life tables* giving the age-specific probability of survival are the same as the actuarial tables used by insurance companies to determine age- and sex-specific rates for life insurance. These life tables can be paired with age- and sex-specific fecundity schedules that summarize the expected reproductive success of individuals in each class. The summation of an individual's age-specific survival probabilities multiplied by the corresponding age-specific fecundity yields the expected lifetime reproductive output of an individual within the population for which the rates were estimated. If this *net reproductive rate* is unity then the population will be stable. If the rate is less than unity then the population is predicted to decline. If the rate is greater than unity, the population is predicted to increase. These predictions make an assumption that the age structure in the population is stable and that the environment is not changing.

In many organisms with indeterminate growth, such as many plants, some sessile marine invertebrates, and many lower vertebrates, survival and fecundity are more closely linked to body size than to age. Models of life histories of these organisms are based on size instead of age and the dynamics of the size–structure of the population can be predicted.

A knowledge of the life history of an organism is essential to assaying the impact of environmental changes on biodiversity and ecosystem functioning. The growth rate and reproductive performance of individuals may be more sensitive indicators than estimates of mortality rates or population density. Monitoring changes in fecundity or body size distributions may also be more easily implemented than assays of mortality rates and population densities.

## Glossary

**Iteroparity** Breeding more than once.

**Monocarpic** Plants that set seed only once and then die.

**Paedomorphic** Retention of larval characters in sexually mature individuals.

**Protandry** Condition in which individuals function as male and then as female.

**Protogeny** Condition in which individuals function as female and then as male.

**Semelparity** Breeding only once and then dying (e.g., annual and biennial plants).

## Bibliography

Calder, W. A., III (1984). "Size, Function, and Life History." Cambridge: Harvard Univ. Press.

Caswell, H. (1989). "Matrix Population Models." Sunderland, MA: Sinauer.

Caughley, G. (1977). "Analysis of Vertebrate Populations." New York: Wiley.

Hutchinson, G. E. (1978). "An Introduction to Population Ecology." New Haven: Yale Univ. Press.

Peters, R. H. (1983). "The Ecological Implications of Body Size." Cambridge: Cambridge Studies in Ecology, Cambridge University Press.

Williams, G. C. (1966). "Adaptation and Natural Selection." Princeton: Princeton Univ. Press.

# Limnology, Inland Aquatic Ecosystems

**Robert G. Wetzel**
*University of Alabama*



Limnology is the study of inland waters—lakes (both freshwater and saline), reservoirs, rivers, streams, wetlands, and groundwater—as ecologically interactive systems. Limnology integrates the functional relationships, growth, adaptation, species composition, nutrient cycles, and biological productivity among inland aquatic communities and ecosystems. The discipline then evaluates how these relationships are regulated by their physical, chemical, and biotic environment.

## I. INTRODUCTION

In the global hydrological cycle, water is conducted continuously from the sea to the atmosphere, to the land, and back to the sea. Almost all the water on the earth's surface is saline in oceans (97.61%), and most of the remainder (2.08%) is polar and glacial ice (Table I). The remainder is mostly groundwater (0.29%) and only 0.009% exists in freshwater lakes. Lakes and groundwaters are temporary storage reservoirs on land with renewal times much shorter than those of the oceans (Table I). Lakes and groundwaters provide the sources from which most of the human population obtains water for agricultural, industrial, and domestic uses. The volume of water flowing from these reservoirs to the sea in rivers is very small (0.00009%) with a mean residence time of about 2 weeks. During high-flow periods, however, surface water often recharges adjacent groundwater aquifers. Return flows from groundwater to rivers occur during periods of low flow or drought, and maintain a base flow in river channels.

Changes in water storage and retention times in lakes are caused by alterations in the balance between inputs from all sources and water losses. Lakes receive water from precipitation directly on the surface, from surface inflows from the drainage basin, and from subsurface groundwater seepage. Lakes lose water by flow from an outlet (drainage lakes), by seepage through the basin walls into the groundwater (seepage lakes), by evaporation, and by evapotranspiration from higher aquatic plants. Saline lakes occur in *closed* basins with no outflow except by evaporation.

Some 40% of the total volume of fresh water is contained in the great lake basins of Siberia, North America, and eastern Africa. Most lakes and reservoirs are very much smaller, however. Lakes are concentrated in the subarctic and temperate regions of the Northern Hemisphere, and reservoirs in the subtemperate and subtropical zones. Most of the millions of lakes and reservoirs are small and relatively shallow, usually less than 15 m in depth (Fig.

**TABLE I**
Water in the Biosphere[a]

| | Volume (thousands of km$^3$) | Percentage | Renewal time |
|---|---|---|---|
| Oceans | 1,370,000 | 97.61 | 3100 years[b] |
| Polar ice, glaciers | 29,000 | 2.08 | 16,000 years |
| Groundwater (actively exchanged)[c] | 4000 | 0.29 | 300 years |
| Freshwater lakes | 125 | 0.009 | 1–100 years[d] |
| Saline lakes | 104 | 0.008 | 10–1000 years[d] |
| Soil and subsoil moisture | 67 | 0.005 | 280 days |
| Rivers | 1.2 | 0.00009 | 12–20 days[e] |
| Atmospheric water vapor | 14 | 0.0009 | 9 days |

[a] From R. G. Wetzel (1983). "Limnology," 2nd ed. Saunders, Philadelphia.
[b] Based on net evaporation from the oceans.
[c] Only the volume of the upper, actively exchanged groundwater is included here.
[d] Renewal times for lakes vary directly with volume and mean depth, and inversely with the rate of discharge. The absolute range for saline lakes is from days to thousands of years.
[e] Twelve days for rivers with relatively small catchment areas of less than 100,000 km$^2$; 20 days for major rivers that drain directly to the sea.

1). As a result, much, often over half, of the sediment occurs within the photic zone and is potentially capable of supporting photosynthesis.

## II. ECOSYSTEM STRUCTURE

The productivity and internal metabolism of aquatic ecosystems are driven and controlled by energy acquired by photosynthesis from solar radiation. Inland waters receive organic products of photosynthesis directly from their aquatic flora and indirectly from their drainage basins as particulate and dissolved organic matter from terrestrial and wetland plants imported by stream water and storm runoff. In addition, solar radiation is absorbed and dissipated as heat. The resulting density stratification affects the thermal structure, water mass stratification, and hydrodynamics of lakes and reservoirs. Heat is distributed and altered by the physical work of wind energy, currents and other water movements, basin morphometry, and water losses. Resulting patterns of density-induced stratification influence physical and chemical properties

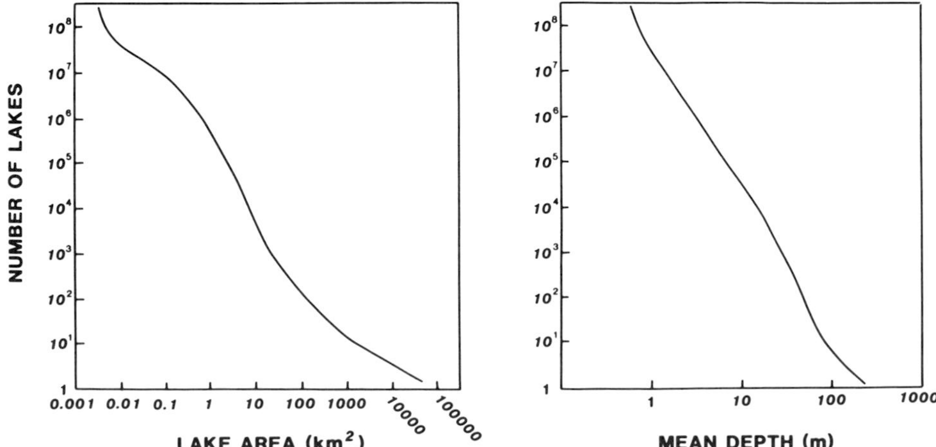


**FIGURE 1** Approximate number of lakes of the world in relation to lake area and approximate mean water depth. [After R. G. Wetzel (1990). *Verh. Int. Verein. Limnol.* **24,** 6–24.]

and cycles of lakes. These characteristics structure the aquatic habitats and have marked attendant effects on all chemical cycles, metabolic rates, and population dynamics and their productivities.

The most abundant chemical species dissolved in inland waters are usually four major cations ($Ca^{2+}$, $Mg^{2+}$, $Na^{+}$, $K^{+}$) and four anions ($HCO_3^-$, $CO_3^{2-}$, $SO_4^{2-}$, $Cl^-$). These species comprise the salinity. The salinity of fresh waters has a world average concentration of about 120 mg/liter, but varies among continents and with the lithology of land masses (Table II).

Of the major constituents of cellular protoplasm of organisms (C, H, O, N, P, and S), the biogeochemical cycles control the availability of phosphorus, nitrogen, and several minor nutrients, and these elements often limit biotic development. Phosphorus, in comparison to other macronutrients required by biota, is least abundant and commonly the first element to limit biological productivity. When phosphorus is in adequate supply, nitrogen availability invariably limits productivity (Table III). Standing inland waters have been categorized into various trophic scales on the basis of ranges of major nutrients and algal productivity (Table III). [*See* BIOGEOCHEMICAL CYCLES.]

Among biological communities, functionally similar organisms can be grouped into *trophic levels* based on similarities in patterns of food production and consumption. Energy is transferred and nutrients are cycled within an overall ecosystem trophic structure. The productivity of each trophic level is the rate at which energy enters the trophic level from the next lower level. Because organisms expend considerable energy for maintenance and because death of an organism routes much energy and nutrients into the detrital pool, only a portion of the energy of one trophic level is available for transfer and use by higher trophic levels. Available energy decreases progressively at higher trophic levels, so that rarely can more than five or six trophic levels be supported. The efficiency of energy transfer from one level to the next is low (5 to 15%), and often decreases as trophic level increases.

Organic matter is either soluble or particulate, and most remains in the water. Oxygen, however, is very insoluble in water, and most diffuses to the atmosphere. When organic matter in lakes is respired by heterotrophic organisms, the amount of organic matter to be oxidized can be much larger than the oxygen available for oxidation. In large lakes, the volume of water with sufficient light to support photosynthesis is small relative to the total volume of water containing dissolved oxygen for respiration. In small lakes, however, the proportion of water supporting photosynthesis is large, and respiration can exhaust oxygen dissolved in the lake water. The exhaustion of dissolved oxygen in large portions of lake ecosystems is exacerbated by the process of eutrophication. Eutrophication is the increased photosynthetic production of organic matter in response to excessive loading of nutrients, particularly phosphorus.

## III. DIVERSITY OF HABITATS

Aquatic ecosystems consist of entire drainage basins. The nutrient and organic matter content of drainage water from the catchment area is modified in each of the terrestrial, stream, and wetland–littoral components, as water moves downgradient to and within the lake or reservoir per se (Fig. 2). Autotrophic photosynthetic productivity is generally low to intermediate in the terrestrial components, highest in the wetland interface region between the land and water, and lowest in the open water (Fig. 2). Similarly in the gradient from land to river channels, the greatest productivity is in the marginal floodplain regions. Autotrophic productivity in river channels is generally low, as in the pelagic regions of lakes. Most of the organic matter of running waters is imported from floodplain and terrestrial sources. [*See* RIVER ECOLOGY.]

The land-water interface region of aquatic ecosystems is always the most productive per unit area along the gradient from land to open water of both lakes and reservoirs. Because most aquatic ecosystems occur in geomorphologically mature terrain of gentle slopes and are small and shallow, the wetland–littoral components usually dominate in productivity and the synthesis of organic matter. The region of greatest productivity is the emergent

**TABLE II**

Mean Composition of Surface Waters of the World (mg/liter)[a]

| | $Ca^{2+}$ | $Mg^{2+}$ | $Na^{+}$ | $K^{+}$ | $CO_3^{2-}$ ($HCO_3^-$) | $SO_4^{2-}$ | $Cl^-$ | $NO_3^-$ | Fe (as $Fe_2O_3$) | $SiO_2$ | Sum |
|---|---|---|---|---|---|---|---|---|---|---|---|
| North America | 21.0 | 5. | 9. | 1.4 | 68. | 20. | 8. | 1. | 0.16 | 9. | 142 |
| South America | 7.2 | 1.5 | 4. | 2. | 31. | 4.8 | 4.9 | 0.7 | 1.4 | 11.9 | 69 |
| Europe | 31.1 | 5.6 | 5.4 | 1.7 | 95. | 24. | 6.9 | 3.7 | 0.8 | 7.5 | 182 |
| Asia | 18.4 | 5.6 | 5.5 | 3.8 | 79. | 8.4 | 8.7 | 0.7 | 0.01 | 11.7 | 142 |
| Africa | 12.5 | 3.8 | 11. | — | 43. | 13.5 | 12.1 | 0.8 | 1.3 | 23.2 | 121 |
| Australia | 3.9[b] | 2.7 | 2.9 | 1.4 | 31.6 | 2.6 | 10. | 0.05 | 0.3 | 3.9 | 59 |
| World | 15. | 4.1 | 6.3 | 2.3 | 58.4 | 11.2 | 7.8 | 1. | 0.67 | 13.1 | 120 |
| Cations (meq) | 0.750 | 0.342 | 0.274 | 0.059 | — | — | — | — | — | — | 1.425 |
| Anions | — | — | — | — | 0.958 | 0.233 | 0.220 | 0.017 | — | — | 1.428 |

[a] From numerous sources cited in R. G. Wetzel (1983). "Limnology," 2nd ed. Saunders, Philadelphia.
[b] Values of calcium are likely less, on the average, than Na and Mg in Australian surface waters.

## *TABLE III*

General Ranges of Photosynthetic Productivity of Phytoplankton and Related Characteristics of Lakes of Different Trophic Categories

| Trophic type | Mean primary productivity (mg C $m^{-2}$ $day^{-1}$) | Phytoplankton density ($cm^3$ $m^{-3}$) | Phytoplankton biomass (mg C $m^{-3}$) | Chlorophyll *a* (mg $m^{-3}$) | Dominant phytoplankton | Light extinction coefficients ($\eta$ $m^{-1}$) | Total organic carbon (mg $liter^{-1}$) | Total P ($\mu$g $liter^{-1}$) | Total N ($\mu$g $liter^{-1}$) | Total inorganic solids (mg $liter^{-1}$) |
|---|---|---|---|---|---|---|---|---|---|---|
| Ultraoligotrophic | <50 | <1 | <50 | 0.01–0.5 | | 0.03–0.8 | | <1–5 | <1–250 | 2–15 |
| Oligotrophic | 50–300 | | 20–100 | 0.3–3 | Chrysophyceae, Cryptophyceae, | 0.05–1.0 | <1–3 | | | |
| Oligomesotrophic | | 1–3 | | | Dinophyceae, Bacillariophyceae | | | 5–10 | 250–600 | 10–200 |
| Mesotrophic | 250–1000 | | 100–300 | 2–15 | | 0.1–2.0 | <1–5 | | | |
| Mesoeutrophic | | 3–5 | | | | | | 10–30 | 500–1100 | 100–500 |
| Eutrophic | >1000 | | >300 | 10–500 | Bacillariophyceae, Cyanophyceae, | 0.5–4.0 | 5–30 | | | |
| Hypereutrophic | | >10 | | | Chlorophyceae, Euglenophyceae | | | 30–>5000 | 500–>15000 | 400–60000 |
| Dystrophic | <50–500 | | <50–200 | 0.1–10 | | 1.0–4.0 | 3–30 | <1–10 | <1–500 | 5–200 |

[a] From R. G. Wetzel (1983). "Limnology," 2nd ed. Saunders, Philadelphia.

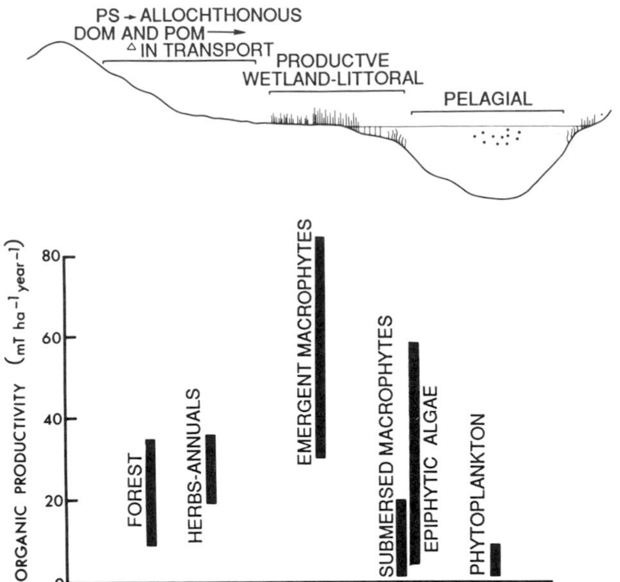


***FIGURE 2*** The lake ecosystem showing the drainage basin with terrestrial photosynthesis (PS) of organic matter, movement of nutrients and dissolved (DOM) and particulate (POM) organic matter in surface water and groundwater flows toward the lake basin, and chemical and biotic alteration of these materials en route, particularly as they pass through the highly productive and metabolically active wetland–littoral zone of the lake per se. [After R. G. Wetzel (1990). *Verh. Int. Verein. Limnol.* **24,** 6–24.]

macrophyte zone. Emergent aquatic plants have a number of structural and physiological adaptations that not only tolerate the hostile reducing anaerobic sediments but exploit the high nutrient and water availability of this habitat. Nutrients entering the zone of emergent macrophytes tend to be assimilated by microflora of the sediments and detritus particles, and are then recycled to the emergent macrophytes. Export from the emergent zone is dominated by dissolved organic compounds released from decomposition of plant detrital material. [*See* AQUATIC WEEDS.]

Submersed macrophytes are limited physiologically by slow diffusion rates of gases and nutrients in water and by reduced availability of light underwater. Internal recycling of resources, both of the gases ($CO_2$, $O_2$) of metabolism and of nutrients, are important to the abilities of submersed plants to function and grow as well as they do under light and gas limitations.

The second most productive component of the wetland–littoral community is the microflora attached to aquatic plants and other surfaces. The surfaces provided by aquatic plants in lakes and rivers can be very large, often exceeding 25 $m^2/m^2$ of bottom. The physiology and growth of attached microflora are intimately coupled to the physical and metabolic dynamics of the living substrata upon which they grow. High sustained growth of attached microflora results from their recycling of essential gases ($CO_2$, $O_2$) and dissolved nutrients. Nutrient uptake is directed primarily to the high net growth of attached microflora and is responsible for the high capacity of wetland–littoral areas to improve the quality of water passing through these communities. [*See* WETLANDS ECOLOGY.]

The wetland–littoral complex, including the marginal floodplains of many rivers, produces the major sources of organic matter and energy of many freshwater ecosystems. Most of the particulate organic matter is decomposed within these interface regions. Organic matter is exported predominantly as dissolved organic matter to the recipient lake or river (Fig. 3).

The deep-water pelagic zone of lakes is least productive along the gradient from land to water, regardless of nutrient availability. Growth of phytoplanktonic algae of the pelagic zone is limited by sparse distribution in a dilute environment where nutrient recycling is restricted by the sinking of senescent phytoplankton below the depth of photosynthesis. When nutrient recycling and availability are increased, greater phytoplankton cell densities attenuate underwater light and reduce the volume of water in which photosynthesis occurs. Despite low productivity per unit area, pelagic productivity can be collectively important in large lakes and for higher trophic levels that depend on this organic matter.

A second trophic level consists of zooplankton (dominated by four major groups of animals: protozoa/protista, rotifers, and the crustaceans cladocera and copepoda) and benthic invertebrates. In the pelagic zone these herbivorous organisms are consumed by small fishes, fry of larger fishes, and predatory zooplankton, which comprise a third trophic level (primary carnivores). A fourth trophic level may consist of medium-sized piscivorous

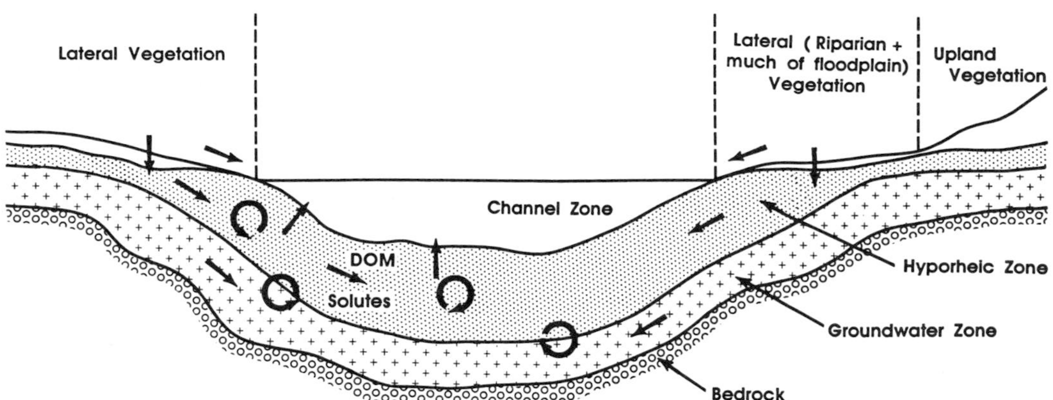


***FIGURE 3*** Lateral and vertical boundaries of flowing water ecosystems. The stream ecosystem boundary is defined as the hyporheic/groundwater interface and thereby includes a substantial volume beneath and lateral to the main channel. Vegetation rooted in the hyporheic zone is therefore part of stream ecosystem production. Arrows indicate flow pathways of dissolved organic matter and inorganic solutes derived from plant detritus within the stream ecosystem. [From R. G. Wetzel and A. K. Ward (1992). *In* "Rivers Handbook, I" (P. Calow and G. E. Petts, eds.), pp. 354–369. Blackwell Scientific, Oxford, England.]

fishes, and the fifth level includes large piscivorous fishes. Higher trophic levels are rare in fresh waters.

The species composition of the higher trophic levels affects the pathways of energy utilization from lower trophic levels. For example, efficiency of consumption of primary production by zooplankton is often appreciably greater in the absence of zooplankton-feeding fishes than in their presence. The community structure of phytoplankton responds variably to grazing impacts in concert with their available resources (light, nutrients, organic constituents) and may or may not be able to compensate for grazing losses in overall primary production.

## IV. ECOSYSTEM FUNCTION: INTERACTIVE REGULATORY MECHANISMS

Most of the organic carbon of aquatic ecosystems, and hence energy available for system operation, exists as dead organic matter, in both dissolved and particulate forms, and is called *detritus*. Dissolved and particulate organic carbon (POC) move with the water, and POC is ultimately deposited at the bottom of static water. Dissolved organic carbon can also sediment if adsorbed to particulate matter or if polymerization occurs. Most heterotrophic decomposition occurs in benthic regions as organic carbon is often largely displaced from sites of production to sites of decomposition.

Although the specific composition of organic matter varies greatly, detritus inevitably carries most of the energy of the ecosystem from its points of photosynthetic origins to places of transformation. Most of that heterotrophic transformation does not occur by digestion in predatory metazoan animals, but rather by microbes and often largely under anaerobic conditions, particularly in sediments. Productivity of the microbes is maximized within the resources available from dead organic matter and nutrients. Microbial heterotrophic utilization is highly dynamic and changes with sufficient rapidity to allow many generation turnovers before predators with much slower turnover rates can respond reproductively. Viral mortality of microbes, as well as ingestion of bacteria by protists (the "microbial loop"), diverts much organic carbon from animal trophic levels. This noningested productivity is of major importance and usually dominates the energy flows within the pelagic region and completely dominates energy flows of the aquatic ecosystems. The microbial heterotrophy is of primary importance to higher trophic levels in feedback processes, both positively (e.g., nutrient recycling and utilization by primary producers) and negatively (e.g., oxygen consumption and production of toxic fermentative metabolic end products). The abundance, distribution, and bacterial decomposition of dissolved and particulate detritus both regulate and stabilize energy metabolism and nutrient availability in aquatic ecosystems.

## V. APPLICATION TO SOCIETAL PROBLEMS

Fresh water is the fulcrum resource upon which agriculture, industry, and domestic life pivot. The reservoirs of water held in natural or artificially created basins contain water for various periods of time. These bodies of water function in a multiplicity of ways in human activities—sources of water, sources of biota for nutrition, sources of aesthetic value to human well-being. These water bodies are also used as purification systems, intentionally or fortuitously, for various waste products of human activities. Many of the purposes and uses to which lakes, reservoirs, and rivers are put are incompatible.

Problems of the alteration of surface waters and groundwaters nearly all focus on (1) introduction of abnormal amounts and/or alien substances or biota or (2) modification of retention or distribution of the water. *Eutrophication* of surface waters results from loading of nutrients, particularly phosphorus, to lakes and reservoirs, which leads to excessive production of certain algae and higher aquatic plants, and often a decrease in biodiversity. Excessive atmospheric loadings of sulfur and nitrogen oxides have resulted in high loadings of strong acids in liquid (rainfall) and particulate (dust) precipitation. The resulting *acidification* of surface waters and groundwaters has radically altered the water chemistry and biota, with a marked reduction in biodiversity. *Pollutants,* such as heavy metals (e.g., mercury, copper, lead) and toxic organic compounds, have contaminated many surface waters and groundwaters to the extent that not only has biodiversity been reduced but many of these water reservoirs cannot be used for human activities. In some cases, introduction of *exotic species* has similarly resulted in competitive exclusion of many endemic species and has also altered chemical conditions of the water. In all of these cases of environmental alterations, paleolimnological analyses of the mineralogy and structure of sediments, their organic and inorganic chemical constituents, and the morphological remains of organisms preserved in the sediments permit interpretations about past states and conditions of the ecosystem.

A great diversity of terrain, geomorphology, climate, and hydrology occurs in every country. The natural diversity is overlain with cultural histories in which humans have systematically and often ingeniously modified hydrological patterns, particularly of surface waters, to accommodate agricultural and water supply needs. Numerous political and human conflicts originated or were incited by the needs for or mismanagement of fresh waters. The present global warming associated with increased loading of carbon dioxide and other pollutant gases to the atmosphere will certainly alter climatic patterns, hydrological regimes, and the availability and quality of fresh waters.

## Glossary

**Benthos** Nonplanktonic animals associated with substrata within sediments or closely above the sediment–water interface.

**Community** Group of interacting populations.

**Detritus** Nonliving organic matter in both soluble and particulate forms.

**Littoral zone** Region of a lake or river between the land and the pelagic zone that is colonized by emergent, floating-leaved, and submersed aquatic plants and their attendant sessile microbiota (periphyton).

**Nekton** Organisms with relatively good swimming powers of locomotion.

**Pelagic zone** Open-water portion of a lake or reservoir beyond the littoral zone.

**Periphyton** Bacteria, fungi, algae, and sessile microfauna growing attached to substrata (sediments, rock, plants, animals, sand); further named in relation to the type of substrata upon which they grow.

**Phytoplankton** Small plant plankton, largely algae.

**Plankton** Small organisms with no or limited powers of locomotion that are suspended in the water and largely dispersed by turbulence and other water movements.

**Population** Defined assemblage of individuals of one species.

**Production** Amount of new organic biomass formed over a period of time, including any losses from respiration, excretion, secretion, injury, death, and grazing.

**Zooplankton** Animal plankton, usually denser than water, that sink by gravity to greater depths.

## Bibliography


Brezonik, P. L. (1994). "Chemical Kinetics and Process Dynamics in Aquatic Systems." Boca Raton, Fla.: Lewis Publishers.

Cooke, G. D., Welch, E. B., Peterson, S. A., and Newroth, P. R. (1993). "Restoration and Management of Lakes and Reservoirs." Boca Raton, Fla.: Lewis Publishers.

Gleick, P. H., ed. (1993). "Water in Crisis: A Guide to the World's Fresh Water [sic] Resources." New York: Oxford University Press.

Hutchinson, G. E. (1957). "A Treatise on Limnology. I. Geography, Physics, and Chemistry"; (1967). "II. Introduction to Lake Biology and the Limnoplankton"; (1975). "III. Limnological Botany"; (1994). "IV. The Zoobenthos." New York: John Wiley & Sons.

Hynes, H. B. N. (1970). "The Ecology of Running Waters." Toronto: University of Toronto Press.

Imberger, J. (1990). Physical limnology. *Adv. Appl. Mechanics* **27,** 303–475.

van der Leeden, F., Troise, F. L., and Todd, D. K. (1990). "The Water Encyclopedia." Chelsea, Mich.: Lewis Publishers.

Ward, J. V., and Stanford, J. A. (1991). Research directions in stream ecology. *Adv. Ecol.* **1,** 121–132.

Wetzel, R. G. (1983). "Limnology," 2nd ed. Philadelphia: Saunders.

Wetzel, R. G. (1990). Land–water interfaces: Metabolic and limnological regulators. *Verh. Int. Verein. Limnol.* **24,** 6–24.

Wetzel, R. G., and Ward, A. K. (1992). Primary production. *In* "Rivers Handbook, I." (P. Calow and G. E. Petts, eds.), pp. 354–369. Oxford, England: Blackwell Scientific.

# Mammalian Reproductive Strategies, Biology of Infanticide

**Glenn Perrigo**
*University of Missouri—Columbia*



Infanticide—defined as the killing of conspecific young—is typically an adaptive rather than pathological behavior. Infanticide occurs widely among mammals and, in most cases, is a violent but extremely effective reproductive strategy that benefits perpetrating individuals. Infanticide in both sexes evolved as a consequence of intraspecific (within-species) competition. Specifically, male mammals compete for the limited resource of females (sexual selection), whereas female mammals, because of the enormous energy burdens of lactation and parental care, compete for limited material assets such as food and shelter (resource competition). When a male kills another male's young, he accelerates ovulation in the victimized mother, mates with her, and soon has his own offspring. When a female kills the young of another female, she lessens competition and usurps more resources for herself and her offspring. Female mammals also kill and cannibalize some of their own young as a means of increasing the viability of those that survive. Physiologically, infanticide is a sexually dimorphic behavior influenced by genetic and phenotypic factors regulated by reproductive hormones. Despite sex differences in behavioral motivation, infanticide enhances reproductive success in both sexes.

## I. INTRODUCTION

Behaviors rarely, if ever, evolve "for the good of the species." Individuals are truly selfish and concerned with their own reproductive success, namely, the propagation of their own genes and *not* those of other members of their species (unless in the special case of kin selection). Our often too humanistic view of a benevolent Mother Nature sometimes make us forget just how hostile and violent the natural world really is.

In many organisms, the threat that offspring will be killed by a member of one's own species is often greater than the threat of predation. In black-tailed prairie dogs (*Cynomys ludovicianus*), for example, over half of the juvenile mortality in one well-studied colony stemmed from killings inflicted by other colony members. This is the phenomenon of *infanticide,* defined here as the killing of prereproductive young of one's own species. Although the deliberate killing of infants may seem like an especially brutal behavior, there are solid biological reasons why this behavior evolved. Likewise, it is now widely accepted that the threat of infanticide has played a pivotal role in the social evolution and life-history strategies of many species.

Infanticide was once thought to represent maladaptive behavior triggered by sociopathological conditions such as crowding. However, when framed in the context of evolutionary or "natural selectionist" thinking, a new and more realistic view of infanticide has recently emerged. Infanticide is now considered a flexible and extremely successful behavioral strategy that enhances reproductive success. Reproductive success means *fitness,* and fitness means the survival and potential spread of an individual's genes in future generations. Field and laboratory studies have documented infanticide among a wide range of organisms, and some studies have actually verified the reproductive benefits that accrue to the perpetrator. [*See* Evolution and Extinction.]

Infanticide occurs widely among vertebrates. This article, however, is limited to the issue of mammalian infanticide and represents but a brief primer on the major ecological and physiological concepts pertaining to infanticide in both male and female mammals. Because the rules of evolution are alike no matter what kind of organism or its level of social organization, the concepts noted herein are equally applicable to infanticide within nonmammals and novertebrates alike. Furthermore, much of the emphasis here is placed on examples from Rodentia (rodents), especially house mice, *Mus domesticus* and *Mus musculus*. (Taxonomic note: worldwide commensal, feral, and laboratory stocks are descendants of *M. domesticus* from Western Europe, whereas *M. musculus* occurs in Eastern Europe and Western Asia.) Small rodents are easily studied and tested in the laboratory and have yielded a wealth of experimental evidence in support of various hypotheses about infanticide. House mice also have a social structure similar to that of many primates and carnivores, and much of what has been learned about the dynamics of infanticide in mice parallels field observations of other mammals.

## II. MALE INFANTICIDE

Two of the most publicized examples of male infanticide occur in African Lions (*Panthera leo*) and primate langurs (*Presbytis entellus*) from India. Lions typically live in a *polygamous* social system in which a dominant male (or several related males) controls a harem of reproductive females. Lionesses do the hunting, while the larger male's main duty is to sire offspring and protect his pride from cub-killing predators, such as hyenas, or defend against infanticidal male lions. Multiple females are sequestered within a pride, so many male lions remain nomadic, being left without access to mates or permanent territories. By far, the greatest threat to a dominant male lion is eviction by marauders attempting to usurp his pride. If a challenger male (or male coalition) succeeds in defeating and driving the dominant male from his territory, one of the first things the usurper male does is kill all the cubs belonging to his vanquished rival. [*See* Sex Ratios; Territoriality.]

Infanticide is not a form of predation, as lion cubs are killed but rarely eaten. Furthermore, the pharmacological evidence from rodents shows that administration of the drug Fluprazine eliminates intermale aggression and infanticide, but does not interfere with predatory behavior. A specific pattern of killing is commonly observed among mammals that commit infanticide, wherein infants are routinely dispatched with swift bites to the head and neck.

The dynamics of infanticide are much the same among langurs, with troop takeovers by alien males often resulting in the death of infants. Male infanticide has been routinely documented in numerous carnivores, such as felids, canids, mustelids, mongooses, and ursids (bears). Male infanticide also occurs in at least nine genera of primates and is ubiquitous among rodents, ranging from rats to muskrats. A peculiar form of male infanticide occurs in sea lions (*Otaria*) when young males raid breeding colonies and abduct pups. This behavior is related to reproduction and sexual motivation because inexperienced males attempt to hone their mating skills by courting and copulating with the abductees. In the process, however, abducted seal pups are often smothered or fatally injured.

## A. Reproductive Ecology of Sex Differences

How does a behavior like infanticide evolve and what are the reproductive benefits for males? The biological basis for infanticide can best be understood by examining the ecology of sex differences, namely, how males and females compete for resources in their environment. *Interspecific* competition (between species) occurs when individuals of different species compete for the same resources such as food, shelter, and territory. In contrast, *intraspecific* competition (within species) occurs when individuals of the same species compete among themselves for resources.

Each sex, however, engages in a different form of intraspecific competition. Females are resource-oriented and routinely compete with other females for material resources, primarily food and quality nest sites. This is because female mammals bear the huge economic burdens of reproduction, principally lactation, and usually provide extensive, time-consuming parental care. Energy is the currency of reproductive success among females, so obtaining sufficient food calories is a female mammal's primary concern. In sharp constrast, reproductive success among male mammals is not so much energy-limited as it is female-limited. Males therefore must compete among themselves for both females and the territory to hold females. Because of these sex differences in resource competition, males and females have different reproductive motivations for committing infanticide.

The special type of intraspecific competition that is so prevalent among males—competition for the limited resource of mates—results in the process of *sexual selection,* defined as differential reproductive success resulting from the ability to obtain mates. Simply stated, some males are better at competing for and securing mates than others. Typical examples of sexual selection pressures include the evolution of elaborate mate-attracting stimuli such as coloration, song, and body size, or dual-purpose structures such as horns and antlers, which also serve as weaponry to fend off rivals and defend territory. Infanticide too is a manifestation of male versus male competition driven by sexual selection pressures.

In lactating females, the neural stimulus of suckling young promotes the secretion of hormones that act to inhibit ovulation. By killing her young, an infanticidal male manipulates the female's physiology by eliminating the stimulus that inhibits her from becoming sexually receptive and ovulating. Thus, when her own young are destroyed she soon ovulates again and typically mates with the infanticidal male. This shortens generation time—the *interbirth interval*—and the infanticidal male benefits in two ways. First, he obtains mating and now has his own reproductive investment. Second, by killing a rival's children he destroys his rival's genes while also eliminating the potential resource competition that could occur between his own offspring and another's offspring.

Shortening the interbirth interval is especially critical for the reproductive success of very short-lived small mammals. In rodents such as house mice, a male gets a chance to mate with a female immediately after parturition when she undergoes a short period of "heat" known as postpartum estrus. If her newborn pups were conceived by another male, the new sire's optimal reproductive strategy is to kill her pups immediately. If he does not kill them and allows the female to continue lactating, she will delay implanting his own fertilized embryos in her uterus for up to 10 days, which expands her interbirth interval up to 30 days as opposed to her normal gestation of only 18–20 days. The average life span of a small mammal in the wild is a matter of days, so most small mammals die or are eaten before they even reach sexual maturity. By accelerating the birth of his own pups by a week, a short-lived male gains a significant time advantage in the race to reproduce. Male mice do, in fact, show a dramatic increase in their motivation to attack and kill newborn pups immediately after copulation, which confirms the evolutionary predictions about an optimized behavioral strategy.

But how do males recognize their own offspring? After all, the accidental killing of one's own offspring is the quickest way to exit the gene pool—behavioral mistakes of this magnitude are eliminated quickly through natural selection. Given the importance of *not* killing one's own offspring, it is not surprising that a variety of physiological mechanisms have evolved to inhibit infanticide in male mammals at the appropriate time, specifically when their own progeny may be present. Male house mice provide the best and most widely studied behavioral examples, having evolved multiple and sometimes redundant mechanisms to ensure that such mistakes do not occur.

### B. A Unique Stimulus–Response System

Some male rodents (*Mus* sp., *Rattus* sp., and probably others, such as *Peromyscus* sp.) possess a unique neural timing system, triggered specifically by the act of ejaculation during mating, that allows a male to track the timing of birth and weaning of his own sired pups. Ejaculation inhibits infanticide; however, there is frequently an extraordinary latency in this response, as a male's infanticidal behavior often does not cease for many days after mating, but virtually always ceases by the time his own sired offspring are born 3 weeks later. When infanticide ceases, most male house mice suddenly express parental behavior similar to that of a newly lactating female. Newborn pups are groomed, retrieved and incubated by a male. But at 50–60 days after mating, male mice often revert to spontaneous infanticide, and this transition back to infanticide happens to coincide with the time when his pups are normally weaned and begin to disperse.

Male mice and rats are thus prone to kill pups both before and after copulation, but the act of ejaculation inhibits them from harming their own progeny during their mate's lactation (Fig. 1). There are no other time-delayed phenomena known in mammals where such dramatic shifts in adaptive behavior are programmed to occur weeks after a specific stimulus such as ejaculation. With regard to the study of behavior, this phenomenon demonstrates that a wide range of potential time-dependent relationships can exist between a classical stimulus and its response, especially when selection pressures are so intense.

Even more physiologically remarkable is that postcopulatory changes in the male's behavior toward pups are correlated with the number of light–dark cycles experienced after mating, suggesting that ejaculation triggers a neural day-counting mechanism that probably synchronizes some sort of synaptic decay (see discussion of the Bruce effect in Section II, E). The use of photoperiodic information as a timekeeper makes perfect ecological sense here, because the daily light cycle provides infallible circadian timing cues, and the use of ejaculation as a neural time-trigger makes perfect physiological sense when one considers the issue of paternity and sex differences. Female mammals are *always* certain of maternity because they sequester fetuses, whereas male mammals can *never* be absolutely certain of paternity. Ejaculation therefore represents a fail-safe stimulus for establishing paternity, at least within the temporal bounds of the female's reproductive cycle.

Whether or not the same neural mechanisms linked to the act of mating exist in larger mammals remains unanswered. Biological rhythms, however, may not be so precise over long periods of time, so photoperiodic precision might be lost whenever much more than a few weeks of gestation are involved. Nevertheless, this does not preclude a male mammal from simply *remembering* his sexual experience. Widespread promiscuity among some female mammals suggests that a male does indeed rely on mating experience to assess his chances of paternity. *Deception* by females is the key here, because some female mammals, especially primates, will allow mating to occur independent from their hormone-driven cycles of fertility. Some female mammals—langurs and lions, for example—actually exhibit pseudo-estrus (sexual receptivity without ovulation). Pseudo-estrous females frequently solicit matings with new males after either a takeover or after a turnover in the social hierarchy when the alpha (dominant) male is replaced. Female copulation with multiple partners probably evolved as a means of obfuscating male paternity and preventing infanticide.

Aside from copulation, other social cues can act either synergistically or independently from the stimulus of ejaculation to suppress infanticide. In wild male *M. musculus* from Israel, ejaculation inhibits infanticide, but contact with a female alone, regardless of copulation, inhibits male infanticide equally well. In sharp contrast, wild male *M. domesticus* from Canada are inhibited by ejaculation, but

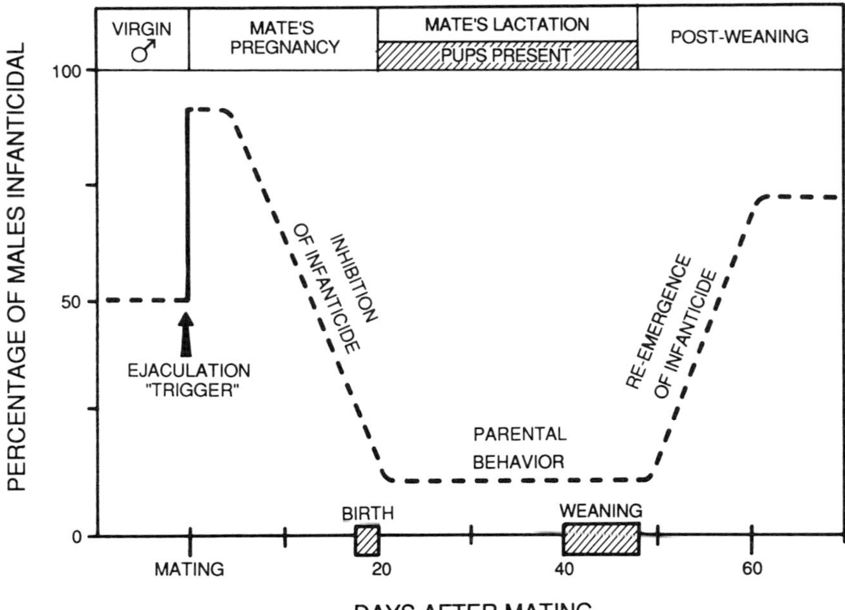


***FIGURE 1*** Ejaculation mediates the time course of changes in a male mouse's behavior toward pups. These data are from a population of CF-1 stock laboratory mice tested over a 2-month period. Males were kept in total social isolation throughout except for brief exposure to a female for copulation. In this particular stock, half of all virgin males are noninfanticidal. However, wild-trapped stocks exhibit an identical response pattern except that most virgin wild males are infanticidal (~90%). Notice the variation in phenotypes. Some males are always parental (10–15%) and some males are always infanticidal (10–15%; see Sections II,D and II,E for further explanation).

contact with females does not in any way inhibit infanticide. Furthermore, in some house mouse populations, ejaculation and female contact are obligatory and both stimuli are required to inhibit male infanticide.

### C. Phenotypic Variation in Infanticide: The Role of Hormones

Some of the behavioral polymorphisms noted in rodents are programmed by *in utero* hormonal events. In mammals that produce multiple offspring, fetuses are positioned randomly in the uterine horns and are therefore exposed to different sex steroid concentrations depending on whether they develop next to same or opposite sex fetuses. Fetuses who develop between two male siblings are exposed to higher concentrations of the male sex steroid testosterone, whereas those who develop between female siblings are exposed to lower concentrations of testosterone. As a consequence, small changes in the dose of sex steroids during late fetal development—the period of sexual differentiation—can differentially organize and sensitize the brains of both males and females. A profound range of phenotypic variation among adult reproductive and behavioral patterns, including adult–infant interactions and infanticide, can thus be traced to intrauterine position. Environmental stress during late gestation can also affect the concentration level of hormones and cause phenotypic variation. Even more germane is the possibility that many common environmental pollutants (particularly organochlorides), even in miniscule concentrations, can mimic the physiological actions of sex steroids, primarily estrogen, and thus alter normal fetal behavioral development.

Not all male mice or rats are infanticidal after mating and some males always seem to behave parentally regardless of sexual experience. Current evidence suggests that the majority of these male phenotypes developed *in utero* between two male siblings. However, there is an inverse relationship between intermale aggression and infanticide in house mice, because fetal exposure to high testosterone concentrations results in adult males who exhibit high levels of intermale aggression. In sharp contrast, some males always kill pups, regardless of how many days have elapsed after mating. The majority of these males apparently developed between two female siblings and, conversely, are usually less aggressive as adults. Among these latter males, mating per se does not seem to inhibit infanticide, but the stimulus of ejaculation *and* female contact are apparently both required to inhibit pup-killing behavior. Fetal hormone exposure also seems to influence timing variation triggered by ejaculation, as well as how some individuals respond to light–dark cues.

### D. Social Structure and Behavioral Variation

The spatial relationships within a typical house mouse deme suggest an ecological answer to the phenotypic variation in behavioral strategies described here. As noted in the Introduction, house mice routinely maintain a social structure like that of many carnivores and primates, with a dominant male defending a territory containing several reproductive females (Fig. 2). Because aggressive males are more likely to hold territory and therefore sire more pups, their optimal strategy should tilt more toward a propensity for noninfanticidal behavior and parenting. In contrast, less aggressive males are more likely to be peripheral or "satellite" males exploiting a "sneaky copulator" strategy similar to that observed in red deer (*Cervus elaphus*). A "sneaky copulator" attempts a quick copulation with an estrous female when the alpha male momentarily leaves his harem unguarded. Mating opportunities may be few for a satellite male, so infanticide may be this phenotype's best strategy for accelerating the birth of pups if he does manage to usurp a postpartum mating. And, as suggested earlier, if this male phenotype replaces the dominant male, prolonged contact with a female after mating will stimulate his transition to parental behavior.

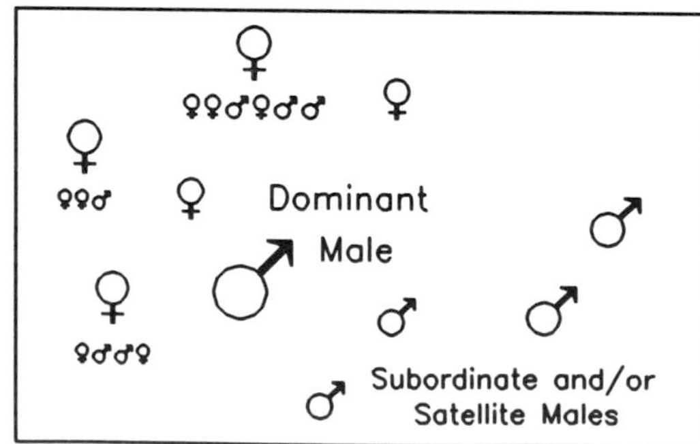


***FIGURE 2*** Social organization of a typical feral mouse microdeme. This diagram could also represent the social structure of various other mammal species, including some carnivores and primates.

Behavioral variation in infanticide is not endemic to rodents, however, because langurs throughout India also show wide population-to-population differences in infanticidal strategies, and other mammals do too. Such variation undoubtedly reflects subtle ecological differences in the social organization and deme structure of different populations. Understanding the multitude of intraspecific strategies of infanticide is one area in need of rigorous field and theoretical work.

### E. Female Counterstrategies against Infanticide

Female promiscuity may be one strategy used to trick males into believing they have offspring present, but females have evolved other counterstrategies to prevent infanticide that also manipulate the sensory system of a male. Chemical communication is the predominant sensory modality in the vast majority of mammals. In rodents, virtually every aspect of their social and reproductive behavior is regulated by pheromones emanating from nearby conspecifics. Although ejaculation can inhibit infanticide, so will chemosensory cues if male Norway rats (*Rattus norvegicus*) are exposed *only* to the odor of a pregnant female. This illustrates the concept of *spatial familiarity*, because in natural situations a male rat would likely remain in the presence of his mate. Her odors thus provide an efficient reinforcing stimulus to inhibit infanticide and promote his transition to parenting. There is deception here too, as her odors will also inhibit infanticide in unmated males who are not the sire. No physical or visual contact is necessary, only her pheromones.

Females rodents can also "remember" the odor of a specific mate for several weeks. This 30- to 50-day memory trace and its decay are mediated by her accessory olfactory system. If the nerve tracts leading from the vomeronasal organ on the roof of the mouth to this area of the brain are severed in either sex, parental behavior is expressed, suggesting that the accessory olfactory system regulates the inhibition and disinhibition of infanticide. The lingering memory of her mate's odor is the primary reason for a well-studied phenomenon, originally discovered in rodents, known as the *Bruce effect*. When a newly inseminated female senses the odor of an alien male—a male who is not the sire—the female is likely to either fail to implant or abort her embryos during the first week of pregnancy.

This physiological response was once thought to be a peculiar laboratory artefact. After all, why would a female curtail her reproductive investment via self-abortion? However, given the concept of spatial familiarity, the odor of an alien male suggests to an inseminated female that her mate has either died or been evicted from his territory. A new stud male would probably kill her newborn pups anyway, so her optimal strategy is to cut her reproductive losses early in pregnancy rather than risk the loss of her entire litter being killed at birth. Some female mice can actually distinguish between infanticidal and noninfanticidal males, for implantation failure is much more likely to occur in the presence of an infanticidal male. Lions, horses, and some primates also have a tendency to abort when confronted with alien males, as typically happens after a male takeover or a shakeup in the dominance hierarchy.

But what about a female mouse who is well into pregnancy and cannot abort her late-term fetuses when a new male takes up residence? In this scenario the female will likely resort to a physical defense of her litter by chasing, biting, and clawing at an attacking male (or female) while trying to drive the intruder from her nest area. Interestingly, her method of attack is either male or female specific. Against a male it is defensive, and her bites are directed specifically at vulnerable areas of his head and abdomen. But if her litter is threatened by another female, her bite pattern is offensive, with frequent strikes at the flanks and sides.

The complex interactions of hormones associated with pregnancy and parturition also trigger adaptive changes in female behavior. The hormone oxytocin, for example, is involved in uterine changes that initiate birth, but it also acts directly on the central nervous system to inhibit infanticide because females tend to be infanticidal throughout pregnancy. Parturition also causes intense postpartum aggression in most female mammals. Despite their elevated aggressiveness, however, when newly parturient females such as house mice have acessible, exposed nests, they are rarely able to prevent infanticide if a female must defend one-on-one against a male. In this situation, the male virtually always wins and destroys her litter. The odds are much more favorable for mammals that nest in burrows or dens, such as hamsters or hyenas, because these structures are more easily defended against intruders.

In a similar vein, some mammals form communal nests where two or more females pool their litters. In house mice, not only do communally raised offspring grow faster, but communal litters also provide for the common defense, because two cooperating females can usually defend against an attacking male. In general, cooperative female defense against infanticide is common among social species. Female lions, for example, will savagely defend the pride's cubs against attack, sometimes successfully.

### F. Subordination and Kin Selection: A Male Counterstrategy

Some male mammals have evolved a counterstrategy to inhibit other males from killing their pups. Most male mammals are aggressive and territorial—male–male encounters often escalate to a fight. In gerbils and mice, if a dominant male wins the fight, the subordinated intruder is typically inhibited from pup-killing behavior. By subordinating his rivals, a dominant male's own offspring are thus safeguarded from attack.

But why would a subordinate, nonbreeding male *not* be infanticidal? For one thing, infanticide is not always the optimal strategy, especially if one considers the concept of *kin selection* and the deme structure of a population. As described earlier (Fig. 2), house mice form social units in which a dominant male controls a territory with several breeding females. Some of these satellite males, typically nonbreeders, might also be nondispersers who were born in the same microdeme, possibly even sired by the dominant male. If a substantial number of genes are shared among animals in a deme, subordination would be beneficial for *inclusive fitness* if it inhibited males from killing close relatives such as full or half-siblings in recently born litters. Kin selection favors this type of social contract, because stability and reproductive success are preserved as long as the dominant breeding male can subordinate his rivals. This type of behavioral interaction is easy to study in small mammals, but one would also predict the same relationships in larger mammals with a similar social structure or higher degree of sociality (e.g., primates and canids). In fact, ritualized subordination and submission is common in domestic dogs and wild canids, such as wolves and coyotes (*Canus lupus* and *C. latrans*).

## III. FEMALE INFANTICIDE

The ecology and physiology of female infanticide are markedly different from those of males. For females, there are two major classes of infanticide: first, a female can kill another female's offspring or, second, under special circumstances, a female may kill and cannibalize her own offspring. The former results directly from resource competition with other females, whereas the latter results from an indirect, internal form of resource competition in which her own physiological systems must vie for the prudent partitioning of limited food calories. Energy needed for survival virtually always takes priority over the energy shunted toward reproduction. Thus, when reproduction conflicts with somatic demands, a female is forced to decide whether the net benefits of current offspring are worth more than her future reproductive potential (i.e., she defers reproduction until energy conditions are more favorable). In a severe energy crunch, offspring may be selectively eliminated and cannibalized, which, despite the loss of some children, greatly increases the viability of the survivors.

### A. Resource Competition: Living in the Material World

Unlike male mammals, who are primarily female-limited, female mammals are resource-limited. The enormous energy costs of lactation (and to a lesser extent pregnancy) make female mammals highly sensitive to the quality and quantity of material resources of their environment. Food and water are absolute requirements. Nevertheless, food energy and the ability to obtain sufficient calories to sustain reproduction are the ultimate constraints on any pregnant or lactating mother. Females have thus evolved fine-tuned energy-related strategies for ensuring that they obtain as much food as possible for their offspring and that they allocate this energy efficiently. Infanticide of conspecifics is one of these energy-capturing mechanisms.

In social and colonial species, females routinely form dominance hierarchies like males. A female's position in the "pecking order" often determines the likelihood that some or all of her offspring may be killed by conspecifics. Canids are well-studied examples of infanticide resulting from resource competition among females. Wild or cape dogs (*Lycaon pictus*) in Africa are cooperative hunters that live in highly structured packs where the dominant female is often responsible for infanticide. Her own selfish reproductive interests are best served by eliminating the offspring of other females because they and their young directly compete with her for resources. If the dominant female has her own young, she will often kill the newborns of other lactating females and usurp their milk resources for her own children by coercing victimized females into sharing nursing duties. Low-ranking females sometimes try to hide their young, but rarely are they successful at preventing infanticide. Thus, subdominant females are often relegated to caring for the offspring of infanticidal females. Again, the physiology of mammalian lactation is the key to understanding the evolution of this behavior.

The neural stimulus of suckling triggers the release of hormones (oxytocin and prolactin) that promote milk production and stimulate maternal behavior. As long as there is a suckling stimulus, a parasitized female will lactate indefinitely and continue to suckle another female's offspring. On the other hand, inclusive fitness benefits still accrue for the parasitized female, for pack members are often closely related (kin selection). Brown hyenas (*Hyaena brunnea*) are also social, cooperative hunters like wild dogs and they too have communal dens where subordinate mothers become "wet nurse" helpers. Infanticide in some primates emulates the pattern of wild dogs. Although it is rare, high-ranking females in a troop have been known to kill infants of lesser status mothers. As an interesting aside, another type of infanticide has been noted frequently among carnivores—*siblicide* occurs when an individual kills a sibling and thus garners more maternal resources for its own use.

The selective killing of offspring in house mice also occurs in the situation of communal nesting, where two females pool their litters and share lactation duties. Sometimes the more dominant female—or sometimes the female who delivers last—may eliminate some of her nestmate's pups to sequester more milk for her own pups. The reason some pups are spared here probably reflects a trade-off for thermoregulatory needs, because a large mass of pups retains heat more efficiently and reduces energy loss (communal pups grow faster). As for the victimized female, she still benefits because some of her own pups survive, plus she retains a partner for the common defense if attacked by an infanticidal male.

The evidence also suggests that in female house mice, as well as in males, different infanticidal strategies are correlated with fetal hormone exposure. Early exposure to slightly higher levels of testosterone appears to create female phenotypes who are much more aggressive as adults than other females and also more likely to be infanticidal. In fact, testosterone injections will cause most female rodents to kill pups.

Female infanticide also occurs for preemptive reasons in seal colonies, namely, to protect a female's own limited energy resources. This is because seal mothers must fatten up and store enough energy reserves to last the entire period of lactation before going ashore and giving birth. Milk is at such a premium in seal colonies that lost, wandering pups who try to suckle from alien mothers are routinely bitten and sometimes killed by lactating mothers to prevent the theft of milk. But seal pups are not always harassed, and some may even be accepted by foster mothers, so there is wide variation in infanticidal strategies among seals as well.

Usurping food energy is not the only factor in resource competition. In some mammals, quality nest sites are equally essential for reproductive success, as they provide both safety and thermal buffering. This seems to be the primary reason for female infanticide in ground squirrels (*Spermophilus* sp.). Marauding females are known to enter another female's nest burrow in her absence and dispatch her litter, often dragging the pups outside before usurping the burrow for their own use. As a general rule, males of colonial rodent species are especially vigilant during the reproductive season in order to fend off intruders of both sexes.

### B. Cannibalizing Your Own Offspring

How can killing and cannibalizing one's own children enhance reproductive success? Reproduction always involves trade-offs, and in many cases the elimination of some children leads to greater fitness among the survivors. (Or greater future reproductive potential, as kangaroos have been known to jettison joeys when pursued. Effectively, this abandonment is infanticide.) The rationale for cannibalization boils down to a simple concept: *energy balance*. In mammals that produce large litters, food consumption during lactation may be triple or even quadruple that of a female's baseline nonreproductive needs.

Although cannibalization is more likely to occur under severe energetic stress, some species—typically large-littered rodents—routinely eliminate a few offspring immediately after birth, even when food is abundant. Most rodents have ten or fewer nipples. Substantially more offspring than this presents a lactating mother with a dilemma because milk cannot be efficiently partitioned among so many young. Thus, new mothers frequently kill and cannibalize some of their own newborns so that litter size is energetically more manageable. This paring of newborns is a normal, organized component of the maternal behavior repertoire of many small mammals, especially hamsters, rats, and mice.

### C. Weaning Strategies: Quality versus Quantity of Offspring

When food stress becomes severe, cannibalization tactics are much more dramatic. Figure 3 shows the sharp contrast in average litter size versus average pup weights at weaning in an experiment where female house mice (*M. domesticus*) and deer mice (*Peromyscus maniculatus*) were challenged to forage harder and harder for less and less food throughout the course of lactation. Deer mouse females tried to wean either five or six pups, regardless of how severe the foraging conditions. As a consequence, their pups became progressively stunted at weaning as females received fewer and fewer calories to partition among the same number of offspring.

Just the opposite occurred in house mice. During the first 10 days of lactation, females routinely cannibalized their own offspring one-by-one. Litter size declined as feeding conditions worsened and progressively more pups within a litter were eliminated. But this ad hoc cannibalization strategy had two beneficial consequences. First, house mouse mothers got a quick boost from recycled calories in the midst of an energy crisis by cannibalizing some of their offspring; and second, their energy partitioning effort was now focused on fewer pups, so survivors received a greater share of milk. As a result, house mouse pups were weaned at similar body weights regardless of the severity of foraging conditions. In summary, Fig. 3 reveals that each species

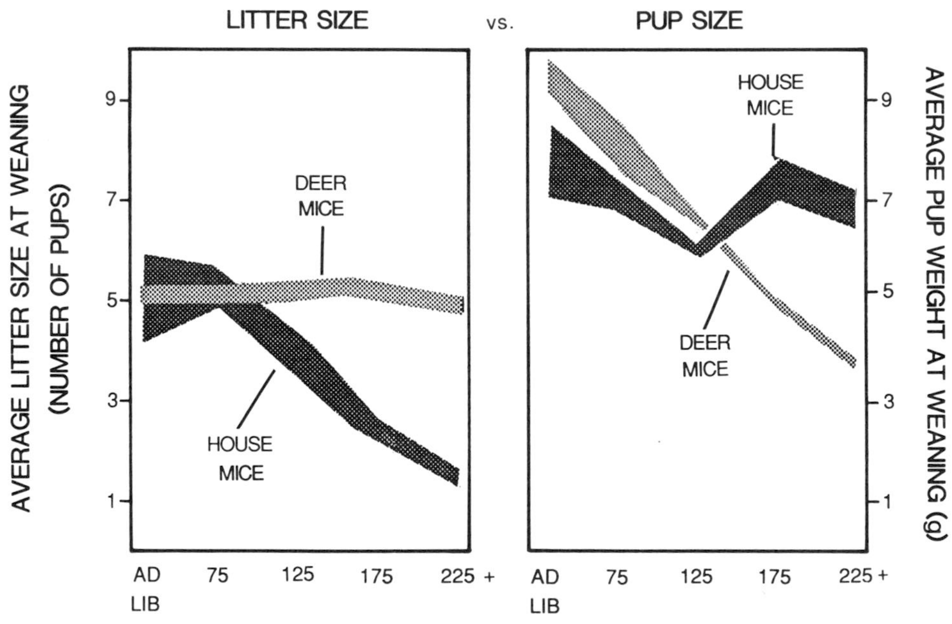


***FIGURE 3*** Litter size versus pup weight at weaning (mean ± sem range after three weeks of lactation) in female deer mice and house mice fed either *ad libitum* food (control) or challenged to run increasingly more revolutions on an activity wheel for each 45-mg food pellet obtained. There were 7 to 10 females of each species in each of the five different foraging conditions tested here. Mothers in each species used a completely opposite strategy to achieve weaning, namely, by manipulating either the quality or the quantity of their offspring (see text for details).

used an inverse solution to reach the same reproductive end point: deer mice defended litter size at the sacrifice of pup quality, whereas house mice defended pup quality at the sacrifice of litter size.

Seasonal breeding, or lack thereof, provides a potential explanation for these contrary strategies. Most deer mice are reproductively quiescent for over half the year, having only a few short months during summer to produce pups. Most of their offspring would not be able to breed until the following year anyway, so this may be why deer mice exhibit such a rigid, "all-or-nothing" commitment to their litters and attempt to wean as many pups as possible, regardless of the local food supply. House mice, on the other hand, kill and cannibalize their own pups whenever it is energetically expedient. House mice are exceptional opportunists and breed year-round as long as there is sufficient food. This species has opted for a strategy of producing a few good-quality offspring and weaning them as soon as possible. House mice are cosmopolitan in their distribution, and this may be a major reason why they gain reproductive footholds so quickly in so many places.

The arguments of seasonality may not always hold true, however, as Syrian hamsters (*Mesocricetus auratus*) are seasonal breeders yet females exhibit a one-by-one pup-killing strategy like that of house mice. Unlike deer mice, however, hamsters can store up substantial amounts of body fat. When food is scarce, hamster pups are cannibalized primarily as a function of their mother's dwindling fat stores and resulting energy deficits. Because cannibalization occurs in so many mammals, a variety of ecological factors probably determine the type of infanticidal strategy that a female uses for maximizing her production of viable offspring.

## IV. HUMAN INFANTICIDE

No overview of mammalian infanticide is complete without the explicit recognition that infanticide occurs in humans. Although it is well documented, there have been extensive and sometimes acrimonious debates among the various social and biological sciences about the relative roles that biological, cultural, and pathological phenomena have played in shaping the evolution of infanticidal behavior in *Homo sapiens*. It must be emphasized, however, that human evolution is intimately intertwined with that of the great apes—notably chimpanzees and gorillas—both species of which are known to commit infanticide. In fact, only since the rise of modern civilization have our ancestors arrived at new types of population structures that have diverged significantly from the basic hunter-gatherer tribal units that humans presumably experienced for most of their evolutionary existence. The pattern of many "primitive" peoples that still exist today emulates that of the primate models described herein (or for that matter, the mouse and lion models). Likewise, anthropologists have reported many such instances of human infanticide, although there are likely many more such instances left unreported, sometimes to protect the academic dogma or to protect indigenous peoples from moral, religious, or political persecution.

This is not meant to oversimplify human infanticide as a purely biological phenomenon. Complex cultural factors obviously play an important role in shaping the wide range of infanticidal strategies observed among different human societies. Infanticide that is not sexually selected, for example, occurs routinely in societies that place an overwhelming cultural bias on the production of sons (infant daughters are thus killed), or in hunter-gatherer societies where appropriate birth spacing is essential to a group's economic and social viability. Nevertheless, recorded history is replete with examples of marauders—from small tribal raiding parties to the armies of warring empires—attacking villages and cities, looting resources, driving out males, abducting reproductive females, and, most significantly, murdering the children of subjugated rivals. This type of intraspecific violence differs only in magnitude when compared to that of typical male takeovers described herein for other mammal species. Furthermore, evolutionary predictions can also be made and tested about human infanticide. As but one example, an examination of murder records in Canada revealed that, just as hypothesized, children were far more likely to be murdered by a step-parent than a biological parent.

## V. GENETICS OF INFANTICIDE

The discovery of a specific infanticide "gene" would represent a Holy Grail of sociobiology. In fact, computer simulations have shown that such a gene would indeed spread like wildfire and rapidly become fixed within a population. However, actual empirical genetic studies of infanticide have been done only in house mice and, not surprisingly, these very few studies have revealed that although there is indeed a genetic substrate for infanticide, it is a highly labile behavior regulated by complex genetic factors. Strain-specific differences in infanticidal behavior are well-documented among house mice. Wild stocks, for example, routinely show a high proportion of infanticidal phenotypes in both sexes (80–90%), but in laboratory inbred stocks, some may show high to intermediate levels of spontaneous infanticide, whereas others exhibit little if any tendency to kill preweaning young.

In one experiment, a reciprocal cross between a wild and a laboratory stock of house mice revealed a peculiar sex-specific pattern of inheritance. Maternal genotype exerted dominance over the infanticidal (and parental) characteristics of male offspring, as hybrid males exhibited the strain-specific phenotype of their mother's stock. But in female offspring, hybrids exhibited the behavioral phenotype of the laboratory stock, regardless of whether derived from a wild-type mother or father. Though such a pattern of inheritance is complex, these sex differences are consistent with the neurobiology and endocrinology of infanticide. Infanticide is a sexually dimorphic behavior regulated by gonadal steroids, and the current physiological evidence suggests that various mouse populations have undergone genetic changes in their sensitivity to steroids within the neural tissues responsible for governing behavior toward pups.

When viewed *in toto,* variation in infanticidal behavior among mammals probably results from an amalgam of polygenic traits influencing a wide range of neural and developmental responses to the organizational and sensitizing effects of sex steroids, and the physiological mechanisms that regulate infanticide within each sex have, to a large degree, evolved somewhat independently of each other. This latter conclusion makes sense in view of the ecology of mammalian sex differences, specifically sexual selection and resource competition, which have resulted in different reproductive motivations for males and females to commit infanticide. In any case, infanticide is a widespread mammalian behavior that enhances reproductive success in both sexes.

## Glossary

**Bruce effect** When a newly inseminated female smells an alien male, the female either fails to implant or aborts her embryos early in pregnancy.

**Competition** When two or more individuals of the same species (intraspecific) or of different species (interspecific) vie for the same resources in the environment (e.g., food, shelter, mates).

**Deme and microdeme** A deme is a population in which all the individuals are capable of randomly interbreeding; a microdeme is simply a descriptive term for a small population unit containing just a few interbreeding individuals (e.g., a lion pride or primate troop).

**Fitness and inclusive fitness** Fitness is the contribution of one individual's genes to the next generation relative to the contribution of other individuals' genes. Fitness is similar to reproductive success, except the latter refers to number of offspring produced. Inclusive fitness is the contribution of an individual's genes plus the added fitness derived from genes shared with close relatives through, for example, kin selection.

**Genotype versus phenotype** Genotype is the actual genetic makeup of an individul, whereas phenotype is the visible characteristics of an individual resulting from genetic and environmental effects. Natural selection operates on phenotypes.

**Infanticide** The killing of prereproductive young of one's own species. In mammals, a more operational definition of infanticide is the killing of conspecific preweaning young.

**Intrauterine position phenomenon** Fetuses are positioned randomly in the uterine horns and therefore are exposed to different sex steroid concentrations (testosterone and estrogen) depending on whether they develop next to same or opposite sex siblings.

**Kin selection** When individuals contribute to the reproductive success of close relatives other than their own offspring.

**Polygamy** When an individual has two or more mates and usually contributes to parental care. Polygyny is when a male has more than one mate. Polyandry is when a female has more than one mate.

**Sexual selection** Differential reproductive success resulting from the ability to obtain mates. Both sexes experience sexual selection, but the most prevalent and exaggerated examples of sexual selection pressures involve males competing for access to females.

## Bibliography

Hausfater, G., and Hrdy, S., eds. (1984). "Infanticide: Comparative and Evolutionary Perspectives." Chicago: Aldine Publishing.

Hrdy, S. B. (1979). Infanticide among animals: A review, classification, and examination of the implications for the reproductive strategies of females. *Ethol. Sociobiol.* **1,** 13–40.

Parmigiani, S., and vom Saal, F. S., eds. (1994). "Infanticide and Parental Care." London: Harwood Academic Publishers.

Perrigo, G., and vom Saal, F. S. (1989). Mating-induced regulation of infanticide in male mice: Fetal programming of a unique stimulus–response. *In* "Ethoexperimental Approaches to the Study of Behaviour" (R. Blanchard, P. Brain, D. Blanchard, and S. Parmigiani, eds.), pp. 320–336. Dordrecht, The Netherlands: Kluwer.

Perrigo, G. (1990). Food, sex, time and effort in a small mammal: Energy allocation strategies for survival and reproduction. *Behaviour* **114,** 191–205.

Perrigo, G., Belvin, L., and vom Saal, F. S. (1992). Time and sex in the male mouse. Temporal regulation of infanticide and parental behavior. *Chronobiol. Int.* **9,** 421–433.

Wade, G., and Schneider, J. (1992). Metabolic fuels and reproduction in female mammals. *Neurosci. Biobehav. Rev.* **16,** 235–272.

# Management of Large Marine Ecosystems

**Kenneth Sherman**
*National Marine Fisheries Service*

**Lewis M. Alexander**
*University of Rhode Island*



Large marine ecosystems (LMEs) are extensive regions of the world ocean, generally greater than 200,000 km$^2$ in extent, characterized by distinct bathymetry, hydrography, productivity, and trophically related populations. Human intervention and climate change are sources of additional variability in the natural productivity of LMEs. Within the nearshore areas and extending seaward around the margins of the global landmasses, LMEs are being subjected to increased stress from toxic effluents, habitat degradation, extensive nutrient loadings, harmful algal blooms, emergent diseases, fallout from aerosol contaminants, and episodic losses of living marine resources from pollution effects and overexploitation. Ninety-five percent of the annual global biomass yields are produced within the boundaries of 49 LMEs. The long-term sustainability of LMEs in producing natural resources for healthy economics appears to be diminishing. A growing awareness that the quality of LMEs is being adversely impacted by multiple driving forces has accelerated efforts to assess, monitor, and manage LMEs to ensure the long-term sustainability of their natural resources by implementation of multisectoral and holistic ecosystem-based management strategies.

## I. BACKGROUND

The coastal areas of the world's oceans are being subjected to increasing stresses from habitat degradation, pollution, and overexploitation of marine resources. In recognition of the need to improve the conditions of coastal ecosystems, the international community reached consensus recently on actions to be implemented for improving the management of marine resources. During the United Nations Conference on Environment and Development (UNCED) in 1992, virtually all coastal nations of the world reached consensus on a declaration on the oceans that recognized the need to: *(1) prevent, reduce, and control degradation of the marine environment so as to maintain and improve its life-support and productive capacities; (2) develop and maintain the potential of marine living resources to meet human nutri-*

*tional needs, as well as social, economic, and developmental goals; and (3) promote the integrated management and sustainable development of coastal areas and the marine environment.* Sustainable development of marine resources is achievable as scientific advice, based on biological, social, and economic considerations, is more widely applied in the development of policies for marine resource use. [*See* CONSERVATION AGREEMENTS, INTERNATIONAL.]

Another important concept to emerge from UNCED is biodiversity. Taken together, biodiversity and sustainability have emerged as the challenges for the 1990s for reversing the present trend toward further degradation of marine ecosystems from overexploitation of resources and growing environmental stresses on coastal areas from around the globe. In this article, we use the term biodiversity to represent the totality of species in a region. In the context of an entire marine ecosystem, biodiversity refers to communities of species and their regional environments. From a management perspective, biodiversity conservation seeks to meet human needs from biological resources, while ensuring the long-term *sustainability* of global biotic wealth, including the existing pool of genetic diversity. In the closing hours of 1993, the United Nations Convention on Biological Diversity, a follow-on to UNCED, was signed by 167 nations. The treaty requires that signatories prepare national strategies to conserve plants, animals, and microorganisms within their national borders, including the habitats (ecosystems) that sustain them. It requires that countries pass laws to protect endangered species, expand protected areas, restore damaged areas, and promote public awareness of the need for conservation and *sustainable use of biological resources*. Sustainable use can be best defined as the use of renewable resources at rates within their capacity for renewal. An economy that is developing in a sustainable manner does so with consideration for balancing the matter and energy of the ecosystem at a level that will allow the system to function and renew itself year after year. [*See* BIODIVERSITY.]

The concept of sustainability as applied to marine resources and ecosystems has been the subject of recent debate in the scientific literature, much of it concerned with the application of science in support of resource management. In the course of the debate on the appropriate balance between science and policy, the ecologist C. S. Holling has recognized that there is presently emerging a new dimension in the application of science to natural resource sustainability issues. It is a science that is multidisciplinary and focused on populations and ecosystems on large spatial scales that are more appropriate to the issue of resource sustainability compared to the more traditional, but nonetheless important, disciplinary and reductionist ecological studies conducted in the absence of any socioeconomic considerations. The definition of sustainability used by Holling[1] and carried forward in this perspective focuses on ecosystem studies in support of "the social and economic development of a region with the goals to invest in the maintenance and restoration of critical ecosystem functions, to synthesize and make accessible knowledge and understanding for economies and to develop and communicate the understanding that provides a foundation of trust for citizens."

In practice, therefore, it would be important to establish institutional arrangements for ensuring that appropriate socioeconomic considerations are exercised in the application of science in support of regimes aimed at the sustainability of renewable resources. Regional examples of this approach to ecosystem sustainability can be found in the objectives of the Convention for the Conservation of Antarctic Marine Living Resources[2] and the ministerial declarations for the protection of the Black Sea[3] and the North Sea.[4] The Black Sea Declaration refers specifically to the objectives of UNCED

---

[1]Holling, C. S. (1993). Investing in research for sustainability. *Ecol. Appl.* **3,** 552–555.

[2]Scully, R. T. (1993). Convention on the Conservation of Antarctic Marine Living Resources. *In* "Stress, Mitigation, and Sustainability of Large Marine Ecosystems." (K. Sherman, L. M. Alexander, and B. D. Gold, eds.), pp. 242–251. Washington, D.C.: AAAS Press.

[3]Hey, E., and Mee, L. D. (1993). Black Sea. The ministerial declaration: An important step. *Environ. Pol. Law* **2315,** 215–217; 235–236.

[4]North Sea Task Force (1991). "Scientific Activities in the Framework of the North Sea Task Force," North Sea Environment Report No. 4. London: North Sea Task Force, Oslo and Paris Commissions, International Council for the Exploration of the Sea.

Agenda 21, Chapter 17, which calls for integrated management and sustainable development of coastal areas, marine environmental protection, sustainable use and conservation of living resources under national jurisdiction, and the need for addressing critical uncertainties for the management of the marine environment and strengthening of international and regional cooperation and coordination.

## II. INSTITUTIONAL FRAMEWORK: LARGE MARINE ECOSYSTEMS

Mitigating actions to reduce stress on marine ecosystems are required to ensure the long-term sustainability of marine resources. The principles adopted by coastal states under the terms of the United Nations Convention for the Law of the Sea (UNCLOS III) have been interpreted as supportive of the management of living marine resources and coastal habitats from an ecosystems perspective. However, at present no single international institution has been empowered to monitor the changing ecological states of marine ecosystems and to reconcile the needs of individual nations with those of the community of nations in taking appropriate mitigation actions. In this regard, the need for a regional approach to implement research, monitoring, and stress mitigation in support of marine resources development and sustainability at less than the global level has been recognized from a strategic perspective. Achievement of UNCED goals will require the implementation of a new paradigm aimed at greater integration of the highly sectorized approach to solving problems of coastal habitat degradation, marine pollution, and the overexploitation of fisheries than has been practiced in ocean monitoring and management by coastal nations during most of this century. It will also require a working partnership between the developed and developing nations of the world. Such an approach should be based on principles of ecology and sustainable development. [*See* MARINE BIOLOGY, HUMAN IMPACTS.]

An ecological framework that may be useful in achieving the UNCED objectives is the large marine ecosystem concept. LMEs are areas that are being subjected to increasing stress from growing exploitation of living marine resources, coastal zone damage, habitat losses, river basin runoff, dumping of urban wastes, and fallout from aerosol contaminants. The LMEs are regions of ocean space encompassing coastal areas from river basins and estuaries out to the seaward boundary of continental shelves and the seaward margins of coastal current systems. They are relatively large regions on the order of 200,000 $km^2$ or larger, characterized by distinct bathymetry, hydrography, productivity, and trophically dependent populations. The theory, measurement, and modeling relevant to monitoring the changing states of LMEs are imbedded in reports on multistable ecosystems and on the pattern formation and spatial diffusion within ecosystems.

From the ecological perspective, the concept that critical processes controlling the structure and function of biological communities can best be addressed on a regional basis has been applied to ocean space in the utilization of marine ecosystems as distinct units for marine research, monitoring, and management. The concept of monitoring and managing renewable resources from an LME perspective has been the topic of a series of national and international symposia and workshops initiated in 1984 and continuing through 1993, wherein the geographic extent of each region is defined on the basis of ecological criteria. The spawning and feeding migrations of fish communities within the LMEs have evolved in response to the distinct bathymetry, hydrography, productivity, and trophodynamics of the systems. Because the spatial dimensions of biological and physical processes directly influencing the success of population renewals within the regions under consideration are large, the term LME is used to characterize them. Several LMEs are in semienclosed seas, including the Black, the Baltic, the Mediterranean, and the Caribbean seas. Within the extent of LMEs, domains or subsystems can be characterized. For example the Adriatic Sea is a subsystem of the Mediterranean Sea LME. In other LMEs, geographic limits are defined by the scope of continental shelves. Among these are the U.S. Northeast Continental Shelf and its four subsystems—the Gulf of Maine, Georges Bank, Southern New England,

and the Mid-Atlantic Bight—and the Icelandic Shelf and the Northwestern Australian Shelf. For LMEs with narrow shelf areas and well-defined currents, the seaward boundaries are limited to the areas affected by coastal currents, rather than relying on the 200-mile Exclusive Economic Zone (EEZ) limits. Among the coastal current LMEs are those within the boundaries of the Humboldt, California, Canary, Kuroshio, and Benguela currents. It is the coastal ecosystems adjacent to the landmasses that are being stressed from habitat degradation, pollution, and overexploitation of marine resources. Nearly 95% of the usable annual global biomass yield of fish and other living marine resources is produced in 49 LMEs identified within, and in some cases extending beyond, the boundaries of the EEZs of coastal nations located around the margins of the ocean basins (Fig. 1). [*See* Continental Shelf Ecosystems.]

## III. RESOURCES AT RISK

Levels of primary production are persistently higher around the margin of the ocean basins than in the open-ocean pelagic areas of the globe. It is within these coastal ocean areas that pollution has its greatest impact on natural productivity cycles, including eutrophication from high nitrogen and phosphorus effluent from estuaries. The presence of toxins in poorly treated sewage discharge, harmful algal blooms, and loss of wetland nursery areas to coastal development are also ecosystem-level problems that need to be addressed.[5] Within several of the coastal LMEs, overfishing has caused biomass flips among the dominant pelagic components of fish communities, resulting in multimillion metric ton losses in potential biomass yield as less-economically desirable species have replaced commercially preferred stocks. The biomass flip, wherein a dominant species rapidly drops to a low level to be succeeded by another species, can generate cascading effects among other important components of the ecosystem, including marine birds, marine mammals, and zooplankton. In addition, recent studies indicate that climate and natural environmental changes are prime driving forces of variability in fish population levels. The growing awareness that biomass yields are being influenced by multiple driving forces in marine ecosystems around the globe has accelerated efforts to broaden monitoring strategies to encompass food chain dynamics and the effects of environmental perturbations and pollution on living marine resources from an ecosystem perspective.

In addition to primary sources of stress, LMEs may also experience secondary and even tertiary forms of stress, although it is often necessary to monitor and study an ecosystem over a considerable period of time to identify with certainty these auxiliary forces for change. Even the identification of a primary stress may at times be difficult. In the case of the Humboldt Current LME, for example, the spectacular decline in anchovy stocks off Peru in the 1950s resulted from a combination of environmentally induced changes in productivity and excessive fishing effort.

Pollution, as a dominant or even secondary source of stress in an LME, is a complex process whose impacts on the composition of the fishery biomass are still imperfectly understood. One source of pollution results from the entry into ocean waters of excessive coastal loading of nutrients—particularly nitrates and phosphates —which come from land-based sources. Rivers also discharge contaminants into the sea, including PCBs, other organochlorines, heavy metals, and toxins.

Environmental and ecosystem degradation may result from man-made changes in the shoreline, such as filling in of wetlands that serve as breeding and nursery grounds for many of the fish stocks of the LME, or dredging and filling in shallow waters immediately offshore—again affecting potential aquatic habitats. Human-induced activities may also interfere with the flow of waters and of nutrients in rivers. For example, Egypt's marine fish catch dropped significantly following the con-

[5]GESAMP [Group of Experts on the Scientific Aspects of Marine Pollution] (1990). "The State of the Marine Environment," UNEP Regional Seas Reports and Studies No. 115. Nairobi: U.N. Environment Program.

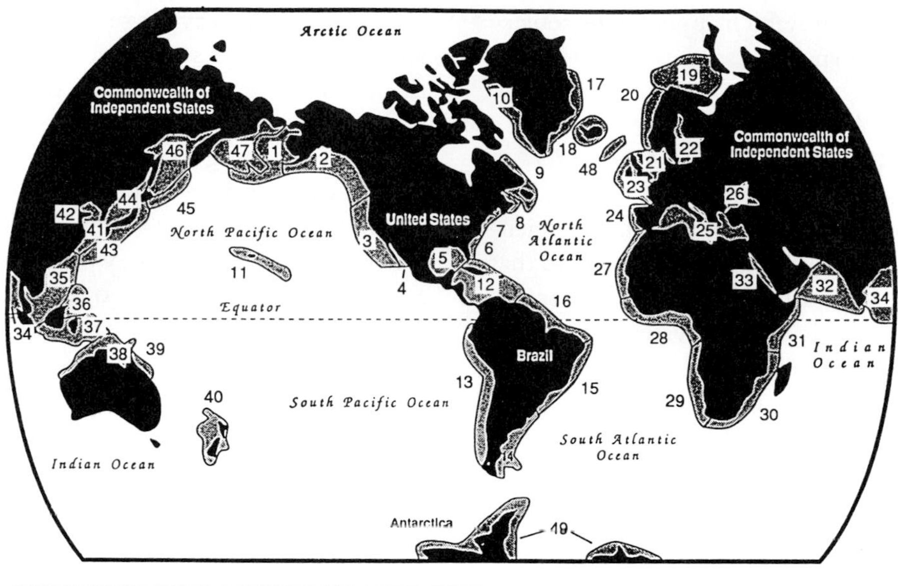


WORLD MAP OF LARGE MARINE ECOSYSTEMS

1. Eastern Bering Sea
2. Gulf of Alaska
3. California Current
4. Gulf of California
5. Gulf of Mexico
6. Southeast U.S. Continental Shelf
7. Northeast U.S. Continental Shelf
8. Scotian Shelf
9. Newfoundland Shelf
10. West Greenland Shelf
11. Insular Pacific--Hawaiian
12. Caribbean Sea
13. Humboldt Current
14. Patagonian Shelf
15. Brazil Current
16. Northeast Brazil Shelf
17. East Greenland Shelf
18. Iceland Shelf
19. Barents Sea
20. Norwegian Shelf
21. North Sea
22. Baltic Sea
23. Celtic-Biscay Shelf
24. Iberian Coastal
25. Mediterranean Sea
26. Black Sea
27. Canary Current
28. Guinea Current
29. Benguela Current
30. Agulhas Current
31. Somali Coastal Current
32. Arabian Sea
33. Red Sea
34. Bay of Bengal
35. South China Sea
36. Sulu-Celebes Seas
37. Indonesian Seas
38. Northern Australian Shelf
39. Great Barrier Reef
40. New Zealand Shelf
41. East China Sea
42. Yellow Sea
43. Kuroshio Current
44. Sea of Japan
45. Oyashio Current
46. Sea of Okhotsk
47. West Bering Sea
48. Faroe Plateau
49. Antarctic

***FIGURE 1*** Boundaries of 49 large marine ecosystems.

struction of the Aswan Dam in the 1960s and the harvest of small pelagics declined from 25,000 to about 3000 tons in the mid-1970s.

The impacts of nutrients vary widely. Nutrients enhance productivity, as illustrated in upwelling ecosystems such as the Humboldt Current off Chile and Peru and the Benguela Ecosystem off South Africa and Mombasa; they have long been noted as regions of high plankton stocks and productivity. Yet too great a level of nutrients produces conditions of eutrophication. In the Black Sea, for example, pollution has resulted in periodic oxygen depletions, the reduction in the number of commercially viable species from 26 to 5, and a massive introduction of jellyfish, which appear to be replacing some of the fish species.

## IV. LMES AND CARRYING CAPACITY

Carrying capacity is a term used by marine scientists to depict the level of biomass and/or yield that can be sustained in the long term in relation to the productivity of a particular ecosystem or region with regard to the diversity, structure, and trophodynamics of populations and communities within the system. In relation to yield, estimates of primary production can serve as an index of the carrying capacity of a marine ecosystem to support a sustainable surplus of fish production for harvesting.

The carrying capacity on a sustainable yield basis for the global ocean has been estimated at between 100 million and 120 million metric tons per year. It should be emphasized, however, that when using the term carrying capacity for marine populations, communities, habitats, and environments within LMEs, it is important to underscore the dynamic aspects of the term because spatial and temporal variations in the carrying capacity of LMEs need to be included in any estimate. The results of a series of studies on the magnitude and trends of changes affecting the carrying capacity and biomass yields of 29 LMEs have been published in several volumes. A summary of the ecosystems investigated, along with volume citations and principal authors of the analyses, is listed in Table I. Examples of the temporal variability in biomass yields are given in Figs. 2 through 5 for several LMEs.

## V. ECOSYSTEM INFORMATION NEEDS

There is a growing awareness among marine scientists, geographers, economists, government representatives, and lawyers of the utility of a more holistic ecosystem approach to resource management. On a global scale the loss of sustained biomass yields from mismanagement and overexploitation has not been fully investigated but is unquestionably very large. Effective management strategies for LMEs will be contingent on the identification of the major driving forces causing large-scale changes in biomass yields. Management of species responding to strong environmental signals will be enhanced by improving the understanding of the physical factors forcing biological changes; whereas in other LMEs, where the prime driving force is overexploitation, options can be explored for implementing adaptive management strategies.

It will also be necessary to conduct supportive research on the processes controlling sustained productivity of LMEs. Within several of the LMEs adjacent to the United States, including the Northeast Shelf, Gulf of Mexico, California Current, and Eastern Bering Sea, important hypotheses concerned with the growing impacts of pollution, overexploitation, and environmental changes on sustained biomass yields are under investigation.[6] By comparing the results of research among the different systems, it should be possible to accelerate an understanding of how the systems respond and recover from stress. The comparisons should allow for narrowing the context of unresolved problems and capitalizing on research efforts under way in the various ecosystems.

Initial efforts to examine changing ecosystem states and relative health within a single ecosystem are under way for four subareas of the U.S. Northeast Continental Shelf Ecosystem—the Gulf of Maine, Georges Bank, Southern New England, and the Mid-Atlantic Bight. Early findings on the structure, function, and productivity of the system have been reported. It appears that the principal driving force in relation to sustainable ecosystem yield is fishing mortality expressed as predation on the fish stocks of the system, and that long-term sustainability of high economic yield species will be dependent on the application of significantly reduced fishing mortality measures and adaptive management strategies. Several alternative management strategies for reducing fishing effort on

[6]Sherman, K. (1992). Large marine ecosystems. *In* "Encyclopedia of Earth System Science" (W. A. Nierenberg, ed.), Vol. 2, pp. 653–673. San Diego, Calif.: Academic Press.

***TABLE I***

List of 29 Large Marine Ecosystems and Subsystems for which Syntheses Relating to Principal, Secondary, or Tertiary Driving Forces Controlling Variability in Biomass Yields Have Been Completed by February 1993

| Large marine ecosystem | Volume No.[a] | Authors |
|---|---|---|
| U.S. Northeast Continental Shelf | 1 | M. Sissenwine |
| | 4 | P. Falkowski |
| U.S. Southeast Continental Shelf | 4 | J. Yoder |
| Gulf of Mexico | 2 | W. J. Richards and M. F. McGowan |
| | 4 | B. E. Brown *et al.* |
| California Current | 1 | A. MacCall |
| | 4 | M. Mullin |
| | 5 | D. Bottom |
| Eastern Bering Shelf | 1 | L. Incze and J. D. Schumacher |
| West Greenland Shelf | 3 | H. Hovgaard and E. Buch |
| Norwegian Sea | 3 | B. Ellersten *et al.* |
| Barents Sea | 2 | H. R. Skjoldal and F. Rey |
| | 4 | V. Borisov |
| North Sea | 1 | N. Daan |
| Baltic Sea | 1 | G. Kullenberg |
| Iberian Coastal | 2 | T. Wyatt and G. Perez-Gandaras |
| Mediterranean–Adriatic Sea | 5 | G. Bombace |
| Canary Current | 5 | C. Bas |
| Gulf of Guinea | 5 | D. Binet and E. Marchal |
| Bengucla Current | 2 | R. J. M. Crawford *et al.* |
| Patagonian Shelf | 5 | A. Bakun |
| Caribbean Sea | 3 | W. J. Richards and J. A. Bohnsack |
| South China Sea–Gulf of Thailand | 2 | T. Piyakarnchana |
| Yellow Sea | 2 | Q. Tang |
| Sea of Okhotsk | 5 | V. V. Kusnetsov |
| Humboldt Current | 5 | J. Alhcit and P. Bernal |
| Indoncsia Seas–Banda Sea | 3 | J. J. Zijlstra and M. A. Baars |
| Bay of Bengal | 5 | S. N. Dwivedi |
| Antaractic Marine | 1,5 | R. T. Scully *et al.* |
| Weddell Sea | 3 | G. Hempel |
| Kuroshio Current | 2 | M. Terazaki |
| Oyashio Current | 2 | T. Minoda |
| Great Barrier Reef | 2 | R. H. Bradbury and C. N. Mundy |
| | 5 | G. Kelleher |
| South China Sea | 5 | D. Pauly and V. Christensen |

[a] Vol. 1: "Variability and Management of Large Marine Ecosystems," AAAS Selected Symposium 99, Westview Press, Boulder, Colo., 1986. Vol. 2: "Biomass Yields and Geography of Large Marine Ecosystems," AAAS Selected Symposium 111, Westview Press, Boulder, Colo., 1989. Vol. 3: "Large Marine Ecosystems: Patterns, Processes, and Yields," AAAS Symposium, AAAS Press, Washington, D.C., 1990. Vol. 4: Food Chains, Yields, Models, and Management of Large Marine Ecosystems," AAAS Symposium, Westview Press, Boulder, Colo., 1991. Vol. 5: Stress, Mitigation, and Sustainability of Large Marine Ecosystems," AAAS Press, Washington, D.C., 1993.

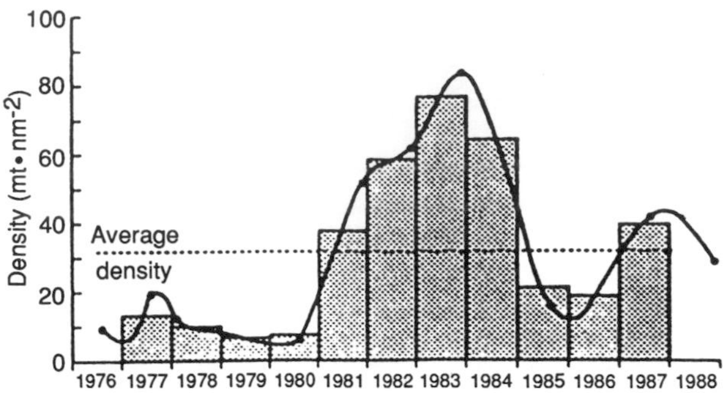


***FIGURE 2*** Sardine biomass from echo-surveys of the northern and central Adriatic Sea during 1976–1988.

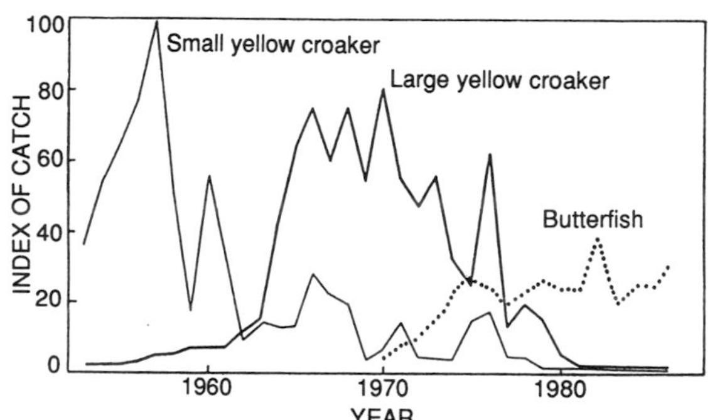


***FIGURE 3*** Changes in annual catch of dominant species of the fishery off Jiangsu in the southern Yellow Sea.

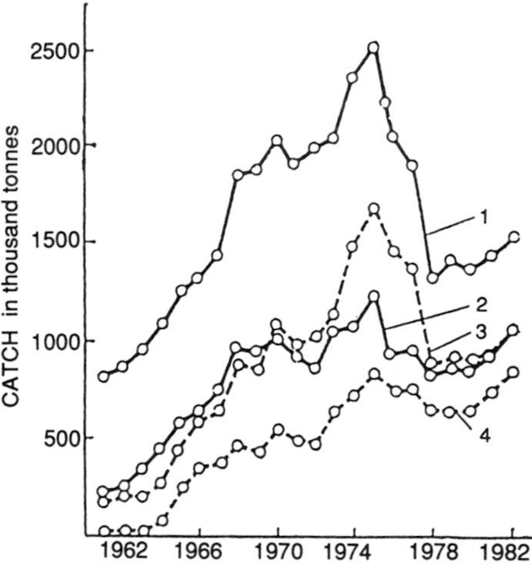


***FIGURE 4*** Fish catch in the Sea of Okhotsk, 1962–1982. (1) Total catch, all countries; (2) USSR catch; (3) total walleye pollock catch, all countries; (4) USSR walleye pollock catch. [From Shuntov, V. P. (1985). "Biological Resources of the Sea of Okhotsk." Moscow: Agropromizdat. In Russian.]

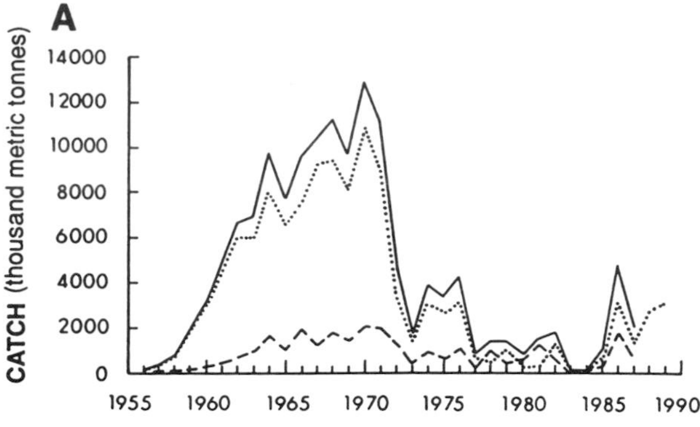


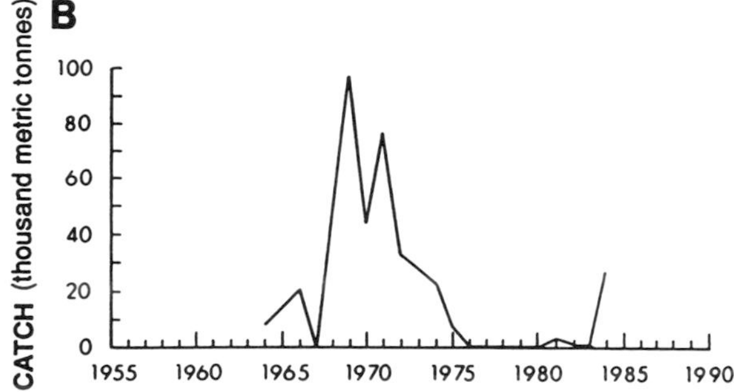


***FIGURE 5*** Catches of anchovy in the Humboldt Current Ecosystem. (A) Anchovy, total (solid line); northern/central stock off Peru (dotted line); southern stock off Peru and northern stock off Chile (dashed line). (B) Stock off Talcahuano in Chile.

the fish stocks of the U.S. Northeast Continental Shelf Ecosystem are under consideration by the New England Fisheries Management Council and the Atlantic States Marine Fisheries Commission. In addition to fisheries management issues and significant biomass flips among dominant species, the Northeast Continental Shelf Ecosystem is also under stress from the increasing frequency of unusual plankton blooms and eutrophication within the nearshore coastal zone resulting from high levels of phosphate and nitrate discharges into drainage basins. Whether the increases in the frequency and extent of nearshore plankton blooms are responsible for the rise in incidence of biotoxin-related shellfish closures, marine mammal mortalities, and the recent discovery of disease-bearing viruses[7] remains an important open question that is the subject of considerable concern to state and federal management agencies.

[7]Epstein, P. R., Ford, T. E., and Colwell, R. R. (1993). Marine ecosystems. *The Lancet* **342,** 1216–1219.

## VI. MANAGING LARGE MARINE ECOSYSTEMS

The concept of managing LMEs raises a whole series of questions. What might be the rationale for a management scheme, and what would be the goals of management? How could agreements be reached among all, or at least some, of the countries bordering an LME both on the goals of management and on the methods to be used in attaining these goals? How much "investment" would be required of the participants—in terms of funds, personnel, and the sharing of decision-making powers in the interests of areawide management? What types of organizational structure would be needed to handle the management effort? These and other issues must be addressed if a successful LME management program is to be put into place.

The term "management" might be defined as the regulation of activities and resources to achieve certain objectives. In the ocean environment, regulations may pertain to both resource and nonresource uses, or, alternatively, to holistic multiple-use aspects of certain geographic regions, such as the North or Black seas, or the Caribbean. Among the desired goals sought from a management system are protecting the health of the marine environment (including the preservation of ecologically sensitive areas), obtaining greater wealth from the sea, accommodating use conflicts, or increasing scientific knowledge of global or regional systems. Any management process involves choices among alternative actions. The more that is known by decision makers about the costs and benefits of each of the alternatives, the greater is the likelihood that successful management programs will be implemented.

### A. Stages in a Management Process

Essentially, there are four stages in a management process. The first is *data acquisition, assessment, and monitoring*—an activity that must continue as long as management is practiced. In the case of LMEs, data are acquired from selected observations of key processes within the ecosystem. From an initial data set, an assessment of the ecosystem state is made, followed by decadal time-series monitoring of key ecosystem components to provide annual assessments of the changing states (health) of the ecosystem. The longer the time series in the monitoring process, and the greater the density of observation stations, the closer analysts can come to understanding the nature of variations of the ecosystem. Obviously, in considering 49 LMEs, the degree of scientific information available to planners and managers varies widely from place to place. At present, for 29 of the LMEs, data acquisition, assessment, and monitoring represent the principal stage in the management process that has been attained.

The second stage is *planning,* involving both the establishment of management objectives and developing a strategy for action in order to achieve these objectives, whether they be in the long- or short-term time frame. A great deal has been written recently about "sustainable development," implying that planners should concentrate on long-term objectives, even at the expense of short-term goals. Planning involves the setting of priorities—often a very difficult task—as well as making commitments to invest (particularly in the sharing of decision-making powers) in the interests of the management plan.

The third process is *implementation* of the management plan, that is, putting management decisions into practice. This may involve considerable costs to one or more of the parties to a management program; if the costs are seen as being unevenly allocated to certain groups, the implementation process itself may be endangered. All manner of trade-offs can be involved as the accommodation of interests is negotiated. Furthermore, the actual impacts of the implementation process may turn out to be completely different from what was anticipated by the program planners.

The final stage is the *feedback system,* where the results of management planning and implementation can be analyzed and evaluated and necessary adjustments made. The system of feedback should be a continuing one during the whole management process.

Remedial actions are required to ensure that the "pollution" of the coastal zone of LMEs is reduced and does not become a principal driving force in any LME. For at least one LME, the Antarctic, a management regime has evolved based on an ecosystem perspective in the adoption and implementation of the aforementioned Convention for the Conservation of Antarctic Marine Living Resources. Other LMEs for which management regimes have been proscribed include the North Sea and the Black Sea. Efforts are also under way to implement ecosystem management within the LMEs of the U.S. EEZ, for example, in the northern California Current Ecosystem.[8] Concerns remain regarding the socioeconomic and political difficulties in management across national boundaries, as in the case of the Sea of Japan Ecosystem where the fishery resources are shared by 5 countries, or the Caribbean Sea Ecosystem where 38 nations share the resources.

A systems approach to the management of LMEs is depicted in Table II. The LMEs represent the link between local events (e.g., fishing, pollution, environmental disturbance) occurring on the daily-to-seasonal temporal scale and their effects on living marine resources, and the more ubiquitous global effects of climate changes on the multidecadal time scale. The regional and temporal focus of season to decade is consistent with the evolved spawning and feeding migrations of the fishes. These migrations are seasonal and occur over hundreds to thousands of kilometers within the unique physical and biological characteristics of the regional LME to which they have adapted. As the fisheries represent most of the usable biomass yield of the LMEs and fish populations consist of several age-classes, it follows that measures of variability in growth, recruitment, and mortality should be conducted over multiyear time scales. Similarly, changes in populations of marine mammals and marine bird species will require multiple-year time-series observations. It is necessary to consider the naturally occurring environmental events and the human-induced perturbations, including coastal pollution, affecting demography of the populations within the ecosystem. Based on scientific inferences of the principal causes of variability in abundance, with due consideration to socioeconomic needs, management options from an ecosystems perspective can be considered for implementation. The final element in the system, with regard to the concept of resource maintenance and sustained yield, is the feedback loop that allows for evaluation of the effects of management actions that consider both fisheries and ecosystem health.

Costs and benefits of any management arrangement, ocean-related or otherwise, are important issues to consider. If the program is a voluntary one (and virtually all ocean-related management programs are), participants must be made to feel that the benefits derived by them through participation outweigh, or at least equal, whatever costs accrue to the individual members. It should be noted that States in a well-defined region, such as a semienclosed sea, may come under pressure to participate in a regional program, even if there appear to be few benefits to the State from such participation. During the mid-1970s, officials of the United Nations Environment Program (UNEP) worked hard to secure the participation of all 18 States bordering the Mediterranean in the Regional Seas Action Plan. Eventually, they succeeded in bringing in 17 of the countries; only Albania held out. Peer pressure might prove to be an important cohesive force in certain LME management efforts.

## B. Parameters of an LME Management Program

Any regional ocean management system has a number of issues that must be considered if a successful program is to be put into place. Among these are the objectives of the management effort, the jurisdictional questions involved, and the nature of the program.

### *1. Objectives of a Management Program*

A primary objective of an LME management program should be the achievement of the long-term sustainability of living marine resources, particularly of commercially valuable species within the framework of reasonable economic benefits. The

[8]Bottom, D. L., Jones, K. K., Rodgers, J. D., and Brown, R. F. (1989). "Management of Living Resources: A Research Plan for the Washington and Oregon Continental Margin," NCRI-T-89-004. Newport, Ore.: National Coastal Resources Research Development Institute.

**TABLE II**
Key Spatial and Temporal Scales and Principal Elements of a Systems Approach to the Research and Management of Large Marine Ecosystems

| | | | | |
|---|---|---|---|---|
| 1. | Spatial–Temporal Scales | | | |
| | | Spatial | Temporal | Unit |
| | 1.1 | **Global** (world ocean) | Millennia–decadal | Pelagic biogeographic |
| | 1.2 | **Regional** (Exclusive Economic Zones) | Decadal–seasonal | Large marine ecosystems |
| | 1.3 | **Local** | Seasonal–daily | Subsystems |
| 2. | Research Elements | | | |
| | 2.1 | Spawning strategies | | |
| | 2.2 | Feeding strategies | | |
| | 2.3 | Productivity, trophodynamics | | |
| | 2.4 | Stock fluctuations/recruitment/mortality | | |
| | 2.5 | Natural variability (hydrography, currents, watermasses, weather) | | |
| | 2.6 | Human perturbations (fishing, waste disposal, petrogenic hydrocarbon impacts, toxic effects, aerosol contaminants, eutrophication effects, pollution effects, viral disease vectors) | | |
| 3. | Management Elements—Options and Advice—International, National, Local | | | |
| | 3.1 | Bioenvironmental and socioeconomic models | | |
| | 3.2 | Management to optimize sustainable fisheries yields | | |
| | 3.3 | Mitigation of pollution stress; improvement of ecosystem "health" | | |
| 4. | Feedback Loop | | | |
| | 4.1 | Evaluation of ecosystem "health" | | |
| | 4.2 | Evaluation of fisheries status | | |
| | 4.3 | Evaluation of management practices | | |

abundance of such species may fluctuate over time as a result of intensive fishing of the stock, environmental changes, or other causes of stress. In the interests of conservation, various regulatory measures may be introduced, such as annual quotas on total catch, increased mesh sizes, closed areas at certain times of the year or closed seasons for fishing over broad regions, limits on the size of the fish that can be retained on board, or restrictions on the number of days that fishing vessels can be at sea.

The results of such measures are not always successful. In the U.S. Northeast Continental Shelf LME, for example, foreign fishing has been virtually excluded since the beginning of 1977,[9] a great deal of fisheries data have been acquired and analyzed over the past several decades, and a plethora of regulations have been imposed on the commercial fisherman. Yet the stocks of gadoids (cod, haddock) and other bottom fish (flounders) have declined to record lows and, for some of these stocks, commercial fishing is close to disappearing. On Georges Bank, the richest fishing ground in the LME, the percentage of spiny dogfish and skates, lower trophic level species, has increased from 25% during the 1960s to 75% in recent years.

A second management objective for LMEs is mitigation through adaptive management. In this case, efforts are made to manipulate the population of certain species in order to improve the conditions of other stocks that are of higher value. It has been argued by some scientists, for example, that extensive harvesting of the spiny dogfish and skates in the Northeast Continental Shelf LME would reduce the pressure on juvenile gadoids; these juveniles are among the prey of the dogfish and skates, and a change in this predator–prey relationship would permit the stocks of cod, haddock, and

[9]Under the terms of the Magnuson Fishery Conservation and Management Act, 1976.

flounder to improve. But why would a commercial fisherman fill his vessel's hold with dogfish and skates in preference to species such as squid, hake, and butterfish, unless some subsidy were offered him, presumably by the U.S. government, for concentrating on the less valuable species? Also, commentators who are knowledgeable about the Northeast Shelf fishery maintain that large-scale ecological tampering might produce results other than those anticipated.

Adaptive management programs can in theory be applied to LMEs where overfishing of preferred stocks is a major driving force for change in the species composition of the fishery biomass. Adaptive management could also be used in the case of coastal pollution, including the filling in of wetlands. Environmental protection and preservation measures affecting these issues could be adopted and enforced. The efforts of the Australians to curb the damage caused by the crown-of-thorns starfish provide an example of attacking the impacts of natural predation. However, adaptive management efforts would have little effect where environmental change is the principal driving force.

A third management objective might be to maximize the economic benefits to the fishermen from the utilization of the resources of the LME. Rather than having "too many boats chasing too few fish" in an open-access fishery, with the result that the average catches per fisherman of commercially valuable species tend to be small and economically unrewarding, the management plan would result in reduction of the total harvesting capacity targeted for the species. Normally, this would mean reducing the number of boats, thereby materially increasing the harvesting potential of the participating boats. There are, however, other measures which might be taken to implement a limited entry system, one of which would be to reduce the number of days that fishing vessels may spend at sea. Assuming that the commercially valuable biomass remains approximately the same over a period of time, reducing effort means that those vessels still in the fishery would have larger and more profitable catches than would be the case in an open-access fishery.

One question to ask in the early stages of a management process is whether the ultimate goal is a fully integrated, or "holistic" plan, which concerns itself with maintaining or restoring the health of the fishery biomass and the health of the ecosystem. A fully integrated plan takes into account other elements of the ecosystem, such as the mitigation of stress on degraded habitats, nonliving resource uses (including offshore installations and undersea cables and pipelines), productivity, coastal and offshore pollution, and interactions with neighboring ecosystems.

Any regional ocean management scheme will initially focus on one or a few specific elements and will evolve, over time, into a more holistic approach developed around four general principles:[10]

1. The ecosystem should be maintained in a desirable state such that
   a. consumptive and nonconsumptive values could be maximized on a continuing basis,
   b. present and future options are ensured, and
   c. the risk of irreversible change or long-term adverse effects as a result of use is minimized.
2. Management decisions should include a safety factor to allow for the fact that knowledge is limited.
3. Measures to conserve a wild living resource should be formulated and applied to avoid wasteful use of other resources.
4. Survey or monitoring, analysis, and assessment should precede planned use and accompany actual use of wild living resources. The results should be made available promptly for critical review.

## 2. Research Needs

In planning a management scheme, consideration should be given to: (1) the incentives for Nations or other entities to enter into a regional management scheme; (2) the costs and benefits of participation, in other words, what is the "investment" in terms of money, personnel, and a reduction in the participating parties freedom of authority; and (3) what are the hoped-for benefits regarding the individual

[10]Holt, S. J., and Talbot, L. M. (1978). New principles for the conservation of wild living resources. *Wild. Monogr.* **59.**

Nation's participation? Unless these incentives are made clear at the outset (they may, of course, change over time) it is difficult to envision a particular country being willing to participate in an LME management plan.

Note should be taken here of two jurisdictional aspects of international LME management opportunities. First is the number of littoral states that might be involved in a regional management scheme. Fourteen of the 49 LMEs are bordered by a single State and 6 others are shared by 2 countries. On the other hand, the Caribbean LME has 34 littoral States as well as the island territories of four non-Caribbean countries. The Mediterranean, as already noted, has 18 bordering States, whereas 10 countries border the South China Sea and 8 States line the Persian Gulf.

Of perhaps equal importance is the status of relationships among the littoral States. For example, in the South China Sea area are Vietnam, the People's Republic of China, and Taiwan; relations among these, as well as with some of the Association of Southeast Asian Nations (ASEAN) countries, are exacerbated by territorial disputes over the ownership of the Spratlys, the Paracels, and other island groups in the South China Sea. Iran, Iraq, Kuwait, and the United Arab Emirates (UAE) are among the countries bordering the Persian Gulf; there is an active dispute between Iran and the UAE over control of three small islands in the mid-Gulf area.

As in the case of other management arrangements between or among governments, there are opposing sets of forces: those that tend toward unity, strengthening the hand of the management authority, and those that oppose or weaken efforts toward joint operations. These forces often exist even in the early stages of a planning process, and their existence should be taken into account by the planners themselves.

One of the most important unifying forces is effective leadership—the presence of persons who are willing and able to promote new concepts of cooperation, risk taking, and commitment in the form of funds and shared authority. These people may be pioneers in cooperative action, but, as time passes, they must also gain a political constituency that can affect decision making within the various participating States. Developing such a constituency generally involves education, through the media, group action, and other means. The political constituency must also eventually include Congressmen or members of other types of legislative groups who are concerned with the obligations that governments incur in the interests of management objectives.

The political process of integration in a regional management program may be helped by the existence of one or more "exogenous" factors, which can affect the perceptions of both the public and the decision makers and create a willingness for cooperation in the realm of ocean management. The event may be a cataclysmic one (a devastating hurricane or massive oil spill) or else a growing awareness of a common problem. The threat of pollution of the coastal marine environment is one of the more likely forms of danger. Such events could serve as strong supports for the introduction or sustaining of a management program.

### *3. Existing Management Systems*

The holistic management of an entire LME is an evolving process, and during the past decade, some progress has been made. As governments move toward the ecosystem management objectives of Agenda 21, the number of reports in the literature describing advances toward ecosystem management is increasing.

For example, in the Yellow Sea LME, the Chinese have successfully introduced large numbers of juvenile fleshy prawns, which are "ideal species for artificial enhancement, since they have lower trophic levels, are short-lived, and have great commercial value."[11] Although the People's Republic of China is the only nation in the region engaged in the prawn enhancement program, two other

[11]Tang, Q. (1989). Changes in the biomass of the Yellow Sea ecosystems. *In* "Biomass Yields and Geography of Large Marine Ecosystems." (K. Sherman, L. M. Alexander, eds.), pp. 7–35. AAAS Selected Symposium 111. Boulder, Colo.: Westview Press, Inc.

Tang, Q. (1993). Effects of long-term physical and biological perturbations on the contemporary biomass yields of the Yellow Sea Ecosystem. *In* "Large Marine Ecosystems: Stress, Mitigation, and Sustainability." (K. Sherman, L. M. Alexander, and B. D. Gold, eds.), pp. 79–93. Washington, DC, AAAS Press.

countries—North and South Korea—also border the Yellow Sea and would be free to harvest these prawn resources east of China's EEZ. Efforts are now under way to establish joint monitoring, assessment, and management of the Yellow Sea Ecosystem by China and South Korea in collaboration with the World Bank, UNEP, and UNDP (the United Nations Development Program).

In the Adriatic Sea, the Italian government has erected eight artificial reefs that provide a refuge from bottom trawling for many organisms such as eggs, juveniles, and adults of pelagic and benthic fish and crustaceans. The results, in terms of unit fishing yields, have been very positive.

On Australia's Great Barrier Reef, a marine park has been established under the aegis of a Marine Park Authority, whose mandate is to ensure "that any use of the reef or associated areas should not threaten the reef's essential ecological characteristics and processes."[12] Here, in a unique marine area under the jurisdiction of one national government, is a pioneering example of a move toward a holistic approach to LME management. The Authority's principal challenges are: (1) to enhance the Great Barrier Reef and its valuable biodiversity as a marine tourist attraction; (2) to achieve a level of sustainable development of fishing and collecting; (3) to understand the crown-of-thorns phenomenon, whereby coral resources are threatened; and (4) to protect the Great Barrier Reef from pollution, particularly from land-based sources. [*See* Coral Reef Ecosystems.]

There are two other areas where holistic management efforts are at least in the planning stage. One involves the Benguela Current LME of the Southeast Atlantic, an LME that South Africa shares with its former territory of Namibia. It is recognized that "fish populations should not be exploited at levels that would unacceptably disrupt the ecological relationships between exploited, dependent, and related species. . . . Under these circumstances, multiple-criteria decision support systems, using methods such as interactive goal programming, are of value because they allow a number of objectives to be evaluated simultaneously."[13] To date these decision support systems have not yet been implemented.

A second planning effort involves the Northern California Current Ecosystem. Here, there is recognition of a need for a Regional Research Plan involving the environmental consequences of oil and gas leasing activities off the coasts of Oregon and southern Washington, the mining of mineral-rich sands, and the possible future dredging of sand and gravel. In response, scientists in the state of Oregon have drafted an Oregon Resources Management Plan, described as "one hopeful step toward integrated management in the Pacific Northwest."[14]

#### *4. LME Management Trends*

The concept of large marine ecosystems as distinct global management entities appears to be gaining acceptance among marine scientists. Steps are being taken in various ecosystems to systematically acquire and analyze data and to develop models of trends in the changing states and health of the various systems.

The 49 large marine ecosystems that have been identified are located around the margins of the ocean basins and extend over the coastlines of several countries. They are in regions of the world oceans that are most affected by overexploitation, pollution, and habitat degradation, and collectively represent target areas for mitigation efforts. A considerable number of international and national organizations, as well as scientists from national marine resource agencies, are prepared to support LME assessment, monitoring, and mitigation activities. Among these organizations are the Global Environment Facility of UNEP, UNDP, and the World Bank—in collaboration with the U.S. National Oceanographic and Atmospheric Administration of

[12]Kelleher, G. (1993). Sustainable development of the Great Barrier Reef as a large marine ecosystem. *In* "Large Marine Ecosystems: Stress, Mitigation, and Sustainability." (K. Sherman, L. M. Alexander, and B. D. Gold, eds.), pp. 272–279. Washington, D. C., AAAS Press.

[13]Crawford, R. J. M., Shannon, L. V., and Shelton, P. A. (1989). Characteristics and management of the Benguela as a large marine ecosystem. *In* "Biomass Yields and Geography of Large Marine Ecosystems." (K. Sherman and L. M. ALexander, eds.), AAAS Selected Symposium 111, pp. 169–219. Boulder, Colo.: Westview Press.

[14]Bottom, D. L., Jones, K. K., Rodgers, J. D., and Brown, R. F. (1993). "Research and management in the Northern California Current Ecosystem." (K. Sherman, L. M. Alexander, and B. D. Gold, eds.), pp. 259–271. Washington, D. C., AAAS Press.

the Department of Commerce, IOC (Intergovernmental Oceanographic Commission), FAO (Food and Agriculture Organization of the U.N.), the U.K.'s Sir Alister Hardy Science Foundation for Ocean Science, and the U.S. National Environmental Research Council. There is strong support for the LME concept among scientists from marine resource agencies in Belgium, Cameroon, China, Denmark, Estonia, Germany, Ivory Coast, Japan, Kenya, South Korea, the Netherlands, Nigeria, Norway, the Philippines, Poland, and Thailand.

## VII. ECOSYSTEM ASSESSMENT AND SUSTAINABLE DEVELOPMENT

In the United States, within the National Marine Fisheries Service of NOAA, greater emphasis has been focused over the past decade on approaching fisheries research from a regional ecosystem perspective in seven of U.S. coastal waters LMEs: (1) the Northeast Continental Shelf, (2) the Southeast Continental Shelf, (3) the Gulf of Mexico, (4) the California Current, (5) the Gulf of Alaska, (6) the Eastern Bering Sea, and (7) the Insular Pacific including the Hawaiian Islands. In 1991, these ecosystems yielded 9.5 million metric tons of fisheries biomass valued at approximately $50 billion to the economy of the United States.[15]

### A. Sampling Systems

The sampling programs providing the biomass assessments within the LMEs have been described in Folio Map 7 produced by the Office of Oceanography and Marine Assessment of NOAA's National Ocean Service. The seven ecosystems under investigation are shown in Fig. 6. Sampling programs supporting biomass estimates in the LMEs are designed to: (1) provide detailed statistical analyses of fish and invertebrate populations constituting the principal yield species of biomass, (2) monitor changes in the principal populations and their environments, and (3) estimate future trends in biomass yields. The information obtained by these programs provides managers with a more complete understanding of the dynamics of marine ecosystems and how these dynamics affect harvestable stocks. Additionally, by tracking components of the ecosystems, these programs can detect changes, natural or human-induced, and warn of events with possible economic repercussions.

Although sampling schemes and efforts vary among programs (depending on habitats, species present, and specific regional concerns), they generally involve systematic collection and analysis of catch statistics and extensive collaborative efforts with other agencies and with academic institutions in the use of NOAA vessels for fisheries-independent bottom and midwater trawl surveys for adults and juveniles; ichthyoplankton surveys for larvae and eggs; measurements of zooplankton standing stock, primary productivity, nutrient concentrations, and important physical parameters (e.g., water temperature, salinity, density, current velocity and direction, air temperature, cloud cover, light conditions); and measurements of contaminants and their effects on living marine resources and coastal ecosystems.

At the shoreward margin of the LMEs, monitoring efforts include the use of mussels, fish, and other biological indicator species to measure pollution effects as part of NOAA's Status and Trends Program, including the bivalve monitoring strategy of "Mussel-Watch" and pathobiological examination of fish. The pilot Environmental Monitoring Assessment Program (EMAP) of the Environmental Protection Agency, focused on the estuarine and nearshore monitoring of contaminants in the water column, substrate, and selected groups of organisms, will be extended to more open waters of LMEs in cooperation with NOAA during 1994. An important component of the associated research to support the monitoring is the definition of routes of exposure to toxic contaminants of selected finfish and shellfish and the assessment of exposure to toxic chemicals by several life-history stages. The routes of bioaccumulation and trophic transfer of contaminants are being assessed and critical life-history stages and selected food

[15]NOAA [National Oceanic and Atmospheric Administration] (1992). "Our Living Oceans. Report on the Status of U.S. Living Marine Resources, 1992," NOAA Tech. Mem. NMFS-F/SP:O-2. Washington, D.C.: NOAA.

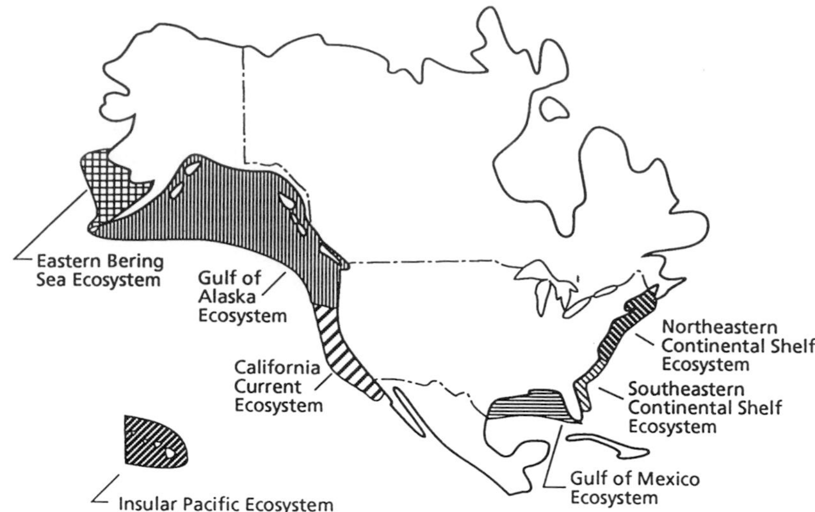


***FIGURE 6*** The seven large marine ecosystems of the United States, including the insular Pacific Hawaiian Islands.

chain organisms are being examined for a variety of parameters that indicate exposure to, and effects of, contaminants. Contaminant-related effects measured include diseases, impaired reproductive capacity, and impaired growth. Many of these effects can be caused by direct exposure to contaminants or by indirect effects, such as those resulting from alterations in prey organisms. The research support program for assessing chemical contaminant exposure and effects in fishing resources and food chain organisms consists of a suite of parameters, including biochemical responses that are clearly linked to contaminant exposure coupled with measurements of organ disease and reproductive status that have been used in previous studies to establish links between exposure and effects. The specific suite of parameters measured will cover the same general responses and thus allow comparable assessment of the physiological status of each species sampled as it relates to chemical contaminant exposure and effects at the individual and population level.

### B. Science and Management

Efforts are under way to place greater focus on the linkage between scientific and societal needs and the utility of long-term, broad-area coastal ocean assessment and monitoring studies aimed at enhancing management practices for the long-term sustainability of marine resources. If the proposition for time-series monitoring of changing ecosystem states is to be realized in this period of shrinking budgets, it would be in the best interests of science and socioeconomic interests to cooperate in the endeavor. The basis for the linkage was emphasized not only in the UNCED declarations on the oceans, but also in a series of recent developments revolving around: (1) global climate change; (2) the legal precedent for international cooperation implicit in the Law of the Sea; (3) a growing interest in marine ecosystems as regional units for marine research, monitoring, and management; (4) the effort of the Intergovernmental Oceanographic Commission to encourage the implementation of a Global Oceanic Observing System; and (5) renewed national interests in improving the health of degraded coastal ecosystems. In the United States, this interest has resulted in the enactment of recent legislation mandating the establishment of a national coastal monitoring program for assessing the changing states of "coastal ecosystem health" and reporting the findings to the U.S. Congress as a recurring biannual responsibility of NOAA and EPA.

A more holistic approach to coastal ecosystems assessment, monitoring, and stress mitigation as a means for fostering international cooperation in achieving sustainability objectives for marine resources between the more developed and less developed countries is progressing. Two international LME programs are in the advanced planning stage, one for the Gulf of Guinea Ecosystem, which brings together five countries of the region—Ivory Coast, Ghana, Nigeria, Benin, and Cameroon. The other program is being developed jointly by marine specialists from China and Korea for the Yellow Sea large marine ecosystem. The first of these projects is scheduled to be implemented in the Gulf of Guinea LME in the spring of 1995.

A comprehensive regional project to assess, monitor, and mitigate stresses on the Black Sea Ecosystem is being supported by the Global Environment Facility. It appears that marine resource managers and scientists are being responsive to management needs by supporting monitoring initiatives related to a more comprehensive systems approach to ensuring the long-term sustainability of the marine resources of Europe, North America, Africa, and Asia.

## Glossary

**Biodiversity** Totality of species in a region. In the context of an entire marine ecosystem, biodiversity refers to communities of species and their regional environments.

**Biomass flip** Process wherein a dominant species in the ecosystem rapidly drops to a low level to be succeeded by another species.

**Carrying capacity** Levels of biomass and/or yield that can be sustained in the long term in relation to the productivity of a particular ecosystem or region with regard to the diversity, structure, and trophodynamics of populations and communities within the system.

**Coastal eutrophication** Excessive levels of nutrients resulting from phosphate and nitrate discharges into river drainage basins and estuaries often resulting in depletion of oxygen and the development of unusual algal blooms that may produce biotoxins harmful to fish and shellfish populations.

**Gadoids** Bottom-living fish, including cod and haddock, now in a severely depleted state along the U.S. Northeast coast.

**Management** Regulation of activities and resources to achieve certain objectives. Among the desired goals sought from a management system are protecting the health of the marine environment and obtaining greater wealth from the sea, while maintaining the ecosystem in a desirable state such that consumptive and nonconsumptive values can be maximized on a continuing basis, present and future options are ensured, and the risk of irreversible change or long-term adverse effects as a result of resource use is minimized.

**Sustainability** Use of renewable resources at rates within their capacity for renewal. An economy that is developing in a sustainable manner does so with consideration for balancing the matter and energy of the ecosystem at a level that will allow the system to function and renew itself year after year.

**Sustainable yield** Carrying capacity of the world ocean on a continuing basis of self-renewal is estimated at between 100 million and 120 million metric tons annually of traditional living marine resources species (finfish, shellfish, algae).

**UNCED** United Nations Conference on the Environment and Development convened in Rio de Janeiro, Brazil, 1992.

**UNEP** United Nations Environment Program.

## Bibliography

Alexander, L. M. (1989). Large marine ecosystems as global management units. *In* "Biomass Yields and Geography of Large Marine Ecosystems" (K. Sherman and L. M. Alexander, eds.), American Association for the Advancement of Science Selected Symposium 111, pp. 339–344. Boulder, Colo.: Westview Press.

Alverson, D. L., Longhurst, A. R., and Gulland, J. A. (1970). How much food from the sea? *Science* **168,** 503–505.

AAAS [American Association for the Advancement of Sciences] (1991). "Food Chains, Yields, Models, and Management of Large Marine Ecosystems." Boulder, Colo.: Westview Press.

AAAS (1993). "Stress, Mitigation, and Sustainability of Large Marine Ecosystems." Washington, D.C.: AAAS Press.

Anthony, V. C. (1993). The state of groundfish resources off the northeastern United States. *Fisheries* **18**(3), 12–17.

Bax, N. J., and Laevastu, T. (1990). Biomass potential of large marine ecosystems: A systems approach. *In* "Large Marine Ecosystems: Patterns, Processes and Yields." (K. Sherman, L. M. Alexander, and B. D. Gold, eds.) pp. 188–205. Washington, D.C.: AAAS Press.

Beddington, J. R. (1984). The response of multispecies systems to perturbations. *In* "Exploitation of Marine Communities," (R. M. May, ed.), pp. 209–255. Berlin: Springer-Verlag.

Belsky, M. H. (1986). Legal constraints and options for total ecosystem management of large marine ecosystems. *In* "Variability and Management of Large Marine Ecosystems" (K. Sherman and L. M. Alexander, eds.), AAAS Selected Symposium 99, pp. 241–261. Boulder, Colo.: Westview Press.

Bottom, D. L., Jones, K. K., Rodgers, J. D., and Brown, R. F. (1989). "Management of Living Resources: A Research Plan for the Washington and Oregon Continental Margin," NCRI-T-89-004. Newport, Ore.: National Coastal Resources Research Development Institute.

Collie, J. S. (1991). Adaptive strategies for management of fisheries resources in large marine ecosystems. *In* "Food Chains, Yields, Models, and Management of Large Marine Ecosystems." (K. Sherman, L. M. Alexander, and B. D. Gold, eds.), pp. 225–242. Boulder, Colo.: Westview Press.

Costanza, R. (1992). Toward an operational definition of ecosystem health. *In* "Ecosystem Health: New Goals for Environmental Management." (R. Costanza, B. G. Norton, and B. D. Haskell, eds.), pp. 239–256. Washington, D.C.: Island Press.

Crawford, R. J. M., Shannon, L. V., and Shelton, P. A. (1989). Characteristics and management of the Benguela as a large marine ecosystem. *In* "Biomass Yields and Geography of Large Marine Ecosystems." (K. Sherman and L. M. Alexander, eds.), AAAS Selected Symposium 111, pp. 169–219. Boulder, Colo.: Westview Press.

Eikeland, P. O. (1992). "Multispecies Management of the Barents Sea Large Marine Ecosystem: A Framework for Discussing Future Challenges." Polhogda, Postboks 326, Fridtjof Nansens vei 17, N-1324 Lysaker, Norway: The Fridtjof Nansen Institute. (R:004-1992; ISBN: 82-7613-030-5; ISSN: 0801-2431).

GESAMP [Group of Experts on the Scientific Aspects of Marine Pollution] (1990). "The State of the Marine Environment," UNEP Regional Seas Reports and Studies No. 115. Nairobi: U.N. Environment Program.

Goldberg, E. D. (1976). "The Health of the Oceans." Paris: UNEC Press.

Hey E., and Mee, L. D. (1993). Black Sea. The ministerial declaration: An important step. *Environ. Pol. Law* **2315,** 215–217; 235–236.

Holling, C. S. (1993). Investing in research for sustainability. *Ecol. Appl.* **3,** 552–555.

Holt, S. J., and Talbot, L. M. (1978). New principles for the conservation of wild living resources. *Wildl. Monogr.* **59.**

Kullenberg, G. (1986). Long-term changes in the Baltic ecosystem. *In* "Variability and Management of Large Marine Ecosystems." (K. Sherman and L. M. Alexander, eds.), AAAS Selected Symposium 99, pp. 19–32. Boulder, Colo.: Westview Press.

Levin, S. (1993). Approaches to forecasting biomass yields in large marine ecosystems. *In* "Large Marine Ecosystems: Stress, Mitigation, and Sustainability." (K. Sherman, L. M. Alexander, and B. D. Gold, eds.), pp. 36–39. Washington, D.C.: AAAS Press.

Lubchenco, J., Olson, A. M., Brubaker, L. B., Carpenter, S. R., Holland, M. M., Hubbell, S. P., Levin, S. A., MacMahon, J. A., Matson, P. A., Melillo, J. M., Mooney, H. A., Peterson, C. H., Pulliam, H. R., Real, L. A., Regal, P. J., and Risser, P. G. (1991). The sustainable biosphere initiative: An ecological research agenda. *Ecology* **72,** 371–412.

Mangel, M., Hofman, R. J., Norse, E. A., and Twiss, J. R., Jr. (1993). Sustainability and ecological research. *Ecol. Appl.* **3**(4), 573–575.

Mee, L. (1992). The Black Sea in crisis: A need for concerted international action. *Ambio* **21**(4), 1278–1286.

Murawski, S. A. (1991). Can we manage our multispecies fisheries? *Fisheries* **16**(5), 5–13.

NSTF [North Sea Task Force] (1991). "Scientific Activities in the Framework of the North Sea Task Force," North Sea Environment Report No. 4. London: North Sea Task Force, Oslo and Paris Commissions, International Council for the Exploration of the Sea.

Pimm, S. L. (1984). The complexity and stability of ecosystems. *Nature* **307,** 321–326.

Rosenberg, A. A., Fogarty, M. J., Sissenwine, M. P., Beddington, J. R., and Shepherd, J. G. (1993). Achieving sustainable use of renewable resources. *Science* **262,** 828–829.

Sherman, K. (1991). The large marine ecosystem concept: A research and management strategy for living marine resources. *Ecol. Appl.* **1**(4), 349–360.

Sissenwine, M. P., and Cohen, E. B. (1991). Resource productivity and fisheries management of the northeast shelf ecosystem. *In* "Food Chains, Yields, Models, and Management of Large Marine Ecosystems" (K. Sherman, L. M. Alexander, and B. D. Gold, eds.). Boulder, Colo.: Westview Press.

Skjoldal, H. R., and Rey, F. (1989). Pelagic production and variability of the Barents Sea ecosystem. *In* "Biomass Yields and Geography of Large Marine Ecosystems." (K. Sherman and L. M. Alexander, eds.), AAAS Selected Symposium 111, pp. 241–286. Boulder, Colo.: Westview Press.

Smayda, T. (1991). Global epidemic of noxious phytoplankton blooms and food chain consequences in large ecosystems. *In* "Food Chains, Yields, Models, and Management of Large Marine Ecosystems." (K. Sherman, L. M. Alexander, and B. D. Gold, eds.), pp. 275–308. Boulder, Colo.: Westview Press.

U.S. Department of Commerce (1988). "Folio Map No. 7. A National Atlas: Health and Use of Coastal Waters, United States of America." Washington, D.C.: U.S. Dept. of Commerce, Office of Oceanography and Marine Assessment.

# Marine Biology, Human Impacts

**T. R. Parsons**
*University of British Columbia*



The greatest human impact on marine biology is the removal of about 90 million tons of fish from the ocean each year by the industrial fisheries of the world; no impact from any pollutant compares with this harvest. Unfortunately the consequences of killing so many fish each year are poorly understood with respect to sustaining the ecology of the sea. Other forms of exploitation of marine biological resources include the use of coral rock for building, destruction of mangrove forests, and the draining of estuarine wetlands for housing, industry, and airports. The effect of these activities can be clearly documented with respect to the loss of habitat for wildlife and the increased exposure of shorelines to the effects of erosion. Though little attempt is being made to understand the ecological effects of world fisheries exploitation, some advances have been made in the management of the more localized effects of extractive development based on shoreline ecology.

## I. INTRODUCTION

Apart from the addition of pollutants to the oceans, humans have affected the balance of life in the sea by a number of other activities. These generally involve the exploitation of a marine resource and may include the removal of fish for food, changes in estuaries to form harbors, changes to coral reefs to accommodate tourists, or the destruction of mangrove habitat for fish farming. Many of these effects have been more devastating on the marine biota than the addition of pollutants. However, because resource utilization is deemed necessary for human survival, the anthropogenic effects of marine resource-based industries on the biota have received less attention than the more obvious and largely avoidable additions of pollutants to the sea. This article discusses examples of resource-based human impacts on the marine biota. [*See* POLLUTION IMPACTS ON MARINE BIOLOGY.]

## II. EXTRACTION OF MARINE RESOURCES

Humans have hunted some marine creatures to extinction, one species being the Steller sea cow (*Hydrodamalis gigas;* Fig. 1). This large relative of manatees and dugongs was discovered in 1741, when a Russian vessel under the command of Captain Vitus Bering was wrecked on the remote Commander Islands in the subarctic North Pacific. As described by Georg Steller, the ship's naturalist, the sea cows were docile, slow-moving mammals that lived in

***FIGURE 1*** An artist's impression, based on existing skeletons and past descriptions, of the extinct Stellar sea cow (*Hydrodamalis gigas*) feeding in a kelp forest.

shallow water and grazed on massive kelp forests. The survivors of the shipwreck were not long in discovering that this huge, gentle creature was easily caught. It had no form of self-defense other than its immense weight of 10 tons. Unfortunately the sea cow tasted much better than the local seals and otters on which the starving crew would have otherwise had to depend for survival. The rescued survivors carried back word about the delightful feast of sea cow that could be had on the Commander Islands, and this report quickly spread to other Russian vessels engaged in the Alaskan fur trade. Thus the fate of this species was sealed. In 1741, there were thought to be about 2000 sea cows on the Commander Islands. This was, however, the last remaining population of these animals, which once ranged all the way from California to the coast of Japan. The gentle nature of this accessible marine mammal probably had attracted the attention of prehistoric humans, and its elimination from all previous habitats would not have been difficult for even the most primitive hunters. By 1768, only 27 years after its rediscovery by humans, the last member of the last population of the Steller sea cow had been killed and eaten. [*See* EVOLUTION AND EXTINCTION.]

Today, the closely related manatees (*Trichechus* sp.) of the tropical Atlantic have been severely depleted through hunting as well as through loss of habitat. However, in some areas, their chances of survival have been increased by habitat protection programs and bans on hunting. Less assurance can be given for the Indo-Pacific dugongs (*Dugong dugong*), which continue to be hunted by local inhabitants in need of a rich source of protein.

The flightless great auk (*Pinguinus impennis*) of the North Atlantic Ocean was also eliminated by humans. This seabird was the original "penguin"; its name was given much later to other flightless birds of the southern hemisphere. A number of "Penguin Islands" still exist off Newfoundland, where the great auk might have been seen by early travelers in the sixteenth and seventeenth centuries. The principal nesting colony of this very large bird (1.0 m in height) was on remote Funk Island, off the northeastern tip of Newfoundland. Eighteenth-century seafarers came to this location to harvest the bird for its soft downy feathers, as well as for meat when needed. There was no fuel on the island and so the giant auks were also burned; their thick layer of fat helped to ignite the fires in which other birds were being scalded to remove their feathers. It is believed that the last of the great auks was killed on Funk Island in 1844.

Today a similar fate may await some marine turtles that are caught for food (both meat and eggs), for their shells, and to produce fertility potions. At one South China Sea site on the east coast of Malaysia, the number of leatherback turtle (*Dermochelys coriacea*) nests declined because of commercial harvesting from nearly 2000 in the 1950s to fewer than 100 in the 1980s. Whereas the leatherback turtle nests in other areas and is not presently an endangered species, other species are declining rapidly. The hawksbill turtle (*Eretmochelys imbricata*) is highly prized for its beautiful shell, which is the true tortoise shell of commerce; this animal is much more severely threatened by hunting in southeast Asia, where its sale boosts the fortunes of local inhabitants. The green turtle (*Chelonia midas*), formerly an abundant species, is the main ingredient for turtle soup. Because their remote nesting sites have become easily accessible to modern travelers, all marine turtle populations can be regarded as potentially endangered. The International Union for the Conservation of Nature (IUCN) recognizes that the limited numbers of green, hawksbill, and leatherback turtles in some specific areas officially classify them as "endangered species." Other turtles are known to breed in only one remaining location; such is the case of the Kemp's ridley turtle (*Lepidochelys kempi*), which breeds near Rancho Neuvo on the east coast of Mexico. There are believed to be fewer than 700 breeding females in the world (or 1% of the estimated original population): they are likely to remain on the endangered species list of IUCN for a long time.

A rather different situation regarding the protection of a marine species occurred in the 1970s with respect to the Canadian harp seal (*Pagophilus groenlandica*). Harp seals are the second-most abundant seal in the world. The three populations that breed in the Gulf of St. Lawrence, on an island off Greenland, and in the White Sea off the Siberian coast probably number over 4 million individuals. However, despite its not being listed as an endangered species, a concerted effort by environmental groups against the killing of the young seal pups for their white fur coats was successful in closing down the harp seal hunt in the 1960s. The primary motive for this closure was not related to conservation, as has been the case in so many other species. Rather, it was the innocent appeal of the young harp seal pup and its violent death that aroused so much public sympathy and forced an end to the hunt.

Perhaps the greatest impact of humans on ocean ecology is not by catching any one particular species, but by the removal of 90 million tons of fish per year by the world's fishing nations. This is the statistic published by the Food and Agricultural Organization (FAO) of the United Nations. However, this figure only represents reported catch of the commercially important species. Those unwanted, unmarketable species that are also caught by the indiscriminate fishing of bottom trawls, purse seines, and driftnets are usually discarded at sea and are not counted as "catch." This includes large quantities of unmarketable fish as well as the much publicized death of propoises and marine birds. For example, several hundred thousand seabirds may be killed annually in driftnets that are set primarily for squid; if this mortality is sustained it could become a significant impact, even for some species of marine birds that may number in the tens of millions. Similarly, over 2 million porpoises are believed to have been killed in a 20-year period since the 1960s by the purse seine tuna fishery. Quite recently, this fishery has been brought under some control that allows the propoises to escape from the purse seine nets. Also unreported and uncounted are the many bottom-dwelling creatures (including turtles) captured in shrimp trawls and the discarding of sharks after collection of their fins, which are used in shark fin soup. Taken together, the unreported and uncounted biomass removed from the ocean would probably be at least equal to the reported catch, making a total mortality of marine creatures of about 200 million tons per year. This is variously estimated to represent between 20 and 50% of the potential total harvestable fish in the oceans. [*See* OCEAN ECOLOGY.]

Because it is often easier to capture the larger fish, many of the species that are harvested are top-level predators in the marine food chain. Ecologically this is similar to removing the lions from the plains of Africa, an action that would surely result in a population explosion of antelopes and other grazers and a decimation of vegetation. The ocean may work in a similar way, or it may have its own response. In any event, there is no form of pollution that can have such a large-scale impact as the actions of this one industry.

In the 1970s, the largest fishery in the world was the anchovy (*Engraulis ringens*) fishery off the coast of Peru. The total harvest in 1970 reached 12 million tons. Overfishing coupled with ocean climate changes led to the collapse of the anchovy population, and it has since yielded only 2 or 3 million tons per year. In retrospect we may ask, "What happened to the ecology of the area when 12 million tons of anchovy were removed in a single year?" No specific studies were conducted to answer this question, but we do know that about 10 million seabirds, which also fed on the anchovy, disappeared at the same time. Further, much of the phytoplankton, on which the adult anchovy fed, was not eaten but simply sank out of the water column to form an anoxic sediment on the seafloor. [*See* MARINE PRODUCTIVITY.]

Another area where biological change was brought about by the intensive extraction of certain species of fish was the North Sea. The catch of herring (*Clupea harengus*) and mackerel (*Scromber scrombus*) in this area started to decline in the period 1965–1975; this was accompanied by a large increase in the abundance of smaller fish that were often the food items of the herring and mackerel. These smaller fish, such as sandlance (*Ammodytes lancea*), became so superabundant that they then formed the basis of new fisheries, yielding in total about twice as much fish from the same area as in the period 1965–1975. Similar changes in the Gulf of Maine are believed to account for the decreased local abundance of right whales (*Eubalaena glacialis*), because the large populations of sandlance decimated the zooplankton on which the right whales also feed.

The problems of overfishing are sometimes compounded by wasteful use of the resource. Much of the Peruvian anchovy harvest was turned into chicken feed, as is the case in the present-day 5-million-ton Chilean sardine fishery. This practice is a waste of high-quality animal protein that could be consumed directly by humans. However, the economics of the world food supply make fish more valuable as poultry feed in America than as human food in Africa or Asia, where it is needed most.

Today the largest fishery in the world is for Alaskan walleye pollock (*Theragra chalcogramma*) in the Bering Sea. In 1986, this fishery was about 6.6 million tons. Recent studies have indicated that the annual removal of this amount of fish has caused decreases in a number of other animal populations that depend on pollock, including sea lions, fur seals, and some marine birds. For example, sea lions in the eastern Bering Sea appear to have declined to one-quarter of their original population during the period 1960 to 1980, which was a period of intensive pollock fishing (Fig. 2). Other marine birds (e.g., auklets) that feed on zooplankton, the food of the pollock, have increased in number because their food has presumably become more abundant with the removal of so much pollock. [*See* FISH CONSERVATION.]

Another documented change in ocean ecology is found in the Antarctic, where hunting reduced the original blue whale (*Balaenoptera musculus*) population from about 228,000 individuals to a population that was estimated in the 1980s to be about 14,000. In the absence of blue whales, there has been a 10-fold increase in the population of crabeater seals (*Lobodon carcinophagus*), a 5-fold increase in minke whales (*Balaenoptera acutorostrata*), and an equivalent increase in some penguin species. All of these animals, like the decimated blue whales, feed on the shrimplike krill (*Euphausia superba*), which became more available to other animals following intensive whaling. This shift in the quantities of krill consumed by different predators from 1900 to 1984 is documented by Dr. R. M. Laws in Table I. More recently, the superabundant krill are becoming a commercial fishery target themselves. If these are depleted, it is unlikely that the blue whale popula-

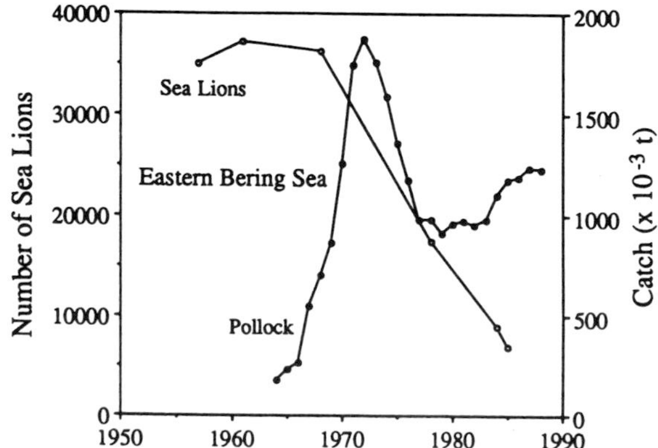


***FIGURE 2*** Changes in the numbers of sea lions and the annual catch of pollock in the eastern Bering Sea. [From A. M. Springer (1992). *Fish Oceanogr.* **1**(1), 87.]

***TABLE I***
**Possible Changes in Patterns of Consumption of Antarctic Krill by the Major Groups of Predators, 1900–1984[a]**

| | Annual consumption (tons × $10^6$) | |
|---|---|---|
| Year | 1900 | 1984 |
| Whales | 190 | 40 |
| Seals | 50 | 130 |
| Birds | 50 | 130 |
| Fish | 100 | 70 |
| Cephalopods | 80 | 100 |
| Total | 470 | 470 |

[a] From R. M. Laws (1985). *Am. Sci.* **73,** 26–40.

tion will ever be able to return to its former levels. [*See* ANTARCTIC MARINE FOOD WEBS; KRILL, ANTARCTIC.]

As the world's fisheries continue to expand, other examples of changes in ocean ecology are emerging in retrospect. For example, there were formerly 26 species of commercial fish in the Black Sea; now only 5 species are fished commercially. A disturbing complexity in this rather confined marine habitat is that there are now huge populations of jellyfish, a phenomenon that has also been recorded from the intensively fished North Sea. Could the removal of fish from the sea result in massive increases in numbers of jellyfish? These invertebrate animals are, after all, the more ancient competitors with fish for zooplankton food. Thus one anthropogenic effect on the oceans could be to cause a reversal in the evolution of marine food chains. There are some specific grounds to support such a hypothesis, but at present the answer remains unknown because it has not been studied.

Fisheries that target invertebrate animals (as opposed to fish) may also disrupt ecological patterns and set up irreversible changes. The Atlantic lobster (*Homarus americanus*) forms the basis for a lucrative fishery along the eastern North American seaboard. Some studies suggest that removal of tons of lobsters from the area has caused an explosion in the population of spiny sea urchins, the principal prey of lobsters. As a result, the urchins eat all the seaweed beds, which are a natural protective cover for young lobsters. Thus a potentially irreversible cycle is begun in which the overharvesting of lobsters destroys the very habitat in which they live. A similar cycle of events can occur between sea otters (*Enhydra lutris*), which eat sea urchins among their prey, and kelp beds, which are a refuge for sea otters as well as being food for sea urchins. Fortunately, however, sea otters are no longer hunted and the few remaining populations that were largely in Alaska have now been reintroduced along other parts of the western coast of North America where they had been eliminated by earlier hunting in the eighteenth and nineteenth centuries.

Too often the disruptive effects of overharvesting have been discovered through hindsight. The most intensive studies of marine harvesting have been supported by fishing companies whose primary interest has been to develop methods that will allow them to continue to catch more fish, more efficiently. There is a need for environmental impact statements on the effect of fishing in changing the whole structure of the marine biota.

## III. CHANGES IN MARINE HABITATS

Marine food chains can also be disrupted in ways that are more subtle than simply destroying fish species. A vital link in the food chain for some migrant birds is often found in an obscure marsh, where they may seek refuge during the winter after spending the summer in the Arctic. The Fraser River estuary in British Columbia, for example, is the winter home to thousands of snow geese (*Anser caerulescens*) arriving from Wrangle Island off the coast of Siberia. In the last 80 years, more than 90% of the marsh grass area used by these birds has been reclaimed for harbors, marinas, housing, industry, and agriculture.

Other birds may fly between the poles during their annual migrations. One such species is the red knot (*Calidris canutus*), which flies 6000 miles between Tierra del Fuego at the southern tip of South America to the northwestern end of Hudson Bay in the Canadian Arctic. When this small bird, which weighs only 75 grams, arrives at its nesting site, it promptly lays a clutch of four eggs. The only way this animal can support such enormous energy demands is by consuming large amounts of food at certain specific staging areas *en route* to the Arctic. One such location is Delaware Bay, on the eastern seaboard of the United States. Here, as many as 100,000 red knots may arrive each year in synchrony with the arrival and spawning of the horseshoe crabs (*Limulus* sp.). The crabs lay prolific numbers of eggs on the beach that are very high in energy and form a major food source for the migrating birds. This site is just one of several important stopover points for red knots. If even one of these important habitats were destroyed by human activities, an essential link would be broken in the migration pattern of about one-third of this bird's world population.

Damage and elimination of coastal marine habitat through the exploitation of its resources have been going on longer than any other anthropogenic impact. As early as the eighteenth century, the princes of the Malabar coast became alarmed at the silting up of local estuaries, which they attributed to intensive deforestation in coastal areas. Along the Arabian coast at the time of Vasco de Gama, there were fringing forests of coastal mangroves. These have virtually disappeared because of the local need for lumber and fodder for camels. Today some attempts are being made to restore these Arabian mangrove forests. However, in other areas such as Bangladesh, so much of the protective barrier of mangrove forest has been destroyed on outer islands to create shrimp farms and paddy fields that storm surges caused by cyclones now enter much farther into the Ganges delta, causing massive destruction and loss of life.

On the Maldive Islands, most of which are only a few meters above sea level, the removal of coral rock (Fig. 3) from the surrounding reefs to use as building material and road construction has created a similar problem. Elimination of part of the protective reef has caused increased erosion of the islands, which are not very large in the first place. It is estimated that on the capital (North Male) island alone, 200,000 $m^3$ of coral rock has been mined in the past 20 years, which represents about one-third of the available coral in shallow water. Even if all mining of coral rock ceased, it would take at least 50 years for the habitat to recover.

Throughout the world, coral reefs are being destroyed and reef ecology disrupted by other activities: corals are removed and sold for ornamental purposes; a lucrative market for aquarium fish has resulted in intensive fishing activity, sometimes using dynamite or poisons; shell collecting and spear fishing have also removed ecologically important species from reefs. In the Philippines, for example, which supplies approximately 70% of the tropical fish for aquaria in North America, it is estimated that 80% of the 33,000 $km^2$ of coral has been damaged, either by poisoning with cyanide or by dynamiting. Shell collecting and spear fishing for large fish in other areas may have caused outbreaks of a starfish, the crown-of-thorns (*Acanthaster planci*), which consumes coral and is normally held in check by certain predatory molluscs and large fish. In Hawaii, parts of the coral reef have been destroyed by silt washed down from housing developments or dredged up in the process of making local marinas. [*See* CORAL REEF ECOSYSTEMS.]

Some alteration in coastal habitats may have caused an increased occurrence of red tides (Fig. 4). The most probable factor is believed to be various forms of eutrophication, which is the addition of nutrients to seawater from either agricultural runoff or sewage. Some red tide organisms are harmless, but others contain a powerful toxin generally referred to as paralytic shellfish poison (PSP). The toxin is produced by certain microscopic algae that are usually red or yellow-brown in color. Shellfish feeding on these algae accumulate the toxin but are not harmed. However, any vertebrate animal (including humans) that feeds on the shellfish may be killed by the toxin.

In summary, although the oceans are much larger than the world's land mass, they have undergone substantial alteration from various resource-based activities of humans. In some cases, the removal of large predators has given rise to a superabundance of smaller organisms; in other cases, loss of habitat could take 50 to 100 years to regrow. In particular, coastal wetlands that have been filled in for airports or housing, or dredged out for harbors, are permanently altered with respect to their original ecological function. Today there are few ocean areas that have not been impacted to some extent by fishermen, tourists, or harbor and real estate development.

**FIGURE 3** Coral mined for building material on the Maldive Islands. (Photo by Barbara E. Brown.)

**FIGURE 4** A red tide in Tokyo Bay. (Photo by Suisan Aviation Company, Tokyo.)

## Glossary

**Endangered species** Small number of a plant or animal species with a low rate of reproduction, living in a few restricted habitats that are susceptible to destruction by humans.

**Overfishing** When the quantity of fish removed by fishing is greater than the amount that can be resupplied by natural growth and reproduction.

**Top-level predators** Animals that have no natural predators (other than humans) and that control the abundance of subordinate species by carnivorous consumption.

## Bibliography

Brown, B. E., and Dunne, R. P. (1988). The environmental impact of coral mining on coral reefs in the maldives. *Environ. Conserv.* **15**(2), 159–165.

Golden, F. (acting editor) (1989). Whither the whales. *Oceanus* **32**(1), 1–144.

Laws, R. M. (1985). The ecology of the Southern Ocean. *Amer. Sci.* **73,** 26–40.

Springer, A. M. (1992). A review: Walleye pollock in the North Pacific—How much difference do they really make? *Fish. Oceanogr.* **1**(1), 80–96.

**Marine Ecosystems**—*See* Management of Large Marine Ecosystems

# Marine Microbial Ecology

**Gerard M. Capriulo**
*State University of New York at Purchase*



Marine microbes, including viruses, bacteria, cyanobacteria, microalgae, and protozoans, have emerged in recent years as central players in both the pelagic and coastal regions of water column and benthic marine ecosystems. They are either directly or indirectly responsible for much marine biomass production as well as respiration, and as such act as both sources of and sinks for organic carbon under different circumstances. They transform various chemical species to more complex or simpler forms, decompose both particulate and dissolved organic matter, and in the process recycle and remineralize important nutrients. They are involved in some of the most fascinating symbiotic associations ever uncovered and are the primary cause of many diseases of both invertebrate and vertebrate organisms, including economically important shellfish, crustacean, and fish species, as well as marine mammals. They greatly affect atmospheric gas composition and thus influence global climate. Indeed, after a quarter century of intensive scrutiny powered by technological advances, marine microbes now stand at the center of modern marine ecosystem paradigms.

## I. INTRODUCTION

Innumerable people, scientists, students and amateurs alike, have peered through the tube of a microscope since it was invented more than 325 years ago. Their efforts have been rewarded with visions of an alien, almost surreal, landscape filled with strange slithering, wriggling, flowing, swimming, attached and stationary, predominantly single-celled creatures of all shapes and sizes. These diverse life-forms, collectively known as microbes (200 $\mu$m or less in size), have roots reaching back some 3.5 billion years and consist of aplastidic, anucleate prokaryotes, including bacteria and cyanobacteria (blue-green algae), and nucleated, plastid-containing eukaryotes that include the slime molds, fungi (e.g., yeasts) and the protists (microalgae, and protozoans including flagellates, amoebae, and ciliates).

Currently microbes occupy center stage in emerging paradigms of marine (and freshwater) ecosystem structure and function. The metabolic activities of microbes result in the decomposition of dissolved and particulate organic matter, the transformation of chemical species of inorganic materials, the production of new organic materials, and the production of large amounts of recycled and remineralized nutrients, which are a primary component of global ocean nutrient cycles and primary production. In different situations, microbes represent either an important sink for or source of carbon, with protists acting as key translators of prokaryote biomass into secondary and higher production (Fig. 1). This production is important to

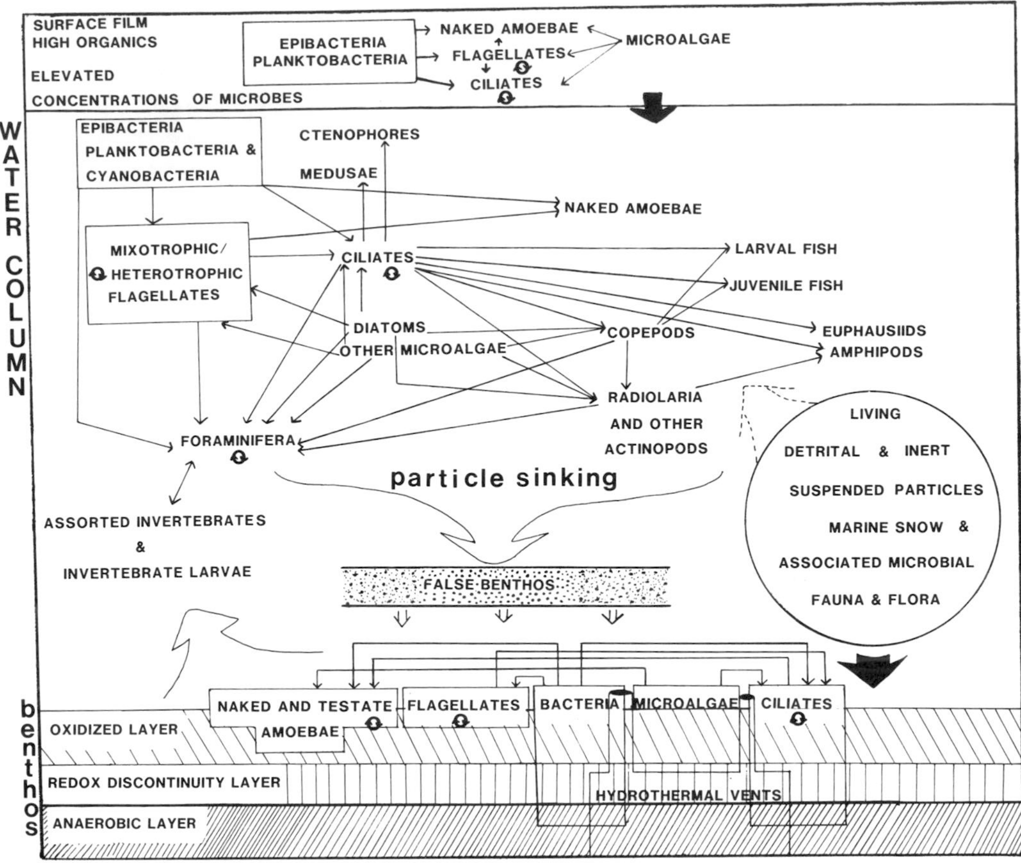


***FIGURE I*** Marine food web overview, indicating trophic relationships among microbes in surface film, water column, and benthic habitats. Arrows denote flow of biomass from prey to predator. Two-headed arrows indicate bidirectional biomass transfer. Closed loop arrows depict within group cycling. [From G. M. Capriulo, ed. (1990). "Ecology of Marine Protozoa." Oxford University Press, Oxford, England.]

the growth of numerous invertebrate and vertebrate marine organisms, including economically important species. The collective metabolic activities of marine microbes also exert a strong influence on the composition of atmospheric gases and therefore on global climate. Among the protists are found some of the most interesting symbiotic associations presently known (parasitism to mutualism), which for certain marine amoebae have fostered the evolution of extremely large size. Study of these symbiotic associations has been instrumental in furthering our knowledge of the evolution of life on earth. [*See* MARINE PRODUCTIVITY.]

## II. THE MICROBIAL PLAYERS

The diversity of microbes that inhabit the world's marine ecosystems is truly awe inspiring. This plethora of species exhibit an equally interesting variety of metabolic pathways and behaviors. It is well beyond the scope of this article to detail such staggering diversity, however, it will provide a digested overview of the key groups that currently appear to play significant roles in marine ecology.

### A. Viruses

Although we have known for some time that viruses are likely agents of infection in various marine invertebrates and vertebrates, it is only recently (since about 1990) that we have come to realize the potentially significant role these organisms play in the world's oceans. Current information suggests that water column concentrations of viruses are typically in the $10^8$ to $10^{11}$ per liter range. A high percentage of these viruses are bacteriophages and cyanobacteriophages. Protist and fungal phages may also be common. It is now believed that viral phages (along with flagellate grazing) are one of the primary controllers of bacterial and cyanobacterial densities in marine systems. Most of these viruses are host specific and cross-infectivity is believed to be low. In addition to these ubiquitous naturally occurring viruses, human and other animal fecal-material-related viruses also are found in coastal marine waters.

## B. Bacteria

Marine prokaryotes consist of two of the three domains of life on earth, namely the Archaea (formerly the Archaebacteria) and the Bacteria (formerly the Eubacteria). The third domain, the Eukarya, contains all the eukaryotic life forms including the slime molds, fungi, algae and protozoans (collectively known as the protists), plants and invertebrate and vertebrate animals. Within the Archaea are included the methanogens, certain extreme halophilic (salt-loving) and hyperthermophilic (heat-loving) prokaryotes, as well as the lone genus *Thermoplasma*. The Bacteria domain includes such forms as the purple photosynthetic bacteria, the green sulfur and nonsulfur bacteria, the cyanobacteria and certain hyperthermophilic bacteria (e.g., the genera *Thermotoga* and *Thermosipho*). To facilitate further discussion in this chapter, members of the Bacteria and Archaea domains will be hereto collectively referred to simply as bacteria. Emphasis will thus be placed on a microbes functional role in marine ecology rather than on systematic relationships.

Marine bacteria are a very versatile group of organisms able to exist at temperatures in excess of 100°C within the plumes of deep-sea hydrothermal vents, to the approximately $-1$°C and below temperatures of polar waters and sea ice. The ability of bacteria to use dilute concentrations of organic matter is temperature dependent, and their metabolic activities appear to be restricted below 2°C. Many bacterial species secrete sticky, mucuslike polymers that tend to aggregate particles in the sea. Nonphotosynthetic bacteria are major consumers of materials in the sea. Water column concentrations of bacteria typically vary in the $10^3$ to $10^6$ per milliliter range, with higher end values typical of coastal waters, especially those with high levels of organic matter (e.g., near salt marshes). More extreme densities are found in heavily eutrophicated areas, for example, around sewage outfalls.

### *1. Phototrophic Bacteria*

Phototrophic bacteria, which are capable of photosynthesis, include both aerobic and anaerobic forms. The anaerobes (also known as anoxyphototrophs) produce ATP but not NADPH in cyclic phosphorylation, typically use $H_2S$ (rather than $H_2O$) as the electron donor for production of the NADPH needed for carbon fixation, and include purple nonsulfur and purple and green sulfur bacteria. The nonsulfur forms frequent inshore waters and bottom sediments, where they live as strict or facultative anaerobes. They are associated with decomposing seaweeds and plant material and can, if necessary, live as complete heterotrophs. The purple and green sulfur bacteria live in sulfur springs, sulfide sands and muds of salt marshes, and the chemocline of anoxic sulfide water basins.

The aerobic (oxyphotobacteria) photosynthetic forms that use $H_2O$ as their electron donor, and that produce both ATP and $NADPH_2$ via photosynthesis, consist of the cyanobacteria (=blue-green algae). Most contain chlorophyll *a* as well as carotenoids and phycobiliproteins. These forms lack chloroplasts and may in fact be the ancestors of chloroplasts found in eukaryotes (see Section III). Their pigments reside in cell membrane-associated thylakoids. Cyanobacteria exist as small (<1 $\mu$m to several $\mu$m) single cells that are ubiquitous in the world's oceans, or as filaments, and are responsible for a significant percentage of global primary production. The filamentous form, *Trichodesmium* (>64 $\mu$m), is a dominant organism of the subtropical and tropical ocean and, in addition to contributing much toward primary production, also carries out nitrogen fixation, converting $N_2$ gas to a form usable by algae and cyanobacteria. The cyanobacteria achieve buoyancy control through the use of gas vacuoles.

### *2. Chemotrophic Bacteria*

Included in chemotrophic bacteria are both chemolithotrophs, which derive energy and reducing power from the oxidation of inorganic compounds such as ammonia, nitrate, and various sulfur compounds, and chemoorganotrophs, which use dissolved organic matter (DOM). The chemolithotrophs include nitrifying, sulfur and iron oxidizing, and hydrogen bacteria. The nitrifying forms are instrumental in the transformation of reduced nitrogen compounds to nitrite and nitrate, nutrients that are used by microalgae, seaweeds, and sea grasses for photosynthesis. Liberated energy is used by these bacteria to fix $CO_2$. They require an environment where ammonia is produced at sufficient levels, generally in regions of animal activity where ammonia is produced as an end product of animal metabolism. Some chemolithotrophs can live as mixotrophs or organotrophs. Among the lithotrophs, the sulfur-oxidizing forms oxidize free sulfur, thiosulfate, and sulfides to sulfate. Colorless sulfur bacteria are widespread in inshore muds and at times compete for substrate with pure inorganic chemical reactions. Forms that contain internal sulfur granules use $H_2S$ released from the decomposition of organic matter and convert it into elemental sulfur.

The chemoorganotrophs (osmotrophic bacteria) are mostly gram-negative forms and include the planktobacteria, which utilize DOM (mono- or polysaccarides), and the osmotrophic epibacteria (Fig. 2), which remain associated with detrital material, living particles, and plant surfaces, either attached or in close zonal proximity. These forms are equipped with enzymes used to break down water column or benthic, small- to large-sized particulate matter (including open wounds and carcasses). Most (but not all) water column bacteria are gram-negative. The gram-negative cell wall holds degradative enzymes much better than does the gram-positive cell wall, which is found more in sediment-associated bacteria. Such protection of degradative enzymes is critical for bacteria living in the generally nutritionally dilute water column.

*Vibrio* spp. are common epibacteria associated with zooplankton and the guts of animals feeding on zooplankton, as well as in sediments high in fecal debris and chitin and in the gut of the shipworm *Toredo,* where they decompose cellulose. Some vibrios are luminescent and free-living in the water column, or are associated with fish guts, decaying fish, and the light organs of squid and certain fish. Other epibacteria, such as *Pseudomonas* spp., act as chemoorganotrophs or facultative chemolithotrophs (using hydrogen or carbon monoxide as an energy source), or as facultative anaerobic denitrifiers. Epibacteria occur as rods or cocci, and some attach via stalks or filaments whereas others glide along surfaces. Some gliding forms construct filaments 10 to 100 $\mu$m in size and are visible to the unaided eye. An example of this is the genus *Beggiatoa,* which is found on mud surfaces rich in $H_2S$, usually associated with decaying seaweeds. Some epibacteria parasitize and lyse other bacteria.

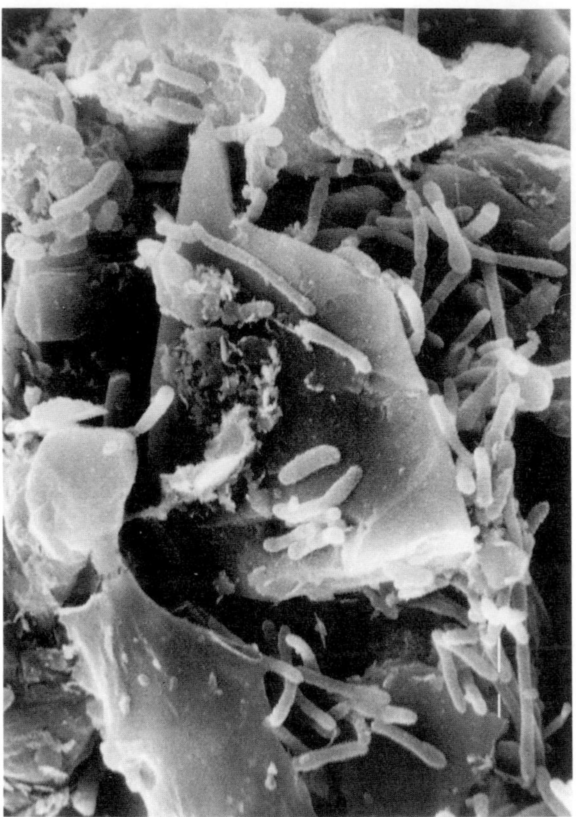

***FIGURE 2*** Marine epibacteria, shown here growing on the primarily mineral-phase lorica of a planktonic tintinnid ciliate.

Obligately, anaerobic bacteria are found in areas where organic matter accumulates. Such areas include water column detrital layers, oxygen minimum layers, sediments in production areas, trenches, semienclosed basins, guts of animals, and parts of salt marshes and estuaries. Most notable among the obligate anaerobes are sulfate reducers, which are widely distributed in benthic marine environments, and methane-producing bacteria, which produce $CH_4$ and $CO_2$ in anaerobic environments such as marshes, guts of animals, and marine snow particles. Methanogens keep the hydrogen ion concentration low, which in turn helps non-methanogenic fermentors, which oxidize organic compounds such as cellulose, proteins, organic acids, and alcohols to acetate and $CO_2$.

## C. Fungi

Marine fungi function as osmotrophs and live on organic-rich surfaces as epiphytes, saprophytes, and pathogens. They include members of the Ascomycetes, Basidiomycetes, Deuteromycetes, Phycomycetes, and Zygomycetes groups. Their role in the marine environment is not yet well understood and many problems exist regarding their taxomony. However, we do know that yeasts dominate nearshore and open ocean faunal communities. They colonize marine muds, open waters, and organic surfaces. Some infect copepods and their eggs, phytoplankton, the egg cases of oyster drills, crabs and crab eggs, sea grasses, the gills of salmonid fish, oysters, clams, barnacles, and red, green, and brown algae. Their collective activities are no doubt important in the ocean's decomposition and nutrient regeneration cycle.

## D. Microalgae

As the primary producers of the world's oceans, the microalgae (including water column phytoplankton as well as shallow-water benthos dwellers) form the base of the marine food web. Diversity in this trophic group is quite high, with densities ranging from about $10^4$ to $10^9$ cells per liter, temporally and spatially. Highest concentrations are associated with coastal, estuarine, and upwelled waters whereas low values typify tropical open waters. A tremendous range in cell sizes is also found, with some species greater than 100 $\mu$m in size whereas others overlap bacteria and cyanobacteria in size, minimally in the 0.5- to 1-$\mu$m range. Some of the smallest microalgae contain only a single chloroplast.

### *1. Diatoms*

Diatoms represent one of the most visible and more dominant groups of microalgae. They are an important food source for microcrustaceans such as the copepods and euphausiids, which in turn are primary food sources for many larval, juvenile, and certain adult fish. Additionally they are an important component of the diets of protozoans such as ciliates, sarcodines, and heterotrophic flagellates, as well as other invertebrates such as the bivalve mollusks. Diatoms exist in a large variety of shapes derived from two basic cell types including radially symmetric (centric) and bilaterally symmetric (pennate) forms, have skeletons (frustules) composed of $SiO_2$ (glass), and occur either as single cells or in chains or colonies (Fig. 3). They are a source not only of cellular particulate carbon but also of released DOM. In fact they release as much as 50% of their photosynthate (organics produced via photosynthesis) back into the water.

This DOM stimulates bacteria and other microbial production and is believed by some to be an attempt by diatoms to cultivate bacteria, some of which produce growth factors, such as vitamins, needed by the diatoms for their own growth. Such activity, which also occurs between certain benthic invertebrates and bacteria, is called microbial gardening.

Diatoms are both oceanic and neritic and live in plankton as well as benthic communities. Some benthic forms secrete mucus that binds sediment on mudflats and changes boundary zone microscale and mesoscale flow dynamics. Diatoms are very abundant in temperate and polar seas, as well as in estuaries, bays, sounds, shelf waters, and upwelling areas. Benthic diatoms also colonize living surfaces such as whales, and some function as parasites, penetrating the skin of their host. Other diatoms are obligate heterotrophs and use things like decaying seaweeds as a source of food, whereas still others are mutualistic symbionts living in various protozoans or higher invertebrates.

### *2. Dinoflagellates*

Dinoflagellates share traits with both the algae and heterotrophic protozoa (Fig. 4). They have an unusual condensed chromosomal arrangement and

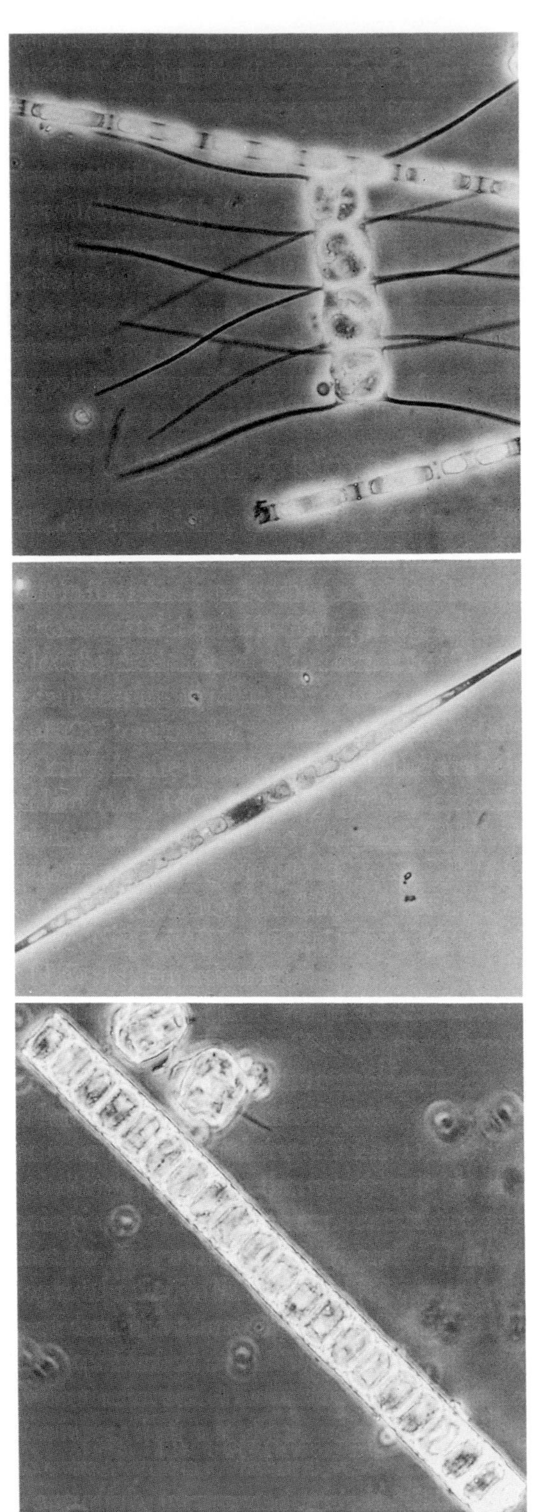

**FIGURE 3** Assortment of chain-forming and unicellular diatoms.

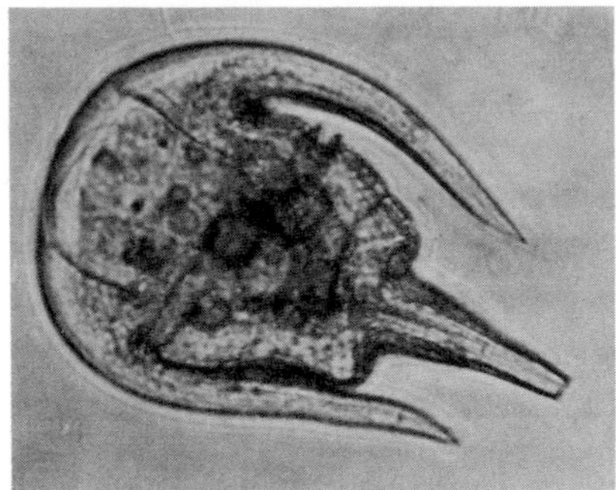

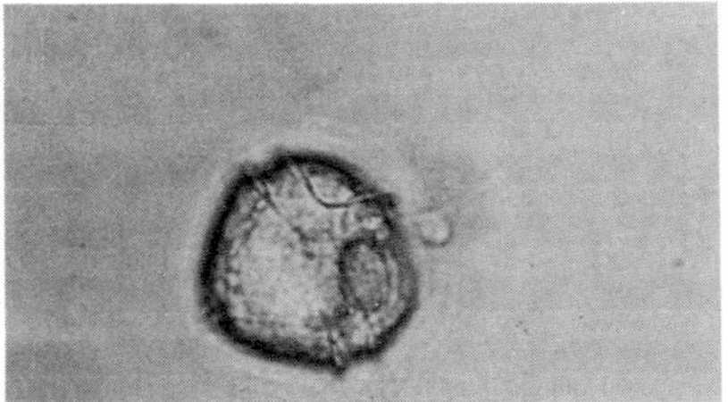

**FIGURE 4** Two examples of thecate dinoflagellates.

distinctive mitochondria. They are equipped with both longitudinal and transverse flagella and with thick (armored) or thin (naked) cell wall plates. Many are apochlorotic and phagotrophic. Certain types (collectively known as zooxanthellae) live as endosymbionts within other protozoa (especially various sarcodines) and higher invertebrates (e.g., hard coals, gorgonians, anemones, nudibranchs, clams). They are found at low to high latitudes in coastal and open ocean waters. In temperate zones they are found in high numbers in the summer and low in winter, usually following the diatom spring bloom. They often dominate subtropical and tropical water plankton. Certain species (e.g., *Gonyaulax tamarensis*) cause toxic red tides, whereas others have evolved to live amid sand grains. Major taxonomic subgroups include the Peridinales, Dinophysales, and Prorocentrales. Dinoflagellates can go into resting cyst (or spore) stages under unfavorable conditions and may await the return of favorable conditions in shallow-water sediments. Excystment rapidly follows restoration of favorable conditions and this may account for unsignaled bloom bursts. Certain species, such as members of the genus *Noctiluca,* have been found to prey on copepod eggs and can reduce the

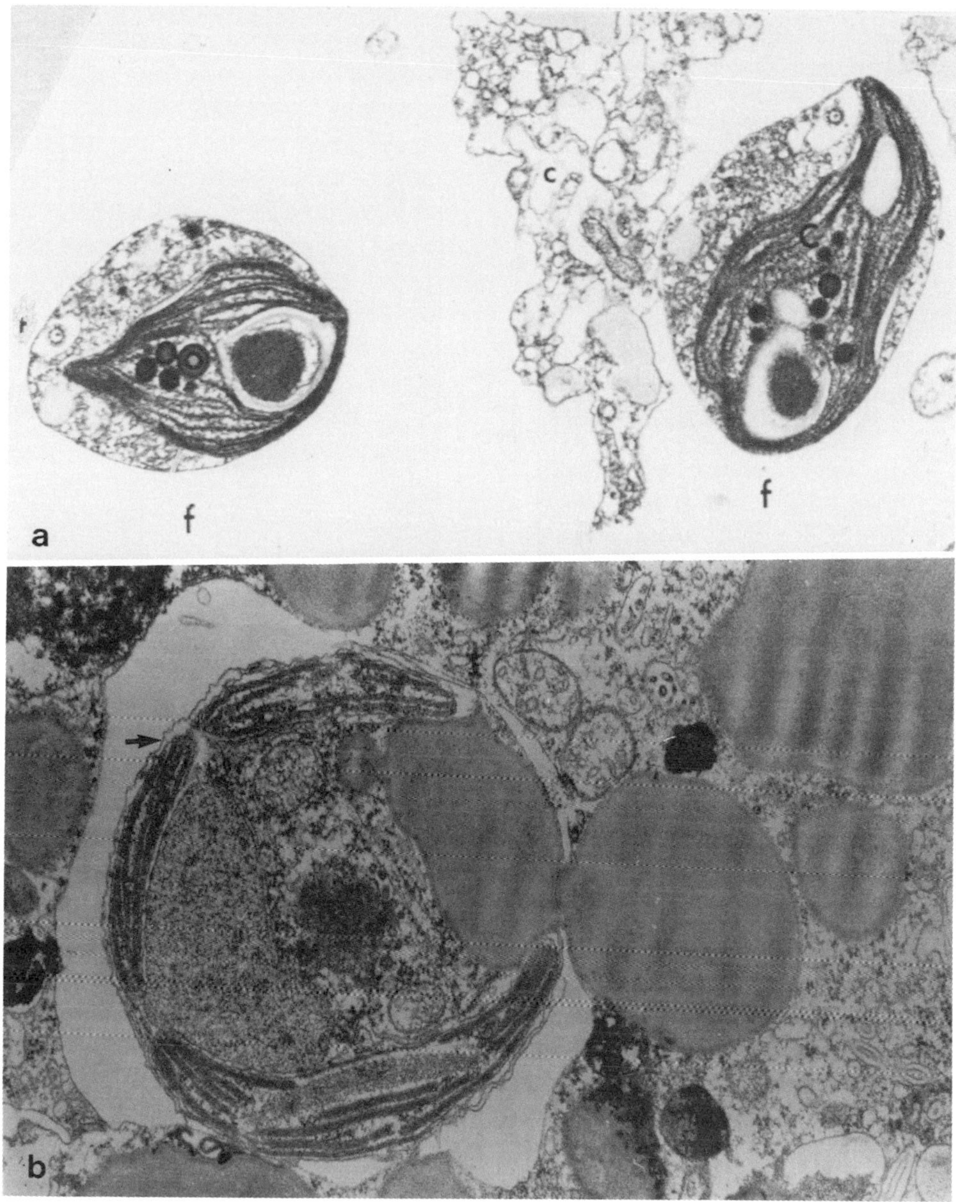


***FIGURE 5*** (a) Two prasinomonad flagellate, f, cells (*Pedinomonas noctilucae*) from a vacuole of the dinoflagellate *Noctiluca scintillans.* (b) Prymnesiomonad zooxanthellae from the planktonic foraminiferan *Globigerinella aequilateralis.* [From G. M. Capriulo, ed. (1990). "Ecology of Marine Protozoa." Oxford University Press, Oxford, England.]

potential size of adult populations by as much as 30%.

### 3. Other Microalgae

Many generally smaller microalgae of different taxonomic affiliations abound in the world's oceans, either as endosymbionts in other organisms (Fig. 5) or as free living cells (Fig. 6). Of particular significance are the haptophytes (= prymnesophytes), which include the coccolithophores (an important component of fossil records) as well as the acrylic acid-containing *Phaeocystis poucheti* (responsible for antibacterial activity, e.g., in the guts of penguins), and toxic chrysochromulinids. The haptophytes

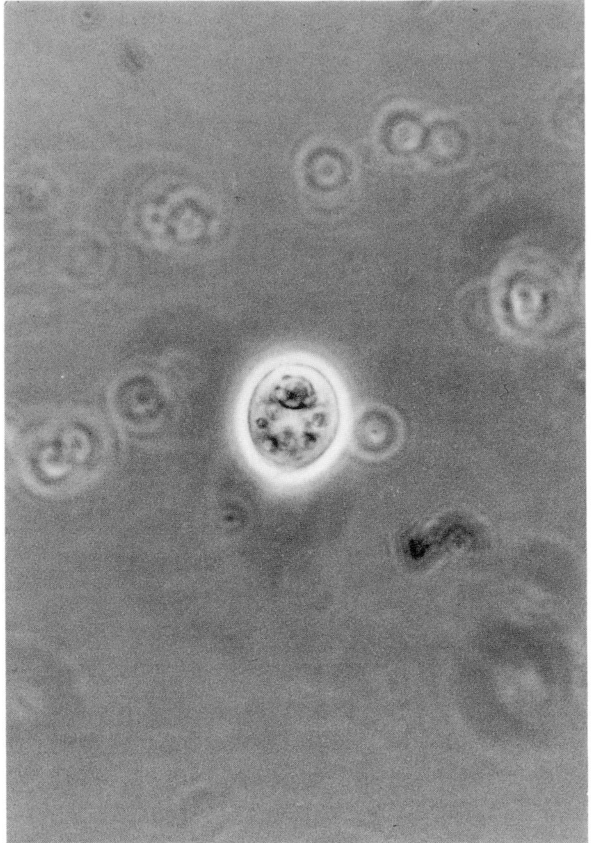

***FIGURE 6*** The marine chlorophyte *Dunaliella tertiolecta*.

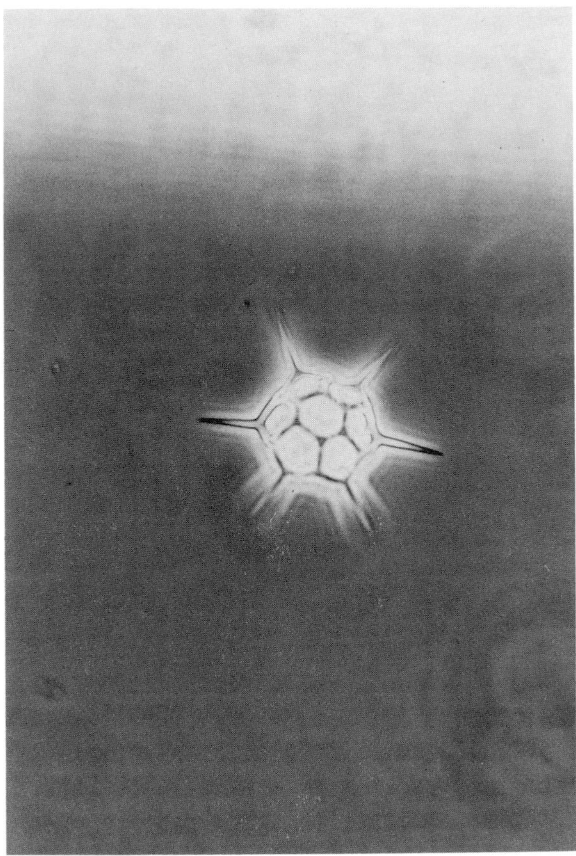

***FIGURE 7*** A typical marine silicoflagellate, showing structure of the silicon-based skeleton.

sport an appendage called the haptonema, which is used for attachment. The prasinophytes are chlorophyll *b* containing algae that possess body scales and flagella (e.g., *Pyramimonas* and *Platymonas*).

Chrysophytes other than diatoms, which are found in significant numbers in marine systems and which also possess skeletons of silicon, are the silicoflagellates (Fig. 7). Cryptophytes are ovoid to pear-shaped forms that have a furrow or groove (e.g., *Rhodomonas, Chroomonas, Cryptomonas*) and often are found in high densities, as are euglenophytes, chlorophytes (e.g., *Chlorella* and *Dunaliella*) (Fig. 6), and rhodophytes.

### E. Protozoans

As was true of the bacteria, fungi, and microalgae, the protozoans are ubiquitous in marine waters ranging from the Arctic to the Antarctic, and occur circumglobally. Different assemblages of protozoans, along with corresponding differences in microalgal and bacterial communities, are found along surface to bottom water and sediment gradients as well as on other spatial (e.g., horizontal) and temporal (e.g., seasonal) scales. Within the protozoan groups, wide-ranging behavioral, nutritional, and growth characteristics can be found. Both chlorophyll-bearing species (claimed in both botanical and protozoological taxonomy) and nonbearing species exist. There are free-living water column and benthic species. Benthic species can live on sediments or within the sediment, in interstitial pore waters. Parasitic, commensalistic, and mutualistic symbiotic relationships abound, and colonial organization is also common. Included in this eclectic group of organisms are phytoflagellates and zooflagellates (including the choanoflagellates and kinetoplastids) (Fig. 8), whose numbers in coastal waters typically

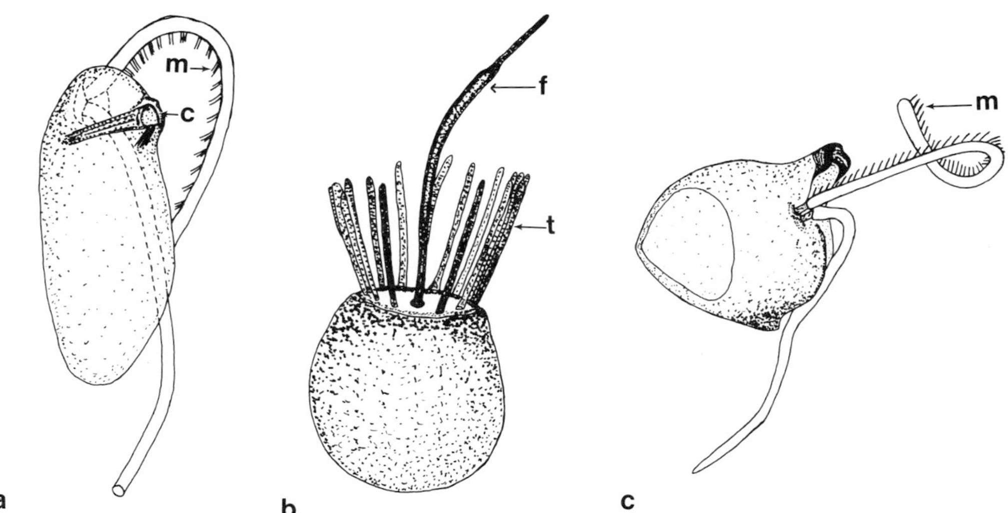


***FIGURE 8*** Three examples of heterotrophic marine zooflagellates: (a) the kinetoplastid *Pleuromonas jaculans;* (b) the choanoflagellate *Monosiga* sp.; (c) the bicoecid *Pseudobodo tremulans*.

reach $10^5$ and $10^4$ per milliliter, respectively, the sarcodines (marine amoebae), including the silicon-skeletoned radiolarians and calcium-based foraminiferans (Fig. 9), the naked amoebae (Fig. 9), and the incredibly diverse ciliates, whose numbers range from several to more than $10^4$ per liter (Fig. 10). The diets of these protozoans include bacteria, microalgae, each other, and metazoan animals. Collectively they play a key role at the lower end of the food web as decomposers, grazers, carnivores, and nutrient regenerators and in biogeochemical cycles.

## III. SERIAL SYMBIOSIS AND THE EVOLUTION OF TROPHIC STRUCTURE

Most experts agree that the first life-forms on earth were anaerobic, heterotrophic prokaryotes. These protocells likely evolved in subsurface marine waters rich in DOM, away from the damaging high UV light intensities of the pre-ozone layer surface earth. In the early stages of life on earth, DOM was circumglobally distributed, in a near limitless supply, in accordance with physical mixing dynamics. The first life-forms were thus inoculated into a global batch culture (the world ocean) with initially little intra- or interspecies competition, and abiotic state variables (e.g., temperature, salinity, pressure, UV light) controlling spatial and temporal growth. At some point DOM became limiting, and competition for resources intensified. Also, diversity of organic chemicals increased as the metabolic activities of existing life-forms diversified and altered the organics in seawater. Such events likely led to increased microbial species diversity. Ability to use existing organics varied by species, and poorer competitors evolved to use the waste products of other organisms or increased their efficiency of uptake of other dissolved materials (lowered their $K_s$ values). Species able to survive at lower nutrient levels, or to metabolize underutilized compounds (e.g., waste products), had the best chance for survival. Additionally, those species able to explore uncolonized marine habitats, such as shallow waters, the air–sea interface layer, and intertidal zones, where the concentration of organics remained high, also had a survival advantage. However, shallow waters, though rich in nutrients, were also characterized by high UV light intensities. Cells with higher pigment content (likely a by-product of evolutionary accident) had a built in shield for some protection, and therefore a competitive edge for utilization of untapped nutrients. Such an edge provided these cells with a competitive release and a new round of exponential growth. Pigment also provided an unanticipated benefit, for

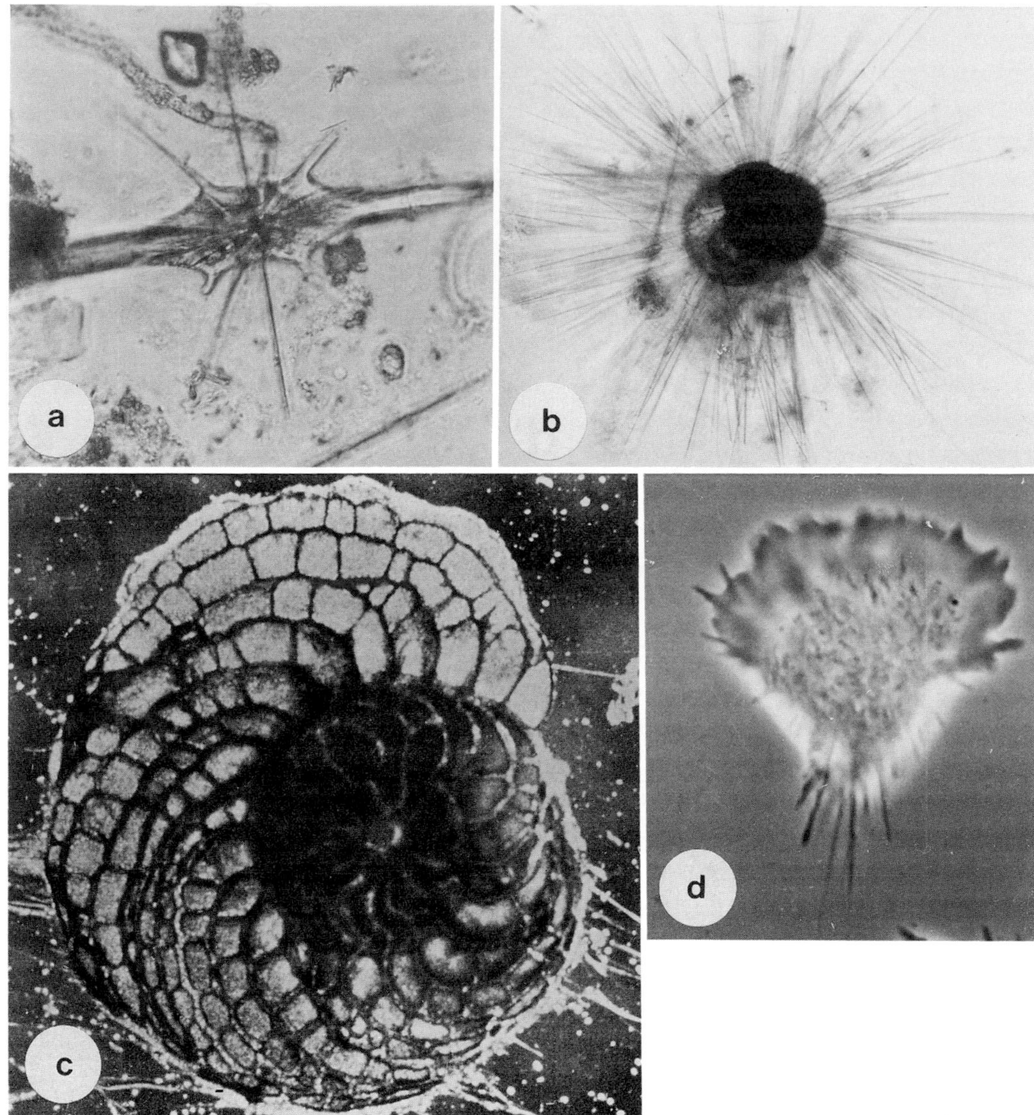


***FIGURE 9*** Testate (a) radiolarian (*Amphilonche* sp.), (b) planktonic foraminiferan, and (c) benthic foraminiferan (*Heterostegina depressa*) and naked (d) *Flabellula hoguae* marine amoebae.

it could garnish electron-reducing power and solar energy for the autofixation of carbon. Thus photosynthesis was born and became heavily favored in competitive clashes. Cells without pigment continued to compete for now low levels of DOM. Some developed the ability to phagocytize (i.e., use other cells and particles for food). Residual DOM from death of cells kept other organisms going on recycled organics. Phagocytizing organisms, with their ability to ingest living, dead, and newly formed (i.e., particle production via bubble bursting in DOM containing surface interface water) particles, experienced competitive release and exponential growth and were heavily favored by natural selection, as were photosynthesizing forms. Microbial species unable to compete became extinct.

At this stage of development three nutrition-gathering modes had become established: DOM and dissolved inorganic matter utilization, photosynthesis (with requirements for some products of other organisms, e.g., vitamins), and phagocytosis. These modes form the basis of nutrition for all current life-forms (i.e., absorptive, photosynthetic, and ingestive modes).

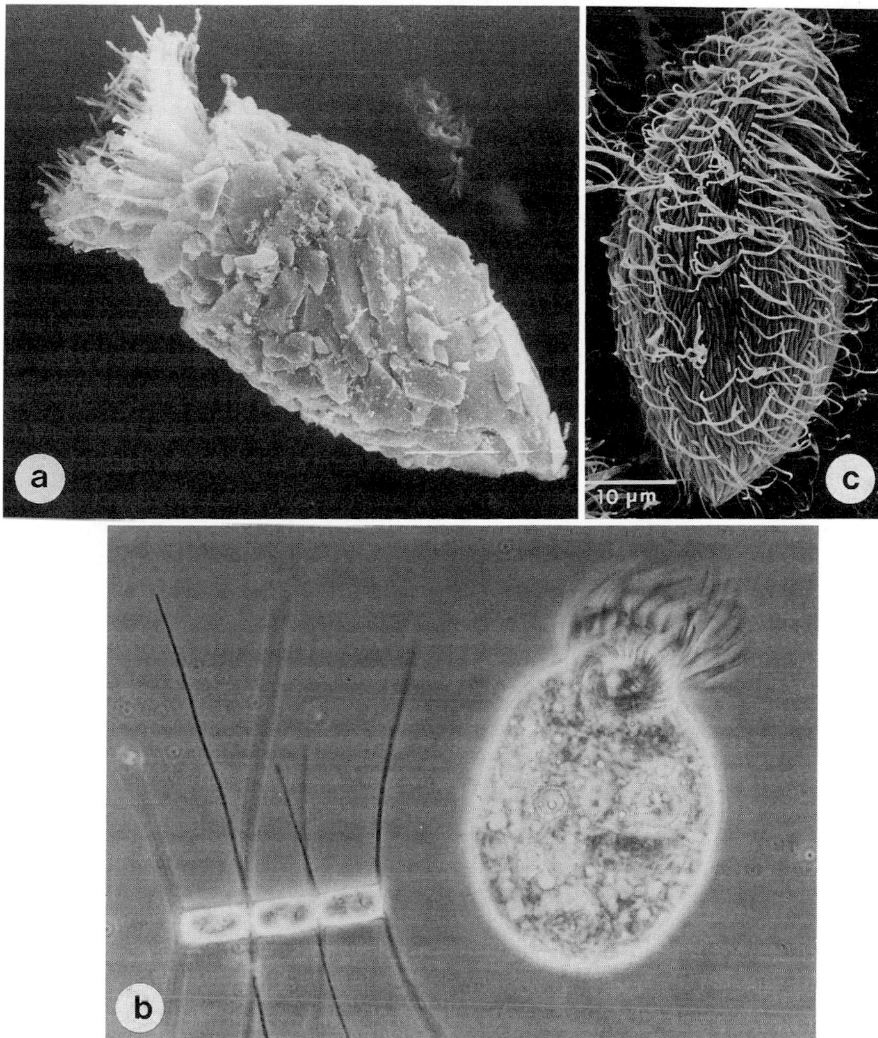


**FIGURE 10** (a) Planktonic loricate, choreotrich tintinnid (*Tintinnopsis* sp.), (b) planktonic naked oligotrich (pictured with the diatom *Chaetoceros* sp.), and (c) benthic scuticociliate (*Cyclidium stercolis*) ciliates.

Photosynthesis was (and still is) restricted to the upper lighted zone of the oceans (upper 150 m) and required inorganic nutrients such as nitrogen, phosphorus, and sulfur. These were provided by the activities of other microbes. In photosynthesis, $H_2O$ serves as the electron donor and free oxygen is released. Oxygen was toxic to existing life-forms. As atmospheric and oceanic concentrations of oxygen increased, most anaerobic microbes were forced to restricted zones in deep- and shallow-water anoxic sediments (e.g., trenches and fjords) or to places in the water column where anoxic conditions persisted (e.g., around organic particles and later in the digestive tracts of metazoans). Many of the photosynthetic bacteria themselves were killed by the toxic effects of oxygen.

Atmospheric oxygen interacted with electrical charges to form ozone, which accumulated into an ozone layer. This layer served (and still does) as a UV shield for the earth. Lower UV radiation resulted in a lowering of mutation rates and a slowing of evolution. Organisms could now more readily colonize shallow waters and so competition in these habitats increased. Some life-forms (both pigmented and nonpigmented) developed the ability to survive short exposures to oxygen. Such tolerance became dependence, as aerobic metabolism developed, with its much higher rate of ATP production as compared to anaerobic metabolism. Aerobic respiration thus led to the next round of competitive release and exponential growth. As anaerobic-microbe-available living space shrank, competition among the anaerobes intensified. Phagotrophic and nonphagotrophic forms dwelled together and osmotrophs were routinely ingested by particle feeders. Aerobic forms occasionally found themselves in anaerobic regions and certain anaerobes could sustain brief exposure to proximal oxygenated habitats. These forms became tolerant of anaerobic (facultative aerobe) and aerobic (facultative anaerobe) conditions, respectively, and came within close contact, particularly within the shallow waters of bays and estuaries where anaerobic muds are found. So aerobic heterotrophs and photosynthetic forms experienced routine contact with phagocytizing anaerobic forms.

Such exposures and interactions are believed to have led to the most fundamentally significant biological events ever to occur on earth. Phagocytizing bacteria, owing to either molecular mistakes or subtle changes in prey cell membrane structure, ingested but failed to digest photosynthetic and/or nonphotosynthetic aerobic cells (of the Bacteria domain) (Fig. 11). These "prey" items now persisted as living cells within cells and maintained their ability to photosynthesize and aerobically metabolize. The anaerobic predator now became a host to aerobic partners. Oxygen was no longer toxic because of the endosymbiont's activities and the "anaerobe" could now routinely journey out of anaerobic habitats and colonize aerobic ones. This fostered a new round of competitive release and exponential growth. The newly evolved partnership "superorganism" experienced enhanced growth and reproduction, more efficient ATP synthesis, and more habitat options. The endosymbionts received protection from predation and nutrients from their host's waste products. These events led to the appearance of the first eukaryotic organisms on earth. Incorporation of both photosynthetic and nonphotosynthetic endosymbionts led to the algae and higher plant taxonomic forms, whereas incorporation of only nonphotosynthetic forms led to protozoan and metazoan lines of descent. The mitochondria of present-day organisms are believed to be descendants of the first heterotrophic aerobic endosymbionts. Similarly, the chloroplasts of algae and higher plants are believed to be descended from the endosymbiotic photosynthetic bacteria.

Present-day protozoa, and some advanced invertebrates as well, routinely ingest algae and leave their prey undigested in a photosynthetically functional state, or digest the prey but retain their fully functional chloroplasts. Also, the chloroplasts and mitochondria of eukaryotic cells contain their own DNA, separate from the "host" cell's DNA. Mitochondria retain a functional set of transfer RNA, bacterial ribosomes, and circular chromosomes characteristic of bacteria. Both mitochondria and chloroplasts divide on their own schedule that is only loosely tied to their "host" cell's schedule. The ancient bacterial host that became a home for endosymbionts is believed to be similar to present-day *Thermoplasma* spp., which live under acidic, hot spring conditions similar to those of the primordial earth and are coated with proteins similar to the histones that form the scaffolding of chromosomes in eukaryotes. Histones are absent from other bacteria.

Motile eukaryotes possess flagella or cilia (undulopodia). Motility was, and is, a major advantage for eukaryotes that could move to regions of higher nutritional value and away from predators if capable of self-directed movement. Such an adaptation undoubtedly led to competitive release and exponential growth. The undulopodia of eukaryotes are composed of tubulin and are ultrastructurally complex and completely different from the simple bacterial flagellum. They have reinforcement microtubular rods with definite arrangements and associated protein complexes, their own DNA, and a resemblance to certain bacteria such as the present-day genus *Spirillum*. *Spirillum* bores into cells and possesses the protein tubulin of similar molecular weight to that found in undulopodia. Partial boring of a form like *Spirillum* into, for example, *Thermoplasma* would provide the host with a beating "tail" and the ability to swim. It is likely that the partially imbedded spirochete even-

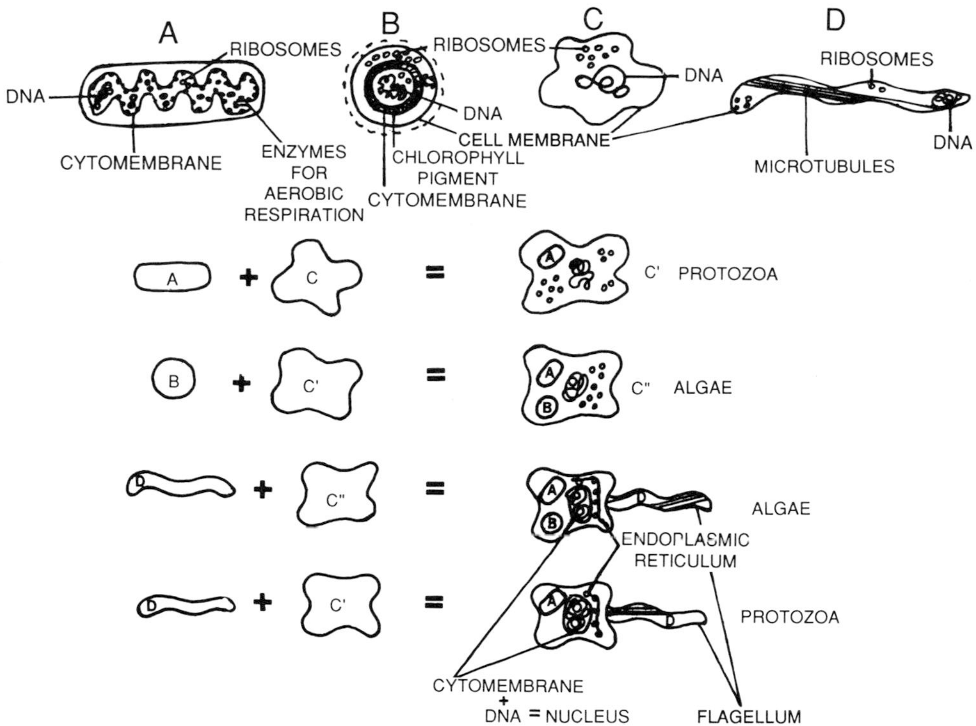


***FIGURE 11*** Schematic representation of serial symbiosis and the evolution of eukaryotic cell lines. A–D represent ancestral prokaryote cell types.

tually transferred some of its DNA and associated proteins to its host, where it formed a coating around the host's DNA. This is believed to be the beginning of the eukaryotic organized nucleus.

Collectively, the foregoing events are called serial symbiosis (Fig. 11). Each of the various events outlined was favored by evolution and natural selection and resulted in competitive release and exponential growth until competition and density dependence again took hold. Such repetitive competition and release cycles form the basis of modern-day trophic structures, and will be discussed in Section V.

## IV. PHYSICAL STRUCTURE OF MARINE ECOSYSTEMS

The world ocean is a nonhomogeneous, nutritionally dilute medium speckled with spatial and temporal heterogeneity. Microbial, primary, and higher productivity is concentrated around coastal zone continental shelf waters and upwelling zones, and their associated sediments. These sediments are rich in inorganic and organic materials and, therefore, life. Deeper waters and sediments also show significant heterogeneity. [*See* OCEAN ECOLOGY.]

Observed chemical and biological variations are driven and configured primarily by physical forces. Differential heating of the earth's surface (caused by the angle of the earth's axis and relative position of the earth with respect to the sun) creates spatial, mainly latitudinal and vertical, variations in atmospheric gases and ocean water densities (due primarily to temperature and salinity differences). Such variations create pressure gradient forces that move both air and water around. Generally speaking, water sinks at the poles and rises at the equator. As water takes in or gives off heat or salt its density changes, and it seeks a physically stable position in the water column. Most of water's properties (e.g., temperature, salinity, and the like) are formed at the air–sea surface and later transported vertically and horizontally, mixing with other water along the way. Around the world there are several distinct sites of large-scale water mass formation. These masses of water are named after the surface site where they originate.

In the area of the Weddell Sea (and to a lesser extent the Ross Sea) adjacent to Antarctica, cold, dense Antarctic bottom water is formed. In the North Atlantic in areas around Labrador and Greenland, North Atlantic Deep water is formed. At several sites in the Atlantic and Indian oceans and a few in Pacific Ocean areas, intermediate water is produced. Because temperature and salinity are conservative properties (i.e., not altered by biological activity), these properties of water change only when water masses of dissimilar properties mix. Circulation of ocean water driven by temperature and salinity (i.e., density) is known as thermohaline circulation and is the principal mover of ocean waters.

Atmospheric circulation is driven by the same physical activities, differing only in speed of movement and degree of mixing owing to its less dense state. Atmospheric circulation patterns dramatically influence ocean circulation. They establish patterns of precipitation and evaporation over both land and water environments and by exerting frictional forces move water around. In turn, land and sea conditions alter atmospheric circulation. In this way, land, sea, and air form an intricate, delicately balanced, mutually influencing set of conditions all driven by the sun's energy. Sea surface temperature, and anomalies therein, have been shown to dramatically affect regional and global climate. The effects of the wind's frictional drag on the oceans is particularly significant under strong wind events such as hurricanes. Such effects set up surface current patterns and Ekman flows, which are generally restricted to the upper 150 m of water.

A third mover of water is tidal circulation, driven by the gravitational effects of the moon, and to a lesser extent the sun, pulling at the waters that envelop a rotating earth. Additionally, water is moved by any large-scale disturbances such as submarine earthquakes. Because water is only minutely compressible, disturbance-generated shock waves are readily transmitted through the water column and result in propagating waves. All the variously produced movements of water are continually undergoing constructive and destructive interferences and are subject to the Coriolis force as well. Such integrated movements are responsible for structuring the global, small to large scale, chemical and biological properties of seawater. The world ocean acts as a giant array of interconnected three-dimensional movements, with various large- and small-scale spin-offs from advective flows routinely occurring (e.g., warm-core and cold-core Gulf Stream and Kuroshio Current rings). Such spin-offs have been found to carry counterclockwise-spinning (cyclonic), nutrient-rich, cold waters into subtropical seas and nutrient-poor, clockwise-spinning (anticyclonic), warmer waters into temperate and boreal waters of the Atlantic and Pacific Oceans. These rings transport heat, momentum, dissolved substances, and weakly swimming organisms around, can form and separate and then be resorbed several times, and have been tracked for up to 3 years.

Structure abounds in the seas. As global circulation is set up, dissolved and particulate inorganic and organic compounds are differentially distributed. These patterns establish other patterns of biological activity and diversity, which are translated up the food web and produce an oceanwide, biological mosaic symphony. The oceans should be viewed in a three-dimensional slice as a series of vertical (and horizontal) density layers with structural dips and rises caused by physical disturbances. There are benthic storms, mid-depth internal waves, and liquid vortices forming and dissipating on any number of temporal and spatial scales. Superimposed on all of this are geological boundaries (e.g., continents, midocean ridge, and islands) that redirect flows and refocus energies of water movement. Such flows reshape coastlines and deposit terrestrial materials imported from rivers and shallow coastlines to new locales. Water masses of different properties may meet head on or run parallel to each other, producing frontal zones important to life-forms. Keeping this dynamic structure as a backdrop, we must now consider the role of microbes in the sea.

## V. ROLE OF MICROBES IN THE SEA

For ease of study purposes and to promote the construction of simplified, stochastic paradigms, scientists have long grouped the inhabitants of the ocean waters into convenient categories. Plankton are thought to be those organisms, large (e.g., jellyfish), medium sized (e.g., microcrustaceans such as krill and copepods), and small (e.g., bacteria, microalgae, and protozoa), whose position in the water is determined primarily by the action of currents. Their swimming abilities, if they possess them, do little to alter their general location, although within a given parcel of water they can maneuver toward or away from food or predators. Nekton are larger organisms (e.g., fish and squid) that can move into and out of currents and therefore determine to a large extent their position within marine ecosystems. This ability does not preclude using currents to aid in directed distance travel or the possibility that entrainment in a strong current could move these organisms against their wills. Benthic organisms are those that primarily inhabit water/lithosphere or other hard-surface interface regions on or within muds, sands, rock, or suspended particles. Within these categories, further breakdown by size is typically employed for additional clarity. Such classifications tend to ignore species compositional differences and rather group organisms into guilds that generally share at least some degree of ecological function and behavioral traits and, therefore, trophic significance. An organism's position on particle surfaces or surface sediment (e.g., epibacteria, epifauna, or epiflora), within sediment (e.g., infauna), or freely suspended in water (e.g., planktobacteria, phytoplankton, zooplankton, microzooplankton) also serves as a convenient identification marker.

Hard and soft substrate benthic habitats represent boundary surfaces that collect particulate debris and have well-developed biological communities, yet other surfaces of all types exist in the ocean realm. The air–sea interface layer, for example, is a zone of enhanced DOM concentrations. This magnified concentration exists in a millimeter-thick surface water film or skin that supports elevated levels of bacteria, protozoa, and other microbes (Fig. 1). At times, microbes of this layer are concentrated 100 to 1000× over those levels in water just below the layer. Within the larger water column itself, density interface layers act as partial barriers, often collecting DOM and particles. These zones form false benthic habitats and support various microbial and larger species more typically found in or on benthic habitats. Such areas support higher density populations of faster-growing bacteria, protista, and their predators and serve as biological oases in an otherwise desert-like deep ocean whose bottom may be miles below. Many planktonic organisms may rely on periodic encounters with such nutritionally enhanced zones for their survival. Internal wave action and disturbances generated by biological activity, such as movements of schools of fish and pods of whales, which often follow density surfaces while foraging or migrating, move organic-rich materials from these surfaces into otherwise nutrient-poor waters above and below them.

These density layers are especially good collectors of flocculent materials known as marine snow. Marine snow is composed of the dead remains of organisms, particularly gelatinous life-forms, discarded larvacean houses, fecal material, mucus particles, and the like. These particles themselves (even away from density layers) form small- to medium-sized oases of concentrated nutrients in nutritionally dilute waters and are heavily colonized by bacteria and other microbes and their predators. Particles of low surface area to volume ratios even develop oxygen gradients, with hypoxic to anoxic conditions developing toward their center. Such water column anoxic zones provide habitat for anaerobic bacteria (e.g., methanogens), which are indispensable in the decomposition cycle and nutrient regeneration.

Studies on the composition of particles in the ocean have demonstrated high microbial activity associated with water column marine snow. The chemical composition of snow particles changes rapidly with depth owing to microbial metabolic activities. Simple carbohydrates and proteins are the first to disappear, with deeper water particles becoming increasingly more refractory with higher concentrations of chitinous and cellulose-based compounds. Many deep benthic invertebrates and microbes have unique enzyme systems tuned to the refractory materials likely to reach the ocean bottom. So, quality of food differs as a function of water column depth, with coastal and other shallow-water systems receiving a higher quality of material, and in greater amounts, owing to the enhanced productivity of these regions. Fecal pellets (and nonpelletized materials) also serve as a source of nutrition for water column and benthic microbes. Epibacteria abound on such material, as do their grazers. Some of the bacteria attach permanently, whereas other species reversibly attach. Nonpermanent attachment and motility allow bacteria to move toward and stay within spheres of diffusing organic matter (e.g., around microalgal cells). During starvation, bacteria undergo cell-surface alterations, form fibrillar structures for adhesion, produce more outer membrane vesicles, increase their hydrophobicity, produce exopolysaccharides, form glycocalyx structures, use stored reserves, lower respiration, stop growing, cease production of nucleic acids, and reduce size to increase their survival potential. Higher up in the protists, a ciliate has been found that lives most of its expected life inside the matrix of fecal pellets. Generally speaking, floating and neutrally buoyant materials of all kinds found in the oceans provide surfaces for microbial attachment or zones of concentrated activity.

Living organisms themselves also provide surfaces for attachment and colonization by marine microbes. Such associations run the gamut from purely parasitic to mutualistic symbioses. A primary example of such a living oasis system can be found in the floating sargassum weed of the subtropical Atlantic Ocean, which concentrates microbial life-forms and serves as a home to larger endemic species such as the sargassum crab and fish. Pathogenic bacteria, fungi, and protists, found on and in coastal water, open water, and benthic organisms, abound in the sea. A good example of this is *Collinia beringensis,* an apostome ciliate that lives in the open blood system of certain krill (euphausiid) species, living on blood nutrients that it draws in pinocytotically. Microbial infections of benthic animals (e.g., gaffkemia in lobsters) are also found.

Luminescent bacteria live mutualistically within special glands of various fish and squid, particularly deeper mesopelagic and abyssopelagic forms. The fish (or squid) provide a protected home and nutrients for their symbionts and use them for signaling or displaying patterns to others for species and mate recognition, for schooling purposes, to attract prey, or to confuse predators. Luminescence is also common in certain dinoflagellates (e.g., *Noctiluca*), which use it to avoid predation by signaling to their predator's predator (known as the burglar alarm theory). Whale skin, in particular, represents a veritable jungle of microbial diversity and abundance. Parasitic diatoms and other protists are living-surface dwellers, causing or making use of wounds as sites of nutrition gathering.

It has been a common belief that most particulate matter settles slowly to the bottom of the ocean, and therefore has been worked on and picked over numerous times before it finally comes to rest in the benthos. It is true that much slow, steady settling does occur. Materials that are fed upon and released as waste products are generally unsuitable as food for some time, until microbial recolonization, which changes the carbon/nitrogen ratio and enhances the overall nutritional value of the material, occurs during the settling process. This reworked material sequentially finds its way to false and ultimately true benthic habitats and associated feeders, with microbes continually "upgrading" the nutritional quality of the recycled material. The microbial biomass itself can be the primary food source as much as or more than the worked on detrital material. Some invertebrate and vertebrate organisms actually cultivate microbes with otherwise low-quality material in a process known as microbial gardening (this is analogous to the phytoplankton cultivation of bacteria discussed earlier).

Not all particles falling through water settle slowly. Obviously large particles, such as the dead remains of large organisms (e.g., whales) fall quickly and serve as major islands of nutrition for microbes and other life-forms. Elaborate benthic communities often form around such bio-islands. However, water column physical forces can also serve as rapid delivery systems, bringing high-quality food to deep sediments. Such is the case with ocean vortices produced by sheer stresses associated with large-scale advective mixing. Such vortices, for example, routinely spin off the Gulf Stream (see earlier discussion). These vortices rapidly transport high-quality (i.e., rich in simple carbohydrates and proteins) particles to deep waters before significant nutritional depletion has occurred, and thus support higher benthic productivity and biodiversity.

### A. Primary Production and Its Fate

Although living and detrital particles that fall through the water are derived from various sources (e.g., algae, protozoa, larger zooplankton, fish), the true source of most of this material is microalgal and cyanobacterial primary production. The highest primary production (and therefore secondary production as well) is centered around coastal waters and upwelling sites. The open ocean is comparatively much less productive per unit volume, although its overall contribution to global production is high owing to the vastness of the open seas (Table I). Obviously, the sediments associated with these regions reflect the productivity of the overlying waters, as modified by horizontal advection.

The largest percentage of global marine primary production is accomplished by nano-and pico-sized algae and cyanobacteria, with larger cyanobacteria (e.g., >64-$\mu$m *Trichodesmium*) responsible for a substantial amount, particularly in subtropical and tropical waters, where they also fix nitrogen. Primary production is high in coral reefs (Table I) (where endosymbiotic zooxanthellae algae live in a mutualistic arrangement within their coral host) as well as in certain waters rich in algal endosymbiont-bearing sarcodines (e.g., foraminifera and radiolaria) or ciliates (e.g., *Mesodinium*). Some ciliates digest algal cells but retain their chloroplasts as functional units.

**TABLE I**
Global Primary Production

| Marine ecosystem | Production (g/m$^{-2}$/yr$^{-1}$) | Net world production ($\times 10^9$ metric tons/yr) |
|---|---|---|
| Algal beds and reefs | 2000 | 1.10 |
| Estuaries | 1800 | 2.40 |
| Upwelling sites | 500 | 0.22 |
| Continental shelf | 360 | 9.60 |
| Open ocean | 127 | 42.00 |
| Total World Ocean | | 55.32 |
| Total Land | | 107.00 |

Large amounts of released dissolved organic matter are associated with this primary production. This DOM is strongly correlated with planktobacterial concentrations which typically range from a low of $10^3$/ml to a more typical for coastal water $10^6$/ml. Higher numbers are found in severely eutrophicated waters. The principal grazers of this biomass are the heterotrophic and mixotrophic flagellates (as well as a few species of planktonic ciliates), which typically range from $10^2$/ml in ocean areas to $10^4$/ml in estuarine systems (Fig. 12). Many of these flagellates are also capable of eating prey larger than bacteria, and in some instances larger than themselves (e.g., microalgae, including chain-forming diatoms). Flagellate ingestion rates vary from 100 to 1000's of bacteria and/or cyanobacteria per day (equaling as much as 100% of the bacteria daily production), with clearance rates (= filtration rates) of 0.0002–0.08 $\mu$l per flagellate per hour and gross growth efficiencies in the ~20–50% range (Table II).

The heterotrophic flagellates are more efficient grazers of bacterial-sized particles than are the ciliates and marine amoebae, which specialize on larger prey such as the zooflagellates, phytoflagellates, and other protistans. Cannibalism is also not uncommon in these forms. A host of other larger organisms also utilize microalgae and protozoans as food, including the bivalve mollusks and macrocrustaceans (e.g., copepods and krill). Thus protozooplanktonic ciliates have a large impact not on bacterial production but rather on primary production (Table II), and are capable of high ingestion and filtration rates, have high gross growth efficiencies, and exhibit community ingestion rates at times equaling those of microcrustacean communities. The water column, spinose, testate marine

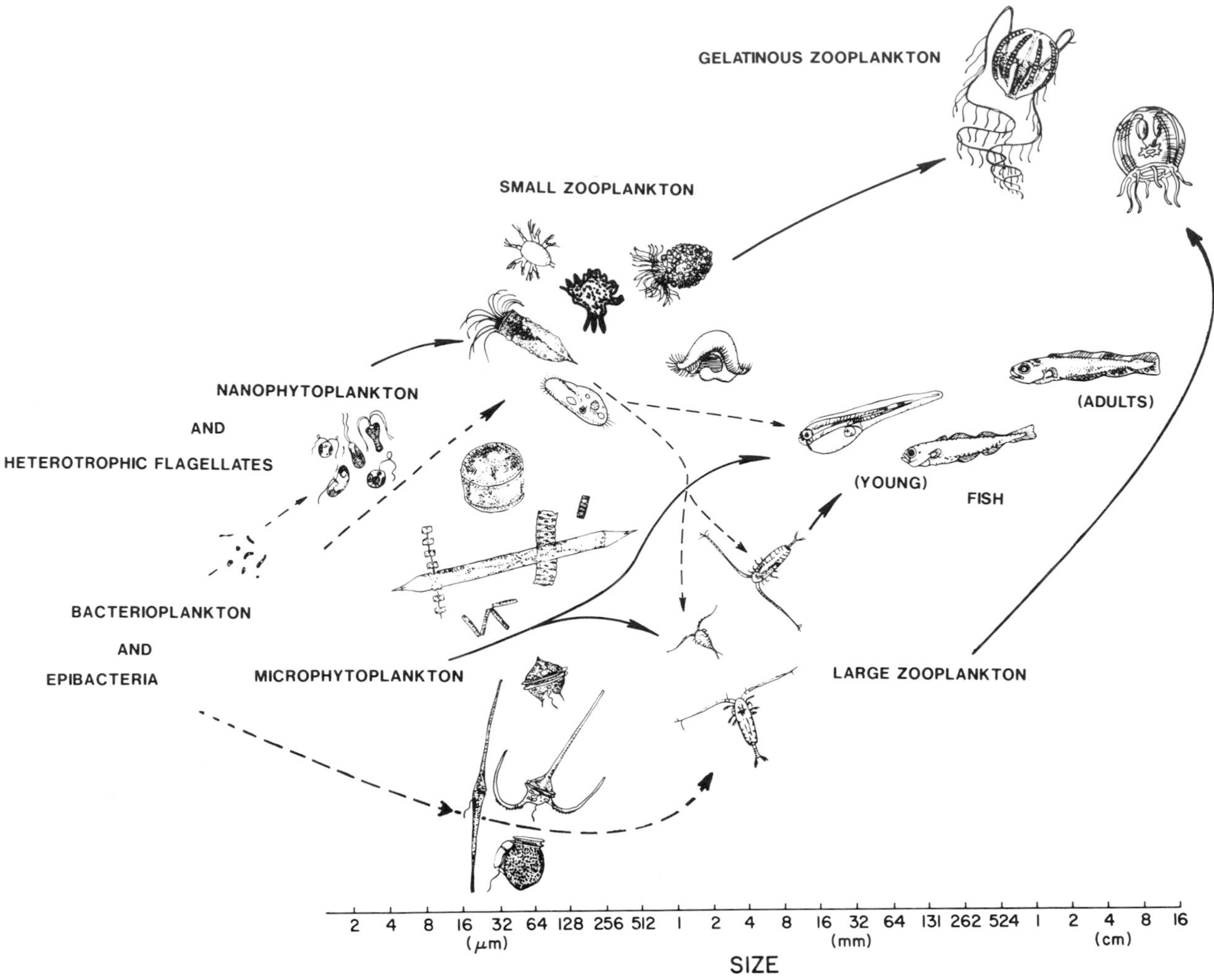


***FIGURE 12*** Planktonic food web model, showing the primary flow of biomass among various predators and prey of different sizes. The solid lines depict past conventional thinking and the dashed lines add to that our current understanding regarding the role of microbes (e.g., bacteria, cyanobacteria, microflagellates, protozoans).

amoebae (e.g., foraminiferans and radiolarians) function primarily as carnivores feeding on microcrustaceans, which they capture via pseudopodial networks and subsequently tear and rip apart by pseudopodial probing and engulfment. Many of these organisms also play host to symbiotic algae, which provide them with nutrition in return for nutrients and protection. Smaller nonspinose forms eat microalgae, some bacteria, and other protists. Colonial radiolarians also exist that reach sizes up to 3 m, although typical colony size is in the centimeter range. These colonial forms feed cooperatively on microcrustaceans and even small fish. Sarcodines such as the foraminiferans are found not only in the plankton but also in benthic environments (especially shallow-water salt marsh and estuarine mud communities). The quantitative impact of these organisms as symbiotic sources of, and grazers on, primary production is not yet well understood.

Certain dinoflagellates also feed on bacterial and microalgal food, whereas others prey on oligotrichous planktonic ciliates. One group feeds via protoplasmic engulfment (veil/pallidium feeding) on chain-forming diatoms (up to 50+ cells long) and other prey. Few data on quantitative importance of such grazing is available.

Although many of the marine protists are able to take up DOM, they rarely can survive solely on this type of nutrition. This is in large part due to

**TABLE II**

**Feeding, Growth Rates and Impact on Bacterial and Phytoplankton Production of Various Marine Protists**

| Organism | Ingestion rate (per protist) | Clearance rate ($\mu$l/protist/hr) | Growth rate (doublings per day) | Gross growth efficiency (%) | Impact on bacterial or phytoplankton production |
|---|---|---|---|---|---|
| Microflagellates | 240–7200 bacteria/day | 0.0002–0.08 | 0.4–9 | 24–49 | 23–100% of daily bacterial production |
| Phagotrophic algae | — | 0.002–0.026 | — | — | Unknown |
| Dinoflagellates | — | 0.4–28 | — | — | Unknown |
| Planktonic ciliates | 0.4–68 ng C/day | 0.04–65 | 0.14–3.4 | 17–76 | 25–60% of yearly primary production |
| Benthic ciliates | 16,000 bacteria/hr<br>10–120% body volume/hr<br>0.24–0.19 × $10^6$ $\mu m^3$/day | 0.003–0.7 | 1.8–4.8 | 10–27 | Unknown |
| Planktonic foraminifera | 0.3–1 copepod/day | — | — | — | Unknown |
| Overall protozoan community | — | — | — | — | 40–100% of daily bacterial production |
| Overall microzooplankton community | — | — | — | — | 8–70% of daily primary production |

the low DOM concentrations found in waters of the world ocean. Under certain enhancement situations (e.g., conditions that exist around marine snow particles, in benthic environments associated with high primary production waters, or around dead animal carcasses), DOM can provide a high percentage of the nutrient requirements of protists (and other animals as well) at least for the short term. However, given the surface area to volume (S/V) ratios of protists, DOM is at best a supplement to their diet and it is the bacteria (and fungi) with high S/V ratios that rely on DOM for nutrition.

A similar argument can be made for bacterial grazers. Because of fluid dynamic viscosity at low Reynolds numbers (i.e., $<1$), which restrict water flow around organisms, the need to keep particle-trapping spaces small to capture bacteria, and the generally low concentrations of bacteria in the water column (e.g., $10^3$–$10^6$/ml), bacteria are a relatively unimportant food source for larger water column protists (e.g., ciliates and sarcodines). Rather, the flagellates with their smaller size and higher S/V are the most efficient and major predators of bacteria in the water column. Some exceptions to this rule are certain planktonic ciliate species that rely mostly on bacteria for food as well as for mucous net feeders such as certain gelatinous zooplankton. The situation is different in the benthos, where DOM and bacteria concentrations are much higher (e.g., $10^8$ bacteria/ml, $10^9$ or more bacteria per gram of dry weight detritus). In these situations a much wider array of predators make use of bacterial biomass. The species of ciliates and sarcodines found in benthic habitats differ greatly from their water column cousins in size, shape, behaviors, and feeding modes.

The quantitative role of protists in benthic habitats has been much less studied than their role in the water column. Densities, species composition, and even presence or absence differ greatly from the oxidized surface layer of sediment, through the redox discontinuity layer, to the anaerobic sulfide layers below (see Fig. 1). Within the oxidized sediment layers of shallow water, where light penetrates to the bottom, microalgae, cyanobacteria, and photosynthetic bacteria are found, as are naked and testate amoebae, flagellates, ciliates, and numerous other kinds of bacteria. Below lighted zones, in aphotic sediments, the algae and photosynthetic bacteria disappear but the other groups remain. Within and below redox layers, bacteria and ciliates persist but other microbial groups become negligible. Strategies and behaviors adapted for living within pore waters or attached to sediment particles are intriguing. For example, the ciliate *Loxodes* contains the enzyme nitrate reductase and employs dissimilatory nitrate respiration (a process otherwise found only in prokaryotes) while also possessing mitochondria and respiring oxygen. This duality allows this ciliate to live near oxic/anoxic boundaries and survive periods of anoxia. Ciliates such as *Trachelorapbis* are elongated with nonciliated ventral surfaces that they use to "glide" along particles. These ciliates possess a fixed cytoproct (mouth) but also can "unzip" sections of their ventral surface to engulf particles. Other forms such as *Euplotes* use fused cilia (cirri) as "stilts" and other cilia as current generators and trapping devices to achieve food particle removal.

The type and concentration of microbes in sediment communities are linked to sediment characteristics. Well-sorted sand sediments develop a rich fauna in their interstitial spaces. Medium-grain-sized sediment is the richest in ciliate fauna. Smaller-sized sediment is dominated by nematodes, and larger-sized sediment by turbellarians, gastrotrichs, and harpacticoid copepods. On silt-sized clays, ciliates, flagellates, and amoebae are present only on flocculent surface layers with ciliates confined to the hypotrich and scuticociliate groups. More than 28 species of ciliates have been found associated with hydrothermal vent systems.

As a general rule, marine microbial trophic relationships can be traced back to serial symbiosis/evolutionary events that established an overall central tendency for the food web, with variations on a theme always present. With the exception of parasitism, microbes (and larger organisms as well) typically eat food of about 10% of their own size. Ingestion of smaller particles usually requires special adaptations such as mucous netting. Eating larger food items requires behavioral modifications related to handling. Nonetheless, some microbes eat food larger than themselves, whereas others concentrate on very small items.

### B. Bacterial Activity

Materials are broken down by bacteria in marine systems either via aerobic respiration or via anaerobic decomposition through fermentation, dissimilatory nitrogenous oxide reduction, dissimilatory sulfate reduction, or methanogenesis (Fig. 13).

Aerobic respiration consists primarily of nitrification processes in which $NH_4^+$ is oxidized into $NO_2^-$ by *Nitrosomonas* spp. and $NO_2^-$ to $NO_3^-$ by *Nitrobacteria* spp. In the anaerobic processes of dissimilatory $NO_x$ reduction, $NO_3^-$ and $NO_2^-$ serve as terminal electron acceptors in place of oxygen, and in the process are converted to reduced nitrogen compounds such as $N_2O$ or $N_2$ gas (denitrification) or ammonia. These reactions require the expenditure of energy. Certain species of bacteria and cyanobacteria fix $N_2$ gas into $NH_4^+$ (e.g., the aerobic bacteria *Azobacter* and cyanobacteria *Trichodesmium,* and the anaerobic bacteria *Clostridium*). The $N_2$ fixation requires anaerobic conditions, which are achieved in oxygenated environments in an oxygen-free heterocyst or in a zone of no oxygen maintained in the center of water column filamentous colonies such as *Trichodesmium,* and the expenditure of energy, usually as carbohydrate. Ammonification is accomplished by algae and microalgae that take up $NO_3^-$ and $NO_2^-$ and

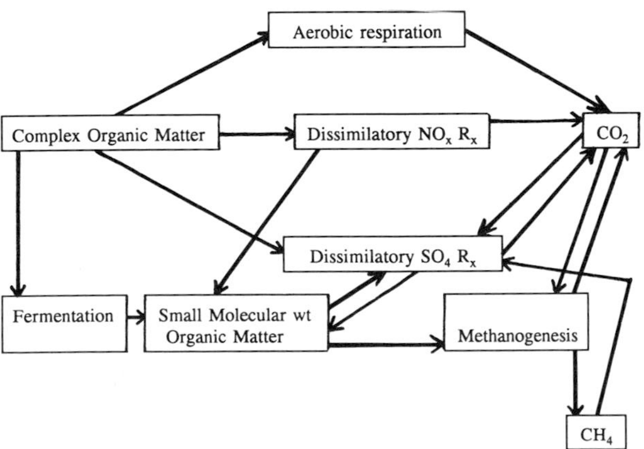


***FIGURE 13*** The pathways of carbon flow resulting from bacterial decomposition. All but the aerobic respiratory pathway are part of the anaerobic decomposition cycle.

convert it to ammonia, through autolysis of cells, via protist and other animal excretion and by bacterial and fungal mineralization. Competition for ammonia between autotrophs and aerobic decomposition bacteria may limit primary production at times.

Dissimilatory sulfate reduction is carried out only by anaerobic bacteria (e.g., *Desulfovibrio*) and is similar to aerobic respiration and dissimilatory nitrogen reduction except that $SO_4^{2-}$ acts as the terminal electron acceptor. Resultant reduced sulfur compounds such as $H_2S$ serve as an energy source for chemoautotrophic organisms. Dissimilatory sulfate reduction occurs as sulfate diffuses down into anoxic waters and sediments, where bacteria use it to oxidize substrates while producing sulfur and $H_2S$. Some sulfide is trapped in the sediment via metal–sulfide complexing and associated precipitation, but most diffuses to aerobic strata, where it is oxidized back to $SO_4^{2-}$ either chemically or biologically. The presence of oxygen inhibits sulfate reduction and sulfate inhibits methane production.

The oxidation of reduced sulfur compounds provides energy to certain bacteria (e.g., *Beggiatoa*) and produces sulfate as an end product. In these bacteria, carbon is provided by other inorganic sources. Anoxygenic photosynthetic green and purple sulfur bacteria use reduced sulfur compounds as electron donors in place of $H_2O$ for photosynthesis. Assimilatory sulfate reduction is carried out by plants and certain microbes that take up sulfate and use it for later incorporation into biomass.

Fermentative pathways, including methanogenesis, are important for decomposition in both oxic and anoxic waters and sediment and occur in concert with oxidative processes. Methanogenesis occurs throughout the ocean, associated with marine snow particles and false and true benthic habitats. Methanogens pick up small-molecular-weight organic matter produced primarily via fermentation and convert it into $CH_4$ while also taking up and releasing $CO_2$. Methanogens provide reduced compounds for growth of other bacteria and lead to *in situ* mineralization, which helps to sustain upper ocean primary production. Methane is also produced at times as a by-product of algal metabolism.

In addition to peaks in heavily organic-detritus-laden habitats such as salt marsh, estuarine and coastal zone sediments, oceanic trenches, and fjords, methane maxima are also found at 50- to 150-m pycnoclines, usually associated with chlorophyll and ATP peaks. The bacterium *Methylomonas pelagica,* found in the Sargasso Sea, uses methane as a carbon and energy source and ammonia and nitrate as a nitrogen source, producing formaldehyde as one by-product of its metabolism. Other mud-dwelling methanogens use formate, hydrogen, methanol, or acetate for their growth.

Within the sediments, organic-detritus-mineralizing bacterial species vary with depth into the sediment as a function of which compounds act as electron acceptors. From the sediment surface down, first oxygen and then nitrate, iron, and manganese oxides, sulfate, and finally $CO_2$ and $CH_4$ act as acceptors. The most abundant acceptor in the sediment is $SO_4^{2-}$.

Overall, as much as 85–95% of the organic carbon produced in the world ocean is recycled in the upper oxygenated water before entering anoxic areas. In coastal systems, such as salt marshes, about half of the aerobic respiration and almost all the anaerobic transformation of organic matter is accomplished by microbial (mostly bacterial) metabolism. Most macrophytic salt marsh and coastal water production is not grazed directly, but rather is utilized via detrital pathways and microbes. Microbes often use DOM and particulate materials not useful to other organisms. In the case of particulates, as discussed earlier they alter C/N ratios (e.g., in estuaries they change C/N ratios from high to lower levels) and enhance particle quality with their own biomass, thus making it available to higher organisms. In this way they introduce usable forms of nitrogen into the sea.

In addition to nitrogenous and sulfur compounds, phosphorous is also very bioactive and is partially regulated by microbes (along with geological and avian influences). Bacteria immobilize $PO_4$, but most mineralization of phosphorus occurs during egestion and excretion by protozoan and metazoan activities, not bacterial ones. However, bacteria indirectly have a considerable influence on the phosphorus cycle. As oxygen is microbially depleted in sediments, dissimilatory sulfate reduction increases, producing $H_2S$. This decreases redox potential, causing a solubilization of inorganic phosphate. If the surface sediments are oxygen rich, phosphate becomes insoluble again, undergoing complex reactions with metals such as iron and manganese and precipitating out of solution. In anaerobic sediments, however, the dissolved phosphate diffuses into the water column and thus becomes available for microalgal use.

In general, bacteria and fungi also aid in the decomposition cycle in marine systems by weakening fibrous, refractory tissues (as do physical/mechanical disturbances) and breaking them into smaller, higher S/V ratio material. Such alteration increases the area available for further bacterial colonization and promotes the grazing activities of protozoans and larger invertebrates, which in turn stimulate additional larger animal activities. All of these activities are functionally related to temperature, pressure, oxygen concentrations, and salinities. Among the autotrophically produced biomass, seaweeds are the fastest to decompose, followed by sea grasses and lastly marsh grasses, which contain higher fiber content.

Much recent marine bacterial research has focused on deep-sea hydrothermal vent systems. These regions are primarily associated with mid-ocean ridge spreading centers, where seawater circulates through newly formed oceanic crust and mixes with magma-associated geochemicals. Such circulation produces modified, superheated seawater that is particularly rich in hydrogen sulfide. This $H_2S$, driven by magma-heat-induced convective currents, mixes with cold, oxygen-rich ocean bottom waters and serves (along with certain other inorganic compounds) as the primary energy source for vent-associated, free-living and symbiotic, chemolithotrophic bacteria. These bacteria form the base of a nonphotosynthetic food web and support the growth of ciliates, mussels, crabs, clams, and pogonophoran worms.

Vent bacterial symbioses are most intriguing. In the vestimentiferan, pogonophoran worm *Riftia,* for example, an organ called the trophosome houses bacterial symbionts in vacuoles of its cells. Hydrogen sulfide, which is toxic to most animals, is utilized by these bacteria as an energy source. Vent bivalves, gutless oligochaete and polychaete worms, nematodes, and turbellarians also house symbiotic $H_2S$-utilizing bacteria. Ecological associations similar to those found at deep-sea vents also exist in certain sulfide- and methane-rich shallow-water sediments, around fluid seeps, and in deep anoxic basins along oceanic margins. In all of these instances chemolithotrophic bacteria oxidize reduced compounds and produce organic compounds that serve as food for other free-living and symbiotically paired organisms. Host organisms often have reduced or no guts. The symbiont bacteria detoxify the $H_2S$ and produce extracellular organic compounds that serve as nutrition for the host or are directly digested for food by the host. These bacteria supply 50 to 100% of their host's nutritional needs. In return, the hosts bridge the gap between oxic and anoxic waters and extend bacterial range while also providing protection from certain predators.

Some vents have giant megaplumes that spew forth hydrothermal fluids. In general, deep-sea hydrothermal vents act as deep ocean oases for sulfide-dependent animals over the open seafloor. Whale carcasses may act as stepping stones for deep-sea species that move from one vent site to another as individual vents become inactive.

Our present view of the oceans and their bioecological structure is radically different from the overly simplistic, homogeneous structure view held just a couple of decades ago. In contrast, we now recognize the sea as an extremely complex system composed of mosaics of subsystems driven primarily by physical forces and boundaries. These forces establish water column and benthic physiochemical characteristics on small to large temporal and spatial (horizontal and vertical) scales, resulting in a four-dimensional time–space realm of shifting pycnoclines (including particle-rich, false benthic zones) and gradients (water column, sediment, on and around floating islands of life, and associated with polar edge ice and ice undersurfaces). Such spatial and temporal variabilities drive algal, bacterial, and cyanobacterial production, which in turn determine patterns of flagellate, protist, and other organismal distributions and production. All of these organisms produce DOM as a product of their metabolic activities and death. Nonphotosynthetic bacteria act as the primary consumers and remineralizers of biologically produced dissolved and particulate materials, which often are first mechanically or biomechanically modified by other events or organisms. Particulates, in small to large sizes and concentrations, are omnipresent in the oceans and are continually sinking at various speeds as they are mechanically, microbially, and generally biologically worked and reworked. The degree of physical and chemical modification that they undergo depends on their residence time in the water column. In addition to production directly or indirectly related to photosynthesis, chemosynthetic production also occurs in certain benthic habitats (e.g., hydrothermal vents, anoxic basins, estuarine sediments), particularly around midocean ridge areas.

Much microbial production remains within the microbial loop, where it is repeatedly recycled. For this reason, many scientists view the microbial component of marine food webs as a sink for carbon rather than as a source of it. Although it is true that a high percentage of marine primary production is consumed by bacteria and other microbes, if one considers the kinds of materials being cycled within the microbial community (i.e., DOM, bacteria, cyanobacteria, pico- to nano-sized algae) a different interpretation emerges.

Dissolved organic matter is generally unsuitable as a primary nutrition source for all but bacteria and certain fungi. Also, small-sized food is the preferred prey of flagellates and, as food size increases somewhat, other protozoa (e.g., ciliates and sarcodines), but not metazoans. So even if at best a small percentage of the originally fixed carbon is transferred up the food web, this is food now packaged in a size and nutritional way to be important to metazoan predators. Also, certain microbes actively aggregate particles via mucuslike secretions or attach to marine snow particles. These aggregates maintain bacterial colonies that enhance the nutritional value of the food. In this way coupling between microbial and metazoan food webs routinely occurs. Another important point to consider is that bacteria and protozoans respond to nutritional pulses via rapid growth and population density increases (i.e., hours to days) as compared to the much slower metazoan response (e.g., copepods exhibit a 4- to 8-week lag in numerical responses). In this way microbes act as water column stabilizers, capturing energy that might otherwise be lost to bacterial decomposition and/or become sequestered in sediments and making it available to higher trophic levels in a nutritionally enhanced state.

So, the microbial loop can be looked at as a stabilizing source of additional nutrition for metazoan-based food webs that rapidly responds to water column and benthic habitat physiochemical changes while also remineralizing nutrients. It acts as both a source of and a sink for carbon, depending on your perspective, and where in the labyrinth of ocean subsystems you hang your hat.

## Acknowledgments

This work was supported in part by a grant from the Long Island Sound Research Fund of the Connecticut Department of Environmental Protection.

## Glossary

**Marine snow** Ubiquitous water column particulate material composed of amorphous, fragile, macroscopic aggregates, often including the dead remains of organisms such as gelatinous life-forms and fecal and mucus material.

**Microbial gardening** Activities carried out by organisms that cause the enhanced growth of microbial organisms that may serve as a source of nutrition.

**Microbial loop** All aspects of aquatic food webs related to bacterial/cyanobacterial and protistan (microalgal and protozoan) interactions.

**Mixotrophic** Gaining nutrition from both autotrophic and heterotrophic activities.

**Serial symbiosis** Series of phagocytoticlike symbiotic events terminating in endosymbiotic associations among prokaryotes that led to the formation of the first eukaryotic cells.

**Undulopodia** Collective term used to refer to eukaryote flagella and cilia.

## Bibliography

Capriulo, G. M., ed. (1990). "Ecology of Marine Protozoa." Oxford, England: Oxford University Press.

*OCEANUS* (1992). Biological Oceanography. **35**(3): 10–61. Woods Hole Oceanographic Institution. Woods Hole, MA.

Ducklow, H. (1983). Production and fate of bacteria in the oceans. *Bioscience* **33,** 494–501.

Margulis, L. (1980). "Symbiosis in Cell Evolution." San Francisco: Freeman.

Reid, P. C., Turley, C., and Burkill, P., eds. (1991). "Protozoa and Their Role in Marine Processes," NATO. ASI Series V. G. 25. New York: Springer-Verlag.

Sieburth, J. McN. (1979). "Sea Microbes." Oxford, England: Oxford University Press.

# Marine Productivity

**R. Perissinotto**
*Rhodes University, South Africa*



From an ecological perspective, productivity is defined as the rate at which new biomass (or weight of living matter) is added to a population or community and is expressed in terms of change in weight or number per unit time. Carbon fixation by photosynthesis (autotrophic or primary production) provides the basic organic matter that is subsequently consumed and recycled through two major pathways (heterotrophic production): the microbial loop and the metazoan food web.

The boundaries between autotrophy and heterotrophy are not always clear and some ciliates and flagellates obtain their metabolic energy from different combinations of the two. These are known as mixotrophs. Similarly, although heterotrophy is conveniently defined in terms of production of herbivorous (secondary productivity) and carnivorous (tertiary productivity) organisms, it is in fact difficult to find instances where some degree of omnivory does not occur.

The trophic status of a marine ecosystem is generally determined by the rate of utilization of inorganic nutrients, which support both the primary and higher production levels. Consequently, marine regions where strong advection and diffusion of primary nutrients occur are called eutrophic and are highly productive, and regions with poor nutrient supply are called oligotrophic and are poorly productive (Table I).

## I. INTRODUCTION

The marine environment plays a critical role in regulating the global carbon cycle. Deep oceanic waters are ≈200% supersaturated with $CO_2$, compared to surface waters in contact with the atmosphere. This difference is the result of a complex system known as the biological pump, which controls the flux of photosynthetically derived organic matter from the sea surface to the deep waters.

The net effect of this system is to reduce the partial pressure of $CO_2$ in the surface waters, causing the drawdown of carbon dioxide from the atmosphere. To increase sequestration of carbon in the deep ocean, it is necessary to increase not only the photosynthetic fixation of carbon, but ultimately the net community production. [*See* GLOBAL CARBON CYCLE.]

When energy flow is considered from the point of view of groups of organisms in similar habitats, then these are referred to as ecological levels. Plankton, nekton, benthos, and fringing communities are the four major ecological levels that dominate the epipelagic (upper layer), mesopelagic (midwater), benthic (seafloor), and intertidal (ocean margin) subsystems, respectively. Plankton and benthos are also conventionally subdivided into functional groups: phyto- for the autotrophs and zoo- for the heterotrophs.

**TABLE I**
Typical Values of Nutrient Concentrations, Phytoplankton Biomass and Production, *f*-Ratio,[a] and Secondary Production for the Different Trophic Regions of the Marine Biosphere.

| Region | Nitrate ($\mu M$) | Chlorophyll*a* ($\mu g \cdot liter^{-1}$) | *f*-ratio (%) | Primary production ($g\ C \cdot m^{-2} \cdot day^{-1}$) | Secondary production ($g\ DW \cdot m^{-2} \cdot year^{-1}$) |
|---|---|---|---|---|---|
| Eutrophic | 12–32 | 1.5–1.8 | 57–82 | 0.3–4.0 | 27–34 |
| Oligotrophic | b.d.–0.05[b] | 0.04–1.0 | 3–21 | 0.05–0.8 | 8–16 |
| High nutrient, low productivity | 7–31 | 0.25–2.2 | 36–48 | 0.1–1.0 | 10–20 |

[a] *f*-ratio = ratio of new-to-total (new + regenerated) production.
[b] b.d. = below detection.

When considering the successive steps of the food pyramid, referred to as trophic levels, the most useful grouping is, however, based on size classes. Marine organisms range in size from meters (fish, squid, and mammals) to less than a micron (bacteria and protists). Similar size classes are designated with a similar prefix (nano-, pico-, etc.), irrespective of the ecological or functional grouping (Fig. 1). The rates of many metabolic processes are determined by the size of the organism. Smaller organisms have a high surface area to volume ratio, which correlates with higher nutrient uptake, growth, and mass-specific respiratory rates, as well as shorter generation times. As a result, biomass and abundance are inversely proportional to size and the larger an organism is, the rarer it is in a given system. [*See* Ocean Ecology.]

Size classes generally correspond well with trophic levels (Fig. 1). The decrease in population biomass with size is then explained in terms of ecological efficiency within the food web. Indeed, because only 10–20% of energy flowing into one trophic level is transferred to the next, there is a progressive decrease in net biomass moving up the food pyramid (Fig. 2). Similarly, temporal changes in the biomass of organisms at different trophic levels are closely dependent on their rates of turnover, which are higher for smaller organisms.

The 10–20% transfer of energy between trophic levels apparently implies that most of the material is lost from the system. However, only a part of this apparent loss is real, in the form of organic carbon respired back to $CO_2$ at each trophic level. The rest, in the form of fecal pellets, organic detritus, and dissolved organic carbon (DOC), is recycled through the food chain, mostly via microbial activity. Also, a complex of bacteria and unicellular microzooplankton operates within the microbial loop, where they use a greater proportion of primary production (as DOC) than is consumed by the larger faunal categories. The relative magnitudes of many of these trophic pathways are still largely unknown and their investigation constitutes one of the most ambitious projects of modern research, the Joint Global Ocean Flux Study (JGOFS).

## II. PHOTOSYNTHETIC PRODUCTION

### A. Phytoplankton Dynamics in the Water Column

Phytoplankton is believed to account for ≈80–90% of net annual marine production. It consists of a mixture of prokaryotic and eukaryotic organisms. Among the prokaryotes, the cyanobacteria (blue-green algae) and the recently discovered prochlorophytes normally account for most of the photosynthetic picoplankton production, especially in oligotrophic waters. The larger phytoplankton (nanoplankton and microplankton; Fig. 1) are, however, entirely composed of eukaryotes. Coccolithophorids (Primnesiophyceae) and autotrophic flagellates usually dominate the nanoplankton, whereas diatoms and dinoflagellates dominate microphytoplankton production. [*See* Marine Microbial Ecology.]

For ecological purposes, phytoplankton is defined as a group of organisms that contain chlorophyll *a* as the main, or one of the main, photosyn-

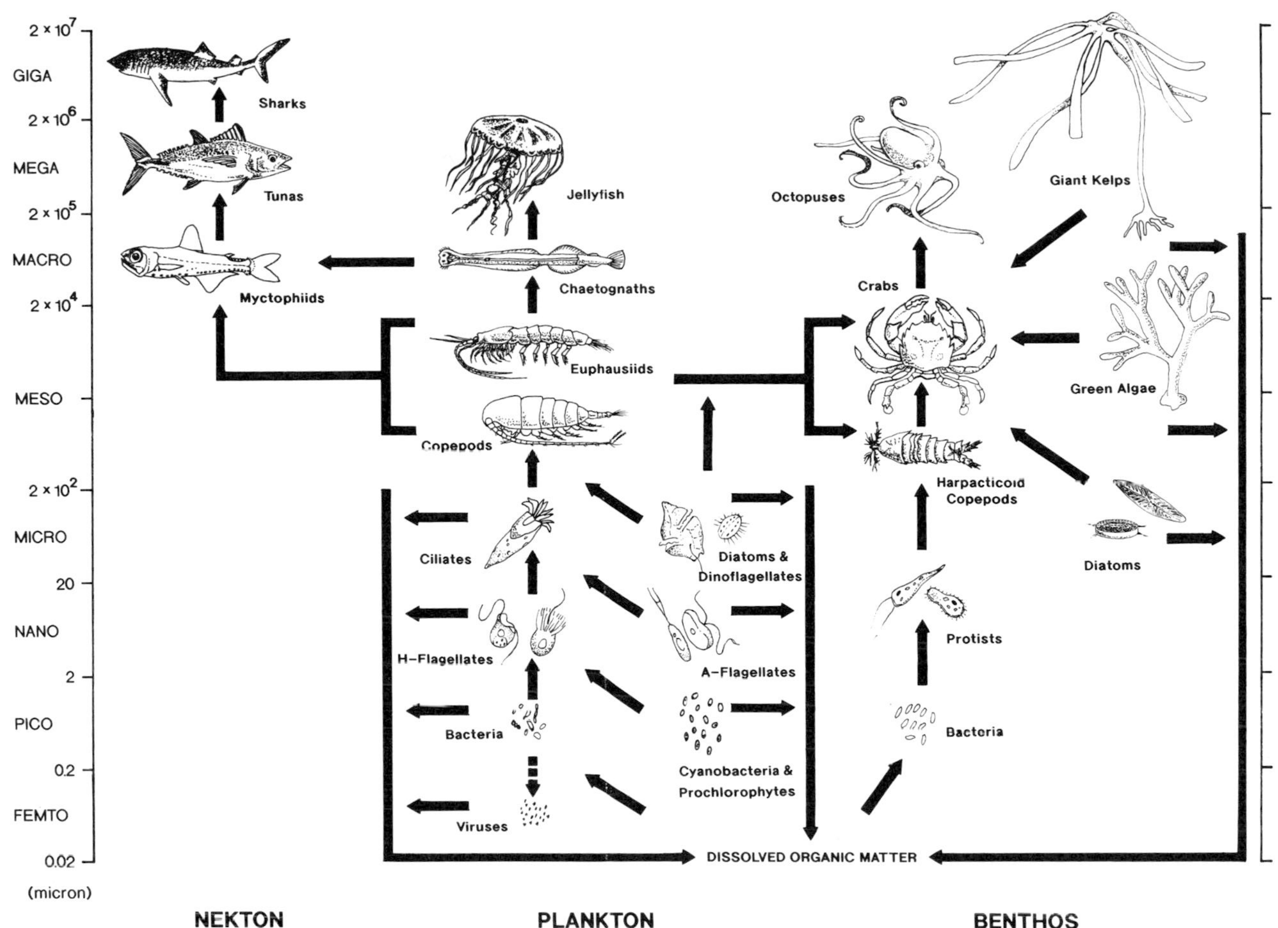


***FIGURE 1*** Size classification and energy flow diagram of the most important components of the pelagic and benthic subsystems. [The plankton compartment has been adapted from T. Fenchel, ed. (1987). "Ecology of Protozoa." Science Tech. Publ., Madison, Wisc., and Springer-Verlag, Berlin.]

thetic pigments. The only exception appears to be the prochlorophyte *Prochlorococcus,* in which chlorophyll *a* is replaced by a divinyl chlorophyll *a* pigment. When chlorophyll *a* constitutes the bulk of the pigments, its concentration provides an index of phytoplankton biomass. This is also used to estimate the photosynthetic capacity ($P^B$) of phytoplankton. This is the ratio of carbon fixation rate over chlorophyll *a,* and is important as an indication of the physiological performance of algal cells.

Phytoplankton production in the euphotic zone of the ocean depends on two different sources of nitrogen supply. One, the autochthonous input, is the *in situ* supply of ammonia, urea, amino acids, and other dissolved organic compounds derived from the excretion of animals and the metabolism of heterotrophic microorganisms. Because the substrates that fuel these activities derive from phytoplankton via the food web, the phytoplankton production resulting from this recycled nitrogen is called "regenerated production." The other principal source of nitrogen is the allochthonous input of nitrate via upwelling and diffusion from deep waters into the euphotic zone. This, along with the secondary sources (sewage, freshwater runoff, and nitrogen fixation), constitutes the basis for "new production."

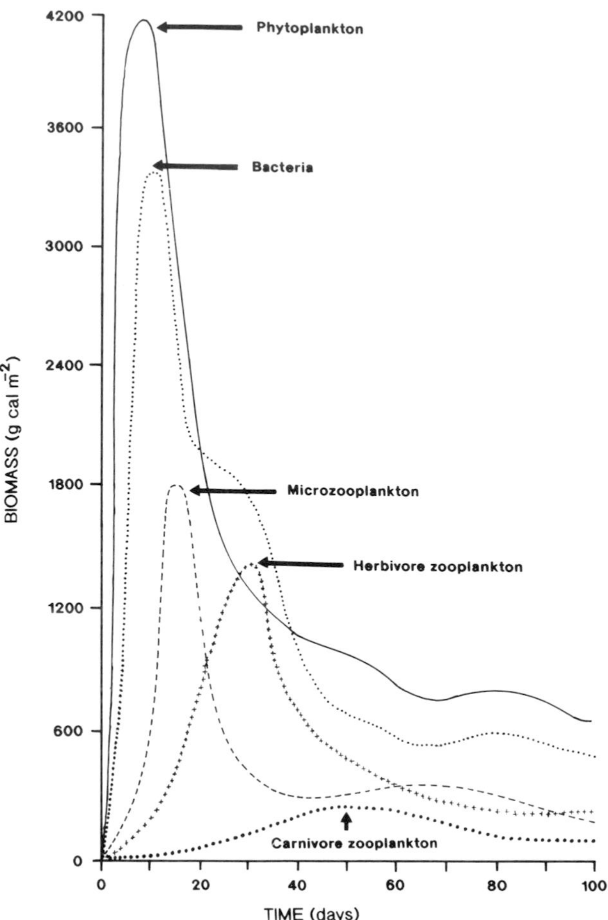


***FIGURE 2*** Time-dependent development of the pelagic trophic system in a tropical upwelling area, after the onset of a phytoplankton bloom. [Adapted from M. E. Vinogradov, V. F. Krapivin, V. V. Menshutkin, B. S. Fleyshman, and E. A. Shuskina (1973). *Oceanology* **13,** 704–717.]

The ratio of the nitrate-based new production to the sum of nitrate- and ammonia-based production (total production) is called the *f*-ratio. On average, the ratio varies in an asymptotic fashion with total production, ranging from ≈0.05 in the oligotrophic oceans to >0.5 in the coastal upwelling regions (Table I). This concept is very important, because under steady-state conditions, the upward transport of new nitrogen (nitrate) is balanced by the removal of nitrogen from the upper mixed layer by sinking organic matter or by predators that leave the area. Corresponding to this is an equivalent amount of carbon that can be exported to the sea sediments from the total production in the euphotic zone, without the production system running down.

Phytoplankton biomass and production vary greatly in different waters (Table I), with highest values found in the coastal upwelling regions of the Americas and Africa. The midoceanic regions have the lowest production rates, and higher levels are found along the equator and in bands lying poleward of midocean gyres. Time-dependent changes in phytoplankton biomass and production are just as marked and important as the spatial variations. Both are induced by a complex interplay of physical, chemical, and biological processes. Of particular importance are the availability of sunlight, macronutrients, and iron, as well as the stability and depth of the upper mixed layer and the grazing pressure exerted by zooplankton.

Phytoplankton photosynthesis is confined to the upper 50- to 100-m layer, called the euphotic zone, which is delimited by the depth of penetration of 1% of the surface irradiance. Algal photosynthesis operates only in the visible range of wavelengths, between 400 and 700 nm. The total irradiance within this wavelength interval is called photosynthetically active radiation (PAR). Light intensity decays exponentially with depth because of absorption and scattering due to the seawater itself (molecular scattering) and the dissolved and suspended matter (particle scattering). The rate of attenuation is measured by the attenuation (or extinction) coefficient, $k$, which varies with the wavelength of light.

Carbon uptake rates can be determined as a function of irradiance using a photosynthesis–irradiance, or P–I, curve (Fig. 3). This may be asymptotic or peaked, with a decrease in photosynthesis at high light intensities due to photoinhibition. The parameters of P–I curves are usually normalized to the biomass of chlorophyll.

Of the macronutrients, nitrogen is considered the primary limiting element in the marine environment. Nitrate controls the rate of new production, whereas ammonia and urea are the major forms fueling regenerated production. Nitrate is a highly oxidated form of nitrogen; both ammonia and urea are reduced forms. Because nitrogen within the phytoplankton cell has to be reduced before being

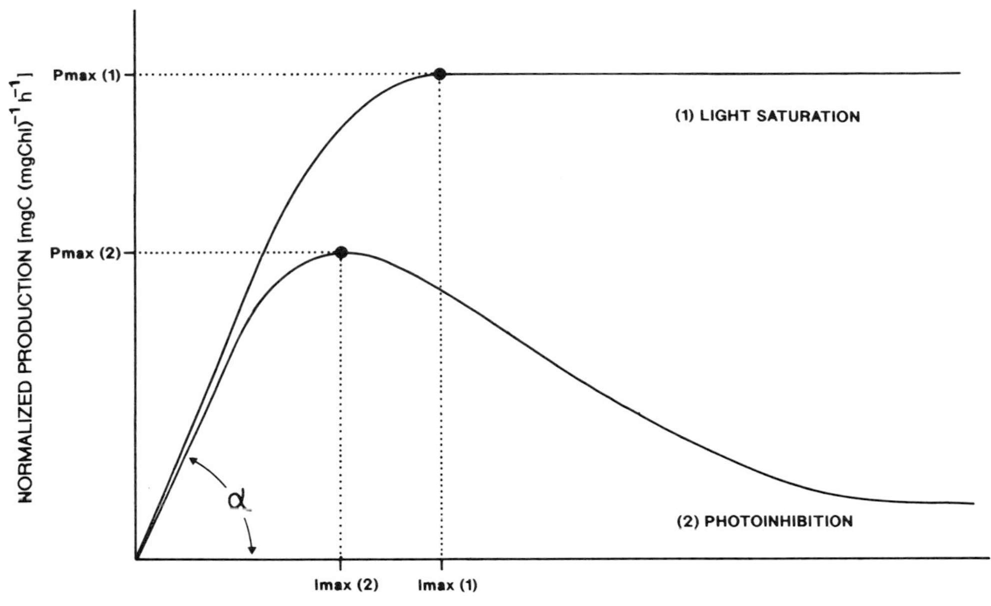


***FIGURE 3*** Models of photosynthesis–light response for (1) light-saturated and (2) photoinhibited phytoplankton assemblages. The light saturation response is more typical of phytoplankton from surface waters, adapted to high irradiances. Conversely, photoinhibition occurs in phytoplankton living at the base of the euphotic zone and adapted to low irradiances.

incorporated into amino acids, it is advantageous to take up urea and ammonia rather than nitrite and nitrate. All phytoplankton size classes show the following order of preferential uptake: $NH_4 >$ urea $> NO_2 > NO_3$. However, there is a tendency for microphytoplankton, particularly chain-forming diatoms, to take up larger proportions of nitrates compared to the nano and pico size classes. The highest levels of new production, leading to highest yields, are therefore likely to occur in the presence of bloom-forming diatoms. The presence of high concentrations of nitrate is in itself not sufficient to result in high productivities, for large oceanic areas have high nutrient concentrations but low primary production (Table I).

The uptake and assimilation of nitrate by phytoplankton in steady-state culture are described by Michaelis–Menten kinetics (a saturation-type curve). The half-saturation constant is approximately 1 $\mu M$ and $NO_3$ concentrations of $\approx 5\ \mu M$ should be sufficient to saturate the uptake and assimilation systems of phytoplankton. Prolonged exposure of cells to concentrations above this threshold may induce a different type of uptake, with nonsaturating linear kinetics, mediated by an increase in enzyme synthesis. This occurs in the phytoplankton of nitrate-rich upwelling systems.

In these areas, phytoplankton growth is maximal and surface nitrates are effectively exhausted after a short bloom phase. There are a few oceanic areas, however, where high nutrient concentrations are found in surface waters throughout the year, and phytoplankton production and biomass are much lower than would be expected from nitrate availability (Table I). These include the Southern Ocean, the equatorial, and the sub-Arctic Pacific.

Recently, a hypothesis based on iron limitation has emerged to explain the paradoxical nature of these regions. Iron as a trace element is required by phytoplankton in very small amounts for the synthesis of chlorophyll and nitrate reduction. In shelf and coastal systems, iron supplies are adequate for high phytoplankton production owing to the resuspension of bottom sediments and runoff from land. However, offshore areas remote from landmasses receive only minor supplies of iron via long-range transport and fallout of atmospheric dust. The Southern Ocean, the equatorial, and the subArctic Pacific are characterized by very low concentrations of dissolved iron. The possibility thus exists that because of iron limitation the enormous amounts of nutrients present there are not utilized. However, although there is some compelling evidence for this, iron enrichment experiments have produced results of controversial interpretation.

Turbulence within the euphotic zone is very important in regulating the photosynthetic activity of phytoplankton. The upper mixed layer is delimited by the depth of the pycnocline, a sharp transition in density often coinciding with the main thermocline. Also, the mixed layer depth ($Z_{mix}$) is negatively correlated to the vertical stability of the water column. For a phytoplankton cell, it is the relationship between $Z_{mix}$ and the euphotic depth ($Z_{eu}$) that regulates the balance of photosynthesis and respiration. When the mixing depth exceeds the euphotic depth, phytoplankton remain in the aphotic zone during part of the day and the carbon balance is shifted toward increasing respiratory losses. If $Z_{mix} < Z_{eu}$, daily production increases and, for a constant $Z_{eu}$, production is higher with a shallow $Z_{mix}$. Similarly, when considering the whole phytoplankton community in the water column, there is a depth at which the vertically integrated rates of photosynthesis and respiration are equal, called the critical depth ($Z_{cr}$). Net production depends on the relationship between $Z_{cr}$ and $Z_{mix}$ and is positive only when $Z_{mix} < Z_{cr}$.

Grazing is responsible for substantial losses in the phytoplankton standing stock. The dynamics of zooplankton grazing will be analyzed in Section IV. What is important here is the level of coupling between autotrophic production and grazing. Tight coupling results in an effective limitation of net production (proximal control), as biomass levels remain constant even in the presence of high specific growth rates of the phytoplankton. This may happen when a lack of environmental variability does not allow the usual temporal separation of grazers and producers.

### B. Benthic and Fringing Macrophyte Communities

Macroalgae and vascular plants are often the most prominent primary producers in shallow, coastal waters. Here, their rates of carbon fixation per unit area may often reach levels an order of magnitude higher than those of the local phytoplankton assemblage.

Marine macrophytes include seaweeds, with a holdfast attached to a rocky bottom, and flowering plants like sea grasses, marsh grasses, and mangroves, which have roots penetrating into the sediments. A minor group of seaweeds are actually free-floating (holopelagic macroalgae) and may be locally important, for example, *Sargassum* in the Sargasso Sea and Gulf Stream. Sea grasses occur primarily in estuarine and lagoonal areas; marsh grasses and mangroves occur in the intertidal range of some temperate and tropical shores. Seaweeds represent by far the largest and most widespread group of marine macrophytes.

Seaweeds are limited to areas where the seafloor is shallow enough to allow sufficient light penetration for photosynthesis. The deepest known macrophyte population is of a crustose coralline rhodophyte, found at 268 m depth in the presence of a photon flux of 0.015–0.025 $\mu$mol · m $^{-2}$ · sec$^{-1}$. For all benthic plants, the key to their productivity is the movement of water around them. Advection and turbulence are important in renewing nutrients and enhancing boundary layer transport. Both photosynthetic rates and nutrient uptake increase with increasing current speed up to an asymptotic threshold ($\approx$5 cm · sec$^{-1}$ in *Macrocystis* kelps). Depending on the turbulence level, kelps within the same population may even develop phenotypic responses in the form of morphological differences between more exposed (narrow and thick blades) and sheltered (wide and corrugated blades) specimens. This is the result of a trade-off between the photosynthetic advantages of a wide, frilly blade morphology and the risk of a plant breaking from increased drag.

Many seaweeds have the capacity to store nitrogen taken from seawater at times of high nutrient and low light availability, and then use the nitrogen to meet growth demands during times of low nutrient but high light availability. Also, unlike phytoplankton, seaweed production may experience both phosphorus limitation in carbonate-rich waters and nitrogen limitation in silicoclastic sediments. Marine macroalgae have much higher C : N atomic ratios than phytoplankton and are grazed very little. Thus, of the total macroalgal production, more is transferred to the higher trophic levels from decomposing plant tissue than from living tissue consumed by grazers. In particular, coastal areas with extended seaweed cover represent an important source of net export of organic matter in the form of plant detritus to deeper, adjoining waters. This provides the substrate that fuels a complex detritus food web as well as enormous amounts of reduced nitrogen, in the form of ammonia and organic nitrogen, to the nitrogen-limited coastal phytoplankton and the energy-limited bacteria.

## III. THE MICROBIAL LOOP

### A. Formation of Dissolved Organic Matter

DOM is defined as the material passing through a membrane filter of pore size 0.2–0.45 $\mu$m, in contrast to the particulate matter (POM), which is retained on the surface of the filter. DOM is regarded as a real trophic level at the base of a food chain, the microbial loop, involving only heterotrophs. The next trophic level is occupied by the bacteria, which in turn pass the energy on to planktonic and benthic protozoans (Fig. 1). From here, the flow of energy and materials joins that of the food chain based on particulate matter (i.e., algae and detritus).

At present, there are three main recognized mechanisms of DOM release, related to the metabolic activities of the algae, macrograzers, and lithic viruses. Phytoplankton cells exude material during photosynthesis at rates of 10–30% of their total primary production. This type of excretion involves a variety of different substances, some of which are early products of photosynthesis. High rates of loss occur under a variety of situations, including exposure to very high irradiances, cell stress or damage, nutrient depletion, light limitation, bacterial infection, and physicochemical shock. Particularly high amounts of DOM are observed during the decaying stage of phytoplankton blooms, which in temperate waters coincides with the seasonal maximum in DOM levels (approximately a month after the onset of the bloom).

Seaweed exudation, along with abrasion, also contributes substantially to the DOM pool, especially in the form of mucilage, which may constitute up to 25% of benthic algal production. In estuarine and coastal waters, the major source of DOM is often represented by excretion from salt marsh macrophytes.

DOM release mediated by zooplankton is due to sloppy feeding, excretion, and egestion. During grazing, zooplankton spill a major portion of the cells' contents into the external organic pool. As much as 5–30% of this may be in the form of DOM. The other major zooplankton contribution is from the excretion of fecal material. Planktonic crustaceans, in particular, produce large fecal pellets surrounded by a pellicle or sheath. The sinking rate of these pellets can be relatively high, in excess of 100 m $\cdot$ day$^{-1}$. Pellets may also reside for long periods in the surface layer, where they lose the pellicle and disintegrate, releasing great amounts of DOM. This happens more easily when the pellets are incorporated into macroscopic, amorphous aggregates larger than 0.5 mm in diameter and known as marine snow. These aggregates consist of a mixture of mucus, pellets, clay mineral particles, moults, and an active microbial community with both autotrophic and heterotrophic components. The sinking rate of marine snow is significantly reduced by the highly porous and flocculent nature of the aggregates.

Another major contribution to the DOM pool appears to come from viral infections. Viruses are even more abundant than bacteria in seawater and it has been recently estimated that they may be responsible for ≈60% and 30% of the total mortality of bacteria and cyanobacteria, respectively. When viruses infect a bacterial host, they cause the host cell to burst or lyse to release the viral progeny. This lysis results in the release of some particulate debris but especially DOM.

### B. Role of Bacteria in the Carbon Cycle

It has been shown with modern fluorescence microscopy that there are $\approx 10^5$–$10^6$ bacterial cells in one milliliter of seawater: two orders of magnitude higher than earlier estimates obtained using counts of colonies grown on agar plates. Bacterial cells are smaller in size as one progresses from estuaries to the continental shelf and to the open ocean. The standing stock of bacterial carbon is generally comparable to phytoplankton biomass and larger than the zooplankton stock (Fig. 2). Net bacterial heterotrophic production is about twice as high as that of the zooplankton and, on a unit volume basis, $\approx$10–30% of primary production. Bacterial assimilation efficiency (the ratio of organic matter assimilated to that decomposed) is only about 50% and so gross production (assimilation plus respiration) is 20–60% of total primary production.

Bacterial growth rates vary widely, with doubling times of a few days in colder, offshore waters and only 0.3–1.5 days in warm, nearshore waters. Bacterial utilization of organic matter is a major pathway for the transfer of energy and matter, indeed, it is probably comparable in magnitude to the link between phytoplankton and macrograzers. This differs substantially from the earlier view that bacteria were the principal agents of mineralization, acting slowly on masses of particulate detritus. Not only are they no longer considered as primary nutrient remineralizers, but it is also clear that they actually compete with phytoplankton in the uptake of inorganic nitrogen (ammonia) under conditions of elevated C:N ratios in the organic substrate.

The rate of supply of DOM as substrate is the major factor controlling heterotrophic bacteria abundance and activity. Because bacteria respond to the release of carbohydrates from phytoplankton, their specific growth rate is usually correlated with the abundance of chlorophyll *a* in seawater. Accordingly, growth rates are maximal during the afternoon and early evening, in coincidence with the accumulation of polymers and particles in the water column. At night, there is a 30–50% decrease in the growth rate.

Within the water column, 80–90% of bacterial growth is due to free-living, planktonic bacteria rather than particle-attached epibacteria. The same applies to biomass, and usually only $\approx$5% of the total bacterial stock is found attached to particles. Thus, although marine snow and other particulate macroaggregates harbor dense bacterial populations and are important sources of DOM, these populations are relatively few in number and contain a trivial fraction of the water-column-integrated bacterial assemblage. Seasonal changes in environmental parameters may affect bacterial growth directly or by changing the rate of DOM release. However, effective dormancy, when phenotypic development is interrupted owing to unfavorable environmental conditions, is not as widespread as previously believed, and may actually occur only at extremely low temperatures.

### C. Protists and the Link with the Metazoan Grazers

The high growth rates of bacteria could potentially produce an enormous accumulation of microbial biomass in the water. However, this does not happen and bacterial abundances remain relatively constant. At steady state, therefore, the high growth rates must be matched by an equally high rate of removal. Most crustacean zooplankton are unable to filter efficiently the less than 1-$\mu$m- sized free-living bacteria, rather flagellate and ciliate protists are the intermediaries that transform this ultrafine organic matter into a particle size range that is readily available to the metazoan grazers (Fig. 1).

Colorless (nonpigmented) flagellates in the size range of 2–20 $\mu$m are currently thought to be the most important group of bacterivores. Bacterial and flagellate abundances are in fact closely coupled and oscillate a few days out of phase, a typical effect of predator–prey interactions (Fig. 2). Flagellates can effectively prey on bacterial concentrations as low as $10^5$ cells$\cdot$ml$^{-1}$ at average rates of 30–200 bacteria per flagellate per hour. As heterotrophic flagellate concentrations in the sea range from $2 \times 10^2$ to $5 \times 10^3$ ml$^{-1}$, up to 250% of the water column may be cleared of bacteria per day by these protozoans.

The colorless flagellates in turn are consumed by larger (>20 $\mu$m) ciliates and mixotrophic dinoflagellates. These larger groups, together with the phytoplankton, are the producing trophic level of the classic metazoan aquatic food web. Therefore, this represents a microbial loop with a three-step food chain, from bacteria to the metazoans via flagellates and ciliates. More recently, however, it has been suggested that direct ciliate bacterivory may short-circuit the loop by making bacteria production available to macrograzers via a simple two-step food chain. Indeed, techniques employing epifluorescence microscopy have demonstrated that small aloricate ciliates (<20 $\mu$m in diameter) often make up a large fraction (up to 50%) of the total heterotrophic nanoplankton biomass. Also, contrary to earlier theories, ciliates can maintain high growth rates when feeding on naturally occurring bacterial concentrations. Occasionally, ciliate grazing can account for 100% of the total protist bacterivory.

Regardless of the number of steps involved in the prokaryotic food chain, ciliates along with other microzooplanktonic organisms represent the actual link between the microbial loop and the metazoan grazers. Ciliates constitute a high-quality, nitrogen-rich food and are an important component in the diet of most carnivorous and omnivorous mesozooplankton, especially copepods, euphausiids, and fish larvae. Thus, the microbial loop operates in parallel with the phytoplankton as a major recycling and repackaging pathway that fuels the classic food chain based on metazoan grazers (Fig. 1).

## IV. THE HIGHER TROPHIC LEVELS

### A. Herbivores and Their Grazing Impact

In the pelagic subsystem, herbivory is due to a variety of organisms mostly in the micro- and mesozooplankton size groups (Fig. 1). Microzooplankton includes both protists (tintinnid and oligotrich ciliates) and metazoans (crustacean nauplii). Unlike most of the mesozooplankton, microzooplanktonic organisms do not undergo significant diurnal migrations and are usually found throughout the euphotic zone, during both the day and night. Protists generally exhibit two layers of maximum concentrations, the surface microlayer and the base of the mixed layer. The deep layer is always denser than the surface, with concentrations often an order of magnitude higher.

Crustacean nauplii also occur throughout the upper mixed layer but tend to be associated mostly with the subsurface chlorophyll maximum. This is positioned near the surface during a phytoplankton bloom and near the bottom of the euphotic zone in a stable, stratified regime.

Mesozooplankton distribution in the water column is completely different from that of the microzooplankton. There is a consistent diurnal cycle characterized by a half-time mismatch between the vertical profiles of the producers and the mesozooplankton consumers. This is due to diel vertical migrations, during which migrants feed in the euphotic zone at night and rest during the day in dark and cold waters, at depths of 200–800 m. Zooplankton vertical migrations have been attributed to several causes, some related to increasing the efficiency of food gathering and digestion. However, it is now widely accepted that diel migration is principally a means of avoiding visual predators, especially epi- and mesopelagic fish.

More closely related to food availability is the seasonal vertical migration that is regarded as a true overwintering strategy. In polar and temperate regions, many zooplanktonic herbivores withdraw in late summer to deep waters (500–1000 m), when food resources become insufficient for growth.

Both protists and metazoans are able to select food particles according to size, morphology, or biochemical characteristics. The range of food particles potentially available to an organism is a function of the morphology of its food-gathering apparatus. Tintinnid ciliates capture and ingest particles smaller than 50% of the oral diameter of their lorica. For copepods and euphausiids, food selection is related to some dimension and morphology of the oral appendages, and often the nature of their diet can be established through an analysis of these parameters. Until recently, attempts to analyze the feeding behavior of crustacean zooplankton, and especially copepods, have been based on the concept that they collect particles using a sieving apparatus composed of setae and setules, through which water would be pushed continuously by the beating of their feeding appendages. However, this mechanism is not consistent with the viscous world of feeding crustaceans, where they operate at very low Reynolds numbers (Re). At low Re of less than 0.1, water flow is laminar and bristled appendages behave like solid paddles rather than open rakes. Particles can neither be scooped up nor left behind because appendages have thick layers of water adhering to them. Indeed, modern high-speed microcinematography has shown that crustacean zooplankton actively capture small parcels of water containing food particles by flinging open and closing specialized appendages.

In the past, grazing estimations have been obtained mostly from bottle incubations of feeding zooplankton. The functional responses obtained with these incubations varied greatly, but were all characterized by the presence of lower and upper feeding thresholds (Fig. 4). These represent food concentrations below or above which the feeding activity of a herbivore ceases or is drastically reduced. However, successive investigations have shown that while a lower threshold is a real feature of grazing, the upper threshold is largely an artifact due to a lack of acclimation of the grazers to the sharp increase in food supply experienced during the bottle incubations. Because of this and other systematic containment errors associated with the bottle method, *in situ* techniques have now been developed. These involve measures of gut contents (using fluorimetry) of freshly caught animals and independent estimates of gut clearance rates that are corrected for chlorophyll breakdown in the gut.

Diurnal variations in grazing show a consistent trend, with high values at night and low values during daytime, irrespective of whether the grazer is a vertical migrator or not. This suggests that vertical migration and feeding activity are distinct processes controlled independently, although both are cued by critical light levels.

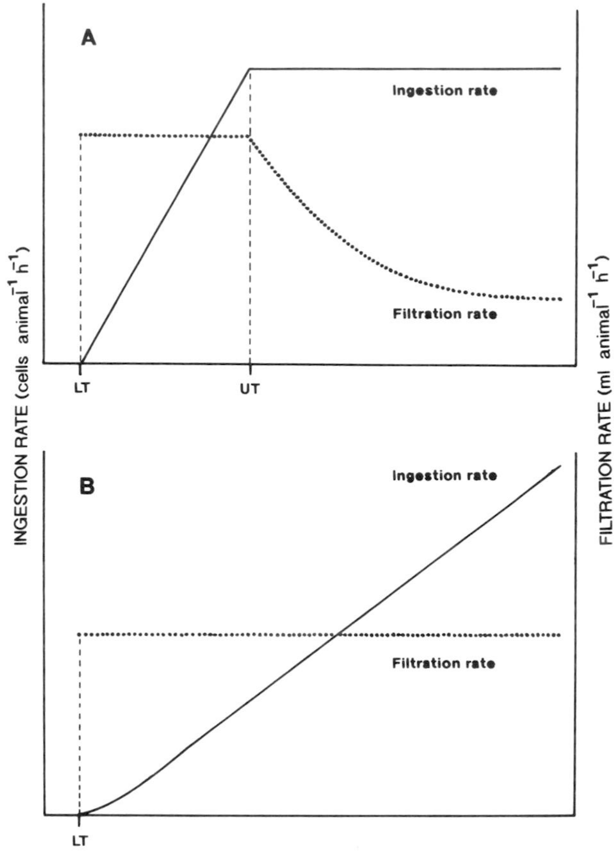


***FIGURE 4*** Functional response curves for zooplankton filtration and ingestion rates versus food concentration. (A) Saturation-type response observed in animals not acclimated to high food levels. (B) Acclimated response obtained when animals are pre-adapted to the highest food concentrations used in the experiment. LT = lower food threshold, below which animals cease grazing; UT = upper threshold, above which animals slow their filtration activities.

Herbivore mesozooplankton rarely consume more than a small fraction of daily primary production, less than 20% in the open ocean and 10–40% in shallow shelf seas. Microzooplankton, however, usually have a higher impact, consuming between 10 and 50% of standing stock, or 25–100% of potential production per day. The extent to which ingested food is assimilated by an organism is known as the assimilation efficiency, and in herbivorous zooplankton it varies between 20 and 80%, decreasing exponentially with food intake.

Benthic herbivory is due to two main functional feeding groups, the filter feeders and the rasping (browsing) grazers. Typical filter feeders are bryozoans, corals, sponges, and bivalve molluscs. Gastropods, sea urchins, and chitons are the main groups that feed by scraping macroalgal substrates. Filter feeders use cilia and setae to create water currents that cause particles to collect on feeding surfaces where mucous bands carry food to the digestive tract. The main supply of particles to these suspension feeders is provided by the downward transport of phytoplankton material. This transport allows coupling between the pelagic and the benthic subsystems. In shallow coastal systems, periodic phytoplankton blooms sink to the seafloor following nutrient depletion. More than half of the total bloom biomass sediments out of the water column in the form of resting spores and living vegetative cells. Even in open oceanic areas, rates of deposition of organic matter are positively correlated with phytoplankton production. Here, sediment trap collections have shown that most of the particulate supply to the benthos is in the form of fecal pellets and macroaggregates of detritus, rather than vegetative cells.

Benthic herbivores that feed by scraping substrata may be classified into three functional groups: epilithic, macroalgal, and epiphytic grazers. Their grazing on live macrophytes has a relatively limited impact, resulting in the consumption of ≈10% of annual net production of coastal vegetation. Some macroalgae contain refractory chemicals, resembling lignins, and compounds that inhibit grazing. However, in some circumstances grazing by sea urchins can severely deplete vegetation, for example, they can entirely eliminate kelp beds.

Epilithic grazers feed directly off hard substrata, scraping microalgae, sporelings of macroalgae, and ephemeral macroalgae. Their impact is strongly dependent on grazer density. Macroalgal grazers, on the other hand, live on and eat perennial macroalgae, and their effects on the community depend not as much on grazer density as on the dynamics of the interaction between grazer and alga (i.e., seasonality of grazing, part of the plant eaten, etc.). Epiphytic grazers live on macrophytes but do not feed directly on their living tissue, instead grazing on epiphytes from their surfaces or feeding on dead, standing tissue.

### B. Predation and the Top of the Food Pyramid

Raptorial predation and carnivory in the marine environment are associated with tertiary production. Because animal prey is easier to digest than plant material, carnivore food assimilation efficiency is generally higher than that for herbivores. Within the zooplankton, chaetognaths, ctenophores, raptorial copepods, and euphausiids have assimilation efficiencies often exceeding 90%, and seldom below 80%. Tentaculate ctenophores may consume up to 1000% of their body weight per day, and chaetognaths have occasionally been reported to remove the entire standing stock of herbivore zooplankton. Many planktonic predators perform diel vertical migrations to track the depth of their prey. Visual predators undertaking diel migrations, especially planktivorous fish, trade off intake against predation risk from larger predators by utilizing a twilight antipredation window, when zooplankton are still visible to them but large predators cannot operate efficiently.

Prey size plays a dominant role in the selection mechanism of a predator, and the ratio of predator to prey size averages about 14 : 1. However, reversals of the size dominance of predators over prey do occur in the feeding of some small pelagic molluscs on jellyfish, and in some small crustaceans that feed off large salps.

Benthic predators may be sedentary, like attached coelenterates, but most are mobile organisms that use visual and chemical senses to perceive prey. Although some of these predators are morphologically and behaviorally adapted to feed on selected prey species, the majority of them are opportunistic feeders, ingesting the most available and vulnerable organisms.

Predation in a community invariably alters the relative abundance of species. This may drive the most vulnerable to extinction but, more important, predation can promote the chances of coexistence of species among which there would otherwise be competitive exclusion (exploiter-mediated coexistence). This has been demonstrated in an epibenthic community where mussels outcompete other invertebrates for space. Starfish remove mussels and so make space available for competitively inferior species such as barnacles. Often, the influence of a predator on the food web can be appreciated only after its extinction. A classic example is provided by the north Pacific sea otter, which preys heavily on sea urchins in kelp forests. Where the otter was exterminated by fur trading on the North American coast, large increases in urchin populations followed, with the result that kelp forests were grazed to near extinction.

In many marine environments the top of the food pyramid is occupied by marine mammals and birds, which may consume up to 40% of the secondary production from herbivorous zooplankton. Because birds and mammals are air-breathing marine predators, a large amount of carbon originally fixed by the phytoplankton would return to the atmosphere via respiration. Recent estimates show that these top predators may in fact transfer into the atmosphere as much as 20–25% of photosynthetically fixed carbon, thereby representing a major leak in the biological carbon pump.

## V. MARINE PRODUCTIVITY AND GLOBAL ECOLOGY

### A. Chemical Pollution and Eutrophication

The marine environment is receiving increased attention as the preferred site for the managed disposal of wastes. This is largely due to the perception that dilution by a large water volume, or containment in sediments for long periods of time at sites remote from humans, makes the ocean a safe disposal system. However, the assimilative capacity of the marine biosphere is very limited and several catastrophic effects of toxic pollution have already been recorded. These have mostly involved three groups of substances: heavy metals, hydrocarbons (both petroleum and chlorinated), and radioisotopes. Their impact is usually much stronger on the higher trophic levels than on the primary producers. [*See* MARINE BIOLOGY, HUMAN IMPACTS.]

Heavy metal contamination is particularly severe in coastal and estuarine systems. The most toxic effects are from lead, mercury, copper, and cadmium. Often, and particularly in polluted estuarine environments, heavy metal accumulation in benthic organisms is in excess of 100-fold over natural, unpolluted levels. Although this makes these organisms unsafe for human consumption, it very seldom results in their elimination or even a dramatic decrease in their productivity because of internal biochemical protection. Long-term (chronic) effects may be more important and may substantially affect the community structure.

Petroleum hydrocarbons have been the object of the most intense investigations because of the enormous amounts ($10^6$–$10^7$ tons) spilled into the oceans every year. Because petroleum oils are natural products, they are readily decomposed by bacteria and invertebrates. There seem to be only few and minor long-lasting (>10 years) effects of petroleum hydrocarbons on the marine environment. The most harmful effects of oil spills are found in populations of marine birds. Large quantities of oil prevent birds from feeding and flying, but even small amounts on their feathers can cause penetration of water to the bird's body, followed by death from exposure.

Chlorinated hydrocarbons are used both in agriculture (DDT, dichlorodiphenyltrichloroethane) and in industry (PCBs, polychlorinated biphenyls). They are very toxic, not readily destroyed during metabolism, and often lethal at concentrations as low as 30–100 ppm. DDT is selectively absorbed by lipids in the fatty tissues of organisms. Because of this property, its concentration increases from the lowest to the highest steps in the food chain as the amount of tissue it is dissolved in becomes a smaller and smaller fraction of the total biomass. This process is known as bioamplification.

Radioactive pollution is associated with the cooling system of nuclear power plants, dumping, and testing of nucler weapons. Exposure of organisms to radioactive materials involves exposure of their protoplasm to the adverse effects of ionizing irradiation. Apart from their immediate toxicity, ionizing irradiations often produce mutations in the genetic material. In general, vertebrates are very sensitive to radiations, crustaceans and plants are less sensitive, and microorganisms are relatively insensitive.

A special type of chemical pollution is represented by the anthropogenic enrichment of nutrients in the marine environment. This is known as cultural eutrophication and is mostly due to wastewater disposal in coastal and estuarine areas. Loading of N and P nutrients is changing the structure and functioning of shallow coastal systems, where phytoplankton and seaweed blooms have become more frequent. However, addition of excess nutrients does not always enhance productivity. In hypereutrophic waters, the inhibiting effects of accumulation of hydrogen sulfide and other byproducts of microbial activity dominate over the growth stimulation by addition of inorganic nutrients and no increase in production occurs. The productivity and species diversity of benthic communities are also negatively affected by eutrophication. This is caused by the sinking and decomposition at the bottom of dense algal blooms and mats, which results in anoxia.

### B. Coupling between Oceans and Climate

Only recently have the large-scale dynamics of the atmosphere been recognized as a primary factor responsible for major changes in oceanic circulation, and thereby in marine productivity. Several types of ocean–atmosphere interactions have had drastic effects on the entire food web, up to the highest trophic levels. The best-studied example is the linkage between El Niño and the Southern Oscillation, or ENSO. The El Niño phenomenon is well known for its devastating effects on the fisheries along the coast of Peru and Chile during 1982–1983. Basically, it consists of extended periods of unusually warm sea surface temperatures (> 29°C) that occur periodically off the coast of South America. The frequency of occurrence is quite irregular, between 2 to 10 years, but the season in which it occurs is always around November–January. The two most recent events were recorded in 1986–1987 and in 1991–1992.

In a normal situation, equatorward winds along the coast blow warm surface waters offshore, to the west. As a result, in coastal waters of the eastern Pacific the surface layer is thinned and the thermocline becomes shallower. This is followed by upwelling of nutrient-rich deep water and a dramatic increase in phytoplankton production, which eventually supports a rich anchovy fishery. During an El Niño event, however, higher atmospheric pressure in the western Pacific results in weaker offshore winds and, therefore, reduced upwelling of nutrient-rich, cold waters. This is why there is persistence of warmer surface waters with no increase in primary production rates. Fisheries are also dramatically reduced and seabirds and mammals, which rely on the fish stock, starve or move away.

El Niño also has a mirror image, for which the name La Niña has been proposed. This consists of a pool of colder (≈24°C) than normal water in the tropical Pacific that has the opposite effect of El Niño. These cold events recur about 4 years apart on average and are the result of strong winds drawing cold water up from several hundred meters beneath the surface.

The meteorological event connected with the warm and cold anomalies in the tropical Pacific has been identified as the Southern Oscillation. This is a large-scale seesaw in atmospheric pressure between the eastern subtropical Pacific and the continental landmasses of Australia and Indonesia. Pressure differences between the two regions drive easterly trade winds, the principal forcing factor of tropical circulation. The trade winds converge in the western side of the Pacific, in regions of high rainfall rates. A drier air mass then returns eastward to the eastern Pacific. El Niño occurs only during periods when the pressure is unusually high in the west and low in the east, because under these conditions the trade winds are weakened throughout the tropical Pacific.

### C. $CO_2$ and Global Climate Change

The powerful linkage between ocean and atmosphere has also caused great concern about the effects that global warming may have on the marine biosphere. In the modern industrial era, human activities have led to a considerable increase in the levels of greenhouse gases. In particular, the atmospheric concentration of $CO_2$ has increased from its preindustrial value of 280 ppm to 355 ppm in 1992. This is mainly due to fossil-fuel burning (≈80%) but also to tropical deforestation and land use (≈20%). Model studies predict that such an increase will result in a global warming of 1 to 2°C within the next century. In the past 65 million years, average surface temperatures have undergone dramatic changes, ranging from up to 10°C warmer to 5°C cooler than they are today. The overall pattern of these climatic extremes has consistently been associated with fluctuations in the concentration of atmospheric $CO_2$.

The oceans play a major role in regulating the global carbon cycle, as the oceanic carbon reservoir is ≈55 times larger than the atmospheric and 20 times larger than the terrestrial stores. The marine biosphere acts as a carbon pump by producing particulate (POC) and dissolved (DOC) organic carbon, which is then exported to deeper waters and decomposed (Fig. 5). This provides a deep source of inorganic carbon and results in a deficit of ≈10% of dissolved $CO_2$ in surface waters compared to the deep ocean. At steady state, this biogenic flux is compensated by an equally large upward transport of dissolved inorganic carbon via upwelling and diffusion. This would produce a zero net transport, except for a relatively small fraction that accumulates in the sediments (Fig. 5). Thus, although the physical processes controlling $CO_2$ air–sea exchange can contribute substantially to the sinking of anthropogenic carbon dioxide (taking up about one-third of the total emissions per year), the biogenic fluxes do not sequester a significant proportion of these emissions.

However, there are, oceanic areas where $CO_2$ removal by the biological pump can potentially increase dramatically. These areas include the equa-

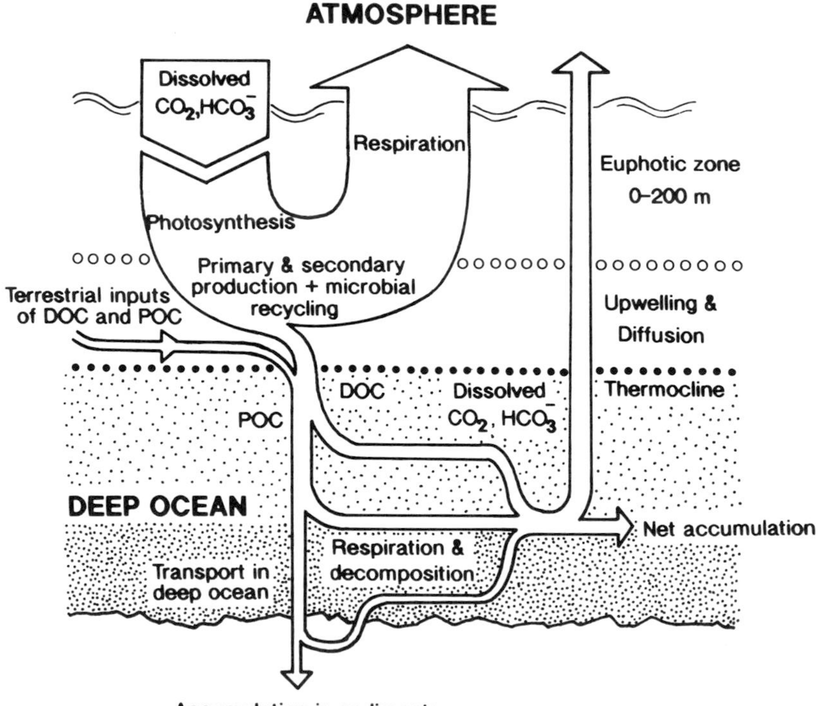


***FIGURE 5*** The biological pathways of the carbon cycle in the marine environment, with emphasis on the sequestration of organic carbon into the ocean interior due to the biological pump. DOC = dissolved organic carbon; POC = particulate organic carbon. [Redrawn from SCOR (1990). "Oceans, Carbon and Climate Change—An Introduction to JGOFS." SCOR, Halifax, Canada.]

torial and sub-Arctic Pacific, but especially the Southern Ocean. Here, very high concentrations of nitrogen and phosphorus nutrients are found in surface waters throughout the year. Yet phytoplankton production is much lower than expected and the biological pump appears to operate with low efficiency. In the case of the Southern Ocean, this situation is known as the Antarctic Paradox. Recent research supports the hypothesis that addition of iron stimulates the growth of phytoplankton in these regions, promoting the utilization of the excess nutrients. It has thus been proposed that the Southern Ocean could be fertilized with iron in an attempt to sequester additional atmospheric $CO_2$ in the deep ocean, where it would be expected to remain over the next few centuries. Model calculations suggest that if this fertilization succeeded in stimulating complete assimilation of nutrients, and was continued for the next 100 years, the buildup of $CO_2$ in the atmosphere could be reduced by 15–25%. However, it must be pointed out that much too little is known about the potential adverse effects that such a manipulation would have on the structure, functioning, and productivity of the marine biosphere as a whole.

## Acknowledgments

I am very grateful to my colleagues C. D. McQuaid, E. A. Pakhomov, and P. W. Froneman for their invaluable comments on the manuscript, and to L. Clennell for assisting with the figure drawings.

## Glossary

**Autotrophic organism** Primary producer that is capable of synthesizing organic matter using inorganic compounds.

**Biological pump** Mechanism of sequestration of organic carbon into the interior of the ocean, based on the production and consumption of biological material by the pelagic subsystem.

**Euphotic zone** Surface water layer delimited by the depth at which there is 1% of the surface illumination.

**Eutrophic region** Highly productive coastal or estuarine area with large nutrient supply owing to horizontal and/or vertical advection.

**Heterotrophic organism** Consumer that is able to utilize only presynthesized organic matter for its metabolic needs.

**New production** Portion of total primary production that is based on the uptake of allochthonous nitrogen nutrients, that is, nitrate ($NO_3$) and nitrite ($NO_2$).

**Oligotrophic region** Oceanic area exhibiting low production rates because of poor nutrient supply.

**Regenerated production** Portion of total primary production resulting from autochthonous, reduced forms of nitrogen such as ammonia, urea, and amino acids.

**Upper mixed layer** Surface water layer that is relatively homogeneous and delimited by the depth of the main pycnocline, or density gradient.

**Upwelling system** Area of high productivity where offshore winds force surface waters away from shore, decreasing thermocline depths so that nutrient-rich deep waters can upwell to the surface.

## Bibliography

Barnes, M., and Gibson, R. N., eds. (1990). "Trophic Relationships in the Marine Environment," Proceedings, 24th European Marine Biology Symposium. Aberdeen, U.K.: Aberdeen University Press.

Barnes, R. S. K., and Mann, K. H., eds. (1991). "Fundamentals of Aquatic Ecology," 2nd ed. Oxford, England: Blackwell Scientific.

Berger, W. H., Smetacek, V. S., and Wefer, G., eds. (1989). "Productivity of the Ocean: Present and Past. Life Sciences Research Report, Vol. 44, Dalhem Workshop Reports." Chichester, England: John Wiley & Sons.

Falkowsky, P. G., and Woodhead, A. D., eds. (1992). "Primary Productivity and Biogeochemical Cycles in the Sea. Environmental Science Research, Vol. 43." New York: Plenum Press.

Jumars, P. A. (1993). "Concepts in Biological Oceanography." Oxford, England: Oxford University Press.

Lalli, C. M., and Parsons, T. R., eds. (1993). "Biological Oceanography: An Introduction." Oxford, England: Pergamon Press.

Longhurst, A. R., and Harrison, W. G. (1989). The biological pump: Profiles of plankton production and consumption in the upper ocean. *Prog. Oceanogr.* **22,** 47–123.

Mann, K. H., and Lazier, J. R. N., eds. (1991). "Dynamics of Marine Ecosystems. Biological–Physical Interactions in the Oceans." Cambridge, Mass.: Blackwell Scientific.

Sherman, K., Alexander, L. M., and Gold, B. D., eds. (1990). "Large Marine Ecosystems: Patterns, Processes and Yields." Washington, D.C.: Amer. Assoc. Advancement of Science.

# Mass Extinction, Biotic and Abiotic

**Michael J. Benton**
*University of Bristol, United Kingdom*



There have been many mass extinctions during Earth's history, events in which 50% or more of species died out during a relatively short time. Studies of individual mass extinctions show a variety of patterns, with some apparently taking place extremely rapidly and others over longer periods of time, some restricted geographically and others worldwide in extent. There is limited evidence for ecological selectivity during mass extinction times: the only genera at risk are those with restricted geographic ranges. Most research has focused on the Cretaceous–Tertiary (KT) event 65 million years ago, which marked the end of the dinosaurs and other dominant reptiles, as well as significant marine groups. Uniquely for the KT event there is abundant evidence of one or more major impacts on the Earth, and these must relate to the extinctions. Evidence for impacts of this sort in association with other major mass extinction events is limited, and it is not clear that there was a single driving mechanism that caused all, or even most, mass extinctions following a periodic cycle.

## I. DEFINITION

The meaning of the phrase "mass extinction" is not entirely clear. The events that are commonly called mass extinctions share many features in common, but differ in others. The common features of mass extinctions may be summarized in three ways:

1. many species became extinct, perhaps more than 30% of the extant biota;
2. the extinct forms span a broad range of ecologies and typically include marine and nonmarine forms, plants and animals, microscopic and large forms; and
3. the extinctions all happened within a short time, and hence relate to a single cause, or cluster of interlinked causes.

None of these factors can be rendered more precisely for a variety of reasons that will be explored in this article. In a qualitative way, paleontologists agree that there have been many mass extinctions in the past, and that these varied greatly in magnitude, but attempts have been made to find a more quantitative definition of which extinctions are truly mass extinctions and which are more localized or ecologically restricted events. [*See* EVOLUTION AND EXTINCTION.]

One approach has been to try to find a statistical definition of mass extinctions. David Raup and Jack Sepkoski of the University of Chicago have argued that times of mass extinctions should be associated with exceptionally high rates of extinction, and that these would stand out clearly from normal "background" extinction rates (species, generic, and familial extinction occur all the time as a normal part of evolution). They calculated a mean rate of disappearance of marine animal families, per million years, for each geologic stage. The stage is a standard subdivision of geologic time, lasting on average 5–6 million years (My) and being identifiable worldwide.

Astonishingly, Raup and Sepkoski found that global mean extinction rates declined through time over the past 600 My, and that a regression line could be plotted (Fig. 1). At the 95% significance level, 5% of points would be expected to lie outside the indicated confidence envelope lying around the regression line, and this was the case. They found statistical outliers for the Late Ordovician, Late Devonian, Permian–Triassic (PTr), Late Triassic, and Cretaceous–Tertiary (KT) intervals, these being the "big five" identified mass extinction events. Hence, they concluded that mass extinction events are qualitatively different from background extinctions in normal times, and that the regression technique provides a way of defining mass extinctions

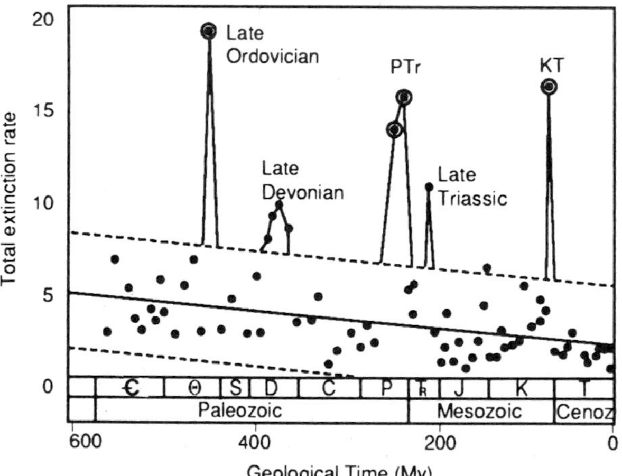


***FIGURE 1*** Total extinction rate, measured as extinctions per million years, for marine invertebrate families. Most points fall either side of a regression line that declines through the last 600 My of the good-quality fossil record. Five sets of statistical outliers, lying above the 95% confidence envelope (enclosed by dashed lines), correspond to the five major mass extinctions. [Based on data in D. M. Raup and J. J. Sepkoski (1982), *Science* **215,** 1501–1502.]

quantitatively. Their technique gave the right answer, but it is statistically flawed in that it is parametric; a regression line can only be drawn if it is assumed that the data are normally distributed, and this is not the case for extinction rate data, where part of the distribution of values, for negative extinction rates, cannot be sampled.

## II. EVENTS

Despite the inappropriateness of this statistical test, paleontologists agree that the five events identified really are mass extinction events, and a number of others have also been identified. Mass extinctions vary in magnitude, and they may usefully be sorted into major, intermediate, and minor events, based on their magnitudes (Fig. 2). The PTr mass extinction is in a class of its own, as it is known that 50% of families disappeared at that time, and this scales to a loss of 80–95% of species. The assumption that a higher proportion of species than families are wiped out is based on the observation that families contain many species, all of which must die for the family to be deemed extinct. Hence, the loss of a family implies the loss of all its constituent species, but many families will survive even if most of their contained species disappear. The "intermediate" mass extinctions (Fig. 2) are associated with losses of 20–30% of families and perhaps 50% of species, whereas the "minor" mass extinctions experienced perhaps 10% family loss and 20–30% species loss. The events shown in Fig. 2 will be reviewed in temporal order.

The *Late Precambrian* event is ill-defined in terms of timing, but such an event clearly occurred about 600 My ago, when earlier metazoan life-forms of the Ediacara type disappeared, and the way was cleared for the dramatic radiation of shelly animals at the beginning of the Cambrian.

A series of mass extinctions occurred during the *Late Cambrian,* perhaps as many as five, which are marked by major changes in trilobite faunas in North America and other parts of the world. Inarticulate brachiopods were also affected. During these events, and just after, animals in the sea became much more diverse, and groups such as artic-

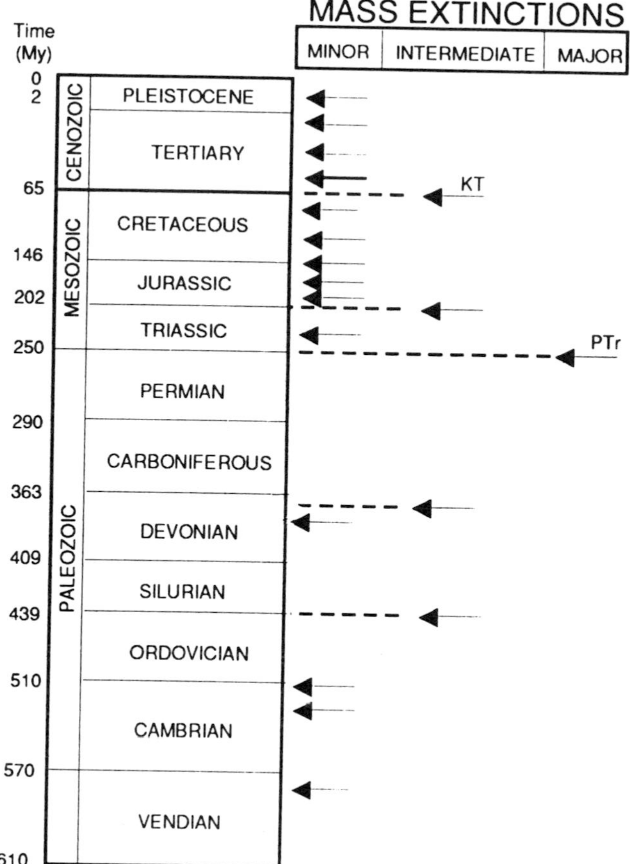


***FIGURE 2*** Mass extinctions through the past 600 My include the enormous Permo-Triassic (PTr) event 250 My ago, which killed twice or three times as many families, genera, and species (50% of families and up to 96% of species) as the "intermediate" events. These were global in extent and involved losses of 20% of families and 70–85% of species. Some of the minor mass extinctions were perhaps global in extent, causing losses of 10% of families and up to 50% of species, but many may have been regional in extent or limited taxonomically or ecologically. Figure is based on various sources.

ulate brachiopods, corals, fishes, gastropods, and cephalopods diversified dramatically.

In the *Late Ordovician,* further substantial turnovers occurred among marine faunas, with extinction of up to 70% of species. All reef-building animals, as well as many families of brachiopods, echinoderms, ostracods, and trilobites, died out. These extinctions are associated with evidence for major climatic changes. Tropical-type reefs and their rich faunas lived around the shores of North America and other landmasses that lay around the equator. Southern continents had, however, drifted over the south pole, and a vast phase of glaciation began. The ice spread north in all directions, cooling the southern oceans, locking water into the ice, and lowering sea levels globally. Polar faunas moved toward the tropics, and warm-water faunas died out as the whole tropical belt disappeared.

The second of the "big five" extinctions occurred during the *Late Devonian,* and this appears to have been a succession of extinction pulses lasting over 10 My in all. The abundant free-swimming cephalopods were decimated, as were the typical armored fishes of the Devonian. Substantial losses also occurred among rugose and tabulate corals, articulate brachiopods, crinoids, stromatoporoids, ostracodes, and trilobites. Causes could be a major cooling phase associated with anoxia on the seabed or massive impacts of extraterrestrial objects.

The largest of all extinction events, the *Permo-Triassic (PTr)* event, is astonishingly one of the least understood. The dramatic changeover in faunas and floras at this time has long been recognized and was used to mark the boundary between the Paleozoic and Mesozoic eras. Most dominant Paleozoic groups in the sea disappeared or were much reduced: rugose and tabulate corals, articulate brachiopods, stenolaemate bryozoans, stalked echinoderms, trilobites, and ammonoids. There were also dramatic changes on land, with widespread extinctions among plants, insects, and tetrapods, which led, in all cases, to dramatic long-term changes in the dominant replacing forms. Causes seem to have been earthbound, perhaps related to the fusion of continents into the supercontinent Pangaea at this time, with associated loss of coastline and shallow seas, global warming, and possible oceanic anoxia (see the following discussion).

The *Late Triassic* events were major, but not so extensive. A marine mass extinction event at the Triassic–Jurassic boundary has long been recognized by the loss of most ammonoids, many families of brachiopods, bivalves, gastropods, and marine reptiles, as well as the final demise of the conodonts. An earlier event, near the beginning of the Late Triassic, also had effects in the sea, with major turnovers among reef faunas, ammonoids, and echinoderms, but it was particularly important on land. There were large-scale changeovers in floras, and many amphibian and reptile groups disappeared, to be followed by the dramatic radiation of the dinosaurs and pterosaurs, as well as the sphenodontids, crocodilians, and mammals. Causes of these events may have been climatic changes associated with the onset of rifting of Pangaea and the opening of the Atlantic, together with drift of continents away from the tropical belt.

Mass extinctions during the *Jurassic* period (Fig. 2) were minor in extent. The early Jurassic and end-Jurassic events at least seem to have been largely restricted to Europe, and to have involved losses of benthic bivalves, gastropods, brachiopods, and free-swimming ammonites as a result of major phases of anoxia. Free-swimming animals were unaffected, and the events are undetectable on land. There is also little evidence for them away from Europe. The mid-Jurassic mass extinction is poorly documented, but may involve losses of cephalopods.

The *Early Cretaceous* mass extinction event is similarly a minor blip in the overall pattern of extinction. The *Cenomanian–Turonian* mass extinction is rather more substantial. Extinctions then occurred particularly among dinoflagellates and foraminifera, as well as cephalopods, echinoids, sponges, bony fishes, and ichthyosaurs, previously important marine reptiles. These disappearances may relate to a major rise in sea level and cooling phase, or to impacts.

The *Cretaceous–Tertiary (KT)* mass extinction is by far the best known, both to the public, because of the loss of the dinosaurs, and to researchers, because of the wealth of excellent geologic sections available for study. As well as the dinosaurs, the pterosaurs, plesiosaurs, mosasaurs, ammonites, belemnites, rudist, trigoniid, and inoceramid bivalves, and most foraminifera disappeared. The postulated causes range from long-term climatic change to instantaneous wipeout following a major extraterrestrial impact. These will be reviewed later.

The *Eocene–Oligocene* event is marked by substantial extinctions among plankton and open-water bony fishes in the sea, and by a major turnover among mammals in Europe at least. The event was rapid, and there is evidence for impact as well as for temperature change.

Later *Tertiary* events are less well defined, which is surprising because the rock and fossil record generally improves toward the present. There was a dramatic extinction among mammals in North America in the mid-Oligocene, and minor losses of plankton in the mid-Miocene, but neither event was large. Planktonic extinctions occurred during the Pliocene, and these may be linked to disappearances of bivalves and gastropods in tropical seas.

The latest extinction event, at the end of the *Pleistocene,* though dramatic in human terms, barely qualifies for inclusion. As the great ice sheets withdrew from Europe and North America, large mammals such as mammoths, mastodons, woolly rhinos, and giant ground sloths died out. Some of the extinctions were related to major climatic changes, and others may have been exacerbated by human hunting activity. The loss of large mammal species was, however, minor in global terms, amounting to a total loss of less than 1% of species.

One mass extinction event that is often ignored, occurring at the *present day,* may pass the numerical test when it is assessed in the future. There are currently some 5–30 million species on Earth, of which only about 2 million have been described. Rates of species loss are hard to calculate: for birds, it is known that about 1% of all 9000 species have gone extinct since 1600, but some 20% of bird species are endangered and could disappear in the next century. Global estimates are based on comparisons with paleontologic examples, from which it is known that background extinction gives a loss of 2–5 families per million years, rising to 15–20 families per million years during mass extinction phases. At present, perhaps several tens of families are being lost per decade, which clearly scales up to a huge mass extinction rate when considered per million years.

## III. PATTERNS

The first defining character of mass extinctions is that many species should have disappeared. This relates to the larger issue of the pattern of mass extinctions, which is tied closely to geological aspects, such as stratigraphy and fossil preservation (see following discussion). It is clearly important, however, to understand patterns of extinction during mass extinction events and patterns of recovery after those events. In broad terms, the scale of extinctions during recognized mass extinction events ranges from losses of 15–50% of families and estimated losses of 35–96% of species overall. In detail, however, these rates are not uniform across all taxa: extinction rates vary from 100% of species within clades that disappear entirely to 0% of species in other clades.

Good-quality fossil records indicate a variety of patterns of extinction. Detailed collecting of planktonic microfossils based on centimeter-by-centimeter sampling up to, and across, crucial mass extinction boundaries offers the best evidence of the patterns of mass extinctions. Regarding the fossil record of foraminifer extinctions in two classic KT sections, some of the patterns reveal rather sudden extinctions, whereas others show a somewhat stepped gradualistic dying-off. Study of the sediments suggests that the sequences are as complete as could ever be achieved, and that the time intervals involved are in the range of 0.5–1.0 My. The proportion of taxa disappearing is high in these cases: some 53–65% of foraminifer species died out in both cases.

Do these superb field examples show catastrophic or gradual patterns of extinction? The problem in interpretation in cases like this becomes one of definition: How sudden is sudden? Paleontologists with different biases can read either conclusion into these fossil records. A "gradualist" would argue that both patterns show a long-term pattern of extinction operating over 0.5 My, and hence far too drawn out to be the result of an instant event. A catastrophist, on the other hand, would see instant extinction occurring in as short a time as 1–1000 years and would argue that the stepped pattern is the result of incomplete preservation, incomplete collecting, or reworking and mixing of sediment by burrowing organisms.

This kind of detailed sampling is not possible for all organisms. The dinosaurs, for example, offer all kinds of practical difficulties. They are preserved in continental sediments, and these are nowhere near as continuously deposited as are sediments in the deep ocean. Specimens are large and rare, so that detailed bed-by-bed sampling is fraught with difficulties. Nevertheless, two teams attempted large-scale controlled field sampling in Montana to establish once and for all whether the dinosaurs had drifted to extinction over 5–10 My, the view of the gradualists, or whether they had survived in full vigor to the last minute of the Cretaceous period, when they were catastrophically wiped out. Needless to say, one team, led by David Archibald of San Diego State University, found evidence for a long-term die-off, and the other team, led by Peter Sheehan of the Milwaukee Public Museum, found evidence for sudden extinction. Each sampling exercise involved teams of dozens of people, logging in one case an estimated 15,000 person-hours of field prospecting and in the other case a total of 150,000 identified specimens. How much more intensive does the program have to be to establish what really happened?

Detailed paleontologic sampling to establish the pattern of mass extinction events is much harder in older rocks. Nevertheless, studies of patterns of marine extinctions in Late Devonian sections (Fig. 3) can be dated to within 0.3 My, and the detailed sequence of events can be discerned. Rates of taxon loss for this event are 21% of families and perhaps 82% of species on a global scale.

In conclusion, many groups seem to have gone extinct geologically rapidly during mass extinction events, perhaps in spans of less than 0.5 My, but that is still far too long an interval for proper biological interpretation as it could encompass gradual and catastrophic kinds of causes.

Following mass extinctions, the recovery time is proportional to the magnitude of the event. Biotic diversity took some 10 My to recover after major extinction events such as the Late Devonian, the Late Triassic, and the KT. Recovery time after the massive PTr event was much longer: it took some 100 My for total global marine familial diversity to recover to preextinction levels.

It is possible to examine the recovery phase after mass extinctions in more detail. One of the most-studied examples is the replacement of terrestrial tetrapod faunas after the KT event. With the dinosaurs and other land animals gone, vertebrate communities were much impoverished. The placental mammals, which had diversified a little during Late Cretaceous times and which ranged in size up to

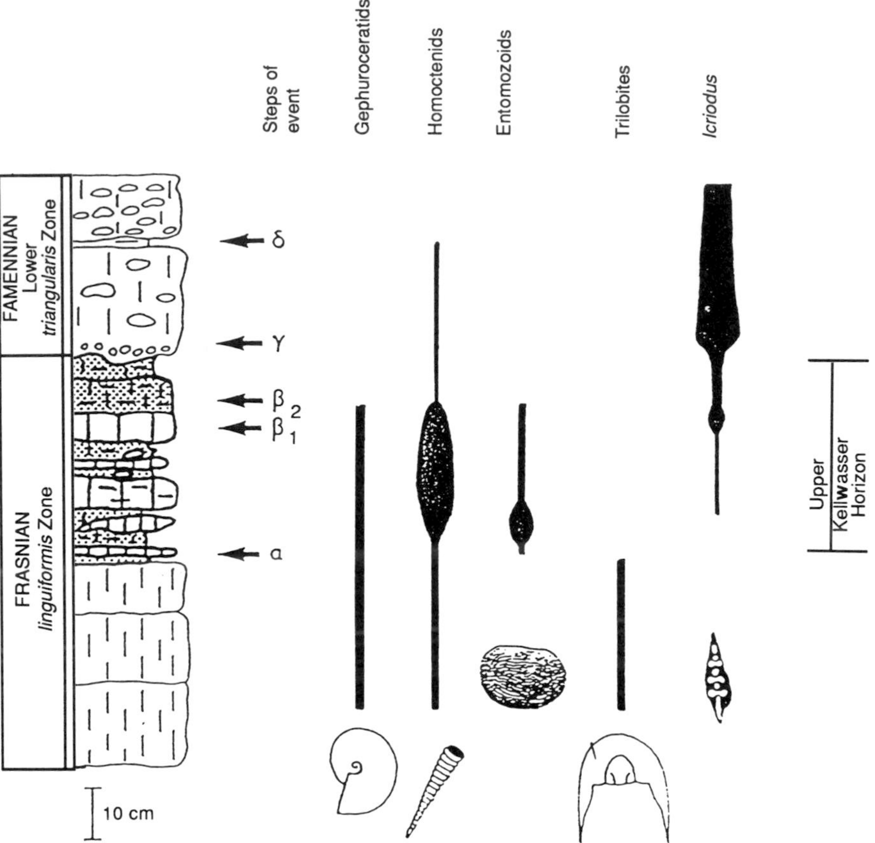


***FIGURE 3*** Precision recording of the Late Devonian mass extinction event in Steinbruch Schmidt, a quarry in Germany. The section, about 1.5 m thick, spans perhaps 0.2–0.3 My and records five pulses of extinction ($\alpha$, $\beta_1$, $\beta_2$, $\gamma$, $\delta$), with progressive losses of trilobites, ammonoids (gephuroceratids), entomozoaceans, and homoctenids, and pulses of extinction of species of the conodont *Icriodus*. [Based on data from E. Schindler (1990). *In* "Extinction Events in Earth History" (E. G. Kauffman and O. Walliser, eds.) pp. 151–159. Springer-Verlag, New York.]

that of a cat, underwent a dramatic radiation (Fig. 4). Within the 10 My of the Paleocene and Early Eocene, twenty major clades evolved, and these included the ancestors of all modern orders, ranging from bats to horses and from rodents to whales. During this initial period, overall ordinal diversity was much greater than it is now: it seems that during the ecologic rebound from a mass extinction, surviving clades may radiate rapidly, and many body forms and ecologic types arise. Half of the dominant placental groups of the Paleocene became extinct soon after, during a phase of filling of ecospace and competition, until a more stable community pattern became established 10 My after the mass extinction.

## IV. SELECTIVITY

The second defining character of mass extinctions mentioned in Section I was that they should be ecologically catholic, that is, that there should be no selectivity. This is a somewhat counterintuitive proposition, because most biologists might predict that large animals, top carnivores, taxa with narrow ecological tolerances, and endemic taxa would be highly extinction-prone. Numerous studies by paleontologists, however, have turned up relatively little evidence for selectivity during mass extinctions. The KT event certainly killed the dinosaurs and some other large reptiles, but a full survey shows that a larger number of microscopic planktonic species died out.

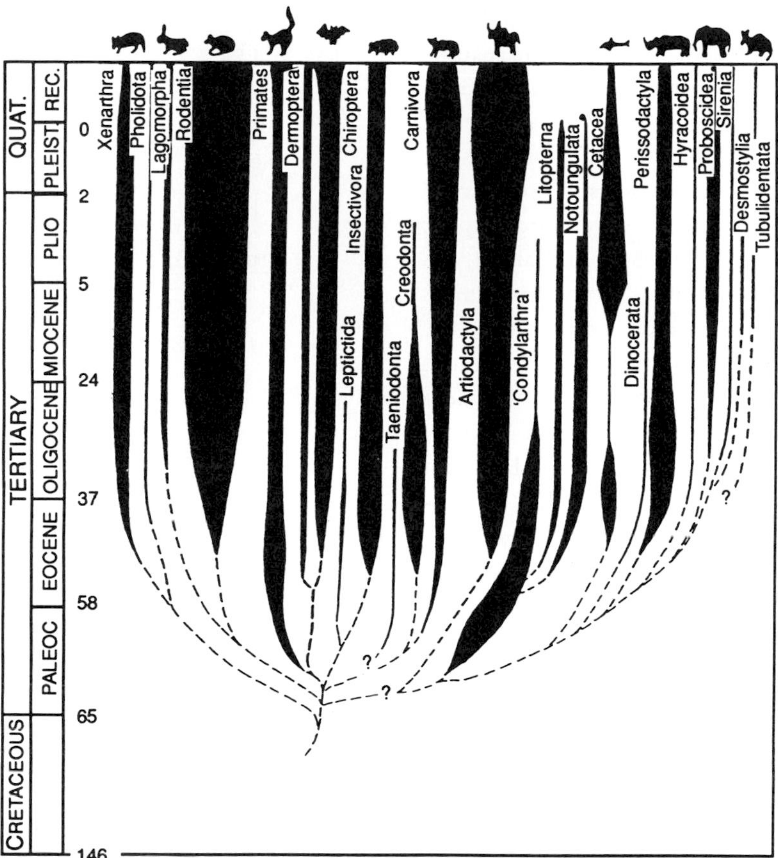


***FIGURE 4*** Radiation of the placental mammals after the KT extinction event. Within the first 10 My after the event (Paleocene, Early Eocene), numerous orders of mammals had arisen, including all modern orders, and some that since went extinct. This shows that it took about 10 My for ecosystems to recover after the devastating loss of the dinosaurs and other taxa during the KT event. [Based on data in P. D. Gingerich (1984). *Univ. Tennessee Stud. Geo.* **8,** 167–181.]

It is not clear that niche breadth was a strong factor either, as whole clades containing generalists and specialists disappeared at the same time. There is also no apparent bias toward extinction of top carnivores.

The only evidence of selectivity during mass extinctions has been against species with limited geographic ranges. David Jablonski of the University of Chicago surveyed all bivalve and mollusc species and genera of the latest Cretaceous and earliest Tertiary in North America and in Europe, and he found that genera that were geographically restricted were selectively killed off when compared to taxa with wider species distributions. So a species less likely to go extinct would seem to be one within a genus that occupies as broad a geographic area as possible. Body size and niche seem to be unimportant.

Geographic realm might also be thought to be significant in selectivity during mass extinction events. It has long been suspected that tropical taxa are more extinction-prone than those with more polar distributions. This idea was based on the observation that some mass extinction events are associated with an episode of cooling. During such a cooling phase, temperate-belt taxa could migrate toward the tropics, tracking their ideal temperature regimes, but the tropical taxa have nowhere to go, and they are squeezed out. A recent study by David Raup and David Jablonski, however, has shown no evidence for latitudinal differences in extinction intensities of bivalves during the KT event.

## V. TIMING

The third defining character of mass extinctions concerns their timing. It is clearly important to know whether a particular extinction event lasted for 5 My or 1 year. At present, stratigraphic resolution is often not good enough to resolve time spans between these two extremes. This is the case even for the much-studied KT event, where opposing experts assert that the extinctions all occurred within a time span of 1 year of 5 My. The dispute arises from problems in fossil sampling and from the inadequacy of dating of mass extinction events (problems in stratigraphy). An assertion that the dinosaurs all died out instantaneously 65 million years ago depends on two kinds of evidence: field observations that dinosaur fossils are relatively abundant through the rocks up to a particular point at which they disappear, and evidence that the disappearance occurs at the same level worldwide.

Fossil sampling is a key issue. Even if a paleontologist can prove that dinosaur fossils suddenly disappear from the rock record at a particular horizon, it cannot simply be assumed that the disappearance is the result of extinction; there may have been an environmental change at that point, and the animals moved elsewhere, or there may have been a substantial hiatus in deposition, or depositional processes may have changed in such a way that bones are no longer buried and preserved. Many aspects of taphonomy (modes of burial and preservation of fossils) must be considered, as well as one of the fundamentals of sedimentary geology, namely, that rocks do not equal time. Many thick rock successions are deposited in a short time as a result of sudden events, such as rock slides, floods, storms, and turbidity flows, whereas other thin parts of a rock sequence may represent vast spans of time. Intensity of study is also important: Peter Ward of University of Washington, Seattle, showed how his records of ammonite extinction at the KT boundary changed with more field collecting (Fig. 5). The pattern that he obtained changed from a rather gradual dying-off to a more catastrophic demise as the result of increased collecting effort.

Stratigraphic issues are also crucial. In dating mass extinctions, the first guide is biostratigraphy, the use of fossils to establish relative ages and to correlate rock units of the same age from continent to continent. The techniques usually work well in

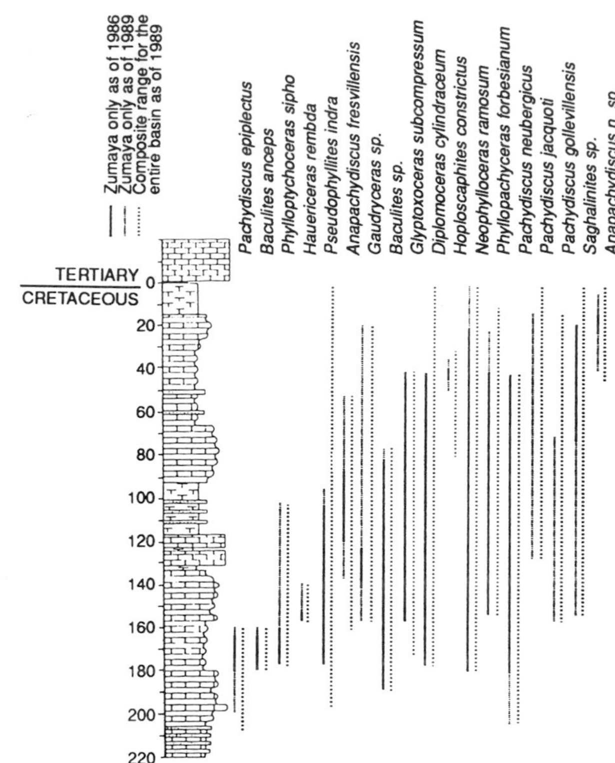


***FIGURE 5*** More collecting makes some paleontologic records more catastrophic. Plot of species ranges of ammonites up to the KT boundary at Zumaya in northern Spain. Ranges established after two field seasons expanded after a further three field seasons, and ranges for the whole basin are even longer. [Based on data from P. D. Ward (1990). *Geol. Soc. Amer. Special Paper* **247,** 425–432.]

defining time bands of 0.5–1 My duration, but they are problematic for shorter intervals. Exact age dating using radiometric techniques may give rather precise dates, but the real error may still be too large for biological purposes. Further, radiometric techniques may only be used on certain rock types, such as volcanic lavas. Other dating techniques have been applied and may become more precise in the future: at present, they are insufficient to guarantee discrimination between instant events and events that lasted for 1 My.

## VI. PERIODICITY

There are many viewpoints on the causes of mass extinctions, but a primary divergence in opinions is between those who seek a single explanation for all mass extinctions and those who believe that each event had its own unique causes. If there was a single cause, it might be sporadic changes in temperature (usually cooling) or in sea level, or periodic impacts on the Earth by asteroids (giant rocks) or comets (balls of ice).

The search for a common cause gained great credence with the discovery in 1984 of an apparently regular spacing between mass extinctions during the last 250 My. Raup and Sepkoski found a regular period of 26 My separating peaks of elevated extinction rates for the record of marine animals (Fig. 6). Initial responses to this proposal were polarized. Many enthusiastic geologists and astronomers accepted the idea, and its clear implication that a regular periodicity in mass extinctions implied a regular astronomically controlled causative mechanism. Numerous models were proposed, all of which involved a regularly repeating cycle that disturbed the Oört comet cloud and sent showers of comets hurtling through the solar system every 26 My. The disturbing factor might have been the eccentric orbit of a sister star of the Sun, dubbed Nemesis (but not yet seen), tilting of the galactic plane, or the effects of a mysterious Planet X that lay beyond Pluto on the edges of the solar system. Geological supporters of periodicity went out to find evidence for massive impacts at mass extinction boundaries to match the physical evidence that had already been established for the KT event.

Critics of periodicity argued that each mass extinction was a unique event and that there was no

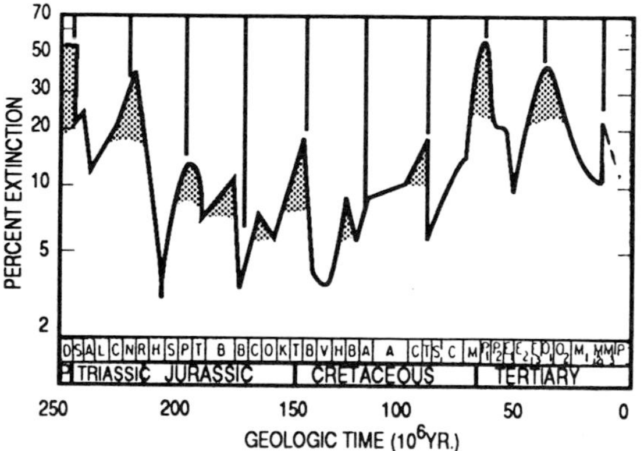


***FIGURE 6*** Periodic extinctions of marine animal families over the past 250 My. Extinction rate is plotted as percentage extinction per My. This is the classic diagram of Raup and Sepkoski. [Based on data from D. M. Raup and J. J. Sepkoski, Jr. (1984). *Proc. Nat. Acad. Sci. U.S.A.* **81,** 801–805.]

linking principle. The 26-My cycle discovered by Raup and Sepkoski was, they argued, a statistical artifact or the result of limited data analysis.

The jury is still out on periodicity. Further statistical analyses have tended to confirm Raup and Sepkoski's original finding, but the search for Nemesis and Planet X has not been successful. Nor has the search for indicators of impact at each of the other identified mass extinctions turned up a great deal of evidence; indicators of impact have been found for only 2 or 3 of the 10 or 11 mass extinction peaks that are elements of the periodic cycle, and the evidence is only really strong for the KT event (see next section). On the other hand, study of ancient impact craters on the Earth has shown that there has been many massive asteroid or comet strikes through geologic time, and these seem to have occurred at similar rates to the rates of extinctions of different magnitudes.

## VII. CAUSES

The key question behind all investigation of mass extinctions is: What was the cause? It is impossible to review the multitudes of causes for mass extinctions that have been proposed in the literature, so this commentary will focus on the two that have attracted the most attention—the PTr event 250 My ago and the KT event 65 My ago.

The PTr event has been surprisingly little studied, especially in view of the magnitude of the extinctions. The reason for the relative lack of attention may be that some of the best sections across this time interval occur in China and Pakistan, but detailed work is at last being done on these. In his review of current knowledge of the PTr event, Doug Erwin of the Smithsonian Institution points out that some data suggest that the event may have been drawn out over the whole of the Late Permian, a span of 10 My or more, whereas others believe that the event was truly rapid (Fig. 7). Postulated models for the extinction include the following: (1) the continents were fusing as Pangaea, and this reduced biogeographic realms on land and in the sea; (2) global temperatures increased; (3) global temperatures decreased; (4) salinity of the sea was

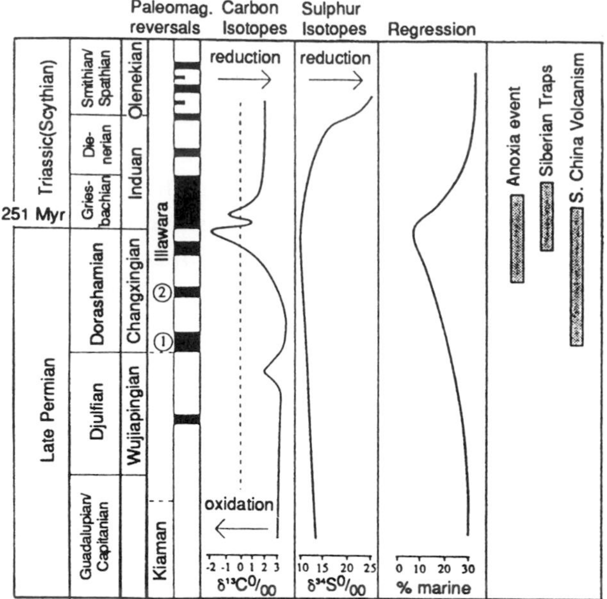


***FIGURE 7*** Physical changes during the Late Permian and across the PTr boundary, which may be related to the mass extinctions: rapid magnetic reversals, isotope changes, marine regression, an anoxic event, and massive volcanism. [Based on data from D. H. Erwin (1994). *Nature* **367,** 231–236.]

reduced; (5) major regression of the sea; (6) seas became anoxic; (7) major volcanic eruptions caused poisoning and excess carbon dioxide; and (8) major impact caused poisoning/excess carbon dioxide/blacking out of the Sun and consequent freezing. Erwin rejects global cooling (3), and notes that there is little evidence for salinity reduction (4), volcanic poisoning (7), and impact (8). He prefers to tie the remaining explanations together and to link them as outcomes of the major Late Permian marine regression and of the outpouring of the Siberian volcanic traps, enormous volumes of lavas that were erupted at this time.

Geologists and paleontologists have devoted astonishing efforts to disentangling what happened 65 My ago during the KT event. Some 500–1000 publications come out each year on this subject, and scarcely a week passes without a report in *Nature* or *Science*. This intensity of research has been maintained since 1980, when Luis Alvarez, of the University of California at Berkeley, and colleagues published their view that the extinctions had been caused by the impact of a 10-km-diameter asteroid on the Earth. The impact caused massive extinctions by throwing up a vast dust cloud that blocked out the sun and prevented photosynthesis, and hence plants died off, followed by herbivores and then carnivores.

There are two key pieces of evidence for the impact hypothesis: an iridium anomaly worldwide at the KT boundary and associated shocked quartz. Iridium is a platinum-group element that is rare on the Earth's crust and reaches the Earth from space in meteorites, at a low average rate of acceration. At the KT boundary, that rate increased dramatically, giving an iridium spike (Fig. 8). Further, several sections have also yielded shocked quartz, grains of quartz bearing crisscrossing lines produced by the pressure of an impact. Other evidence is geochemical and paleontological. A catastrophic extinction is indicated by sudden plankton and other marine extinctions in certain sections and by abrupt shifts in pollen ratios at some KT boundaries. The shifts in pollen ratios show a sudden loss of angiosperm taxa and their replacement by ferns, and then a progressive return to normal floras. This fern spike (Fig. 8), found at many terrestrial KT boundary sections, is interpreted as indicating the aftermath of a catastrophic ash fall: ferns recover first and colonize the new surface, followed eventually by the angiosperms after soils begin to develop. This interpretation has been made by analogy with observed floral changes following major volcanic eruptions.

This "basic" catastrophist model requires that the KT extinction occurred geologically overnight, or at least within a year or so. A variant on this basic model allows for catastrophic extinction in a stepwise manner, involving numerous showers of comets over a span of 1–3 My. Such precision of dating may be impossible: the KT boundary may be identified to the nearest millimeter within any single rock section, and it may be confirmed by the disappearance of the last bones of dinosaurs and by pollen changes on land, and by changes in plankton and invertebrate fossils in the sea. However, other techniques are necessary to correlate such disappearances from place to place around the world and to confirm that they occurred at the same time everywhere.

Radiometric dating gives dates with uncertainties of ±0.5 My at that time period. Magnetostratigraphy may offer more precision. Sporadically, the

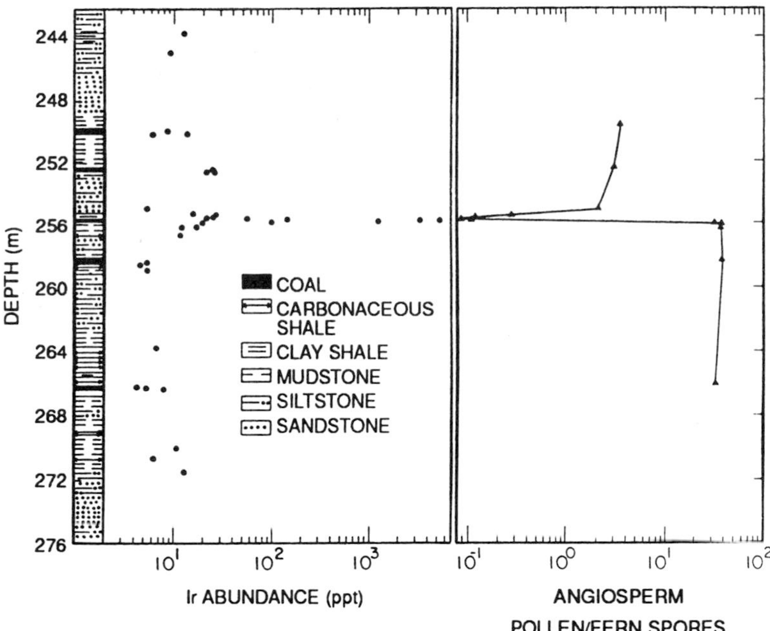


***FIGURE 8*** The iridium spike and fern spike, as recorded in continental sediments in York Canyon, New Mexico. The iridium spike, an enhancement of 10,000 times normal background levels, is generally interpreted as evidence for a massive extraterrestrial impact. The fern spike indicates sudden loss of the angiosperm flora and replacement by ferns, and then partial recovery, a possible indicator of destruction or covering of the soil, perhaps by ash. [Courtesy of C. J. Orth 43.]

north and south poles flip over, and all iron-bearing minerals in rocks that are just being formed acquire the relevant magnetization. In the latest Cretaceous, the Earth's polarity changed eleven times, the KT boundary lying in polarity band 29R (i.e., reversed), which lasted as little as 0.3 My.

The main alternative to the extraterrestrial catastrophist explanation for the KT mass extinction is the gradualist model, which sees declines in many groups of organisms as caused by long-term climatic changes in which the subtropical lush dinosaurian habitats gave way to the strongly seasonal temperate conifer-dominated mammalian habitats. This gradual ecosystem change model has been challenged on the basis of problems in exact correlation of the isolated mammal faunas. The gradualist scenario has been extended to cover all aspects of the KT events on land and in the sea, with evidence from the gradual declines of many groups through the Late Cretaceous. Climatic changes on land are linked to changes in sea level and in the area of warm shallow-water seas.

Recent evidence of the site of impact has strengthened the catastrophist model for the KT event. A putative crater, the Chicxulub Crater, has been identified deep in Late Cretaceous sediments on the Yucatán Peninsula in Central America, and it seems to have produced a range of physical effects in the proximity. A ring of coeval coastline deposits show evidence for tsunami (massive tidal wave) activity, presumably set off by a vast impact into the proto-Caribbean. Further, the KT boundary clays ringing the site also yield abundant shocked quartz and glassy spherules that supposedly geochemically match the bedrock under the crater site. Farther afield, the boundary layer is thinner, there are no tsunami deposits, spherules are smaller or absent, and shocked quartz is less abundant.

Another twist to the KT story is that a third school of thought has gained some ground in recent years, the view that most of the KT phenomena may be explained by volcanic activity. The Deccan Traps in India represent a vast outpouring of lava that occurred over the 2–3 My spanning the KT boundary. Supporters of the volcanic model seek to explain all the physical indicators of catastrophe (iridium, shocked quartz, spherules, and the like) and the biological consequences as the result of the eruption of the Deccan Traps. In some interpretations, the volcanic model explains instantaneous catastrophic extinction, whereas in others it allows a span of 3 My or so for a more gradualistic pattern of dying-off caused by successive eruption episodes.

Thus, the geochemical and petrological data, such as the iridium anomaly, shocked quartz, and glassy spherules, as well as the Chicxulub Crater, give strong evidence for a major impact on Earth 65 My ago, although some aspects might also support a volcanic model. Much of the paleontologic data support the view of instantaneous extinction, but the majority still indicate longer-term extinction over 1–2 My. Key research questions are whether the long-term dying-off is a genuine pattern or whether it is partly an artifact of incomplete fossil collecting and, if the impact occurred, how it actually caused the patterns of extinction found in the fossil record. Available killing models are either biologically unlikely or too catastrophic: recall that a killing scenario must take account of the fact that 75% of families survived the KT event, many of them seemingly entirely unaffected. Whether the two models can be combined so that the long-term declines are explained by gradual changes in sea level and climate and the final disappearances at the KT boundary were the result of impact-induced stresses is hard to tell.

## Glossary

**Ammonites** Extinct cephalopods with coiled shells that lived abundantly as free-swimming carnivores in Jurassic and Cretaceous seas.

**Belemnites** Extinct cephalopods with straight, bullet-shaped internal guards.

**Biostratigraphy** Use of fossils to establish the relative sequence of rocks, and to match rocks of similar age in different parts of the world.

**Correlation** Matching rock ages from place to place within a local basin of deposition, or globally.

**Magnetostratigraphy** Division of geologic time using measurements of the Earth's magnetization as preserved in the rocks. Numerous phases of "normal" (as at present) and "reversed" (poles reversed) magnetization may be distinguished and used for correlation.

**Radiometric dating** Establishment of exact ages by study of unstable radioactive elements and a comparison of proportions of parent and daughter elements where the half-life of breakdown is known.

**Stratigraphy** Study of the sequence and dating of rocks.

**Trilobites** Extinct marine arthropods with three-lobed bodies and many pairs of limbs, which dominated early Paleozoic faunas as carnivores and detritivores.

## Bibliography


Alvarez, W., and Asaro, F. (1990). An extraterrestrial impact. *Sci. Amer.,* October, 44–52.

Benton, M. J. (1990). Scientific methodologies in collision; The history of the study of the extinction of the dinosaurs. *Evol. Biol.* **24,** 371–400.

Benton, M. J. (1995). Diversity and extinction in the history of life. *Science,* in press.

Courtillot, V. E. (1990). A volcanic eruption. *Sci. Amer.* October, 53–60.

Donovan, S. K., ed. (1989). "Mass Extinctions." London: Belhaven Press.

Erwin, D. H. (1993). "The Great Paleozoic Crisis; Life and Death in the Permian." New York: Columbia University Press.

Hildebrand, A. R., and Boynton, W. V. (1991). Cretaceous ground zero. *Natural History* **6,** 46–53.

Jablonski, D. J. (1986). Causes and consequences of mass extinctions; A comparative approach. *In* "Dynamics of Extinction" (D. K. Elliott, ed.), pp. 183–229. New York: John Wiley & sons.

Raup, D. M. (1991). "Extinction; Bad Genes or Bad Luck." New York: Norton.

Raup, D. M., and Sepkoski, J. J., Jr. (1982). Mass extinctions in the marine fossil record. *Science* **215,** 1501–1503.

Raup, D. M., and Sepkoski, J. J., Jr. (1984). Periodicities of extinctions in the geologic past. *Proc. Nat. Acad. Sci. U.S.A.* **81,** 801–805.

Sharpton, V. L., and Ward, P. D., eds. (1990). Global catastrophes in earth history. *Geol. Soc. Amer. Special Paper* **247,** 1–631.

# Modeling Forest Growth

**Harold E. Burkhart**
*Virginia Polytechnic Institute and State University*



Forest management decisions require information about both current and future resource conditions. Inventories taken at one instant in time provide information on current volumes and related statistics. Forests are dynamic biological systems that are continuously changing, and it is necessary to project these changes to obtain relevant information for prudent decision making. Stand dynamics (i.e., the growth, mortality, reproduction, and associated changes in the stand) can be predicted using forest growth models. Growth or increment can be expressed collectively for a stand of trees or for size characteristics of single trees and then summed to obtain unit-area estimates. The summation of increment (cumulative growth) is called yield. Mathematical models of growth and yield are fitted to empirical data by using statistical estimation procedures ("least-squares" or regression analysis). All growth and yield models have a common purpose: to produce estimates of stand characteristics (such as the volume, basal area, and number of trees per unit area) at specified points in time.

## I. INTRODUCTION

Mathematical models of forest growth have been developed for many different purposes. Models aimed at understanding how trees and forest systems function (e.g., physiological and process models) are most useful in research. These models help comprehension, link previously isolated pieces of information, and identify gaps where more work is needed. The benefits of models for understanding come from the development of the model, rather than from its later use.

Forest growth models aimed at prediction have been developed to guide forest management decision making. The purpose of these models is to estimate stand characteristics (e.g., volume, basal area, number of trees per unit area) at specified points in time for stands of varying densities growing on sites of different qualities and having different management treatments (such as different initial spacing, thinning schedules, and fertilizer applications). The scope of this entry is limited to models developed to predict forest growth for management purposes.

## II. APPROACHES TO MODELING GROWTH AND YIELD

Most models of forest growth have been developed for even-aged stand conditions—that is, for stands that have a definite beginning point (birth) with all the trees being essentially the same age. For a given management regime, the factors most closely related to growth and yield of even-aged forest stands are (1) the point in time in stand development, (2) the site quality, and (3) the degree to which the site is occupied. These factors can be expressed quantitatively through the variables of stand age, site index, and stand density, respectively.

Growth is the increase (increment) over a given period of time. Yield is the cumulative production. Thus yield can be regarded as the summation of the annual increments. The growth curve (often referred to as current annual growth or current annual increment) increases up to the inflection point of the yield curve and decreases thereafter. Another important quantity is the mean annual growth or increment, defined as the yield at any given age divided by the total number of years (age) required to achieve that yield. Rotation age (age of final harvest) is sometimes set as the age of maximum mean annual increment because, for a given parcel of land, that is the harvest age that will maximize total wood production from a perpetual series of rotations. One will note that the current annual growth curve crosses the mean annual growth curve at its highest value (Fig. 1).

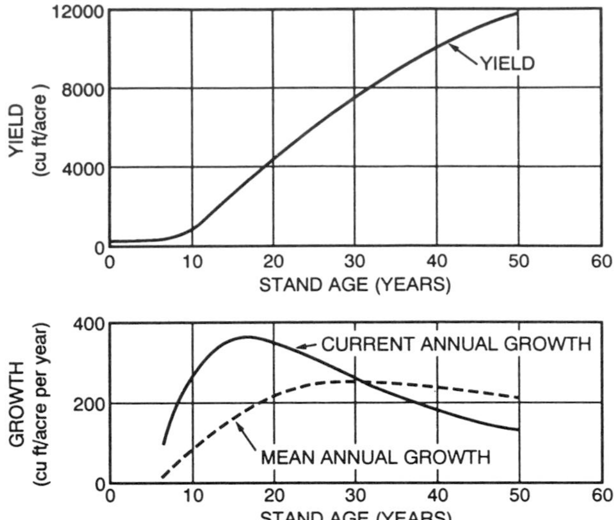


***FIGURE 1*** Relationship between yield, current annual growth, and mean annual growth for even-aged stands with a specified site index and initial stand density. [From T. E. Avery and H. E. Burkhart (1994). "Forest Measurements," 4th ed. McGraw–Hill, New York.]

A wide variety of modeling approaches have been developed to provide quantitative estimates of forest stand dynamics, growth, and yield. These approaches range from whole-stand models to individual-tree models. In whole-stand models the modeling unit is unit-area quantities, such as numbers of trees, basal area, and cubic volume per acre (or hectare). Some whole-stand models provide only aggregate values, whereas others provide procedures for allocating the basal area and volume information by size class. The size-class distribution information is typically developed according to dbh [diameter at breast height, defined as diameter of the stem measured outside the bark at 4.5 feet (1.37 m) above groundline] classes. Models that update size-classes (i.e., dbh class) information directly have also been promulgated. Individual-tree models use trees as the basic modeling unit. The components of tree growth in these models are commonly linked together through a computer program that simulates the growth of each tree and then aggregates these to provide estimates of stand growth and yield. Individual-tree models may be divided into two classes, distance independent and distance dependent, depending on whether or not individual tree locations are required tree attributes.

The basic inputs for forest growth models are stand age, a measure of site quality (typically site index), a measure of site occupancy, crowding, or competition (stand or point density), and a schedule of management treatments. Before describing forest growth models in more detail, site index and measures of stand and point density will be discussed because of their fundamental importance to the development and application of growth and yield models.

## III. SITE INDEX

Of all the commonly applied measures of site quality, tree height in relation to tree age has been found the most practical, consistent, and useful indicator. Theoretically, height growth is sensitive to differences in site quality, little affected by varying stand density levels and species compositions, relatively stable under varying thinning intensities, and strongly correlated with volume. Thus, the average height of the larger trees in the stand at a selected index age, called site index, has been found highly useful for quantifying the wood-producing potential of land.

As generally applied, site index is estimated by determining the average total height and age of dominant and codominant trees in well-stocked, even-aged stands. When these two variables (total height and age) have been ascertained for a given species, they are used for estimating site index (height at a specified index age, such as 25, 50, or 100 years) from a site-index equation.

If one measures a stand that is at an index age, the average height of dominants and codominants is the site index. In most instances, however, stands being measured are less than or greater than the index age. Consequently, an equation is needed to project the dominant stand height to the standard reference age.

Site-index curves are constructed by using regression techniques to fit an equation to measurements of height and age. Height over age for pure, even-aged stands exhibits a generally sigmoid shape (Fig. 2); thus some transformation of the variables is needed if linear regression methods are applied. The most common transformation is

$$\log H_{\rm d} = b_0 + b_1 A^{-1},$$

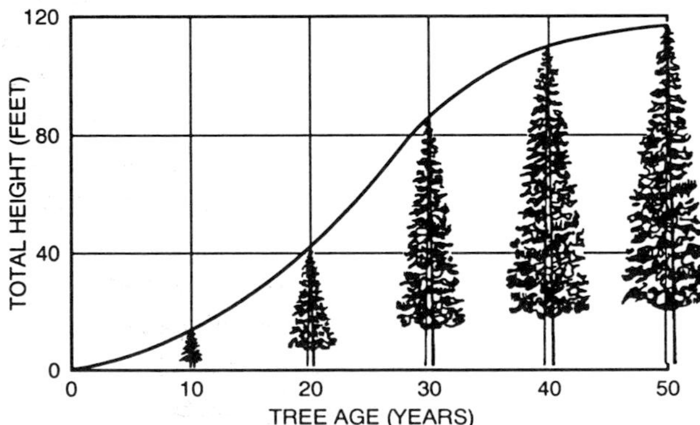


***FIGURE 2*** Cumulative height–growth pattern followed by many coniferous species. [From T. E. Avery and H. E. Burkhart (1994). "Forest Measurements," 4th ed. McGraw–Hill, New York.]

where log $H_{\rm d}$ is the logarithm of the average height of dominant and codominant trees and $A^{-1}$ is the reciprocal of stand age.

The guide curve for a set of anamorphic site-index curves can be established by fitting the model of the logarithm of the height and the reciprocal of age to data from stands of varying site qualities and ages. For anamorphic site-index curves, tree height for different site classes is proportional at all ages. For example, the site-index 50 curve would be one-half of the site-index 100 curve at all ages, including the index age. To avoid bias in the guide curve, it is important that, insofar as possible, all site-index classes of interest be represented approximately equally at all ages. If only poor-quality sites are sampled for the older ages, for example, the guide curve will tend to "flatten" too quickly and bias the entire family of site-index curves.

After the guide curve is estimated, an equation for site index as a function of measured age and height can be constructed by noting that when age is equal to index age $A_{\rm i}$, height is equal to site index $S$, that is,

$$\log S = b_{\rm o} + b_1 A_{\rm i}^{-1}.$$

This imples that

$$b_0 = \log S - b_1 A_{\rm i}^{-1}.$$

Substituting the implied definition for $b_0$ into the original guide-curve equation and algebraically rearranging the terms results in

$$\log S = \log H_{\rm d} - b_1(A^{-1} - A_{\rm i}^{-1}).$$

This form is used to estimate site index (height at index age) when age and height measurements are given. One can estimate height at any age for a specified site index by applying the function as

$$\log H_{\rm d} = \log S + b_1(A^{-1} - A_{\rm i}^{-1}).$$

The height over age curves produced from application of this equation are called anamorphic site-index curves.

The procedures just described will result in height–age curves that have the same shape, regardless of the site-index values (hence the term "anamorphic" site-index curves, meaning one shape). Families of site-index curves that display differing shapes for different site-index curves are called polymorphic. In recent years, polymorphic site-index curves have been constructed for many different species. When available, polymorphic curves generally reflect height–growth trends across a wide range of site qualities more accurately than anamorphic curves and, thus, are generally preferred.

## IV. STAND AND POINT DENSITY

The growth rate achieved by any given forest stand is largely determined by the productive capacity of the site and the degree to which the site is occupied. Site index is the most commonly applied method for quantifying site quality. A large number of methods for quantifying site occupancy or stand density have been proposed and several have been widely applied. The degree to which a forest stand of a given species composition utilizes the resources of a specific site depends on the numbers and sizes of the trees present and on the spatial distribution of the trees.

There are two general categories of density expressions: (1) stand-density measures, which provide an estimate of average crowding, and (2) point-density measures, which provide an estimate of the competitive situation for individual trees.

Stand-density measures, which are aimed at estimating average crowding, are of two types: (1) those that are direct functions of measured stand characteristics, and (2) those that involve comparison of the existing stand to some norm or with some previously established limiting relationship. One obvious measure of density that can be determined by direct observation is the number of trees per unit area. Although number of trees per unit area has been found to be reasonably well correlated with growth in planted stands (where stems typically are more uniformly distributed), it is a less useful measure in naturally reproduced stands. Numbers of stems alone do not include any information on tree size. A measure that incorporates tree size and that has been widely applied is the basal area per unit area, defined as the sum of the stem cross-sectional areas at breast height (4.5 feet or 1.37 m) and expressed as $ft^2$/acre or $m^2$/ha.

Measures of density based on the two component parts of basal area—number of trees per unit area and diameter of the tree of average basal area—have been called stand-density indices. Other measures of stand density that have been applied in forestry include tree-area ratio, crown competition factor, and relative spacing. Most of the commonly used measures of stand density can be shown to be mathematically equivalent. All of these measures can be used to quantify stand density, although none has been found universally superior to number of trees and basal area per unit area as explanatory variables in models developed for the purpose of predicting stand growth. Hence, most models incorporate number of trees or basal area as a predictor. When making predictions of stand growth, the site quality (site index) can be assumed to remain constant over a rotation. By contrast, stand density must be considered as a dynamic component of the stand development prediction system.

Stand-density measures are aimed at providing an estimate of the "average" competition level in stands. Point-density measures attempt to quantify the competition level at a given point or tree in the stand. These competition indices provide an estimate of the degree to which growth resources (e.g., light, water, nutrients, and physical growing space) may be limited by the number, size, and proximity of neighbors. The actual competition processes among trees are much more complex than can be described by a reasonably simple mathematical index. However, these indices have been found useful for predicting tree mortality and growth.

A large number of competition indices have been developed. Three classes or types of competition indices are (1) area-overlap measurement, (2) distance-weighted size ratio indices, and (3) area-available (or polygon) indices.

Area-overlap measures are based on the concept that there is a competition-influence zone around each tree. Typically, this area over which the tree is assumed to compete for site resources is represented by a circle whose radius is a function of tree diameter at breast height. The competitive stress experienced by a given tree is assumed to be a function of the extent to which its competition circle overlaps those of neighboring trees. Various definitions of the area of influence, the measure of overlap, and the use of weights and summing areas of overlap have led to a large number of point-density expressions, although all are conceptually similar. [*See* FORESTS, COMPETITION AND SUCCESSION.]

Competition measures based on distance-weighted size ratios involve the sum of the ratios between the size of each competitor to the subject tree, weighted by a function of the distance between the competing trees. The most common measure of tree size is dbh, but other measures (e.g., height, crown size) have also been employed.

The third general type of point-density measure is based on area available to the subject tree. This approach involves constructing polygons around the subject tree by connecting the perpendicular bisectors of the distance between the subject tree and its competitors. The polygon area for each tree has been called the area potentially available (APA). A basic premise underlying APA is that, within limits, larger APA values should result in a higher survival rate and tree growth (at least in certain dimensions, e.g., dbh).

Comparisons of the effectiveness of various point-density measures for predicting tree growth have been variable, with no index being shown to be universally superior. Some indices seem to be better suited to certain species than others, and within species the performance of a particular index may vary with the stage of stand development and with the management practices followed.

## V. OVERVIEW OF GROWTH AND YIELD MODELS

Although forests are made up of individual trees, it is the growth of stands that is often of primary interest. A stand is defined as a contiguous group of trees sufficiently uniform in species composition, arrangement of age-classes, and condition to be a homogeneous and distinguishable unit. Stands are populations of trees that, as individuals, change in size, die, or are cut. The sum of the changes in individual trees make up stand growth.

### A. Whole-Stand Models

Whole-stand models consist of functions that do not directly account for variation of individual tree size within stands. A yield table generated from a whole-stand function is a tabular presentation of the volume per unit area by age, site index, and other stand characteristics for even-aged stands. Volume is assumed to be dependent on whole-stand attributes such as age, site-index, species, and density.

Variable-density growth and yield models employ multiple regression techniques to derive equations that express the relationship between the dependent variable (yield) and the independent variables of stand and site attributes. No information on volume distribution by size class is provided. An example of a variable-density yield equation is

$$\log Y = b_0 + b_1A^{-1} + b_2S + b_3 \log B,$$

where $\log Y$ denotes logarithm of yield (volume or weight per unit area), $A$ is stand age, $S$ is site index, and $\log B$ is the logarithm of basal area per unit area. The $b_0$–$b_3$ symbolize constants estimated by regression analysis.

Although a wide variety of equation forms have been applied to develop whole-stand growth and yield models, many are variants of the multiple linear regression model just shown. Logarithmic transformation of yield is generally made prior to equation fitting to conform to the assumptions customarily made in linear regression analysis. Furthermore, the use of the logarithm of yield as the dependent variable is a convenient way to mathematically express the interaction of the independent variables in their effect on yield. For example, a unit change in site index has a differential effect on yield, depending on the level of the other independent variables (age and stand basal area). This differential effect—generally called interaction—would not be the case if the dependent variable were yield and interaction terms were not explicitly included.

In most yield analyses, stand age has been expressed as a reciprocal to allow for the "leveling off" (asymptotic) effect of yield with increasing age. Site index is not often transformed prior to fitting, but sometimes logarithmic or reciprocal transformations are employed. In some models, height of the dominant stand has been used in conjunction with age and the variable site index is eliminated. (Note that if any two of the three variables—age, site index, height of the dominant stand—are known, the third can be determined). Use of height rather than site index has the advantage that it is a measured rather than a predicted variable and, thus, more nearly satisfies the assumptions of regression analysis. The measure of stand density is commonly subjected to logarithmic transformation—particularly in models employing basal area—but the exact form in which density is included is quite variable, especially for models that utilize number of trees per unit area as a predictor variable.

When a yield function is available, growth can be estimated by predicting yield at a future time and subtracting the predicted yield at the present time. Alternatively, growth may be the dependent variable when fitting a regression equation to data. Growth and yield equations are called compatible if the growth equation can be derived from the yield equation and vice versa. As previously defined, growth and yield are related biologically but can also be related mathematically by

$$Y = \int_0^A G\,dt,$$

where $Y$ denotes yield, $A$ is stand age, $G$ is instantaneous growth, and $t$ is unit of time.

The distribution of volume by size classes, as well as the overall volume, is needed as input to many forest management decisions. A variety of approaches have been taken to provide the distribution of volume by size classes (generally dbh classes). One widely applied technique for even-aged stands is a diameter-distribution modeling procedure. In this approach, the number of trees per unit area in each diameter class is estimated through the use of a mathematical function that provides the relative frequency of trees by diameters. Mean total tree heights are predicted for trees of given diameters in stands of specified characteristics (i.e., of specified age, site index, and stand density). Volume per diameter class is calculated by substituting the predicted mean tree heights and the diameter-class midpoints into tree volume equations. Yield estimates are obtained by summing the volumes in the diameter classes of interest. Although only total stand values (e.g., age, site index, and number of trees per unit area) are needed as input, detailed stand distributional information is obtainable as output.

A typical dbh distribution for pure, even-aged stands has a single peak (i.e., is unimodal) and is slightly skewed. Curves can be fitted to such diameter distributions by a variety of mathematical functions, but the most popular function used in studies of forest yields is the Weibull function. The parameters of the Weibull function can be estimated (usually by the method of moments or maximum likelihood) for each plot in a data set and then these parameter estimates can be related to overall stand characteristics (e.g., age, site index, and number of trees per unit area). For a given set of stand attributes the predicted Weibull parameters completely characterize the diameter distribution; stand yield estimates are then computed from the diameter-distribution information. This approach to diameter-distribution characterization and yield estimation is sometimes referred to as the "parameter prediction method."

The parameter prediction procedure involves estimating the distribution function parameters for each plot in the data set and then, in a subsequent stage, developing regression equations to relate these parameter estimates to stand characteristics such as age, site index, and number of trees per unit area. Functions for relating the diameter-distribution parameters to stand characteristics typically have not been fully satisfactory. Consequently an alternative method, sometimes called a "parameter recovery method," has been developed and applied. The parameter recovery method consists of forecasting overall stand attributes related to the moments of the distribution (such as average diameter or total basal area) and solving for the parameters of the diameter-distribution model (such as the Weibull function) that will give rise to those overall stand attributes. This approach provides a direct mathematical link between overall stand values and the distribution of those values.

### B. Size-Class Models

An alternative to modeling whole-stand characteristics of forest growth involves dividing the stand into dbh classes and estimating growth and mortality by diameter class. These size-class models are mathematical variants of the stand-table projection method of estimating growth. In forestry terminology, a stand table shows the numbers of trees by dbh class, whereas a stock table displays the volume by dbh class. The stand-table projection method of growth prediction recognizes the structure of a stand, and growth projections are made according to dbh classes. The method is best suited to uneven-aged, low-density, and immature timber stands. In dense or overmature forests where mortality rates are high, stand-table projection may be of questionable value for providing reliable information on net stand growth.

The procedure ordinarily followed in the stand-table projection method of growth prediction may be briefly summarized as follows:

1. A present stand stable showing the number of trees in each dbh class is developed from a conventional inventory.
2. Past periodic growth, by dbh classes, is determined from increment borings or from remeasurements of permanent sample plots.
3. Past diameter growth rates are applied to the present stand table to derive a future stand table showing the predicted number of trees in each dbh class at the end of the growth period. Numbers of trees in each class must then be corrected for expected mortality and predicted in growth (trees growing in the smallest size class of interest).
4. Both present and future stand tables are converted to stock tables by use of an appropriate tree volume equation.
5. Periodic stand growth is obtained as the difference between the volume of the present stand and that of the future stand.

In cases where growth and mortality predictor equations are not available, diameter growth and tree mortality rates are developed from a sample taken in the forest stands of interest. Generalized functions to express stand-table dynamics (changes in numbers of trees by dbh class) and to convert these stand tables to stock tables have been developed for a variety of forest types.

One popular method for modeling changes in forest diameter distributions over time involves the application of Markov chains. A Markov chain is a stochastic process consisting of a sequence of events with a finite number of outcomes. Transition probabilities, which represent the probability of outcome on any given event given the beginning state, are computed. Given an initial state and the probabilities of transitions to all other states, the Markov chain method can be used to develop an estimate of future condition. To apply the Markov process, the Markov and stationarity properties must be satisfied. First, it is assumed that to predict the next state one only needs to know the present state. Second, the transition probabilities between two specific states must remain constant over time. In regard to the dynamics of diameter distributions, these properties imply that (1) the diameter distribution some time in the future depends only upon the distribution now and not on past distributions, and (2) the probability of a tree moving from one diameter class to another in any specific period must remain the same over time regardless of stand conditions. These two fundamental assumptions are reasonable for uneven-aged forests that are maintained in a particular condition through a regular cutting cycle. For even-aged forests one would expect the transition probabilities to vary with stand development. Markov models have been developed for even-aged stands by expressing the transition probabilities as functions of stand density and tree dbh.

Permanent plot data are used to develop the transition probabilities for a specific time step (the measurement interval of the sample plots) when applying the Markov method. The states (outcomes) must be finite, mutually exclusive, and exhaustive. Hence a diameter class interval is usually used such that the possible outcomes are: stay in the same dbh class, move up one dbh class, or move into an "absorbing state" (a state that cannot be changed in the future, e.g., mortality and cut are absorbing states). The present and future stand tables are used to develop present and future stock tables; the difference between the stock table values is an estimate of growth for the time period.

### C. Individial-Tree Models

Approaches to predicting stand growth and yield that use individual trees as the basic unit are referred to as "individual-tree models." The components of tree growth in these models are commonly linked together through a computer program that simulates the growth of each tree and then aggregates these to provide estimates of stand growth and yield. Models based on individual-tree growth provide detailed information about stand dynamics and structure.

Individual-tree models may be divided into two classes, distance independent and distance dependent, depending on whether or not individual tree locations are considered when estimating the growth of each tree. Distance-independent models project tree growth either individually or by size classes, usually as a function of present size and stand-level variables (e.g., age, site index, and basal area per unit area). It is not necessary to know individual-tree locations when applying these models. Typically, distance-independent models consist of three basic components: (1) a diameter–growth equation, (2) a height–growth equation (or a height–diameter relationship to predict heights from dbh values), and (3) a mortality function. Mortality may be stochastically generated (i.e., determined through a random process) or it may be predicted as a function of growth rate. Stand growth is computed from the individual-tree components.

Distance-dependent models that have been developed to date vary in detail but are quite similar in overall concept and structure. Initial stand conditions are input or generated, and each tree is assigned a coordinate location. The growth of each tree is simulated as a function of its attributes, the site quality, and a measure of competition from neighbors. The competition index varies from model to model but in general is a function of the size of the subject tree and the size of and distance to competitors. Most past applications of this modeling approach have employed the area-overlap or distance-weighted size ratio concept when quantifying point density. Tree growth is commonly adjusted by a random component representing genetic and/or microsite variability. Survival is stochastically determined as a function of competition (as indicated by the point density measure) and/or individual tree attributes. Yield estimates are obtained by summing the individual-tree volumes (computed from tree volume equations) and multiplying by appropriate expansion factors.

A wide variety of growth and yield models—ranging from whole-stand models that provide only a specified aggregate stand volume to models with information about individual trees—have been developed. In choosing a growth and yield model, one must be concerned with the stand detail needed for a particular analysis and the efficiency in providing the information required. Obviously, no single growth and yield model can be best for all possible circumstances; rather, users must choose the most appropriate model for their particular application.

## Glossary

**Diameter at breast height (dbh)** Diameter of tree stem measured outside the bark at 4.5 feet (1.37 m) above ground.

**Growth** Increase or increment over a given period of time.

**Site index** Average height of the dominant stand at a specified reference age.

**Stand basal area** Sum of the cross-sectional area of all trees in a stand expressed on a per unit area basis.

**Stand density** Measure of site occupancy, crowding, or competition.

**Stand table** Tabulation of the total number of stems (or average number of stems per unit area) in a stand by dbh classes.

**Stock table** List of total volume of stems (or average volume per unit area) in a stand by dbh classes.
**Tree basal area** Cross-sectional area of tree stem at dbh.
**Yield** Cumulative growth over a given period of time.

## Bibliography

Avery, T. E., and Burkhart, H. E. (1994). "Forest Measurements," 4th ed. New York: McGraw–Hill.

Clutter. J. L., Fortson, J. C., Pienaar, L. V., Brister, G. H., and Bailey, R. L. (1983). "Timber Management: A Quantitative Approach." New York: John Wiley & Sons.

Davis, L. S., and Johnson, K. N. (1987). "Forest Management," 3rd ed. New York: McGraw–Hill.

Dixon, R. K., Meldahl, R. S., Ruark, G. A., Warren, W. G., eds. (1990). "Process Modeling of Forest Growth Responses to Environmental Stress." Portland, Ore.: Timber Press.

Landsberg, J. J. (1986). "Physiological Ecology of Forest Production." London: Academic Press.

Oliver, C. D., and Larson, B. C. (1990). "Forest Stand Dynamics." New York: McGraw–Hill.

Shugart, H. H. (1984). "A Theory of Forest Dynamics." New York: Springer-Verlag.

Wenger, K. F., ed. (1984). "Forestry Handbook," 2nd ed. New York: John Wiley & Sons.

# Molecular Evolution

**T. Ohta**
*National Institute of Genetics, Japan*



Molecular evolution is the discipline that studies the evolution of DNA and RNA sequences and their encoded amino acid sequences. The aim of molecular evolution is to clarify how these sequences that are rich in genetic information change in relation to species divergence. Two main areas of research focus on evolutionary mechanisms and molecular phylogeny. In the first, the rate and pattern of molecular evolution are examined and the roles of natural selection, random genetic drift, and other factors are assessed. As compared with phenotypic evolution, macromolecules are steadily evolving. Through comparative analyses of DNA and protein sequences, the history of genes and organisms can be inferred, that is, molecular phylogeny.

## I. INTRODUCTION

For a long time the study of evolution has been based on morphology—the long neck of a giraffe, the human brain, a bird's wing, and so on. Morphological change in evolution is explained by Darwin's theory of natural selection, but this theory is largely qualitative rather than quantitative. The study of population genetics started more than half a century ago as an attempt to understand evolutionary change quantitatively. Because evolution must take place in all individuals of a species, the changes in gene frequency in the population have been analyzed. However, as long as the interpretations of evolutionary events are based on morphological traits, evolutionary change is very difficult to connect with gene frequency change except in relatively few circumstances. [*See* EVOLUTION AND EXTINCTION.]

The remarkable progress in molecular biology has made it possible to apply population genetics theory to real data. We now know that genetic information is stored in linear sequences of DNA that are stably transmitted from generation to generation, and we can compare linear sequences of DNA or amino acids among species. It is also possible to compare secondary and tertiary structures of proteins and nucleic acids from various sources. Population genetics provides a theoretical foundation for understanding such data. Because natural selection works quantitatively by definition, quantitative treatments based on population genetics are necessary for a full understanding of it. Indeed, for the first time in the course of evolutionary studies, it becomes possible to assess the roles of natural selection and other factors in detail.

Another area of intense study in molecular evolution is the elucidation of the history of genes and organisms, that is, molecular phylogeny. Through comparative studies of genes and proteins, our knowledge of phylogenetic relationships of various species has already greatly increased. The history of genes themselves can also be clarified by sequence analyses of genes and proteins.

## II. POPULATION DYNAMICS OF MUTANT GENES

A basic requirement for understanding the mechanisms of nucleotide or amino acid substitutions in evolution is to distinguish mutations from evolutionary substitutions. Numerous mutations appear in Mendelian populations in every generation, but the majority are lost within a few generations. Thus, those mutations that contribute to evolution are a very small minority of all mutations. It is also necessary to understand the process of frequency increase of mutants in the population in the course of substitution.

At the molecular level, many mutations have a very small effect on the phenotype of the organism, and their behavior is mostly determined by chance, that is, random genetic drift. In 1968, M. Kimura first proposed the neutral theory of molecular evolution, which states that the great majority of mutant substitutions are not caused by positive Darwinian selection, but by random fixation of selectively neutral mutations. For a selectively neutral mutant, the process of frequency change in the population has been theoretically analyzed.

A neutral mutant, if it is ultimately fixed in the population, takes on the average $4N$ generations until this occurs:

$$t_1 = 4N, \tag{1}$$

where $N$ is the effective population size. If $N$ is large, the time is very long. The rate of molecular evolution is measured by averaging the number of substitutions over a very long period of time. It may be expressed as

$$k = \lim_{T \to \infty} \frac{n(T)}{T} \tag{2}$$

where $T$ is the period and $n(T)$ is the number of mutant substitutions in this period. Obviously, for $k$ to be measured accurately, $T \gg N$.

Now consider a locus encoding a protein. Let the mutation rate in terms of base substitutions in this DNA region be $v_g$ per generation, and let $u$ be the probability of fixation of a mutant. Then, in a population of $N$ individuals, the total number of mutations appearing in the population is $2Nv_g$ per generation, and a fraction, $u$, of them spreads through the population, so the rate of substitution in the population per generation becomes

$$k_g = 2Nv_g u. \tag{3}$$

Here $u$ generally depends on the magnitude of natural selection. It should be remembered that, at the molecular level, the number of nucleotide sites of a locus is so large that the probability of having identical mutations more than once is almost nil. Also, the probability of back mutation is negligibly small.

Let us now examine how natural selection influences the rate. If most substitutions are caused by Darwinian natural selection, and the average selective advantage of such substitutions is $s$ with no dominance, the fixation probability is roughly twice the selective advantage, and we have

$$k_g = 4Nsv_g. \tag{4}$$

Hence $k_g$ depends on the product of three parameters, $N$, $s$, and $v_g$. However, when most substitutions are selectively neutral, the fixation probability is equal to the initial frequency, $1/(2N)$, and we have

$$k_g = v_g. \tag{5}$$

In other words, the evolutionary rate is simply equal to the mutation rate, and it is independent of population size.

Deleterious mutations are eliminated from the population by natural selection. However, if the effect is very small, a slightly deleterious mutation can become fixed in the population. The effectiveness of selection is determined by the product of effective population size and selection coefficient, $Ns$. Actual species have various population sizes from very small to very large, and therefore the effectiveness of selection will differ among species. In small populations, a slightly deleterious mutation may behave as a selectively neutral mutation. However, in large populations, selection becomes effective and the mutant allele cannot replace the original allele. Thus, there is a negative correlation between the fixation probability (or evolutionary rate) and population size:

$$k_g \propto 1/N. \tag{6}$$

In 1973, I proposed that a substantial fraction of mutant substitutions at the molecular level are caused by random fixation of very slightly deleterious mutations. This hypothesis seems realistic when one realizes that the effect of an amino acid substitution in a protein often produces only a minor modification of a reaction coefficient. It is also in accord with a recent finding that many enzyme variants in *E. coli* are likely to be mildly deleterious, based on statistical analyses of their frequency distribution. Molecular variants that very slightly disturb the secondary structure of molecules, for example, by opening the stem region of a clover leaf structure of tRNA, may represent a mutant class with mild deleterious effects. In this case, it is likely that a slightly deleterious base substitution is followed by a slightly advantageous (compensatory) substitution. This is supported by the observation that GU or UG pairs, which represent intermediate steps, account for about two-thirds of all non-Watson–Crick pairs in the helical regions of analyzed tRNAs. The model of very slightly deleterious mutation is related to the rate of molecular evolution that will be discussed later. The theory that emphasizes the role of interaction between random genetic drift and selection is called the nearly neutral theory.

## III. RATE OF MOLECULAR EVOLUTION

Since the advent of rapid DNA sequencing techniques, data on the primary structure of genes are accumulating amazingly fast, and now statistical studies of DNA sequences are quite popular. However, before the 1980s, most of the available data on molecular evolution were in the form of amino acid sequences. Hemoglobin $\alpha$ of mammals consists of 141 amino acids, and it is one of the best-studied molecules. If one compares human hemoglobin $\alpha$ with that of gorilla, all the amino acids are identical except one, but 18 amino acids differ between human and horse. Such data on sequence divergence faithfully reflect the phylogenetic relationship. Because we know the approximate time of divergence of mammalian species, it is possible to estimate the rate of amino acid substitution. From a comparison of various mammalian species, the rate of amino acid substitution of hemoglobins $\alpha$ and $\beta$ is roughly $10^{-9}$ per amino acid site per year. It is quite impressive to find that almost the same value is obtained from any two mammalian species. However, there are some variations in the rate of amino acid substitution, if a wider group of organisms are compared. Such variations may be related to functional diversification of gene products in relation to gene duplication or environmental need, as discussed later.

The apparent uniformity of evolutionary rates, as compared with phenotypic evolution, is a most remarkable characteristic of protein evolution and is referred to as the molecular clock. By applying a similar analysis to cytochrome *c* data, the rate seems again to be almost uniform for diverse organisms, including plants, fungi, and mammals. However, the rate is much lower in cytochrome *c* than in hemoglobins, which is thought to be caused by stronger structural constraints on cytochrome *c* than are found on hemoglobins. This is another characteristic of molecular evolution, that is, the stronger the constraint on the molecule, the lower is its rate of evolution. It is now well known that fibrinopeptides have evolved rapidly with little constraint, whereas histone IV has evolved extremely slowly.

For the past 10 years, molecular evolutionary studies have shifted from the analysis of amino acid sequences to that of DNA sequences. Several remarkable features of DNA evolution have emerged. The majority of genomic DNA of higher organisms evolves more rapidly than protein-coding regions, that is, those DNA regions that apparently do not carry genetic information in their primary structure are evolving rapidly. In mammals, the rate of nucleotide substitution in these regions is roughly $5 \times 10^{-9}$ per site per year. The rate of synonymous substitutions in coding regions is slightly lower than this, but is relatively uniform among various genes, whereas the rate of amino acid replacement substitutions differs greatly from gene to gene. These values agree with the results of DNA hybridization studies.

An as yet unsettled problem is whether the rate of DNA evolution depends on generation length. An examination of amino acid sequences seems to have revealed little effect of generation time. DNA hybridization studies, however, indicate that the longer the generation, the lower the DNA evolution rate.

## IV. MECHANISMS OF GENE EVOLUTION

The rate of evolution is different from protein to protein, and the difference reflects the degree of constraint, as explained before. This constraint is directly connected to the function and structure of protein or RNA molecules, that is, the more rigid their function and structure, the lower the rate.

Such evolutionary features may be explained by the simple neutral theory as follows. Let $v_T$ designate the rate of occurrence of new mutations in terms of nucleotide substitutions. Let $f_0$ be the fraction of such mutations that are selectively neutral, and the remaining $(1\text{-}f_0)$ be the fraction of mutations that have a deleterious effect. Then the rate of neutral evolution becomes

$$k = f_0 v_T, \quad (7)$$

because deleterious mutations do not contribute to evolution. Favorable mutations are assumed to be too rare to have any statistical influence. Various degrees of selective constraint may be taken into account by $f_0$, for example, for pseudogenes without constraint, $f_0 = 1$, whereas for amino acid replacement substitutions of histone, IV, $f_0$ is close to zero. Here the question is whether or not mutations may be simply divided into neutral and deleterious classes.

If a substantial fraction of mutations are nearly neutral, such that both random genetic drift and selection influence their behavior, the preceding prediction needs to be modified. The fixation probability of nearly neutral mutations is negatively correlated with the species population size as explained before, and hence the evolutionary rate becomes high in small populations and low in large populations. This prediction is related to one problem of the molecular clock, that is, whether the rate is constant per year or per generation. When the rate of DNA divergence is measured by DNA hybridization, there appears to be a negative correlation between the rate and generation length. However, amino acid sequences seem to evolve chronologically.

In the early 1970s, I tried to explain this seemingly contradictory observation by using population genetics theory. The fundamental idea is that most genomic DNA of higher organisms can freely accumulate base substitutions, that is, most new mutations are selectively neutral. On the other hand, amino acid substitutions are more likely to be influenced by natural selection, that is, many of them may be regarded as nearly neutral, or very slightly deleterious. In general, large organisms tend to have long generation time and small population size, whereas the tendencies are the opposite in small organisms. The generation-time effect and population-size effect would then cancel each other for amino acid substitutions [see equation (6)]. By using DNA sequences, it is possible to examine this prediction. Nucleotide substitutions in protein-coding regions may be divided into synonymous and nonsynonymous substitutions because

Synonymous substitution

0.36 0.18 0.14

Rodentia Artiodactyla Primates

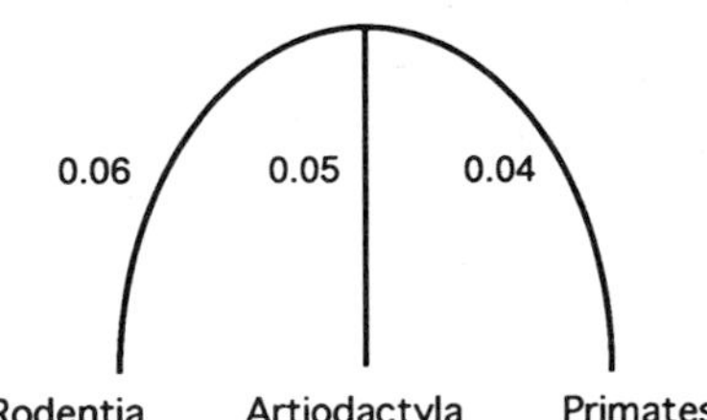


***FIGURE 1*** Star phylogenies of 49 mammalian genes. The numbers beside each branch are the estimated nucleotide substitutions per site.

of redundancy in the genetic code. DNA sequences of 49 mammalian genes, which are thought to be single copy and whose functions were fixed a long time ago, were analyzed. Branch lengths of star phylogenies composed of Rodentia, Artiodactyla, and Primates were estimated separately for synonymous and nonsynonymous substitutions. Figure 1 shows the estimated number of substitutions per site. From the figure, it is clear that the generation-time effect is more conspicuous for synonymous substitutions than for nonsynonymous substitutions, by noting that Rodentia have larger population size and shorter generation time than Primates. In other words, the evolutionary pattern of the 49 genes is in accord with the nearly neutral theory, so that nonsynonymous changes are nearly neutral and synonymous substitutions are neutral.

It has to be noted that these results hold for those genes whose function has been fixed for a long time and does not apply to duplicated genes and others whose functions have been modified in the evolutionary course of the lineages studied. This is because positive correlation is expected between evolutionary rate and population size for beneficial mutant substitutions [see equation (4)]. In the next section, this problem will be discussed in more detail.

## V. ADAPTATION AT THE MOLECULAR LEVEL

So far we have been mainly concerned with selection for keeping the gene function status quo. Once gene function attains a state sufficiently near an optimum, genes are expected to evolve by nearly neutral mutant substitutions. As long as fixed loci are considered, gene functions are usually kept as they are, and on rare occasions an environmental change causes a shift in gene function with the appropriate mutant substitutions. This may occur in protein-coding regions as well as in regulatory regions such as promoters or enhancers. Such substitutions may constitute a major process of adaptation in bacteria.

Adaptation at the molecular level of higher organisms, like mammals, appears to include minor chromosomal changes such as duplication and illegitimate crossing-over. These changes would be acceptable under a genome structure with a large noninformational part, because this part would provide flexibility in gene organization. The *Drosophila* genome seems to be an intermediate between the mammalian and bacterial ones. It is interesting to note that the detailed analyses of DNA polymorphisms in *Drosophila* species reveal many different patterns among loci, suggesting various ways of adaptation. In many cases, the interaction of random drift and selection would be important.

In vertebrates, it is known that multigene families are common. Organismal development is governed by the spatially and temporally regulated expression of various gene families that are the products of hundreds of millions of years of evolution. Thus, the reorganization of genes by duplication or illegitimate recombination is very important for organismal evolution. Comparative studies of gene families show that the genetic material is more versatile than was previously thought, that is, various illegitimate recombinational processes such as unequal crossing-over and gene conversion must have been rather common during evolution. In general, it has been thought that natural selection works to keep genes in status quo for those gene families that were established a long time ago. On the other hand, incipient gene families may be on the way to further progress, in the sense of acquiring more diverse functions, and positive Darwinian selection may be operating.

Is it possible to find incipient gene families in the process of acquiring new functions? There are now several examples of duplicated genes that show accelerated amino acid substitution relative to synonymous substitution. The first example is the emergence of fetal hemoglobin from embryonic hemoglobin in primates. Higher primates have two duplicated hemoglobin $\gamma$ genes and other primates have only one. The $\gamma$ genes of higher primates are turned on in fetal life, whereas the single gene of other primates is turned on in embryonic life and turned off at the beginning of fetal life. Examination of the DNA sequences of the $\gamma$ gene family suggests that the amino acid substitution was accelerated in the duplicated genes in the period when the switch occurred from the embryonic gene to the fetal one phylogenetically. The substitution rate is about 3.5 times the standard rate for hemoglobin genes.

The evolution of the stomach lysozyme of ruminants is another interesting example. An amino acid sequence study has shown that the gene for stomach lysozyme arose by duplication of the gene for nonstomach lysozyme at the time when ruminants diverged from other mammals. It has been shown that, about 30 million years after the divergence of ruminants from other mammals, the rate of amino acid substitution was three times as high as that of ordinary nonstomach lysozyme. Other examples include genes for visual pigment, histocompatibility antigen, immunoglobulin, protease inhibitor, and growth hormone. In genes for immune functions, there seem to be no clear distinctions between incipient and established gene families, and some multigene families belonging to the immunoglobulin superfamily are undergoing continuous reorganization via unequal crossing-over and gene conversion. Such examples reveal remarkable strategies for acquiring enormous diversity in immune reaction, and positive selection must have operated in their origin, even if selective force may have been very weak at the level of individual amino acid sites.

In summary, genes in higher organisms may be said to have been evolving through interactions between natural selection and random genetic drift. In this scenario, not only nucleotide substitution but also illegitimate recombination such as unequal crossing-over and gene conversion are important sources of genetic variability in populations.

## Glossary

**Effective population size** Population size that is equivalent to that of a randomly mating population in terms of random genetic drift. It is roughly equal to the number of reproducing parents in one generation.

**Gene conversion** A nonreciprocal recombination process by which two segments of DNA become identical with each other.

**Molecular clock** Sometimes used to mean the rate of molecular evolution, but more often it means the uniformity of evolutionary rate of nucleotide or amino acid sequences.

**Multigene family** Set of homologous genes having similar or overlapping functions.

**Nearly neutral mutation** Mutation whose behavior in a population is influenced by both random genetic drift and natural selection.

**Neutral mutation** Mutation whose behavior is determined solely by random genetic drift.

**Unequal crossing-over** Crossing-over between two chromosomes or two chromatids at different positions resulting in duplication or deletion of a DNA region.

## Bibliography

Hunkapiller, T., Goverman, J., Koop, B. F., and Hood, L. (1989). Implications of the diversity of the immunoglobulin gene superfamily. *Proc. Cold Spring Harbor Sympos. on Quant. Biol.* **54,** 15–29.

Kimura, M. (1991). The neutral theory of molecular evolution: A review of recent evidence. *Jpn. J. Genet.* **66,** 367–386.

Li, W. -H., and Graur, D. (1991). "Fundamentals of Molecular Evolution." Sunderland, Mass.: Sinauer.

Ohta, T. (1991). Multigene families and the evolution of complexity. *J. Mol. Evol.* **33,** 34–41.

Ohta, T. (1992). The nearly neutral theory of molecular evolution. *Annu. Rev. Ecol. Sys.* **23,** 263–286.

# Nature Preserves


**Joel T. Heinen**
*Florida International University*





Nature preserves of various kinds are currently found in most nations on earth. Ten types of preserves are recognized in international categories, and a number of countries have additional types that may be important for natural, aesthetic, or religious purposes, and may be protected by local, state, provincial, or territorial governments. The broad goals of managing the first five categories of internationally recognized reserves, and many others recognized at lower levels, are to preserve natural habitat for native flora and fauna, and to preserve other types of important natural features such as geological, cultural, or archaeological sites. This article will consider most closely those preserves maintained primarily for their biological interest.


## I. INTRODUCTION

As of 1990, there were 6940 different parks or equivalent reserves internationally recognized by the World Conservation Union, which covered a land area of about 6.5 million square kilometers. For the past century, the number and coverage of parks and reserves grew exponentially until the 1970s, by which time most nations on earth had systems of nature preserves. About 3% of the earth's land area is now protected in some category of national park or equivalent reserves, and the creation of nearshore marine parks has been growing in popularity since the 1970s, especially because of newly recognized threats to ecologically important sites such as coral reefs and mangrove swamps in tropical nations.

Some of these areas are extremely important for national economies as they attract large numbers of nature-based tourists—one of the fastest-growing industries in the world. Tourism is the largest industry in Nepal, for example, and is among the largest in many other developing nations, such as Costa Rica, Kenya, Belize, Ecuador, and Tanzania. Examples of parks and reserves that are particularly important for international nature-based tourism are Amboseli (Kenya), the Serengeti (Tanzania), Tortuguero (Costa Rica), the Galápagos (Ecuador), Chitwan (Nepal), and Volcans (Rwanda), all of which contain large, diverse, and important populations of particularly interesting wildlife and plant species native to those regions.

In addition to these internationally recognized categories, many individual countries have other types of nature preserves that may not be of national or international interest, but may be important locally or regionally for the protection of native wildlife and plants. Many states and provinces and some counties and municipalities in the United States, Canada, and Australia, for example, maintain their own park systems; in these countries and elsewhere there are private reserves that may conserve some important wildlife populations. Examples include reserves managed by conservation organizations such as The Nature Conservancy, private hunting clubs, and ski resorts. Many Asian nations, such as India and Nepal, have sacred groves with important wildlife and plant populations that are locally protected, some of which are rather extensive in area. In short, there are many types of nature preserves in the world and they are ever-growing in importance and popularity for aesthetic, educational, scientific, religious, and economic reasons.

## II. INTERNATIONALLY RECOGNIZED CATEGORIES OF NATIONAL PARKS AND EQUIVALENT RESERVES

The World Conservation Union (IUCN) currently recognizes 10 broad categories of parks and protected areas managed at the national level, all of which can function as nature preserves for the conservation of native fauna and flora. The first eight of these are numbered approximately in decreasing order in terms of the amount of control or protection imposed to preserve important biological, geological, or cultural features within them. Lists of protected areas in Categories I through V are regularly published by IUCN as national parks and equivalent reserves. The primary goal of management of areas designated in Categories I through V is the conservation of natural resources, including native wildlife, plants, and their habitats. Categories VI through VIII have other primary goals, though nature conservation may be an additional or secondary management goal for such reserves. Categories IX and X (Biosphere Reserves and World Heritage Natural Sites) are designated according to international agreements, and such areas may be designated in some national category as well. For example, many world heritage natural sites are also national parks. The following are broad definitions for the 10 categories as recognized internationally. The descriptions are adapted and abbreviated from the IUCN publication "United Nations List of National Parks and Protected Areas." [*See* CONSERVATION AGREEMENTS, INTERNATIONAL.]

### *Category I (Scientific Reserve/Strict Nature Reserve)*

These are defined as areas that contain some outstanding, fragile, or unique ecosystems in which natural processes are allowed to occur without human interference. The areas are maintained primarily for the study of undisturbed natural systems, and tourism and other human uses are generally not permitted. [*See* PARK AND WILDLIFE MANAGEMENT.]

### *Category II (National Park)*

National parks are defined as relatively large areas in which one or several important ecosystems are not materially altered by humans, and where wildlife and plant communities and/or geological sites are found that have special scientific, educational, or recreational interest. Most such sites have more than one special natural feature. Tourism into such areas is promoted for educational, cultural, or recreational purposes. National parks may also be zoned into smaller areas in which some uses are not permitted. For example, some may be subdivided into strict nature zones in which no visitors are allowed, along with tourist zones in which visitation is encouraged.

### *Category III (National Monument)*

National monuments are those areas in which one or more specific natural features are found. They are generally smaller in area than units in Category II, and size is dictated mostly by the specific site or sites to be protected. Management plans usually allow for some tourism, though they are managed to be relatively free of human disturbance. The national monument designation may include sites of natural interest as well as areas of important historic or cultural interest.

### Category IV (Nature Conservation Reserve/Managed Nature Reserve/ Wildlife Sanctuary)

This category is used to designate areas that have important habitat for resident or migratory wildlife of national or global significance. The size of the area designated depends on the habitat requirements of the species to be conserved. The area may require habitat manipulation, and parts may be developed for tourism and public education.

### Category V (Protected Landscape or Seascape)

A wide variety of human-altered natural landscapes occur within this category. These may be areas that contain appealing aesthetic qualities that reflect human–nature interactions, or more natural areas in which recreational uses are high.

### Category VI (Resource Reserve/Interim Conservation Unit)

This general category may be used for relatively large areas, with or without some human habitation, for which appropriate designation has not yet been determined, or is not yet implemented. In many cases, these are poorly studied remote areas in which potential future designations are in need of further study and evaluation.

### Category VII (Natural Biotic Area/ Anthropological Area)

These areas are generally remote and isolated and are maintained to protect the ways of life of traditional human societies, defined as societies in which there is a strong dependence of people on the natural environment for material needs such as food and shelter.

### Category VIII (Multiple Use Management Area/Managed Resource Area)

This category is used for the designation of relatively large areas in which the removal of wood, fodder, wildlife, fish, and other natural products is permitted. They are managed such that removal of natural products is done on a sustained-yield basis.

### Category IX (Biosphere Reserves)

The designation of biosphere reserves is part of the international Man and Biosphere Program. They are designated for a range of purposes, including research, monitoring of ecological processes, and conservation. The purpose of the program is to develop a network of protected representative ecosystems throughout the world in part to study human influences on natural systems. [*See* BIOSPHERE RESERVES.]

### Category X (World Heritage Natural Site)

The designation of world heritage sites is part of the 1972 Convention Concerning the Protection of World Cultural and Natural Heritage. In general, world heritage natural sites contain natural (biological or geological) features that are of outstanding universal value. The primary goal of the convention is to promote international cooperation to protect listed areas.

IUCN lists are designated to include only those areas that are managed by the highest competent authority, generally the national government, and include only areas that are greater than 1000 ha in size. There are several exceptions to these criteria. For example, the national parks in India are managed by the respective forest departments of the state governments, though they are included as Category II reserves, and some state parks in the United States are listed because they are in accordance with general definitions (e.g., Adirondack State Park, New York, is listed as a Category V reserve). Furthermore, the size criterion is lowered in the case of some island preserves.

In addition, IUCN categories may not necessarily correspond to the category designated within a country, as national laws governing the establishment and protection of reserves vary a great deal. For example, many of the national parks in Great Britain are listed under Category V because their management goals and physical features do not correspond to the international definition of a national park (Category II). Few nations recognize all 10 areas within their national legislation, and the national park (Category II) designation is the most common and well known internationally, followed by the wildlife preserve or sanctuary (Category IV) designation.

For the purposes of conservation and ecological monitoring, the IUCN Commission on National Parks and Protected Areas is most concerned with sites listed under Categories I through V, as the management criteria for those listed in Categories VI through VIII are not primarily conservation oriented. The latter designations, however, may in fact be very important for conservation of native wildlife and plants in some cases. For example, several South American nations (e.g., Venezuela) are currently establishing very large Category VIII (Anthropological) reserves for native tribal peoples in remote parts of the Amazon that will function as important nature preserves as well owing to their size and remoteness.

## III. A BRIEF HISTORY OF NATURE PRESERVES

Various types of nature preserves have existed for centuries, at least in some parts of the world, including royal hunting preserves in Europe and South and Southeast Asia. Some of the sacred groves of South Asia are thought to have been under protection for as long as several millennia, especially those of central India. The modern concept of nature preserves of various kinds protected by national governments for the use and enjoyment of all citizens is a much more recent phenomenon. The United States became the first nation to conserve nationally protected areas with the creation of Yellowstone National Park in 1872. Now close to 9000 square kilometers in area, this Category II protected area was inscribed as a World Heritage Natural Site (Category X) in 1978. Canada, Australia, and New Zealand all had national parks of their own by the 1880s. The first national public agency formed specifically to protect such areas was created in Canada in 1911, followed by the United States in 1916.

By the 1930s, the American park system was criticized because many parks in the country were small "mountain top parks" and preserved only scenery without regard for wildlife resources. Conserving animal populations in these areas was sometimes not possible because large mammals regularly moved outside their boundaries. Most modern writers now generally agree that there is a need to conserve large areas and therefore many species and adequately large populations.

Canada and the United States also pioneered several other conservation movements, such as the formation of the world's first migratory bird treaty in 1918 and the creation of the world's first international park in 1932, the Waterton–Glacier International Peace Park. With this came the recognition that many species and ecological systems cross international borders and cannot be effectively conserved within single nations by single sets of legislation. These rudimentary beginnings of international conservation diplomacy have led to many other conventions. The decades of the 1950s and 1960s coincided with high post-World War II population growth rates and new affluence in the West, and eventually a growing awareness about many environmental concerns, including the need for all nations to maintain systems of nature preserves. The First World Conference on National Parks was hosted by the United States in 1962. [*See* CONSERVATION AGREEMENTS, INTERNATIONAL.]

During the 1950s, 1960s, and 1970s, many seminal wildlife studies were also done in developing countries, which led to a greater worldwide awareness of the wealth of natural landscapes and their inhabitants in these nations. Numerous writings on the Serengeti and other sites in East Africa, and major studies of large mammals in Central India, the Himalaya, and China, among others, taught the world of this natural heritage. Many developing countries began to set up their own national legislation and nature preserve systems in the postcolonial era of the 1960s and 1970s. Some, such as India and Kenya, already had parks in place as a result of the colonial British governments, but these were largely for the use of government officials and indigenous royalty for big game hunting and effectively prohibited the rural poor from entering. The postcolonial era in developing nations has seen the awakening of a general consensus that parks and protected areas belong to all citizens. The centennial of the modern movement in the form of the second world conference on national parks was celebrated in 1972, appropriately in Yellowstone National Park.

The third world conference on national parks was renamed "National Parks, Conservation, and People" and took place in Bali, Indonesia, in 1982. The theme of that conference was the role of protected areas for sustainable development, and a majority of participants came from developing nations. Much of the discussion and subsequent papers emanating from the conference focused on the fact that national parks as they were managed in the West were simply inappropriate in the context of developing countries. The fourth world conference on national parks took place in Caracas, Venezuela, in 1992, and followed the themes of conservation and development in the previous conference. The relative success of parks in the United States, Canada, Australia, and New Zealand was due at least partly to the fact that population densities were low in those countries in the first place, and local, indigenous people had been largely removed from their ancestral areas on the North American continent. Such was not the case in much of the developing world, and hence it is now recognized that management plans and zoning rules for developing nations must be designed that are quite different from those used in developed nations.

## IV. DESIGN OF NATURE PRESERVES

Much of the rather large body of literature on the management of nature preserves focuses on ecological and physical aspects of reserve design, of which several will be considered here. Among the most important design criteria is the size of the area. It is generally recognized that larger areas will be more important for the protection of populations of native wildlife and plants because they are more likely to contain many different habitat types that support higher overall biological diversity, as well as larger populations of individual species. The latter can be important to avoid inbreeding, which is a frequent result of small population size.

Despite the generally recognized criterion that larger preserves are better, a good deal of debate has arisen on the topic. Several authors have pointed out, for example, that larger preserves may come at the expense of fewer preserves if nations are only able to protect a finite percentage of their land areas in nature preserves. This can be important for several reasons. For instance, fewer preserves spread across a landscape could jeopardize the goal of preserving intact representations of all ecosystem types. In addition, fewer larger preserves may mean fewer distinct populations of individual species. A catastrophic event such as a hurricane or an epidemic disease could wipe out such species if they occur in one or only a few large preserves. Conversely, the largest species may have inappropriately small population sizes if all reserves in which they are found are below some minimum size criterion; such is one reputed cost of small preserves. Some preserves are notably too small for the protection of important rare species. For example, the only Nepalese reserve to harbor a wild population of Asiatic buffalo (*Bubalus bubalis*) is too small in the sense that all individuals roam outside the preserve boundaries and into agricultural crops during the annual rainy season, when the entire land area of the preserve is prone to flooding. Even some of the largest terrestrial preserves in Africa are thought to be too small to conserve many large mammal species in the future.

This debate, known by the acronym SLOSS (Single Large or Several Small), is still contested on occasion, but most nations are further constrained by political and economic realities and attempt to preserve important natural landscapes of varying sizes where they do occur, and not of the size or number where ecological theory may dictate that they should occur. Many of the wildlife sanctuaries (Category IV reserves) in the United States, for example, were originally wetland areas for which no other land use was feasible. Most researchers now accept that preserve systems should be as comprehensive ecologically as is possible, and each until should be as large as is politically feasible. Even very small preserves that protect entire unique ecosystems can be extremely important for the purposes of conserving rare plants and invertebrates, as well as for conservation education if they occur relatively close to populated areas.

A good deal of debate has also been generated on the positioning of preserves across a landscape. It is generally thought that preserves should be located physically close to other preserves to promote dispersal, and these ideas have generated a rather large body of literature on the importance of corridors to connect nature preserves. If natural corridors allow for the dispersal of animals and plants between different units, and a species becomes extirpated from one area, then other individuals may be more likely to colonize from another area. Colonization of new individuals to an existing population may also be important to reduce inbreeding. There are currently no accepted criteria for how to design natural corridors, and some suggest that corridor systems proposed in some areas may not be beneficial for their intended purpose, and in some cases they may actually have negative effects.

For example, it is likely that relatively thin corridors will be highly disturbed by humans and would only be used by species that are already common in human-altered landscapes. Among mammals, raccoons in North America and common mongooses in Asia come to mind. The more forest-interior dwelling species, such as many small passerine songbirds, are less likely to use such corridors and are a more important focus of conservation because such species in general suffer more from habitat alteration. Lastly, dispersing animals tend to emanate radially from a source, so most individuals are unlikely to use linear corridors even if some do so successfully. Despite this, several new projects are being designed or under implementation in some parts of the United States that propose rather extensive systems of corridors to connect nature preserves. Practically speaking, this may simply be impossible in many cases depending on existing land use in areas between reserves.

Another issue relating to the physical attributes of nature reserves is the shape of the boundary. In general, it is thought that reserves that are more circular in shape may be preferable to similarly sized reserves that are more irregular in shape. Such reserves would tend to minimize both natural and human-induced edge effects. Natural edge effects include the propensity for some predators to use edge areas more intensely for hunting, which can lower reproductive success for prey species nesting in those areas. Also, several studies have shown that microclimate variables such as temperature and humidity can be quite different near the edge of forest reserves, and this can make such areas uninhabitable for forest-interior dwelling species. Human-caused edge effects include a number of factors. Long, irregularly shaped preserve boundaries may make the area more vulnerable to encroachment by poachers, livestock, domestic dogs, introduced weeds, and so on, all of which may jeopardize conservation goals.

As in the case of reserve size, number, and connectedness, most nations are probably not in the position to delineate precisely the optimal shape of nature preserves based on theoretical considerations. Rather, considerations such as historical and current human settlements and uses outside proposed reserve boundaries frequently play the most important role in deciding the physical dimensions and placement of nature preserves.

## V. EMERGING HUMAN CONCERNS

As discussed earlier, the concept of nature preserves has moved in the past century from a largely Western concept of conserving scenery and later wildlife for its own sake for the reception of an elite population, to a concept of conserving natural resources for the many benefits they can provide to people. The people most directly dependent on natural resources from such areas are rural populations of the developing nations. Nationalized preserve systems will therefore be managed differently in most developing countries compared to those in developed countries.

Some of the most critical concerns for managing such areas are frequently unrelated to the ecological theory that has led to suggestions about physical design criteria; park–people interactions and conflict resolution are emerging as extremely important issues. For example, the attitudes of local people who live around Kosi Tappu Wildlife Reserve in southeastern Nepal are generally poor despite that the reserve was the only source of thatch and cane grasses in the region; these grasses are used to construct human dwellings and can be extracted from the preserve on a permit basis for two weeks per year. Furthermore, the reserve was an important source of fish, which sold relatively cheaply in local bazaars, and permission was given on occasion for the removal of minor forest products. In spite of the measurable economic benefits from the reserve, most people expressed negative attitudes and attributed higher costs to crop destruction from wildlife than was observed to occur.

In the case of Nepal, at least some of the negativity expressed by local people may be due to the very fast enactment of comprehensive and prohibitive legislation that removed rights from local people and did not provide alternatives to resource extraction. It is now recognized that management rules need not be so strict, and that some extractive removal of natural products can be allowed on a sustained-yield basis. Furthermore, compensation and alternatives must be provided in preserve management plans in all cases where removal of natural products from the preserve has to be stopped or greatly controlled based on ecological criteria.

In this light, the World Wildlife Fund-US (WWF) began an important initiative called the Wildlands and Human Needs Project in 1985. The mission of the project is to integrate the management of natural areas with rural development schemes to improve the quality of life for the rural poor. M. Wells and K. Brandon have discussed integrated conservation development projects in the context of preserve management in developing nations and outlined some of the failures of the highly centralized approaches used earlier.

Several volumes have been written recently that incorporate economics with nature preserve management. Two in particular present important new approaches to this complex endeavor and consider economic factors that are important and in some cases necessary for justifying conservation activities to planners, politicians, and experts at all levels of government and within bilateral and multilateral aid organizations. J. A. McNeely begins with general background information on economics, biodiversity, and the values of conserving biodiversity. After discussing the strong economic incentives to exploit biological resources, he suggests that new approaches to conservation are needed to alter people's perceptions of which activities are in their own self-interests. Self-interests in this case are defined in economic terms, and he assumes that people need economic incentives to conserve resources. If authorities are to achieve conservation goals, the incentives promoting conservation must be greater than existing economic incentives to exploit resources, which are already strongly in place in many parts of the world.

McNeely then discusses the known values of biological resources and types of incentives and ways to implement them at local, national, and international levels. There are numerous direct and indirect values of conserving biological resources and protected areas (e.g., development of pharmaceuticals and nature-based tourism), and this discussion concludes with some insights about the different types of valuation schemes needed at local versus national levels. Valuation schemes can include estimating economic costs and benefits to people at the local level, as well as estimating foreign currency earnings that result from conservation programs within the national government. It is generally agreed that biotic resources are very undervalued and policy makers must understand the complex relationships between costs and benefits of conservation programs to people at all levels.

McNeely also provides examples of both direct and indirect kinds of incentives: the former includes cash subsidies for reforestation or food in exchange for work in a nature preserve, and the latter includes compensation for crop damage by wildlife for people who live near a nature preserve. There is an enumeration of many different kinds of direct and indirect incentives at local, national, and international levels. As McNeely points out, for incentive systems to work they must be implemented at local levels because it is local people who are likely to suffer from the costs of conservation activities and who have the most access to exploit resources, legally or otherwise. The book also suggests coordination at national and international (donor agency) levels to secure the monetary resources necessary for financing incentives. A broad discussion of potential sources of funding for incentives is provided, as are suggestions that incentives should be supported as much as possible through the existing marketplace. Entry fees, profits from investments, bed taxes for tourists, water use charges, and debt for nature swaps are all possible mechanisms to fund incentives, but all require coordinated governmental policies and, in some cases, international treaties.

Guidelines for ways in which incentives can be included in conservation programs and implemented by central governments, resource management agencies, and donor organizations are also discussed. Some of the guidelines are rather straightforward (e.g., provide initial assessments of available biological resources), but others are rather more difficult (e.g., estimate the contribution of biotic resources to the national economy), though they are presented as a logical series of steps to follow by managers and policy makers. The volume concludes with numerous case studies depicting various kinds of economic incentives in conservation schemes. The first 3 cases give examples of perverse incentives (e.g., incentives that deplete biological resources in Brazil), followed by 13 cases of community-level incentives (e.g., access to grazing and water as an incentive to conserve the Amboseli Ecosystem, Kenya) and 9 cases of national and international incentives (e.g., economic incentives provided to rural communities adjacent to Indian wildlife reserves).

J. A. Dixon and P. B. Sherman look specifically at valuation procedures for the conservation of nature preserves. Their book is divided into two parts: General Issues and Applications. The first chapter is devoted to describing the various costs and benefits associated with nature preserves in general and a discussion of three economically based categories of nature preserves: privately beneficial, socially beneficial, and undetermined beneficial. Preserves in the first category are those in which the benefits can be obtained by individuals or firms: examples are ski resorts or private hunting reserves. In general, privately beneficial areas are small, heavily used, and therefore not as potentially important for conservation. Areas that are socially beneficial are those with much broader and well-defined benefits: they are more common, tend to be larger in area, and include most existing national parks. The third category includes areas that are difficult to assess whether economic benefits are positive or negative. Examples include wilderness areas, wildlife preserves, or remote protected areas where tourism is not an important activity.

A valuable discussion of ways to value benefits depending on the type of resource under consideration is provided and includes issues such as: Does the resource have a market price? Is it known? Can surrogate pricing systems be used if market price is not known? A discussion on the selection of protected areas by a seven-step process with political as well as economic dimensions is given in the third chapter, and the fourth chapter discusses management considerations such as protection, threats to an area, determining allowable uses, and decreasing demands on an area. There is also a description of ways to increase financial returns via the generation of income, followed by a discussion of potential sources of additional funding for nature preserves.

Most of the examples provided in the Applications section are from Thailand, but the procedures are generally applicable. Chapter 5 provides a brief introduction to the protected area system of the country, and Chapter 6 describes the valuation of a national park (Khao Yai) that is very important for tourism as well as for education, watershed protection, and conservation of rare birds and large mammals. Chapter 7 describes a valuation of nonhunting area (Thale-Noi), a protected area designation that in Thai law allows for tourism and some consumptive uses of resources, and a wildlife reserve (Khao Soi Dao) in which tourism is allowed but not promoted. The general principle that emerges from this analysis is that as degree of use decreases, it becomes increasingly difficult to evaluate protected areas economically. The valuation of Khao Soi Dao, for example, concludes that the reserve is undetermined beneficial. There are definite costs to local people (e.g., crop destruction by wildlife) but the benefits, either at the local or national levels, cannot be fully estimated.

Dixon and Sherman also provide examples of national park valuations from other parts of the world, including valuations of Virgin Islands National Park (U.S.A.), Kangaroo Island (Australia), which includes several separate parks, and Korup National Park (The Republic of Cameroon). The main economic benefits in the first two cases result largely from mass-market tourism. The case of Korup is particularly insightful as the valuation shows that several off-site benefits (e.g., improvements in agricultural productivity and fisheries) can be estimated and can provide very convincing arguments for promoting the protection of such preserves. One example found that the value of having large mammal populations in game parks in Kenya was 50 times greater than if the areas were converted to agricultural uses, calculated by estimating average yields and local prices for produce compared to the direct economic values of nature-based tourists coming to these areas.

This approach is important and highly relevant to the present discussion. It goes much farther than other economic analyses that point out only the benefits of protected areas and/or species largely at the national level, because it is crucial to address local interests. However, in this regard, in some cases their book may not go far enough. For example, one potential problem with the valuation regarding game parks in Kenya is that if economic benefits go largely to those other than local people (i.e., international tour operators), and if local people receive direct costs due to the presence of these preserves (e.g., through crop damage), the situation may not be stable despite the obvious national economic importance of the reserves and the wildlife populations in question. Local interests, which can be compensated by local–level incentives, are likely to represent the major key to successful management. Thus, McNeely's work is particularly relevant in this regard.

The estimation of the values of wildlands and wild species within them has gained considerable attention in recent years, and many other volumes could be cited in this context. However, the most important point is that these issues, largely in the realm of the social sciences, are increasingly recognized as the major issues for the effective management of nature preserves worldwide.

## VI. GENERAL DISCUSSION

People have always coexisted with natural areas and resident wildlife populations. Very few places on earth are truly wild in that most had some human habitation and some active forms of management for millennia or longer. Several recent studies on hunter-gatherer societies have shown that some types of human disturbance over time actually enhance biological diversity. Therefore, much of the recent emergence of interest in management scenarios for nature preserves that attempt to foster coexistence has a long historical precedent. In modern times, and with increasing human population densities, industrialization, and general affluence, conflicts between local people and nature preserves are more likely to arise and are more likely to be more extreme. Theory and practice both suggest some mechanisms of resolution that were overlooked in the earlier formulation of national parks and protected areas in many parts of the world.

For example, if reserve management allows local control and sustainable harvest of important resources by relatively small social groups, the situation may be more stable for the long-term persistence of biological resources. In such cases, in the management of local resources, people have an interest in long-term sustainability; thus there may be less need for outside law enforcement that has been strictly imposed in many countries. The removal of control and rights to local residents has in many cases led to protracted conflicts in protected area management. The case of Kosi Tappu, Nepal, is illustrative of this. Despite the fact that the reserve is demonstrably important in the regional economy, and despite corroborative surveys that have shown that benefits are greater and costs are not nearly as great as people believe, the attitude surveys show a great deal of discontent with the reserve. They show that people tend to discount the benefits and exaggerate the costs. Alternative management scenarios that incorporate conservation of wildlife with local development in such cases are needed, as are alternatives for important sources of fodder and fuelwood that are now off-limits owing to the creation of such preserves. This is generally agreed upon by many managers in Nepal and elsewhere.

The types of incentives discussed by McNeely are highly relevant in this context, and many developing nations are beginning to provide seedlings of fuel and fodder trees to local people who live near nature preserves. Nature-based tourism has been shown in many cases to be important, especially when some profits go to local people through employment opportunities. The provision of incentives at the local level is probably the crucial management issue in the protection of many preserves in developing nations if biologically important resources within them are to survive into the next millennium.

In many cases, it may simply not be in the interests of local people to conserve. Even if this is the case, more modern enlightened management schemes have evolved in recent years. For example, India and several other nations have had some success through the use of buffer zones in which extraction of fuelwood and fodder is permitted. These buffer zones surround core areas in which people are not permitted to enter, except for the purposes of tourism. Buffer zones as well as intensive social forestry programs can provide alternatives to local people to remove pressures from the ecologically critical areas contained in the core of nature preserves.

Wells and Brandon describe many integrated conservation development projects (ICDPs) in several developing countries, based on short site visits to each area, interviews with managers and local people, and reviews of project documents. They considered especially projects for which direct links between nature preserves and the communities that surround them were planned. The various projects are extremely different in size, financing, and composition of local communities, so few general conclusions could be drawn from the study and transferred to other preserves. The crucial aspects of reserve design and management should therefore be site specific, given that the geography, biology, and social structures of surrounding communities are variable and individual costs and benefits to local people can be quite different. Incorporating attitudes of local people and using effective comanagement procedures are bound to be more successful in some areas than in all others. All preserve management plans should rely on direct and indirect economic incentives and alternatives where possible. Empowerment and participatory programs are needed in all cases, but the types of programs likely to be useful will vary in specific contexts.

This is not to argue that the physical design criteria discussed by ecologists are unimportant in nature preserve planning and management. What is needed are plans that seek a convergence of interests of conservation of native wildlife and plants with those of local people. In that context, the social sciences have a very large role to play in the design and management of nature preserves. If local conflicts are ameliorated, planning for numerous, appropriately sized reserves with diverse assemblages of species may be more possible in the long term. Social impact assessments and conflict resolution procedures should be built into the management of these areas in the planning stages for the purposes of strengthening the specific conservation goals of such areas.

## Glossary

**Anthropological area** IUCN Category VII preserve, set up primarily to conserve habitat for the maintenance of the culture of peoples living traditional (preindustrial) life-styles. These are also called Natural Biotic Areas.

**Biosphere reserve** IUCN Category IX preserve, designated by the Man and Biosphere Program. The program is attempting to create biosphere reserves in all ecosystems of the world for the purposes of studying long-term effects of humans on natural systems.

**Commission on national parks and protected areas** IUCN commission that has the responsibility of compiling the United Nations List of National Parks and Equivalent Reserves.

**Corridor** In the context of nature preserve management, a corridor is a strip of relatively unaltered habitat that connects two or more preserves, to allow for the dispersal of plants and animals between or among different preserves.

**Integrated conservation development project** Any number of types of rural development projects in developing nations in which conservation of species and natural areas is planned into the overall design.

**IUCN** International Union for the Conservation of Nature and Natural Resources, which has been renamed the World Conservation Union, although it still goes by the abbreviation IUCN. This is a multinational body created in 1948 that compiles lists of rare species and protected natural areas, and provides expertise to national governments on the management of biotic resources.

**Multiple use management area** IUCN Category VIII preserve, used to designate areas in which extraction of natural products (e.g., fish, wildlife, timber) is permitted on a sustained-yield basis.

**National monument** IUCN Category III preserve, which are usually small in area and preserve one or a few natural features of national interest.

**National park** IUCN Category II preserve, which are usually large in area and may contain a high number of outstanding natural features, including wildlife populations and geological sites of national interest. This is the most common nature preserve designation worldwide.

**Protected landscape/seascape** IUCN Category V preserve, which are frequently large in area and altered to some degree by humans. They frequently have heavy recreational uses.

**Resource reserve/interim conservation area** IUCN Category VI preserve, which are designated for relatively remote areas in which further study is warranted to classify the area for future protection.

**SLOSS** Acronym for the Single Large or Several Small debate in preserve design, which is still contested. Several lines of evidence suggest that large preserves are preferable to small reserves, but this may come at the expense of fewer preserves in many nations or states in which only a finite percentage of land area can be set aside.

**Strict nature reserve** IUCN Category I preserve, which are created for scientific study and do not allow recreational uses. This category is not recognized by many nations.

**Wildlife sanctuary** IUCN Category IV preserve, also known as Natural Conservation Reserve or Managed Nature Reserve, which are set up primarily to conserve habitat of migratory or resident waterfowl. This is the second most common type of preserve worldwide.

**World conservation union.** *See* IUCN.

**World heritage natural site** IUCN Category X preserve, used to designate natural areas of outstanding global importance, as determined by the World Heritage Committee. [*See* CONSERVATION AGREEMENTS, INTERNATIONAL.]

**WWF** World Wildlife Fund, now known as the World Wide Fund for Nature. The American affiliate (WWF-US) still goes by the old name.

## Bibliography

Anonymous (1990). "United Nations List of National Parks and Protected Areas." Gland, Switzerland: IUCN Publications.

Dixon, J. A., and Sherman, P. B. (1988). "Economics of Protected Areas: A New Look at Costs and Benefits." Washington, D.C.: Island Press.

Heinen, J. T. (1993a). Population viability and management recommendations for wild water buffalo *Bubalus bubalis* in Kosi Tappu Wildlife Reserve, Nepal. *Biol. Conservation* **65,** 29–34.

Heinen, J. T. (1993b). Park–people relations in Kosi Tappu Wildlife Reserve, Nepal: A socio-economic analysis. *Environ. Conservation* **20**(1), 25–34.

McNeely, J. A. (1988). "Economics and Biological Diversity: Developing and Using Economic Incentives to Conserve Biological Resources." Gland, Switzerland: IUCN Publications.

McNeely, J. A., and Miller, K. R., eds. (1984). "National Parks, Conservation and Development: The Role of Protected Areas in Sustaining Society." Washington, D.C.: Smithsonian Institution Press.

Shafer, C. L. (1990). "Nature Reserves: Island Theory and Conservation Practice." Washington, D.C.: Smithsonian Institution Press.

Simberloff, D., Farr, J. A., Cox, J., and Mehlman, D. W. (1992). Movement corridors: Conservation bargains or poor investments. *Conservation Biol.* **6**(4), 493–504.

Thorsell, J. (1990). "Parks on the Borderline: Experience in Transfrontier Conservation." Gland, Switzerland: IUCN Publications.

Wells, M., and Brandon, K. (1992). "People and Parks: Linking Protected Areas with Local Communities." Washington, D.C.: The World Bank.

Wilcox, B. A. (1980). Insular ecology and conservation. "Conservation Biology: An Evolutionary–Ecological Perspective" (M. E. Soule and B. A. Wilcox, eds.), pp. 95–117. Sunderland, Mass.: Sinauer Associates.

# Nitrogen Cycle Interactions with Global Change Processes

**G. L. Hutchinson**
*U.S. Department of Agriculture*



Ammonia and gaseous N oxides are radiatively, chemically, and/or ecologically important trace atmospheric constituents. Their multiple direct couplings to global change processes that threaten the habitability of our planet make it imperative that sources and sinks regulating their abundance be understood at cellular to global space scales and reaction rate to interannual time scales. It is generally agreed that biological processes in soil/plant systems are an important factor in the global atmospheric budgets of all gaseous N species. Because of the complexity of those processes, however, the magnitude of their budget contributions is poorly constrained and the impact of those contributions poorly defined. Anthropogenic disturbance of some terrestrial N cycles (e.g., by management for maximum economic agricultural production) has degraded their capacity to conserve N against loss to the atmosphere, and the subsequent redeposition of that N sometimes adversely perturbs the N cycles of other terrestrial and aquatic ecosystems in ways just now beginning to be realized.

## I. TERRESTRIAL N CYCLES AS A FUNCTION OF SCALE

Nitrogen is of chemical and geological interest because of its ubiquitous occurrence throughout the lithosphere, the hydrosphere, the atmosphere, and the biosphere. About 7 metric tons of molecular dinitrogen ($N_2$) are contained in the atmosphere over every square meter of the Earth's surface, and even this vast reservoir represents only about 2% of the global inventory of N. Most of the remainder (about 98%) exists in the igneous rocks of the Earth's crust and mantle, whereas amounts contained in the hydrosphere, the biosphere, coal deposits, ocean sediments, and terrestrial humus amount to only traces by comparison. [*See* Biogeochemical Cycles.]

Nitrogen has even greater biological interest because, except for C, H, and O, no other element approaches its near omnipresence in the structural (e.g., membranes), metabolic (e.g., enzymes), and genetic/reproductive (e.g., nucleic acids) systems of living organisms. Because of its immense biological importance, the relative abundance of N on our planet seems a propitious circumstance indeed. However, most organisms' demand for N cannot be met by $N_2$, which is so plentiful, but instead must ultimately be satisfied by ammonium ($NH_4^+$), the relatively scarce, fully reduced form of N. It is ironic that the same properties that make $N_2$ the most stable diatomic molecule known, and thus contribute to its abundance, also make the molecule very difficult to convert to a form that can participate in the oxidation–reduction schemes that are the basic feature of biological N transformations.

Some of those transformations are shown in Fig. 1, which summarizes the N cycle in soil and shows some of the ways it interacts with the planetary cycle of N. It is interesting to note that the term "N cycle" is really a misnomer in both cases; that is, although there is overall counterclockwise movement throughout the diagram, any given N atom usually does not follow a circular pathway to return to its original form, but instead moves randomly from one form to another. It is also important to realize that the interpretation of any such diagram requires specifying its temporal and spatial dimensions. For example, the ammonia ($NH_3$) volatilized from animal excrement is often pictured as a loss of N from soil systems. However, most of this $NH_3$ is redeposited on soil, water, and plant surfaces downwind from the site of volatilization, so if the temporal and spatial dimensions of the N cycle diagram are expanded to include this process (as was intended in Fig. 1), then the conclusion is that this N is really not lost, but instead redistributed. It is generally true that expanding the time and space dimensions of the system under study causes many of its N source/sink relationships to look more and more like N transfer processes.

At global/annual space and time scales, the exogenous addition of N to terrestrial ecosystems is limited primarily to biological $N_2$ fixation by symbiotic and nonsymbiotic microorganisms and (over the last few decades) industrial $N_2$ fixation by chemical processes. Some combined N is also added to soil in various atmospheric wet and dry deposition processes, but except for the small amount formed by electrical discharge during thunderstorm activity, most of this N represents simply the return of dust, $NH_3$, and gaseous N oxides recently volatilized from the planetary surface. The added combined N is assimilated by microbes and plants and eventually finds it way back to the soil as waste or decaying plant and animal residues. Much of it is then again released in ammoniacal form and subsequently absorbed and assimilated by other microbes and plants. Such intracyclical reuse of N continues as generation succeeds generation, but may be interrupted if the $NH_4^+$ in soil is oxidized to nitrite ($NO_2^-$) or nitrate ($NO_3^-$). These oxidized forms of N may also be recovered and reduced again to $NH_4^+$ by still other plants and microbes, but because of their susceptibility to leaching and denitrification, some may be lost from the terrestrial environment.

There is little doubt that in the geologic past, the combined N gained through biological $N_2$ fixation on the planetary surface was balanced by denitrification. If it were not so, fixation over eons of time would have resulted in the pedosphere or hydrosphere, rather than the atmosphere, serving as the principal reservoir for the N released from primary rocks and minerals. Other soil N loss mechanisms, such as crop removal, leaching, erosion, and volatilization of $NH_3$ or $NO_x$ [nitric oxide (NO) + nitrogen dioxide ($NO_2$)], can be important factors in the N budget of a particular ecosystem, but as pointed out, they result principally in transporting N from one terrestrial or aquatic ecosystem to another rather than returning it to the atmospheric reservoir of $N_2$.

The rapidly increasing worldwide rate of industrial $N_2$ fixation is soon likely to overtake and exceed its biological counterpart, thus raising the question of whether denitrification will increase commensurately. Although the supply of atmospheric $N_2$ would last millions of years at present fixation rates, the question has much more immediate significance because gaseous N species involved

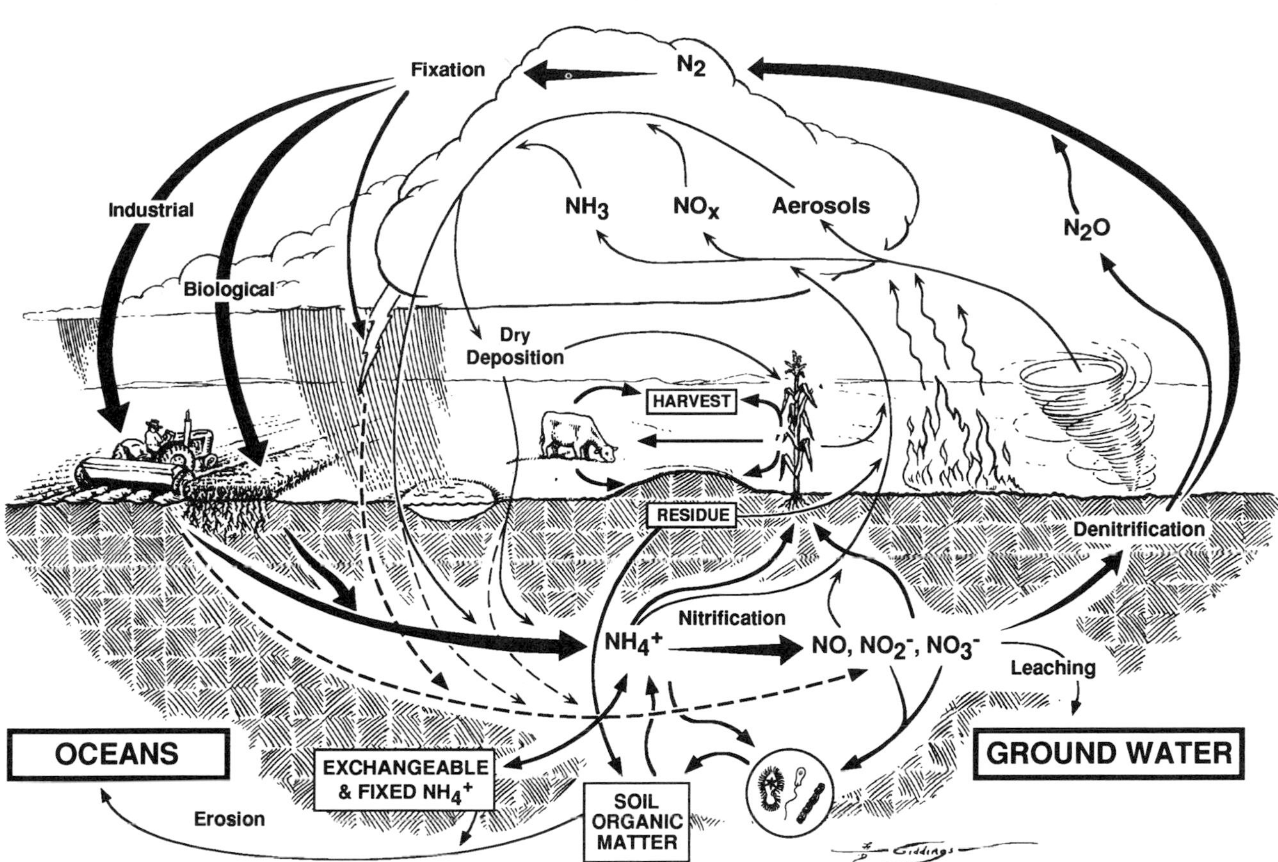


*FIGURE 1* The N cycle as a function of scale. Heavy tapered arrows represent $N_2$ fixation, oxidation of the resulting $NH_4^+$ to $NO_3^-$, and its return to the atmospheric $N_2$ reservoir by denitrification over large time and space scales. In the lower central portion of the diagram, medium arrows with solid heads describe the local N cycle at short time and space scales. The lightest arrows depict intermediate-scale N redistribution, including transfer within the terrestrial environment via gaseous N exchange (open arrowheads) and from terrestrial to aquatic ecosystems (solid arrowheads). Arrows with dashed lines represent $NO_3^-$ additions to soil that bypass the $NH_4^+$ pool.

in the global change issues presently occupying headlines in both the popular and scientific press are produced and/or consumed by denitrification and its prerequisite process, nitrification, both of which are described in greater detail in Section IV. To provide context for that and Section V concerning assessment and prediction of soil gaseous N exchange, I will first summarize the relation of $NH_3$, $NO_x$, and nitrous oxide ($N_2O$) to contemporary global change issues (Section II) and then briefly describe the global atmospheric budgets of the three gases (Section III). Finally, in Section VI, I will revisit the preceding discussion of soil and planetary N cycles to speculate on prospective permutations in their interactions with global change processes. [*See* NITROUS OXIDE BUDGET.]

## II. RELATION OF $NH_3$, $NO_x$, AND $N_2O$ TO GLOBAL CHANGE

Gaseous N compounds have multiple critical influences on the chemistry and physics of the atmosphere and, therefore, global change processes (Fig. 2). Briefly, $N_2O$ is a radiatively active gas that contributes 5–6% of the anthropogenic forcing of the global energy balance, and thus the potential for climate warming. In addition, its persistence allows time for $N_2O$ emitted at the Earth's surface to be transported to the upper atmosphere, where about 5% is oxidized to NO while the remainder is photolyzed to $N_2$. Before being redeposited to the planetary surface, the NO thus produced catalyzes a complex set of

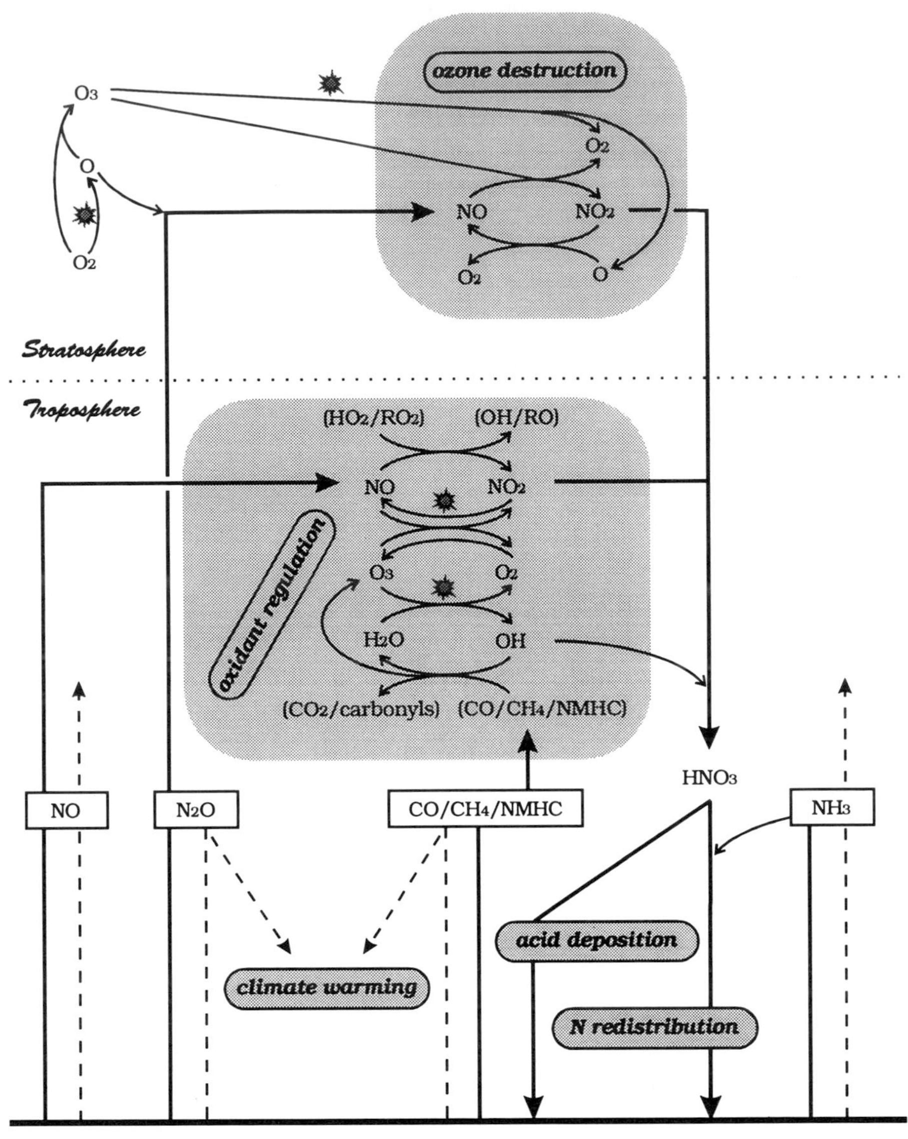


***FIGURE 2*** Relation of gaseous N species to the five shaded global change processes. Rectangles indicate emissions from soil and plant surfaces; arrows with light, heavy, and dashed lines indicate reaction, transport, and surface reradiation, respectively. Briefly, $N_2O$ is radiatively active and thus contributes to *climate warming;* following transport to the stratosphere, it is oxidized to NO, which catalyzes the *destruction of $O_3$* produced there via $O_2$ photolysis. NO is not radiatively active, but plays a key role in tropospheric *oxidant regulation*. NO, $NO_2$, and $O_3$ form a photostationary state in sunlight, whereas the alternative oxidation of NO by peroxy radicals results in net $O_3$ production. Photolysis of tropospheric $O_3$ produces the OH responsible for oxidizing the greenhouse gases CO, $CH_4$, and NMHC via reactions that regenerate $O_3$ if they occur in the presence of adequate NO. OH also oxidizes $NO_2$ to $HNO_3$, the wet and dry removal of which contributes to *acid deposition* and *N redistribution*. $NH_3$ is neither photochemically nor radiatively important, but is a uniquely significant acid neutralization agent and contributes to N redistribution.

gas-phase and heterogeneous reactions that results in the destruction of stratospheric ozone ($O_3$), which absorbs biologically harmful solar ultraviolet radiation before it reaches life-forms on the planetary surface. NO emitted at the surface is too short-lived to permit sufficient upward transport to interact with stratospheric $O_3$, and the gas is not itself radiatively active. However, it influences the concentrations of important greenhouse gases by regulating the level of tropospheric oxidants involved in removal of carbon monoxide (CO), methane ($CH_4$), nonmethane hydrocarbons (NMHC), and other gases. The rapid oxidation of NO by $O_3$ and subsequent photodissociation of $NO_2$ results in the three gases forming a photostationary state in sunlight, thereby regulating the rates of many important tropospheric photochemical reactions. Some of the $NO_2$ is eventually further oxidized by hydroxyl radical (OH) and removed from the atmosphere as nitric acid ($HNO_3$), which is the fastest-growing component of acidic deposition—another global change issue not presently in the limelight, but one that will appear again. [*See* GREENHOUSE GASES IN THE EARTH'S ATMOSPHERE.]

Although it is neither an important greenhouse gas nor an active participant in stratospheric or tropospheric photochemistry, $NH_3$ is nonetheless crucial because it is the dominant gaseous base in the atmosphere and the principal neutralizing agent for not only the $HNO_3$ produced by $NO_x$ oxidation, but also sulfuric acid ($H_2SO_4$) and other acids of both natural and anthropogenic origin. Because of its influence on the composition and pH of aerosols and cloud water, $NH_3$ is an important regulator of regional air quality and acid deposition patterns. In addition, the emission, transport, and subsequent redeposition of both $NH_3$ and $NO_x$ accomplishes substantial N redistribution both within and among natural and disturbed ecosystems, which can, in turn, initiate quite complex interactions among various ecosystem processes. For example, the deposition of atmospheric $NH_3$ and $NO_x$ from anthropogenic sources has been credited in some cases with enhanced production and in other cases with a general decline of temperate forests, depending on the amounts involved and the balance of other nutrients and growth factors.

## III. SOURCES, SINKS, AND ACCUMULATION OF ATMOSPHERIC COMBINED N

Because of the critical relations of gaseous N compounds to contemporary environmental issues and other biospheric and atmospheric processes, it is essential to identify and characterize source and sink terms in the global budgets of these important trace atmospheric constituents. The data in Table I were adapted from budgets published by other authors, but I have grouped some source and sink terms to facilitate comparisons of their importance among gases. Of the sources of $NO_x$ listed in the table, estimated magnitudes of all but the biogenic soil source term have changed little since they were described in several review articles that appeared in the mid-1980s. The most recent estimate of the soil source magnitude is about 20 Tg N per year (1 teragram equals $10^{12}$ g), computed by summing the products of the mean measured $NO_x$ emission rate, area, and length of the growing season for each of the Earth's major biomes for which these data were available. The large estimated flux from savanna was based on measurements from only three sites all in the same Latin American country, so it is subject to the possibility of substantial revision once more savanna sites are studied. Total emissions from undisturbed land is also subject to certain upward revision because nearly half the Earth's land surface is covered by biomes for which no $NO_x$ exchange estimates were available, including deserts, semideserts, polar deserts, peatland, mixed forest, taiga, and tundra. It is generally agreed that most of the $NO_x$ emitted by soil is NO, with direct soil emission of $NO_2$ accounting for substantially less than 10% of the total. Table I values for biogenic soil NO emissions and for energy-related combustion $NO_x$ emissions are comparable and substantially exceed the only two remaining sizable sources (i.e., lightning and biomass burning), so the soil source may represent as much as 40% of the total budget. The mean resi-

***TABLE I***
Global Atmospheric Budgets of $NO_x$, $N_2O$, and $NH_3$[a]

| Source or sink | $NO_x$ | $N_2O$ | $NH_3$ |
|---|---|---|---|
| | (Tg N/year) | | |
| Fossil fuel combustion | 21 | <0.1 | 2 |
| Biomass burning | 8 | 0.4 | 5 |
| Sea surface | <1 | 2 | 13 |
| Domestic animal waste | — | — | 32 |
| Human excrement | — | — | 4 |
| Lightning | 8 | — | — |
| $NH_3$ oxidation by OH | 1 | — | — |
| Stratospheric input | 0.5 | — | — |
| Soil emissions | | | |
| Cultivated land | 7.2 | 1.2 | 9 |
| Tropical forest | 2.6 | 3.4 | — |
| Savanna | 7.7 | — | 6 |
| Other undisturbed soils | 2.7 | 1 | 4 |
| Total sources[b] | 59 | 8 | 75 |
| Wet deposition | 12–42 | — | 46 |
| Dry deposition | 12–22 | — | 10 |
| Stratospheric photolysis | — | 10.5 | — |
| $NH_3$ oxidation by OH | — | — | 1 |
| Atmospheric accumulation | — | 3.5 | — |
| Total sinks[b] | 59 | 14 | 57 |

[a] Adapted from J.S. Levine, ed. (1991). "Global Biomass Burning: Atmospheric, Climatic and Biospheric Implications." MIT Press, Cambridge, Mass.; J. E. Rogers and W. B. Whitman, eds. (1991). "Microbial Production and Consumption of Greenhouse Gases: Methane, Nitrogen Oxides, and Halomethanes." American Society of Microbiologists. Washington, D.C.; W. H. Schlesinger and A. E. Hartley (1992). *Biogeochem.* **15,** 191–211; and references therein. The data represent "best estimates" provided by the authors or (except for $NO_x$ sinks) the logarithmic mean when ranges were given. A dash indicates insignificant or unavailable terms. Tg = teragram ($10^{12}$ g).

[b] It is accepted that wet and dry $NO_x$ deposition should total the sum of $NO_x$ sources and that the apparent difference between total $NH_3$ sources and sinks represents uncertainties in identified budget terms, not atmospheric accumulation.

dence time of $NO_x$ in the atmosphere ranges from a few hours in summer to times at least an order of magnitude longer in winter, and its concentration varies from only tens of pptv [parts per trillion ($10^{12}$) by volume] or less in the clean remote marine troposphere to tens of ppbv [parts per billion ($10^9$) by volume] and occasionally even more in urban areas or near industrial point sources.

The relatively long estimated atmospheric lifetime of $N_2O$ (150 years) and the smaller number of important natural sources of this gas make its global budget somewhat simpler, but no more certain (Table I). Following recent downsizing of the combustion source and tropical terrestrial source estimates in the budget, the sum of known $N_2O$ sources amounts to only about half the sum of its stratospheric photochemical sink and the observed 0.2–0.3% annual growth rate in its tropospheric concentration (currently about 320 ppbv). Moreover, adopting a more recent smaller estimate of the $N_2O$ atmospheric lifetime (110 years) implies an even larger global sink strength, so the unknown source needed to balance the budget may be quite large indeed. Contributions of the Earth's major biomes to the total soil source of $N_2O$ are not known with any certainty, although it is generally believed that humid tropical forests dominate. Studies of the impact of deforestation on these systems have yielded ambiguous results, with some indicating up to threefold enhancement and others a net reduction in $N_2O$ emissions integrated over time. Immense variability in the measurements of $N_2O$ emission from temperate and boreal forests, grasslands, and agricultural lands hampers quantifying or even ranking their contributions to the global budget of the gas.

The global atmospheric $NH_3$ budget is larger than that for the gaseous N oxides, and there is good evidence that it is increasing. Ammonia's abundance (reported to range from about 0.06 $\mu g/m^3$ in the remote marine atmosphere to greater than 300 $\mu g/m^3$ downwind from strong terrestrial point sources) is exceeded among N-containing gases only by $N_2$ and $N_2O$. Its mean atmospheric lifetime (about 10 days) is also intermediate between $NO_x$ and $N_2O$, but much closer to that of $NO_x$. There is general agreement that domestic animal waste is the principal source of atmospheric $NH_3$, but losses from the ocean surface, biomass burning, agricultural fertilizers, and undisturbed soil/plant systems may also contribute significantly. The estimated strength of the latter source is poorly constrained because soil/plant systems can both absorb and release gaseous $NH_3$, because scavenging of $NH_3$ by acidic sulfate and nitrate aerosols confounds the interpretation of concentration and flux measurements, and because the sensitive automated techniques required to establish a large data base have not been readily available. The major sinks for atmospheric $NH_3$ are foliar sorption and conversion to $NH_4^+$ via acid neutralization reactions, followed by particulate deposition or washout of dissolved $NH_4^+$ in precipitation.

## IV. PRODUCTION, CONSUMPTION, AND TRANSPORT OF GASEOUS N IN SOIL

### A. Ammonification, Immobilization, and Volatilization of $NH_3$

Because $NH_4^+$ is central to soil microbial as well as higher plant and animal metabolism, and because N is the nutrient most frequently limiting to primary productivity, it is likely not fortuitous that many natural ecosystems have developed very efficient means for conserving and reusing N (Fig. 1). Most of the $NH_4^+$ released by decomposer microorganisms or added via $N_2$ fixation and atmospheric deposition processes is exchangeably bound to the negatively charged surfaces of soil particles, nonexchangeably fixed in clay lattice structures, absorbed and assimilated by plant roots, or immobilized in various soil organic fractions. However, excess $NH_4^+$ may be volatilized as $NH_3$ following deprotonation, or oxidized by soil nitrifying microorganisms to species more susceptible to loss from the terrestrial environment. It is generally true that when the N supply exceeds biological N demand, $NO_3^-$ begins to accumulate, and the N cycle becomes more "leaky." [*See* RHIZOSPHERE ECOPHYSIOLOGY.]

Although the supply of $NH_4^+$ in unfertilized soil is determined primarily by the balance of biologically mediated production versus consumption processes, its exchange across the soil–atmosphere interface is governed principally by physical/chemical principles. At sufficiently high pH ($pK_a = 9.27$ at 25°C), a significant fraction of total ammoniacal N in the soil solution exists in deprotonated form, which equilibrates according to Henry's Law with gaseous $NH_3$ in the soil's air-filled pore space. That $NH_3$ will be transported upward through the surface–atmosphere boundary if the equilibrium vapor pressure of $NH_3$ in soil air exceeds its partial pressure above the surface; in the opposite case, $NH_3$ will be removed from the atmosphere. Additions of $NH_4^+$-releasing fertilizers and/or animal excreta enhance $NH_3$ volatilization by increasing the $NH_3$ content of the soil solution. For the same reason the volatilization rate is enhanced by increasing temperature, decreasing cation exchange capacity (CEC), reduced biological N demand, and increasing pH. Because $NH_3$ transport through soil generally occurs primarily by diffusion, its surface–atmosphere exchange rate is also influenced by factors that regulate gas diffusion in soil. Proximal and increasingly distal biotic and abiotic controls on soil $NH_3$ exchange are summarized in Fig. 3.

In some cases the $NH_3$ volatilized from soil may be partially reclaimed by foliar absorption before it escapes the overlying plant canopy. In an extreme example, less than 10% of the $NH_3$ lost from soil beneath an ungrazed Australian pasture escaped through the top of its dense grass/legume canopy. In other cases, losses from plant leaves may add to the total surface–atmosphere $NH_3$ exchange rate. Both the magnitude and direction of $NH_3$ exchange across the surface of a plant leaf is controlled by its "$NH_3$ compensation point," which reflects the equilibrium vapor pressure of the gas in its substomatal cavities.

### B. Nitrification

Nitrification in soil is accomplished primarily by a few genera of aerobic chemoautotrophic bacteria—*Nitrosomonas* and *Nitrosospira,* which oxidize $NH_4^+$ to $NO_2^-$, and *Nitrobacter,* which converts $NO_2^-$ to $NO_3^-$. Although the biochemical pathway of these transformations is not clearly established, there is abundant evidence that both NO and $N_2O$ are usually included among the products. Studies using specific inhibitors have demonstrated that both the $N_2O$ and NO produced by this process are a direct result of the activity of those organisms responsible for its first step, that is, oxidation of $NH_4^+$ to $NO_2^-$. Recent evidence suggests that production of $N_2O$ by $NH_4^+$ oxidizers results from a reductive process in which the bacteria use product $NO_2^-$ as an electron acceptor, especially when $O_2$

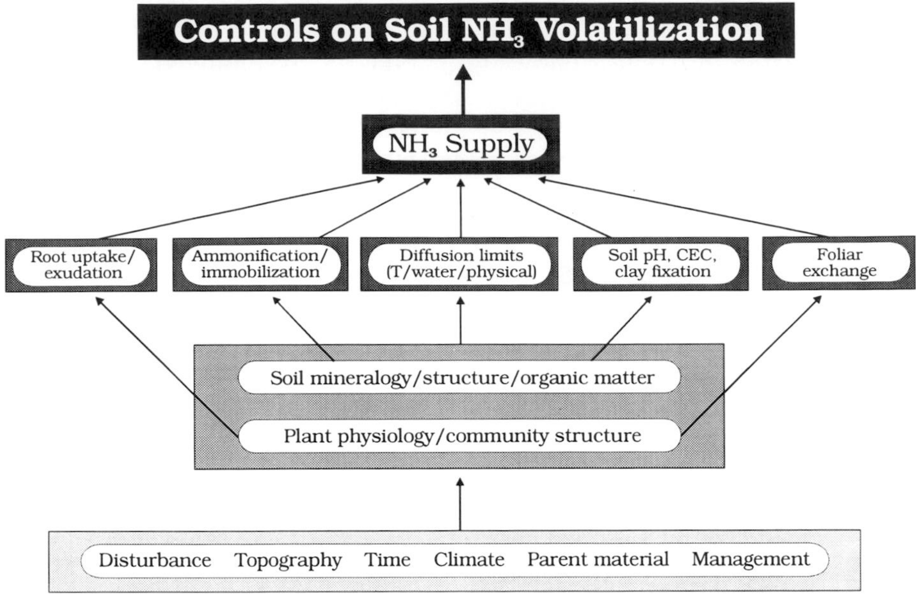


***FIGURE 3*** Schematic diagram showing the dependence of $NH_3$ volatilization on proximal environmental controls (darkest shaded rectangles), which are in turn regulated by increasingly distal factors (increasingly lighter shaded rectangles).

is limiting. Our knowledge of nitrifier physiology is not sufficient to predict whether their *in situ* production of NO also results from $NO_2^-$ reduction, or if it represents decomposition of an enzyme-bound intermediate in the oxidation pathway from $NH_4^+$ to $NO_2^-$.

The total amounts of NO and $N_2O$ generated by chemoautotrophic nitrifiers are regulated by two separate, but interdependent, sets of controllers—those that establish the overall rate of $NH_4^+$ oxidation and those that determine the $NO:N_2O:NO_3^-$ ratio of oxidation products. Nitrifying bacteria are widely distributed in soil, and their requirement for $O_2$ is satisfied in all but a few very anaerobic environments (e.g., sediments, bogs, sludge), so $NH_4^+$ availability is the factor that most frequently limits the overall process rate. Regulation of these two proximal controllers by increasingly distal factors is summarized in Fig. 4. The $N_2O$ yield of nitrification is typically less than 1%, often much less, particularly in well-aerated soil; NO yields are often in the range 1–4% of the $NH_4^+$ oxidized, but values as high as 10% and as low as 0.1% have also been reported. Because the $N_2O$ yield is generally accepted to be more sensitive to $O_2$ availability, the $NO:N_2O$ ratio of nitrification products, normally on the order of 10 to 20 in well-aerated environments, decreases along with $O_2$ partial pressure.

## C. Denitrification

Denitrification is defined here as the respiratory reduction of $NO_3^-$ or $NO_2^-$ to gaseous NO, $N_2O$, or $N_2$ that is coupled to electron transport phosphorylation. Although the identity of N compounds involved in the biochemical pathway of denitrification is well established, the nature of NO's relation to the process, as well as the mechanism for formation of the N–N bond during reduction of $NO_2^-$ to $N_2O$, are the subjects of considerable current controversy. Unlike the narrow species diversity of organisms responsible for nitrification in soil, denitrification capacity is common to several taxonomically and physiologically different bacterial groups. Denitrifiers are basically aerobic bacteria with the alternative capacity to reduce N oxides when $O_2$ becomes limiting. They are so widely distributed that denitrifying activity in a given habitat is usually assumed to be restricted not by lack of enzyme, but instead by other factors required for the process to occur—availability of

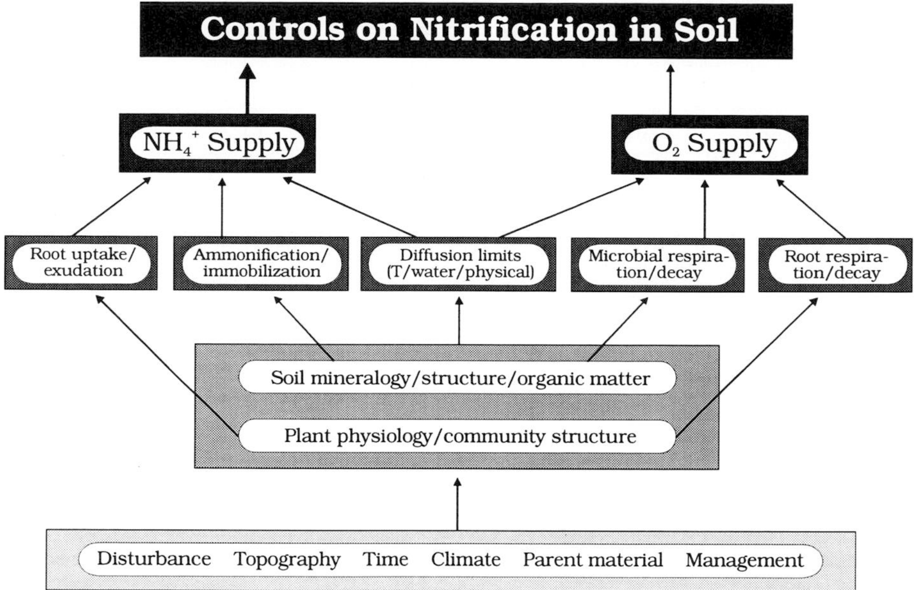


***FIGURE 4*** Schematic diagram showing the relative importance of the two principal proximal environmental controls on nitrification in soil (darkest shaded rectangles) and their regulation by increasingly distal factors (increasingly lighter shaded rectangles).

suitable reductant (usually organic C), restricted $O_2$ availability, and presence of N oxides. The relative importance of these three denitrification controllers varies among habitats, but for soil and other habitats exposed to the atmosphere, $O_2$ availability is nearly always the most critical. Regulation of these three proximal process controllers by increasingly distal factors is summarized in Fig. 5.

Dinitrogen and $N_2O$ are the usual end products of denitrification in soil, and that fraction of the process interrupted at $N_2O$ ranges from almost none to the preponderance of N reduced, depending principally on the relative availability of oxidant versus reductant. When the availability of oxidant overshadows the supply of reductant, then substrate N oxide may be incompletely reduced, resulting in a larger $N_2O : N_2$ ratio of end products. Conversely, when the overall rate of denitrification is limited by the supply of oxidant, most of the N oxide is converted to $N_2$. NO is often an important denitrification product in unsaturated soil incubated under an artificial $O_2$-free atmosphere in the laboratory, but not in the natural environment where the process generally occurs only when soil water content is high enough to restrict $O_2$ availability. In the latter case, the concomitant increase in time required for NO diffusion to the soil surface, combined with its instability toward further reduction, allows very little of this gas to escape.

### D. Abiotic Processes

Abiotic production of $N_2O$, and particularly NO, occurs primarily through a set of reactions collectively termed chemodenitrification. The most important of these reactions is the disproportionation of nitrous acid known to occur in acid soils, especially those high in organic matter content. Although this reaction has not been demonstrated in neutral or alkaline soils in the laboratory, the required accumulation of $NO_2^-$ and low pH may occur at microsites in undisturbed soils as a result of solute concentration in thin water films during freezing or drying, or because of proximity to a colony of $NH_4^+$ oxidizers, etc. Despite its potential importance in these isolated instances, chemodenitrification is generally believed to be responsible for a much smaller fraction of soil NO and $N_2O$ production than either nitrification or denitrification. The possibility that abiotic processes contribute significantly to net $NH_3$ production/consumption in soil is even more remote.

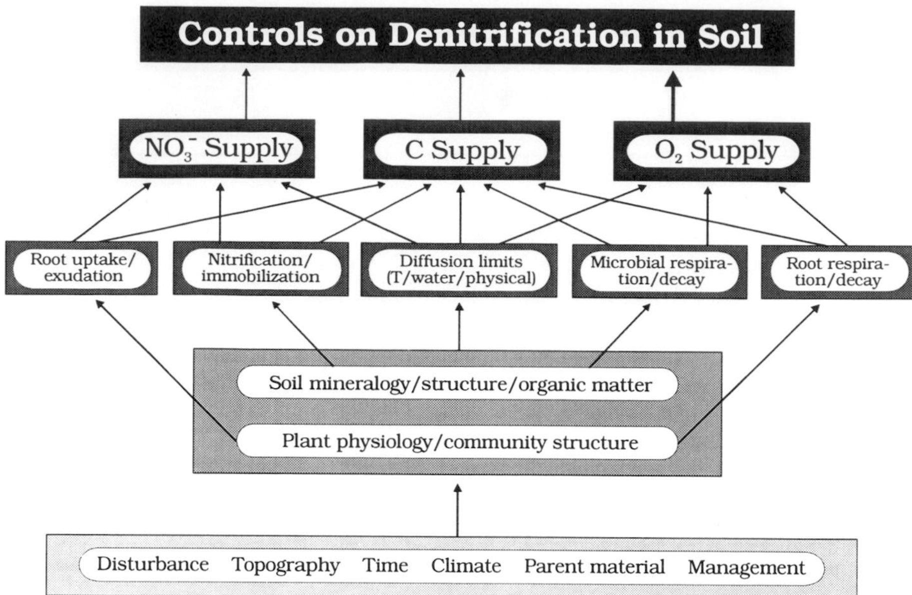


***FIGURE 5*** Schematic diagram showing the relative importance of the three principal proximal environmental controls on denitrification in soil (darkest shaded rectangles) and their regulation by increasingly distal factors (increasingly lighter shaded rectangles).

## V. ASSESSMENT AND PREDICTION OF SOIL–ATMOSPHERE GASEOUS N EXCHANGE

Despite our rapidly increasing understanding of cellular and local-scale processes responsible for the production, consumption, and transport of gaseous N species in soil, applying that knowledge at landscape, regional, or global space scales and seasonal to interannual time scales remains troublesome. Much of the difficulty reflects scaling problems associated with the need to draw inferences about large-scale exchanges from the highly variable fluxes typically measured over small areas and short times. Budget calculations like those in Table I compare best estimates of the magnitudes and distributions of known sources with known sinks, and then attempt to reconcile the net source of each gas with the magnitude, distribution, and trend in its atmospheric concentration. This approach has already resulted in greater understanding of the relative importance of known sources and sinks and established the potential for the existence of additional unknown sources and sinks. Biogenic soil source terms in the budgets were derived by dividing the Earth's surface into a few major biomes and then characterizing each by the product of its area and the mean exchange rate from randomly selected sites assumed to be representative of that biome. However, the criteria for establishing that a particular site (or measurement time) is representative are not straightforward, which renders the entire approach hopelessly inadequate for achieving a predictive understanding of the immense variability in soil $NH_3$, $NO_x$, and $N_2O$ exchange rates across time and space domains larger than the scale of available measurements. [*See* SOIL ECOSYSTEMS.]

Accordingly, the long-range goal of additional field measurements should be to capture the exchange rates in terms of their basic physical, chemical, and biological controllers, so that dependence of the flux on these controllers can be described by simulation models parameterized by variables observable at the scales of interest. Observed patterns in the measurements of surface–atmosphere $NH_3$, $NO_x$, and $N_2O$ exchange, combined with rapidly expanding knowledge of the processes responsible for their production, consumption, and transport in soil, suggest that the minimum set of predictor variables must characterize dependence of the fluxes on soil temperature, soil N availability, and soil water content, all of which are discussed in greater detail in the following.

### A. Dependence on Soil Temperature

The soil exchange rates of gaseous N species generally increase approximately twofold for each 10°C rise in temperature over the range 15 to 35°C, which matches the temperature dependence of most microbial processes, including ammonificaton, nitrification, and denitrification. The tendency for soil NO emissions to decline rapidly at temperatures greater than about 35°C may result partially from soil desiccation and its consequent reduction in the mobility of dissolved nutrients, but is probably due primarily to the failure of chemoautotrophic nitrifiers to grow above 40°C. Heterotrophic organisms responsible for ammonification and denitrification have higher tolerance to elevated temperature, so the soil exchange rates of $NH_3$ and $N_2O$ exhibit less sensitivity to high temperature. In fact, the Arrhenius equation that describes this exponential temperature dependence generally applies only over the range 15 to 35°C for nitrification, but over the much broader range 15 to 75°C for denitrification. Changes in temperature below about 15°C typically have much greater effect on the rates of all biological processes than changes above this threshold and are not well characterized by the Arrhenius equation. However, the potential for significant production of gaseous N compounds at low temperature cannot be discounted because the responsible organisms are known to possess a significant capacity for adaptation to extreme climates.

In addition to its effect on the microbial production of gaseous N compounds in soil, temperature also has a strong influence on the physical and chemical parameters that regulate their transport and subsequent exchange with the atmosphere. Unfortunately, these parameters (e.g., diffusion coefficients, solubilities) also depend on soil texture, soil water content, the composition of aqueous and nonaqueous soil phases, and other factors, which significantly complicates achieving a predictive understanding of the net effect of temperature on gaseous N exchange.

### B. Dependence on Soil N Availability

It is elementary that the availability of organic and inorganic N in soils should strongly influence their rates of gaseous N exchange. For example, the potential for $NH_3$ volatilization grows with the size of the soil $NH_4^+$ pool, if other controlling factors are favorable (Fig. 3). Because analogous relations for soil NO and $N_2O$ exchange are less direct, many authors have tried to characterize their dependence on soil N availability via correlations with a net nitrification assay, or net mineralization assay, or nitrification/denitrification substrate pool sizes (i.e., soil $NH_4^+$ and $NO_3^-$, respectively). Unfortunately, microbial process assays are labor-intensive and condition dependent, and concentration measurements may not accurately reflect the prevailing rate of substrate supply to the responsible organisms, because the turnover rates of soil inorganic N pools are sometimes very rapid (i.e., one day or less). All reported correlations tend to be site specific or study specific, and no single predictive parameter or suite of parameters has emerged that accurately reflects the effect of N availability on soil gaseous N exchange across all gases and sites.

Failure to find common predictors probably reflects that very different processes are involved (i.e., oxidative versus reductive, chemical versus biological, etc.), that process-limiting factors other than N availability may be more important at some sites than others, and that the scale chosen for investigation influences the nature of the predictors likely to be found useful. For example, soil $NO_3^-$ concentation was a good predictor of NO emissions in a comparison of hardwood forests with fertilized corn fields, but it did not account for substantial variation within each location. At the latter scale, modest topographic gradients or local-scale effects of crop residues may be important contributors to the observed variability.

### C. Dependence on Soil Water Content

Except for its universal requirement by all life processes, soil water's most important effect on gaseous N exchange results from its strong influence on both gas-phase and solution-phase diffusive transport rates. Higher water contents increase the ratio of water-filled to air-filled soil pore space and result in thicker water films lining the remaining air-filled pores, thus enhancing the transport of species in solution, but retarding that of the gases in soil air. Thus, for aerobic microbial processes like nitrification, the overall process rate is probably limited by solution-phase substrate diffusion through thin water films in dry soil, and by gas-phase $O_2$ diffusion in wet soil. The optimum soil water content for most aerobic processes is about 60% water-filled pore space (WFPS), but it is important to realize that this value applies to overall process rates rather than the production of specific end products. For example, the optimal WFPS for $N_2O$ production by nitrifiers may be somewhat higher than for $NH_4^+$ oxidation, because $N_2O$ is apparently produced by these bacteria only in response to incipient $O_2$ deficiency. Because it is an anaerobic process, the optimum WFPS for denitrification exceeds 60%, but maximum production of $N_2O$ by the responsible bacteria may occur at a WFPS that is somewhat lower than the optimal value for $NO_3^-$ reduction, because the $N_2O : N_2$ ratio of denitrification products is determined by the relative availabilities of oxidant versus reductant.

Separate from these transport-related effects of soil water content, a large burst of gaseous N emissions often occurs immediately following wetting of very dry soil. Emission rates during such an event may be up to three orders of magnitude higher than preceding or following rates, so the quantity of soil N lost during its short duration may exceed the total amount emitted during the comparatively long periods between times that the soil dries enough to support another emissions burst following the next addition of water. Subsequent irrigation or precipitation events may further increase emissions above the levels measured from dry soil, but the amount of increase is small compared to that observed for the initial watering of very dry soil. Results from a soil incubation study in my laboratory revealed that a second burst of NO similar to the one that followed initial wetting occurred only where desiccation reduced the evolution rate to near zero prior to rewetting. Reasons for the unusually large response of gaseous N emissions to wetting of dry soil remain unclear.

### D. Example Soil Exchange Models

Despite the difficulties associated with parameterizing the dependence of gaseous N exchange on its controller variables, recent literature documents progress toward this goal. For example, in my ongoing small-plot research at Akron, Colorado, the observed variability in NO and $N_2O$ exchange rates was successfully simulated by the product of a constant that reflected soil N availability, an exponential function of soil temperature, and a dual-slope linear function of soil WFPS, thereby accounting for all three of the principal environmental controls on soil gaseous N exchange. Evidence supporting the potential of this approach includes that order-of-magnitude differences between sets of fluxes measured on sampling dates with widely divergent conditions were reduced to less than a factor of 2 after normalization to the same temperature and percentage WFPS, and the remaining differences between sets were correlated with measured changes in various soil N availability indices. Better understanding of the model constant's dependence on the sizes and/or transformation rates of identifiable soil N pools and an effective method of adjusting the effect of precipitation for soil water content prior to wetting are needed before this approach can be extended over larger areal and temporal domains.

Other recent modeling efforts take advantage of known differences in N availability among land use types, the empirical temperature dependence noted in large data sets, and relations of the three proximal gaseous N exchange controllers to various surrogate variables (e.g., basic climate data, soil physical properties, and agricultural management practices). Compared to simple linear extrapolation of point measurements from randomly selected sites assumed to be representative, all such modeling approaches promise to (1) provide the basis for more precise inferences about regional/global and seasonal/annual scale phenomena using either ground-based or remotely sensed data; (2) provide information concerning the timing and distribution of emission events, which in turn helps to define the nature and dynamics of source ecosystems; and (3) permit the prediction of how future natural and anthropogenic disturbances might influence the soil source of atmospheric N species.

## VI. SUMMARY AND PERSPECTIVE

Gaseous $NH_3$, $NO_x$, and $N_2O$ are radiatively, chemically, and/or ecologically important trace atmospheric constituents that are directly or indirectly related to acid deposition, global warming, stratospheric $O_3$ depletion, groundwater contamination, deforestation, and biomass burning, which include most of the major environmental issues facing society today. Interactions between these global change processes and the terrestrial N cycles diagrammed in Fig. 1 are numerous, complex, bidirectional, and generally increase in scope with every anthropogenic perturbation of N cycle components. For example, a sudden increase in the N supply of a pristine ecosystem (e.g., by application of an $NH_4^+$-releasing fertilizer to enhance productivity) augments the site's $NH_3$ volatilization potential and often stimulates the decomposition of soil organic matter, which eventually results in smaller microbial biomass, greater wind and water erosion potential, diminished native soil fertility, and devalued soil physical characteristics (e.g., aeration, permeability, water-holding capacity). The larger supply of $NH_4^+$ also supports greater activity of soil nitrifying microorganisms, which not only produce NO and $N_2O$ directly but also increase the relative importance in the local N cycle of $NO_2^-$ and $NO_3^-$ that are subject to both leaching and denitrification, a potential source of additional gaseous N oxides.

Expanding the time and space scales over which the anthropogenically disturbed N cycle is viewed magnifies the number and complexity of its interactions with global change processes. The $N_2O$ generated by microbial processes is relatively inert in the troposphere and not readily redeposited to the surface, but the opposite is true of $NH_3$ and $NO_x$, whose downwind deposition may transfer the influence of the disturbance far beyond the area impacted directly. Chronic low-level deposition of gaseous and particulate $NH_3$ and $NO_x$ to temperate forest ecosystems, for example, initially increases their primary productivity without perturbing their relatively conservative N cycles. However, as N availability eventually begins to overwhelm C availability to soil microorganisms and the availability of water and other nutrients to higher plants, changes similar to those described for the directly perturbed area begin to appear. Because close coupling of C and N dynamics in most pristine ecosystems is an essential feature of their capacity to conserve N, it is generally true that small changes in the supply of either element cause disproportionately large changes in their N source/sink characteristics, as well as other N-dependent ecosystem functions. Examples of the latter include (1) the capacity for microbial oxidation of atmospheric $CH_4$ in aerobic temperate zone forest and grassland soils is substantially reduced by relatively small additions of N, and (2) carbonyl sulfide (COS) and carbon disulfide ($CS_2$), which exert important influences on the Earth's heat budget as well as on the chemistry of the stratosphere, are produced mainly via soil biological processes that are unusually sensitive to inorganic N availability.

At still larger time and space scales, the multiple, complex, bidirectional linkages of terrestrial N cycles to atmospheric processes and properties that control the habitability of our planet raise critical questions regarding the potential fate of the rapidly expanding quantity of N fixed by industrial processes. In the geologically recent past, fixation of atmospheric $N_2$ was apparently balanced by denitrification, but there is little evidence to suggest that the rate of this process in intensively managed agricultural soils has increased enough to match inputs of industrially fixed N fertilizer over the last few decades. In fact, it is generally agreed that denitrification in agricultural soils returns to the atmosphere only a small fraction of the difference between the quantity of N fertilizer applied and that removed in harvested products.

It is possible that excess N in agricultural ecosystems is exported to the oceans for denitrification, but the best available evidence does not unequivocally establish even the direction of the sum of net $N_2$ and $N_2O$ fluxes across the air–sea interface. The N might also still be in transit to the oceans, existing as $NO_3^-$ below the crop rooting zone of agricultural soils and in the groundwater beneath them; or it could be denitrified there, although the low level of energy materials in most subsoils and groundwaters is likely to limit the rate of this process. Nonetheless, it is noteworthy that neither of the methods most often used to quantify denitrification in soil (acetylene blockage and isotopic techniques) would include $N_2$ from this source.

Alternatively, terrestrial N cycling processes may become spatially decoupled on landscape and regional scales. For example, when supplied with N from external sources, habitats particularly suitable for organisms that produce N gases might exhibit N exchange rates much larger than could be supported by their internal rates of N processing. An obvious example is a riparian system adjacent to heavily fertilized agricultural land. Other examples do not require hypothesizing such intense source activity. Denitrification rates typical of all the Earth's uncultivated land (90% of the total) likely represent a greater fraction of total N inputs than for cultivated land, so excess industrially fixed N might be denitrified there following redistribution via the emission and subsequent downwind redeposition of $NH_3$ and $NO_x$. Interestingly, total upward and downward global fluxes of the two gases approach that of recent total $N_2$ fixation rates, at least within the uncertainty with which these fluxes of atmospheric N are known.

Our present capability to address questions raised by the foregoing speculation, as well as to predict other biosphere–atmosphere interactions at global/annual scales, is severely limited by immense variability in the spatial and temporal distributions of measured $NH_3$, $NO_x$, and $N_2O$ exchange rates and the difficulty associated with quantifying soil $N_2$ emissions in the presence of a high background atmospheric concentration of this gas. Because direct monitoring of the exchange rates at large scales is improbable with existing or foreseeable technologies, development of applicable trace gas exchange models driven by variables observable at these scales is imperative. This goal requires not only better understanding of the interacting physical, chemical, and biological processes that regulate gaseous N exchange across the surface–atmosphere interface, but also acquisition of comprehensive data sets against which proposed models can be tested. Thus, there is also urgent need for additional measurement programs that must include intercalibrations and intercomparisons of both sampling techniques and analytical procedures, and reports of which must describe explicitly the sample allocations and data analyses employed to characterize the exchange rates. In addition, integrating and extrapolating these data over regional-to-global space scales and seasonal-to-interannual time scales require improved and updated data bases of climatological, ecological, and land use information, although it is generally true that even existing data bases cannot be adequately utilized without additional flux measurements and better simulation models. Despite the difficulty of these challenges, the recent advances summarized in this article promise additional rapid progress toward the overall triumvirate goals of understanding the controls on gaseous N production, consumption, and transport at all scales, integrating that knowledge with the rapidly expanding data base of exchange measurements to explain their existing temporal and spatial distributions, and then predicting how future natural and anthropogenic disturbances might influence those distributions.

## Glossary

**Ammonification** Microbial process resulting in release of ammonium from N-containing organic compounds.

**Chemoautotroph** Organism capable of using carbon dioxide or carbonates as the sole source of C for cell biosynthesis, and of deriving energy from the oxidation of reduced inorganic or organic substrates.

**Chemodenitrification** Conversion of nitrite or nitrate to molecular dinitrogen or a gaseous N oxide via nonbiological processes.

**Denitrification** Conversion of nitrite or nitrate to molecular dinitrogen or a gaseous N oxide via respiratory reduction coupled to electron transport phosphorylation.

**Heterotroph** Organism capable of deriving C and energy for cell biosynthesis via the utilization of organic compounds.

**Immobilization** Conversion of an element from inorganic to organic form in microbial or plant tissues.

**Nitrification** Biological oxidation of ammonium to nitrite and nitrate, or a biologically induced increase in the oxidation state of N.

**Stratosphere** Region of the atmosphere extending from about 12 or 15 to 50 km above the Earth's surface. It contains only about 15% of total atmospheric mass, but 90% of the total amount of ozone. Stratospheric temperature increases with height owing to absorption of incoming solar radiation by ozone.

**Troposphere** Lower portion of the atmosphere in which temperature usually decreases with height and most weather phenomena occur (clouds, precipitation, etc.). It extends 12–15 km above the Earth's surface and contains about 85% of total atmospheric mass.

## Bibliography

Andreae, M. O., and Schimel, D. S., eds. (1989). "Exchange of Trace Gases between Terrestrial Ecosystems and the Atmosphere." Chichester, England: Wiley & Sons.

Bouwman, A. F., ed. (1989). "Soils and the Greenhouse Effect." Chichester, England: Wiley & Sons.

Harper, L. A., Mosier, A. R., Duxbury, J. M., and Rolston, D. E., eds. (1993). "Agricultural Ecosystem Effects on Trace Gases and Global Climate Change," ASA Spec. Publ. No. 55. Madison, Wisc.: American Society of Agronomy.

Houghton, J. T., Jenkins, G. J., and Ephraums, J. J., eds. (1990). "Climate Change, the IPCC Scientific Assessment." Cambridge, England: Cambridge University Press.

Langford, A. O., Fehsenfeld, F. C., Zachariassen, J., and Schimel, D. S. (1992). Gaseous ammonia fluxes and background concentrations in terrestrial ecosystems of the United States. *Global Biogeochem. Cycles* **6,** 456–483.

Levine, J. S., ed. (1991). "Global Biomass Burning: Atmospheric, Climatic, and Biospheric Implications." Cambridge, Mass.: MIT Press.

Oremland, R. S., ed. (1993). "Biogeochemistry of Global Change: Radiatively Active Trace Gases." New York: Chapman & Hall.

Rogers, J. E., and Whitman, W. B., eds. (1991). "Microbial Production and Consumption of Greenhouse Gases: Methane, Nitrogen Oxides, and Halomethanes." Washington, D.C.: American Society of Microbiologists.

Schlesinger, W. H., and Hartley, A. E. (1992). A global budget for atmospheric $NH_3$. *Biogeochem.* **15,** 191–211.

Williams, E. J., Hutchinson, G. L., and Fehsenfeld, F. C. (1992). $NO_x$ and $N_2O$ emissions from soil. *Global Biogeochem. Cycles* **6,** 351–388.

# Nitrous Oxide Budget

**R. J. Haynes**
*New Zealand Institute for Crop and Food Research*



Nitrous oxide ($N_2O$) is a gas implicated in the greenhouse effect and is also involved in catalytic destruction of the ozone layer. Of global $N_2O$ emissions, 22–28% originate from oceans, 46–55% from soils, 15–28% from fertilizer applications to agricultural soils, and 2–4% from fossil fuel and biomass burning. The major processes involved in $N_2O$ emissions from both oceans and soils are biological denitrification carried out by heterotrophic bacteria and autotrophic nitrification carried out by chemoautotrophic bacteria belonging to the family Nitrobacteraceae. These biological processes are regulated by factors such as $O_2$ partial pressure, temperature, pH, and, in the case of denitrification, levels of available organic matter. Nitrous oxide is relatively inert in the troposphere and is able to move comparatively unimpaired into the stratosphere, where it is destroyed by photochemical decomposition reactions.

## I. INTRODUCTION

It is estimated that the rate of N input to the biosphere through the combined use of industrially fixed N fertilizer and cultivated leguminous crops is currently about double the total rate of $N_2$ fixation by biological and other processes that occurred before human intervention. As a consequence of these increasing N inputs, the rates of those processes that return fixed N back to the atmosphere (including $N_2O$ emission) are also increasing. Nitrous oxide is currently present in the troposphere at an ambient concentration of 310 ppbv (parts per billion by volume) (amounting to about 1500 Tg N in total; 1 teragram = $10^{12}$ g) but appears to be increasing at a rate of about 0.25% per annum (i.e., 2.5–4.5 Tg N/year).

Nitrous oxide is a very effective "greenhouse" gas and has the potential for thermal absorption of 250–290 times that of $CO_2$. The estimated contribution of $N_2O$ to greenhouse warming over the past 100 years is about 5%. Nitrous oxide is also a gas involved in photochemical reactions that catalyze the destruction of the stratospheric ozone layer. Because of these environmental effects, the increases in global $N_2O$ emissions in the last two decades have caused considerable discussion. Agriculture has been the main contributor to increased losses. [*See* GREENHOUSE GASES IN THE EARTH'S ATMOSPHERE.]

Two major biological processes are the source of most $N_2O$ evolved from soils, surface waters, sediments, and oceans. Autotrophic nitrification, in which ammonium ($NH_4^+$)-N is converted to nitrate ($NO_3^-$)-N with the release of some $N_2O$, is thought to be primarily responsible for low $N_2O$ fluxes emitted by most soils under aerobic conditions. Biological denitrification is a process in which aerobic bacteria grow in the absence of $O_2$ while reducing $NO_3^-$ and nitrite ($NO_2^-$) to the gaseous products $N_2O$ and dinitrogen ($N_2$). Dinitrogen is normally the major product formed. This process is thought to be responsible for pulses of $N_2O$ that often come from soils after rainfall or irrigation.

## II. GLOBAL BUDGET

An estimated global budget for $N_2O$ is shown in Table I. About 20 to 30% of total global $N_2O$ emissions originate from the oceans through the processes of denitrification and nitrification. An average $N_2O$ flux of about 2 Tg N/year from oceans is relatively small compared to the stratospheric sink of 7–13 Tg N/year. The global flux of N ($N_2O$ plus $N_2$) originating from denitrification and nitrification in oceans is estimated to be approximately 50–80 Tg N/year so that $N_2O$ losses account for only about 3% of the total flux [*See* NITROGEN CYCLE INTERACTIONS WITH GLOBAL CHANGE PROCESSES.]

***TABLE I***
**Estimated Global Budget for Nitrous Oxide**

| Source or sink | Range (Tg/year) |
|---|---|
| Source | |
| Oceans | 1.5–2.5 |
| Unfertilized soils | 2.9–5.2 |
| Fertilizer | 0.8–3.2 |
| Biomass | 0.02–0.2 |
| Fossil fuel burning | 0.2–0.3 |
| Total | 5.3–11.4 |
| Sink | |
| Photolysis in stratosphere | 7–13 |

Soils are the dominant source of $N_2O$ and contribute 45–55% of total emissions. In forest soils of temperate and boreal zones, N availability is suboptimal and nitrification is a relatively minor process. As a result, estimated emissions of 0.7–1.5 Tg N/year are small. However, in recent times there has been a substantial input of N through anthropogenic deposition of nitrogen oxides ($NO_x$) and ammonia ($NH_3$). Farm livestock and fertilizer use are the main sources of atmospheric $NH_3$ that is deposited onto forest land. The increasing abundance of $NH_4^+$ in soils of these systems is likely to result in an increased rate of nitrification and denitrification and increased $N_2O$ emissions. Nitrous oxide emissions from tropical forest soils (2.2–3.7 Tg N/year) are generally higher than those from temperate and boreal regions, reflecting the higher N availability in these soils.

Agricultural operations are a significant direct contributor to $N_2O$ emissions. Applications of N fertilizers lead to enhanced $N_2O$ emissions and increasing fertilizer rates generally result in greater emissions. The reason for this is that fertilizer additions result in increased levels of substrate (i.e., $NH_4^+$, $NO_2^-$, and $NO_3^-$) in the soil for the processes of nitrification and denitrification. Emissions of $N_2O$ can also be enhanced by addition of organic materials having a high N content (e.g., animal and green manures) to soils. Based on total production rates of different types of fertilizers, the global loss of fertilizer-derived N as $N_2O$ is estimated as 0.5–2%. A similar amount may be emitted through nitrification and denitrification from fertilizer-derived N that has been lost to ground and surface waters via leaching and runoff. The total loss rate becomes 1–4%. The total global fertilizer N consumption in 1990 was about 80 Tg and the total $N_2O$ emission rate due to fertilizer applications amounted to 0.8–3.2 Tg N/year with an average of about 1.7 Tg N/year. The total flux of $N_2O$ from soils and fertilizers (i.e., 3.7–8.4 Tg N/year) amounts to only about 3–5% of total denitrification plus nitrification losses ($N_2$ plus $N_2O$) estimated from these sources (i.e., 105–185 Tg N/year).

Small quantities of $N_2O$ are also emitted into the atmosphere by anthropogenic activities such as fossil fuel and biomass burning. Until recently it had been assumed that the combustion of fossil fuels represented the dominant source of the increase in atmospheric $N_2O$ during the twentieth century. However, discovery of errors in estimates and measurements have led to a major downward revision of estimates.

## III. BIOLOGICAL DENITRIFICATION

Biological denitrification is a respiratory process that is carried out by a limited number of normally aerobic bacteria whereby they can grow in the absence of $O_2$ while reducing $NO_3^-$ or $NO_2^-$ to $N_2$ and/or $N_2O$. The majority of these bacteria are heterotrophs and obtain their energy and cellular C from organic substrates. In the absence of $O_2$, $NO_3^-$ or oxides derived from it take the place of $O_2$ for the metabolism of organic matter and the generation of adenosine triphosphate. That is, $NO_3^-$ and its derived oxides serve as electron acceptors during the oxidation of organic substrate and a more reduced N oxide or $N_2$ is produced.

Denitrifying bacteria are biochemically as well as taxonomically diverse and over 20 genera have been reported to denitrify. However, only a few genera are dominant in soil, marine, freshwater, and sediment environments. *Pseudomonas* spp. are the most numerous in all environments and *Alcaligenes* is commonly the second most numerous population. The presence of denitrifiers in surface soils may be regarded as ubiquitous as their density frequently exceeds $10^6$ per gram of soil and even higher concentrations are present in the volume of soil in close proximity to plant roots (the rhizosphere). Possession of a denitrifying pathway may well help free-living bacteria survive anaerobic conditions in soil.

A few $N_2$-fixing organisms are known to have the ability to denitrify under anaerobic conditions. These include a considerable number of strains of *Azospirillum,* which are commonly associated with roots of tropical grain and forage grasses. Several strains of *Rhizobium* also possess denitrifying capabilities in their free-living state and under anaerobic conditions can produce $N_2$ or $N_2O$.

The pathway of denitrification is

$$\underset{(+5)}{NO_3^-} \rightarrow \underset{(+3)}{NO_2^-} \rightarrow \underset{(+2)}{[NO]} \rightarrow \underset{(+1)}{N_2O} \rightarrow \underset{(0)}{N_2}$$

The oxidation state of nitrogen in each of the species is indicated in brackets. The specific enzymes responsible for the reductions are $NO_3^-$ reductase, $NO_2^-$ reductase, NO reductase, and $N_2O$ reductase. Not all denitrifying organisms are capable of synthesizing each of the enzymes necessary for the complete reduction of $NO_3^-$ to $N_2$. Although NO behaves as an intermediate, there remains considerable debate about whether it should be regarded as a true intermediate.

The necessary conditions for the occurrence of biological denitrification are the simultaneous presence of $NO_3^-$ (or a more reduced nitrogen oxide), a readily metabolizable carbonaceous substrate, and a low partial pressure of $O_2$ in the soil.

## IV. AUTOTROPHIC NITRIFICATION

Nitrification is the process by which ammonium ($NH_4^+$) is oxidized to nitrate ($NO_3^-$) via nitrite ($NO_2^-$). It is an oxidative process mediated by chemoautotrophic bacteria belonging to the family Nitrobacteraceae. Organisms in this family derive their energy from the oxidation of either $NH_4^+$ or $NO_2^-$. A list of currently recognized genera of Nitrobacteraceae is shown in Table II.

Chemoautotrophic nitrifiers are aerobes that synthesize their cell constituents from $CO_2$. The

**TABLE II**
Members of the Family Nitrobacteraceae

| Oxidation | Genus | Species | Habitat |
|---|---|---|---|
| Ammonium to nitrate | *Nitrosomonas* | *europa* | Soil, water, sewage |
| | *Nitrosolobus* | *multiformis* | Soil |
| | *Nitrosovibrio* | *tenuis* | Soil |
| | *Nitrosopira* | *briensis* | Soil |
| | *Nitrosococcus* | *nitrosus* | Soil |
| | *Nitrosococcus* | *oceanus* | Marine |
| | *Nitrosococcus* | *mobilis* | Marine |
| Nitrite to nitrate | *Nitrobacter* | *winogradskyi* | Soil, water |
| | *Nitrospina* | *gracilis* | Marine |
| | *Nitrococcus* | *mobilis* | Marine |

driving force for $CO_2$ reduction is the production of adenosine triphosphate during the oxidation of $NH_4^+$ to $NO_2^-$ or $NO_2^-$ to $NO_3^-$. The $NH_4^+$ oxidizers use $NH_4^+$ as an oxidizable substrate in an energy-yielding reaction to generate $NO_2^-$:

$$NH_4^+ + 1.5O_2 \rightarrow NO_2^- + H_2O + 2H^+$$

The $NO_2^-$ formed is oxidized further by $NO_2^-$ oxidizers in another energy-yielding reaction:

$$NO_2^- + 0.5O_2 \rightarrow NO_3^-$$

Both types of organisms occur together in soils so that $NO_2^-$ rarely appears in any quantity except under conditions of low pH, where $NO_2^-$ oxidizers are inhibited.

The biochemical pathway of nitrification is shown in Fig. 1. Both $N_2O$ and NO can be produced during the process. Conversion of $NH_4^+$ to $NO_2^-$ occurs through a complex multistep route in which $N_2O$ is formed by a separate reduction side reaction when the $NH_4^+$-oxidizing bacteria use $NO_2^-$ as an electron acceptor under conditions where $O_2$ is limiting. The ability of nitrifying organisms to reduce $NO_2^-$ to $N_2O$ is probably a mechanism to avoid the accumulation of potentially toxic levels of $NO_2^-$ in the immediate environment. The ratio of $N_2O : NO_3^-$ produced during nitrification is increased with decreasing $O_2$ concentrations. The ratio also increases with increasing acidity, although the mechanism responsible is as yet unknown. Although increasing acidity and decreasing $O_2$ availability increase the relative proportion of $N_2O$ produced, these factors also tend to retard rates of $NH_4^+$ oxidation so the net effect on $N_2O$ production is unclear. The ratio of $N_2O : NO_3^-$ produced via nitrification is usually below 1 : 100.

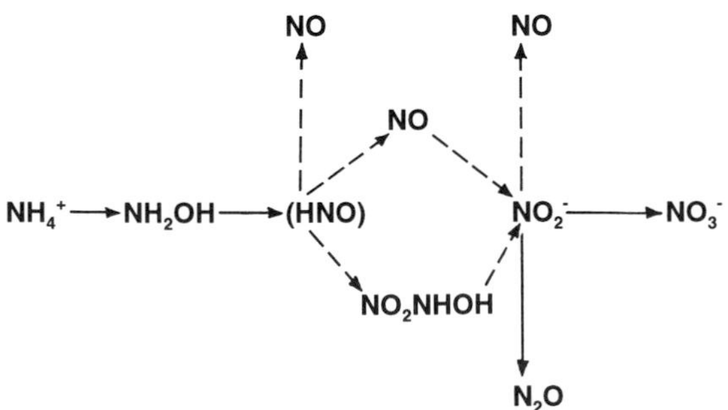


***FIGURE 1*** Pathways of chemoautotrophic nitrification. Unconfirmed pathways are indicated by dashed lines.

## V. OTHER BIOLOGICAL PROCESSES

In addition to denitrification and autotrophic nitrification, at least three other biological processes are known to yield $N_2O$ as a transient by-product. Nondenitrifying fungi and bacteria are able to respire $NO_3^-$ anaerobically as far as $NO_2^-$, but when growing fermentatively they can further dissimilate $NO_2^-$ to $NH_4^+$. Nitrous oxide is produced as a minor product. This pathway may well be a contributing source of $N_2O$ from systems that suffer prolonged anaerobic periods (e.g., in sediments and rice paddy fields). It could also be an important process in acidic coniferous forest soils, where fungal activity predominates.

Assimilatory $NO_3^-$ reduction is a second process whereby $N_2O$ can be evolved. When the microbial biomass is using $NO_3^-$ as its N source it must reduce the $NO_3^-$ to $NH_4^+$ before N assimilation can take place. During this reduction process, which occurs under aerobic conditions, $N_2O$ is produced as a by-product. In addition, certain assimilatory nitrate-reducing yeasts have been shown to be able to produce $N_2O$.

## VI. NONBIOLOGICAL PROCESSES

Chemodenitrification is a term commonly used to describe various chemical reactions of $NO_2^-$ ions within soils that result in the emission of a variety of nitrogenous gases (e.g., $N_2$, NO, $NO_2$, and $N_2$). As an $N_2$-producing process, it gains in importance whenever $NO_2^-$ accumulates in soil. Conditions that favor $NO_2^-$ accumulation in soils include an alkaline pH, where nitrification of $NO_2^-$ to $NO_3^-$ is inhibited, and acidic conditions, where nitrous acid ($HNO_2$) can form more readily.

Under acidic conditions (pH < 4.9) and a redox potential of 0 to 200 mV, $HNO_2$ dismutates chemically to form the nitrogenous gases $NO_2$ and/or NO:

$$3HNO_2 \rightarrow 2NO + HNO_3 + H_2O$$

or

$$HNO_2 \rightarrow NO + NO_2 + H_2O$$

The NO and $NO_2$ produced during these processes can be further reduced chemically by organic constituents to $N_2$ and $N_2O$. With increasing pH, the $HNO_2$ levels decline and $N_2O$ production through $HNO_2$ dismutation also declines.

Chemical reactions involving hydroxylamine ($NH_2OH$), an intermediate formed during nitrification, may also produce $N_2O$. The chemical reaction between $NH_2OH$ and $NO_2^-$ may be responsible for some production of $N_2O$ in well aerated as well as anaerobic soils:

$$NH_2 + NO_2 \rightarrow N_2O + 2H_2O$$

Nitrous oxide may also be formed by decomposition of organic matter oxides with nitrous acid:

$$R_2C = NOH + HNO_2 \rightarrow R_2C = O + N_2O + H_2O$$

The reaction between $HNO_2$ and phenolic constituents and with compounds containing free amino groups may also be responsible for some $N_2O$ emissions from soils.

## VII. FACTORS CONTROLLING $N_2O$ FLUXES

### A. Aeration and Moisture

The activity and synthesis of all the N oxide reductase enzyme systems involved in denitrification are repressed by $O_2$. Thus, factors that reduce the $O_2$ partial pressures in soils are positively correlated with $N_2O$ production by biological denitrification. Soil moisture is an important factor, because with increasing moisture content, air-filled pores become water-filled and denitrifying organisms switch from $O_2$ to $NO_3$ (or another N oxide) as a terminal electron acceptor. The latter enzymes in the denitrifying sequence are more sensitive to $O_2$ than are the earlier reductases. As a consequence, the $N_2O : N_2$ ratio in the product gases tends to be higher in the early stages of anaerobiosis but declines rapidly if anaerobiosis is protracted. Following rainfall or irrigation, brief pulses of $N_2O$ emissions are commonly recorded that are usually attributed to denitrification.

Nitrous oxide production through autotrophic nitrification is also stimulated by moisture addition at low to moderate soil moisture levels where the soil remains aerobic. However, denitrification can also occur in well-structured aerobic soils due to the occurrence of anaerobic microsites. Anaerobic pockets in soils may often be localized areas of intense respiratory activity where $O_2$ demand exceeds supply.

Though denitrification is favored in flooded soils, the highly anaerobic conditions strongly favor the $N_2$ rather than $N_2O$ as an end product. Thus flooded soils generally contribute less $N_2O$ to the atmosphere than drained soils.

An oxygen concentration of 0.1 ml/liter is generally considered the upper limit in which denitrification can occur in seawater. Anoxic microsites may be particularly important sites of denitrification in waters with oxygen contents of 0.1–0.2 ml/liter. Below an oxygen concentration of about 0.3 ml/liter nitrification begins to be limited in seawater and the $N_2O : N_2$ ratio emitted increases. At oxygen concentrations of 0.1 to 0.3 ml/liter nitrous oxide can apparently be produced simultaneously in marine environments by nitrification and denitrification.

### B. Organic Matter Levels

The availability of organic matter is an important factor moderating both the rate and extent of denitrification. The dependence on C availability results in higher denitrifying potentials being generally found in surface soils than in subsoils. Denitrification rates are usually positively correlated with the water-soluble C content of soils. Large fluxes of $N_2O$ are known to originate from organic peat soils. Additions of organic materials (e.g., crop residues, organic manures, waste effluents, and sewage sludge) generally increase total denitrification and $N_2O$ losses from soils. The ratio of $N_2O:N_2$ emitted may, however, tend to be reduced by addition of organic materials. With increasing available C there is generally a more complete reduction of $NO_3^-$ and therefore less $N_2O$ production relative to that of $N_2$. [*See* Global Carbon Cycle.]

Large concentrations of organic C are rare in seawater and organic C content has been implicated in controlling denitrification in seawater.

### C. Nitrifiable Nitrate-N

In soils, the availability of $NO_3^-$ at the sites of denitrification can be an important limiting factor so that denitrification can be dependent on $NO_3^-$ concentrations up to about 40–100 $\mu$g N/g soil. Above these levels, denitrification is often not affected by $NO_3^-$ concentration.

Soil $NO_3^-$ concentrations can also affect the ratio of $N_2O:N_2$ in the product gases. If the availability of oxidant (i.e., $NO_3^-$ or $NO_2^-$) greatly exceeds the availability of reductant (i.e., organic C), then the oxidant may be completely used and $N_2O$ is likely to be produced. Thus, conditions of low C availability and high $NO_3^-$ or $NO_2^-$ availability favor a high $N_2O:N_2$ emission ratio.

The rate of $N_2O$ loss via nitrification will be influenced by the amount of $NH_4^+$ rather than $NO_3^-$ containing fertilizers. In some studies, $N_2O$ emissions have been found to be greater from $NH_4^+$ rather than $NO_3^-$ containing fertilizers, suggesting that nitrification can be the major source of $N_2O$ in some situations. Nitrate concentrations are low in most marine environments (i.e., 0–50 $\mu M$) and consequently the diffusion of $NO_3^-$ to the cell surfaces of denitrifiers usually regulates the denitrification rate.

### D. Temperature

The optimum temperature for denitrification is about 25°C and there is a rapid decrease in the rate below 5°C. Denitrification will proceed up to about 65°C and the minimum temperature for the process may vary from −2 to 10°C. For nitrification the optimum temperature is between 30 and 35°C; below 5°C and above about 40°C the activity is low. Under field conditions, $N_2O$ release rates can sometimes show a diurnal pattern that closely follows the diurnal fluctuations in temperature.

As temperatures decline below about 15°C, the $N_2O:N_2$ ratio of the denitrification products increases. However, this effect seems to have little effect on seasonal $N_2O$ fluxes because the effect of seasonally varying temperatures on denitrification is often confounded by other factors such as variations in soil moisture content and simultaneous generation of $N_2O$ via nitrificaton.

The temperature in most denitrification zones in oceans range from 0 to 15°C, which is a somewhat narrower temperature range than that found in most soils. Seasonal temperature changes in most marine environments are small and temperature therefore probably has little influence on seasonal rates of $N_2O$ production.

### E. pH

Optimum pH for both biological denitrification and autotrophic nitrification is in the range 7.0 to 8.0. The $N_2O$ reductase enzyme system is more sensitive than the other reductase systems to low pH so that the $N_2O:N_2$ ratio emitted increases as pH decreases. At pH 4.0, $N_2O$ may be the major product of denitrification.

There are few areas in the ocean where pH values of less than 7.3 are encountered, thus $N_2$ is favored as the major end product of denitrification in seawater.

## VIII. STRATOSPHERIC SINK

A major concern regarding increased $N_2O$ emissions is that the rate of reactions in the stratosphere that lead to the destruction of the ozone ($O_3$) layer may be increased. The stratosphere ozone layer shields the biosphere from harmful UV radiation and also influences the vertical temperature profile in the atmosphere and thus earth surface temperatures. [*See* Atmospheric Ozone and the Biological Impact of Solar Ultraviolet Radiation.]

Nitrous oxide has relatively low water solubility and is therefore not removed to any marked extent by tropospheric precipitation. It is also essentially inert with the troposphere and is able to move relatively unimpeded into the stratosphere. Atmospheric destruction of $N_2O$ occurs through photochemical decomposition reactions in the stratosphere:

$$N_2O + h\nu \rightarrow N_2 + O \quad (1)$$
$$N_2O + O('D) \rightarrow N_2 + O_2 \quad (2)$$
$$N_2O + O('D) \rightarrow 2NO \quad (3)$$

The electronically excited O('D) atom is produced by photolysis of stratospheric ozone. Approximately 10% of the stratospheric $N_2O$ is thought to be converted to NO by reaction (3), whereupon the NO formed can catalyze stratospheric ozone destruction by the reaction pathway

$$NO + O_3 \rightarrow NO_2 + O_2$$
$$NO_2 + O \rightarrow NO + O_2$$

The net result is $O_3 + O \rightarrow 2O_2$. Stratospheric destruction of $N_2O$ has been estimated at 8.4–9.4 Tg N/year, corresponding to an average residence time of about 170 years. The relatively long lifetime of $N_2O$ means that changes in its atmospheric concentration have a long-term effect.

## Glossary

**Biological denitrification** Process in which gaseous nitrogenous products ($N_2$ and $N_2O$) are formed from $NO_3^-$ and/or $NO_2^-$ through the actions of certain heterotrophic bacteria, which under anaerobic conditions use $NO_3^-$ or $NO_2^-$ as a terminal electron acceptor for respiratory metabolism.

**Chemoautotrophic nitrification** Oxidation of ammonium to nitrite and thence nitrate mediated by aerobic bacteria that derive their energy solely through this oxidation and their C by assimilation of $CO_2$.

**Ozone** Molecule containing three oxygen atoms ($O_3$) that is found in trace quantities in the atmosphere and that has the ability to absorb ultraviolet radiation with wavelengths between 240 and 320 nm.

**Ozone layer** Layer in the lower stratosphere between altitudes of 15 and 30km, where ozone is at its highest concentration.

**Stratosphere** Layer of the atmosphere that extends from about 10–16 km to about 50 km.

**Troposphere** Lowest layer of the atmosphere extending from the earth's surface to a height of about 10–16 km.

## Bibliography

Aulakh, M. S., Doran, J. W., and Mosier, A. R. (1992). Soil denitrification—Significance, measurement, and effects of management. *Adv. Soil Sci.* **18,** 1–57.

Bolle, H. J., Seiler, W., and Bolin, B. (1986). Other greenhouse gases and aerosols. *In* "The Greenhouse Effect, Climatic Change, and Ecosystems" (B. Bolin, B. R. Döös, J. Jäger, and R. A. Warrick, eds.), Scope 29, pp. 157–204. Chichester, England: Wiley & Sons.

Bouwman, A. F. (1990). Exchange of greenhouse gases between terrestrial ecosystems and the atmosphere. *In* "Soils and the Greenhouse Effect" (A. F. Bouwman, ed.), pp. 61–127. Chichester, England: Wiley & Sons.

Eicher, M. J. (1990). Nitrous oxide emissions from fertilized soils: Summary of available data. *J. Environ. Quality* **19,** 272–280.

Goering, J. J. (1985). Marine denitrificaton. *In* "Denitrification in the Nitrogen Cycle" (H. L. Golterman, ed.), pp. 191–224. New York: Plenum Press.

Jones, J. G. (1985). Denitrification in freshwaters. *In* "Denitrification of the Nitrogen Cycle" (H. L. Golterman, ed.), pp. 225–239. New York: Plenum Press.

Smith, K. A., and Arah, J. R. M. (1990). Losses of nitrogen by denitrification and emissions of nitrogen oxides from soils. *Proc. Fertilizer Society,* No. 299.

Watson, R., Rodhe, H., Oeschger, H., and Siegenthaler, U. (1990). Greenhouse gases and aerosols. *In* "Climatic Change: The IPCC Scientific Assessment" (J. T. Houghton, G. J. Jenkins, and J. J. Ephraums, eds.), pp. 7–40. Cambridge, England: Cambridge University Press.

# Northern Polar Ecosystems

**L. C. Bliss**
*University of Washington*



## I. INTRODUCTION

Northern polar regions, the Arctic, are defined in different ways. Geographers and astronomers often consider the Arctic Circle as the dividing line, whereas most biologists use the combination of tree line and the 10°C mean July isotherm as the southern limit (Fig. 1). The circumpolar Arctic covers about 5% of the earth's surface—roughly 360,000 km$^2$ in Alaska, 2,480,000 km$^2$ in Canada, 2,167,000 km$^2$ in Greenland and Iceland, and 2,560,000 km$^2$ in Russia and Scandinavia. Over these vast areas there are considerable differences in climate, soils, ice cover, topography, and variable sizes of floras (cryptogam and vascular), and faunas (invertebrates to mammals). There are often small pockets of trees extending beyond tree line along river terraces and south-facing slopes. Throughout these cold-dominated lands the soils are permanently frozen (permafrost) to a depth of 100–650+ m. Only the upper portion (20–60 cm, but 100–300 cm along rivers and steep slopes) thaws in summer, and it is the active layer.

The circumpolar Arctic is divided into the Low Arctic and High Arctic (Fig. 1) based on ecological characteristics (Table I). Tundras dominate the Low Arctic, with polar semideserts and polar deserts dominating the High Arctic. Tundra is a generic term that includes vegetation types ranging from tall shrubs (2–5 m) to dwarf shrubs (5–50 cm), graminoids, and mosses. In most sites there is a total plant cover of 80–100%. Woody plants are an important component in all but the wettest tundra sites. Species richness of birds, fishes, and mammals is much greater in the Low Arctic than farther north (Table I). The semideserts of the High Arctic are dominated by mosses and lichens rather than vascular plants—the latter accounting for only 5–25% of cover. The number of species of birds and mammals is much less than in the Low Arctic. The polar deserts cover vast areas with typically a total plant cover of only 1–3%. Woody or semi-woody species (*Dryas*) are not a part of these landscapes, but are present in the semidesert landscapes.

## II. PHYSICAL ENVIRONMENT

### A. Climate

The Arctic is erroneously considered a region of perpetual ice and snow, for the summers are snow-free but short (1.5–4 months) while the winters are long and snow covered. These northern lands are characterized by continuous light in summer and darkness in winter. The limited number of arctic

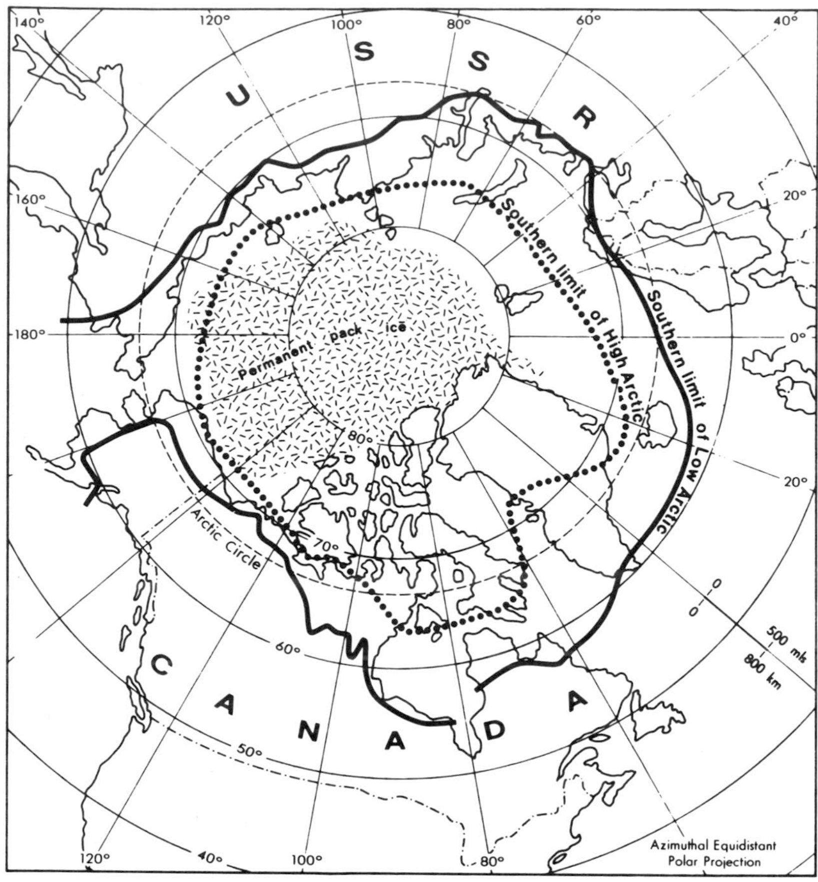


***FIGURE 1*** Circumpolar arctic showing the southern boundaries of the Low and High Arctic and the area of continuous permafrost. [From L. C. Bliss (1979). *Can. J. Bot.* **57,** 2167–2178.]

weather stations and the often short duration of their records have hindered our understanding of the dynamics of arctic climates. The use of radiosonde data and satellite imagery since the 1950s has greatly facilitated our knowledge of these lands.

In summer, arctic air masses seldom extend south of 50–60°N and generally lie near the arctic tundra–forest transition. In winter, with continuous snow cover across Canada, Scandinavia, Russia, the northern United States, and China, and into the southern republics of the former Soviet Union, modified arctic air often extends to low latitudes (25–30°N).

Because the Arctic has been defined as that region north of the climatic tree line, this poses the question: what sets the northern limits for tree growth? For North America, the accepted understanding is that tree line is determined by the relative position of the Arctic Front (leading edge of polar air masses) in summer and the magnitude of the mean annual net radiation. Net radiation ($R_n$) represents the energy available at ground level and on plant and animal surfaces for heat-transfer processes:

$$R_n = H + LE + G + P,$$

where H = sensible heat flux, LE = latent heat flux (evaporation and transpiration), G = soil heat flux, and P = photosynthesis (a very minor component). Net radiation generally accounts for 45–50% of global (shortwave) incoming radiation. Net radiation averages about 650–800 MJ $m^{-2}/yr^{-1}$ across Alaska and Canada at tree line. It is assumed that similar relationships occur across Russia in controlling tree line. In the Arctic, a major portion

***TABLE I***

**Physical and Biological Characteristics of the Low and High Arctic of North America**

| Characteristics | Low Arctic | High Arctic |
|---|---|---|
| Environmental | | |
| Length of growing season (mo.) | 3–4 | 1.5–2.5 |
| Mean July temperature (°C) | 4–12 | 2–8 |
| Mean January temperature (°C) | −20 to −30 | −25 to −35 |
| Accumulated degree days above (°C) | 600 to 1400 | 150 to 600 |
| Mean precipitation June–August (mm) | 35–200 | 25–100 |
| Mean annual precipitation (mm) | 120–800 | 60–500 |
| Active layer depth (cm) | | |
| fine-textured soils | 20–60 | 20–60 |
| coarse-textured soils | 100–200+ | 70–150 |
| soil temperature at −10 cm (°C) | 5–12 | 2–8 |
| Soil pH | Mostly 5–6.5 | Mostly 6–8 |
| Organic layer (cm) | | |
| lowlands | 50–300+ | 5–50 |
| uplands | 2–20 | 0–2 |
| Biological | | |
| Total plant cover (%) | | |
| tundra | 80–100 | 80–100 |
| polar semidesert | 20–80 | 20–80 |
| polar desert | 1–5 | 1–5 |
| Vascular plant flora (# species) | 700–800 | 350–400 |
| Vascular plant growth forms | Graminoid and low woody shrubs are common | Cushion, rosette, and graminoid are common |
| Large land mammals (# species) | 4–8 | 1–6 |
| Small land mammals (# species) | 15–30 | 5–12 |
| Nesting birds (# species) | 30–100 | 2–25 |
| Freshwater fishes (# species) | 10–25 | 1–9 |

of net radiation is utilized in thawing snow and ice each spring.

Polar high-pressure systems form over northern Canada and Russia in winter, resulting in outward flow of cold, dry air (anticyclones). Low-pressure systems (cyclones) that form over the Gulf of Alaska, the Norwegian Sea, and the Barents Sea seldom penetrate farther north in winter. In summer the Arctic has a more uniform sea-level pressure system with a weaker zonal flow, and cyclonic systems are able to penetrate the Arctic with increased precipitation. These summer storms are more frequent between Alaska and eastern Siberia, over northern Baffin Island and western Greenland, and in the European Russian Arctic. Inversions (warm air aloft and cold air at the surface) occur on a large scale over the Polar Ocean, especially in winter.

Mountain systems play a major role in controlling weather systems on a continental basis. The Aleutian low-pressure systems are cut off from the Polar Ocean by mountain ranges in Alaska and eastern Siberia. The cyclonic storms that reach Greenland are diverted northward along the west coast. The Icelandic lows often reach the Polar Ocean because there are no barriers, whereas the cyclonic systems that reach the European Russian Arctic are often blocked by the Polar Ural Mountains. The intense polar air masses swing south along the east side of the Rocky Mountains, reducing their impact on the Pacific Northwest Coast, and the Ural Mountains in Russia partially block cold, dry air from penetrating into western Russia.

Another important aspect of arctic climatology is the impact of downslope winds (katabatic winds) from adjacent uplands. These winds in the eastern Canadian Arctic, Greenland, and the Polar Urals provide both heating as the air settles and contracts and strong winds that may average 15–25 m sec$^{-1}$ for several hours. If these winds occur in late spring, they can remove 10–30 cm of snow in a few hours, with important ecological impacts, for example, exposing microtines to a cold, bare surface with inadequate covers before their summer burrows have thawed.

The Canadian Arctic is a huge region with the Arctic Archipelago contributing 58% of the land area. In summer a shallow low-pressure system frequently develops over northern Quebec and southern Baffin Island, resulting in a high degree of cyclonic activity and cloudy, mild, wet weather. The northwestern arctic islands are dominated by cold, dry air masses most of the time, resulting in low stratus clouds but limited precipitation. Mean annual temperature in the northwestern islands averages −16° to −18°C and precipitation is very low (<100–150 mm $yr^{-1}$). Snow cover lasts from late August through mid-June. The northeastern islands (Axel Heiberg and Ellesmere) are characterized by high mountains, large ice caps with glaciers, and again low precipitation (100–150 mm) with many clear days, the result of the Greenland high-pressure system. Mean annual temperature averages −18° to −19°C. The east central and southeastern islands are dominated by an arctic maritime climate with cold winters, cool summers, and high levels of coastal fog. Precipitation averages 300–600 mm over much of the mountainous terrain and coastal areas; mean annual temperature averages −13° to 15°C. The arctic mainland of Canada has a continental climate, dominated by cold polar air masses in winter but with warmer summers. Mean annual temperatures are not as low (−9° to −12°C), but the annual range is greater. Snow cover is present September through early June.

In arctic Alaska, the Brooks Range effectively blocks some of the coldest arctic air in winter and the cyclonic storms that develop over the Gulf of Alaska and the Bering Sea bring more winter precipitation to the west coast. Spring comes earlier to western and northern arctic Alaska because of the penetration of mild Pacific air. Springtime temperatures are often 15°C higher than in the Canadian Arctic. Along the northern coast of Alaska, mean annual temperature averages −12°C and precipitation averages 100–180 mm, whereas inland at Umiat mean annual temperature is −12°C and precipitation averages 150 mm. Climatic data for representative Low and High Arctic stations are presented in Table II.

### B. Soils

Arctic soils are generally less well developed than in temperate regions because of low temperatures and the youthfulness of these lands since deglaciation. Early in the development of arctic soil science it was believed that the processes of soil formation were very different from those of temperate regions. It is now recognized that the same processes occur in the Arctic as elsewhere, but at reduced rates. The process of *Podzolization* occurs in the Low Arctic, where there are lichen–heath plant communities and where acid parent material predominates. Organic acids are effective in mobilization and transport of Fe and Al into the B horizon. A weakly leached E horizon and a dark humus, iron-rich B horizon develop. [*See* SOIL ECOSYSTEMS.]

*Brunification,* resulting in Arctic Brown soils, is found in Alaska, the Canadian High Arctic, and the mountains of Siberia, where there are lichen–dryas rather than lichen–heath communities. This soil-forming process is based on the bonding of iron, clay, and humus within the upper mineral soil. Plants that produce a mull type of humus, which is organic material incorporated with mineral matter and a parent material rich in iron and clay, seem necessary and thus these soils are usually found associated with igneous rocks. They can occur on carbonate parent materials provided decarbonation occurs.

*Melanization* is a soil-forming process in which soils develop from carbonates that are high in silicates and where the vegetation is herbaceous, including dryas. Rendzina soils with a mollic horizon are found in the Brooks Range, on pingos, and high-center polygons in northern Alaska and in the Hudson Bay area of Québec, Canada.

Soils formed under the process of *Gleization* are restricted to poorly drained sites of slopes and level uplands. Vegetation is often dominated by cottongrass tussock tundra with abundant heath

**TABLE II**
Climatic Data for Selected Stations in the Low and High Arctic[a]

| Station (latitude) | Temperature (°C) Mean monthly July | Aug. | Mean annual | Total degree days >0°C | Precipitation (mm) Mean monthly July | Aug. | Annual |
|---|---|---|---|---|---|---|---|
| Low Arctic | | | | | | | |
| Baker Lake, N.W.T. (65°) | 10.8 | 9.8 | −12.2 | 1251 | 36 | 34 | 213 |
| Umiat, Alaska (69°) | 11.7 | 7.3 | −11.7 | 993 | 24 | 20 | 119 |
| Godthaab, Greenland (64°) | 7.6 | 6.9 | −0.7 | 809 | 59 | 69 | 515 |
| Umanak, Greenland (71°) | 7.8 | 7.0 | −4.0 | 682 | 12 | 12 | 201 |
| Khatanga, Russia (72°) | 12.2 | 8.9 | −12.8 | 838 | 40 | 45 | 228 |
| Ustye, Russia (76°) | 4.4 | 2.8 | −13.3 | 244 | 38 | 50 | 162 |
| High Arctic | | | | | | | |
| Barrow, Alaska (71°) | 3.8 | 2.2 | −12.1 | 288 | 22 | 20 | 100 |
| Resolute, N.W.T. (75°) | 4.3 | 2.8 | −16.4 | 222 | 26 | 30 | 136 |
| Eureka, N.W.T. (80°) | 5.5 | 3.6 | −19.4 | 318 | 13 | 9 | 58 |
| Scoresbysund, Greenland (70°) | 4.7 | 3.7 | −6.7 | 333 | 38 | 33 | 428 |
| Nord, Greenland (81°) | 4.2 | 1.6 | −11.1 | 202 | 12 | 19 | 204 |
| Cape Zhelaniya, Russia (76°) | 1.7 | 2.2 | −8.3 | 134 | 25 | 38 | 128 |
| Cape Chelyuskin, Russia (78°) | 1.7 | 1.1 | −13.9 | 117 | 28 | 28 | 113 |

[a] Data from various sources; generally 10-yr averages.

shrubs, or by wetland sedges and grasses. These soils are common in the Low Arctic of Alaska, Canada, and Russia, but also occur in wetlands of the High Arctic. A thick organic horizon often occurs at the surface, the result of decreased decomposition; the B horizon is often dull-colored or mottled with Fe and Al.

Organic soils, formed under the process of *Paludification,* occur where the soils are constantly wet. Peat accumulation by the abundant growth of sedges and mosses results in soils that are usually acid (pH 4.5–6.0). These soils are common on the Coastal Plain Province of northern Alaska, the Yukon Territory, and throughout the Northwest Territories, though less common in the High Arctic. These soils are also common in the lowlands of arctic Russia.

*Calcification* is a common soil-forming process in the polar deserts of the High Arctic in Canada, northern Greenland, Svalbard, and the arctic archipelagos of northern Russia, where limestones and dolomites predominate. These soils develop where there is little if any plant cover, and consequently there is little incorporated organic matter, and always low levels of N and P. A schematic diagram of the relative importance of these soil-forming processes is given in Fig. 2. Table III presents chemical characteristics of major soils within the Lowand High Arctic, and Figs. 3 and

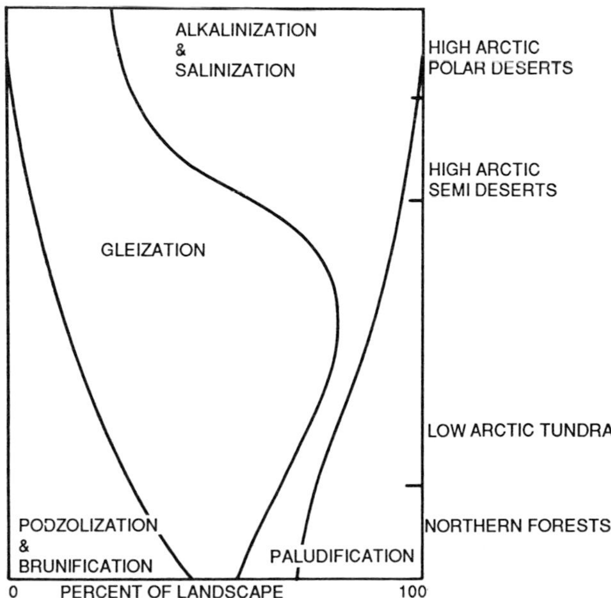


**FIGURE 2** Generalized diagram of the relative importance of different soil-forming processes within the Arctic.

**TABLE III**
Chemical Characteristics of Major Genetic Soils within the Arctic

| Horizon | Depth (cm) | pH | Exchangeable cations (meq $kg^{-1}$ soil) Na | K | Ca | Mg | N (%) | Organic matter (%) |
|---|---|---|---|---|---|---|---|---|
| Arctic Brown (Pergelic Cryochrept), N. Alaska | | | | | | | | |
| A1 | 0–20 | 4.3 | 2.6 | 2.0 | 11 | 7.0 | 0.4 | 11.5 |
| BW1 | 20–46 | 6.1 | 1.5 | 2.0 | 19 | 0.0 | 0.0 | 0.7 |
| BW2 | 46–69 | 7.2 | 1.5 | 2.0 | — | — | 0.0 | 0.4 |
| C | 69+ | 8.2 | 0.3 | 1.0 | — | — | 0.0 | 0.6 |
| Tundra (Pergelic Cryaquept), N. Alaska | | | | | | | | |
| A1 | 0–10 | 4.6 | 9.0 | 1.0 | 31 | — | — | 7.1 |
| E | 10–30 | 4.6 | 9.0 | 1.0 | 47 | — | — | 10.0 |
| Bg | 30–76 | 5.2 | 7.0 | 2.0 | 272 | — | — | 32.0 |
| C | 76–114 | 6.4 | 6.0 | 2.0 | 156 | — | — | 5.7 |
| Arctic Brown (Pergelic Cryochrept), Truelove Lowland | | | | | | | | |
| Ak | 0–8 | 7.1 | 1.0 | 1.0 | 590 | 190 | 0.81 | 8.4 |
| Bwk | 8–23 | 7.7 | | | | | 0.10 | 1.4 |
| Ck | 23–61 | 7.8 | | | | | 0.20 | 0.0 |
| Tundra (Pergelic Cryaquept), Truelove Lowland | | | | | | | | |
| Oi | 0–23 | 6.2 | 1.0 | 2.0 | 490 | 150 | 2.24 | 26.1 |
| Cgk | 23–28 | 7.6 | — | — | — | — | 0.04 | 0.0 |

4 present diagrams of soil profiles in relation to landscapes.

### C. Permafrost and Patterned Ground

One of the most important features of the Arctic is the presence of permanently frozen ground (permafrost) and the resultant conspicuous features called patterned ground. Permafrost describes soil, rock, and seafloor conditions where the temperature remains below 0°C for two or more years. Although ice is generally an important component of frozen ground, dry permafrost occurs in coarse gravels and frozen rock where there is little or no

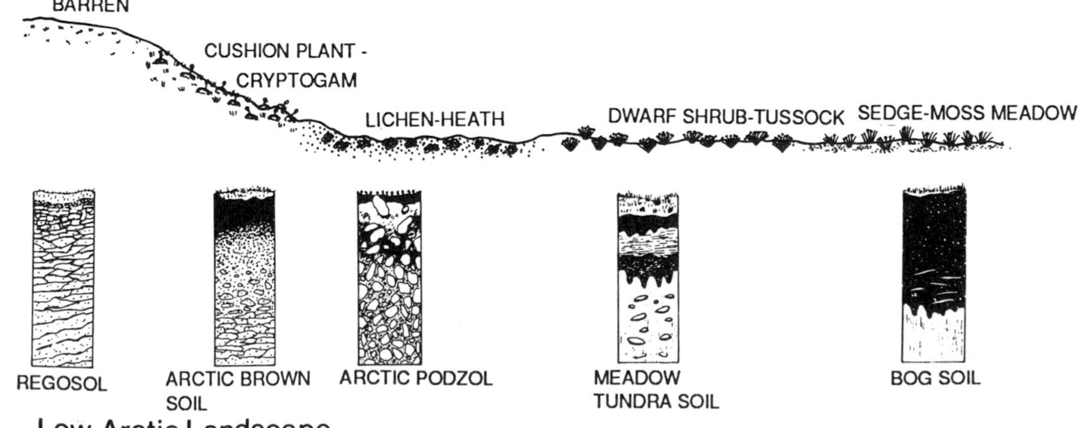


**FIGURE 3** Generalized diagram of soil profiles on a landscape basis in the Low Arctic. [Based on J. C. F. Tedrow (1977). "Soils of the Polar Landscape." Rutgers Univ. Press, New Brunswick, NJ; F. C. Ugolini (1986). *Quart. Res.* **26,** 100–120.]

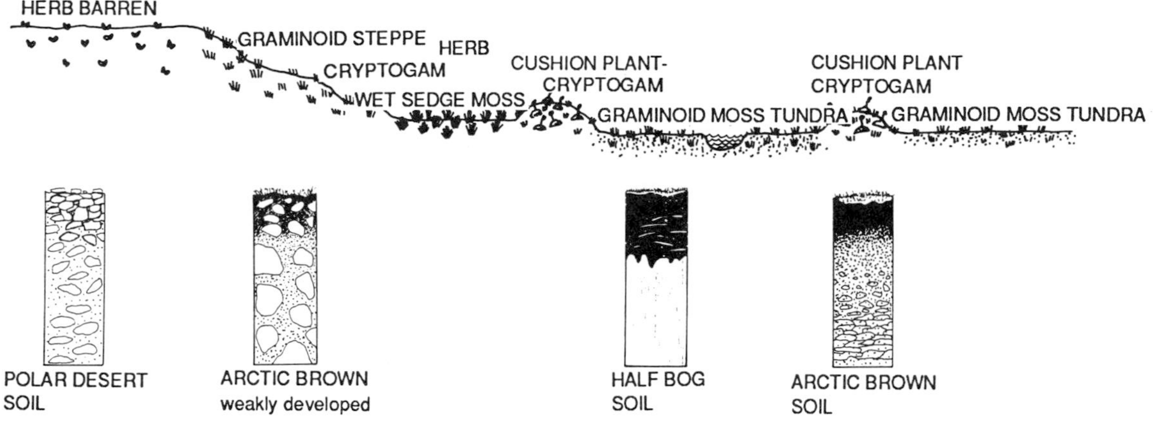


***FIGURE 4*** Generalized diagram of soil profiles on a landscape basis in the High Arctic. [Based on J. C. F. Tedrow (1977). "Soils of the Polar Landscape." Rutgers Univ. Press, New Brunswick, NJ; B. D. Walker and T. W. Peters (1977). "Truelove Lowland, Devon Island, Canada: A High Arctic Ecosystem." Univ. Alberta Press, Edmonton, Canada; F. C. Ugolini (1986). *Quart. Res.* **26,** 100–120.]

ice. The amount of ice decreases with depth except where massive ice lenses and ice wedges occur (Figs. 5 and 6). In both Canada and Russia, continuous and discontinuous permafrost underlies about one-half of the total land. About 20% of the world's land is underlain with permafrost.

Continuous permafrost underlies the Arctic and the northern portion of the taiga (boreal forest), whereas discontinuous permafrost is confined to the boreal forest. There is no permafrost under large lakes and rivers or in deeper ocean waters. Smaller lakes have a thawed basin underneath with permafrost below. Permafrost depends on mean annual temperature, soil and rock thermal conductivity, proximity to the sea, and topographic position. Permafrost thickness is about 320 m at Barrow, Alaska, 500–600 m in the Mackenzie River Delta region, and 400–600 m on Devon Island in the High Arctic. Mean annual temperature has risen 2°C or more near the top of the permafrost table in at least some parts of the Arctic in the last 30–100 yr, a possible indication of global warming.

In summer, arctic soils thaw to a depth of 20–60 cm where peats and fine-textured soils occur, and to about 100–300 cm in sands and gravels. The depth of this thawed layer or active layer depends on plant cover, soil texture, water-holding capacity, and the amount of ice present. About 60–80% of the annual thaw occurs within 3–4 weeks of

***FIGURE 5*** Massive ice lenses near Tuktoyaktuk, N.W.T., Canada.

***FIGURE 6*** Ice wedge near Tuktoyaktuk, N.W.T., Canada.

snowmelt, slowing later in the season as net radiation is reduced following the summer solstice. Refreezing occurs from the top down and the bottom upward as air temperature and net radiation drop in August and September. As a result, an unfrozen layer, under pressure at depth and often water-saturated, is squeezed to the surface, forming "frost boils."

Ice-rich permafrost often results in interesting landforms that can break the monotony of subdued landscapes. Low ice-cored mounds, often 1–2 m in height, are called *palsas*. The most spectacular features are the conical hills, with solid ice cores, called *pingos*. There are about 1450 pingos in the Mackenzie Delta region of Canada and they also occur in Russia, Greenland, Banks Island in the Canadian High Arctic, and Alaska. These hills, often 10–50 m in height, have relatively warm and dry soils and thus are preferred habitat for arctic ground squirrels, arctic foxes, and various vascular plants requiring warmer and drier soils.

Land features resulting from freeze–thaw cycles and permafrost are called patterned ground. These features, including circles, polygons, stone nets, earth hummocks, solifluction steps, and stripes, are found in both alpine and arctic environments. Circles, stripes, and polygons range in size from 10–25 cm to 5–10 m across, with a few 50–100 m in diameter. Circles and polygons occur on level areas to slopes of only 1–2°, whereas elongated polygons and circles may occur on slopes of 2–7° and solifluction steps and stripes on slopes >7°. Patterned ground features may be nonsorted (lacking a border of stones) or sorted (having a stone border). If movement of soil and rocks occurs after formation, there will be a few lichens and mosses on the surfaces, as opposed to fossil patterned ground where cryptogams and vascular plants are common on rocks and soil.

Circles or soil boils (spot medallions in the Russian literature) often have vascular plants on their margin with open soils of silts and clays in the center. Organic soils may develop along the margins where plants have helped stabilize the soils. These soils frequently contain desiccation-cracked polygons.

Ice-wedge polygons are the most common feature in the Arctic; nonsorted large polygons (>1 m) are common in lowlands. These polygons have an ice wedge coincident with their border and the border is raised or depressed with respect to the polygon center, depending on whether the ice wedge is growing or melting. Within wetlands, ice-wedge polygons have either raised or depressed centers; the latter melt each summer, forming a small pond with algae, zooplankton, and often fairy shrimp. Raised-center polygons are often quite dry in summer and often are only minimally snow covered in winter. This results in cushion plants and dwarf heath shrubs, lichens, and mosses covering the tops, with sedges and grasses in the wet troughs between adjacent polygons (Fig. 7).

Soil or earth hummocks, 20–50 cm across, are common throughout the Arctic on gentle slopes or level areas. These hummocks are usually covered with plants, with considerable amounts of organic matter incorporated in the soils.

Solifluction steps and sorted or nonsorted stripes occur on slopes of usually 7–25°. The steps often contain different species than the riser, the latter with dwarf or medium-sized shrubs and the step (tread) with cushion plants. In the High Arctic, these features may lack any plant cover. The downslope movement of saturated soil and stones over a permafrost table is called gelifluction or solifluction. The rates of movement average 1–5 cm $yr^{-1}$.

**FIGURE 7** Raised-center polygons with cushion plants on the tops and sedges in the troughs with ice wedges below the surface.

## III. VEGETATION, PLANT PRODUCTION, AND PLANT ADAPTATIONS

### A. Plant Community Patterns

As indicated earlier, the Arctic covers a huge circumpolar area, with a considerable range in climate, soils, and topography. Consequently there is a large range in plant community types, from shrublands to sedge–moss meadows or mires in tundras of the Low Arctic to polar semideserts and polar deserts in the High Arctic. Tundra as used here refers to the vegetation of the Low Arctic where plant cover is complete or nearly so and where woody species, sedges, grasses, mosses, and lichens predominate.

#### *1. Low Arctic*

#### a. Tundras

The geomorphology of the North American Arctic and that of Russia is very different. The Brooks Range in Alaska and the Richardson Mountains in the Yukon Territory limit the southern extent of the tundra. Much of the Canadian mainland Arctic was glaciated by the Wisconsinan glaciation; consequently there is limited soil and this restricts vegetation and wildlife. The rest of the North American Arctic (49%) lies within the Canadian Arctic Archipelago. The ocean between these islands creates a major barrier for plants and animals moving farther north. There are also considerable changes in the climate from the mainland to the scattered islands. The eastern islands are mountainous (200–2000 m) and the western islands have low relief. In contrast, most of the Russian Arctic is continental, with less than 10% within small groups of islands. The Taimyr Peninsula is a huge landmass that extends to 78°N, broken only by the Byrranga Mountains, reaching elevations of 200–1000 m. These landmass differences result in gradual changes in vegetation and wildlife northward in the Taimyr compared with the mosaic pattern of biota, temperature, and precipitation in the Canadian High Arctic. [*See* TUNDRA, ARCTIC AND SUB-ARCTIC.]

***i. Shrub Tundras*** North of the forest-tundra in Alaska, Yukon Territory, and the Mackenzie River Delta, shrub communities occupy steep river and lake banks and floodplains of the larger rivers. These habitats have deeper winter snow that protects the shoots from ice and snow abrasion and desiccation. Major species include *Betula nana, B. glandulosa, Salix glauca, S. alaxensis, S. pulchra,* and *S. lanata*. An understory of dwarf shrubs of heath species—*Ledum palustre, Vaccinium uliginosum, V. vitis-idaea, Empetrum hermaphroditum, Arctostaphylos alpina, A. rubra,* and *Cassiope tetragona,* and species of *Carex* and *Eriophorum*—and an abundance of mosses and lichens provide a continuous cover.

Much of the mainland eastern and central Canadian Arctic contains little shrub tundra because of limited winter snow and strong winds. Shrub communities of *Betula nana, B. glandulosa,* and *Salix glauca* extend to 62°N on the colder eastern coast, but to 74°N along fiords of western Greenland—the result of warmer ocean waters in Davis Strait.

Shrub vegetation dominates in the southern part of the Russian Low Arctic, with taller shrubs of *Salix lanata* (1–3 m) along river and lake banks with lower-statured shrubs (0.25–0.5 m) of *Betula nana* in the west and *B. exilis* in the east, along with *Salix lanata, S. glauca, S. phylicifolia,* and *S. pulchra*. A ground cover of *Carex ensifolia,* the same dwarf heath shrubs listed for North America, and *Dryas octopetala* and *D. punctata* predominate along with abundant lichens and mosses (Fig. 8). From the Ural Mountains to the Chukotka Peninsula there are patches of the shrub *Alnus fruticosa*. In all of these tundras there is a continuous ground cover of mosses and lichens.

***ii. Sedge–Dwarf Shrub Tundras*** Communities of *Carex ensifolia* ssp. *arctisibirica,* the common dwarf heath shrubs listed the foregoing, and *Dryas punctata* dominate large areas in Siberia north of the shrub tundra. Only prostrate clumps of *Betula nana, B. exilis, S. reptans,* and *S. pulchra* occur in the interfluves with small thickets of *Salix lanata* (50–80 cm tall) along valleys. *Dryas* mats are common on fell-fields, along with herbs such as *Oxytropis middendorffii, O. nigrescens,* and several species of *Pedicularis*. There are true mesic meadows of grasses and mixed forbs on the Taimyr Peninsula;

**FIGURE 8** Shrub tundra with *Salix lanata, Betula nana, Vaccinium uliginosum, V. vitis idaea, Carex ensifolia* spp. *arctisibirica,* and *Eriophorum vaginatum* on the Yamal Peninsula, Russia. [From L. C. Bliss and N. D. Matveyeva (1992). "Physiological Ecology of Arctic Plants: Implications for Climate Change." Academic Press, NY.]

these meadows are very species rich and have no counterparts in the North American Arctic.

***iii. Tussock–Dwarf Shrub Tundras*** Tundras dominated by the tussock sedge *Eriophorum vaginatum* with *Carex bigelowii* and *C. lugens* in North America and *C. consimilis* in Siberia, along with the common heath shrubs *Vaccinium, Ledum, Empetrum,* and *Arctostaphylos,* and an abundance of mosses and lichens dominate large areas in Alaska (Fig. 9), the Yukon Territory, and across eastern Siberia on imperfectly drained soils. *Eriophorum* is much less common in the sandy soils of the Yamal and Gydan peninsulas, although all the other species are present along with scattered low shrubs of *Betula exilis, Salix pulchra,* and *S. glauca. Betula nana* and the two willow species are common in the Foothill Province of Alaska and the Yukon Territory. To the east, where the soils are thin due to Wisconsinan glaciation, these communities are less common.

**FIGURE 9** Tussock-dwarf shrub tundra in northern Alaska dominated by dwarf shrubs of various heath species, tussocks of *Eriophorum vaginatum* with *Carex lugens* and abundant lichens and mosses. [From L. C. Bliss and N. D. Matveyeva (1992). "Physiological Ecology of Arctic Plants: Implications for Climate Change." Academic Press, NY.]

***iv. Mires*** Wetland communities are limited in the mountains of northern Alaska and the Yukon Territory, but gain importance in the Foothill Province and dominate on the Coastal Plain Province (Fig. 10). Sedge-dominated mires occur in lowlands across the Canadian Shield, but in limited extent, as they are in Greenland. Across the Russian Arctic, mires are a common feature, especially in the central part of the Yamal Peninsula and in lowlands of the Yana, Indigirka, and Kolyma river basins.

Peats are often thick (1–5 m) in these poorly drained lands and the active layer is shallow (15–25 cm) in these cold and nutrient-poor soils. In Alaska and western Canada the dominant sedges are *Carex aquatilis, C. membranacea, C. chordorrhiza, C. rariflora, C. rotundata, Eriophorum angustifolium,* and *E. scheuchzeri*. The grasses *Arctophila fulva, Dupontia fisheri,* and *Arctagrostis latifolia* occur in a gradient from shallow ponds to medium-drained up-

**FIGURE 10** Sedge-dominated mire near Prudhoe Bay, Alaska. Dominant sedges include *Carex aquatilis, Eriophorum angustifolium,* and *E. scheuchzeri.*

lands. *Menyanthes trifoliata* and *Equisetum limnosum* are found in shallow pond waters (50–60 cm deep), whereas *Potentilla palustris* and *Hippuris vulgaris* are found in very shallow waters (20–30 cm). Mosses are abundant but lichens are minor in these wetlands. Depressed-center and raised-center polygons with massive ice wedges are common.

In the Siberian lowlands the dominant sedges are *Carex stans, C. chordorrhiza, C. rariflora, Eriophorum angustifolium*, and *E. medium*, along with *Caltha arctica* and *Comarum palustre*. Mosses are abundant, especially species of *Sphagnum*.

***v. Salt Marshes*** Coastal salt marshes are a minor feature of arctic vegetation, yet where sands and silts occur and shoreline ice push is minor, small marshes occur, dominated by several species of *Puccinellia, Carex, Cochlearia officinalis*, and *Stellaria humifusa*. *Puccinellia phryganodes* marshes dominate along Hudson Bay, where they are heavily grazed by lesser snow geese. These animals play an important role in nutrient release from their feces and thus help maintain the productivity of these lands, provided the goose populations are not too large.

### 2. High Arctic

#### a. Tundra

***i. Mires*** In the High Arctic, mires, which are restricted to lowlands, are structurally and floristically related to those of the Low Arctic. Less than 6% of the Canadian Arctic Archipelago consists of these communities, yet they are the major habitat of muskoxen and the breeding grounds for shorebirds and waterfowl. The dominant graminoid is *Carex stans* with lesser amounts of *C. membranacea, Eriophorum scheuchzeri, E. triste, Dupontia fisheri*, and *Alopecurus alpinus*. Mosses are also abundant, but species of *Sphagnum* are absent. With a shorter and colder growing season, peats are shallow (10–20 cm).

Similar mires occur in Siberia with the same species as in North America, but with *Carex ensifolia* and *Dryas punctata* on the raised rims of low-center polygons. Small moss–peat hummocks develop in some places with several species of *Sphagnum* and other mosses.

#### b. Polar Semideserts

***i. Cushion Plant–Cryptogam*** In the uplands and mountains of Alaska and the Yukon (Low Arctic), mats of *Dryas* with associated species of *Draba, Saxifraga, Salix arctica*, and *Carex rupestris* dominate. This same vegetation predominates in the southern Canadian arctic islands (Fig. 11) and to a lesser extent in the northern islands and the dry, continental valleys of Greenland. These communities occur on well-drained uplands that have limited snow cover (2–10 cm) in winter and a relatively deep active layer in summer (50–100 cm). Mosses and lichens are abundant.

In the Russian High Arctic, similar communities are found near the northern part of the Taimyr and Gydan peninsulas, on a narrow strip of mainland along the Arctic Ocean in eastern Siberia, and on the islands of Novaya Zemlya, Novosibirskie Ostrova, and Wrangel Island. Mats of *Dryas punctata* occur in central Siberia and *D. punctata* and *D. integrifolia* in Chukotka. Species of *Draba* and *Saxifraga* are common as in North America, along with *Salix polaris, S. arctica*, and *Luzula confusa*.

***ii. Cryptogam–Herb*** Communities dominated by bryophytes and lichens, but with scattered vascular plants of *Alopecurus alpinus, Luzula confusa*, and *L. nivalis* along with several species each of *Draba, Saxifraga, Minuartia, Papaver radicatum*, and

**FIGURE 11** Cushion plant-lichen semidesert landscape, northern Banks Island, N.W.T. Dominant species include *Dryas integrifolia, Salix arctica, Saxifraga oppositifolia*, and *Cassiope tetragona* in slight depressions.

*Cerastium alpinum*, occur on sandy to clay loam soils in the central and western Queen Elizabeth Islands (Fig. 12). The summers are exceedingly short and heavy cloud cover is common, although precipitation is minimal most summers.

Communities of similar structure and floristics occur at the tip of the Taimyr Peninsula and on some of the islands of the Franz Josef Land Archipelago and the Severnaya Zemlya Archipelago. Species of *Draba, Saxifraga, Phippsia algida, Papaver polare, Cerastium arcticum*, and *Stellaria edwardsii* are the more common species along with diverse mosses and lichens. Vascular plant cover here and in North American communities averages 10–20%, but the cover of mosses and lichens is much greater (30–50%).

#### c. Polar Deserts

In Canada, huge areas in the Queen Elizabeth Islands are nearly devoid of plants, both vascular and cryptograms. Many of these lands are in uplands, but they also occur at and near sea level in areas where soil drainage is great early in summer, where the soils are of fine texture, and where soil churning is common. Vascular plant cover averages 1–3% with very few lichens and mosses, except below snowbanks where meltwaters provide surface runoff much of the summer. Here, cryptogamic crusts of cyanobacteria, crustose lichens, and a few mosses and fungi maintain a moist surface and one with higher levels of nitrogen. In these sites more vascular plants are found, but typically fewer than 15–20 species and a cover of only 5–10% (Fig. 13). The most common plants are *Draba corymbosa, D. subcapitata, Papaver radicatum, Saxifraga oppositifolia, S. caespitosa, Puccinellia angustata, Phippsia algida*, and *Minuartia rubella*.

In Russia, polar deserts are restricted to the larger arctic islands of Novaya Zemlya, Severnaya Zemlya, Zemlya Franz-Josef, and some smaller islands such as Ostrova De Longa. Similar areas occur in northern Greenland. The species of vascular plants, lichens, and mosses are similar to those cited for northern Canada.

### B. Plant Production and Carbon Reserves

The Arctic is often viewed as a land of little biological productivity and, compared with forest and many grasslands, this is true. The previous section describes in general terms the major plant communities or vegetation types based on plant physiognomy and dominant species. However, this gives no idea of areal extent of these types: net annual production, carbon reserves, and major environmental factors that control production and carbon storage. Arctic lands that are devoid of ice total $5.60 \times 10^6$ km$^2$, with Eurasia contributing 45%,

***FIGURE 12*** Cryptogam-herb-dominated landscape, King Christian Island, N.W.T. Dominant vascular species include *Luzula confusa, Papaver radicatum, Alopecurus alpinus*, and several species each of *Draba* and *Saxifraga*.

***FIGURE 13*** Polar desert landscape on the plateau (330 m) of Devon Island. Within the cryptogamic crust are scattered plants of *Papaver radicatum, Draba corymbosa, Saxifraga oppositifolia, S. caespitosa*, and *Cerastium alpinum*. [From L. C. Bliss and N. D. Matveyeva (1992). "Physiological Ecology of Arctic Plants: Implications for Climate Change." Academic Press, NY.]

Canada 42%, Greenland and Iceland 6.5%, and Alaska 6.5% (Table IV). Of the major vegetation types, tall shrub tundra contributes the smallest area, but the largest net annual production (1000 g $m^{-2}$) and a large standing crop, yet little soil organic matter. These tallest communities are confined to river terraces, riparian zones along streams, and steep slopes where winter snows are deep, soils thaw more deeply in summer, and habitats where soil nutrients are often greater. The low shrub communities cover much larger areas, have less net annual production and standing crop, but contribute significantly larger carbon reserves (10.6%). These latter communities occupy well-drained soils with winter snow that covers the shrubs.

The tussock (cottongrass)–dwarf shrub communities are found on imperfectly drained soils of level to rolling upland terrain, habitats generally lower in soil nutrients. These communities account for 13% of total arctic carbon reserves. Mires dominated by sedges and mosses in lowlands of the Low and High Arctic total 1.012 × $10^6$ $km^2$. They occupy the largest land area and produce less net production per year, but they store large amounts of carbon in standing crop and in soil organic matter, 70.2% of the total arctic carbon reserves. These communities occupy wet lands of low nutrient reserves, for the soils are generally acid.

**TABLE IV**
Aerial Extent, Standing Crop, and Net Annual Production of Major Vegetation Types in the Arctic

| Vegetation type | Aerial extent ($X10^6$ $km^2$) | Standing crop ($Kg\ m^{-2}$) Plants | Standing crop ($Kg\ m^{-2}$) Soil OM | Net annual production ($g\ m^{-2}$) |
|---|---|---|---|---|
| Low Arctic | | | | |
| Tall shrub | 0.174 | 5.800[a] | 0.40[a] | 1000[a] |
| Low shrub | 1.282 | 3.100[a] | 1.00[a] | 375[a] |
| Tussock–dwarf shrub | 0.922 | 7.400 | 1.00[a] | 225[a] |
| Mire | 0.880 | 4.560 | 38.75 | 220 |
| Semidesert | 0.358 | 1.470 | 2.54 | 45 |
| High Arctic | | | | |
| Mire | 0.132 | 2.360 | 21.00 | 140 |
| Semidesert | 1.005 | 1.155 | 1.03 | 35 |
| Desert | 0.847 | 0.024 | 0.02 | 1 |
| Ice caps | 1.967 | | | |

[a] Denotes estimate.

Semideserts of the Alaskan and Canadian mountains and uplands of the Canadian Shield in the Low Arctic and the uplands in the arctic islands of the High Arctic are dominated by cushion plants, herbs, and cryptogams. They occupy lands with a much shorter and colder growing season. Soils are nutrient-poor and cold. These communities occupy 24% of the land, but their low net production, low standing crop, and soil organic matter result in their contribution of only 6% to the carbon reserves (Table IV). Polar deserts that occupy the most stressful and shortest growing seasons of all contribute almost no carbon reserve, for carbon reserves are only 46 g $m^{-2}$ compared with 2.19 kg $m^{-2}$ within semideserts.

The Low Arctic comprises 65% of the ice-free area but 91% of the carbon reserves aboveground and belowground in the Arctic. The High Arctic, with its low summer temperatures, short growing season, and low nutrient content of soils, makes a very small contribution to total carbon reserves. Thus, it is somewhat surprising that these lands support as much wildlife as they do.

### C. Plant Adaptations

Plants, like animals of polar regions, have received much attention in recent years because of the stressful environments in which they live. The combination of low soil and air temperatures, low nutrient availability, soil churning, considerable ranges in soil moisture, and a short growing season limit growth and sexual reproduction.

#### *1. Reproduction and Seedling Establishment*

Annuals and short-lived perennials are rare in these environments, with most species being long-lived. Grasses, sedges, and rushes that have been aged live 20–70 years, with some living 90–190 years. Basal stems of shrubs and massive root and rhizome systems of sedges may live 200–400 years. Most graminoids depend on tillering and shrub species on branch layering, yet most species also flower and fruit abundantly and produce some viable seed. However, seedlings of most species are rarely found in undisturbed vegetation.

Seeds of many species ripen in late fall and are dispersed in winter or spring over snow. Seed dormancy mechanisms include chilling, after-ripening, light requirements, and seed coat inhibition as in many temperate region species. Germination percentages are often low and the germination process is often slow, requiring 2–3 weeks for full germination. The optimum temperature for germination is generally 15–30°C, yet some seeds germinate at <10°C. These temperature requirements seem higher than what one might expect, but if germination occurs only in warmer summers, this would favor a higher percentage establishment in those rare summers.

### *2. Plant Growth*

Flower and leaf buds overwinter in an advanced stage, thus permitting rapid growth once there are a few hours per day of temperatures above freezing. Leaf expansion and shoot elongation of shrubs are generally completed within 2–3 weeks. Graminoids initiate leaves throughout the growing season, even into early autumn. An important adaptation of many arctic species is the ability to carry some green leaves over winter, which permits them to initiate photosynthesis as soon as snow melts or becomes thin.

Root growth generally begins as soon as soils warm above 2–4°C, but lateral roots grow later in summer, often after shoot growth has slowed or stopped. Some studies have shown that decreased root growth is correlated with reduced photoperiod. Maximization of shoot growth early in the growing season and root growth later reduces competition for carbohydrates and nutrients within the plant and enables both shoot and root development to occur under the most favorable environmental conditions.

Arctic plants with evergreen or wintergreen (leaves function 1+ year) leaves have lower growth rates and lower rates of photosynthesis and transpiration compared with graminoids and deciduous forbs. In general, species with relatively rapid growth rates occur on nutrient-rich sites and species with slow growth rates dominate on nutrient-poor sites. Various studies have concluded that plant growth rates and plant community net production are more closely correlated with availability of soil nutrients, soil moisture, soil temperature, and aeration than with air temperature. Much of this can be explained by the fact that evergreen- and wintergreen-leaved (shrubs and forbs) species have low rates of photosynthesis and respiration, yet can maintain relatively high tissue nutrient concentrations in sites with low available soil nutrients. Deciduous shrubs, graminoids, and forbs grow more rapidly and have higher rates of photosynthesis and respiration, but often have lower tissue nutrient concentrations on sites with higher available nutrients.

Although shoots and roots of arctic plants do grow at low temperatures (0° to 5°C), various studies have shown that the use of small greenhouses greatly enhances plant growth and flowering. Different species respond differently to temperature rise and, in time (2 to 4 years), it appears that nutrients become more limiting, unless rates of decomposition also increase. More recent research points to nutrient limitations as a major constraint to plant production. Low nutrient availability results from low air and soil temperatures, wet soils, and a short growing season, all of which impede decomposition and nutrient mineralization. Current studies under way in the Arctic using greenhouses of various sizes and kinds may provide insight into how species and plant and animal communities will respond to global temperature rise and nutrient availability, as influenced by decomposition rates, in the next century.

The rapid growth of arctic plants following snowmelt results from the translocation of root and rhizome carbohydrates and nutrient reserves into shoot primordia. Once photosynthesis is initiated in graminoid shoots, they become self-sufficient in carbohydrates to continue growth. Root systems of some graminoids appear to be long-lived and to have rather low turnover rates. Many high arctic forbs have greatly reduced root systems, yet maintain wintergreen leaves that may act as sites for limited carbohydrate storage. New shoot growth appears to be delayed in some of these species compared with the graminoids.

Lipid concentration is often quite high in arctic species, leading some researchers to postulate that high concentrations relate to membranes necessary for high metabolic capacity. Others postulate that lipids aid in the production of resins in antiherbivore defense.

### *3. Photosynthesis*

Photosynthetic rates in arctic species are quite similar to those of plants in temperate regions, although maximum rates are reached at lower temperatures (10° to 15°C), positive carbon gain in some species occurs at 0° to 4°C, photosynthesis has rather large $Q_{10}$ responses (2–3), they have relatively high mesophyll resistance ($r_m$) to compensate for lower water use efficiency, and they have rather high chlorophyll levels. All known arctic plants utilize the $C_3$ pathway. The relatively low temperature optimum for photosynthesis seems to result from high concentrations of the enzyme ribulose bisphosphate (RuBP), which permits higher photosynthetic rates at lower temperatures. Rates of respiration are also high in arctic plants, including photorespiration. Photosynthetic rates are highest in forbs and some deciduous shrubs, intermediate in graminoids, and lowest in evergreen shrubs and wintergreen-leaved rosette species.

Rates of photosynthesis are much lower in mosses and lichens than in vascular plants. The evergreen leaves of bryophyte and lichen thalli enable them to become photosynthetically active as soon as temperature, light, and moisture become favorable. Temperature optima are low and there is little resistance to water loss; thus as these plants dry, photosynthetic rates drop rapidly. Data from arctic and alpine environments show that most positive net photosynthesis in lichens occurs on days with rain or fog. Temperature acclimation for photosynthesis occurs in lichens as it does in vascular plants. In contrast with vascular plants, lichens can carry on positive rates of photosynthesis at −3° to −6°C.

### *4. Water Relations*

The Arctic is often viewed as a wet landscape in which water would not be limiting, and much of the Low Arctic fits this description. Yet even there, on sunny days with moderate wind, water stress occurs. Plants growing in medium- to well-drained habitats are subject to greater water stress. Water stress is measured as a negative water potential; the lower the water content of plant tissue, the lower or more negative is the water potential.

Plants growing in wet sedge–moss mires, the edges of ponds, and along polygon troughs generally develop leaf water potentials ($\psi_l$) of −0.2 to −0.8 MPa. Species more typical of better-drained soils of polygon tops develop $\psi_l$ of −0.8 to −1.4 MPa. These measurements suggest that water stress may limit photosynthesis by inducing stomatal closure, and thus influences where these species grow. A major reason that water stress is limited in wetland species is that soil and root temperatures influence stomatal conductance and this is influenced by root membrane permeability in cold soils.

Soil moisture conditions are somewhat different in the High Arctic, where plants maintain lower water potentials, for soil moisture is often more limiting. Many habitats have coarse-textured soils that are well drained in midsummer, although the soils are saturated at the bottom of the active layer. *Dryas integrifolia* dominates many landscapes where soil water potential ranges from −1.5 to −2.0 MPa within the rooting zone and can drop to −4.0 MPa for short periods of time. Photosynthesis dropped to near zero at leaf water potentials of −2.0 to −3.0 MPa.

In the northwestern Queen Elizabeth Islands, most summers are cloudy and moist, but short dry periods do occur in some years. In cool, wet summers, $\psi_s$ did not drop below −0.03 MPa at 0–5 cm soil depth, and not below −0.03 to −0.6 MPa at −5 to −10 cm. *Luzula confusa,* a species of moist soils, showed water stress at −0.5 MPa and photosynthesis dropped to 25% of maximum at $\psi_l$ of −0.7 MPa. Two grasses, *Phippsia algida* and *Puccinellia vaginata,* showed little drought stress when $\psi_l$ often ranged from −1.5 to −2.5 MPa.

Water relations also help explain the distribution of male and female plants of *Salix arctica.* Female plants are more common in wet habitats with their colder but more nutrient-rich soils (6 : 1), whereas male plants predominate in more xeric, lower nutrient, and higher soil temperature habitats (1.5 : 1).

In wet habitats, female plants maintain higher rates of leaf conductance and lower $\psi_l$ than male plants in these colder soils. The reverse is true in the dryer and warmer soils, where male plants maintain higher leaf conductance and lower $\psi_l$.

Within the polar deserts, many areas in the eastern islands contain silty loam soils developed from dolomite. These soils become supersaturated in spring, yet the surface soils (−1 to 2 cm) dry considerably by midsummer. Soils at depth have plenty of water and adult plants seldom if ever undergo significant water stress. However, the lack of plants in many sites appears to result from this surface drying that inhibits seedling establishment and from the abundant soil churning in autumn with refreezing. Most vascular plants are confined to snowflush sites, where there is a cryptogamic crust that holds moisture and prevents soil churning, and where cyanobacteria fix nitrogen, one of the most limiting factors to plant growth in these most extreme environments.

### 5. Nutrients and Nitrogen Fixation

As discussed earlier, arctic soils are infertile, with especially low levels of N and P. Many species have all or some of the following characteristics to compensate for these limited resources. These include: (1) low absorption rates of nutrients per plant; (2) species with high root : shoot ratio; (3) leaves evergreen or at least wintergreen; (4) low growth rates, yet shoots maintain relatively high concentrations of nutrients; (5) seasonal storage of nutrients and carbohydrates belowground; and (6) reduced rates of leaching and increased translocation from senescing leaves. Tundra species are adapted to absorb phosphorus more efficiently at low soil temperatures than are temperate species. Deciduous species accumulate biomass, N, and P in new leaves and twigs until midseason, then senesce, whereas evergreen species accumulate N and P aboveground throughout the growing season. Most evergreen species store minimal levels of nutrients and carbohydrates in stems and roots. *Carex stans* in the High Arctic reduces its total amount of N, P, and K before peak biomass is achieved. However, the green shoots of graminoids that remain over winter have high levels of nutrients and these are important for winter grazing by caribou, muskoxen, and brown lemming. There are few examples of soils with toxic levels of Al, Fe, or Mg, although high levels of S and Na occur in some locations.

Cyanobacteria are the major nitrogen fixers in arctic ecosystems, although soil bacteria and legumes contribute to total fixation in some locations. Soil moisture and temperature are major factors controlling rates of fixation, though P availability is of secondary importance. Fixation rates are highly variable in space; the highest rates occur on marine algae in brackish-water lagoons, pond margins, and wet sedge–moss mires. The lowest rates occur on well-drained beach ridges, rock outcrops, and barren areas in the polar deserts, where there are no cryptogamic crusts. Although most arctic ecosystems fix at least some nitrogen, the rates are often low and in general they provide about 10% of annual uptake by vascular plants. Nutrient release from decomposition must be a major source for plant growth, especially because most species have relatively high tissue concentrations of nitrogen.

### 6. Pollination Biology

Many arctic species have large, colorful flowers and thus are attractive to insect pollinators as well as serving as basking sites. The parabolic-shaped, solar-tracking flowers of *Dryas* and *Papaver* may be 8–10°C warmer than adjacent air. These flowers serve as basking sites for flies and midges. Empidid and syrphid flies are the most important pollinators at Lake Hazen on Ellesmere Island. Bumblebees pollinate the flowers of *Pedicularis* species and those of other species. These higher flower temperatures also speed maturation of fruits and seeds, an important adaptation where the growing season is so short.

## IV. ANIMAL ADAPTATIONS AND POPULATIONS

Arctic animals, like plants, are adapted to taking advantage of a short growing season where growth and production are limited by available food and by low temperatures, but are aided by long periods of continuous light for feeding. With continuous light for most or all of the growing season, one might expect that animal activity would lack a diurnal rhythm. For many groups of organisms this is not the case, although the resting period of light-active songbirds shifts into evening hours. The hatching from insect pupae is linked to time of day and small mammals are more active at "night," whereas caribou and muskoxen are less active at "night." The synchronization of their biological clock with rotation of the earth appears to relate to azimuth position of the sun in relation to landmarks and the spectral composition of light (diurnal shifts in red : far red ratios). The very limited data on lemming, Peary caribou, and muskoxen indicate that animal activity is greater in "daytime" during the long winter darkness or semilight period. Temperature and the accumulation of heat energy are as critical for animals as for plants. This is even more pronounced in winter, for the plants are dormant and usually buried in snow, while the resident birds and mammals remain active, except for hibernation of ground squirrels and denning of polar and grizzly bears.

A major difference from temperate regions is the reduced diurnal amplitude of temperature (both summer and winter), yet there is still a tremendous range of mean daily or monthly temperature and insolation from summer to winter (Fig. 14), resulting in very great seasonal changes in climate. Animals, like plants, are not very well adapted to low temperatures; they just tolerate them by low rates of growth and physiological activity. On the

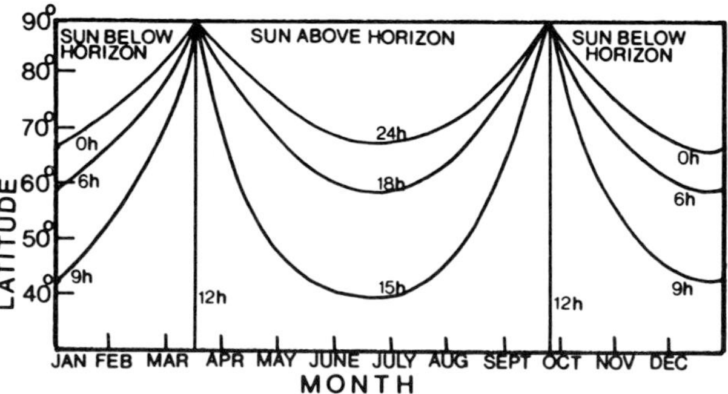


***FIGURE 14*** Duration of daylength at latitudes 40° to 90°N. latitude.

other hand, many organisms respond quite rapidly to increased temperature, for the $Q_{10}$ of many metabolic processes of plants, invertebrates, and vertebrates is between 2 and 4, rather than the 1 or 2 more characteristic of temperate regions. The $Q_{10}$ values for 16 invertebrate species on Devon Island averaged 4 at temperatures of 2° to 12°C, but the range was 1.1 to 27.3. In general, the $Q_{10}$ for most species decreased with increasing temperature. Higher $Q_{10}$ for both aquatic and terrestrial invertebrates may help these organisms maintain adequate metabolic rates at the low prevailing temperatures.

Growth rates of invertebrates are exceedingly slow in the Arctic and therefore maximum life span of various organisms is measured in years, compared with weeks or months in temperate regions. Again, data from Devon Island indicate that Enchytraeidae and Tardigrada worms in wet sedge–moss meadows have a life span of 2.8 and 1.3 yr, respectively. Collembola average 7 yr, Diptera 4.3 yr, Symphyta Hymenoptera 7 yr, and the moth *Gynaephora groenlandica* 10–14 yr.

Life spans are related to temperature. For example, the meadow soils at −1 cm averaged 313°C accumulated degree days (three summers) and the elevated, warmer, and drier raised beach ridges averaged 624°C. The estimated life span for these same invertebrate groups were accordingly 50% shorter on the raised beach ridges compared with the cold and wetter soils of the meadows. The slow growth rates of arctic char (*Salvelinus alpinus*) also demonstrate this relationship and the importance of nutrient regime relative to the drainage basin. Water temperature in these lakes range from 0° to 5°C. At Char Lake on Cornwallis Island, the estimated age of arctic char is 17+ yr at 20 cm and 10 yr at 20 cm for char in two lakes on the Truelove Lowland, Devon Island. Growth rates are higher on the Truelove Lowland because of higher biological production and lake metabolism, the result of more lush vegetation in the watersheds of these lakes compared with the polar desert watershed on Cornwallis Island.

The mechanisms for maintenance of warmth by arctic mammals and birds has interested biologists over the years. P. F. Scholander and his colleagues at Barrow, Alaska, studied the insulation effectiveness of fur thickness in winter for a series of arctic mammals (Fig. 15). Small arctic mammals (lemmings—*Dicrostonyx, Lemmus;* ground squirrels—*Spermophilus;* weasels—*Mustela*) seek sheltered burrows or nests in winter, for their fur is thin with correspondingly limited insulation. In contrast, the large mammals (wolf—*Canis;* muskox—*Ovibos;* caribou, reindeer—*Rangifer;* fox—*Alopex;* polar bear—*Ursus*) have thick fur and appear to be undisturbed by winter cold except during storms with severe windchill. The implications of these findings are that small mammals must maintain higher metabolic rates at lower temperatures than is necessary for thick-furred mammals (Fig. 16). The reduced surface area : volume ratio of large animals (Bergmann's rule) reduces not only heat loss, but also loss of water by evaporation. According to this rule, body size tends to increase toward colder climates, which results in greater heat conservation. Reduction of heat and water loss are important adaptations for animals that must obtain water from snow in winter, an energy-demanding process.

Although snow and ice are abundant in the Arctic, there is no comparable flora and fauna adapted to glacier snow and ice as there is in many alpine environments of temperate alpine regions, where green algae, Collembola, and Tardigrada are quite common.

Winter snow plays a very important role for arctic mammals and birds. Species of *Lagopus* commonly dig themselves into the snow, especially during storms. Lemmings, voles, and ermines spend most of the winter at the base of the snow in winter nests, developing runways for feeding and hunting as well as for defecation sites. The bottom layer of snow is the warmest and it develops interconnected latticelike structures that facilitates animal movement. This layer of snow is often called *pukak.* Snow depths of 30 cm at Barrow in early March resulted in 10° to 20°C difference in temperature between the air–snow interface and the ground surface, and comparable snow depths on Devon Island resulted in a 12°C temperature difference. Most winter nests of lemmings are found where snow depths of 60–150 cm are found

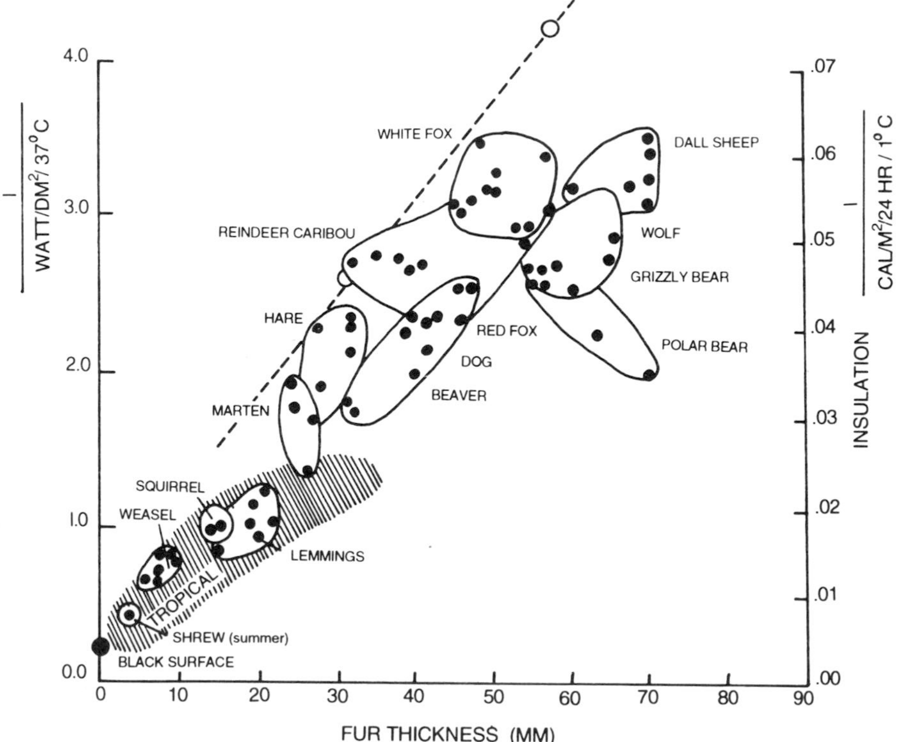


***FIGURE 15*** Winter fur thickness of northern animals. [From L. Irving (1972). "Arctic Life of Birds and Mammals." Springer-Verlag, NY.]

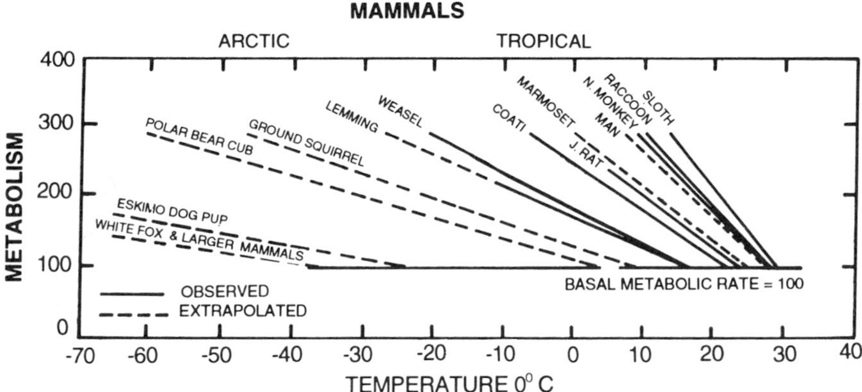


***FIGURE 16*** Metabolic rates for arctic mammals compared with a few tropical species. [From L. Irving (1972). "Arctic Life of Birds and Mammals." Springer-Verlag, NY.]

along gullies and the transition between raised beach ridges and sedge–moss meadows. The birthing sites for arctic mammals are quite variable. Ground squirrels, weasels, wolves, and foxes deliver their young in nests in underground burrows. Lemmings, voles, and shrews build winter nests at the snow–ground interface, whereas caribou and muskoxen give birth on the open tundra when the ground is still snow covered in the High Arctic. Hares build nests, typically among rocks.

Polar bears on land and seals at sea each construct snow caves where they can stay much warmer, especially in spring when young are born. Another adaptation of mammals is the reduction in length of snouts, ears, and legs to reduce loss of body heat. Poikilothermic animals maintain body fluids with antifreeze such as glycoproteins in marine fishes and other glycerine-based compounds in invertebrates. Adipose tissue (blubber) is an important insulation layer for seals, walruses, and whales that spend their entire life in cold waters.

### A. Invertebrates

Invertebrates are the least known major group of tundra animals, yet they typically account for a greater biomass than the vertebrates. The majority of invertebrates live belowground and are involved with decomposition, rather than as aboveground herbivores and carnivores. Less than 10% of invertebrate herbivore, decomposer, and microbial feeder biomass is converted to carnivore production. The soil fauna is dominated by Collembola, Acarina, Nematoda, Enchytraeidae, and larvae of Diptera. Earthworms are present only in the Siberian Arctic, and Lepidoptera, Trichoptera, and Coleoptera are relatively minor families, especially in the High Arctic. Major groups of invertebrates absent from the High Arctic include Mollusca, Myriapoda, Isopoda, and the insect orders Thysanura, Orthoptera, Plecoptera, Odonata, Thysanoptera, Formicidae, and Heteroptera.

In a summary of the International Biological Program (IBP) data, it is apparent that species of Diptera dominate the insect biota from Alaska, Canada, and Russia. Species of Collembola are also common to all sites, for they, like the larvae of many Diptera, feed on fungi and soil particulate matter. More than 60 species of arthropods are common to these three tundra sites, although the species richness data (Table V) indicate that the Tareya, Taimyr Peninsula, site in Siberia has a more diverse fauna, again indicating the importance of the huge landmass of Siberia that served as a center for arthropod evolution, as it did for mammal and plant evolution. Species richness remains relatively high on the Truelove Lowland, Canada, in spite of its northern latitude and its island location. The presence of a high arctic oasis with its relatively diverse habitats and the abundance of soil-dwelling species accounts for its richness.

**TABLE V**
Numbers of Invertebrate Species in Select Taxa at Three Arctic Sites[a]

| Taxon | Tareya Taimyr (37° 30'N) | Point Barrow (71°N) | Truelove Lowland (75° 33'N) |
|---|---|---|---|
| Protozoa | — | 20[b] | 13[b] |
| Nematoda | 162[b] | — | 18[c] |
| Oligochaeta | | | |
| Enchytraeidae | 10[b] | 15 | 7 |
| Lumbricidae | 1 | 1 | 0 |
| Mollusa | 3 | 0 | 0 |
| Araneida | 8 | 36 | 8 |
| Acarina | 50[b] | 60 | 19 |
| Insecta | | | |
| Collembola | 62 | 43 | 23 |
| Plecoptera | 3 | 1 | 0 |
| Heteroptera | 4 | 2 | 0 |
| Trichoptera | 6 | 5 | 1 |
| Lepidoptera | 38 | 5 | 13 |
| Coleoptera | 43[b] | 15 | 3 |
| Diptera | 178[b] | 79[b] | 64[b] |
| Insect orders | 12 | 10 | 7 |

[a] From J. K. Ryan (1980). "Tundra Ecosystems: A Comparative Analysis." Cambridge Univ. Press, NY.
[b] More known.
[c] Other data source.

On the Truelove Lowland, Protozoa accounted for 70% of the invertebrate energy flow within the wet sedge–moss mires, followed by Enchytraeidae, Nematoda, Diptera, and Crustacea. On the beach ridges with drier soils, Nematoda accounted for 60% of energy flow, followed by Enchytraeidae, Collembola, Diptera, and Acarina. Although some of the soil invertebrates are herbivores (Lepidoptera, Hymenoptera, Symphyta), a larger number are carnivores (Araneida, portions of Protozoa, Nematoda, Tardigrada, Crustacea, Acarina, Collembola, Diptera, and Hymenoptera), but the most important groups are detritus feeders (Protozoa, Rotifera, Nematoda, Enchytraeidae, Tardigrada, Crustacea, Acarina, Collembola, and Diptera). It is assumed that these same groups of invertebrates perform similar roles in other terrestrial arctic ecosystems, but the data are not available.

Few if any arctic insects have been studied in as much detail as *Gynaephora groenlandica*. This species has a large larva size that enhances experimentation and requires a relatively long time to complete its life cycle (10 to 14 yr). The shorter time estimate resulted from a model based on energy budget analyses, whereas the longer time was based on number of years to complete the six larval instars. Adult eggs to instar II and instar II to III each takes one year, but three years elapse for each succeeding instar and from VI to pupa. The preferred foods are *Salix arctica* and *Oxyria digyna* leaves and flowers of *Saxifraga oppositifolia* and *Dryas integrifolia*. Feeding, molting, and pupation all occur early in the season (middle to late June). At Alexandra Fiord, Ellesmere Island, N.W.T. (78°53'N), where these feeding studies were conducted, the larvae spend 20% of the time feeding, 15% moving about, and 60% basking, where body temperature can reach 30°C. By maintaining most of their biological activity near the summer solstice, the larvae maximize the use of incoming solar radiation, an excellent adaptation where energy is limiting. In winter their freeze tolerance is a product of glycerol syntheses from D-glucose. The presence of glycerol is important for the winter survival of many invertebrates. Parasitism by a wasp (*Hyposoter*) and a fly (*Exorista*) account for a large proportion (56%) of the summer mortality; predation by birds is minor (3%). Winter mortality is also relatively low (13%).

### B. Birds

One of the most interesting groups of arctic animals is the birds. Many arctic areas are near the ocean, consequently the avian fauna is often a mixture of marine and terrestrial species, especially in the High Arctic. A few species, typically 3–5, are winter residents, while migrant species swell the numbers in summer to 25–50+ species in the Low Arctic and to 15–25 species in the more lush areas of the High Arctic. On a circumpolar basis, there are about 140 breeding species plus visitors. [*See* BIRD COMMUNITIES.]

With the increase in temperature and photoperiod in spring, there is a rapid rise in the arrival of migratory species. Snow Buntings (*Plectrophenax nivalis*) and Lapland Longspur (*Calcarius lapponicus*) are two of the early arrivals, often feeding on spiders and seeds until summer insects are available with snowmelt and warming of soils. At Anaktuvuk Pass, Alaska (68°N), an average of 14 days elapse between arrival to egg laying for 28 species. Although the growing season is short, it does not appear that the time required for nesting or the time spent as nestlings is shortened in arctic populations compared to more temperate populations of the same or closely related species. The complex reproductive procedures for physiology and behavior must be initiated by individuals on their winter grounds and continue during migration. The number of eggs laid by sandpipers (*Scolopacidea*) and plovers (*Charadrudae*) is 4 in all climates and arctic loons (*Gaviidae*) lay 1–2 eggs as elsewhere. Oldsquaw ducks (*Clagula hyemalis*) lay 10–12 eggs, a large expenditure of energy for these northern latitudes. For some closely related groups of species, clutch size increases from low to high latitudes. The duration of incubation and nestlings staying in the nest are highly variable. For seabirds along the Barents Sea, the range is from 25 days in the Common Eider (*Somateria spectabilis*), 63 days for the Kittiwake (*Rissa tridactyla*), to 95 days for the Green Cormorant (*Phatacrocorax aristatelis*).

There are limited data on densities of adult birds (Table VI), but these data show great variation for different sites. Densities of breeding pairs and the number of species present are significantly different from the Low to the High Arctic and also populations vary in relation to the diversity and productivity of habitats. One of the longest records of population numbers is the 16-yr record for the Truelove Lowland, Devon Island, where summer density of adults has ranged from a low of 16.5 $km^{-2}$ in 1971 to a high of 46.7 $km^{-2}$ in 1983 and averaged 31.1 $km^{-2}$ for the period. Earlier studies recorded a mean density of 10.3 breeding pairs $km^{-2}$ in the early 1970s. Populations of Snow Buntings and Lapland Longspurs varied synchronously, but fluctuations were asynchronous in Lesser Golden Plovers (*Plarialis dominica*) and Black-bellied Plovers (*P. squartarola*); Arctic Terns (*Sterna paradisaea*) have become rare to nonexistent since 1989. Typically in the Arctic, 5 or fewer species make up 80–90% of the breeding population. On the Truelove Lowland, 6 or 7 species typically account for 80% of the regularly breeding populations of 18 species and the 15 species of occasional breeders.

Of the 140 species that breed in the Arctic, 28% are waders, 22% are seabirds, and 21% are waterfowl. Thus the presence of lakes and nearness to the ocean explain why populations are larger than might be expected at these northern latitudes. Waders are an important group, especially on the coastal tundra of Alaska, where the summer population was estimated to be 5,500,000 in the 1970s. Migrating flocks of waders (Knots and Turnstones) and Brant Geese (*Branta bernicla*) stop in Iceland in May to feed and accumulate large amounts of fat prior to flying on to northwest Greenland and the Canadian Arctic Archipelago for nesting. In a similar man-

**TABLE VI**
Breeding Bird Numbers and Population Density for Various Arctic Sites

| Location | Latitude (°N) | No. of species: Regular | No. of species: Occasional | Total species | Density (pairs per 100 ha) | Area (ha) |
|---|---|---|---|---|---|---|
| Frobisher Bay, N.W.T. | 63 | — | — | — | 20 | 72 |
| Mackenzie River Delta, N.W.T. | 69 | 92 | — | — | 207 | 25 |
| Igloolik Island, N.W.T. | 69 | 26 | — | — | 29 | — |
| Interior Arctic Alaska | 68–69 | 84 | — | — | — | — |
| Prudhoe Bay, Alaska | 70 | | | | 72 | 100 |
| Barrow, Alaska | 70 | 22 | — | 151 | 52 | 35 |
| Tareya, Russia | 73 | 31 | — | 61 | 140 | — |
| Truelove Lowland, N.W.T. | 75 | 18 | 10 | — | 10 | 4300 |
| Alexandra Fiord, N.W.T. | 79 | 10 | — | 24 | 13 | 1200 |
| Lake Hazen, N.W.T. | 82 | 14 | 4 | 18 | 5 | 2230 |

ner, Knots and Brant Geese migrate in early June from the Wadden Sea, across Scandinavia and Finland, to nest on the Taimyr Peninsula, Russia, a distance of 4000 km. These species and others migrate in flocks, often in V-formations, which promotes energy saving. The energy saving that results from flying in a flock, compared with a single bird, is estimated to be 20–40%. Flock size greater than 30 birds results in a minimal additional increase in energy saving.

The waders also comprise the largest group of birds in the Eurasian tundra, where the total species number about 100. Here as elsewhere, many species are insectivores. Red Phalaropes (*Phalaropes fulicarius*) are excellent swimmers, scooping up invertebrates, including chironomid larvae, from the surface. Species of *Calidris* typically feed in mires (wet sedge–moss meadows), where they feed on Collembola, tipulid larvae, and other invertebrates. The plovers (*Charadriidae*), Knot (*Calidris canutus*), Sanderling (*C. alba*), and Curlew Sandpiper (*C. testacea*) feed on beetles, spiders, crane flies, and Diptera. Although there are relatively few songbird (passerine) species, some of the most abundant species in the tundra and polar semideserts include the Lapland Longspur, Shore Lark (*Eremophila alpestris*), and Snow Bunting. Their diet includes seeds and insects, the latter picked from the soil rather than from plants. The Wheatear (*Oenanthe oenanthe*) is common in the Low Arctic but less so in the High Arctic. This species nests in ravines, stony slopes, and feeds on larger insects—crane flies, beetles, and even bumblebees.

At least nine species of the gull order, Laridae, nest in the Arctic, including some of the best-known low and high arctic species. These include the Arctic Tern, Glaucous Gull (*Larus hyperboreus*), and three species of jaegers (skuas). The jaegers feed on bird eggs, small rodents, and some of the larger insects. The true gulls along with Arctic Terns are insectivores, the latter adapted to taking insects above water surfaces.

Among the ducks, Oldsquaw (*Clangula hyemalis*) and the eider ducks feed on freshwater invertebrates in summer but from the sea in winter, as do the jaegers. Geese (*Anser* spp.) are herbivores, as are the ptarmigan. Geese feed on the rhizomes of sedges (*Carex, Eriophorum*) and grasses (*Arctophila fulva, Puccinellia*), and where they gather in large numbers they can convert sedge–moss mires to areas of moss. Lesser Snow Geese, which inhabit the eastern Canadian Arctic, are estimated to number 2 million and have been increasing in recent years. These increases may result from changes in agricultural and conservation practices in winter and spring staging areas. On their nesting grounds, densities can exceed 2000 $km^{-2}$. The goslings grow rapidly from an initial weight of 80 g to 1500 g in less than 7 weeks. Although geese may consume about 90% of the graminoids available, their high rates of defecation return large amounts of nitrogen to the soil, which stimulates regrowth of the plants; this is a very important feedback mechanism that helps to maintain *Puccinellia phryganodes* and *Carex subspathacea* inspite of exceptionally heavy grazing. However, after snow melts and before plant growth begins, snow geese grub rhizomes and roots. Where grazing is intense, there are small to large patches nearly devoid of vegetation. This has recently led to intensive mudflats formerly occupied by salt marsh. The Willow and Rock Ptarmigan (*Lagopus lagopus, L. mutus*) feed on shrubs, mostly species of *Salix, Alnus,* and *Dryas*. Ducks, ptarmigan, and jaegers consume large amounts of berries in autumn.

In the Eurasian tundra, there are 10 species of predators, including the Rough-legged Buzzard (*Buteo lagopus*), White-tailed Eagle (*Haliaeetus albicilla*), Hen Harrier (*Circus cyaneus*), Peregrine Falcon (*Falco peregrinus*), Gyrfalcon (*F. rusticolus*), Snowy Owl (*Nyctea scandiaca*), Short-eared Owl (*Asio flammeus*), and the three species of jaeger. All but the buzzard, eagle, and harrier are common in the North American Arctic. The jaegers, Snowy Owl, and the Peregrine Falcon all nest in the High Arctic despite the limited available food—birds, rodents, and, for the jaegers, insects and some plants in addition to birds and lemmings. The Snowy Owl and Peregrine Falcon exert a considerable influence on rodents, consuming large numbers of lemmings during a "lemming high." When lemmings are more abundant, clutch size averages two eggs for jaegers, but in a lemming low some nests contain only one egg. Clutch size for Snowy

Owls ranges from four to nine, with a larger clutch size when lemmings are more abundant.

### C. Mammals

About 60 species of mammals are found in the Arctic, but only 42 are regular inhabitants. Of these regular inhabitants, 71% are small mammals (rodents—*Dicrostonyx, Lemmus, Microtus;* insectivores—*Sorex;* and lagomorphs—*Lepus*). These animals are often present in such large numbers that their total weight and grazing influence are greater than that of the large herbivores (caribou—*Rangifer,* or muskox—*Ovibos*). These are the only important large arctic herbivores, but after ice retreat 8000–12,000 yr B.P., there were many more species (great bison—*Bison;* horse—*Equus;* wooly mammoth—*Elephas;* moose—*Alces;* moose stag—*Cervalces;* muskox—*Ovibos;* woodland muskox—*Symbox;* caribou and reindeer—*Rangifer;* wapiti—*Cervus;* saiga antelope—*Saiga;* yak—*Bos;* and sheep—*Ovis*). Only sheep, caribou, reindeer, muskox, and moose have survived to the present.

The large herbivores of the modern Arctic are ruminants and the fossil record shows that all but the arctic horse were ruminants or ruminantlike. Although there is an abundance of food available during the growing season, the nutritional quality of vegetation is less than that of seeds and fruits. Thus the ability to utilize cellulose was an important step in the evolution of grazers and browsers. Anaerobic microorganisms in the ruminant gut, which synthesize the enzyme cellulase, enable these animals to ferment and utilize about 65% of the cellulose consumed. The volatile fatty acids produced account for 50–60% of the digestible energy intake and, along with the B vitamins and essential amino acids produced, give a selective advantage to ruminants. These mammals also conserve nitrogen by recycling urea nitrogen to the rumen by saliva or by diffusion of urea from blood through the rumen wall. These are important adaptations for winter survival where water and forage are limited resources.

The native ungulates of the Arctic, that is, dall sheep (*Ovis dalli*), moose (*Alces alces*), muskoxen (*Ovibos moschatus*), caribou, and reindeer (*Rangifer tarandus*), have evolved physiological, nutritional, and behavioral adaptations that enable them to live in these stressful environments. These adaptations include time of breeding, ability to maintain heat balance in winter on a reduced food intake, but ability to recycle nutrients, and in summer the ability to maximize food intake, growth, and fat deposition.

Seasonal changes in metabolic rates and protein requirements of caribou, reindeer, and muskoxen follow patterns of food availability. In late September and October, the metabolic and protein requirements of females of these species decline. Following the rut, metabolic requirements remain high for males, then decline in December. Metabolic and nutrient requirements remain low in winter, rising in females just before calving. Males and nonpregnant females lag 2–3 weeks in these metabolic shifts. These changes in metabolic rates are believed to relate to changes in appetite and are interpreted as mechanisms for lowering energy requirements in winter when there is a food shortage.

The largest energy demands come after the young are born. Calving in caribou and reindeer comes as the snow melts and when there is new nutrient-rich forage available, especially new shoots of *Eriophorum*. However, Peary caribou and muskoxen give birth before snowmelt in the High Arctic, which places a greater energy drain on these lactating mothers. Caribou and muskox milk is rich, enabling calves to grow rapidly, for barren-ground caribou calves must be ready to migrate 500 to 1000 km in autumn, and Peary and Svalbard caribou and muskox calves must be able to remain in the Arctic, where winter conditions are severe.

Caribou from the Alaskan, Canadian, Russian, and Greenland Arctic feed selectively on grasses, sedges, woody plants, forbs, and lichens. *In vitro* digestibilities are higher for lichens than for forbs in winter for Peary caribou. The mouths of caribou and reindeer are narrow, permitting them to be more selective in feeding than muskoxen with their wider mouths. Selective grazing by caribou permits them to maintain increased digestibility of forage and to increase their daily weight gain in summer. Their performance is influenced by selecting plants high in nutrients and avoiding species higher in secondary compounds. In winter, caribou can successfully paw through snow 50–60 cm in depth, provided it is not too crusted. Snow hardness greater than 25–29 kg $cm^{-2}$ prevents caribou and reindeer from cratering, whereas muskoxen avoid a snow hardness of only 10 kg $cm^{-2}$. Of greater importance are the occasional autumn icings in the High Arctic. Early snow may melt and refreeze to form an ice layer that encases much of the vegetation. Large die-offs of muskoxen and caribou can result from these conditions, as in 1973–1974.

Measurements of forage quality in the High Arctic indicate that protein levels in summer and winter are higher for sedges than is typical for similar species in more temperate latitudes. This gives these mammals a nutritional advantage. Mortality for muskoxen and Peary caribou often comes in spring when the animals have utilized much of their fat reserves in surviving the 9+ months of winter. These animals crater through 20–50 cm of snow for food and it has been observed in Alaska and the Canadian High Arctic that less energy is utilized in seeking food if animals aggregate to crater and feed. Male muskox often displace females at crater sites in winter grounds and during wolf attacks. Rumen contents of Peary caribou indicate their variable winter diet, for they eat whatever is available. This contrasts with barren-ground caribou and reindeer, which consume large proportions of lichens, graminoids, and mosses. In winter, muskoxen consume graminoids and some willow in the High Arctic, but more woody species in the Low Arctic.

Densities of large herbivores on the tundra are low (0.1–3.4 animals $km^{-2}$), but can be much higher on select sites during calving and winter feeding. In summer, these ungulates can spread over a large landscape, feeding as they slowly move. As an example, muskox on five lowlands in northeastern Devon Island averaged 0.3–1.4 $km^{-2}$ in the early 1970s. The Truelove Lowland constitutes 15% of the total usable range and on an annual basis, the population spent 17% of the potential muskox days there. However, only about 3–8% of the population utilized the lowland in June through November, but 15–37% of the population utilized the area in December through April, the time when successful cratering for forage was most critical. Densities thus ranged from 0–0.5 animals $km^{-2}$ in summer to 1.9–2.6 animals $km^{-2}$ in winter.

Muskoxen have increased their numbers in northern Greenland from a low around 1960 to a high in the mid-1980s. The population low probably resulted from ice crust formation in late autumn or winter resulting in their inability to obtain adequate forage. Northward, where weather conditions are more stable, snow depth becomes more critical in terms of local winter migration. At present, animal density north of Scoresbysund and Jamison Land averages 0.5–1.0 $km^{-2}$, but only 0.3 $km^{-2}$ in northeastern Peary Land. In summer, muskoxen forage opportunistically on sedges and willows. The quality of forage is high in summer (21–28% crude protein) and *in vitro* degestibility is also high (60–75%). At Kap Kobenbaven, animals spend <50% of daily activity feeding and slightly less time (48%) at Jamison Land (71°N). In contrast, animals feed only 35% of the day in northeast Alaska (69°N) and on Melville Island, N.W.T. (75°N). This appears to relate to the decreased growing season northward and also to reduced diversity of vegetation.

Arctic hare (*Lepus arcticus*—Canada, Greenland; *L. americanus*—Alaska; *L. timidus*—Eurasia) are present throughout the Low and High Arctic, but only in small numbers (<1 $km^{-2}$), except in some areas where mats of *Salix arctica* are abundant as on Ellesmere Island, where herds of 150–200+ animals occur. Their food includes grasses, sedges, saxifrages, mountain sorrel, and willows. Where hares are more common, clumps of *Salix* are often consumed in winter, leaving only a few cut stems and a few very dry fecal pellets.

Mating of hares occurs during April and May and the young are born in June; the average litter is four to five leverets. They are born in a nest of mosses and grasses, usually in the protection of rocks or shrubs. By September they are fully grown and their fur color changes from light brown or gray to their permanent white coat. They remain active all winter, often huddled together for shelter during storms. Compared with other lagomorphs, these animals are larger at maturity (4–4.5 kg). Excited animals stand on tiptoes and turn their heads to catch the slightest sound, then bound away on their hind legs in kangaroo style.

Arctic ground squirrels (*Spermophilus parryi*) are the largest members of this genus and the most northern, though they are confined to mainland North America. The closely related Asian species, *S. undulatus,* is limited to northeastern Siberia. Permafrost is a major limiting factor in the distribution of these species. Typical habitats include riverbanks, lake shores, pingos, moraines, and stabilized sand dunes. The colony area is usually honeycombed with burrows, which are used for many years. Hibernation dens are dug as side tunnels off the main burrows and are lined with grasses, moss, lichens, leaves, and animal hair. Arctic ground squirrels spend 6–7 months in hibernation in their dens 0.5–0.75 m below the surface. Body temperature often drops from 36.4° to 17°C during hibernation, a mechanism that greatly extends their energy reserves. Adults appear in late April to early May before snow melts. Mating occurs in May; the young are born 25 days later and by September they are nearly full grown. The average litter size is 5–10. A study in northwestern Alaska reported a territorial population density of 1.2 $ha^{-1}$, but a total population density including young of 6–7 $ha^{-1}$ in June–July and dropping to 3–3.5 $ha^{-1}$ in October in a 2-year period.

Unlike most tundra rodents, ground squirrels maintain a relatively constant population level as a result of two types of groupings. The breeding colonies are territorial and seem to limit predation by foxes and grizzly bears via extensive vocalization. The refugee populations, driven from the colonies, are nonbreeding, live in habitats only periodically suitable for squirrels, and are subject to heavy predation, which is in part the result of being solitary and nonvocal. There is no predation during hibernation.

These animals consume a wide variety of plants, including mushrooms, grasses, forbs, and woody shoots of willow and birch. In the fall, seeds and fruits are consumed and with leaves are stored for consumption the following spring before new vegetation is available.

When one thinks of arctic land mammals, lemmings usually come to mind. Brown lemming (*Lemmus sibericus, L. lemmus*) and collared lemming (*Dicrostonyx groenlandicus*) are circumpolar species, although the latter species occurs in both the Low and High Arctic. As with voles, these species remain active all winter. Their nests are usually under snowbanks, where there is greater insulation under deep snow. Their runways at the ground surface are visible in spring as the snow and ice melt. Brown lemming are more typical of wet sedge–moss communities, where they feed on grasses, sedges, and to a lesser extent forbs. In winter they consume the green shoots of graminoids and mosses along with bark and twigs of *Salix* and *Betula*. In contrast, collared lemmings feed on dicotyledons more than monocotyledons and they are more typical of better-drained soils.

Brown lemming build up large populations every 3–5 yr, then crash or populations move out to occupy other habitats. Under ideal conditions (mild winter, early and deep snow) both species breed in winter, although the most productive months are June–August in the brown lemming and March–September in the collared lemming. At peak densities, *Dicrostonyx* number about 125 $ha^{-1}$ (Siberia) and *Lemmus* about 155 $ha^{-1}$ (Barrow, Alaska). In the High Arctic of Devon Island, N.W.T., a lemming high numbered 7–10 $ha^{-1}$ and a low of $<1$ $ha^{-1}$. The rapid buildup in populations results from a short gestation period of about 3 weeks, a litter size of 7.3 in *Lemmus* and 4.5 in *Dicrostonyx,* and two to four litters per year. The rapid buildup in populations is a product of high reproductive capacity, deep winter snow, high quality and quantity of available food, and low predation. Peak lemming populations build up over winter when available forage is at a minimum. Because brown lemmings consume only green shoots of graminoids and mosses in winter, large amounts of dead, cut shoot material are evident at snowmelt. By leaving massive amounts of cut material, disrupting the moss layer, and depositing feces, lemmings increase decomposition rates, thus contributing to nutrient release and subsequent stimulation of plant growth.

Microtines, including the red-backed vole (*Clethrionomys rutilis*) and the root vole (*Microtus oeconomus*), play an important role in arctic food chains. They are a major food of foxes, ermines, snowy owls, and the three species of jaegers.

Arctic fox (*Alopex lagopus*) inhabit all of the circumpolar Arctic and extend south into the alpine and forest–tundra. They also follow polar bear on the frozen polar seas to consume remains of seals. The winter fur color is white or blue to pearly gray, but changes in summer to a yellowish white with areas of brown to black. Arctic foxes with their acute olfactory sense can detect lemmings in their nests under the snow and in summer burrows. Foxes feed primarily on lemmings and voles, but also rob eggs and fledglings of ground-nesting birds. In the High Arctic, foxes play a very significant role in limiting successful reproduction of birds. They also feed on ground squirrels and young hare as well as caching lemmings, eggs, and fledglings for later consumption.

Except during the breeding season, foxes are solitary animals. Breeding occurs in mid-February through April, with the young born between mid-May and mid-June. Litter size averages 6.4 but is larger if lemmings are abundant and is lower to none during a lemming low.

The arctic wolf (*Canis lupus arctos*) is limited in numbers as a result of limited prey and hunting. Wolf packs are still present on Ellesmere, Melville, Banks, and Victoria islands, as well as in the Brooks Range and hills to the north. Wolves are also found in large parts of Siberia. Wolves follow caribou and reindeer herds, selecting young and weakened animals, the result of old age and disease. They also take a few muskoxen, but again usually young animals. In addition to taking big game, they successfully hunt lemming, ground-nesting birds, and arctic hare. It is estimated that a single adult wolf kills 1.5 large animals per month. Wolves appear to mate for life and the adults typically hunt in pairs. Breeding is in late February to mid-March, with pups arriving in May. Litter size averages 7 (5–14), with the young born in dens initially dug by badgers, foxes, or arctic ground squirrels. By August or early September the pups join adults in hunting.

Grizzly bears (*Ursus arctos*) prefer open areas and at present are found in alpine tundra, subalpine forests of the Rocky Mountains, and into the Arctic. The European brown bear occurs throughout Europe and Asia, except in the far north of Siberia. These bears are omnivorous, feeding on new grass, roots, and bulbs of herbs upon emergence from hibernation. In summer they feed on berries and dig up lemmings, arctic ground squirrels, insects, fungi, and roots. They also capture fish as well as take the occasional caribou and mountain sheep. Sows breed every second year in late June to early July. Cubs are born during hibernation from mid-January to early March. Average litter size is two, but one to four cubs may be born. Young cubs stay with their mother the first summer and are weaned at 4–5 months. These solitary animals have huge home ranges and thus are generally rare, although they do congregate along rivers to fish.

## V. ECOSYSTEMS

### A. Evolution of Arctic Ecosystems

In early Tertiary times (ca. 55 million yr B.P.), the fossil records from southern Alaska indicate that tropical, subtropical, and warm temperate forests predominated. Even in Miocene time (22 million yr B.P.) there were conifer–deciduous forests on Devon Island, at what is now 75°N, and similar mixed conifer–deciduous forests dominated in Central Alaska in the Middle Miocene (15 to 10 million yr B.P.). These northern lands were farther south due to plate tectonics and the north pole was not in its present location during the early and mid-Tertiary. Cooling that began in the Oligocene continued in the Pliocene Epoch (6 to 3 million yr B.P.) on the Seward Peninsula, exemplified by coniferous forests and insect groups similar to present-day forests in southern British Columbia and northern Washington.

The western islands of the Canadian Arctic Archipelago also contain fossil evidence of late Tertiary (6 to 4 million yr B.P.) forests of mixed conifers and deciduous species from Banks Island to Ellef Ringnes Island. The fossils from Meighen Island (80°N) indicate that a forest–tundra vegetation and insect fauna occurred in Miocene to early Pliocene time. A forest–tundra vegetation with heathlands has been reported from Peary Land, Greenland (82°N), at the Pliocene–Pleistocene transition (2.5 to 2 million yr B.P.). There is little evidence of arctic ecosystems until late Pliocene as these northern lands cooled.

There is also evidence for the evolution of an arctic fauna and flora on the highlands of Central Asia. With major climatic cooling of polar regions and northward drift of the tectonic plates, there was rapid evolution of plants and animals adapted to arctic and alpine environments. The fossil record shows there have been four periods of major mammal migration between Asia and North America. The evidence indicates that a much larger number of genera and species evolved in Eurasia and spread to North America than the reverse. These migrations of mammals and plants occurred over the Bering Land Bridge, called Beringia. Beringia is central to understanding present-day arctic and alpine ecosystems in terms of migration of people as well as animals and plants. Mammal remains from reworked sediments along rivers reveal the presence of mammoth, bison, horse, caribou, mountain sheep, steppe antelope, muskox, and sabertoothed tiger in Beringia sometime 40,000 to 13,000 yr B.P. However, there is little if any evidence that these species all occurred together in any one place. With this diversity of species, the landscapes must have supported a diversity of plant communities, rather than the simplified grassland–large mammal ecosystem suggested by some authors.

## B. Terrestrial Ecosystems

### 1. Primary Production

Detailed studies of terrestrial ecosystems are available from Barrow, Alaska, and the Truelove Lowland, Devon Island, N.W.T. Both sites are dominated by wet sedge–moss meadows (mires). Although Devon Island is farther north, solar radiation is greater because of more clear days in summer. The growing season averages 50 days on the Truelove Lowland and 70 days at Barrow. Soils at both sites are organics (Histic Pergelic Cryaquepts) in wet meadows and Arctic Brown soils (Pergelic Cryochrepts) on the gravelly raised beaches. Evapotranspiration and precipitation are generally in balance each summer; thus soils recharge with water each autumn prior to refreezing. The active layer in the poorly drained organic soils is 20–30 cm, but 80–120 cm in the well-drained Arctic Brown soils of the raised beaches.

The vascular plant floras (139 and 96 spp; respectively) are relatively small at Barrow and the Truelove Lowland, but relatively large numbers of bryophytes (122 and 162 spp; respectively) and lichens (70 and 120 spp; respectively) are found. *Carex aquatilis, Dupontia fisheri, Eriophorum angustifolium, E. russeolum,* and numerous mosses dominate the wetter sites, with *Arctophila fulva* in ponds at Barrow. *Carex bigelowii, Poa arctica, Luzula arctica, Arctagrostis latifolia,* and species of *Saxifraga* and *Salix* occupy the high-center polygons and raised beaches.

*Carex stans,* with lesser amounts of *C. membranacea, E. angustifolium,* and *Arctagrostis latifolia,* dominates the wet meadows on the Truelove Lowland. *Dryas integrifolia, Salix arctica, Saxifraga oppositifolia, Carex rupestris,* and *C. nardina,* along with numerous species of lichens, dominate the crests and slopes of the raised beaches.

Data on standing crop and net plant production show that graminoids dominate the wet meadows and that bryophytes are major components. Cyanobacteria are important in the wet meadows owing to their role in nitrogen fixation, averaging 70 mg N $m^{-2}$ $yr^{-1}$ at Barrow and 120–380 mg N $m^{-2}$ $yr^{-1}$ on the Truelove Lowland. The raised beaches, with their cushion plants, scattered graminoids, and lichens, have much lower standing crop and net production (Table VII). Because of their drier soils and limited cyanobacteria, nitrogen fixation is also much lower in these habitats (7–30 mg N $m^{-2}$ $yr^{-1}$).

### 2. Secondary Production

The dominant grazers in the Arctic are mammals, to a lesser degree birds, and aboveground invertebrates are of little importance. As stated earlier, most arctic invertebrates live belowground. At Barrow, the brown lemming (*Lemmus sibericus*) is the dominant grazer, with lesser numbers of collared lemming (*Dicrostonyx groenlandicus*). The former species feeds on graminoids and mosses and the latter mostly on forbs and dwarf shrubs in drier sites. Populations of *Lemmus* have fluctuated from 150 to 225 animals $ha^{-1}$ in a high, but only 1 to 5

**TABLE VII**
Biomass and Net Production (g $m^{-2}$) for the Barrow, Alaska, and Devon Island, N.W.T., Ecosystems[a]

| Component | Barrow, Alaska sedge–moss | | Truelove Lowland sedge–moss | | Devon Island Cushion plant, crest–slope | |
|---|---|---|---|---|---|---|
| | Biomass | Net production | Biomass | Net production | Biomass | Net production |
| Carnivores | T | T | T | 0.9 | T | 0.6 |
| Herbivores | 0.2 | 1.1 | 2.7 | 0.7 | T | T |
| Vascular plants | | | | | | |
| Aboveground | 47 | 45 | 86 | 44 | 104 | 15 |
| Belowground | 996 | 142 | 1085 | 104 | 54 | 3 |
| Bryophytes | 40 | 22 | 908 | 33 | 30 | 2 |
| Lichens | 5 | T | 0 | 0 | 49 | 2 |
| Algae | — | — | 4 | 4 | <1 | T |
| Total plants | 1088 | 209 | 2083 | 185 | 237 | 22 |
| Detritivores | 20.8 | 30.1 | 21.9 | 19.2 | 6.1 | 2.9 |
| Total ecosystem | 1109 | 240 | 2108 | 206 | 243 | 25 |

[a] From F. S. Chapin III, P. C. Miller, W. D. Billings, and P. I. Coyne (1980). "An Arctic Ecosystem: The Coastal Tundra at Barrow Alaska." Dowden, Hutchinson, and Ross, Stroudsburg, PA; Miller, P. C., P. J. Webber, W. C. Oechel, and L. L. Tieszen (1980). "An Arctic Ecosystem: The Coastal Tundra at Barrow, Alaska." Dowden, Hutchinson and Ross. Stroudsburg, PA; and L. C. Bliss (1977). "Truelove Lowland, Devon Island, Canada: A High Arctic Ecosystem." Univ. Alberta Press, Edmonton, Canada.

animals $ha^{-1}$ in a lemming low. The cycles, which average 3 to 6 yr, increase in numbers in the spring of winters with deep snow, high-quality forage, and low winter predator density (*Nyctea scandiaca*—snowy owl; *Alopex lagopus*—fox; *Mustela nivalis*—weasel). The sharp declines in population are attributed to overgrazing, with a drop in nutrient content of graminoids and a rise in predator populations. Winter reproduction is lower when snows are of shallow depth.

Collared lemming is the only microtine species in the High Arctic and its numbers are always low (0.1 to 10 animals $ha^{-1}$). Most breeding occurs from April to June when temperature and light levels are higher, but snows are deeper. Lemmings are the dominant grazers in the drier habitats, where they feed on *Dryas, Salix,* and *Saxifraga oppositifolia,* their preferred foods.

Muskoxen (*Ovibos moschatus*) are the dominant grazer within the sedge–moss mires on the Truelove Lowland, although they remove only 1–2% of the total aboveground forage per year in the mires. Large herbivores are minor at Barrow, but do increase near Prudhoe Bay, Alaska. Willow Ptarmigan (*Lagopus lagopus*), arctic ground squirrels (*Spermophilus parryii*), and the two species of lemming are relatively minor grazers compared with caribou at Prudhoe Bay.

### 3. Detrital Herbivores

At Barrow and the Truelove Lowland, 95–99% of the annual plant production enters the pool of dead organic matter and is available to invertebrate detritivores and microbivores. At Barrow the nematodes and springtails are most numerous, but owing to their small size, their biomass is not as great as that of the enchytraeid worms. The enchytraeid worms are mostly fungal and bacteria feeders; the nematodes are herbivores on plant roots as well as microbial feeders on bacteria, algae, and fungi. The Collembola (springtails) are detrital and microbial feeders, but some are herbivores and others are carnivores.

On the Truelove Lowland the enchytraeid worms dominate the biomass within the cushion plant systems of the raised beaches and in the wet sedge–moss mires. Protozoa are most productive in the mires due to their small size but short life cycles. Other important groups are nematodes, enchytraeids, and dipterans. On the raised beaches the nematodes are most productive, followed by enchytraeids, springtails, and dipterans. All the invertebrates assimilated about 5% of the net primary plant production, four times that of all the vertebrates (small and large mammals and birds).

### 4. Tertiary Production

As with temperate regions, carnivores are a minor component of these systems in terms of biomass and net production, yet there are relatively many species, especially during a lemming high at Barrow. Important predators include Snowy Owls, three species of jaeger, Glaucous Gulls, and Short-eared Owls of the avian fauna. Least weasel (*Mustela nivalis*), short-tailed weasel (ermine) (*M. erminea*), and arctic fox are the major mammal carnivores. Important insectivores include four species of sandpipers and the passerine species Lapland Longspur and Snow Bunting.

Avian predators are more abundant than mammalian predators at the Truelove Lowland, again indicating the greater biomass of invertebrates compared with vertebrate herbivores that serve as a food source. Of the 15 avian species that regularly breed on the Lowland, 13 species are predators. Arctic fox and short-tailed weasel are the only predator mammals of importance here, for wolves are seldom seen. Foxes depredated 42% of the lemming nests and 63% of all observed bird nests. Weasels depredated only 12% of winter lemming nests.

The larvae of crane flies are the most important invertebrate carnivore group at Barrow; other predators include species of spiders and beetles. Spiders and portions of the Nematoda, Tardigrada, Collembola, Diptera, and wasps are the main detrital carnivores on the Truelove Lowland.

### 5. Decomposers

Rates of decomposition are influenced by soil temperature and moisture, oxygen levels, pH, organic substance, and the kinds and numbers of microbes present. Soil invertebrates and protozoa indirectly influence decomposition rates by directly feeding on organic matter fragments and by grazing on bacteria and fungal hyphae. Substrates with low molecular weight that are water and/or alcohol soluble (simple sugars, citric and succinic acids) decompose more rapidly and by more organisms than substances with a large molecular weight (starch, cellulose, hemicellulose, lignin, and pectin). At Barrow about 25% of aboveground plant products are more soluble and this fraction contains most of the leaf nitrogen, phosphorus, and potassium.

Bacteria are more numerous in the wetter soils of the mires at Barrow and the Truelove Lowland, whereas fungi are more abundant in the warmer and drier soils of raised beaches and tops of polygons. Wet acid soils are not very favorable for nitrogen-fixing bacteria, but denitrifiers are common. The microbial flora declines rapidly with soil depth and shifts to a dominance of anaerobic bacteria.

The most abundant fungi are sterile forms, Basidiomycetes, and Fungi Imperfecti at both locations. The abundance of Basidiomycetes at Barrow reflects their role as mycorrhizal symbionts (endomycorrhizae), but these groups are not common at the Truelove Lowland. Ectomycorrhizal and mycorrhizallike structures are present on vascular plant species of raised beaches and sedge–moss mires on the Truelove Lowland, but they peak only in middle to late season, probably due to warmer soils.

### 6. Biomass and Net Production

At Barrow, microbial biomass ranged from 12 to 20 g $m^{-2}$; the lower values were found in the drier microhabitats of polygonal rims and the higher values were in wet meadows. On the Truelove Lowland, microbial annual production was 17 g $m^{-2}$ in the sedge–moss mires and only 2 g $m^{-2}$ on the drier raised beaches. Bacterial species of *Arthrobacter, Bacillus,* and *Pseudomonas* are able to utilize low levels of carbon and tolerate low temperatures, indicating that these species are well adapted to arctic conditions.

### 7. Nutrient Budgets

The shallow active layer (20–40 cm) at most sites, cold soils, and limited rates of mineralization of organic matter result in very limited pools of soil nutrients, especially N and P. At Barrow the inputs of N by fixation and precipitation account for only 5% of the nitrogen that cycles annually through the vegetation (2 g $m^{-2}$). Thus the limited pool of nitrogen available to plants results from litter decomposition. On the Truelove Lowland, fixation and precipitation account for about 13% of the nitrogen utilized per year (2.4 g $m^{-2}$). At both of these arctic ecosystems, annual input of nitrogen accounts for only about 0.01% of the total nitrogen stored in each system. The turnover time for nitrogen in the living and dead plant material is 11 yr at Barrow and 16 yr in the mires on the Truelove Lowland.

The cycling of phosphorus is even more limiting within these systems. On the Truelove Lowland the annual input accounts for only 2% of the annual amount cycling through the vegetation and only 0.005% of the total P within the system. At Barrow the comparable figures for wet mire are 0.6% and 0.002%. The turnover time for P is estimated to be 225 yr at Barrow and 410 yr at the Devon Island site. All of this indicates that these nutrient-deficient systems place severe constraints on the vegetation; species must be nutrient-conservative.

These ecosystems add little carbon per year to the large accumulation of carbon in the soils. This indicates that decomposition is slow and that these ecosystems appear to be in or near a steady state regarding carbon accumulation. This pattern could change significantly if global change results in warmer and possibly drier arctic summers.

#### *8. Energy Flow*

Energy flow models have been prepared for both the raised beach systems with their cushion plants and lemmings and the wet sedge–moss mires with muskoxen on the Truelove Lowland. Only models for the Devon Island and Barrow mires are presented here (Figs. 17 and 18). Although the species are quite different on the beach ridges, the basic function is very similar. Plants contributed 99% of the standing crop and 90% of the net annual production. The microflora, consisting mostly of bacteria, accounted for 0.9% of the standing crop, and the saprovores (Collembola, Crustacea, Diptera, Enchytraeidae, Rotifera, Tardigrada) that feed on the microflora and partially decomposed organic matter accounted for 0.1%. The muskoxen and birds, so conspicuous on the landscape, account for only a very small fraction of the standing crop and net annual production. The same can be said for the abundant but small soil invertebrates (Fig. 17).

The structure and function of the Barrow system is very similar (Fig. 18), with plants accounting for 98.7% of the standing crop, the microflora 1.1%, and the vertebrates and invertebrates an additional 0.2%. In both the Barrow and Truelove Lowland systems, most of the energy is diverted to the detritus component. Herbivores and carnivores are of great interest to people and they are the most appropriate trophic levels for humans to hunt. Yet they account for such a small fraction of the biomass and net annual production within these ecosystems that their hunting must be highly regulated if they are to continue to function as a part of these northern systems.

### C. Aquatic Ecosystems

Several arctic ponds and lakes have been studied, but rivers and streams have received little attention. The aquatic studies have included ponds at Barrow, Toolik Lake in the foothills, near the Brooks Range in Alaska, Char Lake on Cornwallis Island, the lakes on the Truelove Lowland, and a series of lakes northwest of Hudson Bay, Canada. The Kuparuk River in Alaska has been studied in terms of trophic structure and life history of key species.

Arctic lakes and ponds are species-poor in phytoplankton and zooplankton, a product of low water temperatures (seldom above 2° to 5°C, an ice-free surface for only a few weeks, and very low nutrient regimes). Ponds and shallow lakes freeze solid and thus support no fish. However, their shallow water permits emergent macrophytes to develop, resulting in much higher levels of primary production. The ponds at Barrow contained 45 species of benthic algae, mostly green algae and cyanobacteria. Phytoplankton included green and golden-brown algae, numbering 105 species. Their populations do not build up owing to the heavy grazing by Crustacea and two species of fairy shrimp. The lack of fish permits the presence of large species of zooplankton not found in lakes with fish and results in only two trophic levels. The high production of

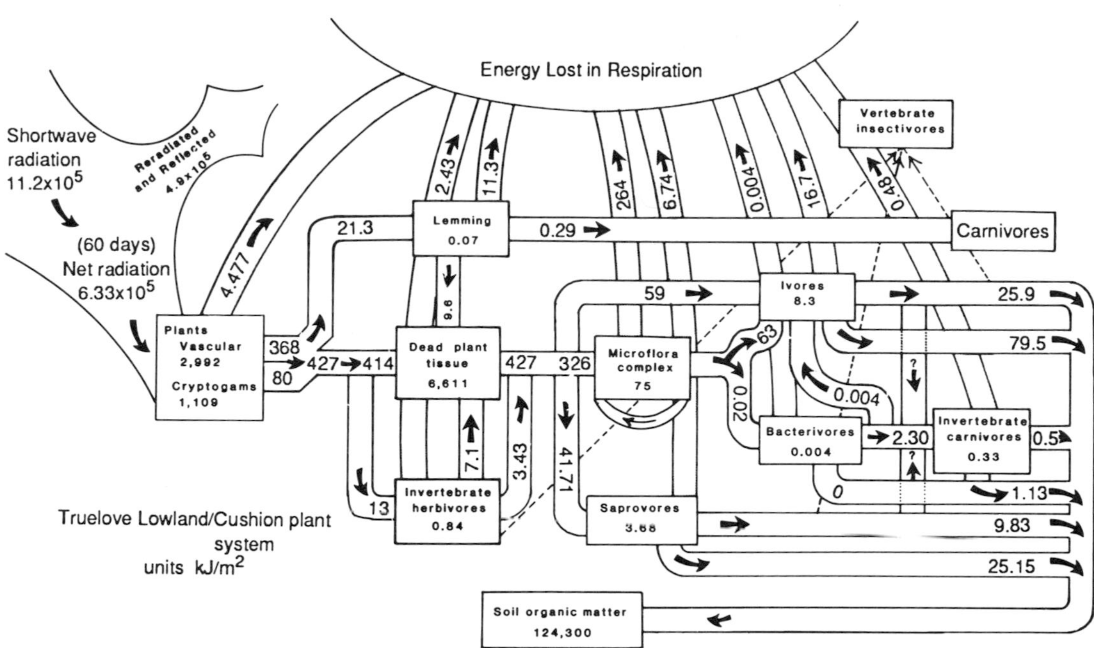


*FIGURE 17* Energy flow within a hummocky sedge-moss mire on the Truelove Lowland, Devon Island. Biomass (boxes) and energy flow (arrows) are expressed in kJ $m^{-2}$ [From L. C. Bliss (1986). "Grazing Research at Northern Latitudes." Plenum Pub. Corp. NY; D. W. A. Whitfield (1977). "Truelove Lowland, Devon Island, Canada: A High Arctic Ecosystem." Univ. Alberta Press. Edmonton, Canada.]

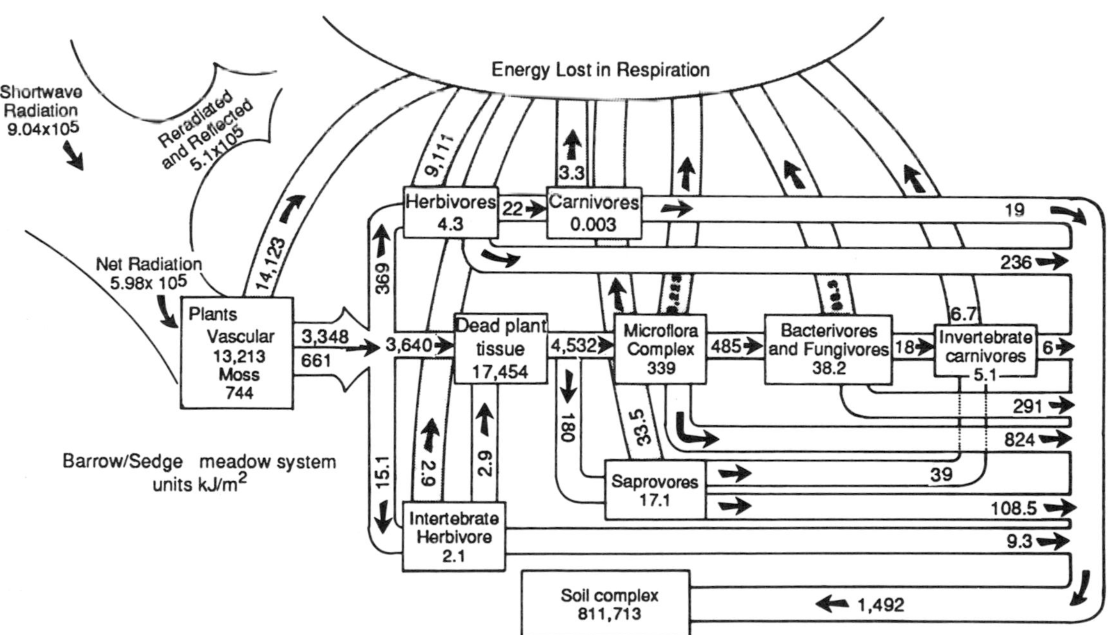


*FIGURE 18* Energy flow within a sedge-moss mire at Barrow, Alaska. Biomass (boxes) and energy flow (arrows) are expressed in kJ $m^{-2}$. [From L. C. Bliss (1986). "Grazing Research at Northern Latitudes." Plenum Pub. Corp. NY; S. F. Mac Lean (1980). "An Arctic Ecosystem: The Coastal Tundra at Barrow, Alaska." Dowden, Hutchinson and Ross, Inc., Stroudsburg, PA; F. S. Chapin, P. C. Miller, W. D. Billings, and P. I. Coyne (1980). "An Arctic Ecosystem: The Coastal Tundra at Barrow, Alaska." Dowden, Hutchinson and Ross, Inc. Stroudsburg, PA.]

macrophytes and benthic algae but low production by planktonic algae shift organic carbon to sediments and into the detritus compartment. The grazing food chain in these ponds is unimportant compared with the detritus food chain (Fig. 19). Net annual production of benthic algae averaged 14 g C $m^{-2}$ $yr^{-1}$ whereas phytoplankton averaged only 1 g C $m^{-2}$ $yr^{-1}$ in these Barrow ponds.

Biologically, arctic lakes are quite simple in terms of species richness and carbon production. Lakes in the Low Arctic often have 3 to 5 species of fish, but lakes in the High Arctic seldom contain more than one species. Char Lake on Cornwallis Island contains only arctic char, 2 species of zooplankton, and 6 species of midges, yet about 100 species of benthic algae and 21 species of nematodes. The lakes on the Truelove Lowland also contain only arctic char, landlocked populations as elsewhere in the High Arctic. Where the surrounding watershed is barren of vegetation, as at Char Lake, Cornwallis Island, and Loon Lake on The Truelove Lowland, arctic char grow more slowly and the lakes have a lower annual metabolism than lakes where the watershed contains more lush vegetation. Lakes with an outlet to the sea permit sea-run char and these lakes and rivers contain char that are larger. Eggs are typically laid in autumn, and hatch the following March or April, and the fry reach the ocean that summer. Landlocked populations reach sexual maturity in 6 to 17 years, whereas anadromous populations are mature in 4 to 7 years; egg production is much greater in anadromous populations. There are only three trophic levels in these lakes.

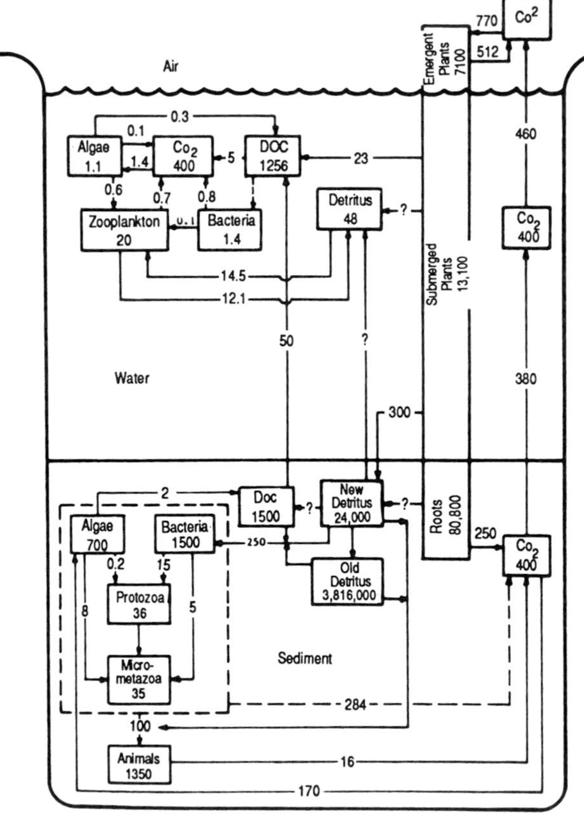


*FIGURE 19* Energy flow within ponds, Barrow, Alaska. Biomass (boxes) and energy flow (arrows) expressed in kJ $m^{-2}$. [From J. E. Hobbie (1980). "Limnology of Tundra Ponds." Dowden, Hutchinson and Ross, Inc., Stroudsburg, PA.]

There are at least 34 species of fish in the arctic rivers of North America and Russia, but only 9 species extend into the southern islands of the Canadian Archipelago, and only 3 species are found in the Queen Elizabeth Islands: arctic char (*Salvelinus alpinus*), ninespine stickleback (*Pungitius pungitius*), and fourhorn sculpin (*Myorocephalus quadricornis*). These same species are the only freshwater fishes of Greenland.

The upper reach of the Kuparuk River, Alaska, has been studied in terms of the influence of adding phosphoric acid on biological productivity. Algal biomass increased dramatically in years 1 and 2, but only slightly in years 3 and 4. Growth rates of the herbivore grazers (mayfly nymphs of *Baetis* and the chironomid *Orthocladus rivularum*) and a filter feeder (*Prosimulium*) increased. The more abundant invertebrates increased growth rates and fecundity of the only fish species, arctic grayling (*Thymallus arcticus*). This food chain comprising three trophic levels is a bottom-up control, but with a top-down feedback with the potential increase in grayling population in time. It has also been determined that the colonization cycle hypothesis holds for this river section. The river does not become biologically depleted of mayflies, for although the nymphs drift at least 2.1 km downstream, 30 to 50% of the adult *Baetis* fly an average 1.6–1.9 km upstream to lay their eggs, thereby maintaining the population and the river ecosystem.

### D. Marine Ecosystems

The Arctic Ocean is the smallest ocean in the world, covering about $12 \times 10^9$ km$^2$. This enclosed ocean, in contrast with the Antarctic Ocean, has limited water exchange with other oceans, is mostly ice covered, has a large input of fresh water from massive rivers, has low salt content and strong salinity gradients, and has a broad continental shelf. Shores of this ocean lie near 70°N. Also in contrast with the Antarctic, the Arctic Ocean is divided by groups of islands into the Greenland Sea, the Barents, Kara, Laptev, East Siberian, and Chukchi seas north of Russia, and the Beaufort Sea north of Alaska and Canada. The continental shelf lies about 150 m below the surface, averaging 200 km in width north of the Canadian islands and Alaska, but is wider north of Russia (400–1500 m). The central basin or abyss is 3000–3600 m deep; it is 5450 m at its deepest point.

As with the land masses, the Arctic Ocean is characterized by long, dark, and cold winters (−30° to −40°C), and the summers are short and cold (2° to −1°C) but with months of continuous light. Marine waters enter the Arctic Ocean mainly through Fram Strait, the passage between Svalbard and Greenland. Various parts of the Arctic Ocean differ in temperature and salinity. Surface waters (0–50 m) are coldest (−1.0° to −1.5°C) and are less saline (<31‰). The holocline separates the cold, saline waters (31–34‰) of the abyss. The nutrient levels of the Chuckchi, East Siberian, and western Beaufort seas are higher than those of the Barents and Kara seas. This reflects the more nutrient-rich waters of the North Pacific compared with North Atlantic waters.

#### 1. Phytoplankton and Macroalgae

The Arctic Ocean is not only the smallest but also the most oligotrophic of the world's oceans. There is a single bloom of phytoplankton per year, reaching a peak during summer (Fig. 20). The growth of ice algae generally precedes that of the phytoplankton. Historically it was believed that diatoms were the major producers, but now it is recognized that green algae and pyrophyta are also important.

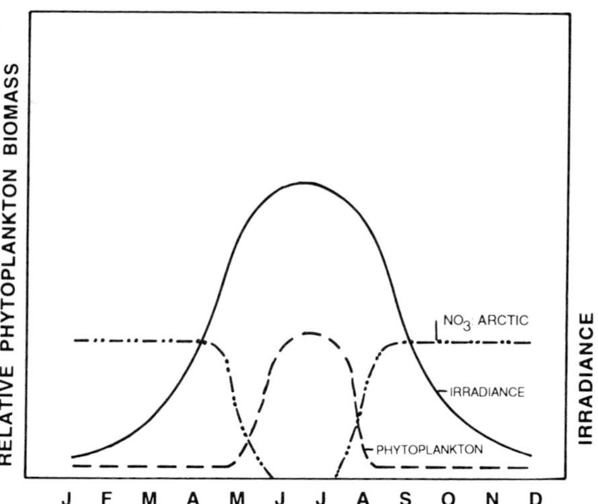


***FIGURE 20*** Bloom of phytoplankton in relation to irradiance and nitrate concentrations. [From W. O. Smith and E. Sakshaug (1980). "Polar Oceanography Part B, Chemistry, Biology and Geology." Academic Press, NY.]

In general, algal blooms occur at the edge of sea ice and are restricted to the upper 40–50 m of the water column. Algal blooms are initiated in April in the Barents Sea, but not until August off Ellesmere Island. About 200 species of phytoplankton have been reported in the Barents Sea [*See* OCEAN ECOLOGY; MARINE PRODUCTIVITY.]

Algal growth rates have been calculated as 0.76 dbl day$^{-1}$ (doubling per day) at −1.8°C and as 1.03 dbl day$^{-1}$ at 3°C, resulting in overall estimates of 0.31–0.75 dbl day$^{-1}$ for the Arctic Ocean. These growth rates result in primary production estimates of 225 mg C m$^{-2}$ day$^{-1}$ or 27 g C m$^{-2}$ yr$^{-1}$ in the continental shelf waters, but only 75 mg C m$^{-2}$ day$^{-1}$ or 9 g C m$^{-2}$ day$^{-1}$ in the open ocean. In the warmer waters of the Bering Sea, net production is estimated to be 89 mg C m$^{-2}$ day$^{-1}$ at the ice margin, and only 6 mg C m$^{-2}$ yr$^{-1}$ under ice and in open water. Nutrients are generally depleted, especially nitrogen, after a bloom (Fig. 20). Generally there is a deep vertical mixing of waters, with maximum nutrient content prior to a bloom. As light increases in spring, net production occurs and biomass increases (Fig. 20). As with terrestrial arctic systems, there is a greater uptake of $NH_4$ than $NO_3$ by the phytoplankton.

The limited data on macroalgae indicate a rich flora of 175 species in the Canadian arctic waters.

Of these, 160 have their affinity in the Atlantic, 3 in the Pacific Ocean, and 12 are endemic. Species of *Fucus* and *Laminaria* are important in the total biomass.

### 2. Zooplankton

Production of zooplankton is governed by the low temperature of ocean waters, extremes in solar radiation, and cycles in pelagic primary production. Copepods of the genera *Calanus* and *Metridia* are most frequently found, although *Microcalanus pygmaeus* and *Oithona similis* are also present. Studies from ice islands show that different species reach maxima at different depths.

Life-history studies at Ice Island T-3 in the Canadian Basin show that *Calanus hyperboreus* has a 3-yr life cycle: year one—egg to copepod stages II and III; year two—stages IV and V; and year three—adult and spawning. Gravid females are near the surface prior to the phytoplankton bloom, then they slowly descend from 100 m to 300 m. Males are always fewer than females and they are most abundant in deep waters (>400 m). Biomass of 1 mg $m^{-3}$ occurs in July–August at a depth of 50–100 m but decreases by two- to fourfold at depths of 200–1000 m. There appears to be little or no diel migration of organisms, only ontogenetic migrations.

Invertebrates are more abundant in the Chukchi Sea than the Beaufort Sea. The dominant copepods of the relatively shallow Kara, Laptev, and East Siberian seas include *Drepanopus bungei, Limnocalanus grimaldi, Oithona similis,* and *Senecella calanoides.* Medusae and copepods dominate in the warmer 5–10 m depth waters of the Beaufort Sea. The euphausiid genera *Euchausia* and *Thysanoessa,* so abundant in production of krill in the Antarctic, are minor in the Arctic Ocean. However, studies of *Thysanoessa* and the common *Calanus hyperboreus* show that lipid content increased from 29% in June to 64% of body weight in August after the phytoplankton bloom. It is believed that these high lipid levels provide the metabolic substrate used by *Calanus* in overwintering. The synthesis of wax esters can be transformed through the food web to whale blubber and to myctopid and bathypelagic fish.

### 3. Upper Trophic Levels

Arctic marine food webs are generally believed to have only three or four trophic levels from algae to top predators, however, a recent study from Lancaster Sound, between Baffin and Somerset islands to the south and Devon Island to the north, illustrates a more complex food chain (Fig. 21). There are few detailed studies on arctic marine mammals, marine birds having received more attention. The abundance of surface ice year-round is a major inhibitor for predators, consequently the number of species is limited but their body size is large (baleen and toothed whales, seals, walrus, and polar bears). Open waters in winter (polynyas) thus become very important, for algal blooms are greater at the ice margin, a substrate in which micronekton and macronekton seek refuge from predators. These polynya also permit seabirds and whales to feed in winter, rather than being forced to move farther south.

Within Lancaster Sound, phytoplankton are the major primary producers. Ice algae and kelp add lesser amounts of net production. Of the 116 species of macro-zooplankton sampled, 83% were calanoid copepods, 5% chaetognatha, 5% gastropods, and 4% amphipods. Four species of copepods, *Pseudocalanus acuspes, Calanus hyperboreus, C. glacialis,* and *Metridia longa,* account for nearly all the energy flow. Arctic cod feed primarily on large *Calanus* and amphipods. The abundant bivalve *Mya truncata* is a major component of the benthic community in the upper 15 m of water. Arctic cod (*Boreogadus saida*) occur in two types of distribution, as dispersed individuals and in schools, with an estimated upper number of 75,000 tonnes of biomass. It is estimated that about 148,000 tonnes of cod are consumed annually by whales (46,000 tonnes), seals (78,000 tonnes), and seabirds (23,300 tonnes). Arctic cod are central to the maintenance of this ecosystem (Fig. 21).

About one million seabirds nest along the rocky coasts of eastern Lancaster Sound and western Baffin Bay and several million nonbreeding birds spend part of each summer there. The major nesting species include Northern Fulmars (*Fulmarus glacialis*), Black-legged Kittiwakes (*Rissia tridatyla*),

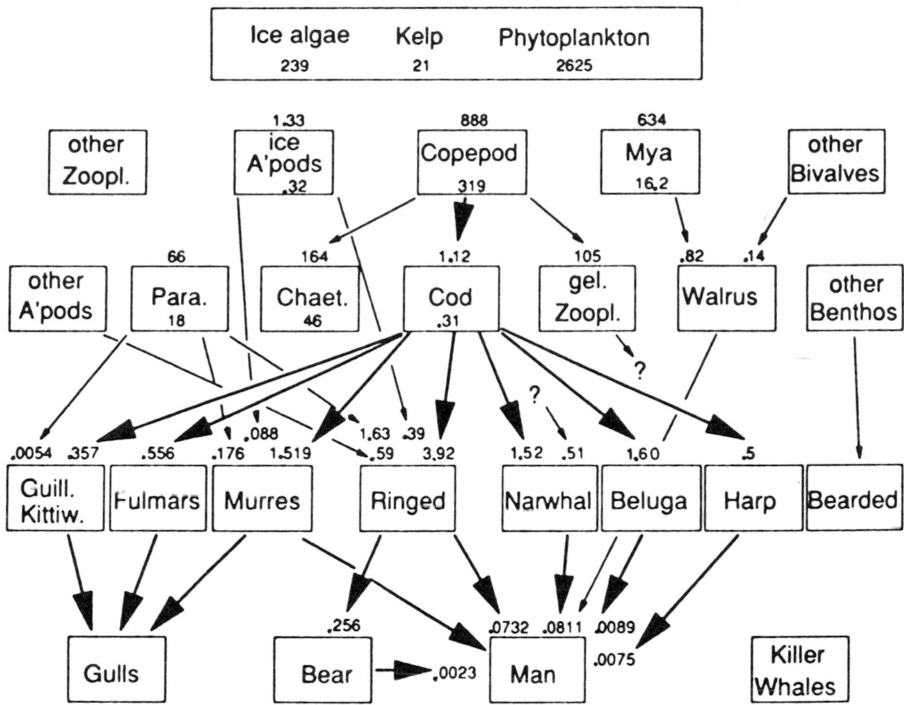


***FIGURE 21*** Trophic structure of a marine ecosystem, Lancaster Sound, N.W.T. Energy flow is given in kJ m $^{-2}$ yr$^{-1}$; numbers above boxes are ingestion, numbers in boxes are calculated net production (growth), and numbers for humans are kill rates. Note the importance of cod within this ecosystem. [From H. E. Welch (1992). *Arct. Alp. Res.* **23,** 11–23.]

Thick-billed Murres (*Uria lomiva*), Dovekies (*Alle alle*), and Black Guillemots (*Cepphus grylle*). These species nest from late June to early July. Fulmars, murres, and guillemots feed along fast ice edges and therefore are uncommon where pack ice is extensive. Only one seabird rookery is not within bird flight range of a polynya in the North American Arctic, indicating the importance of these open waters to the food chain. Oldsquaw and eider ducks nest on islands in lakes or near lakes, yet they are benthic feeders in shallow ocean and fresh waters.

Of the arctic seas, the Barents Sea appears to be one of the richest biologically. The phytoplankton number about 200 species and are fed upon by relatively few species of zooplankton. The number of fish species is also large, with Arctic cod, haddock (*Melanoghrammus aeglefinus*), and Barents sea capelin (*Mallotus villosus*) being some of the most important species.

The top carnivore in these marine systems is the polar bear (*Ursus arcticus*), especially in the marine ecosystems of the High Arctic. This circumpolar animal is not an endangered species, numbering about 10,000 individuals. Their common habitat is the broken edge of the polar ice pack and not multiyear ice, unless there are leads where they can hunt seals. They are solitary animals with acute smell but poor eyesight. They pay little attention to sounds for they are used to hearing cracking ice.

These animals consume mostly ringed seals but they also take harbor and bearded seals and young walrus. When eating seals, they consume skin and blubber first and meat last. In summer they float south on ice floes, come ashore, and work their way north along coasts. Many come ashore near Churchill, Manitoba, where they consume grasses, mushrooms, berries, and roots prior to freezeup. The lack of meat at this time of year results in their being hungrier and thus more dangerous to people.

Breeding occurs every other year in April and May. In autumn, pregnant females excavate dens in snowbanks on coastal hillsides or along pressure ridges at sea. They hibernate 160–170 days, during which time they typically give birth to two cubs.

Cubs stay with their mother the first summer and winter, but are driven off the spring of the second year. Young animals hibernate about 105 days and adult males only 50–60 days. Denning sites are common in eastern Lancaster Sound, especially the south coast of Devon Island, west coast of Banks Island, the Yukon coast, and Wrangle Island.

## VI. ROLE OF INDIGENOUS PEOPLE

Based on the limited detailed ecosystem studies in the Arctic, it is evident that primary production is much greater in terrestrial than in aquatic and marine ecosystems. However, the terrestrial systems are less able to support year-round production of large mammals, unless caribou are domesticated (reindeer), managed, and followed to their winter ranges within the northern boreal forests. In contrast, marine systems support five rather than three trophic levels and the fourth level (seals, walrus, whales, seabirds) provides more food at more dependable locations than do most terrestrial systems.

It is no accident that most indigenous peoples in the Arctic have lived along coastlines or major rivers or lakes, where they can harvest mammals from the sea and fish from the rivers or lakes. Caribou, muskox, and waterfowl are important sources of additional protein, supplemented with plant roots and berries in summer. Records from Alaska at the time Europeans first arrived indicate that adults consumed 3000 g of meat per day when available.

There are presently approximately 140,000 Eskimos in the Arctic. They prefer to be called by their native names, the Inupiat and Yupik in Alaska, Inuit in Canada, and Yuit in Russia. Indigenous peoples probably reached arctic Russia about 7000 to 8000 yr B.P., and North America via the Bering Land Bridge about 5000 yr B.P. They then spread rapidly through the Canadian Arctic to Greenland about 4000 yr B.P. The lives of these people began to change in the 1800s with the arrival of European explorers and whalers, with more significant changes in life-style beginning in the early to mid-1900s. Today most people live in settlements, although they still hunt and fish at considerable distances from their permanent homes. Much of their food is now purchased from stores, for the land and waters cannot support their rapid increase in population since the 1940s and 1950s.

## Glossary

**Active layer** That portion of the soil that thaws each summer. Depth generally ranges from 10 to 30 cm with peaty soils and 50 to 150 cm with sands and gravels.

**Arctic** Those lands that are climatically north of tree line circumpolar.

**Biomass** Living mass of plants and/or animals within a given area, usually expressed in grams or kilograms per unit area.

**Ecosystem** Community of plants, animals, and environment that is studied in terms of structure and function (e.g., energy flow and nutrient cycling).

**High arctic** Those arctic lands where there is a limited cover of flowering plants in which woody species are very minor, and where mammal, avian, and fish species are also very limited. These lands are mostly north of 70°N in North America.

**Low arctic** Those arctic lands where there is a nearly complete cover of flowering plants in which woody species are very important along with a rich fauna of mammals, birds, and fishes.

**Mire** Wet lands where the soils are relatively nutrient rich and nearer neutrality, and where graminoids predominate.

**Patterned ground** Result of soil churning that produces polygonal or striped patterns of rocks, soil polygons or stripes, and terraced slopes.

**Permafrost** Ground that remains frozen for two or more years. Continuous permafrost covers arctic lands but does not extend below large lakes or large rivers. Permafrost extends into the arctic seas.

**Polar desert** Those lands at high elevation in the Low Arctic and throughout much of the High Arctic where total plant cover averages 1–5%. The flora and fauna are greatly decreased in number of species compared with the Low Arctic and semideserts.

**Polar semidesert** Those lands in mountains of the Low Arctic and throughout the High Arctic where total plant cover averages 20–80% and where species richness of plants and animals is intermediate between tundra and polar desert.

**Trophic level** Food level within an ecosystem such as plants (producers), herbivores, carnivores (consumers), and decomposers.

**Tundra** Those lands within the Arctic, mostly Low Arctic, where plant cover approaches 100% and where plant and animal species richness is relatively high.

## Bibliography

Alexandrova, V. D. (1980). "The Arctic and Antarctic: Their Division into Geobotanical Areas." New York: Cambridge University Press.

Bliss, L. C., ed. (1977). "Truelove Lowland, Devon Island, Canada: A High Arctic Ecosystem." Edmonton, Alberta, Canada: University of Alberta Press.

Bliss, L. C., Heal, O. W., and Moore, J. J., eds. (1980). "Tundra Ecosystems: A Comparative Analysis." New York: Cambridge University Press.

Brown, J., Miller, P. C., Tieszen, L. L., and Bunnell, F. L., eds. (1980). "An Arctic Ecosystem. The Coastal Tundra at Barrow Alaska." Stroudsburg, PA.: Dowden, Hutchinson and Ross, Inc.

Chapin, F. S., III, Jefferies, R. L., Reynolds, J. F., Shaver, G. R., and Svoboda, J., eds. (1992). "Arctic Ecosystems in a Changing Climate: An Ecophysiological Perspective." New York: Academic Press.

Chernov, Y. I. (1985). "The Living Tundra." New York: Cambridge University Press.

Hobbie, J. E., ed. (1980). "Limnology of Tundra Ponds." Stroudsburg, PA.: Dowden, Hutchinson and Ross.

Hobbie, J. E. (1984). Polar limnology. *In* "Lakes and Resources. Ecosystems of the World" (F. B. Taub, ed.), Vol. 23, pp. 63–105. Amsterdam: Elsevier.

Irving, L. (1972). "Arctic Life of Birds and Mammals." New York: Springer-Verlag.

Postma, H., and Zijlstra, J. J., eds. (1988). "Continental Shelves. Ecosystems of the World," Vol. 27. Amsterdam: Elsevier.

Remmert, H. (1980). "Arctic Animal Ecology." New York: Springer-Verlag.

Smith, W. O., Jr., ed. (1990). "Polar Oceanography. Part B. Chemistry, Biology and Geology." New York: Academic Press.

# Nutrient Cycling in Forests

**Peter M. Attiwill**
*University of Melbourne*



Nutrient cycling in forests is an all-embracing term that includes the range of processes that govern the availability of nutrients for the growth of forest trees, interactions between plant and soil in the uptake and return of nutrients, microbial interactions in which nutrients are transformed between organic and inorganic forms, and the balance between inputs and outputs of nutrients. Growth of forests is sustained through processes of nutrient cycling; by these processes, nutrients are accumulated in surface soils where they are again available for uptake, and losses of nutrients from the ecosystem are minimized.

## I. INTRODUCTION

A continuous supply of nutrients in inorganic form is essential for the growth of plants. In modern ecology, plants are treated as part of the ecosystem—the whole complex of all the biotic and abiotic aspects of an area. The study of ecosystems, then, is the study of the processes by which energy and materials essential for functioning (e.g., water and nutrients) or detrimental to functioning (e.g., pollutants) enter, leave, and are transferred within the ecosystem. Nutrient cycling is a term used to cover all the pathways and processes by which nutrients enter, leave, and are transferred within forest ecosystems. Nutrient cycling is both a determinant and a consequence of forest growth. The study of nutrient cycling in forests therefore covers many topics, and it has produced an extensive literature.

The study of nutrient cycling is generally traced to the work of E. Ebermayer in 1876. At that time in middle Europe, litter from the forests—the twigs and leaves and other fine materials shed from the trees and lying on the forest floor—was gathered by raking and used as bedding or litter in the barns where farm animals spent the winter. After the winter, the litter, now enriched with animal droppings, was spread over agricultural fields. The nutrients in the litter that otherwise would have been available to the trees were thereby removed, and Ebermayer showed that this removal reduced the growth rate of the forest.

Ebermayer's work stimulated a large amount of work, particularly in Europe, on the distribution of nutrients in forests. By the 1960s there was a mass of data on nutrient pools for a whole range of forests around the world, but it was recognized that this information on its own was static; since the 1960s, there has been a concentration of studies on the dynamic *processes* of nutrient cycling. It is only through the knowledge of processes that we can begin to answer questions such as: How can nutrient supply for forest growth be sustained? How can long-term nutrient supply be assessed and what are the effects on supply of disturbances such

Encyclopedia of Environmental Biology
Volume 2

as wildfire and logging? How are the nutrients that have been cycled during forest growth conserved within the ecosystem, and does disturbance of the ecosystem result in a loss of nutrients?

The thirteen elements known to be essential for the inorganic nutrition of plants vary in their chemistry, in their geological abundance, in their biochemical roles, and in their mobility in soils, plants, and microorganisms. The study of nutrient cycling in forests therefore covers many fields, including soil chemistry, plant physiology and biochemistry, soil microbiology, geochemistry, and hydrology. Most of our knowledge, however, centers on the two macroelements nitrogen and phosphorus, which are most critical to the sustained productivity of forests. In contrast, our knowledge of the cycling of the trace elements—like copper, zinc, and molybdenum—is poor. [*See* SOIL ECOSYSTEMS.]

## II. THE COMPONENTS OF NUTRIENT CYCLING

Diagrammatic and schematic models of nutrient cycling (Fig. 1) can be considered for a number of components:

- inputs of nutrients in rain, dust, by biological fixation, and by weathering of parent rock;
- outputs of nutrients in drainage waters, in gaseous forms, and in harvested materials;
- uptake of nutrients from soil by plants;
- internal redistribution of nutrients within the plant;
- return of nutrients from plant to soil by leaching, in litterfall, in root turnover, and by mortality and treefall;
- decomposition of the forest floor, incorporation of organic matter in soil, and mineralization of organically bound nutrients to simple ion, available once more for uptake.

These components will be discussed in the following sections; some of the major points are summarized for the nutrients nitrogen, phosphorus, calcium, magnesium, and potassium in Table I.

### A. Inputs of Nutrients

#### *1. Inputs in Rain and Dust*

Nutrients are contained within both rainwater and dust, but where rainfall is collected in open gauges (as is most common), there is no discrimination between the two and the term "bulk precipitation" is sometimes used. Inputs in dust can be locally important (e.g., along roadsides) but more generally it can be assumed that there is an equivalent output.

The source of nutrients in rainwater is aerosols in the atmosphere. Aerosols are small particles that act as condensation nuclei for water vapor or that are washed out of the atmosphere by falling rain. The origins of aerosols are oceanic, terrestrial, extra-terrestrial, biogenic, industrial, and volcanic. However, the principal nuclei for condensation come from sea spray. Sodium is therefore the dominant ion in rainwater near coasts, and the concentration ratio of sodium ion to other ions decreases with increasing distance from the coast. At inland sites, calcium and magnesium from dust may become dominant. The annual input of nutrients to forests in rainwater is therefore highly variable depending on location (Table II). At some locations, the annual input of some nutrients (e.g., sulfur) is more than enough to meet the demands of plants. Even small inputs of sodium (a nonessential element for plants) at dry inland sites can accumulate in the subsoil over thousands of years, to emerge to the surface when evapotranspiration is altered by land-clearing for agriculture; this is the problem of "dry-land salting." In some rain shadow areas of the world, there is so little iodine in rain that local people have suffered from goitre; this has been simply overcome with iodized salt.

There is debate in the literature on the extent to which aerosols impact and adhere to tree leaves, later to be washed off in rain and thereby contributing to nutrient input. There is no doubt that impaction is of local importance, particularly in

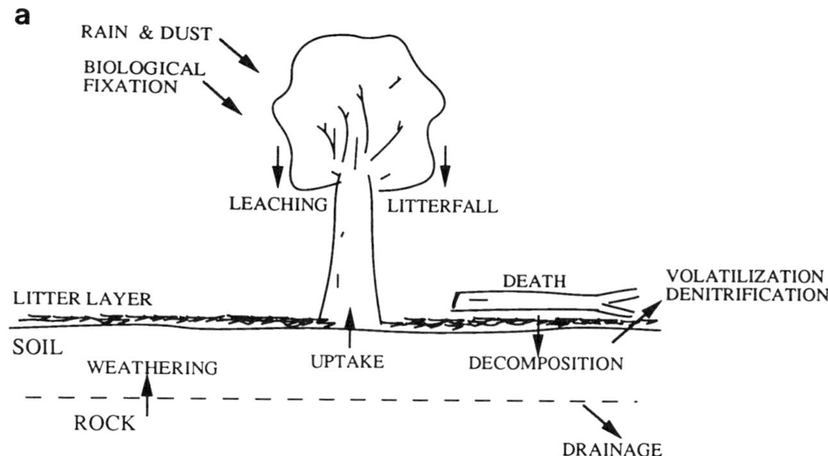


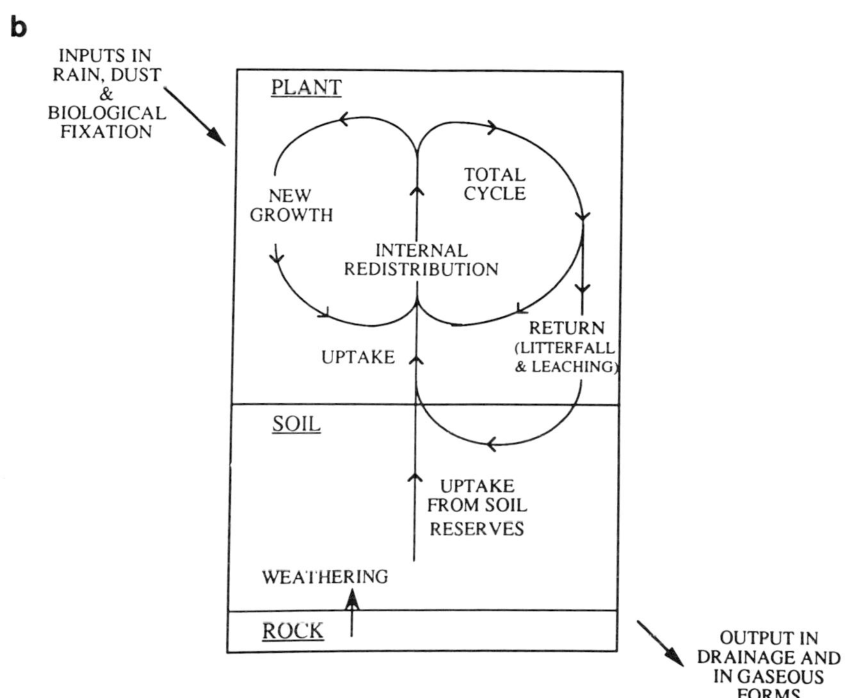


***FIGURE 1*** Diagrammatic (a) and schematic (b) representations of the cycling of nutrients in a forest [From P. M. Attiwill and M. A. Adams (1993). *New Phytologist* **124,** 561–582.]

coastal areas as shown by the salt-pruning of coastal vegetation. In general, however, the evidence for aerosol impaction making a significant contribution to nutrient input is not strong.

### *2. Fixation of Atmospheric Nitrogen*

Inputs of nitrogen are of special interest because the rocks from which soil has formed contain little or no nitrogen. All the nitrogen in a forest ecosystem therefore represents the accumulated inputs over time of nitrogen in rain and by biological fixation of atmospheric nitrogen. To put this into perspective, the atmosphere contains 78% by volume of the gas nitrogen ($N_2$), a form of nitrogen that plants cannot use for nutrition. At the global scale, the annual input from the atmosphere to the biosphere (excluding the use of fertilizers) is 150 megatons (Mt). Inputs in rain account for about

**TABLE I**
Summary of the Cycling of Major Elements in Temperate Forests

| Element | Role in nutrition | Form taken up by plants | Typical concentration in tree leaves (mg/kg dry weight) | Mobility in plants | Typical amount in aboveground parts of trees (kg/ha) | Main cycling pathway | Release during litter decomposition | Mobility in soil |
|---|---|---|---|---|---|---|---|---|
| Nitrogen | Constituent of amino acids, proteins, nucleic acids, etc. | $No_3^-$ $NH_4^+$ | 10–20 | Moderate | 200–500 | Litterfall> redistribution> leaching | Immoblized until late stages | $NO_3^-$: in solution and highly mobile; $NH_4^+$: cation exchange and relatively immobile |
| Phosphorus | Constituent of sugar phosphates, nucleic acids, key role in ATP–ADP reactions | $H_2PO_4^-$ | 0.7–2.0 | Highly mobile | 10–50 | Redistribution> litterfall> leaching | Immobilized until late stages; can be leached in some forest types | Adsorbed, highly immobile |
| Calcium | Structural, especially in middle lamella of cell walls | $Ca^{2+}$ | 2.0–10.0 | Mostly immobile | 100–400 | Litterfall> leaching> redistribution | Released at rate of litter decomposition | Cation exchange and relatively immobile |
| Magnesium | Central atom in chlorophyll molecule | $Mg^{2+}$ | 2.0–5.0 | Moderate | 20–200 | Litterfall> redistribution> leaching | Released faster than litter decomposition | Cation exchange and relatively immobile |
| Potassium | Enzyme reactions, osmotic relations | $K^+$ | 2.0–15.0 | Mobile | 100–300 | Leaching> redistribution> litterfall | Leached early during decomposition | Cation exchange and relatively immobile |

**TABLE II**
The Input of Nutrients in Rainfall to Forests

| Nutrient | Amount (kg/ha/year) |
|---|---|
| Nitrogen | 3–20 |
| Phosphorus | Trace–0.5 |
| Calcium | 1–17 |
| Magnesium | 1–10 |
| Potassium | 1–20 |
| Sodium | 1–150 |
| Chlorine | 1–345 |
| Sulfur | 2–300 |

20 Mt/year and the remaining 130 Mt/year is due to biological fixation. [*See* NITROGEN CYCLE INTERACTIONS WITH GLOBAL CHANGE PROCESSES.]

Biological $N_2$ fixation is the process whereby microorganisms use energy derived from photosynthesis to reduce atmospheric nitrogen to ammonia:

$$N_2 + 3H_2 \rightarrow 2NH_3$$

All the microorganisms that are capable of $N_2$ fixation are prokaryotic, that is, they lack a true nucleus, and the enzyme responsible for the reduction is structurally and functionally similar in all $N_2$-fixing microorganisms. The enzyme nitrogenase contains two proteins, the larger of which includes molybdenum. Apart from these similarities, $N_2$-fixing microorganisms vary widely. Some are free-living and nonsymbiotic (e.g., some bacteria and blue-green algae), some are symbiotic, and some are difficult to classify (e.g., microorganisms on the leaves of some tropical trees fix $N_2$ but the association with the tree is not truly symbiotic). The symbiotic organisms are perhaps the best known; they include the bacteria *Rhizobium,* which forms nodules in the roots of leguminous plants (e.g., species of *Acacia* and many other shrubs in Australian forests); an actinomycete that forms nodules in the roots of the tree species *Alnus* and *Casuarina;* lichens, which are associations between a fungus and an alga; the fern *Azolla,* which has an algal symbiont that can fix large amounts of nitrogen (e.g., in rice paddies), and many others.

In agriculture, the management of $N_2$ fixation is highly advanced. Inputs of 100–200 kg/ha/year are routine, and alfalfa has been reported at 600 kg/ha/year. Rates of this order have been recorded for *Alnus* in North America, but for many other forests of the world, rates of 2–5 and at the maximum 10 kg/ha/year are common. The major difficulty is that there is no simple and routine method for the reliable measurement of the rate of $N_2$ fixation. Chemical analysis of the soil is of little help. Suppose a forest soil contains 10 t/ha of nitrogen; a rate of $N_2$ fixation of 10 kg/ha/year adds only 0.1% to the total, far too small to detect within the normal variability of soil analysis. Increasingly, estimates of nitrogen fixation are based on the measurement of the natural enrichment of $^{15}N$, but this method may not be universally applicable and is certainly not routine.

### 3. Rock Weathering

The nutrients that are available to forest trees, with the exception of nitrogen as just outlined, come from the chemical weathering of minerals contained in the rock from which soil has formed. This is not an input into the ecosystem of which, of course, the parent rock is a part, but rather it is an input of nutrients from unavailable forms into the pool that is available to plants. Estimates of the rates of rock weathering have come from comparisons of elemental concentrations in igneous rock and in weathered rock, from measurement of the weathering rinds on glacial, granitic boulders, and from mass balance studies, where for a given nutrient,

$$X_w = (X_o - X_i) + X_{p,s},$$

where $X_w$ = mass of nutrient release by rock weathering;
$X_o$ = mass of nutrient in streamflow;
$X_i$ = mass of nutrient in rainfall; and
$X_{p,s}$ = change of mass of nutrient in plants and soil.

A rule-of-thumb is that igneous rocks weather at the rate of 1 cm per 1000 years. Estimates of nutrient inputs by this rule are compared with the mass of nutrients in a forest in Table III.

## B. Outputs of Nutrients

### 1. Output in Drainage Waters

The water balance of an ecosystem is given by

$$P = I + D + U + ET + S$$

where $P$ = precipitation (rain and snow);
$I$ = interception by vegetation, and then evaporated;
$D$ = discharge to streams;
$U$ = deep seepage;
$ET$ = evapotranspiration (evaporation plus transpiration by plants); and
$S$ = change in soil moisture.

The water balance can be studied within defined, water-tight catchments. Precipitation can be accurately measured. The rate of discharge, or streamflow, is measured in specially calibrated V-notch weirs, constructed so that all the water flowing from the catchment passes through the weir. The water balance can then be calculated provided that deep seepage *(U)* is negligible. The catchments must therefore be carefully chosen.

As water percolates through the forest soil on its way to streams, the concentration of ions in solution reaches equilibrium through adsorption and exchange. Concentrations of ions in streamwater are the result of these processes. The sampling of streamwater for chemical analysis is relatively simple because concentration of most ions is largely independent of the rate of streamflow, the consequence of the strong buffering capacity of forest soils. The concentration of phosphorus (as orthophosphate ion, $H_2PO_4^-$) in streamwater is generally very small ($10^{-6}$ mol/litre or less) because phosphate is strongly adsorbed by soil colloids. Concentrations of the more mobile ions like nitrate ($NO_3^-$) are variable as $NO_3^-$ is not held by soil colloids. If a forest soil produces no $NO_3^-$, then disturbance of the ecosystem (by natural treefall, logging, or fire) temporarily increases the loss of $NO_3^-$ in streamwater. In deciduous forests of the Northern Hemisphere, the strong biological control of nitrogen cycling is shown by an increase of $NO_3^-$ in streamwater in winter when the trees are not taking up nutrients. In contrast, the concentration of $Na^+$—not an essential element for plants—in streamflow (generally on the order of $10^{-4}$ to $10^{-3}$ mol/liter) is not biologically regulated, and the amount of $Na^+$ leaving a given catchment in streamflow is primarily a function of the water balance.

**TABLE III**
Amounts of Nutrients Released from Weathering of a Granite Compared with the Amount of Nutrients in a Eucalypt Forest

| | Amount of nutrient (kg/ha) | | | |
|---|---|---|---|---|
| | P | Ca | Mg | K |
| Released during 50 years of rock weathering | 7 | 170 | 60 | 440 |
| In aboveground stand of 50-year-old eucalypt forest | 27 | 310 | 190 | 230 |

### 2. Other Losses

Nitrogen behaves differently from the other nutrients in that it can be lost in gaseous form, either as ammonia ($NH_3$) where pH is high, or as nitrogen ($N_2$) or oxides of nitrogen ($NO_2$, $NO$, $N_2O$) where facultative anaerobic bacteria reduce $NO_3^-$ in the process of *denitrification*. *Volatilization*—the conversion of $NH_3$ to $N_2$—is not common in forest soils, which are mostly acidic, but may be significant following heavy applications of nitrogenous fertilizer. Rates of denitrification can be considerable under reducing conditions (e.g., flooded soils) and following the application of N fertilizers. Although populations of denitrifiers are ubiquitous in soils, rates of denitrification are not easy to measure in the field. Under undisturbed forest conditions, the few studies of denitrification have found low to insignificant rates and it is generally ignored in nutrient cycling studies (see earlier comments on the difficulties of measuring small changes in N balance).

Wildfires are a common disturbance in many forest ecosystems. Nitrogen and sulfur in organic forms are oxidized and volatilized at relatively low temperatures (>300°C or so) and these temperatures are reached during wildfire in the litter layers and upper few centimeters of the soil. The other nutrients are more refractory; however, they too can be lost in significant amounts in ash and other particles carried off by turbulence and convection. Fire increases the available supply of most nutrients in the soil and may therefore lead to an increase in the rate of loss of nutrients in streamflow (see earlier discussion). Finally, any major disturbance such as wildfire leads to the likelihood of loss of topsoil by erosion. The nutrients that have been cycled by forest growth are concentrated in the topsoil and losses by erosion can produce localized areas of major nutrient depletion.

### C. Uptake of Nutrients by Trees

Most of the work on the uptake of nutrients by forest trees has been based on a balance-sheet approach in which the pools of nutrients in various components of the forest are measured. The methodology for determining dynamic estimates of nutrient fluxes has not been developed—the results of glasshouse experiments on nutrient uptake with 1- or 2-year-old plants, for example, are of little value in considering the long-term (100 years or more) events in a forest. A further limitation in most balance-sheet studies is that they are based only on the aboveground parts of trees because the physical and technological difficulties of assessing both structure and function of the root systems of trees are enormous.

The root system has two functions; first, it is the anchoring structure that supports the vast mass of stem, branches, and leaves above ground. This part of the root system is large and structurally and functionally highly developed. The second function is in the acquisition of nutrients and water for forest growth, and this is achieved through fine roots and their relatively simple root hairs, both of which are short-lived. The root system of most forests is associated with a specific fungus or group of fungi, and this association is called a *mycorrhiza*. The mycorrhizal association stimulates intensive branching of the short, lateral roots and this, together with the prolific hyphae of the fungus, results in a very considerable increase in the surface area effective in nutrient uptake. Furthermore, it seems likely that a variety of organic acids and catalytic enzymes are excreted into the soil by mycorrhizas and that these acids solubilize sources of nutrients that would be otherwise unavailable to the plants. However, we know little about root physiology and turnover in most forest ecosystems. [*See* RHIZOSPHERE ECOPHYSIOLOGY.]

The annual demand for a given nutrient for the functioning and growth of a forest is met in part by internal redistribution, in part by return from plant to soil, and in part by uptake from soil reserves (Fig. 1). The relative importance of these three processes varies with developmental stage, as shown for a *Eucalyptus* forest in Fig. 2. In the first years, all the photosynthetic gains go toward increasing short-lived and physiologically active tissues—foliage, metabolic transport systems, and fine roots. At this time, uptake of nutrients from soil reserves is at a maximum. When leaf area and transport systems in the stem (phloem and xylem) reach their maximum development, most of the photosynthetic gains are stored in the stem as physi-

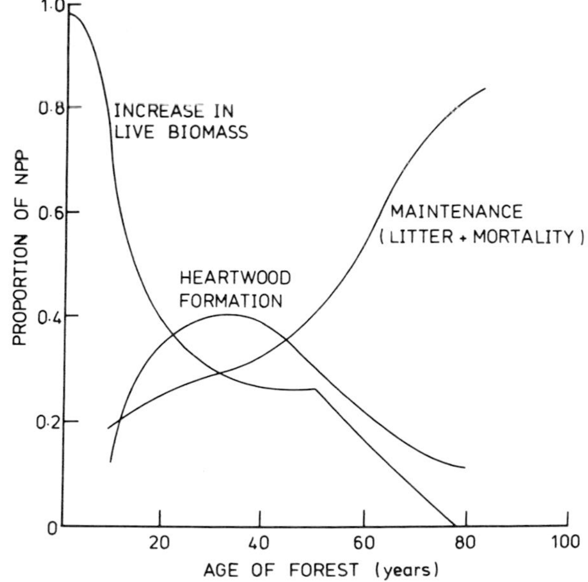


**FIGURE 2** Changes in the allocation of annual net primary production (NPP) to various growth processes in a forest *(Eucalyptus obliqua* in Australia). [From P. M. Attiwill and G. W. Leeper (1987). "Forest Soils and Nutrient Cycles."Melbourne University Press, Melbourne, Australia.]

ologically inactive heartwood. Then as heartwood accumulates, the stem reaches a structural limit; the forest is then said to be mature, net biomass increment is close to zero, and most of the annual photosynthetic gain is shed from the trees as litter.

The pattern of nutrient cycling therefore changes markedly throughout the life of the forest. In the first years, the trees obtain all of their nutrients by uptake from soil reserves. As the forest matures, increasing proportions of nutrient supply come from internal redistribution and from nutrients released from decomposing litter and roots.

### D. Internal Redistribution of Nutrients within Trees

The various nutrients differ greatly in their mobility within plants. For example, much of the calcium is bound within the middle lamella of cell walls, where it is immobile. In contrast, potassium is nonstructural and highly mobile, moving in ionic form. Phosphorus as orthophosphate, $H_2PO_4^-$, is involved in energy transfers in which organic compounds are phosphorylated and dephosphorylated, and it too is highly mobile.

Trees are a mixture of physiologically active and inactive tissues. As new, physiologically active tissues develop and create a demand, the mobile nutrients (e.g., phosphorus, nitrogen, potassium) are *translocated* from aging and senescing tissues. For example, the mobile nutrients are translocated from senescing leaves before abscission, the mobile nutrients are translocated from aging sapwood before it becomes physiologically inactive (heartwood), and translocation during fine-root turnover meets much of the nutrient demand for the production of new roots.

Although translocation is a physiological process, the difficulties of studying it directly in structures as large as forest trees are enormous. Most estimates of translocation are therefore based on the results of translocation, and this result is a *redistribution* of nutrients. Even redistribution is difficult to study where age of the tissues cannot be determined accurately and where variation between tissues of the same kind is large. In deciduous trees, the day of leaf initiation is known with precision, and in pines the position of a needle within the annual whorl provides a good estimate of age. In evergreen hardwoods, however, sampling leaves of a given age can be extremely difficult.

The magnitude of internal redistribution of nutrients probably differs between deciduous and evergreen trees, and for the latter between angiosperms and gymnosperms. It is also influenced by soil fertility and by interactions between soil fertility and climatic regulation of productivity. Although most estimates of redistribution are based on measurements of the change in the *total* amount of a given nutrient with time, much of the current work aims at identifying the *forms* in which the given nutrient is translocated. This work should greatly improve our understanding of the adaptation and functioning of forest ecosystems.

### E. Return of Nutrients from Trees to Soil

#### *1. Foliar Leaching by Rain*

Some of the rain *(R)* that falls onto a forest is intercepted *(I)* by the tree crowns from where it evaporates. The rain that falls through the canopy to the forest floor is called throughfall *(T)* and that which flows down the tree stems is called stemflow *(S)*, so that

$$R = I + (T + S).$$

Both $T$ and $S$ vary widely with species and location, depending on foliage type and foliage area index, type of bark, and amount and intensity of rainfall. Throughfall is commonly 70–90% of rainfall but may be as little as 50% in pines. Stemflow is commonly around 5% of rainfall but may be as little as 2% in forests with fibrous-barked stems that absorb a lot of water.

Both throughfall and stemflow are enriched in chemical composition relative to rainfall. Though part of this enrichment may be due to dissolution of aerosols impacted on the tree crowns (see Section II, A,1), most evidence shows that enrichment is due to leaching. In laboratory experiments, inorganic ions and a range of carbohydrates, amino acids, and organic acids are readily leachable from leaves. Furthermore, ionic ratios (e.g., $Na^+/K^+$) in throughfall are markedly different from those in rainfall; because the ionic composition of rainfall is itself due to aerosols, we would expect no difference in ionic ratios if the chemical enrichment of throughfall was also due to aerosols. For example, the ratio $Na^+/K^+$ is much lower in throughfall relative to that in rainfall, showing that $K^+$ has been leached from leaves in greater amounts than has $Na^+$. In fact, most studies of nutrient cycling world wide show that, of all the nutrients, $K^+$ is leached in the greatest amount. In contrast, $H_2PO_4^-$ is leached only in trace amounts.

For some species, pH is greater in rainfall than in throughfall, whereas for others it is less; thus for some species cations must be exchanged on the leaf surface. Leaching therefore involves both cation exchange and diffusion across a concentration gradient. The pH of stemflow may be as low as 3, and this is due to organic acids and to organic residues produced by decomposition in the moist stem bark.

### 2. *Litterfall and the Return of Nutrients from Plant to Soil*

At maturity, a large proportion of the annual net primary production (NPP) is shed as litter (Fig. 2) and the rate of litterfall therefore provides an index of productivity. The rate of litterfall ranges from 1 t/ha/year in forests of lower productivity (drier or colder areas) to 10 t/ha/year in warmer and wetter forests. The annual return of nutrients from plant to soil includes up to 200 kg/ha of nitrogen and 10 kg/ha of phosphorus. Tree litter is particularly rich in calcium that is not translocated from plant tissues prior to litterfall.

In deciduous forests of the Northern Hemisphere, all the leaves fall as litter in autumn, the end of the growing season. In evergreen forests, litterfall is more or less continuous throughout the year, but with a maximum in summer. Litter is not homogeneous. About half of the litterfall in a eucalypt forest, for example, is leaves that have senesced and died and the other half is a mixture of more woody components—twigs, bark, and fruit capsules. The death and fall of large trees is also part of the return of nutrients to the soil. In some forests, these trees may lie on the forest floor in a slowly decomposing state for centuries, and so the time delay between treefall and the release of nutrients in an available form may be very great.

## F. Decomposition and Mineralization

### 1. *Decomposition of the Litter Layers*

Litter falls onto the forest floor—the litter layer. The litter layer forms a *mull* (transition between organic layers and mineral soil is not obvious) in some forests or a *mor* (transition is obvious) in others. The litter layer is a zone of comminution (breaking down to smaller pieces) and decomposition through the digestive and oxidative activities of bacteria, fungi, and a range of micro- and macrofauna. Through these activities, carbon in organic matter is oxidized to the gas carbon dioxide, and the various elements essential to plants are mineralized—transformed from various organic combinations to their inorganic form of simple ion, available once more for uptake by plants.

Litter as it is shed from trees is a mixture of organic compounds ranging from less decomposable (e.g., lignin), through moderately decomposable (e.g., cellulose), to easily decomposable (e.g., soluble sugars). During decomposition, the most readily available substrates are attacked first, and the least available become concentrated. Thus decomposition involves the eventual formation of a residue that is completely and indefinitely stable.

At maturity, the weight of the litter layer $(Q)$ is constant with time and annual litterfall $(A)$ is equal to decomposition, and so we can calculate an annual decomposition constant:

$$k = A/Q \text{ per year.}$$

This is the proportion of litter that is decomposed each year. Decay therefore follows an exponential pattern.

The change in weight of the litter layer with time $(t)$ is

$$dQ/dt = -kQ$$

and if at $t = 0$ the weight of the litter layer is $Q_o$, then

$$Q = Q_0 e^{-kt}$$

and the half-life is

$$t_{0.5} = 0.693/k \text{ years.}$$

Values of $k$ vary from 0.01 ($t_{0.5}$ about 70 years) in the cool wet forests of the Northern Hemisphere, where deep litter layers develop, to 2 ($t_{0.5}$ about 0.35 year) in the warm wet rain forests of the tropics, where there is almost no accumulation of litter on the forest floor.

Forest litter is of relatively poor nutritional quality as assessed by agricultural standards. Organic matter is a source of carbon and hence energy for decomposers, but these decomposers also need nutrients. For example, organic manures rich in nitrogen and with a carbon to nitrogen ratio (C/N) less than 10 are of high quality as a green manure for agriculture. Nitrogen is in excess of the requirements of microorganisms to metabolize and oxidize the carbon supply, and this nitrogen is *mineralized*. However, at C/N ratios greater than 20, carbon is plentiful but nitrogen is scarce; because heterotrophic microorganisms are strong competitors for nitrogen, they *immobilize*—incorporate in their own biomass—the small amount of nitrogen that is available from litter and soil. Plants are weaker competitors for nutrients, and under these conditions they may run into nitrogen deficiency.

The C/N ratio of forest litter may be as great as 70 or more, and the C/P ratio as great as 1000–2000. During the first stages of litter decomposition, therefore, nitrogen (and in many forests, phosphorus) is immobilized. In eucalypt forest, for example, the total quantity of these nutrients in the decomposing litter increases (Fig. 3), as the decomposers must use N and P from external sources in their metabolism of the carbon supply. In contrast, potassium is not limiting to the growth of the microbial population; it is released from organic matter, most probably by simple leaching, at a faster rate than the rate of decomposition. The rate of release of calcium falls in between the extremes and generally follows the rate of decomposition quite closely.

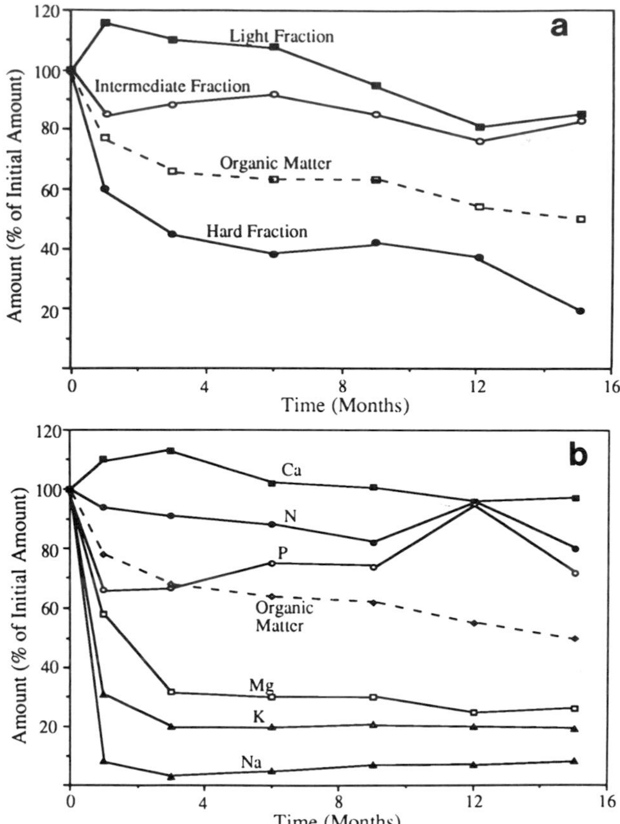


***FIGURE 3*** Changes in the amounts of empirically defined organic fractions (a) and elements (b) in decomposing leaf litter. [From J. Maheswaran and P. M. Attiwill (1987). *Can. J. Bot.* **65,** 2601–2606.]

The question remains: How do forest trees compete successfully for nitrogen and other nutrients in an environment of sustained and intense decomposition and immobilization? The incorporation of 10 t/ha of forest litter each year with a C/N ratio of 70 in an agricultural field would cause severe nitrogen deficiency and decreased productivity of the crop plants. A likely answer is that, through the mycorrhizal association between root and fungus, the tree becomes heterotrophic and therefore a stronger competitor for nitrogen. The role of mycorrhizas in phosphorus uptake by forest trees is more definitive. Mycorrhizas increase the affinity of the root for $H_2PO_4^-$ and may solubilize phosphorus compounds in the soil by the release of organic acids and phosphatase enzymes.

### 2. Decomposition of Soil Organic Matter

Soil organic matter is heterogeneous, and the major group of compounds is called "humic acid." In the most general sense, we can say that humic acid is a colloid, that it has a molecular weight that varies around 200,000, and that it includes compounds derived from the oxidation and breakdown of lignin and nitrogen compounds. We can think of soil organic matter as covering a range of stability classes—from classes that are readily decomposable and therefore short-lived (weeks) to classes that are physically and chemically stabilized, highly resistant to decomposition, and therefore long-lived (hundreds to thousands of years).

The most readily decomposable class is of most interest to short-term nutrient cycling. These are the compounds that will be rapidly decomposed by microorganisms, mineralizing nutrients to a form once again available to plants. The resistant class—by far the greater—is chemically important because it carries a negative charge and so holds supplies of exchangeable cations ($Ca^{2+}$, $Mg^{2+}$, $K^+$, $Na^+$, $NH_4^+$), and it also plays a major role in determining the physical structure of the soil, especially in binding soil particles as aggregates that are water-stable. The resistant class, however, plays little role in nutrient cycling.

Suppose the surface 10 cm of a forest soil contains 10% organic matter. If the bulk density of the soil is 1 $Mg/m^{-3}$, the total amount of carbon in the soil is 100 t/ha. Suppose litterfall is 7 t/ha/year, about 3.5 t/ha/year of carbon. Suppose further that *all* of this carbon is incorporated into the soil (in fact, this is false because much of the carbon in the litter layer is respired and hence lost as carbon dioxide; it will be seen, however, that the supposition only strengthens the developing argument). If carbon in litterfall is in equilibrium with carbon in soil organic matter, then $r$, the proportion of soil organic matter decaying each year, is 0.035 per year and $t_{0.5}$ is 20 years, much too long for our model of nutrient cycling (Fig. 1). If, for example, nutrient cycling in organic matter operates on a time scale of $t_{0.5}$ = 0.5 year (i.e., 6 months), then the input of 3.5 t/ha/year of carbon must be in equilibrium with only 2.5 t/ha of carbon in soil organic matter (or only 2.5% of total soil carbon). Note that the smaller the input becomes due to loss of carbon respired as carbon dioxide from the litter layer, the smaller the equilibrium amount of carbon in the soil must be.

These calculations illustrate the importance of differentiating between a small pool of soil carbon with a rapid turnover time and a much larger pool of soil carbon that is stable within the time frame of nutrient cycling. Unfortunately there are no chemical methods for fractionating soil organic matter into these components, and we must rely on conceptual computer models to describe the turnover of nutrients through organic matter. Methods based on the use of stable and radioactive isotopes to determine nutrient fluxes, and on $^{13}C$-nuclear magnetic resonance (NMR) to determine organic matter composition are becoming well-developed, although their use is limited to laboratories with specialized scientists and relatively large budgets.

### 3. Mineralization of Nutrients from Soil Organic Matter

Most interest in the mineralization of nutrients from soil organic matter has focused on the element nitrogen. Generally more than 95% of the total nitrogen in a forest soil is in organic forms that must be mineralized to $NH_4^+$ or $NO_3^-$ before the nitrogen is available for plant uptake.

The first stage is the breakdown of organic matter by heterotrophic microorganisms to give ammonium ion, $NH_4^+$, this process being called *ammonification*. This $NH_4^+$ is relatively immobile in the soil as most soils have a negatively charged surface and hence hold $NH_4^+$ through electrostatic forces. $NH_4^+$ produced by ammonification is taken up again by microbes in their further metabolism of carbon, is taken up by a plant root, or, where nitrogen is in excess of the needs of microbes, is used by autotrophic bacteria which oxidize it to nitrate ($NO_3^-$) via nitrite ($NO_2^-$). This latter process is called *nitrification;* the anion $NO_3^-$ is not held by the negatively charged soil colloid and so it is highly mobile within the soil solution. It is taken up by a root, may be lost by denitrification (see earlier), or may be lost from the ecosystem in drainage waters.

The process of mineralization is therfore dynamic and constantly changing. The availability of nitrogen relative to carbon, together with water availability and temperature, determines the balance between production and uptake of $NH_4^+$ by microbes. Where production exceeds uptake by microoganisms, net mineralization is positive; where uptake exceeds production, net mineralization is negative and nitrogen is *immobilized* by the microbes. In turn, the net rate of ammonification together with water availability and temperature determines the rate of nitrification. The concentration of $NO_3^-$-N in soil therefore shows very large changes with time.

In the great majority of boreal and temperate forests of the world, the C/N ratio in forest soils is relatively high (>20) and ammonification is dominant. Most forest trees therefore take up nitrogen in the form of $NH_4^+$. However, in warm temperate forests and tropical rain forests, turnover of nitrogen is rapid enough to allow nitrification.

The rate of mineralization in agricultural soils is generally measured by incubating soil samples either anaerobically (under water) or aerobically in the laboratory. For forest soils, *in situ* (i.e., in the field) incubations of soils are increasingly used. These methods involve the containment of soil samples in a core or in a plastic bag. Because tree roots are severed and the uptake of nitrogen is thereby prevented, the increase in inorganic N after a period gives a measure of the rate of nitrogen mineralization. Ion-exchange resins are sometimes used to capture the nitrogen as it is mineralized. All the methods of measuring the rate of mineralization can be at best regarded as giving an index of the rate. However, these indexes are realistic, as shown by the good agreement between the measured rate and the rate of turnover of nitrogen in litterfall (Fig. 4).

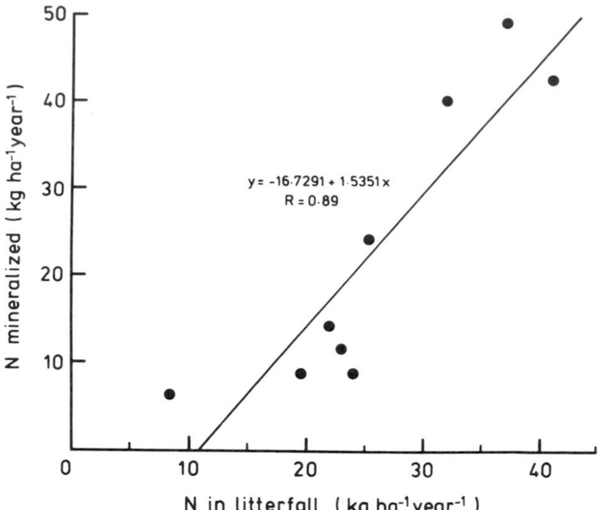


**FIGURE 4** The relationship between the rate of mineralization of nitrogen measured *in situ* and the turnover of nitrogen in litterfall. [From M. A. Adams, P. J. Polglase, P. M. Attiwill, and C. J. Weston (1989). *Soil Biol. Biochem.* **21,** 423–429.]

Measuring the rate of mineralization of phosphorus is a far more complex problem. Unlike nitrogen, virtually all the phosphorus in a soil is derived from minerals in the parent rock. With continuing forest growth and nutrient cycling, much of this inorganic P becomes bound in organic forms in the surface layers of the soil. However, the emphasis in studies of the chemistry of soil phosphorus has been on the set of inorganic equilibria that determine the concentration of phosphorus in the soil solution available for uptake by plants (Fig. 5). This set of equilibria is dominated by the extremely low solubility of inorganic phosphorus compounds and by reactions at the surface of the soil colloid by which phosphorus is *adsorbed*. In acidic forest soils, adsorption is dominated by exchange between $H_2PO_4^-$ and $OH^-$ on the surfaces of colloidal, hydrated oxides of iron and aluminum (Fig. 5). Exceedingly insoluble iron and aluminum phosphates are then precipitated where the concentrations (or, strictly, the activities) of $H_2PO_4^-$, $Fe^{3+}$, and $Al^{3+}$ exceed solubility products.

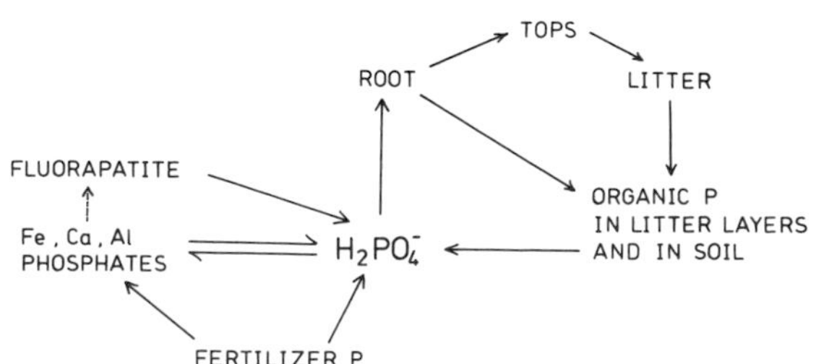


**FIGURE 5** The cycle of phosphorus in a forest [From P. M. Attiwill and G. W. Leeper (1987). "Forest Soils and Nutrient Cycles." Melbourne University Press, Melbourne, Australia.]

In many forest soils, 50% or more of the total phosphorus is in organic form, the result of organic turnover and nutrient cycling. As for nitrogen, this phosphorus in organic form must be mineralized to simple ion, $H_2PO_4^-$, before it is available to plants. Thus nutrient cycling is completed and the availability of phosphorus is sustained (the right-hand side of the phosphorus cycle, Fig. 5). The difficulty in measuring the rate of mineralization of phosphorus is that as soon as $H_2PO_4^-$ is produced, there is a reactive surface to adsorb it (the left-hand side of the phosphorus cycle, Fig. 5).

For these reasons, scientists have used a number of methods that do not aim to measure the rate of mineralization directly but rather to give indexes of its importance. These methods include the measurement of the activity of soil phosphatase enzymes, of the amount of phosphorus in the microbial biomass, of the concentration in soil of various empirically defined fractions of organic phosphorus, and of the mineralization of phosphorus from known organic substrates labeled with the $^{32}P$ isotope. In accord with the foregoing discussion, the availability of phosphorus in a forest soil is determined by both organic and inorganic equilibria, and the importance of organic equilibria as indexed by the activity of phosphatase enzymes increases with forest age (Fig. 6). In forest soils where inorganic equilibria are relatively weak, organic equilibria may entirely dominate in surface soil (Fig. 7).

## III. THE STABILITY OF NUTRIENT CYCLING

The recovery of forests after disturbance has most generally been studied in terms of species composition, diversity, and productivity. Recent studies, however, have included the stability of nutrient cycling and the recovery of nutrient cycling processes.

Nutrient cycling in the undisturbed forest is conservative, or "tight"—losses from the ecosystem are only a small proportion (1–10%) of the large quantities of nutrient that are cycled from plant to soil. As the forest grows to maturity and beyond, the rate of biomass accumulation decreases and becomes negative when the trees lose more biomass through decay than they gain through NPP. At this stage, trees die or are blown down. Gaps are created where litterfall decreases and soil temperature and soil moisture increase; under these conditions, the rate of mineralization increases. Because nutrient uptake has been disturbed, the mineralized nutrients accumulate in the soil to the extent that

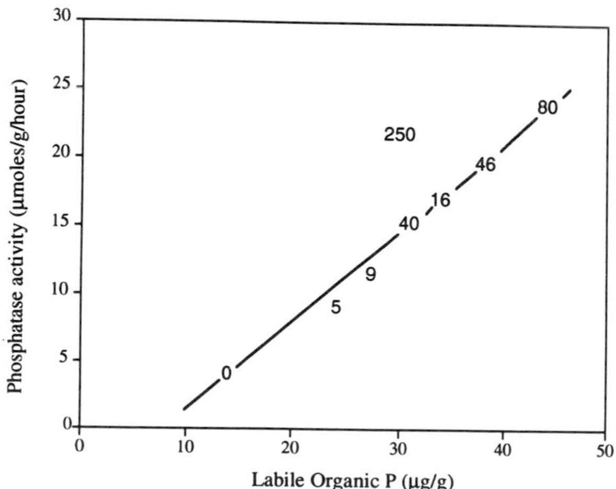


***FIGURE 6*** Relationship between phosphatase activity in surface soil and age of *Eucalyptus regnans* forest. [From P. J. Polglase, P. M. Attiwill, and M. A. Adams (1992). *Plant and Soil* **142,** 177–185.]

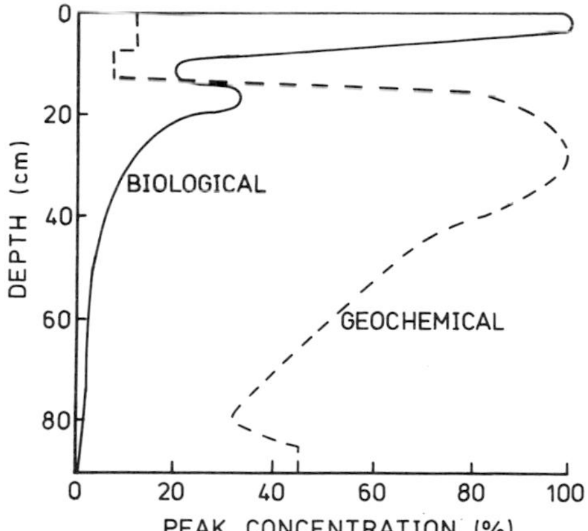


***FIGURE 7*** The relative distributions of biological agents (fine roots, bacteria, and fungi) versus geochemical agents (iron and aluminum hydrated oxides) contributing to phosphorus retention in northeastern hardwood forest of the United States. [From T. Wood, F. H. Bormann, and G. K. Voigt (1984). *Science* **223,** 391–393. Copyright 1984 by the AAAS.]

they may be leached from the ecosystem in streams. Here, the conservative nature of nutrient cycling has been disrupted and the ecosystem has become "leaky." What then happens in the gaps depends on the particular growth requirements of the forest. In some forests, seedlings establish beneath the mature forest ("advanced growth") and are liberated when a gap is created. In some forests, new plants establish within the gap either from seed or vegetatively from root or shoot suckers. In some forests, disturbance must be of stand-replacing magnitude before new plants can establish, and the most common stand-replacing disturbance is fire.

The general picture of nutrient stability following disturbance (Fig. 8) accounts for the varying behavior of elements in the soil–plant system, but the scale of change depends on the magnitude and intensity of disturbance. Following disturbance, net biomass increment decreases to zero or less and the output in steamwater of the most soluble ($NO_3^-$, $K^+$) and relatively soluble $Mg^{2+}$, $Ca^{2+}$) nutrients increases (curve A in Fig. 8c). As the forest regenerates, conservative nutrient cycling is reestablished and the output of nutrients in streamwater rapidly decreases.

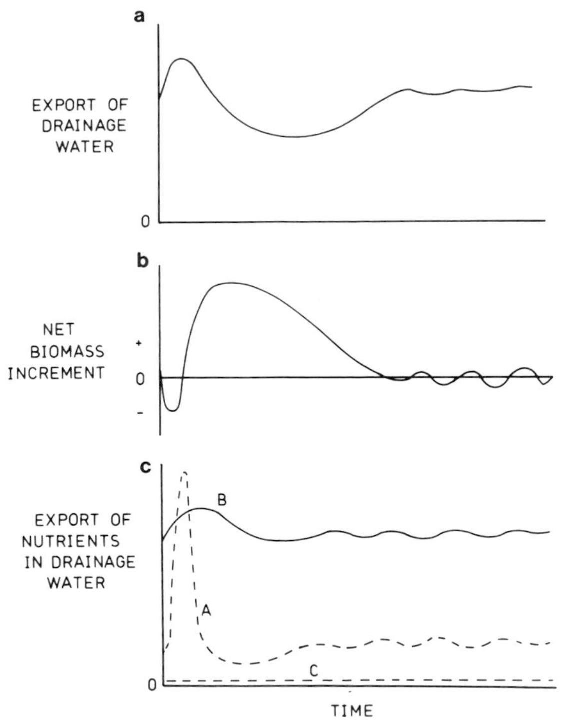


***FIGURE 8*** Changes in (a) export of drainage water, (b) net biomass increment, and (c) export of nutrient in drainage water with time after disturbance and regeneration of a forest. In (c) three curves are shown: (A) relatively soluble nutrients, (B) relatively soluble but nonessential elements, and (C) relatively insoluble nutrients.

The stage immediately after disturbance is therefore most critical in two ways. First, the ecosystem is most susceptible to nutrient loss and, second, the output of nutrients may result in eutrophication of lakes and streams. The risk of both is increased following fire of high intensity, as fire mineralizes large amounts of nutrients both by the oxidation of a large amount of organic matter and by chemical reactions in the soil. Many forest types are both adapted to fire and dependent on large-scale fires for their regeneration. Regeneration is stimulated by fire, and rapid rates of immobilization of nutrients both by plants and by the microbial population feeding on the rich supply of mineralized nutrients restrict the loss of nutrients from the ecosystem in the critical postfire period.

The output of phosphorus is always small (curve C in Fig. 8c) for soils that have significant adsorption surfaces. However, it should be mentioned that all soils are not of this kind. Sandy soils have a low colloidal content, geochemical equilibria are not of the strength shown in Fig. 7, and the leaching of phosphorus when biological equilibria are perturbed by forest clearing can contribute to serious problems of eutrophication.

Finally, biological control of nutrient cycling is clearly seen by a comparison of the behavior of soluble, nonessential elements (e.g., Na, Cl, curve B in Fig. 8c) with that of soluble, essential elements (curve A in Fig. 8C). The output of the former is more or less constant, roughly equal to input, and unaffected by changes in forest growth.

All of this illustrates the conservative nature of nutrient cycling and sustained productivity in the undisturbed forest. Harvesting the forest for its timber products results in relatively little nutrient depletion because only tissues low in nutrients (especially heartwood) are removed, and then only once every 100 years or so under sustainable forest management. In contrast, agriculture removes the most nutrient-rich tissues (grain, foliage) every year and much agriculture depends on regular additions of fertilizers. Sustainable agricultural practices now increasingly encourage a recognition of nutrient cycling through the better management of organic residues after cropping to give an increase in soil organic matter, thereby leading to improved fertility, water-holding capacity, and stability of the soils.

## Glossary

**Adsorption** Colloidal particles carry a charge on their surface and so attract ions of opposite charge; such ions are said to be *adsorbed*. In soils, $H_2PO_4^-$ is adsorbed on the surfaces of hydrated oxides of iron and aluminum.

**Ammonification** Production of $NH_4^+$ in the breakdown of organic matter by heterotrophic microorganisms: the first step in the process of mineralization of nitrogen.

**Bulk density** Weight of soil per unit volume. Bulk density varies from 0.8 to 1.0 $Mg/m^{-3}$ for friable surface soils of good structure and from 1.3 to 1.6 $Mg/m^{-3}$ for compacted surface soils and subsoils.

**Denitrification** Process in soils in which facultative anaerobic bacteria use $NO_3^-$ as a terminal electron acceptor instead of O, this reduction giving $N_2O$ and $N_2$, which are then lost from the soil.

**Ecosystem** Whole complex of biotic and abiotic aspects of an area.

**Essential nutrients** Elements known to be essential for the physiological functioning of plants are (a) *major elements* (> 30 $\mu$mol per g dry weight)—nitrogen, phosphorus, calcium, magnesium, potassium, sulfur—and (b) *minor* or *trace elements* (< 3 $\mu$mol per g dry weight)—boron, chlorine, copper, iron, manganese, molybdenum, zinc.

**Eutrophication** Process that makes a water body nutritionally richer.

**Interception** Rain that wets the canopy of a forest, from where it evaporates back to the atmosphere.

**Mineralization** Process of making mineral—the transformation of an element from an organically bound form to an inorganic form.

**Mycorrhiza** Symbiotic relationship between plant roots and a fungus.

**Net primary production** Difference between photosynthetic gain and respiratory loss of plants (the primary producers) in an ecosystem.

**Nitrification** Production of $NO_3^-$ from $NH_4^+$ by autotrophic bacteria in soil: the second step in the mineralization of nitrogen. The reaction proceeds via $NO_2^-$, which is short-lived and never appears in quantity.

**Stemflow** Rain that passes through the canopy of the forest and reaches the forest floor by flowing down the stems of trees.

**Throughfall** Rain that passes through the canopy of the forest, directly to the forest floor.

## Bibliography


Aber, J. D., and Melillo, J. M. (1991). "Terrestrial Ecosystems." Philadelphia: Saunders College Publishing.

Adams, M. A., Polglase, P. J., Attiwill, P. M., and Weston, C. J. (1989). *In situ* studies of nitrogen mineralization and uptake in forest soils: Some comments on methodology. *Soil Biol. Biochem.* **21,** 423–429.

Attiwill, P. M., and Adams, M. A. (1993). Nutrient cycling in forests. *New Phytol.* **124,** 561–582.

Attiwill, P. M., and Leeper, G. W. (1987). "Forest Soils and Nutrient Cycles." Melbourne, Victoria, Australia: Melbourne University Press.

Likens, G. E., Bormann, F. H., Pierce, R. S., Eaton, J. S., and Johnson, N. M. (1977). "Biogeochemistry of a Forested Ecosystem." New York: Springer-Verlag.

Maheswaran, J., and Attiwill, P. M. (1987). Loss of organic matter, elements, and organic fractions in decomposing *Eucalyptus microcarpa* leaf litter. *Can J. Bot.* **65,** 2601–2606.

Polglase, P. J., Attiwill, P. M., and Adams, M. A. (1992). Nitrogen and phosphorus cycling in relation to stand age of *Eucalyptus regnans* F. Muell. III. Phosphatase activity and pools of labile soil P. *Plant and Soil* **142,** 177–185.

Postgate, J. R. (1987). "Nitrogen Fixation," 2nd ed. London: Edward Arnold.

Stevenson, S. J., and Elliott, E. T. (1989). Methodologies for assessing the quantity and quality of soil organic matter. *In* "Dynamic of Soil Organic Matter in Tropical Ecosystems." D. C. Coleman, J. M. Oades, and G. Uehara, (Eds.) pp. 173–199. Honolulu: NifTAL Project and University of Hawaii Press.

Vogt, K. A., Grier, C. C., and Vogt, D. J. (1986). Production, turnover and nutrient dynamics of above and below ground detritus of world forests. *Adv. Ecol. Res.* **15,** 303–377.

Wood, T., Bormann, F. H., and Voigt, G. K. (1984). Phosphorus cycling in a northern hardwood forest: Biological and chemical control. *Science* **223,** 391–393.

# Nutrient Cycling in Tropical Forests

**Carl F. Jordan**
*University of Georgia*



A nutrient is a chemical element such as nitrogen, phosphorus, or potassium that is essential for soil fertility. When a nutrient is taken up by a plant, either through roots or leaves, it is used in metabolic processes that result in growth and reproduction. When the plant is eaten, the nutrients are ingested by the animal. When the plant dies, or when the animal that ate the plant dies, the nutrients are released by the process of decomposition and are taken up by microbes such as bacteria or fungi. From the microbes, the nutrients may move into roots of plants or into the mineral soil. From the soil, the nutrients may circulate back into the roots or may be lost to the ecosystem through processes such as leaching or volatilization. The pathway followed by a nutrient as it moves through the plants, animals, and soil of an ecosystem (Fig. 1) is called a nutrient cycle.

## I. INTRODUCTION

In contrast to temperate and high-latitude ecosystems where biological process slow down or stop during the winter, biological processes often occur year-round in tropical rain forest ecosystems. Therefore in tropical rain forests there is frequently a greater demand for nutrients than in other ecosystems. The greater demand for nutrients is met in one of two ways: a high rate of nutrient cycling or a high efficiency of nutrient cycling. High rates and/or high efficiencies cause nutrient cycles in tropical forests to be quantitatively different, and perhaps even qualitatively different than cycles in other ecosystems. The characteristics of a nutrient cycle in a particular tropical forest ecosystem have important implications for the diversity and productivity of the biota and for the sustainable management of that ecosystem. [*See* NUTRIENT CYCLING IN FORESTS.]

## II. CLIMATE OF TROPICAL FORESTS

The climate of tropical forest regions is often thought to be extremely hot and wet. While the climate *is* hot, it is often not as hot as temperate zone continental areas such as the Great Plains during the summer. And while it can be wet, it also can be quite dry for weeks or even months at a time. In some regions, such as parts of the Caribbean coast of Central America, the climate is wet year-round, but along a westward transect, a dry season becomes more and more pronounced, and in some

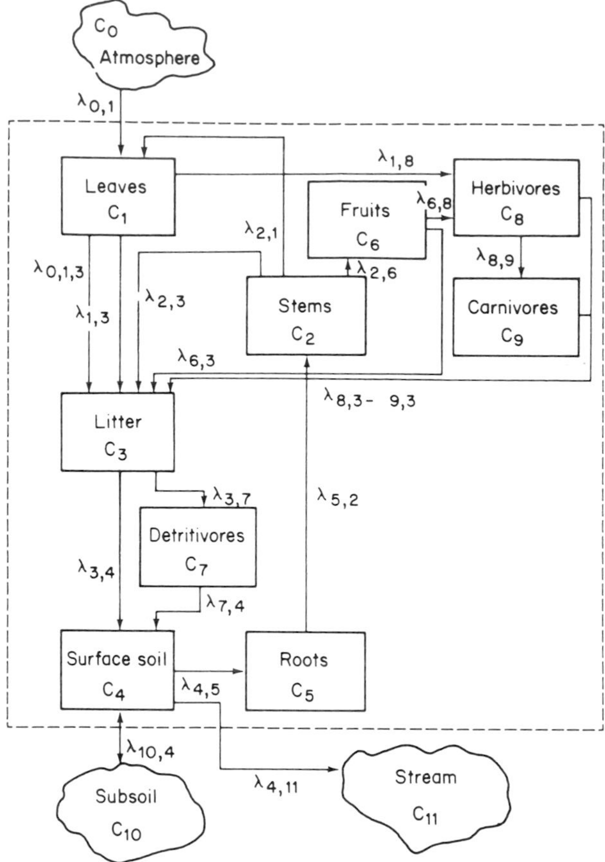


***FIGURE 1*** The cycle of a nonvolatile nutrient such as potassium in a tropical forest ecosystem. Compartments are numbered. Flows between compartments such as leaves and litter are identified by a λ, with subscripts indicating the compartment of origin and the compartment of destination. [Adapted from Golley, (1975).]

areas rain may fall only in June through August. Where there is insufficient rain for forest to develop, the vegetation may be savanna or desert.

The factor that sets apart the climate of tropical forests is that it is hot (or at least well above the freezing point) all year-round. These continually high temperatures have important effects on ecosystem process rates. Constant high temperatures can mean that the growing season for plants never ceases. A continual growing season results in high annual rates of plant growth and of nutrient uptake. The continuous production of leaves in this climate results in continuous food availability for herbivores, and their populations can quickly explode. Consequently, in tropical forests there is potential for high annual rates of nutrient movement through food chains. High annual growth also means high annual rates of nutrient return to the soil through leaf fall and through the death and defecation of herbivores and predators. [*See* SOIL ECOSYSTEMS.]

## A. Photosynthesis and Primary Production

Photosynthesis is the process through which carbon is taken from the atmosphere and fixed in carbon-containing compounds in leaves. Measurements of photosynthesis are often given as milligrams of carbon fixed per leaf per minute. Average rates of photosynthesis of leaves in a tropical rain forest scarcely differ from average rates of leaves in a deciduous temperate forest during the summer. [*See* GLOBAL CARBON CYCLE.]

Primary production is the process through which the carbon fixed during photosynthesis is converted into leaves and wood, similar to the concept of "growth," except growth often is expressed in terms of volume or height increment whereas production is expressed in terms of increment of dry weight. Measurements of primary production are often given as weight gain (kilograms or pounds) per unit area (hectare or acre) per growing season or per year. Average rates of net primary production can be two or more times higher per hectare per year in the wet tropics than at higher latitudes, not because photosynthesis is any faster, but because the number of days during which photosynthesis takes place during the year is greater.

## B. Decomposition

Decomposition is the process through which litter (dead leaves and tree trunks) and carcasses of animals are broken down and eventually converted to the carbon, hydrogen, oxygen, and nutrient elements from which they were constituted. Like photosynthesis, decomposition, when measured on a short-term rate (minutes, hours, or days), does not differ much between tropical and temperate forests during the summer. But on a yearly basis, rates of decomposition are higher in the tropics than at higher latitudes. Therefore, over the long term, the rate at which nutrients are released into the soil is highest in the tropical rain forest environment.

### C. Soil Weathering

All living organs respire, that is, give off carbon dioxide when complex carbon-bearing compounds are broken down with the subsequent release of energy for growth or metabolism. When roots give off carbon dioxide into the soil, it reacts with soil water to form carbonic acid. This acid can percolate down to the bedrock underlying the soil, decompose the minerals, and release nutrients such as calcium and phosphorus which then become available for uptake by plant roots. This process is called weathering. Because roots respire year-round in the wet tropics, weathering can take place year-round. In regions where bedrock is recently uplifted and fresh, year-round weathering can result in a relatively large supply of nutrients. However, in areas such as the central Amazon Basin that have had no important geological activity for hundreds of millions of years, the bedrock has been weathered very deeply and is devoid of most nutrient elements. [*See* RHIZOSPHERE ECOPHYSIOLOGY.]

### D. The Nitrogen Cycle

The nitrogen cycle is different from the cycles just described in that the atmosphere constitutes an important part of the cycle. Nitrogen is taken from the atmosphere, often by microbes living symbiotically with roots of species in the legume family of plants. Once in the plant, nitrogen is used in proteins and other compounds. When a leaf or a plant dies, it falls to the ground as litter. Some of the nitrogen is converted into ammonium and nitrate, and moves through the soil to roots. However, some nitrogen is released from litter into the atmosphere by bacteria called denitrifiers. Because nitrogen fixation and denitrification occur year-round in the tropics, the yearly rate of nitrogen cycling can be relatively high.

### E. Nutrient Cycling

Because primary production, decomposition, soil weathering, and other processes such as nitrogen fixation are generally greater in tropical forests than in forests at higher latitudes, it has been hypothesized that rates of nutrient cycling are more rapid in tropical forests than in temperate forests. It has been difficult to compare data to determine if this is true because of differences in methodologies in the studies that have been accomplished. Nevertheless, an early attempt at comparison (Table I) suggested that indeed nutrient cycling was more rapid in the tropics.

For simplification, the forest ecosystems compared were broken into four major categories: soil, wood, canopy (living leaves), and litter (mainly dead leaves on top of the soil). For each category ("compartment" in the jargon of systems analysis), the turnover time was determined by dividing the total stock of a nutrient in that compartment by the rate at which the nutrient left that compartment. Then the turnover times for the four categories were summed to give "total cycle time." The total cycle time for calcium does indeed appear to be shorter in the tropical systems, indicating a higher rate of nutrient cycling, at least for this particular nutrient element.

Although nutrient cycles appear to be faster in some tropical forests, there are exceptions. For example, the total cycle time of calcium in a rain forest on a low-fertility soil of the Amazon region of Venezuela was 45.8 years, much higher than that of the tropical forests in Table I.

The study showed that the high demand for nutrients in this forest did not result in a high rate of nutrient cycling, but rather in a high efficiency of nutrient use. The nutrient-use efficiency was estimated by a parameter called the "cycling index," a measure of the amount recycled in an ecosystem compared to the total amount recycled plus amount lost. Thus a cycling index of 0.9 means that 90% of the nutrient moving through that ecosystem is recycled whereas 10% is lost. The cycling index for calcium in the Venezuelan forest is 0.79, compared to values of 0.31 for the Puerto Rican rain forest, 0.67 for a hardwood forest in Tennessee, and 0.76 for a Douglas fir forest in the northwestern United States.

This result means that tropical rain forests cope with high nutrient demand in one of two ways:

**TABLE I**
Compartmental Turnover and Cycling Times for Calcium

| Location of ecosystem | Type of ecosystem | Soil | Wood | Canopy | Litter | Total cycle time (years) |
|---|---|---|---|---|---|---|
| Puerto Rico | Rain forest | 3.0 | 6.4 | 0.9 | 0.2 | 10.5 |
| Ghana | Moist forest | 8.2 | 6.8 | 1.5 | 0.2 | 16.7 |
| England | Pine plantation | 11.2 | 6.1 | 0.8 | 3.4 | 21.5 |
| Belgium | Oak–ash | 184.9 | 21.5 | 0.4 | 0.6 | 207.4 |
| Northwest United States | Douglas fir | 57.4 | 20.2 | 5.5 | 10.2 | 93.3 |
| Northeast United States | Northern hardwoods | 14.0 | 10.8 | 0.8 | 34.8 | 60.4 |

When soils are fertile, the nutrients are recycled quickly. When soils are infertile, the nutrients are recycled efficiently.

## III. NUTRIENT CYCLES ON RICH AND POOR SOILS

There is a large variability in process rates and nutrient cycling between tropical forests. Some of the variability is because of differences in the length of the dry season. However, in forests that have a dry season of less than 4 or 5 months per year, the nature of the soil and its fertility often is the factor that has the greatest influence on the character of the nutrient cycle. The importance of soil fertility on the structure and function on tropical rain or moist forests is illustrated in Table II. The forests are arranged from nutrient-poor tropical forests on the left along a gradient of increasing soil fertility to nutrient-rich tropical forests on the right.

The most nutrient-poor forest is Amazon caatinga, a scrubby forest growing on coarse, heavily leached sands called podzols. The Oxisol forest is typical of many forests in the lowland tropics growing on very old soils formerly called "lateritic." Because of their great age, most of the nutrients have been leached away. The Puerto Rican rain forest is on basaltic soils of volcanic origin, but is relatively old. The forests in Ivory Coast and Malaysia are on somewhat more fertile soils. The Dipterocarp forest in Malaysia has many trees of the family Diptercarpaceae, very tall, straight trees of high commercial value. The Costa Rican soils are alluvial, but are of relatively recent volcanic origin, and the Panama site is on soil derived from rock high in calcium.

### Structure and Function of Forests

The root systems of forests on poor soils are relatively large compared to root systems on richer soils (parameter 1, Table II). A large root biomass seems to be an adaptation to the low-nutrient environment. In infertile soils where nutrients are dilute, a relatively large biomass is required to gather sufficient nutrients to support growth.

In contrast to roots, the aboveground biomass (trunks, branches, parameter 2) is greater in forests on fertile sites. With more nutrients available in the soil, trees can get larger faster. The root : shoot ratio, parameter 3, is an index that incorporates both above- and belowground biomass.

In forests on the soils lowest in nutrients, a "mat" of fine roots (parameter 4) often occurs on top of the soil surface. The roots penetrate the accumulation of decomposing leaves and wood on the soil surface and appear to attach themselves to the litter. Often what happens is that a fungus called a mycorrhiza grows on the very fine root hairs and penetrates the decomposing leaves and wood. Nutrients released from the leaves and wood can move directly from the litter into the roots. This so-called "direct" recycling of nutrients is an efficient means of preventing nutrient loss through leaching or volatilization and can be an important adaptation where nutrients are scarce. The mat of roots and litter is also rich in bacteria and other types of fungi. Although these do not penetrate the decomposing leaves and wood, they take up nutrients as soon as

***TABLE II***

Ecosystem Characteristics of Tropical Rain Forests

| Parameter | Amazon caatinga, San Carlos, Venezuela | Oxisol forest, San Carlos, Venezuela | Lower montane rain forest El Verde, Puerto Rico | Evergreen forest, Ivory Coast | Dipterocarp forest, Pasoh, Malaysia | Lowland rain forest La Selva, Costa Rica | Moist forest, Panama |
|---|---|---|---|---|---|---|---|
| 1. Root biomass (tons/hectare) | 132 | 56 | 72.3 | 49 | 20.5 | 14.4 | 11.2 |
| 2. Aboveground biomass (tons/hectare) | 268 | 264 | 228 | 513 | 475 | 382 | 326 |
| 3. Root/shoot ratio | 0.49 | 0.21 | 0.32 | 0.10 | 0.04 | 0.04 | 0.03 |
| 4. Percentage of roots in mat on soil surface | 26 | 20 | 0 | 0 | 0 | 0 | 0 |
| 5. Specific leaf area ($cm^2/g$) | 47 | 65 | 61 | — | 88 | 139 | 160 |
| 6. Leaf area index | 5.1 | 6.4 | 6.6 | — | 7.3 | — | 16 |
| 7. Leaf biomass (tons/hectare) | 10.8 | 9.8 | 10.8 | | 8.3 | | 10.4 |
| 8. Leaf litter production (tons/hectare/year) | 4.95 | 5.87 | 5.47 | 8.19 | 6.30 | 7.83 | 11.3 |
| 9. Leaf turnover time (years) | 2.2 | 1.7 | 2.0 | — | 1.3 | — | 0.9 |
| 10. Wood production (tons/hectare/year) | 3.93 | 4.93 | 4.86 | 4.0 | 6.4 | — | — |
| 11. Time for 95% leaf litter decomposition (years) | 3.95 | 5.77 | 1.09 | 0.91 | 0.91 | 0.86 | 0.94 |
| 12. Phosphorus concentration in leaf litter (parts per million) | 380 | 138 | 200 | 732 | 304 | 494 | 758 |
| 13. Nitrogen concentration (% weight of leaf litter) | 0.7 | 1.1 | — | 1.6 | 1.2 | 1.9 | — |

[From Jordan (1985)]

they are released from the litter. Then when these microbes die, the nutrients become available for uptake by roots.

Specific leaf area (parameter 5) is the ratio of leaf area to leaf weight. It is an index that indicates how tough and resistant a leaf may be to herbivory and decomposition. A leaf with a low specific leaf area means that it is relatively thick. Thick leaves are often more leathery, or "sclerophyllous," and consequently are more difficult for insects and other herbivores to chew and for microbes to break down. Such leaves may represent an adaptation to the low nutrient environment. Where nutrients are "expensive" in terms of the resources needed to gather them (such as large root biomass, parameter 1), it is an advantage for a tree to produce leaves that are tough and difficult to chew as a defense against herbivores. In contrast, on soils where nutrients are plentiful, it may be easier for a tree to replace any leaves that are eaten than to defend all leaves since replacements are "cheap."

"Leaf area index" (parameter 6) is simply the number of layers of leaves in the forest. The leaf area index is higher in forests on richer soils. However, the total leaf biomass in tropical forests does not vary significantly between forest types (parameter 7). This is because forests tend to have more layers of thin leaves on rich soils whereas they have fewer layers of thick leaves on poor soils.

Although the leaf *biomass* varies little between forests, the *rate* of leaf production (as measured by leaf litter production, parameter 8) is greater in forests on rich soil than on poor soil. Like leaf production, wood production (parameter 10) increases along a gradient of increasing soil fertility. Like most plants, tropical trees produce more leaves and wood when soil is fertile.

The "turnover time" of leaves is the average length of time that a leaf remains on a tree. On nutrient-poor soils, it is approximately 2 years (parameter 9), but on the richest soils, it can be less than a year. Turnover time is another reflection of the durability of the leaves. Tough leaves with a high specific leaf area last longer than relatively thin leaves. This toughness is also reflected in the length of time required for the leaves to decompose (parameter 11). In nutrient-poor sites, it requires 4 or 5 years for 95% of the leaf biomass to decompose, whereas it takes less than a year on richer sites.

It is not enough to know that nutrients are critical in a particular forest. It is also important to know specifically which nutrient may be limiting. One indication that a particular nutrient may be limiting is a relatively low concentration of that nutrient in comparison with concentrations in leaves from other forests. Phosphorus and nitrogen are two of the most common limiting elements in the tropics. Concentrations of these nutrients in the forests of Table II are given in lines 12 and 13. Phosphorus is lowest in the leaves of the Oxisol forest, suggesting that in this ecosystem, phosphorus could be critical. For nitrogen, the lowest concentration is in the caatinga forest. Physiological studies of plants in that environment have indicated that in fact, nitrogen is a limiting factor there.

## IV. NUTRIENT "STRESS" IN TROPICAL FORESTS

### A. Are Some Tropical Forests "Nutrient Stressed?"

Because nutrients seem to be in short supply in some tropical forests, the question often asked is whether tropical forests are under nutrient stress. To answer that question, one must know what symptoms a stressed forest would exhibit. One symptom would be a relatively low biomass. Another would be relatively low primary production. If the aboveground biomass and leaf litter production are considered, the Venezuelan and Puerto Rican forests in Table II might be considered stressed. The relatively low phosphorus concentration in leaf litter of the Puerto Rican and the Venezuelan Oxisol forests and the low nitrogen in the caatinga forests suggest that low levels of these nutrients could be a cause of stress.

It can be argued, however, that these forests are not really under nutrient stress since none of the individual trees exhibit symptoms of stress such as leaf yellowing. The trees have adapted to the low nutrient condition and are perfectly healthy. Pouring fertilizer onto these forests generally would not increase the biomass or productivity of the forest (it might, however, result in a replacement of existing trees with other faster-growing trees).

Whether tropical forests are under nutrient stress is an academic question until the forests are cut down and replaced by plantations, ranches, or agricultural fields. Then the question of nutrient status of the soils and its relationship to plant growth take on very important practical implications. For example, although the trees growing on the Oxisol site in Venezuela (Table II) may not be individually stressed in the physiological sense, an annual crop planted at the site will quickly show signs of stress due to phosphorus deficiency.

## B. Geography of Nutrient Stress

### *1. Old, Highly Weathered Lowlands*

Large areas of the lowland or flat tropics such as the central Amazon Basin, the plateau covering parts of peninsular Southeast Asia, and much of India and Africa are covered with soils that are very old and highly weathered. Their great age, coupled with the intense weathering that has occurred because of the hot and often humid climate, has resulted in a process called laterization. In this process, a large proportion of the nutrients have been lost through leaching over geological millennia. However, high concentrations of iron and aluminum remain in the soil. Where iron is present in its highest concentrations, the soil can be removed, shaped, and dried to form bricks. This type of hardened soil has been called "laterite." While deforestation in these regions can cause the soil to become hard and impermeable, in reality it rarely occurs because the vigorous growth of vegetation forms a cover that prevents the soil from drying.

Although leaching has removed most of the nutrients from the soil, there still is enough nutrient input to maintain forest cover in most regions. The nutrients are contained in atmospheric dust deposited on the forests or are dissolved in the rain. Actual nutrient stress only begins when these forests are cleared and agriculture is attempted. [*See* FOREST CLEAR-CUTTING, SOIL RESPONSE.]

Immediately after the tropical forest on such sites is cut and burned, there are sufficient nutrients in the soil for one or two agricultural crops. The nutrients come from the ash of the burned trees or from decomposing litter and trunks that remained unburned. After a few years, nutrients such as potassium and calcium are leached down through the soil. Nitrogen, held primarily in soil organic matter, disappears as the litter and soil organic matter decomposes. However, phosphorus is often the most critically limiting element. Under undisturbed forests, phosphorus is kept soluble by organic acids released from decomposing litter. But when the forest is destroyed and the litter and soil organic matter decompose, the phosphorus reacts with the iron and aluminum present in the soil and becomes immobilized. [*See* BIOMASS BURNING.]

In the past when shifting cultivators who farmed such soils saw a decline in crop yield, they usually abandoned their fields and moved on to clear another area. Slowly the forest regrew on the abandoned area, and as it grew, it accumulated nutrients, mostly brought in from the atmosphere. One study carried out on a series of abandoned sites of shifting cultivation near San Carlos, Venezuela, indicated that it could take well over 100 years for a site to regain its fertility (Fig. 2).

Now, however, with rapidly increasing populations and decreasing land availability, it is often not

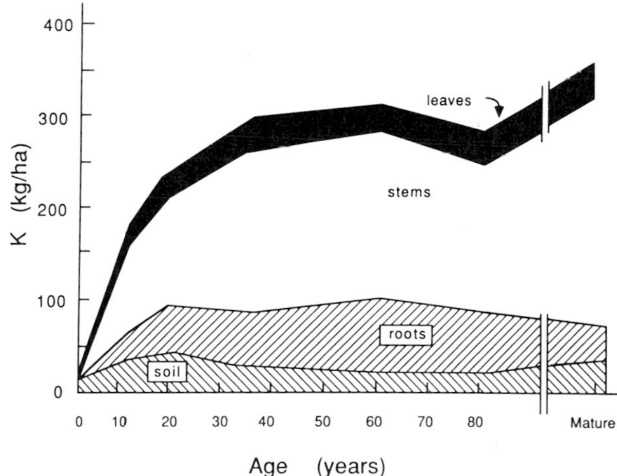


***FIGURE 2*** Stocks of potassium as a function of age in a Venezuelan rain forest. Individual components are stacked on top of each other. The stock of each component is indicated by the thickness of the respective shaded, clear, or hatched areas. The uppermost line represents the total in ecosystem. [Adapted from Jordan (1987).]

possible for the tropical farmer to move on to new virgin sites. This sets the stage for hunger, poverty, land conflicts, and a host of other problems plaguing tropical developing countries.

Theoretically of course, it is possible to add fertilizers to nutrient-depleted soils of the tropics. In some areas, large commercial farmers do add fertilizers. However, in many regions of tropical rain forest, fertilizers are prohibitively expensive. In remote locations, it is more expensive to ship in fertilizers than to ship in food.

Small-scale agriculture without fertilizers becomes impossible after a few years. However, pastures can be extended for a much longer time. After a few years when the grass production declines, ranchers commonly burn the pasture. Burning releases nutrients tied up in the dead aboveground grass litter and in the weeds and shrubs that have invaded the pasture, which renders these nutrients available for new grass growth. The first burning may increase production for a while, but the period of good production is a year or less because nutrient stocks in the pasture grasses and weeds are low. Frequently there is a second, third, and fourth burn, each resulting in a new pulse of nutrients, but each pulse is smaller than the previous one, resulting in an even shorter period of good production. Finally the ecosystem is so depleted of nutrients that there are none left to mine and the pasture has to be abandoned.

Establishment of new vegetation is very difficult. In some cases, prolonged use of pastures and consequent erosion have converted the vegetation into a shrub-like heath. For example, in the region near Paragominas in the eastern part of the Amazon Basin, hundreds of square miles are covered by impenetrable thicket. It is highly unlikely that this region will ever return to forest. Lack of nutrients and soil erosion are only part of the problem. The other part of the problem is that there is no seed source for future forests because the original forest was completely destroyed. But even if there were seeds, it would be almost impossible for seedlings to become established in the thick weeds that cover abandoned pastures.

Pastures are sometimes fertilized, and such treatment extends the time that active grazing is possible (Fig. 3). However, the amount of fertilizer required continually increases and fertilization soon becomes economically infeasible.

### *2. Sandy Soils*

"Podzolization" is a soil-forming process in which iron and aluminum are leached out of the upper soil horizon and are deposited and concentrated in a horizon below the rooting zone. At one time, the process of podzolization was thought to occur only in soils of cold regions. Recent observations that it also occurs on sandy soils of the tropics has drawn attention to the forests that grow on these soils.

Although there may be a thick layer of undecomposed humus on the surface of podzol sands, there is very little organic matter in the soil itself. Because of the sandy texture and the lack of organic matter, podzol sands are the most nutrient limited of all tropical soils. Nitrogen deficiency often is a dominant symptom, but limitations of other nutrients are also frequently important. The scrubby Amazon caatinga forest in Venezuela (the first forest in Table II) grows on a sandy podzol.

Because of the coarse nature of the sand, water quickly drains through and carries with it partially decayed humus. This incompletely decomposed organic matter gives the characteristic tea color to the "blackwater" rivers in many parts of the world.

In Southeast Asia, forests on podzol sands are called "Kerangas." Although agriculture is impossible, and even forestry is not desirable because the trees are small and crooked, Kerangas forests are sometimes utilized for firewood and charcoal. Once destroyed, Kerangas forests require extremely long periods to regenerate because of lack of nutrients.

### *3. Volcanic Soils*

Many soils, especially in Central America and some of the Pacific islands, are derived from volcanic lava. These soils are generally much younger and therefore richer in nutrients than the lateritic soils of the lowlands. Many important tropical fruit industries such as pineapple, cocoa, and banana are located on volcanic soils or on alluvial soils derived from volcanic rock.

Although these soils are richer in nutrients, nevertheless phosphorus often is a problem when origi-

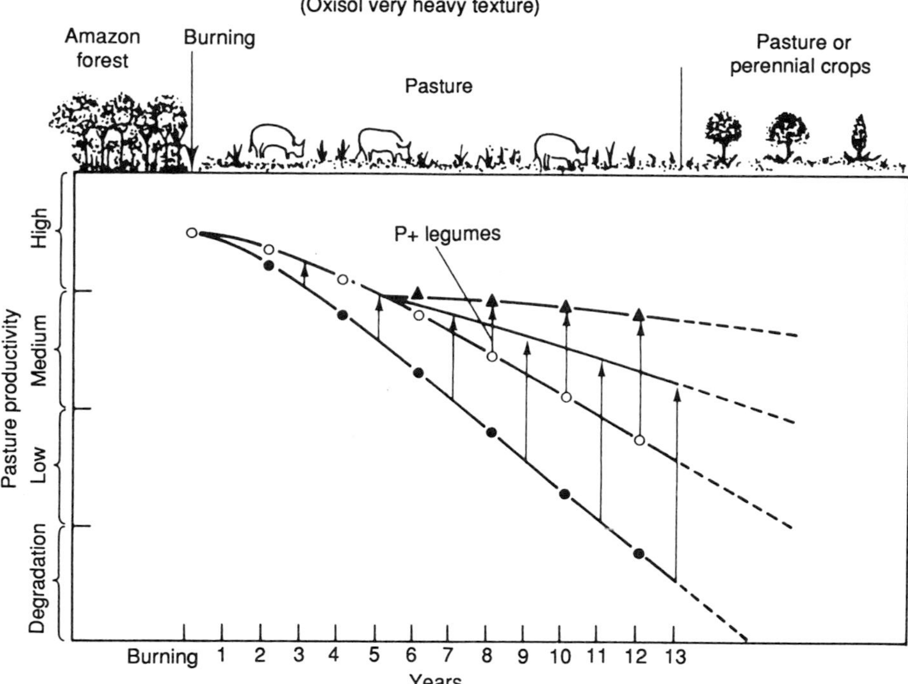


***FIGURE 3*** Dynamics of nutrients in an Amazonian pasture. Black dots indicate productivity at overgrazing. Clear dots represent grazing at "optimal" stocking without fertilizers. Arrows show the improvement in the system resulting from the addition of phosphorus fertilizers and leguminous shrubs. [From Serrao *et al.* (1978).]

nal forests are removed and replaced by agriculture. Volcanic soils, like lateritic soils, are high in aluminum oxides with the potential to immobilize phosphorus and render it less available to crop plants. Shifting cultivation on volcanic soils, like shifting cultivation on lowland lateritic soils, is possible only for a few years. However, there is an important difference between the two soil types.

Weeds can be an important detriment to agriculture on lateritic soils, but they may not be as important to continuous agriculture as on volcanic soils. Following cutting and burning of forests on volcanic soils, the nutrient dynamics are similar to those on lateritic soils: there is an initial increase in the soil, followed by a gradual decrease. However, stocks of nutrients on volcanic soils are relatively high. As a result, weedy shrubs and trees invade the site. Some weeds are excellent competitors for nutrients, especially phosphorus. It is not entirely clear why wild, weedy species can often outcompete crop species for phosphorus and other nutrients. It may be related to a specially adapted root-mycorrhizae association or it may be due to unique organic acids secreted by the roots of some weedy species. Regardless, weeds often become so aggressive after a few years of cultivation that cropping becomes extremely difficult. Sometimes the only solution is to abandon the plot and allow natural reforestation to take place. After a number of years, the trees shade out the weedy shrubs and grasses and cultivation again is possible following cutting of the trees. The length of time required for regeneration of a site on volcanic soils could be as little as a decade or less.

On some Pacific islands, indigenous cultivators have worked out a scheme whereby annual crops are cultivated for 1 year, perennial crops are interplanted among the annuals and harvested for 2 or 3 years, and certain woody species are planted among the perennials. After 10 years, the trees can be cut and another crop of annuals can be planted.

Erosion is frequently a problem on volcanic islands. On the island of Mindinao in the Philippines, islanders have evolved a system of cultivation that is effective in preventing erosion. "Hedges" consisting of fast-growing leguminous trees and grasses are planted along the contour of mountainous slopes. Individual hedges are about 4 m apart. Once the hedges are established, cultivation for corn, upland rice, or other crops can be carried out on the terrace between the hedges. The hedges prevent erosion. They are pruned back several times a year to prevent shading of the crops. The litter is used for cattle fodder or is left on the soil as mulch to improve soil quality.

### *4. Mountainous Soils*

Some tropical mountains are formed by geological uplift. Like volcanic soils, they are relatively rich because of their young age. However, they are often relatively thin, and erosion is frequently severe when they are cleared and converted to agriculture. However, like volcanic soils, they can be used for production if care is used to prevent erosion. For example, experiments in Peru have demonstrated that logging is possible without erosion, if the logging is carried out in bands about 50 m wide and if alternate bands of forest are left intact until the logged strip begins to regrow.

Soils in "cloud forests" on top of some tropical mountains have attracted scientific interest. Cloud forests are forests that are almost always immersed in clouds. Typically, the trees are short, crooked, and heavily festooned with mosses and lichens because of the continual mist. The growth form of the trees is similar to those on podzol sands that suffer from nutrient deficiency. Soils are covered with a thick layer of undecomposed humus. Scientists have theorized that these forests are in fact suffering from nutrient shortage because of the extremely slow decomposition of the humus on the soil surface and the consequent slow release and recycling of the nutrients. Slow decomposition is caused by cool temperatures and by the saturated condition of the humus, which results in relatively slow anaerobic bacterial activity.

### *5. Alluvial Soils*

Some of the best soils in the tropics are alluvial, such as the "varzeas" or flood plains along the Amazon. Tropical rivers such as the Amazon often show seasonality. During the rainy season, the Amazon carries a heavy load of sediment, carried from the Andes. The waters overflow the banks and cover the varzea. Then with the approach of the dry season, the waters drop and the silt is deposited in the varzea, enriching the soil.

Crops are planted as the river recedes and must be harvested before it rises again. The time available for growth depends on the seasonality and the exact location within the varzea.

### *6. Mid-elevation Tropical Valleys*

Some of the best soils in the tropics are in mid-elevation valleys of mountain ranges. They are composed of young alluvial soils washed down from higher elevations. Temperatures are moderate, so the intense weathering that leads to phosphorus fixation in the lowlands is not predominant. Neither are conditions so cool and moist that nitrogen immobilization is a problem.

Because of the pleasant conditions, such valleys were frequently the first areas settled by colonists and the first areas deforested for agriculture. As a result, there are few original forests left in tropical valleys. Many agricultural experiment stations have been located in such valleys, and their results have led to the impression that the tropics are ideal for agriculture.

## V. AGRICULTURE IN THE TROPICS

Early scientists who explored tropical rain forests were impressed by the large stature of the trees and the apparent frenzy of vegetative growth. Back in Europe, forests of such character would indicate soils that would be excellent for agriculture. As a result, these scientists recommended large investments in temperate-style agriculture for the tropics.

These recommendations have often not led to the hoped for results. In general, large-scale monocultures of annuals or of pastures and plantations have not been economically successful in the tropics and have often been ecologically damaging. There are a variety of reasons for the dismal results. For one thing, pests and weeds are more of a problem in tropical rain forest areas because of the continual growing season which allows an exponential growth of pest populations. At higher latitudes, freezing temperatures kill back pest populations, but no such control exists in the wet tropics. Second, logistics are often very difficult in less-developed tropical countries, and bureaucratic red tape frequently ties up business ventures.

But often forgotten in the excuses for failed development projects in the tropics is the fact that soils of the wet tropics are usually not very fertile. Native forests are able to survive and to achieve a large stature as a result of adaptations to the low-nutrient environment. Herbivore-resistant leaves and efficient recycling of nutrients from decomposing litter to roots are two such adaptations. When the forest is cut down and converted to agriculture and pastures, these nutrient-conserving mechanisms are destroyed.

Despite the frequent failure of large-scale temperate-style development in the tropics, the belief still persists that the tropics are a good place for agriculture. That belief has been strengthened in recent decades by the success of the "green revolution." The green revolution is the name given to an international effort in agriculture, in which grains with high productive potential were bred and distributed to farmers in regions of the tropics. However, in order to attain the high potential yield, the farmer had to use lots of fertilizers. Those large farmers who were able to afford fertilizers generally profited greatly from the green revolution, and some countries such as India increased their self-sufficiency in agriculture. However, the green revolution was of little help to the small farmer who could not afford the fertilizers and the pesticides and herbicides that also were frequently needed.

For the small farmer, or for regions where chemicals are not practical from a logistic or environmental point of view, another approach besides the capital-intensive, chemical-intensive methods of the green revolutionists must be tried. One idea that has attracted a lot of scientific interest is something called "agroforestry."

### A. Agroforestry

Agroforestry is a very general concept covering a wide spectrum of approaches to crop and tree production. On one end of the spectrum are "home gardens," small plots surrounding homesteads in which a wide variety of trees, fruits, vegetables, grains, and animals are raised for domestic use and consumption (Fig. 4). On the other end of the spectrum are large-scale, government or industry sponsored "forest village" projects, covering hundreds of square miles. These projects were initiated to produce high quality trees for export or for industry, but to resolve local social problems, peasant farmers are allowed to cultivate subsistence crops between the seedlings of the forest plantation for the first few years. In between these two extremes are a variety of cultivation schemes used by peasant farmers, businessman farmers, government foresters, and landowners concerned with the welfare of the local populations.

The common feature of all agroforestry schemes is that the cultivation plot contains a variety of crops, both in space and in time, and that some of the crops are trees. The trees may be used for fruit or nut products, for extracts such as gums and oil, for firewood and charcoal, or for timber. The key to successful combinations is finding species that will interact with a minimum of competition. Figure 5 gives an example for a coconut plantation. In the early years, there is ample space between the seedlings for sun-demanding annual crops such as grains or short-lived tree crops like papaya. After about 8 years, the canopy of the palms closes. Intercropping is not possible until about 20 years, when the canopy becomes high enough that there is room for shade-adapted species such as cocoa.

Agroforestry is not a new practice. In fact, the earliest hunters and gatherers in the wet tropics evolved a form of cultivation that is very similar to present-day agroforestry. However, for centuries, it was looked down upon by Europeans as being primitive. Only in recent decades with the gradual realization that large-scale monocultures of annual plants are usually not appropriate for the wet tropics has there arisen an interest in agroforestry among agricultural scientists. They discovered that despite the fact that it was practiced by primitive peoples, agroforestry is in fact highly sophisticated in the sense that it is closely attuned to the ecological characteristics of the region: the climate; seasonal-

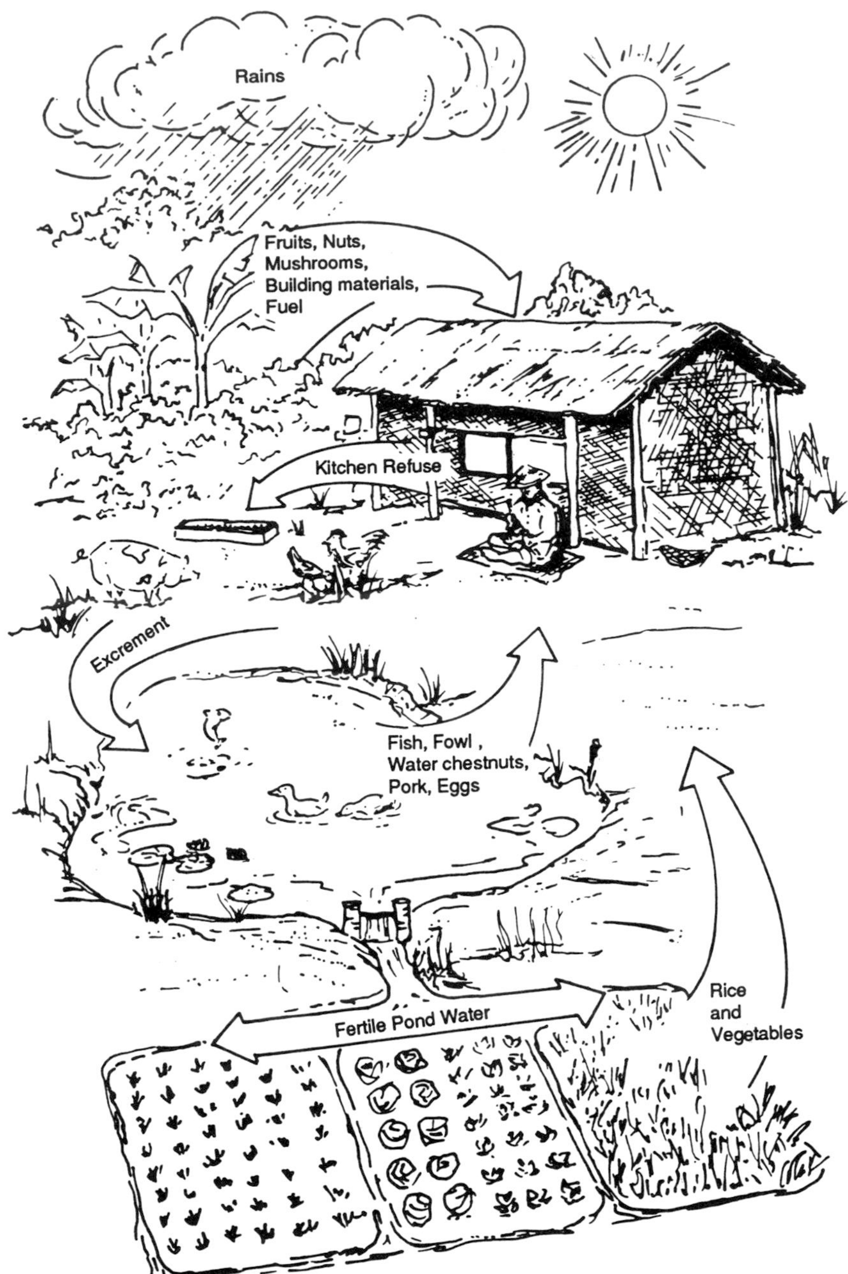


***FIGURE 4*** Hypothetical polycultural agroecosystem.

ity; rainfall; the topography, whether the site was on a mountainside or on the shore of an island; and the nutrient cycling characteristics of the soil.

## B. Trees and Soil Fertility in Agroforestry Systems

Trees are important for maintaining soil fertility in agroforestry systems. Trees in natural forests are the principle storage reservoir for calcium and potassium. When trees are cut and burned, these nutrients are leached away, but in agroforestry systems, the trees conserve these nutrients. Their roots take up these elements from deep soil horizons and supply them to the crops via litter when the trees shed their leaves.

Leguminous trees are especially important in agroforestry systems. These trees have the ability

***FIGURE 5*** The growth phases of coconut palm indicating possibilities for crop combinations. [From Nair (1979).]

to capture atmospheric nitrogen, in symbiosis with nitrogen-fixing microbes. As a result, their leaves are especially high in nitrogen, an important nutrient for crops such as coffee. The traditional method of growing coffee in Mexico is to have an overstory of nitrogen-fixing trees and an understory of coffee trees. Nitrogen removed when the coffee is harvested is replaced by the root activity of the overstory legumes.

Trees are also important in recycling phosphorus. When tropical soils are denuded, phosphorus often becomes immobilized in the soil. However, trees and the mycorrhizal fungi associated with their roots keep the phosphorus cycling through the ecosystem.

Given the role that trees perform in the nutrient cycle of tropics, it is clear why it can be important to incorporate trees into the agricultural systems of these regions.

## VI. CONCLUSION

It is not a coincidence that the natural vegetation in the wet tropics is usually forest. The only type of vegetation that can persevere indefinitely in the continually hot and wet climate is a forest community because of its ability to conserve and recycle nutrient elements. When the forest is cut, loss of nutrients begins, and the rate at which the ecosystem becomes unproductive depends on the severity of the disturbance. Nutrient loss following conversion of tropical forests to agriculture can be compensated for by the addition of fertilizers. However, fertilizers can be prohibitively expensive in many tropical regions.

Agricultural practices that incorporate trees into the system often are more practical in tropical rain forest regions. While crop productivity in agroforestry systems may not be as high as in intensively managed monocultures, the crop production is more sustainable, in the sense that it is less dependent on external sources of fertilizers and chemicals to control insects and weeds.

Nutrient cycles differ between regions in the tropics. On more fertile soils they are rapid whereas on poor soils they are conservative. In both cases, rain forests are poor choices for agricultural development, at least in the sense of large-scale monocultures of grains. However, sustainable agriculture in the tropics is possible if the farmer works with nature instead of against it. By selecting combinations of species that complement each other in their abilities to recycle nutrients and to supply food and fodder for subsistence and for export, continuous agricultural production is possible in the tropics. Designing systems suitable for the social, economic, and political realities of tropical regions pose a challenge to ecologists, agronomists, foresters, anthropologists, and all those concerned with natural resources of the tropics.

## Glossary

**Alluvial** Soil deposited by a river when it falls at the beginning of a dry season.

**Anaerobic** Without oxygen; is often applied to the type of decomposition that occurs in the absence of oxygen.
**Biomass** Weight of an organism or all the organisms in an ecosystem.
**Carnivore** An animal that eats other animals.
**Decomposer** Microbes such as bacteria and fungi that break down organic debris, such as dead leaves and wood.
**Detritivore** Organisms that feed on dead plants or animals.
**Ecosystem** Living biota and the soil and water that occupy a delimited area such as a forest or lake.
**Food chain** Plants, the animals that eat the plants, the animals that eat those animals, and the decomposers that eventually eat all organic material.
**Herbivore** An animal (including insects) that eats plants exclusively.
**Humus** Partially decayed organic material on the forest floor.
**Leguminous** A plant belonging to the legume family; many of these plants have a symbiotic relationship with microbes that fix atmospheric nitrogen.
**Litter** Dead organic material such as leaves and trunks on the forest floor.
**Microbe** A microscopic organism such as bacteria.
**Mycorrhiza** A type of fungus that lives in symbiosis with the roots of many plants.
**Organic** A compound containing carbon; includes all living material and material that once was living.
**Primary production** Growth of plants, as measured by an increase in weight.
**Shifting cultivation** Practice of cutting the forest, farming the soil for a few years, then abandoning the site and moving on when crop growth declines.
**Slash and burn** Agriculture following cutting and burning of the forest.
**Volatile nutrient** A nutrient such as nitrogen that can exist in a gaseous state.

## Bibliography


Anderson, A. B. (1990). "Alternatives to Deforestation: Steps toward Sustainable Use of the Amazon River Forest." New York: Columbia Univ. Press.

Golley, F. B. (1975). "Mineral Cycling in a Tropical Moist Forest Ecosystem." Athens, Georgia: University of Georgia Press.

Gomez-Pompa, A., Whitmore, T. C., and Hadley, M., eds. (1991). Rain forest regeneration and management. *In* "Man and the Biosphere Series," Vol. 6. UNESCO, and The Parthenon Publishing Group, Carnforth, UK.

Jordan, C. F. (1985). "Nutrient Cycling in Tropical Forest Ecosystems." Chichester: Wiley.

Jordan, C. F. (1989). An Amazonian rain forest: The structure and function of a nutrient stressed ecosystem and the impact of slash and burn agriculture. *In* "Man and the Biosphere Series," Vol. 2. UNESCO, and The Parthenon Publishing Group, Carnforth, UK.

MacDicken, K. G., and Vergara, N. T. (1990). "Agroforestry: Classification and Management." New York: Wiley.

Nair, P. K. R. (1979). Agroforestry Research: A retrospective and prospective appraisal. *In* "International Cooperation in Agroforestry." (T. Chandler and D. Spurgeon, Eds.), pp. 275–296. Nairobi, Kenya: Proc. Conf. Internat. Cooperation in Agroforestry.

Proctor, J., ed. (1989). "Mineral Nutrients in Tropical Forest and Savanna Ecosystems." Oxford: Blackwell.

Serrao, E. A., Falesi, I. C., de Veiga, JK. B., and Teixeira Neto, J. F. (1978). Productivity of cultivated pastures on low fertility soils in the Amazon of Brazil. *In* "Pasture Production in Acid Soils of the Tropics," (P. A. Sanchez and L. E. Tergas, Eds.), pp. 195–225. Cali, Colombia: Semin. Proc. Cintro Internacional de Agricultura Tropical.

ISBN 0-12-226732-X